# Lecture Notes in Networks and Systems

Volume 1356

The series “Lecture Notes in Networks and Systems” publishes the latest developments in Networks and Systems—quickly, informally and with high quality. Original research reported in proceedings and post-proceedings represents the core of LNNS.

Volumes published in LNNS embrace all aspects and subfields of, as well as new challenges in, Networks and Systems.

The series contains proceedings and edited volumes in systems and networks, spanning the areas of Cyber-Physical Systems, Autonomous Systems, Sensor Networks, Control Systems, Energy Systems, Automotive Systems, Biological Systems, Vehicular Networking and Connected Vehicles, Aerospace Systems, Automation, Manufacturing, Smart Grids, Nonlinear Systems, Power Systems, Robotics, Social Systems, Economic Systems and other. Of particular value to both the contributors and the readership are the short publication timeframe and the world-wide distribution and exposure which enable both a wide and rapid dissemination of research output.

The series covers the theory, applications, and perspectives on the state of the art and future developments relevant to systems and networks, decision making, control, complex processes and related areas, as embedded in the fields of interdisciplinary and applied sciences, engineering, computer science, physics, economics, social, and life sciences, as well as the paradigms and methodologies behind them.

Indexed by SCOPUS, EI Compendex, INSPEC, WTI Frankfurt eG, zbMATH, SCImago.

All books published in the series are submitted for consideration in Web of Science.

For proposals from Asia please contact Aninda Bose (aninda.bose@springer.com).

K. Reddy Madhavi · Nagaraj Ramrao ·
Kishore Kumar · K. Srujan Raju ·
Mathini Sellathurai
Editors

# Proceedings of Sixth International Conference on Computer and Communication Technologies

IC3T 2024, Volume 1

Springer

*Editors*
K. Reddy Madhavi
School of Computing
Mohan Babu University
Tirupati, Andhra Pradesh, India

Kishore Kumar
Department of Electronics
and Communication Engineering
National Institute of Technology
Warangal, Telangana, India

Mathini Sellathurai
Department of Science and Engineering
Heriot-Watt University
Edinburgh, UK

Nagaraj Ramrao
Department of Electronics
and Communication Engineering
Mohan Babu University
Tirupati, Andhra Pradesh, India

K. Srujan Raju
Department of Computer Science
and Engineering
CMR Technical Campus
Hyderabad, Telangana, India

ISSN 2367-3370 ISSN 2367-3389 (electronic)
Lecture Notes in Networks and Systems
ISBN 978-981-96-5237-2 ISBN 978-981-96-5238-9 (eBook)
https://doi.org/10.1007/978-981-96-5238-9



This Springer imprint is published by the registered company Springer Nature Singapore Pte Ltd.
The registered company address is: 152 Beach Road, #21-01/04 Gateway East, Singapore 189721, Singapore

# Committees

## International Advisory Committee

Dr. Rangaraj M. Rangayyan, Professor, Emeritus of Electrical and Computer Engineering, University of Calgary, Alberta, Canada
Dr. Ajay K. Katangur, Professor and Head, Computer Science, MSU
Dr. Junfeng Que, Professor, Clayton State University, Atlanta, USA
Dr. Anitesh Barua, Professor, UT, USA
Dr. Mathini Sellathurai, Professor, Heriot-Watt University, UK
Dr. Joao Manuel R. S. Tavares, Professor, University of Porto, Portugal
Dr. Kun-Lin Hsieh, NTU, Taiwan
Dr. Ahamad J. Rusumdar, KIT, Germany
Dr. V. R. Chirumamilla, EUT, Netherland
Dr. Halis Altun, MU, Turkey
Dr. P. N. Suganthan, NTU, Singapore
Dr. Mohammad Reza Daliri, Professor, Iran
Dr. Boka Kumsa, Wollega University, Ethiopia
Dr. Joy Iong-Zong Chen Ph.D., Da-Yeh University, Taiwan
Dr. Siddharth S., Ph.D., University of Essex, UK
Dr. Ashok. R. Patel, University of Massachusetts, USA
Dr. Bhuvan Unhelkar, University of South Florida, USA
Dr. M. S. Mekala, Yeungnam University, Gyeongsan, Korea
Dr. Asadi Sreenivasulu, University of Newcastle, Australia
Dr. Srinivas Nowduri, Professor of CSI, Pueblo, Colorado, USA
Dr. Shiva Kaleru, Juniper Networks, USA
Dr. M. Madhusudhana Subramanyam, King Khalid University, Kingdom of Saudi Arabia
Dr. Siva Ram Rajeyyagari, Dean, IT and E-Learning, Shaqra University, Shaqra
Dr. Mohammad Gouse Galety, Samarkand International University of Technology, Samarkand, Uzbekistan

## National Advisory Committee

Dr. D. Surya Narayana, Professor, Energy Science Engineering, IIT Mumbai
Dr. N. V. S. Narasimha Sarma, Director, IIIT Trichy
Dr. Tushar Kanti Bera, National Institute of Technology, Durgapur
Dr. K. Chandrasekaran, NIT Karnataka, Surathkal
Dr. Annappa, NIT Karnataka, Surathkal
Dr. Devender Gurjat, NIT Silchar
Dr. K. Jairam Naik, NIT Raipur
Dr. G. C. Nandi, IIIT, Allahabad, Jhalwa, Uttar Pradesh
Dr. J. Girija, Bangalore Institute of Technology, Bengaluru
Indra Riyanto, Ph.D., Universitas Budi Luhur, Indonesia
Dr. D. Ganesh Gopal, UIET, Chandigarh University, Mohali, Punjab
Dr. B. Balamurugan, Galgotias University, Greater Noida, Uttar Pradesh
Dr. Thirunavukkarasu K., Karnavati University, Uvarsad, Gandhinagar, Gujarat
Dr. Valli Kumari, Professor, Andhra University, Visakhapatnam
Dr. Brojo Kishore Mishra, GIET University, Gunupur, Odisha
Dr. G. S. Pradeep Ghantasala, Chitkara University, Punjab, India
Dr. A. Ananda Rao, JNTUA, Ananthapuramu
Nazori A. Z., Ph.D., Universitas Budi Luhur, Indonesia
Dr. S. Viswanadha Raju, JNTUHCEJ, Jagtial, Karimnagar
P. Thiyagarajan, Ph.D., Central University of Tamil Nadu, Thiruvarur, Tamil Nadu
Dr. G. Madhavi, JNTU, Vizianagaram
Dr. Thulasi Bikku, Amrita Vishwa Vidyalayam
Dr. Chandra Mouli V. S. A., BVC Engineering College, Odalarevu
Dr. S. Nagini, VNR VJIET, Hyderabad
Dr. Bramhananda Reddy, VNR VJIET, Hyderabad
Dr. M. Sudhakar, Chistats, Pune
Dr. B. V. A. N. S. S. Prabhakar Rao, VIT Chennai
Manish Prateek, Ph.D., Swami Rama Himalayan University, Dehradun
Dr. Nagendra Panini Challa, VIT AP, Amaravati
Dr. Beebi Naseeba, VIT AP, Amaravati
Dr. Ramasree, Aditya University, Surampalem
Dr. M. RudraKumar, MGIT, Hyderabad
Dr. K. P. N. V. Satyasree, Usha Rama College of Engineering
Dr. Samreen Fiza, Presidency University, Bengaluru
Dr. Tirumala Vasu, Presidency University, Bengaluru
Gopal Chaudhary, Ph.D., Bharati Vidyapeeth's College of Engineering, New Delhi
Deepika Koundal, Ph.D., University of Petroleum and Energy Studies, Dehradun
Neha Sharma, Ph.D., Chandigarh University, Chandigarh

## Chief Patrons

Dr. M. Mohan Babu, Chancellor, MBU
Mr. Vishnu Manchu, Pro-Chancellor, MBU

## Patrons

Mr. Vinay Maheswari, Executive Director, MBU
Dr. Nagaraj Ramrao, Vice-Chancellor, MBU
Dr. K. Saradhi, Registrar, MBU

## Advisors

Mr. Vikas Singh, Chief Growth and Strategy Officer, MBU
Dr. Avireni Srinivasulu, Dean, Research and Innovation, MBU
Dr. B. M. Satish, Dean, School of Engineering and Technology, MBU
Dr. B. Narendra Kumar Rao, HOD-CSE (AI&ML), School of Computing, MBU

## Editorial Board

## Conveners

Dr. K. Reddy Madhavi, Professor CSE (AI&ML), School of Computing, Mohan Babu University
Dr. J. Avanija, Professor, CSE (AI&ML), School of Computing, Mohan Babu University

## Co-Conveners

Dr. I. Siva, Professor, School of Engineering, Mohan Babu University
Dr. Subba Rao Polamuri, CSE, Aditya University, Surampalem, Andhra Pradesh, India

## Coordinators

Dr. P. Dhana Lakshmi, CSE (AI&ML), School of Computing, Mohan Babu University
Mr. N. Bala Krishna, CSE (AI&ML), School of Computing, Mohan Babu University

## Organizing Secretaries

Ms. K. Ghamya, Assistant Professor, CSE (AI&ML), School of Computing, Mohan Babu University
Ms. K. Sandhya, Assistant Professor, CSE (AI&ML), School of Computing, Mohan Babu University

## Organizing Committee

Department of AI&ML, School of Computing, Mohan Babu University

## Technical Review Committee

Dr. Ankit Agarwal, IIT Madras
Dr. C. Rohith Bhat, Saveetha School of Engineering (SIMATS)

Dr. Ravinder, Chandigarh University
Dr. Voruganti Naresh Kumar, CMR Technical Campus, Hyderabad
Sapana Amol Kolambe, Pimpri Chinchwad College of Engineering Nigdi, Pune
P. Anjaneyulu, Aditya Institute of Technology and Management, Tekkali, Andhra Pradesh
Dr. Suresh Kumar, Geeta University, Panipat
Dr. Suraya Mubeen, CMR Technical Campus, Hyderabad
Uppalapati Srilakshmi, Sridevi Women's Engineering College
Rashi Saxena, GRIET, Hyderabad
M. R. Ramesh, Indira Gandhi Centre for Atomic Research, Kalpakkam
Dr. Ashok Matani, Government College of Engineering, Jalgaon
Ma. Eliza A. Sano, Cavite State University-Silang Campus
Dr. R. Jeevitha, PSGR Krishnammal College for Women
Dr. G. Madhukar, CMR Technical Campus, Hyderabad
Dr. Jonnadula Narasimharao, CMR Technical Campus, Hyderabad
Srikanth Chintakindi, Cloud Software Group
Raghi K. R., St. Joseph's College of Engineering
Dr. A. Naveen, Don Bosco College (Co-Ed), Yelagiri Hills, Tamil Nadu
Dr. J. Senthil Kumar, Kalaignarkarunanidhi Institute of Technology
Mrs. Mamta Sanjay Koban, AISSMS Institute of Information Technology, Pune
Dr. S. Amuthameena, SRM Institute of Science and Technology, Tiruchirappalli
Sreekanth Yalavarthy, HCL Technologies, USA
Dr. Prithviraj Singh Solanki, Shivalik College of Engineering, Dehradun
Boudhayan Bhattacharya, Brainware University
Suryaprakash Nalluri, Citigroup
Dr. Anita G. Khandizod, Symbiosis Skills and Professional University
Rambabu Inaganti, Smith&Nephew, USA
Muruganantham Balakrishnan, SRM Institute of Science and Technology
Dr. V. Ceronmani Sharmila, SRM Institute of Technology and Science, Kattankulathur
Dr. K. Bharath Kukar, CMR Technical Campus, Hyderabad
Sadhana Singh, ABES Institute of Technology, Ghaziabad
Dr. K. Kalaiselvi, Saveetha College of Liberal Arts and Sciences, SIMATS
Pragya Pandey, ABES Institute of Technology
Dr. Raghu N., Jain University
Dr. Lanke Pallavi, B. V. Raju Institute of Technology
Vikram D., Rathinam Technical Campus
Harsha Vikas Patil, MIT ACSC, Alandi, Pune
Dr. Santhosh Kumar Allemki, Gurunanak Institute of Technology
Dr. Jyoti A. Khubchandani, Vishwakarma Institute of Technology
Dr. Payal Khubchandani, Vishwakarma Institute of Technology, Pune
Dr. Saroja Kumar Rout, Vardhaman College of Engineering, Hyderabad
Dr. Pavithra K., Alliance University
Madhusmita Sahu, C.V. Raman Global University, Bhubaneswar
Dr. P. Bharath Kumar Chowdary, VNR VJIET, Hyderabad

Dr. G. N. R. Prasad, Chaitanya Bharathi Institute of Technology
K. Lova Raju, Vignan's Foundation for Science, Technology and Research
Arun Pandiyan Perumal, Illinois Institute of Technology
Hemant Kumar Saini, Amity University
Dr. P. AnnanNaidu, GMR Institute of Technology, Rajam, Vizianagaram
Dr. Ravikumar Ch., Chaitanya Bharathi Institute of Technology
Harshavardhan Nerella, MassMutual, USA
Praveen Borra, Florida Atlantic University
Prasad Gandham, Microsoft
Dr. Samta Jain Goyal, Amity University
Dr. Ch. N. Santhosh Kumar, Anurag Engineering College, Kodad
K. Ravikanth, Aurora University
Mr. K. Sreekanth, NNRG (Autonomous), Hyderabad
Ashish Kumar Dass, NIST University
G. V. Subbarao, KLEF
Hemant Kumar Saini, Amity University
Dr. Pradnya Zode, Yeshwantrao Chavan College of Engineering
Dr. Aswini J., Saveetha Engineering College
Dr. Raghu N., Jain University
Dr. S. K. Khaja Shareef, KLH University
Avinash Khanderi, Walmart Inc.
Sapan Bharadwaj Bonala, AVCO Consulting Inc.
Sukanya Konatam, Vialto Partners, Texas, USA
Kiran Kumar Kakkireni, Equinox IT Solutions LLC
Nikhil Gunda, Digital Connect, Hyderabad

# Preface

DST SERB Sponsored 6th International Conference on Computer and Communication Technologies (IC3T) 2024, organized by the Department of Artificial Intelligence and Machine Learning, School of Computing, Mohan Babu University, located in Tirupati, Andhra Pradesh, India, is a State Private University established by the Sree Vidyanikethan Educational Trust under the patronage of Dr. M. Mohan Babu, a philanthropist, former Rajya Sabha Member, producer, and actor. The University advocates for excellence in diverse disciplines, including Engineering, Pharmacy, Liberal Arts, Science, Agriculture, Art and Design, Media Studies, Film Academy, and Paramedical and Health Sciences. Featuring a 5-star rated campus, an advanced curriculum, skilled guidance, and over 65 clubs. MB School of Engineering and MB School of Computing (formerly Sree Vidyanikethan Engineering College) have attained significant accreditations and rankings, which include: GSAAA-International Accreditation, NBA, NAAC, SII Green Rankings, MHW Rankings, AICTE-CII Platinum Survey, NIRF, Institution's Innovation Council Ratings. Boasting a robust placement record, including a remarkable package of Rs.60 Lakhs offered by Google, the university also actively mentors other engineering colleges under AICTE Margdarshan Scheme To foster Research, Innovation and Entrepreneurship, MBU has DST-STI Hub worth of 3.61 Crores, AICTE-IDEA Lab worth of 1.1 Crores and V-Hub, Innovation Centre, Recognized as one of the best-performing institutions from South Zone by Institution Innovation Council (IIC), Bagged several laurels in Hackathons organized by MHRD and AICTE. The university has successfully implemented World Bank 6 Crores project TEQIP-II as one of the best-performing institutions at the National Level, meeting all the key deliverables. The institute got sanctions projects worth Rs.2.86 Crores are funded by agencies like DST, INUP, ISRO-RESPOND, and UGC for faculty research projects, AICTE grant of Rs.97.58 Lakhs for laboratory development under Modernization and Removal of Obsolescence Scheme (MODROBS), Rs.82.5 Lakhs for Staff Development Programs, and Rs.8 Lakhs for establishment of Entrepreneurship Development Cell (EDC) and received Rs.1.4 Lakhs from EDII, Ahmedabad, to organize Entrepreneurship Awareness Camps under DST-NIMAT Project. Implemented Pradhan Mantri Kaushal Vikas Yojana for Technical Institutions (PMKVY-TI) through AICTE with a sanctioned amount of Rs.31.54 Lakhs to offer technical training to unemployed youth. Sanction of DST grant of Rs.20 Lakhs under FIST Programme for promotion of Research. Establishment of Atmospheric Research Laboratory, implementing a DST, Government of India Project with a fund of Rs.20 Lakhs. AICTE sanctioned Skill and Personality Development Programme Centre (SPDPC) for SC/ST students with a grant of Rs.15.3 Lakhs. Provision of facilities worth Rs.15 Lakhs to support Research and Development by INUP, IIT Bombay. Member of professional societies such as IEEE with 16 societies, ISOI, CSI, ACM, SAE India, IWS, ISHRAE, IE, ASCE, IGS, SCRS and Institutional Memberships: EdCIL, CII, IIEC, CISCO, IUCEE and ISTE, Establishment of National MEMS Design Centre under the aegis of NPMASS and IISc, Bangalore, JNTUA recognized research centers in EEE, ECE, and CSE Departments. The university also offers Industry Collaboration programs with prominent companies like IBM and SAP.

IC3T 2024 aims to provide a cohesive forum for advanced, multidisciplinary research focused on the design of smart computing, information systems, and electronic systems. The theme centers on diverse innovation paradigms in system knowledge, intelligence, and sustainability that can be utilized to offer practical solutions to a range of societal, environmental, and industrial challenges. The scope encompasses the implementation of emerging computational and knowledge transfer methodologies, enhancing solutions across various fields including Computer Science and Engineering, Electronics and Communication Engineering, Electrical and Electronics Engineering, and Electronics and Instrumentation Engineering. IC3T 2024 witnessed the participation of nearly 200 delegates both online and offline from various institutions. We are pleased to share that the technical committee received total of 911 quality manuscripts from main track and five special sessions include—Industry-Driven Machine Learning, Deep Learning with Explainable AI by Dr. Malla Sudhakara and Dr. Asadi Srinivasulu—University of Newcastle, Australia, Recent Trends in Cognitive Systems by Dr. Nagendra Panini Challa, Dr. B. Naseeba, VIT AP, Amaravathi, Security Solutions for Integrated IoT Ecosystems by Dr. Siva Ram Rajeyyagari, Shaqra University, Shaqra and Dr. M. Rudra Kumar, MGIT Hyderabad, Smart Innovations in Mezzanine Technologies by Dr. S. Nagini, and Dr. A. Brahmananda Reddy, VNRVJIET, Hyderabad, Advancements in AI-Driven Communication Technologies: Bridging Innovation and Practical Applications (AADCT) by Dr. S Rama Sree and Dr. P. Subba Rao, Aditya University, Surampalem, Andhra Pradesh. Manuscripts received from all over India and Abroad Countries like USA, Ireland, China, Australia, United Arab Emirates, Veitnam, Saudi Arabia, Uzbekistan, and from various industries from abroad like Salesforce, Massmutual, HCL Technologies, Juniper Networks, and Smith&Nephew. An exclusive editorial team is constituted by accounting national and international experts to publish the warrant articles. After rigorous review process, as per the Springer guidelines, 109 articles, of these, 59 papers in Volume I have been chosen for publication in the Scopus Indexed Springer Book Series on Lecture Notes in Networks and Systems (LNNS).

The event was applauded for the 4 plenary talks by International experts 2-offiline sessions by Shri. Aninda Bose, Springer Nature, London and Dr. Sam Goundar, RMIT University, Australia, 2-online sessions by Dr. Asadi Srinivasulu, University of Newcastle, Australia and Dr. Siva Ram Rajeyyagari, Shaqra University, Shaqra, 9 session tracks: 4-offline, 4-online, and 1-blended, each track with two session chairs, besides the Inaugural and Valedictory sessions.

We are indebted to the technical committee members and reviewers for providing timely and quality reviews. We take this opportunity to thank the management of Mohan Babu University for providing financial, infrastructure support, and other resources to organize the prestigious conference. We wish to submit our sincere thanks to the Science and Engineering Research Board, Department of Science and Technology (DST), Government of India for their kind financial support to host this conference which elevated this as a benchmark for our technical followers and colleagues. Finally, we would like to thank all the volunteers for their tireless efforts for a long time in every step of the journey for the success of this conference, which inspired organizers and motivated others.

Tirupati, India — Dr. K. Reddy Madhavi
Tirupati, India — Dr. Nagaraj Ramrao
Warangal, India — Dr. Kishore Kumar
Hyderabad, India — Dr. K. Srujan Raju
Edinburgh, UK — Dr. Mathini Sellathurai

# Contents

# About the Editors

**Dr. K. Reddy Madhavi** received Ph.D. from JNTU, Anantapur, M.Tech. from JNTU Hyderabad, and B.Tech. from SV University, Tirupati, and currently working as a professor, School of Computing, Mohan Babu University, Tirupati. She has filed and published 21 national and international patents and few of them granted too. To her profile, she has more than 70 papers in international journals/conferences indexed in SCI/Scopus/Web of Science/Google Scholar and five edited/authored books. She has undergone a course on design thinking in collaboration with IUCEE EPICS Purdue University, USA. She is also an innovation ambassador associated with IIC, MHRD, Government of India. She is the convener of Springer International Conferences, ICDIC 2021, ICIHCNN-2022, and other conferences NCKITS-2022, NCKITS-2023, and ICATAS 2023.

**Dr. Nagaraj Ramrao,** an eminent academician, researcher, and an able administrator, is the Vice-Chancellor of Mohan Babu University, Tirupati, Andhra Pradesh. He holds a Ph.D. from Visvesvaraya Technological University. During his career spanning more than 30 years, he served as the vice-chancellor of Kalasalingam Academy of Research and Education, Srivilliputhur, Tamil Nadu, director of Dhirubhai Ambani Institute of Information and Communication Technology, (DA-IICT), Gandhinagar, Gujarat, mentor director of the Indian Institute of Information Technology (IIIT)-Vadodara, Gujarat, director and principal of the Oxford College of Engineering, Bengaluru, where he also served as a professor and head of the ECE Department. Earlier, he was a professor and director of the R. V. Center for Cognitive Technologies, a R&D Center, ECE Department, R. V. College of Engineering, Bengaluru, and an assistant engineer at Karnataka Power Corporation Ltd. and also at MSRIT, Bengaluru.

**Dr. Kishore Kumar** earned his Ph.D. in signal processing in 2004. He headed the Department of Electronics and Communication Engineering at NIT Warangal from 2015 to 2017. He was also the head of the Institute Computer Center, NIT Warangal, from 2018 to 2020. He was a visiting professor at ESIGELEC University in France in 2009, a visiting professor at TALLINN University, Estonia, for a semester for an MS in Computer Science, and a visiting professor at the Asian Institute of Technology, Bangkok, in 2019 under the Secondment of Indian Faculty by MHRD, Government of India. He worked as a technical officer in the Cabinet Secretariat, Prime Minister's Office, Government of India. He is a member of the Central Council Board (CCB) and a member of the Andhra Pradesh State Council for Higher Education (APSCHE).

**Dr. K. Srujan Raju** is currently working as the dean of Student Welfare and Heading Department of Computer Science and Engineering and Information Technology at CMR Technical Campus. He obtained his Doctorate in Computer Science in the area of Network Security. He has more than 20 years of experience in academics and research. His research interest areas include computer networks, information security, data mining, cognitive radio networks and image processing, and other programming languages. He was invited as a session chair, keynote speaker, technical program committee member, track manager, and a reviewer for many national and international conferences also appointed as subject Expert by CEPTAM DRDO–Delhi and CDAC.

**Dr. Mathini Sellathurai** is currently the dean of Science and Engineering and the head of the Signal Processing for Intelligent Systems and Communications Research Group, Heriot-Watt University, Edinburgh, UK, and leading research in signal processing for radar and wireless communications networks. She has five years of industrial research experience. She held positions with Bell-Laboratories, New Jersey, USA, and with the Canadian (Government) Communications Research Centre, Ottawa Canada. She was an associate editorship for the *IEEE Transactions on Signal Processing* (2005–2018) and *IEEE Signal Processing for Communications* Technical Committee member (2013–2018). She was an organizer for the IEEE International Workshop on Cognitive Wireless Systems, IIT Delhi, India, 2009, 2010, and 2013, and the general chair of the 2016 IEEE Workshop on Signal Processing Advances in Wireless Communications (SPAWC), Edinburgh, UK.

# Dynamic Resource Allocation in Cloud Using Reinforcement Learning

**R. Sivasubramanian, Thayabba Khatoon Mohammed, C. N. Rajalakshmi, and Harshavardhan Nerella**

**Abstract** This research suggests an actor-critic approach for cloud computing resource allocation that is based on deep reinforcement learning. The suggested approach generates the allocation strategy using an actor network and assesses the allocation's quality using a critic network. A deep reinforcement learning technique is used to train the actor and critic networks in order to maximize the allocation strategy. A simulation-based experimental investiGenetic Algorithm is used to assess the suggested approach, and the findings demonstrate that it performs better overall in cost, efficient energy and resource utilization than a number of the current allocation approaches. In an attempt to lower energy usage, some methods for workloads management or virtual machines have been already, but, those methods frequently are unable to account for the states of the server. In order to achieve logical task allocation, this study suggests the Dynamic Workload Allocation Scheme (DWAS), which continuously captures the dynamicity of server conditions and accounts for the impact of various workloads on energy consumption. The goal is to guarantee workload quality of service (QoS) with less computational of physical servers. This technique computes the predicted cumulative return over time using a dynamic algorithm based on reinforcement learning (RL) to discover the optimal task allocation in terms of energy efficiency. Through simulation, we find that, in comparison with

R. Sivasubramanian (✉)
Department of AIML, Malla Reddy University, Hyderabad, India
e-mail: sivar2000@gmail.com

T. K. Mohammed
Department of Computer Science and Engineering, VNR Vignana Jyothi Institute of Engineering and Technology, Hyderabad, Telangana, India
e-mail: thayyabakhatoon_md@vnrvjiet.in

C. N. Rajalakshmi
Department of CSE, Malla Reddy University, Hyderabad, Telangana, India
e-mail: c.n_rajalakshmi@mallareddyuniversity.ac.in

H. Nerella
Cloud Engineer, MassMutual, Austin, TX, USA

 
K. R. Madhavi et al. (eds.), *Proceedings of Sixth International Conference on Computer and Communication Technologies*, Lecture Notes in Networks and Systems 1356,
https://doi.org/10.1007/978-981-96-5238-9_1

the current baseline allocation methods, the proposed DWAS can lower the workload around 15%. The report also discusses potential applications of the suggested technology in cloud computing and makes recommendations for further research.

**Keywords** Resource allocation · Load balancing · Reinforcement learning · Dynamic workload allocation scheme

## 1 Introduction

Reinforcement learning (RL) methods have been utilized in cloud computing. Here are a few instances of RL algorithms used to solve this issue: QT-Education: This model-free reinforcement learning technique iteratively updates QT-values depending on the rewards from various states and actions in order to learn an optimal action-value function. It has been used to address cloud computing's dynamic load balancing issue. Deep Reinforcement Learning: To learn more sophisticated policies, this entails merging deep neural networks with reinforcement learning. In cloud computing, it is used for balancing the load to maximize the distribution of virtual computers among physically existing servers. DWAS: This variation of RL algorithms learns a policy and estimates its value by fusing policy-based and value-based techniques (like QT-learning). Priority-based Scheduling: In this method, tasks are ranked in order of significance after being given a priority based on their specifications (such as the resources they'll need and how long they should take). Most valued priority appointments are drafted first, while lower priority appointments are booked next. But this easy-to-use method might only work well in certain circumstances. Genetic Algorithms: In this method, a schedule that maximizes a certain objective function is evolved through the use of a genetic algorithm [1, 4]. In order to produce the next generation of schedules, the proposed method generates a cluster of schedules, assesses each one's fitness using the main procedure and delivers most fit methods. Although it could take a lot of iterations to reach an ideal solution, this method is useful for correcting complicated problems.

## 2 Literature Survey

Nevertheless, an incorrect policy gradient and excessive variance result from the linear functional form that is utilized in the traditional actor-critic to derive the action-value function. To solve this problem, an advantage actor-critic was created by [8, 10]. This gives values as responses and be modified by the actor. Actor makes selections based on random chance. On the other hand, the policy gradient variance can be greatly decreased with the help of the advantage functions. Function approximation error occurs when a neural network unable to accurately deliver the correct shore up function when approximating the function's value or strategy procedure in a policy-value system. We were successful in creating a double-deep QT-network-based reinforcement learning technique. In order to counteract state-space burst and quickly determine the optimal operational loading strategy, they resorted to a DNN-based amount. The authors suggest a DL method works on the SARSA structure in order to optimize computational dumping efficiency[11, 13]. In the authors' simulation tests, a number of scheduling strategies, including a genetic algorithm and the first-come, first-served algorithm, are contrasted and compared with the proposed system General Algorithm. The suggested method outperforms the competing algorithms when makespan, or overall execution time, and resource utilization are taken into account imply that, especially in highly continuous action spaces, function rough error can lead to unstable and subpar appearance in actor-critic techniques. In response, they present the TD3 policy, which makes use of trine crucial adjustments to raise the function approximation's accuracy. An actor that makes recommendations for asset distribution and a critic that assesses decisions' credibility make up a policy-value deep RL algorithm. Both the critic and the actor are represented as neural networks, and they are taught by combining reinforcement learning with supervised learning [3]. An environment simulation of a cloud data center is used to assess the algorithm. According to the results, it performs better in terms of workload completion time and resource consumption than traditional resource allocation algorithms. According to the paper's conclusion, the suggested strategy may enhance the cloud data center's resource allocation's flexibility and efficiency [12]. Power allocation model for residential homes was discussed [2]. After allocating data in the cloud, the security and recovery was discussed [5, 9]. Enhancement of the resource allocation was that re-allocation of resources was elaborated[6]. In case of geographically distributed environments, routing method was discussed to find the shortest path [7].

## 3 Proposed Work

The suggested method makes use of RL to determine the best resource allocation policy given the available resources and the workload at hand. Two parts make up this dynamic algorithm: recommendations for asset distribution by actor and the decisions' quality will be diagonized by critic**.** Both critic as well as the actor are built as neural networks that are taught via reinforcement learning combined with supervised learning. Figure 1 is an environment simulation of a cloud data center that is used to assess the algorithm. The outcomes demonstrate that it performs better in terms of workload completion time and resource consumption than conventional resource allocation algorithms. For any given state-action pair (nst, pat), the optimal QT-value is determined by adding the estimated addition of the instant value (IRt $++$ ) where fall of maximum QT-value of next state (nst $++$ ) and actions (pat $++$ ) that will be pulled from it. Future rewards weight is determined by the discount factor ($\alpha$).

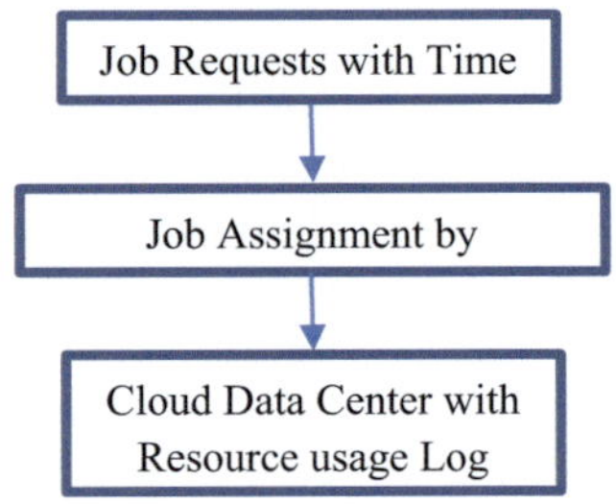


**Fig. 1** Scheduler in dynamic allocation

$$Q * (\mathrm{nst}, \mathrm{pat}) = \sum [\mathrm{IRt} + 1 + \beta \mathrm{max\_pat} + 1Q * (\mathrm{nst} + 1, \mathrm{pat} + 1)|\mathrm{nst}, \mathrm{pat}] \quad (1)$$

The equation for the ideal QT-score in LR. This offers one method that iteratively determines the best QT-score for a specific state-action combination by taking into account the best QT-scores of the subsequent state and every action that may be taken from it. The optimal QT-score of the current state-action (nst, pat) is represented by $Q$ * (nst, pat) in Eq. (1). This QT-value indicates the predicted total cost that can be achieved by adhering to the better policy proceed from here and forwards the action. The instant reward received following action pat in state nst is denoted by IRt + 1. This is the prize that was awarded at timestamp ($t$ + 1) in question. The reduction factor, denoted as α(alpha), establishes the value assigned to further rewards in comparison with immediate Genetic Algorithmins. It is a number between 0 and 1 that is meant to make sure the agent takes into account both the possibility of future rewards being delayed and the uncertainty around them while making decisions max_pat + 1. The highest QT-score for the upcoming state nst + 1 and possible actions pat + 1 that can be considered from it is $Q$ * (nst + 1, pat + 1) (Fig. 2).

Figure 3, this is the anticipated reward that can be attained by continuing to pursue the best course of action. The max executor opts largest QT, which, in accordance

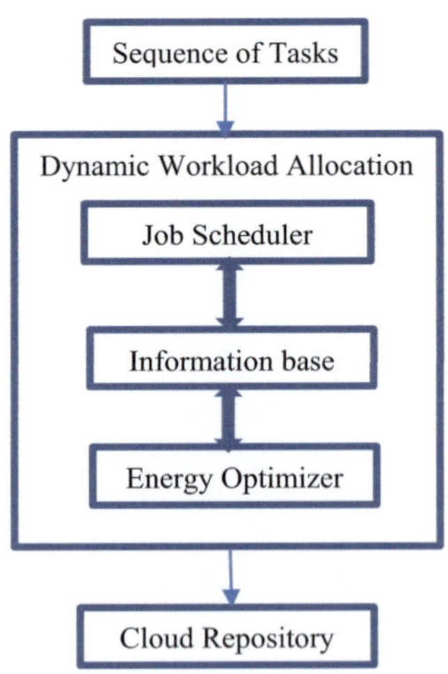


**Fig. 2** Sequence of actions in resource allocation

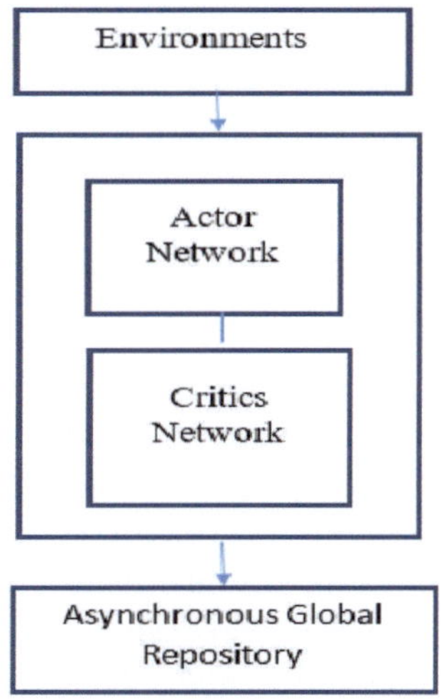


**Fig. 3** Scheduling based on environment

with the existing policy, indicates the optimal course of action for the following state. The expectation operator, Σ[…], determines the estimated score of the expression enclosed in brackets based on the state/action pair (nst, pat) that is currently in effect. In summary, the assumed sum of reward for a state & action combination is expressed by the Bellman equation for the best QT-value by means of the instant price, the highest original added values from the subsequent state, and all feasible values. This equation, which is essential to reinforcement learning, is utilized in numerous QT-learning methods to improve the QT-scores as well as enhance the broker's policy in the long run. Next, self-adaptive systems are introduced, which have the ability to track their actions and adjust to changing circumstances in order to achieve performance targets.

**Algorithm 1:** Active task assignment scheme

| |
|---|
| *Initialization:*<br>Initialize the weights for the actor n/w $\theta$ and the other network |
| Set the learning rates $\gamma n$ and $\gamma m$, along with the Twin Delayed bug factor $\beta$ |
| *Training Loop:* |
| For every training epoch $n = 1,2,\ldots,N$: |
| *Episode Loop* |
| Choose an action according to the current policy |
| $at =$ choose of action($st$) |
| *Execute Action:* |
| Take action $at$, observe price $Rt$ and subsequent state $st++$ |
| *TD Error Calculation:* |
| Calculate the TD error in the critic network: |

(continued)

(continued)

| |
|---|
| TD_error $= rt + \beta \cdot V\pi\theta\ (st + 1) - V\pi\theta\ (st)$ |
| *Policy Slope Calculation:* |
| Calculate the strategy deviation in the actor network using: $J(\theta) = \mathrm{K}\pi\theta - \log\pi\theta\ (st,at)\cdot\delta\pi\theta$<br>*Modify State:* |
| Move to the next state: $st = st + 1$ |
| *End of Episode:* |
| Repeat steps 4–8 until the episode terminates |
| *End of Epoch:* |
| Repeat steps 3–9 until the specified number of epochs $N$ is reached |

To compare the status of current policy with the expected value, the advantage function is used to update the policy gradient. Figure 2 is sequence of actions in reinforcement learning technique called an actor-critic algorithm that is used to concurrently optimize a method and estimate its score. Allocating computational resources (such as virtual machine) to jobs in this method will minimize total running time and maximize asset utilization by adhering to specific limitations which is the scope of computational heavy load balancing in cloud computing.

*Algorithm can be understood as follows:*

This algorithm is a variant of the actor-critic method in reinforcement learning. The actor network selects actions based on a policy, while the critic network evaluates those actions to provide feedback. The TD error helps to update the critic network, ensuring it accurately predicts future rewards. The actor network updates its policy to maximize the expected reward, considering the feedback from the critic network. The Twin Delayed bug factor ($\beta$\beta$\beta$) helps to balance immediate and future rewards in the TD error calculation. This approach iteratively improves both networks to achieve better performance in the given environment.

*Strategy updates 2*

$$\forall\varphi = \omega\sigma\forall\_(\mathrm{lkgk}, \varphi) \tag{2}$$

adjusts the policy weights ($\varphi$) of the actor in accordance with the policy $\pi$, taking into account, likelihood of choosing action pat in state $k\_t$ and the advantage estimate ($\tau$).

*Assessment function update 3*

$$\forall \mathrm{Z}(\mathrm{pat}) = \beta(\delta\_t - V(\mathrm{pat})) \tag{3}$$

updates ($\delta\_t$) and ($\alpha$) to update the value function ($Z$) of critic for a state pat.

*Total reward 4*

$$\text{Rx} = \sum \text{arir\_ib} \tag{4}$$

The final reduced reward acquired over a series of action steps is calculated using this equation, in which I is the discount factor and ri is the instantaneous reward received at $\gamma$.

*Probability of acceptance 5*

$$\text{Probx} = (1(1 + \exp(-f))) \tag{5}$$

Calculates the likelihood that a virtual machine instance will accept a task assignment, where $x$ is the sum of attributes that indicate how well-suited the sample value is for the task.

*Resource utilization 6*

$$Z = \sum a/(m * Z\_\text{max})\text{ab} - 1 \tag{6}$$

The resource usage in a cloud system is calculated using this equation, where b is available resources, $Z\_max$ is the highest resource consumption, and $a\_i$ is the usage of asset $I$.

## 4 Simulation Setup in Cloud Environment

A cloud environment is created and configured using Cloudsim, so that the effectiveness of the recommended approach can be evaluated there. The recommended deep reinforcement learning method is employed in a cloud environment created by Cloudsim and is implemented in Python. We used Google's task events dataset to create user tasks. For our simulations, we used four virtual machines, each with a terabyte of storage and 16 GB of RAM. A virtual computer needs anything from 5 to 100 GB of storage space in order to complete a task. From the Google Task dataset, nine user tasks are generated. The first set of charges is scheduled at random, and the performance of each virtual system is assessed so that RL can explore and learn about its environment. Each virtual machine's average processing speed is calculated in meGenetic Algorithmbytes per second using the simulation results that are acquired when random scheduling is implemented. We separated the prizes we could give them into four categories using four virtual machines. If the computation is carried out on the strongest virtual machine, it is worth + 2. When the assignment is given to the next best virtual system, the action gets a plus one. If the worker's job is assigned to minimum productive VM, the operation will be penalized by a factor of − 2. If the action selects the virtual system with the next fastest processing speed, an A-1penalty is imposed. A VM's effort is rewarded in the suggested deep RL-based scheduling system. Virtual machines are thus provided with incentives. Since the choice to employ a virtual machine is linked with rewards and penalties rather than the task itself, the scheduling process is non-preemptive.

## 5 Simulation Results

The performance of various methods for a few tasks pertaining to resource allocation in a cloud setup is compared in Table 1. Latent period, CPU usage, output, and job end time are the four metrics shown in the table. Here some of the algorithms are compared with the proposed method. According to the table's figures, it appears that DSOS and GENETIC ALGORITHM have the fastest reaction times, with GENETIC ALGORITHM having the longest downtime at 240.0 ms. In terms of CPU utilization, the highest value is 0.380 for WOA, and the highest throughput numbers are 0.327 and 0.380 for PSO and DQL. The majority of algorithms have comparatively short task completion times, with third method (in the table) having the smallest value at 4/100. With a reaction in 0.51 s, CPU usage of 334 ms, output of 31 ms, and job end time of 36 ms, DWAS performs reasonably well when compared to the other algorithms in Table 1.

Figure 4 represents the various parameters like algorithm task completion, time (ms) and throughput for various methods like genetic algorithm, DSOSA, MSDEA, PSOA, WOWA, DQLA, and DWAS. From the above graph it is clear that the algorithm proposed is optimal among others.

## 6 Conclusion

By efficiently allocating resources and distributing workloads over several servers, the suggested approach seeks to reduce the overall execution time of processes. We'll be expanding the suggested model soon and incorporating a precedence order in combination. The state score, action score, and reward function need to be reconsidered in view of the relative importance of jobs. Methods can be experienced to understand allotment principles that take into account the jobs' precedence ranking by studying things like how high-priority careers alter their rewards. The suggested agent outperforms all other current algorithms in terms of performance. DSOS and MSDE may perform better since, as the amount of data increases, their search areas get quite large, consuming a lot of CPU power and significantly degrading the performance of other algorithms. The performance is marginally better than throughput and task completion when the quantized levels are increased, but not as much as with the suggested DWAS. The brain finds it difficult to interpret actions and rewards as a result of the noise produced by the quantization process. DWAS determines the optimal path of action. The results of the suggested model are 23% better than

**Table 1** Performance metrics for various algorithms

| Algorithm | Latent period | CPU usage | Output | Job end time |
|---|---|---|---|---|
| Genetic Alg | 240 | 0.051 | 0.067 | 0.064 |
| DSOSA | 241 | 0.042 | 0.054 | 0.051 |
| MSDEA | 310 | 0.041 | 0.048 | 0.040 |
| PSOA | 340 | 0.370 | 0.380 | 0.045 |
| WOWA | 410 | 0.380 | 0.380 | 0.450 |
| DQLA | 461 | 0.341 | 0.327 | 0.384 |
| DWAS | 410 | 0.334 | 0.316 | 0.360 |

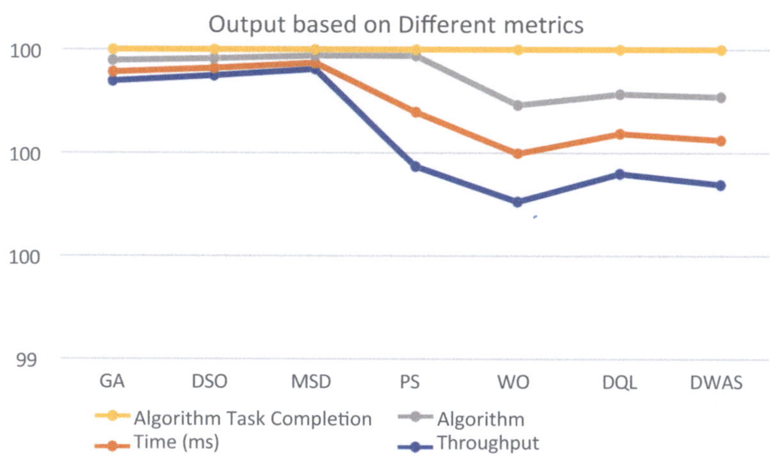


**Fig. 4** Output based on different parameters

those of previous studies. In terms of indexing LB values, DQL outperforms another method by 15%. The task completion time is on average succinct, and each algorithm scores employed in this experiment revealed that the reaction time could raise not > 10%. Throughput had to decrease by no more than 23% in comparison with other algorithms. Finally, DWAS learning reduces the usage of the CPU from 38 to 12%, although all other methods can utilize up to 46% of it. In order to facilitate collaboration on computing activities, we want to establish an edge cloud computing network system in our future work. We will also examine ways to simplify the computation and communication aspects of the training process. Instead of using local training data, we will attempt to employ merged learning-based reinforcement learning, which just requires real data that comes in to the data corner in order to disclose sample arguments.

## References


1. Al-Habob AA, Dobre OA, Armada AG, Muhaidat S (2020) Task scheduling for mobile edge computing using genetic algorithm and conflict graphs. IEEE Trans Veh Technol 69(8):8805–8819
2. Mohemmed TK et al (2022) Power allocation model for residential homes using AI-based IoT. Meas Sens 24:100461
3. Ferratti GM, Sacomano Neto M, Candido SEA (2021) Controversies in an information technology startup: a critical actor-network analysis of the entrepreneurial process. Technol Soc 66:101623. https://doi.org/10.1016/J.TECHSOC.2021.101623
4. Li C, Zheng P, Yin Y, Wang B, Wang L (2023) Deep reinforcement learning in smart manufacturing: a review and prospects. CIRP J Manuf Sci Technol 40:75–101. https://doi.org/10.1016/J.CIRPJ.2022.11.003
5. Sathyanarayana et al S (2022) Secure data sharing in cloud computing using revokable blockchain ıdentity encrytion (RIE). Purakala 31(2):1395–1403
6. Sivasubramanian R et al (2019) Systematic Grid Reallocation Through Reliable λ Process", IEEE Trans. 978-1-4673-9223-5/15
7. Sivasubramanian R et al (2018) Robust routing strategy for efficient reallocation in multi-grid environment. Int J Eng Technol 7 3(12):7086–7090
8. Sun J, Zhu Z, Li Y, Chai H, Qi G, Wang H, Hu YH (2019) An integrated critic-actor neural network for reinforcement learning with application of DERs control in grid frequency regulation. Int J Electr Power Energy Syst 111 286 299. https://doi.org/10.1016/J.IJEPES.2019.04.011
9. Wang X, Zhang L, Liu Y, Zhao C, Wang K (2022) Solving task scheduling problems in cloud manufacturing via attention mechanism and deep reinforcement learning. J Manuf Syst 65:452–468. https://doi.org/10.1016/j.jmsy.2022.08.013
10. Zhang W, Chen Q, Yan J, Zhang S, Xu J (2021) A novel asynchronous deep reinforcement learning model with adaptive early forecasting method and reward incentive mechanism for short-term load forecasting. Energy 236:121492. https://doi.org/10.1016/J.ENERGY.2021.121492
11. Zhou L, Zhang L, Ren L, Wang J (2019) Real-time scheduling of cloud manufacturing services based on dynamic data-driven simulation. IEEE Trans Ind Inform 15(9):5042–5051. https://doi.org/10.1109/TII.2019.2894111
12. Fernandez-Gauna B, Graña M, Osa-Amilibia JL, Larrucea X (2022) Actor-critic continuous state reinforcement learning for wind-turbine control robust optimisation. Inf Sci 591:365–380. https://doi.org/10.1016/J.INS.2022.01.047

13. Serrano-Guerrero X, Briceño-León M, Clairand JM, Escrivá-Escrivá G (2021) A new interval prediction methodology for short-term electric load forecasting based on pattern recognition. Appl Energy 297. https://doi.org/10.1016/j.apenergy.2021.117173

# IoT-Based Secure Health Monitoring System for Neonatal Incubators

**M. Munisankar and T. Ramashri**

**Abstract** Newborn babies who are in incubators require continuous monitoring since they cannot express their pain. Because of the lack of a continuous monitoring system for neonatal incubators, there is no information about the child's health status, which creates havoc for medical staff and treatment will be delayed. Every year, about 20 million unseasonable and low birth weight infants are born in underdeveloped nations, with roughly 4 million of them being bones. These fatalities are mostly caused by the availability of incubators or the unreliability of conventional incubators. Also, while telemedicine has emerged in pastoral regions, there is a shortage of healthcare personnel, making it unsuitable for introduction healthcare. As a result, traditional incubators and corrective approaches require telemedicine contact and feedback. From existing models, only a few parameters are monitored, which does not fully fulfill the secure functioning of an incubator. In the present paper, a secure model with additional monitoring parameters is designed. In this paper, an intelligent monitoring system for neonatal incubators that includes multicomponent tracking and real-time data using various sensors is integrated into a microcontroller. The baby's health and the surroundings in the incubator are regularly monitored. Medical staff receives information about the patient's condition through the ThingSpeak cloud, web server, Pushover application, and Excel spreadsheets, which they can use to process and analyze the child's current condition. Parents also benefit from the ability to monitor their children remotely.

**Keywords** Arduino UNO · ESP8266 · ThingSpeak · Heartbeat monitoring · $SpO_2$ · Ambient temperature and humidity · Body temperature · Neonatal incubator

M. Munisankar (✉) · T. Ramashri
Department of Electronics and Communication Engineering, SV University, Tirupati, India
e-mail: munisankar404@gmail.com



K. R. Madhavi et al. (eds.), *Proceedings of Sixth International Conference on Computer and Communication Technologies*, Lecture Notes in Networks and Systems 1356,
https://doi.org/10.1007/978-981-96-5238-9_2

# 1 Introduction

The proposed infant incubator is a device consisting of a rigid, box-shaped housing that allows one to keep a newborn baby in a regulated environment while it receives treatment [1]. In the newborn incubator, the temperature, relative humidity, and oxygen concentration remain constant. The incubators are used in conjunction with other equipment and methods to ensure the best possible conditions and continuous monitoring of children in need of special care. A baby may need to stay in an incubator for various reasons including premature birth, breathing problems, infections, jaundice, low birth weight, etc.

# 2 Literature Review

Generally, the condition that occurs in a premature baby when it stops breathing or its heart rate decreases is called Apnea. The heart rate monitor proposed is an infrared sensor that persistently tallies the number of beats over a period of one minute and sends this data to the controller and after that to the cloud [2]. If the newborn's body temperature is not regulated, the risk of hypothermia is high [3]. Premature babies have high skin permeability, which is why it is so important to maintain an adequate temperature. The data collected by IoT devices can help doctors determine the best treatment process for newborns and achieve the expected results. It is important to consider not only temperature, humidity, heart rate, etc., but also other factors that may affect the child's health, such as the factor of skin contact with family members/ guests and staff, which directly contributes to the cause of illness and irregularities. The baby's temperature, because its skin, is even more sensitive and can become inflamed. The goal of using an incubator is to create a very stable environment in which a low-weight infant can maintain a constant body temperature. When building the incubator, the humidity in the room must be taken into account because the temperature sensitivity in the room is relatively simple. The trans epidermal weight loss (TEWL) occurs in a newborn when optimal humidity is not maintained. In this paper, a new IoT-based secure health monitoring system for neonatal incubators is proposed.

This paper proposed an effective system for monitoring newborns in incubators. Here are some of the report's key contributions:

- The central real-time monitoring system for premature babies provides real-time data on newborn skin temperature and box temperature, ensures the consistency and quality of recorded temperature data, and provides clinical advice on diagnosis and treatment to doctors.
- Maintains normal humidity and temperature in the incubator to provide newborns with a properly controlled environment.

- Medical personnel can remotely monitor the physical condition of the newborn in the incubator, reducing the risk of injury caused by the abnormal physical condition of the newborn through the effective LED system, which displays different alarm conditions depending on the premature baby and the normal condition of the child.
- Parents of premature babies can also remotely monitor their baby's progress in real-time, reducing stress, minimizing the frequency of NICU visits, and improving the quality of the NICU environment.

Continuously monitoring children's health parameters in real time, triggering the LED system for situational evaluation, and sending emergency alerts when needed.

The system collects and stores real-time health data, enabling better tracking and analysis of a child's well-being. Tracked parameters can be recorded and represented in Excel spreadsheets for detailed review, while cloud storage ensures secure long-term data retention for medical evaluations. A web server provides live monitoring with dashboard visualization, and integrated Pushover notifications deliver instant alerts in critical situations.

Share this data with your child's parents: An effective visual display of the data is easily accessible to parents providing real-time data about their children and allowing them to monitor their child's health.

## 3 System Development and Working Principle

The basic concept of the proposed system is a continuous health monitoring system of the newborn incubator. The three main features of our system are data storage and processing. Mentioned above, the Arduino Uno collects numerous sensor signals, including $SpO_2$, heart rate, room temperature, humidity, and body temperature and tries to restore important features to detect and identify potential heart diseases in a newborn and to combat various diseases and emergencies that may occur due to fluctuations in any of the parameters more drastic than the safe limit. Accurate identification and diagnosis of potential diseases and threats often require specific historical data. Therefore, a cloud database like ThingSpeak cloud is created to graphically store the sensor node's sensor data for each user. The Excel spreadsheet also stores data for the required period. A web server-based data visualization system has been implemented using Python. In particular, a web GUI is created for only accessed clients to get to the information in real time. Then the real-time data of heart rate, $SpO_2$, ambient temperature, humidity, and body temperature [4] is displayed. The same data is uploaded to the ThingSpeak cloud and displayed on the website in real time, powered by the Python-based Tkinter package [5].

Disease Identification and Reporting: Our solution is designed to provide a user-friendly data access interface. Figures 1 and 2 are the block diagram and sensor node diagram of this model. Whenever the baby is placed in the incubator, the Arduino Uno collects the child's heart rate, body temperature, etc. and sends the data to the server.

The system is designed for three main user groups: hospital medical staff (doctors and nurses), family members, and affiliated healthcare workers, all of whom have access to the server for relevant data. Minor fluctuations in heart rate, body temperature, and other vital signs can pose serious risks to a child's health. This system is designed to safeguard children by allowing real-time monitoring and analysis of their health data, ensuring that timely interventions can be made when necessary. Therefore, it is important that the patient's health status can be monitored and understood. The Pushover app allows you to identify suspicious and abnormal sensor readings and send notifications to identified users such as family members and doctors. Parents can use this information to monitor their children's health, and doctors can use it to diagnose possible illnesses. If an unusual condition is detected, an alert is sent via the app to the relevant affected party, such as doctors or family members.

The integrated comprehensive LED alarm system makes it easy to understand the scenario and take appropriate next steps. Figures 2 and 3 show the block diagram and developed sensor node and incubator box. Below you will find a short description of our system:

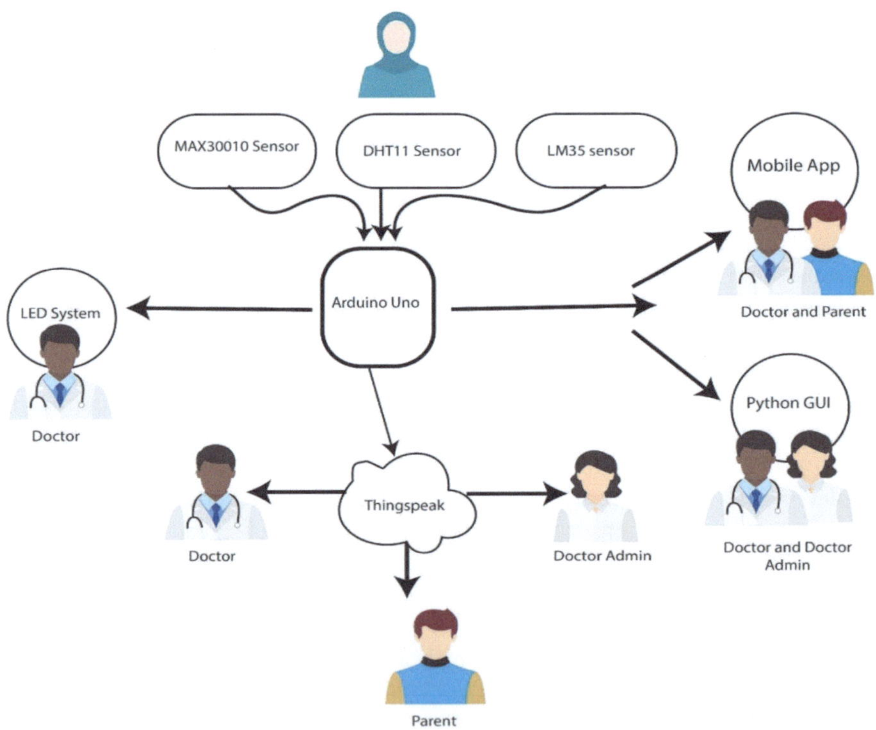


**Fig. 1** Sensor node diagram

**Fig. 2** Developed sensor node and incubator box

## *3.1 Sensors*

DHT11 to monitor the appropriate environment for the newborn: Low humidity raises incubator temperature and oxygen demand for premature babies [4]. We need to maintain a steady temperature in a reasonably compact region without damaging the infant in the incubator [1]. Lower humidity leads to problems such as hypothermia and dehydration in newborns. When humidity is low, there is a rise in imperceptible water loss, which increases heat evaporation and moisture loss. Relatively high humidity levels are also not suitable for children as they increase the risk of germs and bacteria multiplying [1]. The website contains information about the temperature, number of pulses, and all other parameters measured by the sensors instantly in real-time. ThingSpeak stores data and generates charts that can be viewed on your computer or mobile app for better data visualization. The data is analyzed by the LED system. This includes determining the best range of safe parameters to detect medical emergencies and predict possible infections/diseases that could pose a danger to the child. The Pushover app gives you timely notifications so you can prevent deaths and provide priority care.

The upgrade includes a total of five parameters by integrating both digital and analog sensors: a $SpO_2$ sensor for measuring blood oxygen levels and a DHT11 sensor for monitoring ambient temperature. This enhances health monitoring and environmental control capabilities. A Python-based GUI web server is integrated to display real-time data values and facilitate rapid diagnostics, streamlining decision-making for medical staff. It has separate conditions and loops for both mature and premature babies and immediately alarms the medical facilities in case of any danger when the parameters exceed their usual safe limits [2]. In the case of premature babies, the conditions have been made more sensitive as minor aberrations in the parameters may harm the baby and could lead to fatal consequences.

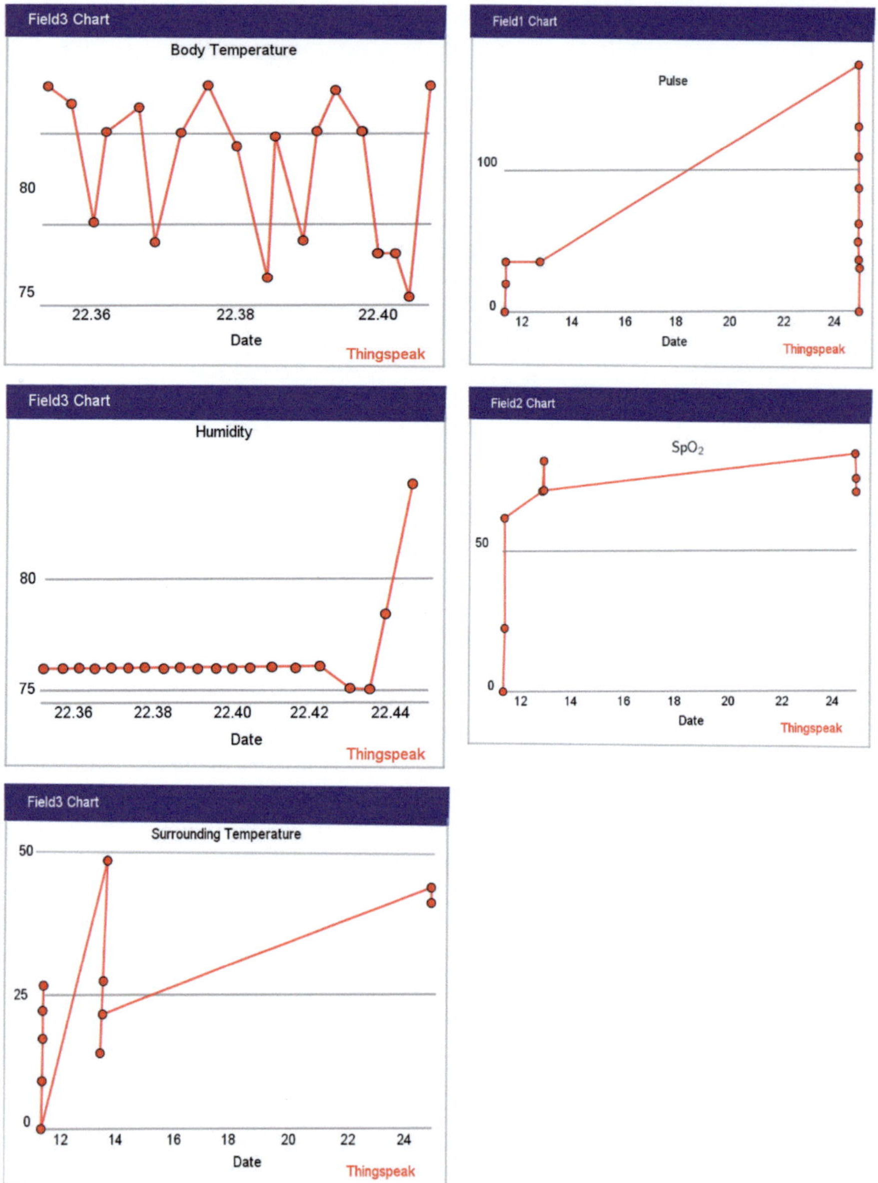


**Fig. 3** Snapshot of the ThingSpeak channel for data visualization: heart rate (pulse), $SpO_2$, ambient temperature, ambient humidity, and body temperature

# 4 Results and Discussion

Initially, the hardware was implemented for heart rate and $SpO_2$ monitoring, body temperature monitoring, and ambient monitoring on an individual basis. The collected results for each bodily health metric were studied by modifying behavioral or environmental conditions to better understand how the sensors perform in response to certain changes. The analysis of the signals acquired and how they were integrated to transfer the data to the cloud storage for graphical representation on the Android application.

## 4.1 ThingSpeak

We use ThingSpeak to be able to see any sudden changes in the patient's health status quickly and graphically. It allows only authorized persons to access data about infants. As a result, information is secure and cannot be misplaced since it is always upgraded to a cloud. A graphical representation of the patient's information which includes heartbeat, body temperature, ECG, body position, EMG and $SpO_2$, is shown in Fig. 4, which can be viewed by logging onto the ThingSpeak server using a personal computer.

Figure 3 shows a screenshot of the continually acquired data in the ThingSpeak. Any medical staff member with an internet connection can simply monitor a specific patient as well as the ambient state in which the infant is now present. The patient data is secured since accessing it requires passing employing a password-protected framework, which implies that as it were authorized faculty and guardians may screen it.

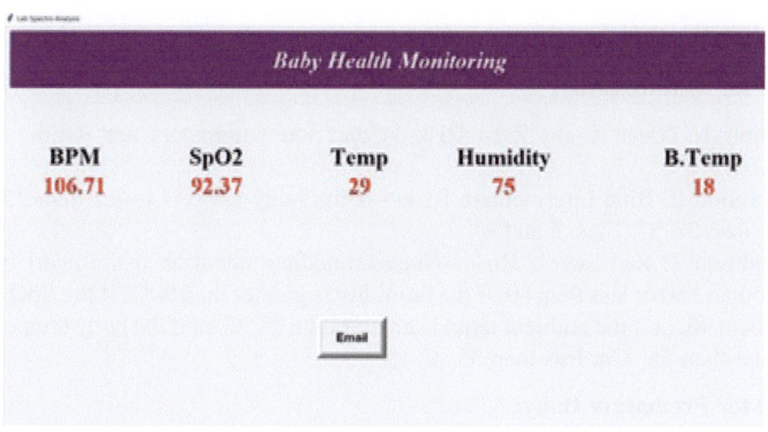


**Fig. 4** Python-based GUI which makes real-time updates

### 4.2 Web Server Using Python-Based GUI

We used a Python package named Tkinter to create a web server that displays the instantaneous data of each parameter, on a real-time basis [6]. This web server contains the 'email' button which sends an Excel sheet of the stored data to the desired personnel for further analysis.

### 4.3 Excel Sheet

As demonstrated in the accompanying figure, an Excel sheet includes real-time data on body temperature, ambient temperature and humidity, heart rate, and $SpO_2$ [1]. Furthermore, doctors can prescribe for a patient by easily monitoring his health status trends and evaluating past-present recordings. This is a one-of-a-kind function that allows the doctor to keep track of the baby's progress and assess it over a set of times.

### 4.4 LED System

The system has two modes, one for a premature baby and the other for a normal baby. The Arduino-based system uses pin D12 as a mode selector with two operational states:

Case 1 (Mature infant mode): Activated when D12 is left floating (no connection) or reads HIGH (with internal pullup enabled).

Case 2 (Premature infant mode): Triggered when D12 is pulled LOW (e.g., via a resistor to GND)

**Case 1 for Mature Baby:**

Condition 1: Green Light Zero Risk—Condition parameters are within a safe threshold.

Condition 2: Blue Intermediate Risk—If the body temp is lesser than 35°C or greater than 38 °C (Figs. 5 and 6).

Condition 3: Red Severe Risk—Need immediate attention if the heart bpm is greater than 150 or less than 80, if the humidity is greater than 90%, if the $SpO_2$ level is less than 90, or if the ambient temp is greater than 38 °C, or if the body temperature is greater than 38 °Cor less than 35 °C

**Case 2 for Premature Baby:**

Glowing of an orange LED continuously indicates the presence of a premature baby in the incubator.

**Fig. 5** LED system for mature baby indicating emergency

**Fig. 6** LED system for premature baby indicating emergency

Condition 1: Green Light Zero—Risk condition parameters are within the safe threshold.

Condition 2: Red Severe Risk—Need immediate attention For a premature baby, any significant change from normal health conditions can be fatal; therefore, we have removed the intermediate level of alert and added only a second emergency level. The threshold therefore for this is narrower than mature baby (if the heart bpm is greater than 150 or less than 80, or if the humidity is greater than 90%, or if the $SpO_2$ level is less than 90, or if the ambient temp is greater than 38 °C Celsius, or if the body temperature is greater than 38 °C or less than 35 °C).

If continuous monitoring is needed, the device must be attached to the baby's hand and Ardinuo must be on 24/7 for the baby's temperature, and for also surrounding temperature, the child may feel suffocated as it is attached to the baby's hand and the network must be there. The website must have access to the hospital staff and parents only.

## Discussion

The previous existing methods to check the neonatal babies condition who are in incubator are Bluetooth, GSM, and RF which have some limitations but the proposed method in this paper could overcome those limitations; the comparison is written in the following Table 1

**Table 1** Difference between existing methods and proposed method

| Bluetooth, GSM, RF methods | Proposed method |
|---|---|
| 1. Range of access is fixed, for Bluetooth distance of 20 m at a frequency of 2.5 GH, GSM only message sending is possible at a frequency of 900 MHz to 2.4 Ghz, and RF is only possible for a particular frequency; it can send information for few kilo meters only | 1. The range of access is not fixed in this method so the access can be from anywhere at any frequency via app connected through WiFi or mobile data |
| 2. As the range of access is fixed, only appointed doctors were able to access and no other doctor or parents can be able access at a time | 2. As the range of access is not fixed, multiple users can be accessed at the same time |
| 3. High risk level is not indicated unless an appointed doctor is within the range of access. It may take some time to identify risk and start the immediate treatment | 3. High risk level can be shown for multiple users who are connected so with in the short span of time, the risk is identified and can start the immediate treatment |

## 5 Conclusion

The proposed system keeps track of the infant's heart rate as well as the temperature and humidity of the environment. Temperature monitoring is carried out to ensure that the neonate's surroundings are appropriate. As the body temperature rises, temperature monitoring will aid in the detection of many other internal disorders such as infections, the common cold, and pneumonia, which all have a similar sign of fever. The system combines numerous functions into a single device, analyzes data from sensors, and interacts with the device according to circumstances specified by conventional clinical research. The led system in this method also helps to identify the risk level and these systems indicators are also connected to the app. The led system is divided into two cases for mature and premature babies, based on this led system the cause is identified and accurate treatment is done for the babies as the hospital staff, appointed doctors and parents are connected via app so it would be easier to check the baby's condition continuously. It can also aid parents, especially those in rural areas, by reducing their number of visits to the hospital and stress by being timely updated with their child's health progress.

## References

1. Chuenpirom T, Srirussamee K, Pintavirooj C (2022) Remote temperature monitoring for infant incubator using thermal camera. In: 2022 14th biomedical engineering international conference (BMEiCON), 2022
2. Koli M, Ladge P, Prasad B, Boria R, Balur NJ (2018) Intelligent Baby Incubator. In: 2018 Second international conference on electronics, communication and aerospace technology (ICECA), 2018
3. Shalannanda W, Zakia I, Sutanto E, Fahmi F (2020) Design of hardware module of IoT-based infant incubator monitoring system. In: Proceedings of 2020 6th International Conference on wireless and Telematics (ICWT). IEEE, pp 1–6
4. Pradeep Kumar S, Samson VRR, Bharath Sai U, Malleswara Rao PLSD, Kedar Eswar K (2017) Smart health monitoring system of a patient through IoT. In: 2017 International conference on I-SMAC (IoT in social, mobile, analytics and cloud) (I-SMAC), 2017
5. Lamidi L, Hamzah T, Triwiyanto T, Nuristadarro N, Elfredo D (2022) Baby incubator monitoring center using wi-fi network for data transmission. J Biomimetics Biomater Biomed Eng 55:275–287
6. Shahed MT, Azim Meem AF, Habib MS, Se, GP, Hossain MS, Uddin MS (2023) An IoT enabled cost effective smart healthcare for real-time COVID- 19 patient early identification and monitoring system using smartphone. Int J Sens Wirel Commun Control 13(2):71–81

# Threat Analysis Using Social Media Patterns

**Kartik Palsetty and Pranali Dhawas**

**Abstract** Threat analysis is the one of the crucial issue arising now days. This study digs into social media threats analysis specifically focuses on twitter data. It includes threat identification, network analysis, sentimental analysis to classify threats into the categories and its evaluation. Sentimental analysis is generally use to access the emotions behind tweets while threat is use to identify the harmfulness of tweet. Whereas network analysis is use to find the spread of tweet with its location and origin. This study provides a complete evaluation of threat severity by combining outcomes of sentiment, network, and threat findings of social media tweets. This investigation concludes the categorization of threats for preemptive risk management with visual representations that enables recognition for categorization.

**Keywords** Social media threat analysis · Twitter data analysis · Sentiment analysis · Threat detection · Network analysis · Threat severity assessment

## 1 Introduction

As social media has become integral to daily life, it has also introduced various dangers, such as cyberbullying, hate speech, misinformation, and cyber-attacks. The study aims to create a comprehensive framework for analyzing social media threats through three main components: sentiment analysis, threat detection, and network analysis. Sentiment analysis evaluates the emotional tone of tweets, threat detection identifies harmful content, and network analysis examines the spread and impact of these threats. The seeks to combine insights from these areas to assess threat severity and develop strategies for risk mitigation and online security enhancement. It also addresses ethical concerns, especially regarding data privacy, and acknowledges the need for an adaptable and scalable framework due to the evolving nature of social media. This study is grounded in practical application, using real-world Twitter data

K. Palsetty · P. Dhawas (✉)
G H Raisoni College of Engineering, Nagpur, India
e-mail: pranalidhawas10@gmail.com

 
K. R. Madhavi et al. (eds.), *Proceedings of Sixth International Conference on Computer and Communication Technologies*, Lecture Notes in Networks and Systems 1356,
https://doi.org/10.1007/978-981-96-5238-9_3

to test the framework's effectiveness. The ultimate goal is to contribute to a safer digital environment, promoting a healthier online ecosystem.

This study focused on the analysis the twitter data to predict the online threats. The task was divided into three parts: first task is to use sentiment analysis to evaluate the emotional tone of tweets and understand user sentiment on social media. The second task involves refining methods for detecting harmful content that could pose threats to individuals and organizations. The third task utilizes network analysis to study how these threats spread within the platform, assessing their influence and reach. In the conclusion, research aims to develop a methodology for evaluating threat severity by integrating findings from sentiment analysis, threat detection, and network analysis to provide a comprehensive view of social media threats.

## 2 Related Work

Crisis management, public health communication, and threat detection have been extensively studied by using social media which imitates its growing implication in the digital age. Exploration in this field highlights the position of leveraging social media platforms to spread timely and precise information, as well as to sense emerging risks.

Discovers the dissemination [1] of health information on social media during the Zika outbreak, revealing an information gap between public concerns and the posts provided by health authorities. This study underscores the need for better alignment in hazard communication to address the public's information needs effectively. It was examined [2] the initial detection of evolving suicide risks using social media data. This literature is based on the Blue Whale Challenge which demonstrates the possible of social media to provide early warnings, allowing extra rapid public health replies to online threats. In [3] analyzes natural disasters from social media usage during natural disasters to discover the communication patterns on Twitter vary depending on the various types of disaster. Their study suggests that disaster communication approaches should be personalized to the specific context and behavior of social media users. Analyzed crime rate [4] calculations using social media data which is predominantly focused on violent crime. According to their study, crime hot spots, challenging the traditional reliance can be shifted on residential population data by using social media as a substitution for mobile populations. Major consequences for this findings are crime prevention and source sharing. This signifies that more dynamic data origins should be merged into crime analysis. It was specified the detection of threats [5] in social media content. They proposed a model that extract triggered keywords and contextual cues spontaneously rank the threat possibilities of tweets. This study uses a Dutch tweets which proves effective in classifying threatening content, and enhanced the effectiveness of security monitoring efforts.

As above literature mentioned various aspects like disaster management, security, and threat and crime detection, public health like suicide risks by using social media data. It sounds for improved communication strategies, the usage of progressive data analytics, and the development of automated tools to monitor and respond to emerging risks in real-time.

## 3 Text Analysis Literature Using NLP

The management and the understanding of large amount of data, in any form like structured or unstructured twitter data is extensively analyzed using natural language processing in this study. NLP combines language processing, computer science, and artificial intelligence to enable computers to interpret human languages, with applications such as Google Translate and Grammarly [6]. The study focuses on sentiment analysis of Twitter data to extract public opinion from tweets.

Data extraction is performed through the Twitter API, requiring specific credentials. The extracted data undergoes exploratory data analysis to address class imbalances in sentiment categories (positive, negative, neutral) using up-scaling and downscaling techniques [9]. A training model is then developed for sentiment analysis. Preparing tweets for analysis involves several data pre-processing [7] and cleaning steps: creating a corpus, converting text to lowercase, removing stop words and punctuation, excluding URLs and usernames, trimming spaces, and applying text stemming. These steps ensure efficient processing and focus solely on textual content for prediction. Consistency is achieved by converting all text to lowercase, removing stop words and punctuation, and applying stemming to reduce words to their root forms, with care taken to avoid over-stemming. Table 2 illustrates the stemming process (Tables 1).

This paper evaluates various stemming algorithms, including Potter's, Lovins, Dawson, and NGram, ultimately choosing Potter's stemming algorithm. For analyzing word frequency in tweets, it employs a Document Term Matrix (DTM), which represents tweets as rows and words as columns, with each cell indicating the frequency of a word in a tweet. To enhance model training efficiency, the paper

**Table 1** Example of a corpus of tweets

| Corpus row id | Corpus text |
|---|---|
| 1. cb774db0d1 | Sooo SAD |
| 2. 088c60f138 | Bullying me |
| 3. 9642c003ef | Leave me alone |

**Table 2** Example of stemming

| Words | Root word |
|---|---|
| Consult, consultant, consulting, consulted, consultancy | Consult |
| Likes, likely, liked | Like |

**Table 3** Confusion matrix

| | Class negative (actual) | Class positive (actual) |
|---|---|---|
| Class negative (Predicted) | TN | FN |
| Class positive (Predicted) | FP | TP |

introduces a Sparse Document Term Matrix (SDTM) that excludes infrequent words, reducing the number of columns without losing accuracy.

The classification and regression tree (CART) is a decision tree algorithm which is particularly used for data classification. CART is specialized in handling the versatile task in classification and regressions. CART follows top-down approach with gini impurity matrix which helps to reduce classification errors and helps to efficient partitioning data.

The 'Area Under Curve (AUC)' of an ROC curve helps to measure the AUC score. AUC Score plots the true positive rates (TPR) with respect to false positive rate (FPR). True positive rates can be defined as TP with respect to TP & FN whereas false positive rate as FP with respect to FPP & TN. The AUC score ranges between 0 and 1, where 0 signifies inaccurate predication and 1 with accurate prediction (3).

The evaluation and the performance of the classifier model on test data with some known values is done by the very crucial tool confusion matrix. Here the predicted class values are arranged in roes and the actual class value are arranged in columns to form a confusion matrix, which includes true positive (TP), false positive (FP), true negative (TN), and false negative (FN). The comparison between the actual and the predicted values will depict the classifier model performance with the help of confusion matrix which will use to calculate accuracy and precision of the model. The term accuracy is the measurement of ratio between correctly predicted true value and true negative instances by the total predictions, whereas the precision is the ration of true positive prediction with the sum of true and false positive predictions.

$$\text{Accuracy} = (\text{TP} + \text{TN})/(\text{TP} + \text{FP} + \text{TN} + \text{FN})\text{and}$$

$$\text{Precision} = \text{TP}/(\text{TP} + \text{FP})$$

## 4 Data Source

The Twitter Developers Portal provides access to various datasets and tools via the Twitter API, which requires appropriate credentials and permissions. These datasets support diverse research purposes such as sentiment analysis, threat detection, and network analysis [8]. For this study, the dataset from the portal was fundamental, offering valuable insights into real-world Twitter interactions, thus serving as the cornerstone for the social media threat analysis research [10].

## 5 Methodology of the Experiment

The study encompasses six stages: Data collection from Twitter Developers, Preprocessing and data cleaning, Document-term matrix generation, Sparse Document Term Matrix (SDTM), Classifier model and Model evaluation [6]. The flowcharts depicting these stages can be observed in Fig. 1.

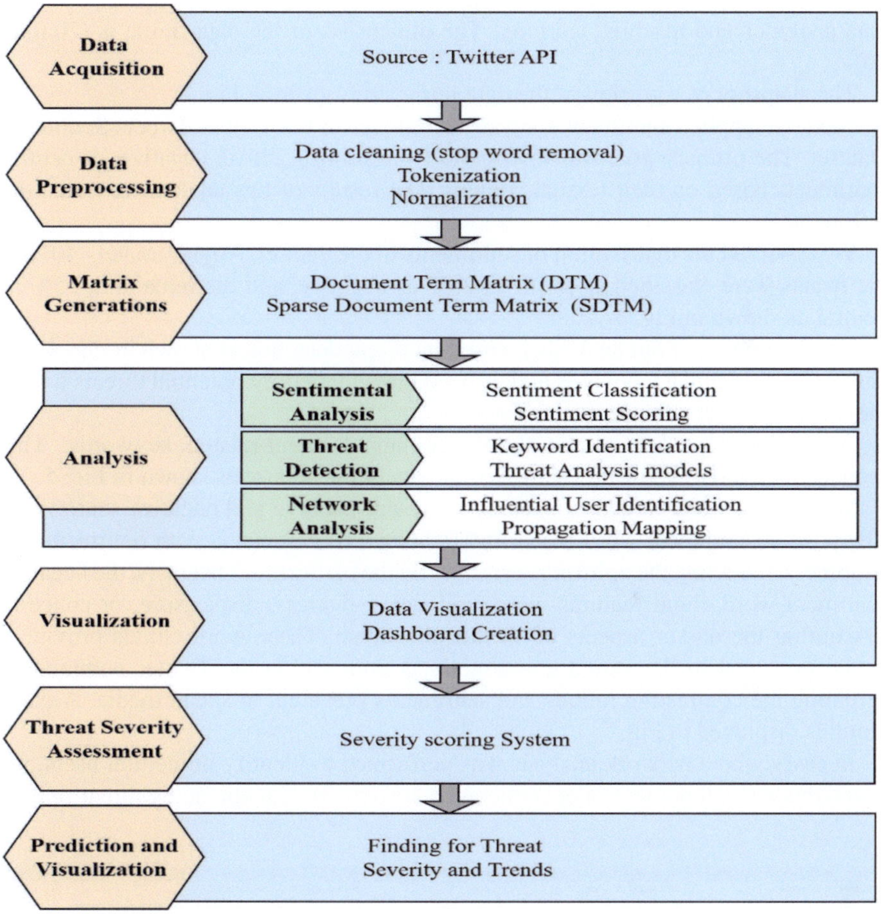


**Fig. 1** Methodology for threat analysis using social media patterns

## 6 Results and Discussion

The results and discussion section of the paper involves presenting and analyzing the findings obtained from the Twitter threat analysis with sentiment analysis. It aims to provide insights into the effectiveness of the threat detection system and the significance of sentiment analysis in identifying potential threats. The study described in this paper was carried out using the python programming language. The dataset used in the experiment was obtained from the Twitter developers portal for data analytics and machine learning. The dimension of the data-frame is (20,050, 26).

The snapshot of a portion of the data retrieved is given in Fig. 2:

Next, we applied sentiment analysis techniques to the Twitter Tweets Sentiment Dataset. The primary goal was to classify tweets into positive, negative, or neutral sentiments based on their textual content. The results of this analysis revealed the following.

We observed the distribution of sentiments in the dataset. Approximately 40% of the tweets were classified as positive, 25% as negative, and the remaining 35% as neutral as shown in Fig. 3.

The next phase of our analysis focused on threat detection as shown in Fig. 4. We defined a set of threat keywords and utilized them to identify potential threats within the dataset. The results were as follows.

Approximately 25% of the tweets contained threat-related keywords. This indicated a relatively low prevalence of threats in the dataset as shown in Fig. 5.

We created word clouds to visually represent positive and negative sentiments. The positive sentiment word cloud highlights words associated with optimism and happiness, reflecting the uplifting narratives on the platform. Conversely, the negative sentiment word cloud features words indicating distress, displeasure, or concern, illustrating the darker aspects of Twitter discourse. These word clouds provide a visual overview of the varied emotional spectrum within the Twitter community, revealing the contrasting moods and sentiments prevalent in social media. A word cloud is displayed in Fig. 5.

In cases where network analysis was performed to identify influential users, we found that a small number of accounts were responsible for spreading threat-related

| unit_id | golden | unit_state | trusted_judgments | last_judgment_at | gender | gender_confidence | profile_yn | profile_yn_confidence | created | description |
|---|---|---|---|---|---|---|---|---|---|---|
| 815719226 | FALSE | finalized | 3 | 2015-10-26T23:24:00 | male | 1 | TRUE | 1 | 2013-12-05T01:48:00 | i sing my own rhythm. |
| 815719227 | FALSE | finalized | 3 | 2015-10-26T23:30:00 | male | 1 | TRUE | 1 | 2012-10-01T13:51:00 | I'm the author of novels filled with family drama and romance. |
| 815719228 | FALSE | finalized | 3 | 2015-10-26T23:33:00 | male | 0.6625 | TRUE | 1 | 2014-11-28T11:30:00 | louis whining and squealing and all |
| 815719229 | FALSE | finalized | 3 | 2015-10-26T23:10:00 | male | 1 | TRUE | 1 | 2009-06-11T22:39:00 | Mobile guy. 49ers, Shazam, Google, Kleiner Perkins, Yahoo!, Sprint PCS, AirTouc |
| 815719230 | FALSE | finalized | 3 | 2015-10-27T01:15:00 | female | 1 | TRUE | 1 | 2014-04-16T13:23:00 | Ricky Wilson The Best FRONTMAN/Kaiser Chiefs The Best BAND Xxxx Thank yo |
| 815719231 | FALSE | finalized | 3 | 2015-10-27T01:47:00 | female | 1 | TRUE | 1 | 2010-03-11T18:14:00 | you don't know me. |
| 815719232 | FALSE | finalized | 3 | 2015-10-27T01:57:00 | brand | 1 | TRUE | 1 | 2008-04-24T13:03:00 | A global marketplace for images, videos and music. Sharing photos, inspiration, ( |
| 815719233 | FALSE | finalized | 3 | 2015-10-26T23:48:00 | male | 1 | TRUE | 1 | 2012-12-03T21:54:00 | The secret of getting ahead is getting started. |
| 815719234 | FALSE | finalized | 3 | 2015-10-27T01:52:00 | female | 1 | TRUE | 1 | 2015-09-08T04:50:00 | Pll Fan // Crazy about MCD // Ramen is bae |
| 815719235 | FALSE | finalized | 3 | 2015-10-27T01:49:00 | female | 1 | TRUE | 1 | 2011-05-13T03:32:00 | Renaissance art historian, University of Nottingham; fuelled by Haribo, partial to c |
| 815719236 | FALSE | finalized | 3 | 2015-10-26T23:17:00 | brand | 0.7002 | TRUE | 1 | 2011-11-16T17:14:00 | Clean food that tastes great while providing energy & nutrients! No guilt granola, \ |
| 815719237 | FALSE | finalized | 3 | 2015-10-26T22:33:00 | brand | 1 | TRUE | 1 | 2015-02-22T20:06:00 | highly extraordinary auctions |

**Fig. 2** Snapshot of the data

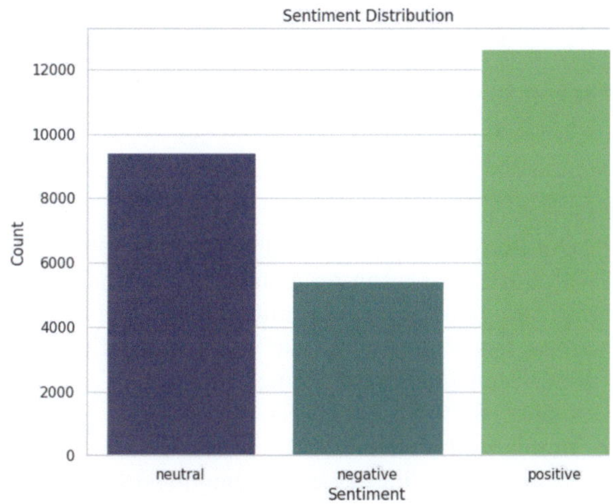


**Fig. 3** Sentiment distribution

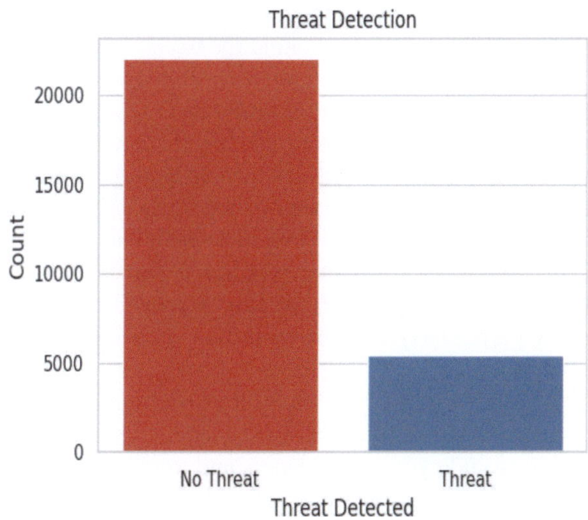


**Fig. 4** Threat detection

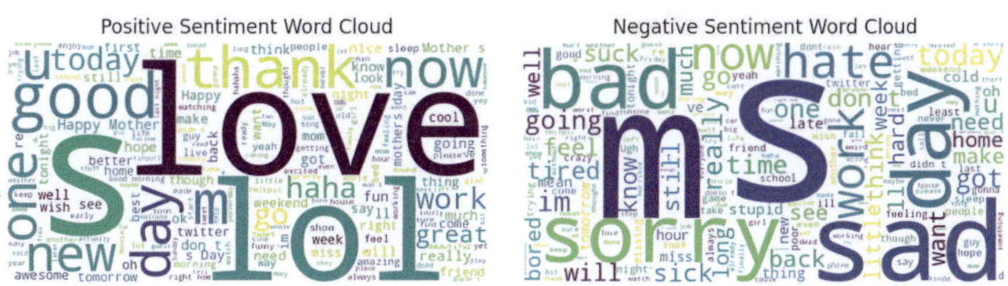


**Fig. 5** Word cloud for positive and negative sentiment words

```
Number of nodes (users): 20071
Number of edges (interactions): 619
```

**Fig. 6** Network analysis

content. Understanding the network dynamics helped us recognize potential sources of threats as shown in Fig. 6.

Our study utilized network analysis to investigate the complex interactions among Twitter users, focusing on identifying key individuals who significantly influence information dissemination, conversations, and trends. As per analysis of these interactions, this study found that significant influences which act a bridge between various segments of the Twitter community and facilitate the spread of ideas are getting holded by certain users.

In this study, we used 10,000 tweets for application of sentimental analysis to understand the various aspects of emotions of tweeter data. In this study, the sentiment score for threat analysis is ranges between − 1 and 1, where − 1 represents extreme negativity and 1 represents utmost positivity, which categorizes the diversity of the aspects of the emotion s behind the tweets of users.

As we analyses the large amount of dataset, this study exposed a very diverse sentiments among twitter users and there thoughtful deep insight into their collective mood. As above mentioned sentiment score range of − 1 to 1, this study offers a subtle view point toward the various aspects of the emotions behind the tweets of user for better understanding of the threat. The social media world relates with the influence which is very important for better understanding information dissemination and community formation. Figures 7 and 8 show the exploration of impact and specific users with their account, activities, within the platform ecosystem. This study particularly focuses on twitter data and especially the influence analysis of user activities.

Our network analysis identifies central nodes—highly influential individuals on Twitter—and examines how their content spreads. By analyzing influence scores, we pinpoint key opinion leaders and trendsetters, revealing the mechanisms behind information dissemination and threat propagation. Our research also investigates the temporal evolution of threat severity and sentiment scores within the Twitter dataset. Through longitudinal analysis, we gain insights into the dynamics of threats and sentiment changes over time, identifying patterns and trends that correlate with variations in sentiment.

This temporal perspective helps us unravel the intricate relationship between public sentiment and the emergence of potential threats. Our findings not only illuminate how the Twitter community responds to evolving events but also contribute to the development of predictive models that can forecast shifts in sentiment and impending threats, facilitating more proactive risk management in the realm of social media.

In our research, we implement a robust threat classification system that synergizes sentiment scores and threat severity scores to predict whether a given tweet poses a potential threat or not. This study developed an fine-tuned model that offers precise

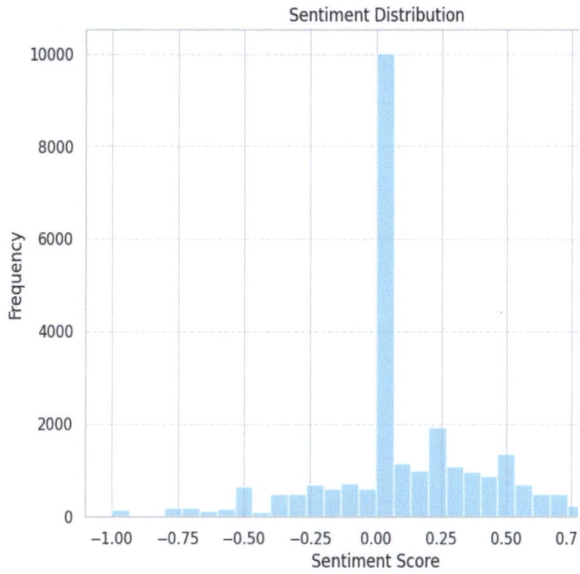


**Fig. 7** Sentiment score

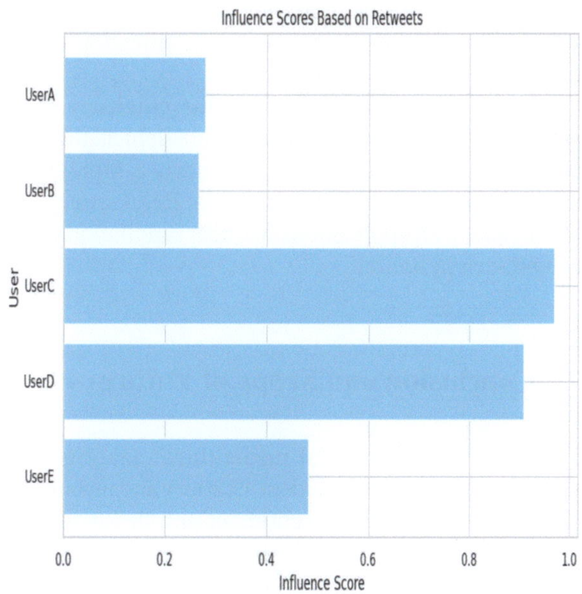


**Fig. 8** Influence score

threat classification by using leveraging machine learning algorithms and the amalgamation of these two critical metrics. The understanding of the emotional context behind the tweets id given by sentimental score, which allow us to measure the contents in categories like predominately negative or might warrant further scrutiny.

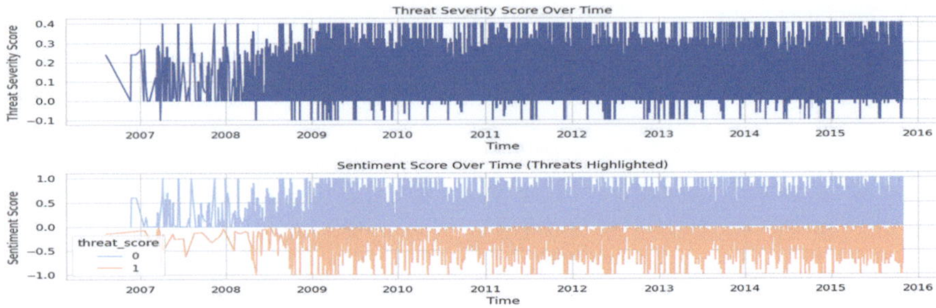


**Fig. 9** Threat severity score and sentiment score

Figure 9 depicts the threat severity score which accessed the potential harm implied by contextual indicators and keywords.

This study conducted using linear regression and random forest regression on threat analysis for twitter data. These machine leaning algorithms are used to predict the 'retweet_count' based on 'threat_severity_score' and 'sentiment_score,' to evaluated the relationship between tweets significance and its variables. The Random Forest Regression, an ensemble technique, was particularly useful in capturing complex patterns. We classified threats using a predefined threshold for 'threat_severity_score' and equated these classifications with the actual threats, visualizing the results in scatter plots.

Figures 10 and 11 showcased the efficiency of the models for identifying the potential threat in tweets. These figure depicts the significance of threat severity and sentiments in predicting threat significance, highlighting the welfares of collaborative methods, particularly Random Forest Regression, in management complicated real-world data, with applications in threat calculation and social media monitoring which can be seen in figures.

## 7 Conclusion and Scope of Future

This study explores social media threat analysis using Twitter data by combining sentiment analysis, threat identification, and machine learning regression techniques. The results indicate that sentiment and threat severity scores, along with machine learning models, particularly Random Forest Regression, are effective for assessing threats on Twitter. The research highlights the significance of incorporating emotional context and real-time assessment for better threat detection.

Future work could enhance threat severity scoring by adding more features and advanced machine learning algorithms, improving accuracy. Incorporating real-time data and dynamic sentiment analysis models would aid in timely threat detection. Expanding the use of Natural Language Processing (NLP) for multilingual assessment would broaden the project's global applicability.

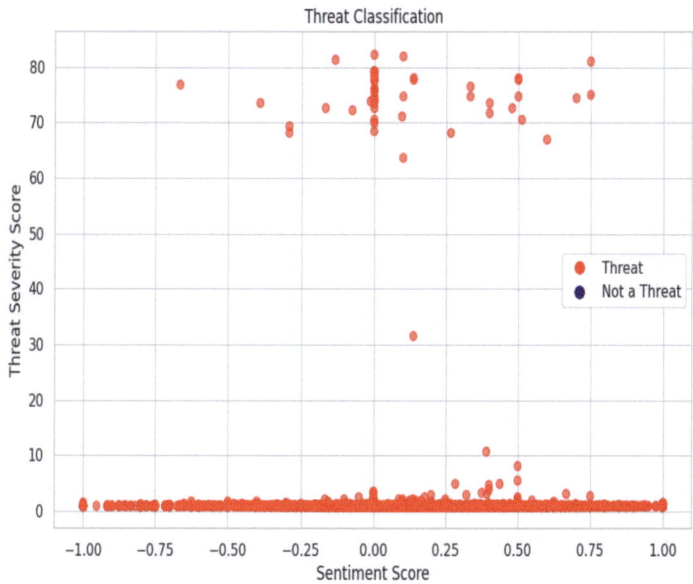


**Fig. 10** Threat classification

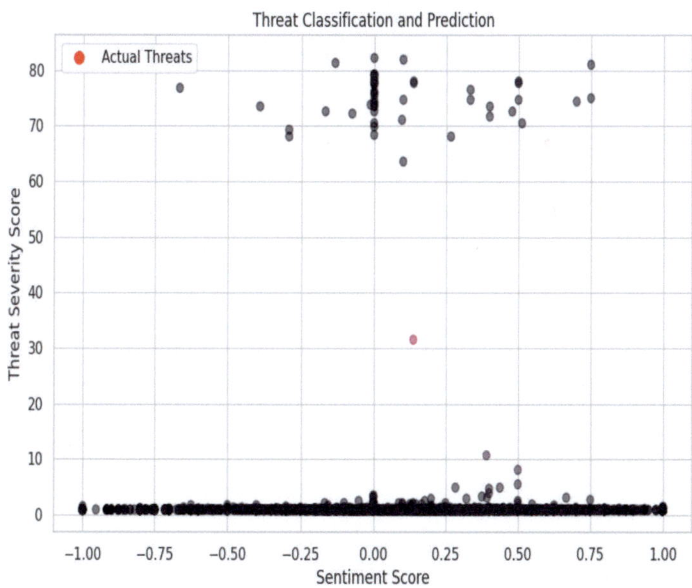


**Fig. 11** Threat prediction

## References


1. Gui X, Wang Y, Kou Y, Reynolds TL, Chen Y, Mei Q, Zheng K (2018) Understanding the patterns of health information dissemination on social media during the Zika outbreak. In: AMIA annual symposium proceedings 2018, 16 Apr 2017, pp. 820–829. PMID: 29854148; PMCID: PMC5977662
2. Sumner SA, Galik S, Mathieu J, Ward M, Kiley T, Batholow B, Dingwall A, Mork P (2019) Temporal and geographic patterns of social media posts about an emerging suicide game. J Adolesc Health 65(1). https://doi.org/10.1016/j.jadohealth.2018.12.025.
3. Niles MT, Emery BF, Reagan AJ, Dodds PS, Danforth CM (2019) Social media usage patterns during natural hazards. PLoS ONE 14(2):e0210484. https://doi.org/10.1371/journal.pone.0210484
4. Malleson N, Andresen MA (2014) The impact of using social media data in crime rate calculations: shifting hot spots and changing spatial patterns. Cartogr Geogr Inf Sci 42(2):112–121. https://doi.org/10.1080/15230406.2014.905756
5. Spitters M, Eendebak PT, Worm DTH, Bouma H (2014) Threat detection in tweets with trigger patterns and contextual cues. In: 2014 IEEE joint intelligence and security informatics conference, The Hague, Netherlands, pp. 216–219. https://doi.org/10.1109/JISIC.2014.39.
6. Dhawas P, Dhore A, Bhagat D, Pawar RD, Kukade A, Kalbande K (2024) Big data preprocessing, techniques, integration, transformation, normalisation, cleaning, discretization, and binning, pp. 159–182. https://doi.org/10.4018/979-8-3693-0413-6.ch00
7. Dhawas P, Ramteke MA, Thakur A, Polshetwar PV, Salunkhe RV, Bhagat D (2024) Big data analysis techniques, pp 183–208. https://doi.org/10.4018/979-8-3693-0413-6.ch007
8. Bouchama F, Kamal M (2021) Enhancing cyber threat detection through machine learning-based behavioral modeling of network traffic patterns. Int J Bus Intell Big Data Anal 4(9):1–9
9. Jakka GJ (2021) Extracting Malware threat patterns on a mobile platform. University of the Cumberlands
10. Imran M, Castillo C, Diaz F, Vieweg S (2015) Processing social media messages in mass emergency: a survey. ACM Comput Surv (CSUR) 47(4):1–38

# Design and Analysis of 9 T Single-Ended Low-Power High-Speed SRAM

**D. B. Shilpa and Kishan Singh**

**Abstract** The current work proposes an innovative design for a 9 T CMOS SRAM device to increase stability and lower power usage. To ensure the operation of our suggested 9 T SRAM circuit, we compared our results with previously available data. The suggested cell has similar metrics to the intended 9 T SRAM cell, which uses the same design ideas and technology. The recommended cell's stability was assessed using N-curve measurements. The recommended 9 T SRAM cell conducts write operations via bit line (BL) charging and discharging, resulting in reduced power usage. When compared to normal SRAM cells, the proposed SRAM circuitry requires at least 92% less reading power. Furthermore, the recommended SRAM's write power usage was reduced by a minimum of 97% as compared to standard SRAM designs. The recommended SRAM unit reduces reading time by at least 63% when compared to existing SRAM cells. In addition, power consumption during data writing to the specified SRAM was lowered by at least 96% when compared to traditional architectures.

**Keywords** CMOS · SRAM · Delay · Power usage

## 1 Introduction

Portable mobile gadgets along with personal electronic assistants are becoming widespread and are transforming several aspects of our everyday life. The primary factor driving the growth of the handheld computing industry is the need for multimedia computation, namely, in video and image-based operations. Video applications need extensive use of integrated memory access, leading to substantial power usage

D. B. Shilpa (✉)
Research Scholar, VTU Belagavi, Guru Nanak Dev Engineering College, Bidar, India
e-mail: shilpabdarvesh@gmail.com

K. Singh (✉)
Professor and Head, ECED, Guru Nanak Dev Engineering College, Bidar, India
e-mail: kishanskrish@gmail.com

 
K. R. Madhavi et al. (eds.), *Proceedings of Sixth International Conference on Computer and Communication Technologies*, Lecture Notes in Networks and Systems 1356,
https://doi.org/10.1007/978-981-96-5238-9_4

thereby restricting the duration of the battery's lifespan. The importance of power absorption is becoming crucial owing to the enhanced connectivity and operating rates, besides the rapid expansion of battery-operated products [1].

Minimal power design has grown into a popular term, and designing for minimal power consumption has long been a crucial part of System on Chips. The main explanation for the growing significance of low-power architecture nowadays remains the escalation of leakage current when device dimensions decrease. Low-power architecture is mandatory for the implementation of battery-operated devices. Portable electronics are constrained by their tiny size, making it impractical to include bigger batteries. Additionally, the low capacity of tiny batteries results in restricted battery life, so limiting their use. Hence, the use of a low-power design is of the utmost importance to prolong the lifespan of the battery. Nowadays is crucial to prioritize the decrease of power usage and the area within digital gadgets, since they mostly consist of memory. This is essential for enhancing system reliability, effectiveness, and dependability [2, 3].

Reducing the operating voltage has become an efficient method for reducing dynamic power usage since it has a quadratic connection. Ensuring a good yield of SRAM with scaled technologies remains more difficult owing to the susceptibility of SRAM bit cells to processing changes caused by the usage of minimal-sized components. The quantity of integrated SRAM in contemporary Systems on-chips (SoC) rises to fulfil the performance demands of each technological advancement [4, 5]. As per the International Technology Roadmap for Semiconductors (ITRS), it is anticipated that the memory capacity will occupy 90% of the total area of the semiconductor. The aforementioned pattern is additionally persisting for real-time video Systems on Chip (SoC). There are two crucial factors to consider in SRAM cell structure: (i) the size of the cell and (ii) the cell's stability. Power dissipation within memories is becoming a crucial factor in modern technology architecture. With the advancement of technological scaling, an increasing number of devices are being included in the system, leading to a proportional rise in leakage power. The effects of reducing voltages and decreasing chip size on power usage, leakage power, and data stability within an SRAM unit are substantial [6].

The ensuing portions of this work are arranged as follows: Sect. 2 presents a literature review of current SRAM designs. Section 3 describes the proposed 9 T SRAM module and analyzes its effectiveness. Section 4 explores the experimental results, comparing the cell to current structures. Section 5 gives the corresponding results.

## 2 Literature Survey

The design of the circuit described in [7] efficiently addresses read/write stability issues regardless of process changes through the use of the read-coupled technique. In addition, the low power methodology outlined in [7] enables the development of a device which significantly diminishes hold power. Sachdeva et al. [8] investigated the use of 12 T SRAM cells that utilize asymmetrical single-ending read and write architectures. Despite the implementation of the power-gated writing assistance process, the system remains susceptible to reading fluctuations characterized by substantial nonuniformity within the subarray.

Lakshmi et al. [9] suggested that the constraints of the 6 T SRAM cell be mitigated by implementing an 8 T SRAM unit. A reduction in SNM throughout reading activities could suggest increased writing capability. As a consequence, circuit architects are given greater freedom for optimization because the processes are entirely distinct.

An SRAM circuitry was developed by Singh et al. [10] utilizing two separate accessing transistors. The threshold voltage method increases the proportion of current which is switched on or off by employing transistors along the reading path.

The existing design maintains a high level of efficiency as a result of the reduced cell leakage and increased impedance. This facilitates an increase in the number of SRAM cells contained across a single column. Chakraborty et al. [11] centered their efforts on the optimization of performance statistics, such as power along with latency, that have previously proven to be crucial when it comes to the IC market.

Pal et al. [12] conducted a study that compared the proposed architecture with various SRAM designs, including 7 T, FD8T, & 9 T bit cells. The study investigated metrics for performance and reliability, especially evaluating the performance of the solutions in response to process changes. The lower dynamic power accomplishment may be ascribed to the use of a single-bit line. On the other hand, the reduction of leakage power is accomplished by arranging transistors in a series in the route where discharge occurs. Xue et al. [13] found that computer memory can store both information and instructions. It might possess either a temporary or lasting characteristic. Unlike RAM, serial access memory does not provide for rapid access to both write and read activities. The utilization of reading buffer circuits [14], sleep NMOS transistors [15], and voltage swinging control [16] represent three additional approaches being explored in the development of power-efficient SRAM architectures that prioritize reliability. Single-ended writing/reading architectures are particularly useful for reducing power usage due to switching as well as leakage, in addition to reducing chip size requirements.

Nevertheless, this approach is impeded by the prolonged duration required for write and read access. Moreover, the capacity to sense and read data is much diminished in single-ended approaches when there are variations in the manufacturing procedure, as a result of the scaling of the supply voltage. Nonetheless, carrying out this assistance requires a significant chip footprint, and is being considered a tradeoff considering the current requirement for more memory and fewer dimensions. Conversely, A. Sachdeva et al. [17] unveiled an innovative SRAM device called 11 T SRAM, which possesses the capability to process data using differential signals for reading and single-ended signals for writing.

This eliminates the need for additional circuitry to enable both operations. The suggested SRAM has read buffers to alleviate the ATS problem. This technology reduces the demand for extra power use in voltage conversions and showcases the capacity to adapt to changes in DRV. An asymmetrical write and single-ended read configurations were investigated by Sachdeva et al. [18] in their study of 12 T cells. Notwithstanding its implementation, the write-assist approach remains vulnerable to read disturbance and demonstrates significant WSNM.

By utilizing a feedback-based optimization strategy, the MST11T SRAM circuit proposed in [19] improves WSNM; however, its single-ended design and large number of commands produce additional latency.

In response to the limitations inherent in 6 T SRAM technology, Lakshmi et al. [20] developed an enhanced 8 T SRAM memory device. An opportunity to improve one's writing skills is provided by RSNM. Additional prospects exist for circuit architects to enhance the performance of their designs. The differentiating SRAM design intended to be described in reference [21] was implemented to improve both RSNM as well as WSNM.

This results in reduced power consumption as it reduces operational variability. Nonetheless, a read-disturbance complaint remains. As the number of transistors connected to a single-bit line increases, its effectiveness and operational features also deteriorate. The existing SRAM memories, which have a reduced power use, are recommended for addressing the read-disturbance concern.

Navaneetha et al. [22] argue that to optimize the performance of FinFET circuits, the overall VLSI sector should proactively anticipate diverse threats. Outstanding stability is a characteristic feature of each of the SRAM storage cells mentioned above.

As a method to decrease power consumption and architectural footprint, single-ended SRAM configurations were suggested in reference [23]. The bit line toggling behavior of the SRAM is less than 0.5, resulting in energy conservation.

Xue et al. [24] discovered that semiconductor memory can be utilized for saving instructions and information. This could have been a short or extended duration.

## 3 Proposed 9 T SRAM Architecture

This paper presents a unique 9 T SRAM cell, construction aimed at reducing both current leakage along with power usage. The diagram illustrating the design of the 9 T SRAM cell using 45 nm technology is seen in Fig. 1. The suggested SRAM cell consists of a writing access transistor (NM3) that is regulated by the Writing Word Line (WWL), also a reading access transistor (NM5) that is regulated by the Reading Word Line (RWL). The NM6 transistor is used as a threshold voltage transistor. It is attached to the CNTRL input and remains in the ON state throughout both operations of reading and writing. During the hold state, the transistor will be deactivated, resulting in a decrease in the amount of current that leaks. Throughout the writing process, the word line (WWL) transitions into a high state, while the read word line (RWL) and bitline bar (BLB) continue to stay at a logic zero level. As a result, the reading access transistor (NM5) has been turned off. To write a '1' onto the storage device, the bit line (BL) has been precharged to a sufficiently high voltage, and then a '1' is applied via the write accessing transistor (NM3). Similarly, when writing '0' into the cell, the BL (bit line) gets drained. As an outcome, the proposed cell utilizes only a one-bit line (BL) to execute the write operation, which reduces both power consumption and leakage power. The Read Word Line (RWL) progresses from zero toward a high value during the reading process, whereas the Write Word Line (WWL) remains at a logically low level.

As a result, the writing accessing transistor gets turned off. Before the reading operation, the BL & BLB lines are pre-charged to a logic high voltage level. If we presume that '1' has been kept on the left side and '0' gets stored on the right side, therefore the BL (bit line) is drained via NM4 and NM5 (transistors). As the 'M6' has been turned off, there is currently no way to release the BLB. Therefore, BLB is regarded as very valuable. If the value '1' is stored on the right side, the BLB (battery load bus) is drained via the 'NM2' and 'NM5' components. As the 'NM4' is disconnected, there is no available route to release the BL. Therefore, it can sustain a high value. By entirely isolating the storage nodes from their bit lines (BL) throughout the reading operation, the stability of the system rises dramatically.

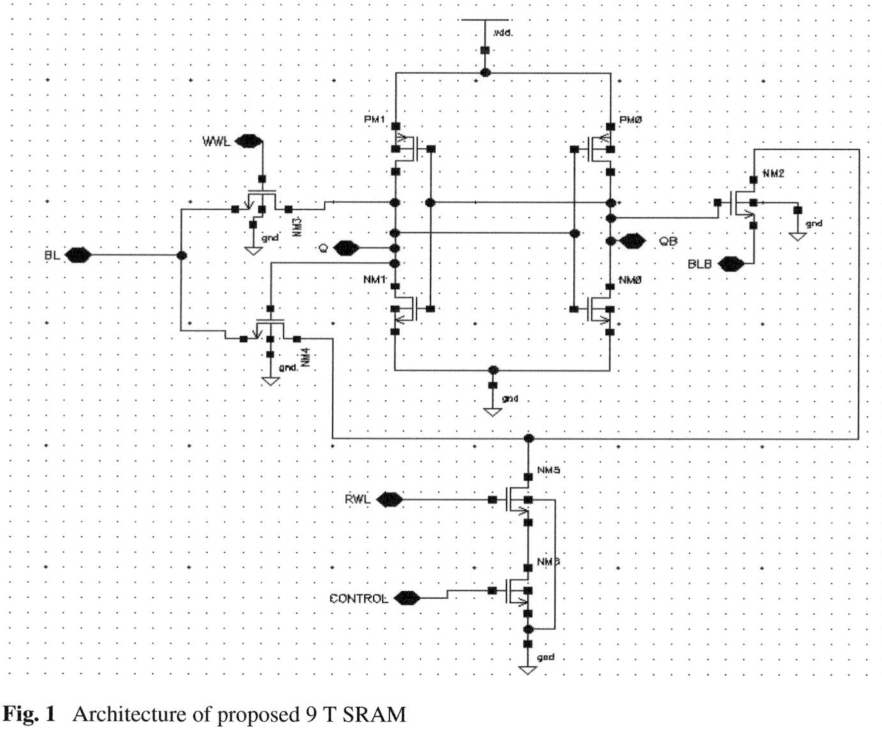


**Fig. 1** Architecture of proposed 9 T SRAM

## 4 Simulation Results

The required 9 T SRAM cell was designed and developed, and simulation findings were obtained using 45 nm CMOS technology in CADENCE virtuoso tools. In addition, the suggested SRAM cells' efficiency in comparison with existing architectures was evaluated by employing a 45 nm technology node. Numerous computations were performed to determine their power usage, latency, process variability tolerance, and stability effectiveness. Using a 1 V power supply, the room temperature remained at 27 °C. This study was conducted for writing and reading considering the different cells under investigation. Inter-die variations often affect the threshold voltage. These factors could be having an impact on energy utilization, delay time, as well as reliability. To determine the level of resistance to change, this research thoroughly examined and analyzed a variety of factors. The transient analysis of the proposed SRAM is carried out and corresponding waveforms of writing and reading modes are depicted in Figs. 2 and 3, respectively.

The comparative findings of many significant attributes comprising all the existing SRAM modules are displayed in Table 1. The computerized simulation results for the traditional and recommended SRAM cell configurations are presented in Table 1. The operating temperature of the cell is maintained at 27 °C, the power supply voltage is set at 1 V, and the transient evaluation time is 1000 ns. The SRAM circuitry that has been proposed exhibits enhanced reliability, decreased latency periods, and increased performance. The power consumption associated with conventional SRAM cells has risen drastically as a result of their bit line discharge activities. The proposed SRAM cell configuration reduces power usage by decreasing discharge activity associated with writing and reading bit lines. When compared to standard SRAM cells, the recommended SRAM circuitry uses at least 92% less reading power.

In addition, the power consumption while writing the recommended SRAM had been greatly lowered by a minimum of 97% compared to traditional designs. The reading time of the proposed SRAM unit remains at least 63% faster than that of the

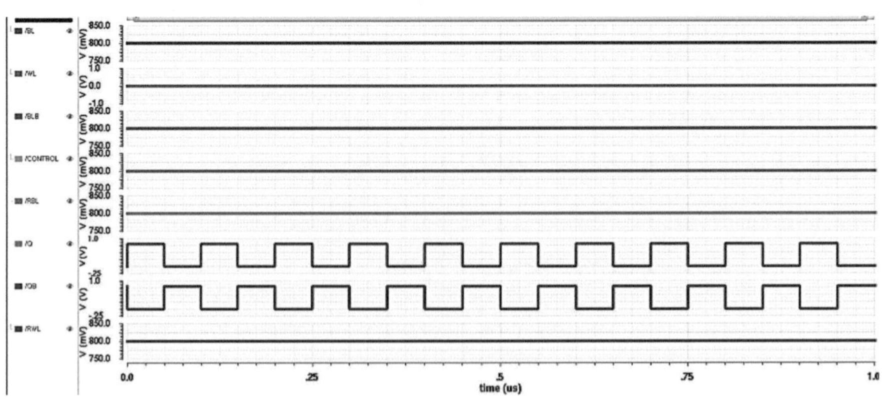


**Fig. 2** Transient simulation of write mode operation of suggested 9 T cell

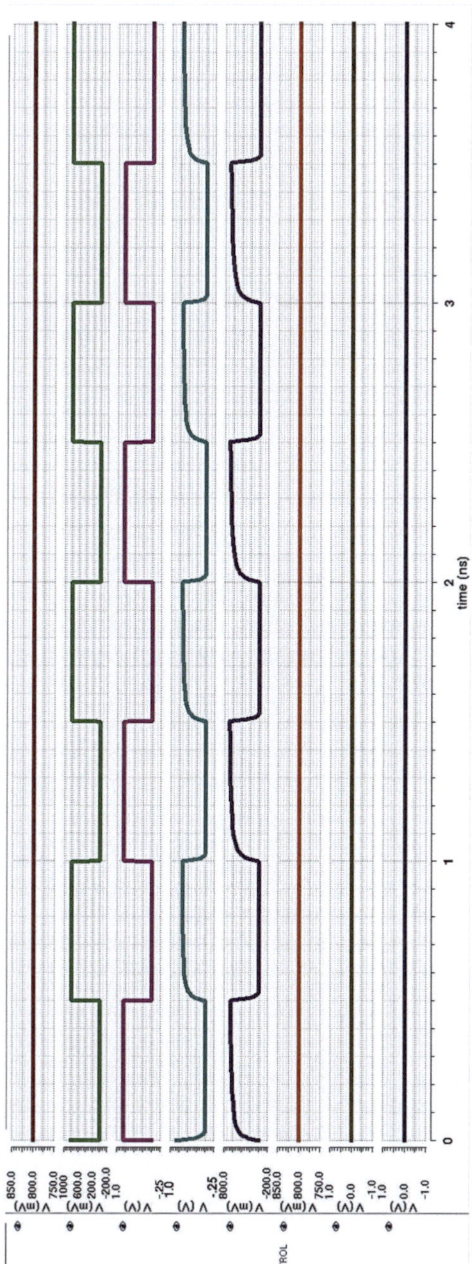


**Fig. 3** Transient simulation of reading mode operation of suggested 9 T cell

**Table 1** Performance analysis of the suggested 9 T SRAM circuit against current SRAM designs

| Reference | [25] | [25] | [25] | [26] | [26] | [33] | Proposed SRAM |
|---|---|---|---|---|---|---|---|
| Number of transistors | 6 | 8 | 10 | 10 | 11 | 11 | 9 |
| CMOS technological node (nm) | 45 | 45 | 45 | 45 | 45 | 45 | 45 |
| Write mode power (uW) | 1.31 | 1.66 | 1.4 | 25.5u | 0.54 | 0.44 | 0.011 |
| Read more power (uW) | 1.13 | 1.51 | 1.23 | 1.47 | 0.61 | 0.46 | 0.037 |
| Leakage Power (uW) | 0.39 | 0.53 | 0.83 | 0.99 | 0.5 | 0.27 | 0.002 |
| Write delay (ps) | 347.7 | 471.3 | 531.6 | 431.9 | 349.2 | 339.3 | 14.79 |
| Read delay (ps) | 229.9 | 161.8 | 279.8 | 249.6 | 140.4 | 141.3 | 51.5 |
| WSNM (mV) | 301 | 29 | 349 | 275 | 481 | 485 | 440 |
| RSNM (mV) | 151 | 159 | 189 | 321 | 419 | 431 | 495 |

SRAM cells presently in use. In addition, power consumption during data writing to the stated SRAM was reduced by a minimum of 96% in comparison with conventional designs.

To evaluate the reliability of the recommended SRAM in comparison to pre-existing cells, the SNM has been put into effect. The term 'SNM' denotes the minimal amount of direct current (DC) disturbance required to alter the configuration of a bit that is stored within a node. When in standby, read, or write mode, the SNM can often be measured and denoted by the symbols." HSNM, RSNM, and WSNM. Increasing the SNM value during every operation strengthens the SRAM cell. The SNR has been established utilizing the conventional butterfly curve method.

The RSNM has the capability of undergoing evaluation at a potential voltage of 1 V under the subsequent conditions: RWL is coupled to VDD; bit lines BL along with BLB have been attached to VDD; WWL has been grounded; and RBL was pre-charged. This configuration results in an enhanced RSNM.

An RSNM diagram has been produced to determine the condition of balance of the cell, which was supposed to be achieved by varying the VDD from 0 to 1.0 V. SRAM RSNM, as illustrated in Fig. 4. WWL connects to VDD, whereas bit lines BL and BLB are assigned values ranging from 0 to 1 across the writing operations.

The RWL was effectively connected to the ground. The WSNM within Fig. 5 corresponds to the proposed SRAM. HSNM denotes the minimum voltage that an SRAM cell needs to preserve data accuracy at all times while in dormancy. Frequently, measurements are carried out in a standby state, with BL and BLB connected to VDD, while WWL along RWL is grounded.

Monte Carlo analysis of Read and write mode power consumption are depicted in Figs. 6 and 7, respectively, followed by read and write mode analysis are depicted in Figs. 8 and 9, respectively. The sigma value is chosen as 3 with 200 iterations for each plot.

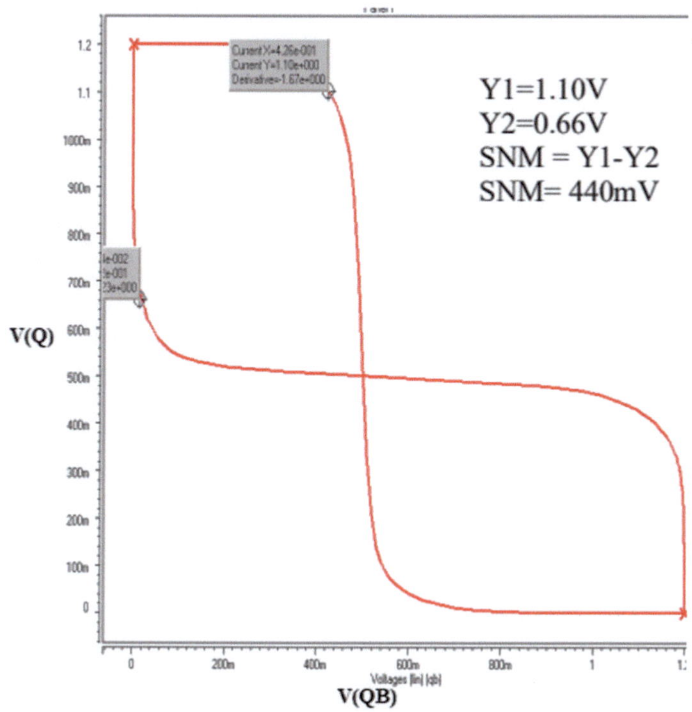


**Fig. 4** SNM of the proposed SRAM in read mode

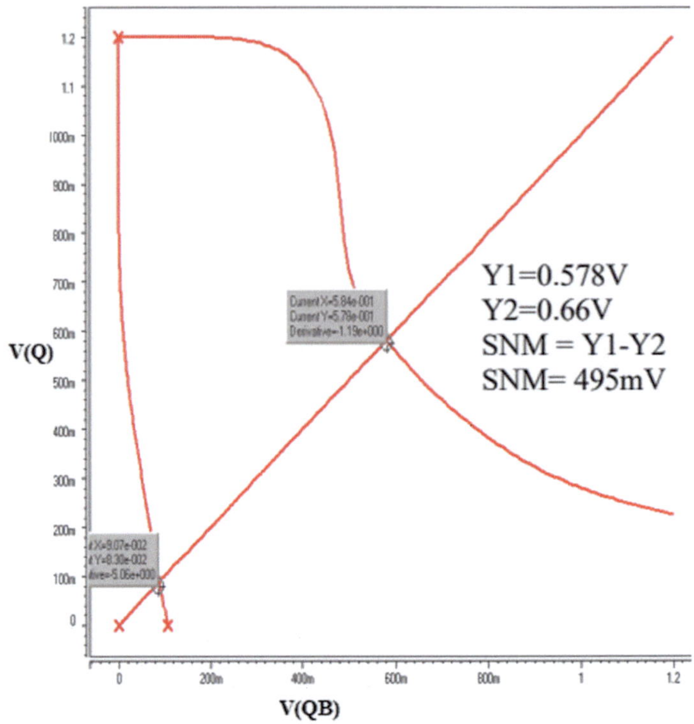


**Fig. 5** SNM of the proposed SRAM in write mode

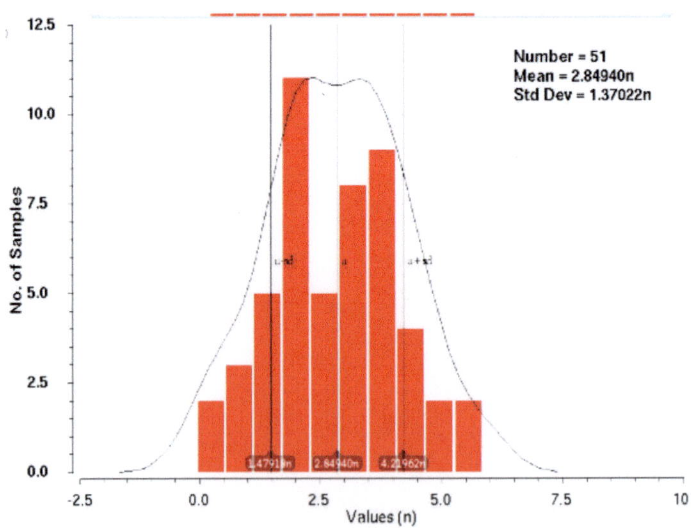


**Fig. 6** Monte Carlo simulation of power consumption in write mode

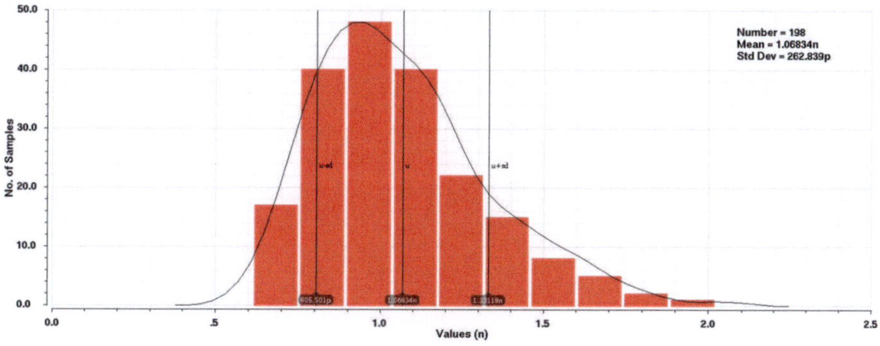


**Fig. 7** Monte Carlo simulation consumption in read mode

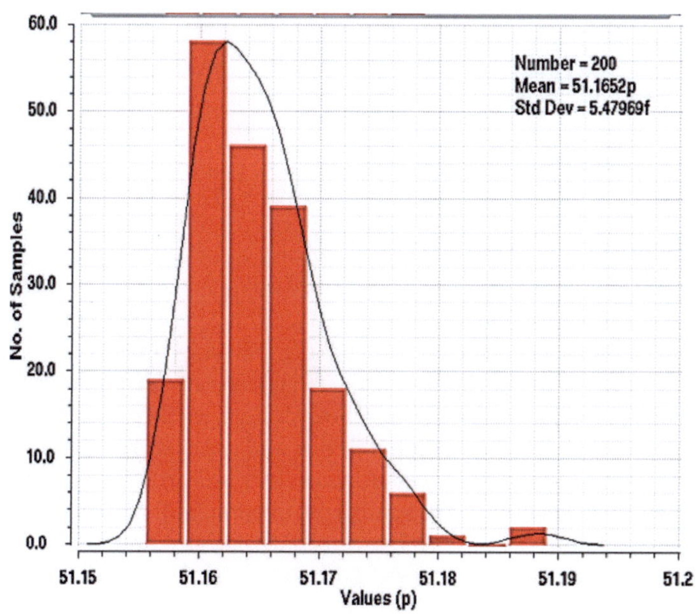


**Fig. 8** Monte Carlo simulation of delay in read mode

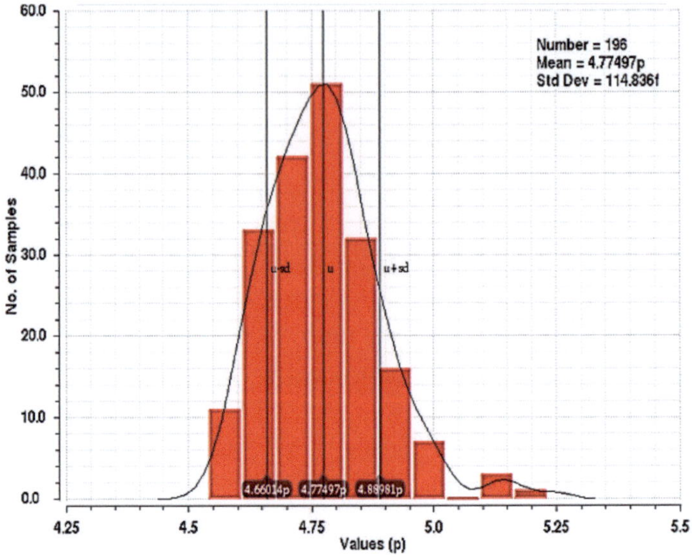


**Fig. 9** Monte Carlo simulation of delay in write mode

## 5 Conclusion

An innovative 9 T SRAM device was introduced, wherein the writing process is executed upon a bit line to decrease power usage. For substantially equivalent data, the recommended SRAM device has become appropriate for cryptography, video streaming applications and also for any SoC design, as SRAM being a memory element is found in almost every digital device and applications. Stability parameters were analyzed utilizing the N-curve, which provides insights into both reading stability and writing capability. When compared to normal SRAM cells, the proposed SRAM circuitry requires at least 92% less reading power. Furthermore, the recommended SRAM's write power usage was reduced by a minimum of 97% as compared to standard SRAM designs. The recommended SRAM unit reduces reading time by at least 63% when compared to existing SRAM cells. In addition, power consumption during data writing to the specified SRAM was lowered by at least 96% when compared to traditional architectures.

## References

1. Calhoun BH, Chandrakasan AP (2006) Static noise margin variation for sub-threshold SRAM in 65-nm CMOS. IEEE J Solid State Circ 41(7):1673–1679

2. Goel A, Sharma RK, Gupta AK (2012) Process variations aware area efficient negative bit-line voltage scheme for improving write ability of SRAM in nanometer technologies. IET Circuits Devices Syst., , 6, (1), p. 45
3. Calhoun BH, Chandrakasan AP (2007) A 256-kb 65-nm Sub-threshold SRAM design for ultra-low-voltage operation. IEEE J Solid-State Circuits 42(3):680–688
4. Tu M-H, Lin J-Y, Tsai M-C et al (2012) A single-ended disturb-free 9T subthreshold SRAM with cross-point data-aware write word-line structure, negative bit-line, and adaptive read operation timing tracing. IEEE J Solid State Circ 47(6):1469–1482
5. Islam A, Hasan M (2012) Leakage characterization of 10T SRAM cell. IEEE Trans Electron Devices 59(3):631–638
6. Gupta, S., Gupta, K., Pandey, N.: A 32-nm subthreshold 7t SRAM bit cell with read assist. IEEE Trans. Very Large Scale Integr Syst 25(12):3473–3483 (2017)
7. Sharma V et al (2019) A robust, ultra low-power, data-dependent-power-supplied 11t SRAM cell with expanded read/write stabilities for internet-of-things applications. Analog Integr Circ Signal Process 98(2):331–346
8. Sachdeva A, Tomar VK (2020) A Schmitt-trigger based low read power 12T SRAM cell. Analog Integr Circ Sig Process 105(2):275–295
9. Lakshmi TV, Kamaraju M (2021) A review on SRAM memory design using FinFET technology. Int J Syst Dyn Appl 11:1–21
10. Singh A, Sharma Y, Sharma A, Pandey A (2019) A novel 20nm FinFET based 10T SRAM cell design for improved performance. In: International symposium on VLSI design and test. Springer, Singapore, pp 523–531
11. Chakraborty A, Singh Tomar R, Sharma M (2023) Optimization of low power 12 T SRAM bit cell using FinFET in 32 nm technology. Mater Today Proc 80:226–232
12. Pal S, Bose S, Wing-Hung K, Islam A (2020) A highly stable reliable SRAM cell design for low power applications. Microelectron Reliab 105. Article ID 113503
13. Xue X, Sai Kumar A, Khalaf OI et al (2023) Design and performance analysis of 32 × 32 memory array SRAM for low-power applications. Electronics 12(4):834
14. Upadhyay P et al (2015) A design of low swing and multi-threshold voltage based low power 12t SRAM cell. Comput Electr Eng 45:108–121
15. Limachia MJ, Thakker RA, Kothari NJ (2018) Characterization of a novel 10t SRAM cell with improved data stability and delay performance for 20-nm tri-gated FinFET technology. Circ World 44(4):187–194
16. Gavaskar K, Ragupathy U (2019) Low power self-controllable voltage level and low swing logic based 11t SRAM cell for high-speed CMOS circuits. Analog Integr Circ Signal Process 100(1):61–77
17. Sachdeva A, Tomar V (2020) Design of a stable low power 11-t static random access memory cell. J Circ Syst Comput 2050206
18. Sachdeva A, Tomar VK (2020) A schmitt-trigger based low read power 12T SRAM cell. Analog Integr Circuits Signal Process 105(2):275–295. https://doi.org/10.1007/s10470-020-01718-619
19. Abbasian E, Mani E, Gholipour M, Karamimanesh M, Sahid M, Zaidi A (2022) A schmitt-trigger-based low-voltage 11 T SRAM cell for low-leakage in 7-nm FinFET technology. Circuits Syst Signal Process 41(6):3081–3105. https://doi.org/10.1007/s00034-021-01950-z20
20. Lakshmi TV, Kamaraju M (2022) A review on SRAM memory design using FinFET technology. Int J Syst Dyn Appl 11(6):1–21. https://doi.org/10.4018/IJSDA.302665
21. Sharma P, Gupta S, Gupta K, Pandey N (2020) A low power subthreshold schmitt trigger based 12T SRAM bit cell with process variation-tolerant write-ability. Microelectron J 97:104703. https://doi.org/10.1016/j.mejo.2020.104703
22. Navaneetha A, Bikshalu K (2023) Reliability analysis of FinFET based high performance circuits. Electronics 12(6):1407. https://doi.org/10.3390/electronics12061407
23. Lorenzo R, Paily R (2022) Half-selection disturbance free 8T low leakage SRAM Cell. Circuit Theory Apps 50(5):1557–1575. https://doi.org/10.1002/cta.3232

24. Xue X, Sai Kumar AS, Khalaf OI, Somineni RP, Abdulsahib GM, Sujith A, Dhanuja T, Vinay MVS (2023) Design and performance analysis of 32 × 32 memory array SRAM for low-power applications. Electronics 12(4):834. https://doi.org/10.3390/electronics12040834
25. Joshi S, Alabawi U (2017) Comparative analysis of 6T, 7T, 8T, 9T, and 10T realistic CNTFET based SRAM. J Nanotechnol 2017:4575013
26. Ashish Sachdeva, V.K. Tomar," Design of 10T SRAM cell with improved read performance and expanded write margin", IET Circuits Devices Syst. 2021;15:42–64.

# AlexNet-Based Deep Learning Model for Fruit Image Classification

**K. G. Suma, Gurram Sunitha, Gorantla Sreeram, Kariyavula Meghana, Duvvuru Deeraj, Kodavath Supriya, and Asadi Srinivasulu**

**Abstract** Various applications of image processing in agriculture are extensively explored by researchers. Fruit image classification plays a key role in agricultural and food industries. Accurate identification of fruits is important for quality control, sorting, and inventory management. Automating fruit classification requires advanced techniques to handle different shapes, sizes, and colors. Deep learning has become a powerful tool for this task. It extracts complex patterns from images. Among deep learning models, AlexNet is proven to be effective in image recognition tasks. Its layered architecture efficiently captures visual features and handles large datasets, making it ideal for fruit classification problem. This research proposes an AlexNet-based deep learning model for fruit image classification. The model is designed to enhance classification accuracy by leveraging AlexNet's strength in feature extraction. A fruit image dataset comprising 100 different fruit images with over 90 K images has been used to train the AlexNet model. Other deep models such as LeNet-5, SqueezeNet, and VGG16 have also been investigated for fruit image classification. Experimental results showed that AlexNet has achieved a recall value of 0.964. Hence, it is excellent at identifying most of the fruit images. The accuracy of 0.974 indicates that AlexNet has correctly classified almost all images. The precision value 0.940 indicates that AlexNet is also good at minimizing false positives.

**Keywords** Agriculture · Food industry · Fruit sorting · Deep learning · AlexNet

K. G. Suma
School of CSE, VIT-AP, Amaravati, India

G. Sunitha (✉)
Department of CSE, School of Computing, Mohan Babu University, Tirupati, India
e-mail: gurramsunitha@gmail.com

G. Sreeram · K. Meghana · D. Deeraj · K. Supriya
Department of CSE, Sree Vidyanikethan Engineering College, Tirupati, India

A. Srinivasulu
University of Newcastle, ATC Building, Callaghan, NSW, Australia

 
K. R. Madhavi et al. (eds.), *Proceedings of Sixth International Conference on Computer and Communication Technologies*, Lecture Notes in Networks and Systems 1356,
https://doi.org/10.1007/978-981-96-5238-9_5

## 1 Introduction

Fruit image classification plays a key role in agricultural and food industries [1]. Accurate identification of fruits is important for quality control, sorting, and inventory management. Traditional methods rely on manual processes, which are prone to errors and time-consuming. Automated systems offer a reliable alternative, ensuring faster and more consistent classification.

Automating fruit classification requires advanced techniques to handle different shapes, sizes, and colors [2]. Deep learning has become a powerful tool for this task. It extracts complex patterns from images [3]. It improves classification accuracy compared to conventional methods. Among deep learning models, AlexNet is proven to be effective in image recognition tasks [4]. Its layered architecture efficiently captures visual features and handles large datasets, making it ideal for image classification problems.

This research proposes an AlexNet-based deep learning model for fruit image classification. The model is designed to enhance classification accuracy by leveraging AlexNet's strength in feature extraction. The study focuses on training the model with a diverse fruit dataset and evaluating its performance across key metrics.

## 2 Literature Survey

Various applications of image processing in agriculture are extensively explored by researchers. The potential of image processing to improve agricultural productivity has been proven. Advancements in technology and methods have enhanced accuracy in identifying various horticultural products in agricultural fields.

Various automated systems have been developed for fruit grading and classification using computer vision [5]. Various techniques and algorithms provided enhanced accuracy and efficiency in fruit sorting. Deep learning has been proven to effective in improving classification performance in agricultural applications [6].

Analyzing visual features is important for automated fruit classification [7]. Different feature extraction techniques and classification algorithms have been assessed for their effectiveness in identifying various fruits. The importance of selecting appropriate visual features to improve classification accuracy has been highlighted.

Vegetable classification using the AlexNet deep model has been investigated [8]. The authors demonstrated how high-performance classification can be achieved through image analysis, thereby showcasing the model's effectiveness in accurately identifying various vegetables. Research has been performed on optimizing crop classification in precision agriculture by utilizing AlexNet and high-resolution UAV imagery [9]. This study highlights the integration of advanced imaging techniques with deep learning to enhance crop identification accuracy in agricultural practices. An intelligent system has been developed for mango fruit grade classification

[10]. It combines AlexNet-SPP with a Mask R-CNN-based segmentation algorithm, enhancing classification accuracy. Various studies explored deep learning applications for fruit recognition from images [11]. Various methods and their effectiveness in classifying different fruit types have been investigated.

Convolutional neural networks have been applied to fruit image processing [12]. Various CNN architectures have been thoroughly investigated, and their impact on improving classification performance has been recorded. Pure-CNN, a framework for classifying fruit images, emphasizes the architecture's efficiency and performance in fruit recognition tasks [13]. One of the research works focused on an efficient image classification technique for weather-degraded fruit images [14]. Deep learning methods were proven to improve recognition accuracy in challenging conditions. A multi-fused convolutional neural network model has been designed for fruit image classification [15]. It aimed to enhance classification performance through advanced techniques.

## 3 Proposed Methodology

The architecture of AlexNet deep model is as follows [4]. It contains five convolutional layers (Conv1-Conv5), three max pooling layers (Maxpool1-Maxpool3) and three fully connected layers (FC1-FC3).

**Input Layer**. Let $I$ be the input image with the dimensions $W \times H \times C$, where $W$ is the width, $H$ is the height, and $C$ is the number of channels.

**Conv1**. It extracts low-level features using $N_1$ filters of size $F_1 \times F_1 \times C$. The output feature map $O_1$ is computed using the ReLU activation function $O_1 = \sigma(W_1 * I + b_1)$ where $W_1$ represents weights, $b_1$ is bias term, and $*$ denotes convolution operation.

**Maxpool1**. This layer downsamples the spatial dimensions of the feature map.

$$O_{\text{pool1}} = \text{MaxPool}(O_1, \text{pool\_size} = P_1, \text{stride} = S_1) \tag{1}$$

**Conv2**. It extracts higher-level features using $N_2$ filters of size $F_2 \times F_2 \times N_1$. The output feature map $O_2$ is computed using the ReLU activation function.

$$O_2 = \sigma\left(W_2 * O_{\text{pool1}} + b_2\right) \tag{2}$$

**Maxpool2**. This layer downsamples the spatial dimensions.

$$O_{\text{pool2}} = \text{MaxPool}(O_2, \text{pool\_size} = P_2, \text{stride} = S_1) \tag{3}$$

**Conv3**. It extracts further higher-level features using $N_3$ filters of size $F_3 \times F_3 \times N_2$. The output feature map $O_3$ is computed using ReLU.

$$O_3 = \sigma\left(W_3 * O_{\text{pool2}} + b_3\right) \tag{4}$$

**Conv4**. It extracts further higher-level features using $N_4$ filters of size $F_4 \times F_4 \times N_3$. The output feature map $O_4$ is computed using ReLU.

$$O_4 = \sigma(W_4 * O_3 + b_4) \tag{5}$$

**Conv5**. It extracts further higher-level features using $N_5$ filters of size $F_5 \times F_5 \times N_4$. The output feature map $O_5$ is computed using ReLU.

$$O_5 = \sigma(W_5 * O_4 + b_5) \tag{6}$$

**Maxpool3**. This layer downsamples the spatial dimensions.

$$O_{\text{pool3}} = \text{MaxPool}(O_5, \text{pool\_size} = P_3, \text{stride} = S_3) \tag{7}$$

**Flattening Layer**. This layer flattens the output from Maxpool3.

$$O_{\text{flat}} = \text{Flatten}\left(O_{\text{pool3}}\right) \tag{8}$$

**FC1**. This layer performs computation.

$$O_{\text{fc1}} = \sigma(W_{\text{fc1}}.O_{\text{flat}} + b_{\text{fc1}}) \tag{9}$$

**FC2**. This layer performs computation.

$$O_{\text{fc2}} = \sigma(W_{\text{fc2}}.O_{\text{fc1}} + b_{\text{fc2}}) \tag{10}$$

**FC3**. This layer contains $C_{\text{output}}$ nodes and performs computation.

$$O_{\text{output}} = \sigma\left(W_{\text{output}}.O_{\text{fc2}} + b_{\text{output}}\right) \tag{11}$$

**Softmax**. This layer performs computation,

$$\text{softmax}\left(O_{\text{output}}\right) = \frac{e^{O_{\text{output}}}}{\sum_{i=1}^{C_{\text{output}}} e_i^{O_{\text{output}}}} \tag{12}$$

**Loss Calculation**. Using cross-entropy loss function to compute the error between predicted and true labels.

$$L = -\frac{1}{N}\sum_{i=1}^{N} y_i \log y_i \tag{13}$$

**Backpropagation and Weight Update**. Calculate gradients using backpropagation and update weights using an optimization algorithm.

$$W_{\text{new}} = W_{\text{old}} - \alpha \nabla L \tag{14}$$

## 4 Experimentation Results and Discussions

The dataset used for this research comprises thousands of high-resolution fruit images, each labeled with its corresponding fruit variety, including common fruits like apples, bananas, oranges, and more [16]. These images exhibit consistent lighting conditions and backgrounds to facilitate effective learning. Dataset captures a wide range of fruit shapes, colors, and orientations, ensuring robustness and generalization of the trained model. Figure 1 presents sample training images from fruit dataset. Figure 1 presents sample fruit classification results of AlexNet.

To prepare the dataset for training, images are preprocessed by resizing them to the input size required by the AlexNet architecture and normalizing pixel values. By using this comprehensive fruit dataset and the powerful features of the AlexNet architecture, the resulting deep model achieves high accuracy in fruit classification tasks. Table 1 presents the configuration of AlexNet for fruit image classification.

Table 2 presents the hyperparameter setup for training AlexNet for fruit classification. Figure 2 presents sample predictions of AlexNet model. Figures 3, 4 and 5 present the performance curves (accuracy and loss) of deep models AlexNet, SqueezeNet, LeNet-5 for fruit classification, respectively. Table 3 presents experimental results of the performance of various deep learning models in fruit classification task. LeNet-5 performed decently with relatively balanced values across all metrics. However, given that it is a shallow model, its performance is lower than

(a) Ackee (b) Barbadine (c) Grapes

**Fig. 1** Sample training images from fruit dataset

**Table 1** AlexNet configuration for fruit image classification [4]

| Layer | Type | Input size | Filter size / stride / padding | Output size | Details |
|---|---|---|---|---|---|
| 1 | Convolutional | 227 × 227 × 3 | 11 × 11 / 4 / 0 | 55 × 55 × 96 | 96 filters, ReLU |
| 2 | Max pooling | 55 × 55 × 96 | 3 × 3 / 2 / 0 | 27 × 27 × 96 | Max pooling |
| 3 | Convolutional | 27 × 27 × 96 | 5 × 5 / 1 / 2 | 27 × 27 × 256 | 256 filters, ReLU |
| 4 | Max pooling | 27 × 27 × 256 | 3 × 3 / 2 / 0 | 13 × 13 × 256 | Max pooling |
| 5 | Convolutional | 13 × 13 × 256 | 3 × 3 / 1 / 1 | 13 × 13 × 384 | 384 filters, ReLU |
| 6 | Convolutional | 13 × 13 × 384 | 3 × 3 / 1 / 1 | 13 × 13 × 384 | 384 filters, ReLU |
| 7 | Convolutional | 13 × 13 × 384 | 3 × 3 / 1 / 1 | 13 × 13 × 256 | 256 filters, ReLU |
| 8 | Max pooling | 13 × 13 × 256 | 3 × 3 / 2 / 0 | 6 × 6 × 256 | Max pooling |
| 9 | Fully connected | 6 × 6 × 256 (Flattened) | – | 4096 | ReLU |
| 10 | Fully connected | 4096 | – | 4096 | ReLU |
| 11 | Fully connected | 4096 | – | 100 | ReLU + Softmax for classification |

**Table 2** Hyperparameter setup for training AlexNet for fruit classification

| Hyperparameter | Configuration |
|---|---|
| Batch size | 32 |
| Loss function | Cross-entropy loss |
| Optimizer | SGD with momentum = 0.9 |
| Learning rate | 0.001 |
| Momentum | 0.9 |
| Learning rate scheduler | StepLR with step size of 7 and gamma of 0.1 |
| Number of epochs | 50 |

the other deeper models. The recall, precision, and F1-score are respectable, but it struggles to reach the higher performance levels of the more modern architectures. SqueezeNet showed the weakest performance among the four models. This is due to its smaller architecture, which trades off accuracy for model efficiency. The low recall and precision values suggest that SqueezeNet struggled to correctly classify certain fruit images, which impacted the overall performance. This model prioritizes speed and smaller size over high classification accuracy. VGG16 performs very well,

(a) Correct Classification as Custard Apple (b) Correct Classification as Cluster Fig (c) Misclassification as Black Berry

**Fig. 2** Sample fruit classification predictions by AlexNet

with high accuracy and recall. Its F1-score indicates that it maintains a good balance between precision and recall. This is expected given the depth of the model and its ability to learn detailed features.

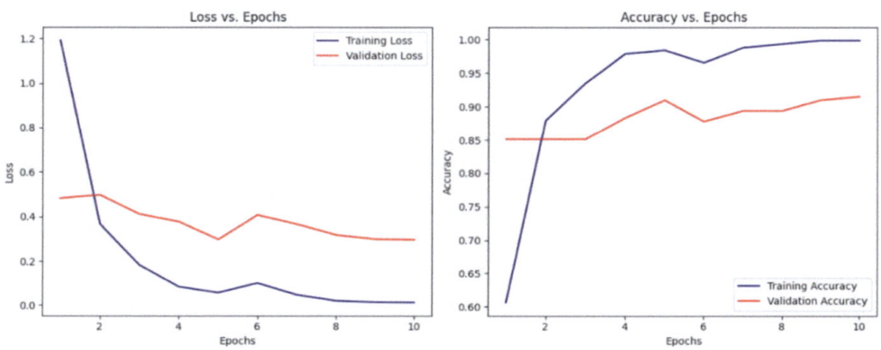


**Fig. 3** Performance curves (accuracy and loss) of AlexNet

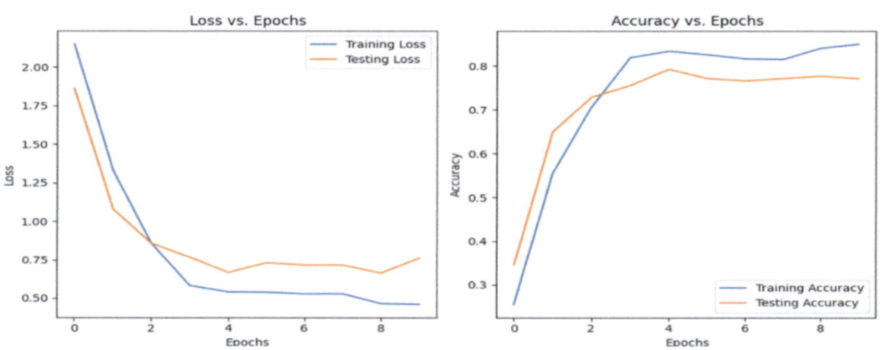


**Fig. 4** Performance curves (accuracy and loss) of SqueezeNet

**Fig. 5** Performance curves (accuracy and loss) of LeNet-5

**Table 3** Experimental results of investigating various deep models for fruit classification

| Deep model | Recall | Accuracy | F1-score | Precision |
|---|---|---|---|---|
| LeNet-5 | 0.825 | 0.855 | 0.844 | 0.828 |
| SqueezeNet | 0.771 | 0.771 | 0.773 | 0.779 |
| VGG16 | 0.932 | 0.945 | 0.912 | 0.924 |
| AlexNet model | 0.964 | 0.974 | 0.955 | 0.940 |

VGG16 is highly effective at classifying fruit images but requires more computational resources than LeNet-5 or SqueezeNet. AlexNet achieved highest performance across all metrics. With a recall of 0.964, it is excellent at identifying most of the fruit images. The accuracy of 0.974 indicates that it has correctly classified almost all images. The high F1-score suggests a good balance between precision and recall, and the precision value (0.940) indicates that it is also good at minimizing false positives.

## 5 Conclusion

Various applications of image processing in agriculture are extensively explored by researchers. Fruit image classification plays a key role in agricultural and food industries. Accurate identification of fruits is important for quality control, sorting, and inventory management. Automating fruit classification requires advanced techniques to handle different shapes, sizes, and colors. Deep learning has become a powerful tool for this task. It extracts complex patterns from images. Among deep learning models, AlexNet is proven to be effective in image recognition tasks. Its layered architecture efficiently captures visual features and handles large datasets, making it ideal for fruit classification problem. This research proposes an AlexNet-based deep learning model for fruit image classification. The model is designed to enhance classification accuracy by leveraging AlexNet's strength in feature extraction. A fruit image dataset comprising 100 different fruit images with over 90 K images has been used to train the AlexNet model. Other deep models such as LeNet-5, SqueezeNet, VGG16 have also been investigated for fruit image classification. Experimental results showed that AlexNet has achieved a recall value of 0.964. Hence, it is excellent at identifying most of the fruit images. The accuracy of 0.974 indicates that AlexNet has correctly classified almost all images. The precision value 0.940 indicates that AlexNet is also good at minimizing false positives.

## References


1. Vibhute A, Bodhe SK (2012) Applications of image processing in agriculture: a survey. Int J Comput Appl 52(2):34–40
2. Tripathi MK, Maktedar DD (2020) A role of computer vision in fruits and vegetables among various horticulture products of agriculture fields: a survey. Inf Process Agric 7(2):183–203
3. Ukwuoma CC, Zhiguang Q, Bin Heyat MB, Ali L, Almaspoor Z, Monday HN (2022) Recent advancements in fruit detection and classification using deep learning techniques. Math Probl Eng 2022(1):9210947
4. Krizhevsky A, Sutskever I, Hinton GE (2012) ImageNet classification with deep convolutional neural networks. Adv Neural Inf Process Syst 25
5. Kumar A, Gill GS (2015) Automatic fruit grading and classification system using computer vision: A review. In: 2015 Second international conference on advances in computing and communication engineering. IEEE, pp 598–603
6. Chaudhari D, Waghmare S (2022) Machine vision based fruit classification and grading—a review. In: ICCCE 2021: proceedings of the 4th international conference on communications and cyber physical engineering. Springer Nature Singapore, Singapore, pp 775–781
7. Ghazal S, Qureshi WS, Khan US, Iqbal J, Rashid N, Tiwana MI (2021) Analysis of visual features and classifiers for fruit classification problem. Comput Electron Agric 187:106267
8. Zhu L, Li Z, Li C, Wu J, Yue J (2018) High performance vegetable classification from images based on AlexNet deep learning model. Int J Agric Biol Eng 11(4):217–223
9. Ajayi OG, Iwendi E, Adetunji OO (2024) Optimizing crop classification in precision agriculture using AlexNet and high resolution UAV imagery. Technol Agron 4(1)
10. Yeh JF, Lin KM, Lin CY, Kang JC (2023) Intelligent mango fruit grade classification using AlexNet-SPP with Mask R-CNN-based segmentation algorithm. IEEE Trans AgriFood Electron 1(1):41–49
11. Gill HS, Murugesan G, Khehra BS, Sajja GS, Gupta G, Bhatt A (2022) Fruit recognition from images using deep learning applications. Multimed Tools Appl 81(23):33269–33290
12. Naranjo-Torres J, Mora M, Hernández-García R, Barrientos RJ, Fredes C, Valenzuela A (2020) A review of convolutional neural network applied to fruit image processing. Appl Sci 10(10):3443
13. Kausar A, Sharif M, Park J, Shin DR (2018) Pure-CNN: a framework for fruit images classification. In: 2018 International conference on computational science and computational intelligence (CSCI). IEEE, pp 404–408
14. Singh Gill H, Singh Khehra B (2020) Efficient image classification technique for weather degraded fruit images. IET Image Proc 14(14):3463–3470
15. Sinha BB, Dhanalakshmi R (2024) A multi-fused convolutional neural network model for fruit image classification. Int J Cogn Comput Eng 5:416–424
16. Fruits 100 Dataset From Kaggle. https://www.kaggle.com/datasets/marquis03/fruits-100

# Impact of Hidden Markov Models on Real World Scenarios

**Gayam Srinivasa Reddy, Prem Kumar Deepak, Mudigallu Lashmi Narayana Charan, Vemula Vasuki Rohinidevi, Gurram Sunitha, Javvaji Venkatarao, and Asadi Srinivasulu**

**Abstract** Machine learning in current world presents itself across various domains where data plays major role. It involves process from data collection to decision-making. This paper mainly focuses on the application of Hidden Markov Models in machine learning, aiming on investigating forward and Viterbi algorithms. Forward algorithm focuses on determining the probability of an observed sequence. On the other hand, Viterbi algorithm predicts the most likely sequence of hidden states based on the observed sequence. With the help of practical examples, this research demonstrates how HMMs can model scenarios where future outcomes are influenced by hidden factors. Beyond this, the potential of these algorithms extends to broader fields, including artificial intelligence, where these algorithms can be used to predict and control the unnatural behavior of artificial intelligence based systems by finding the hidden factors leading to unimaginable outcomes. As AI continues to grow across various domains or industries, the ability to predict, assess, and control

G. S. Reddy · G. Sunitha (✉)
Department of Computer Science and Engineering, School of Computing, Mohan Babu University, Tirupati, India
e-mail: gurramsunitha@gmail.com

P. K. Deepak
Department of Computer Science and Engineering, School of Computing, Vel Tech Rangarajan Dr. Sagunthala R&D Institute of Science and Technology, Chennai, India

M. L. N. Charan
Department of Electrical and Electronics Engineering, School of Engineering, Mohan Babu University, Tirupati, India

V. V. Rohinidevi
Department of Artificial Intelligence and Data Science, Chaitanya Bharathi Institute of Technology, Hyderabad, India

J. Venkatarao
CSE, K L E F Deemed To Be University, Green Fields, Vaddeswaram, Andhra Pradesh, India

A. Srinivasulu
University of Newcastle, ATC Building, Callaghan, NSW, Australia

 
K. R. Madhavi et al. (eds.), *Proceedings of Sixth International Conference on Computer and Communication Technologies*, Lecture Notes in Networks and Systems 1356,
https://doi.org/10.1007/978-981-96-5238-9_6

such behaviors using HMMs becomes increasingly valuable. This research outlines the advantages, limitations, and future directions for HMMs in real world machine learning applications.

**Keywords** State · Hidden states · Observable states · Transition probabilities · Emission probabilities · Forward algorithm · Viterbi algorithm

## 1 Introduction

In machine learning and artificial intelligence, designing complex sequential data and predicting accurate hidden processes remains a big challenge through various domains like healthcare, natural language processing, and robotics. Traditional methods like Brute-force algorithms frequently fail due to their inefficiency in capturing temporary dependencies and hidden states in sequential data observed on storing dynamically the calculated values. This paper focuses on Hidden Markov Models (HMMs) and algorithms used to predict this hidden state sequences, mainly forward and Viterbi algorithms. HMMs are good in handling uncertainties in observable data, where hidden processes are not directly observable but can be predicted through probability models. Forward algorithm is efficient in computing the probability of observed sequences, while the Viterbi algorithm identifies the most likely hidden state path. With help of these algorithms practical applications such as disease progression prediction in healthcare, error detection in AI systems, etc. can be addressed. This research work illustrates how the combination of HMMs with these algorithms represents a significant advancement in accurately modeling and predicting outcomes in complex as well as majorly in data driven environments.

## 2 Literature Review

Forney et al. describe the Viterbi algorithm, which is foundational in decoding of hidden state sequences widely used in tasks like speech recognition, etc. [1]. Jelinek et al. focuses on the significance of probability methods used in speech recognition, using HMMs to manage discrepancies in observed data [2]. The work of Baum et al. gives the theoretical groundwork for HMMs by introducing the forward and backward algorithm and statistical inference techniques for finite state Markov chains [3, 4]. Their contributions are essential for training HMMs. Dempster et al. demonstrate the Em algorithm, which complements HMMs training by using a method for maximum likelihood prediction with incomplete data [5]. Eisner, J. presents an interactive spreadsheet designed to teach the forward–backward algorithm. The tool aims to simplify learning in NLP and computational linguistics contexts [6]. In computational biology, Sankoff et al. investigated sequence arrangement under deletion and insertion conditions [7]. Vintsyuk et al. introduced dynamic programming for speech bias, both aligning with HMMs capabilities in sequence modeling [8]. Tamposis et al. explained the semi supervised learning of HMMs in biological problems with observable sequences of data, advancing their models in medical field [9]. HMMs are applied in areas such as speech recognition, bioinformatics, and financial modeling [10]. Advantages of using HMMs in handling sequential data and probabilistic state transitions has been proven [10]. A systematic review of HMMs and their diverse applications has been performed covering their use in fields like robotics, genomics, and natural language processing [11]. The Viterbi algorithm and its application in Indonesian speech recognition systems has been studied and demonstrated to improve accuracy in recognizing speech patterns [12]. An adaptive HMM combined with the Viterbi algorithm has been designed for segmenting both ordinary and medical images [13]. The study focuses on enhancing image segmentation accuracy through adaptive modeling techniques. An HMM has been implemented to predict financial market regimes [14]. The study explored the use of HMM for analyzing and forecasting market conditions. A hand gesture recognition system is developed using HMM for computer vision applications. The study explored how HMMs enhance gesture recognition accuracy in vision-based system [15].

## 3 Proposed Methodology

This research mainly focuses on HMMs to analyze the observed data with help of the forward and Viterbi algorithms. This consists of the following steps:

**Data Preparation**: Observed data set, that is data from healthcare records or sequence from natural language processing problems, are processed priorly to fit into the model.

**Model Initialization**: HMMs data, including transition probabilities, emission probabilities, and initial state probabilities, are initialized. Initialization can be done randomly or through techniques or using domain specific concepts.

**Forward Algorithm**: It is used to calculate probability of observed sequence. This involves continuously calculating the probability of observations and helps in evaluating the model fitted with data (Fig. 1).

**Viterbi Algorithm**: The Viterbi algorithm is used to predict the most happening sequence of hidden states from data fitted in model. This can be achieved using dynamic programming to find the most probable path by backtracking with maximum probability of hidden states.

**Model Evaluation**: The efficiency of HMM is determined by using metrics like accuracy and precision depending on the problem model trying to solve. Cross-validation, testing on hidden data makes best of the model.

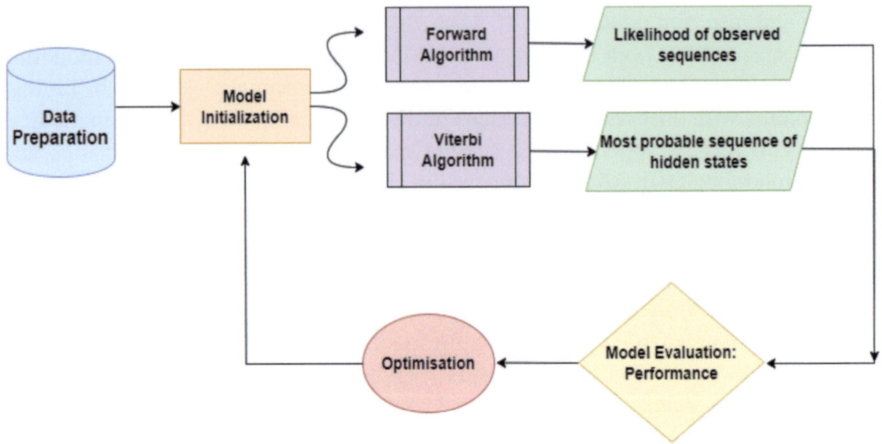


**Fig. 1** Architecture of proposed model

**Optimization**: Optimization of model involves observed data tuning. Optimization is performed to improve model performance. This can involve multiple ways to improve the accuracy and precision of transition and emission probabilities.

Basic Terminologies in Hidden Markov Model are as follows:

**Hidden States(X)**: These are the unobservable states of the system. Let $X = \{X_0, X_1, X_2..., X_{N-1}\}$ represent the set of hidden states, where $N$ is the total number of hidden states.

**Observations (Y)**: These are the observable states generated by the hidden states. Let $Y = \{Y_1, Y_1, ..., Y_t\}$ represent the set of possible observations set, where t is the total number of possible observations.

**Transition Probabilities (A)**: It is a matrix represent the probabilities of transitioning from one hidden state to another state. Let $A = [a_{ij}]$ denote the transition probability matrix, where $a_{ij}$ represents the probability of transitioning from state $X_i$ to state $X_j$. The rows of this matrix sum equal to 1.

**Emission Probabilities (B)**: It is a matrix representing the probabilities of observing each possible outcome from each hidden state. Let $B = [b_{jk}]$ denote the emission probability matrix, where $b_{jk}$ represents the probability of observing outcome $Y_k$ given hidden state $X_j$. The rows of this matrix sum up to 1.

**Initial State Probabilities ($\pi$)**: This is a vector representing the probabilities of starting in each hidden state. Let $\pi = [\pi_1, \pi_2, \pi_3,.., \pi_N]$ represent the initial state probability vector, where $\pi_i$ represents the probability of starting in state $X_i$.

In HMM, a transition state diagram is a graphical representation of the transitions between hidden states over time. When modeling sequential data, HMMs are probabilistic models that are frequently utilized. In these models, the underlying system is thought to be a Markov process with hidden states influencing observable outcomes.

Using the data in Fig. 2, a transition state diagram is created (Fig. 3). This diagram will include "cloudy" and "sunny" as hidden states, and "happy" and "sad" (represented by emojis) as observable states. From the report, it can be summarized that cloudy days (☁): 10, sunny days (☀): 15. Now probabilities can be calculated as follows.

$P$(cloudy) $= P(X_0) =$ Number of cloudy days / Total days $= 10 / 25 = 0.4$; $P$(sunny) $= P(X_1)$ Number of sunny days / Total days $= 15 / 25 = 0.6$; So, $\pi = [0.4, 0.6]$.

**Fig. 2** Data illustrates the co-relation between a person's mood and the weather over a period of 25 days

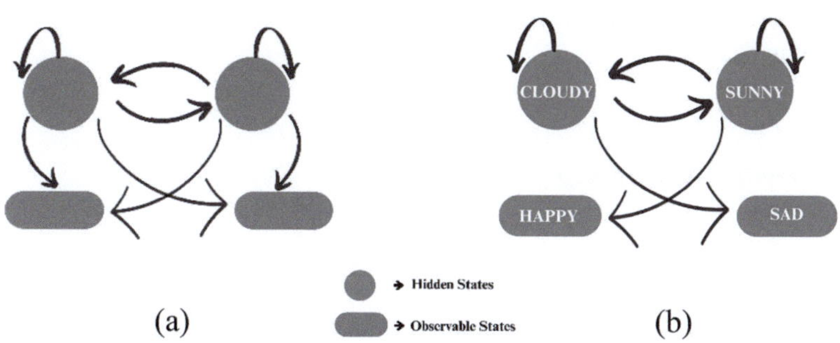


**Fig. 3** Transition state diagram with two hidden and two observable states

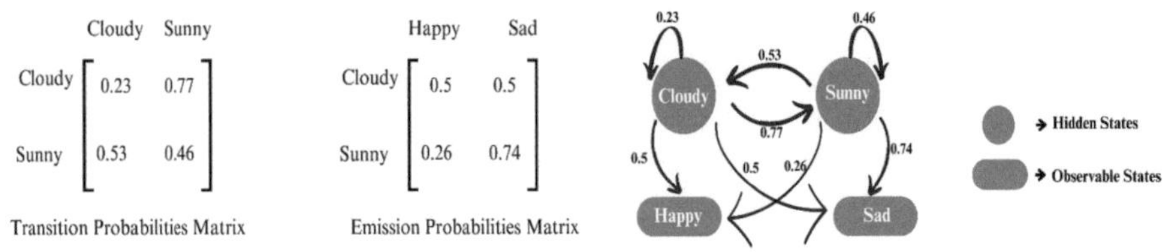


**Fig. 4** Probability matrices and transition state diagram

For the transition probabilities, the number of changes in weather from one kind to another is to be considered when compared to the total number of instances of the original type of weather.

P (sunny to cloudy) = (No.of ☀ to ☁ transitions) / (No.of ☀ days)= 8 / 15 ≈ 0.53
P (sunny to sunny) = (No.of ☀ to ☀ transitions) / (No.of ☀ days) = 7 / 15 ≈ 0.46
P (cloudy to cloudy) = (No.of ☁ to ☁ transitions) / (No.of ☁ days)=2 / 9 ≈ 0.23
P (cloudy to sunny) = No.of ☁ to ☀ transitions / No.of ☁ days = 7 / 9 ≈ 0.77

Here the total likelihood of transitions from a cloudy day is not equal to 1 due to data could not define 26th day weather. It may be cloudy or sunny. That is why the 25th day weather is being neglected. These probabilities are based on the data provided in the image and are empirical probabilities that describe the likelihood of the weather conditions and transitions observed during these 25 days.

According to Baye's Theorem: $P\ (X \mid Y) = P\ (X \text{ and } Y) / P\ (Y)$

$P\ (X \text{ and } Y) =$ (No.of $X$ and $Y$ days) / (Total No.of days).

$P$ (Cloudy|Happy) $= P(X_0|Y_0) = P$ (Cloudy and Happy)/$P$ (Cloudy) $= 0.2 / 0.4 = 0.5$

$P$ (Cloudy | Sad) $= P(X_0|Y_1) = P$ (Cloudy and Sad)/$P$ (Cloudy) $= 0.2 / 0.4 = 0.5$

$P$ (Sunny |Happy) $= P(X_1|Y_0) = P$ (Sunny and Happy)/ $P$ (Sunny) $= 0.16 / 0.6 \approx 0.26$.

$P$ (Sunny | Sad) $=$ P$(X_1|Y_1) = P$ (Sunny and Sad) /$P$ (Sunny) $= 0.44 / 0.6 \approx 0.74$.

With reference from above values, transition and emission probability matrices are shown in Fig. 4.

## 3.1 *Forward Algorithm*

The forward algorithm serves as a valuable tool for determining the likelihood of a sequence of observable states occurring, denoted as $P\ (Y = Y_1, Y_2, Y_3, \ldots, Y_t)$. By computing, the chance that the observed sequence will occur based on this probability can be determined. The forward algorithm determines the probability of a series of observations given an HMM model. The general approach for forward algorithm is as follows.

**Step 1. Initialization**: Calculate forward probability at $t = 1$.

$$\alpha_1(X_i) = \pi[X_i] \times P(X_i|Y_i) for 0 \leq i \leq N-1, (t=1) \quad (1)$$

**Step 2. Recursion**: For each time step $t = 2, 3,.., t$, compute the forward probabilities for each state $X_j$. ($Y_t$ is observable state at $t$-time step).

$$\alpha_t(X_j) = \sum_{i=0}^{N-1} \alpha_{t-1}(X_i) \times P(X_i|X_j) \times P(X_j|Y_t), \forall t > 1 \quad (2)$$

**Step 3. Final Probability**: Calculating final probability of observed sequence by sum of the forward probabilities at final time step $t$ where $N$ is the number of hidden states.

$$P(Y_1, Y_2, \ldots, Y_t) = \sum_{i=0}^{N-1} \alpha_t(X_i) \quad (3)$$

where $\alpha_t$ $(X_j)$ is forward probability for each state $X_j$ at $t$-time step, $P(X_i \mid X_j)$ is the likelihood of a transition from state $X_j$ to state $X_i$, $P$ $(X_j| Y_t)$ is the likelihood of emission from state $Y_t$ that result $X_j$, $\Pi[X_i]$ is the initial state probability at time step $t = 1$.

So, the probability of the entire observation sequence can then be calculated as the total likelihood of existing in each concealed state at the previous time step.

$$P(Y) = \sum_{i=0}^{N-1} \alpha_t(X_i) \quad (4)$$

For instance, let us take an observation sequence $Y = \{\text{Happy, Sad, Happy}\}$ comprising of two observable and hidden states each. Assuming the hidden states to be $X = \{\text{Cloudy, Sunny}\}$, the forward algorithm can be employed to determine the likelihood that a sequence will occur. Considering the observable states Happy as $Y_0$ and *Sad* as $Y_1$, hidden states Cloudy as $X_0$, Sunny as $X_1$. Assuming observed sequence be $Y = \{Y_0, Y_1, Y_0\}$. $Y_t$ denotes observed state at time $t$. From above sequence $Y_1 = Y_0$, $Y_2 = Y_1$, $Y_3 = Y_0$. To find the probability of happening the sequence, forward algorithm is used to find $P(Y = Y_0, Y_1, Y_0)$.

$$P(Y_1, Y_2, \ldots, Y_t) = \sum_{i=0}^{N-1} \alpha_t(X_i) \quad (5)$$

$$P(Y = Y_0, Y_1, Y_0) = \alpha_3(X_0) + \alpha_3(X_1) \quad (6)$$

At the first-time step $t = 1$, the forward probabilities are initialized for all states. This is done by multiplying the initial state probabilities with the probability of observing the first event given each state.

$$\alpha_1(X_0) = \pi[X_0] * P(X_0|Y_0) = 0.4 * 0.5 = 0.2$$

$$\alpha_1(X_1) = \pi[X_1] * P(X_1|Y_0) = 0.6 * 0.26 \approx 0.16$$

The forward probabilities for each state are determined during each subsequent time step $t > 1$ by adding up the forward probabilities of all previous states at time $t-1$, multiplying by the transition probabilities from those states to the current state, and then multiplying the observation probability of the current event given the current state.

$$\alpha_t(X_j) = \sum_{i=0}^{N-1} \alpha_{t-1}(X_i) * P(X_i|X_j) * P(X_j|Y_t) \tag{7}$$

$$\alpha_2(X_0) = \sum_{i=0}^{N-1} \alpha_1(X_i) * P(X_i|X_0) * P(X_0|Y_2) \tag{8}$$

$$\alpha_2(X_1) = \sum_{i=0}^{N-1} \alpha_1(X_i) * P(X_i|X_1) * P(X_1|Y_2) \tag{9}$$

Therefore, by using the forward algorithm the probability of happening of sequence $Y$ is $\approx$ 0 089. [$P\ (Y_0, Y_1, Y_0) \approx 0.089$].

## 3.2 *Viterbi Algorithm*

The most probable path, or most likely chain of hidden states, in HMM can be found using the dynamic programming Viterbi algorithm given a series of observations. Recognition of speech, bioinformatics and natural language processing are just a few of the areas in which it is especially helpful. Let us consider, $T$ as the number of observations or time steps, N reflects the HMM's maximum permitted state count, $Y = \{Y_1, Y_2, \ldots, Y_M\}$ the sequence of observed symbols, $X = \{X_1, X_2, X_3 \ldots, X_N\}$ as the set of possible states, $A[i][j]$ denoting the likelihood of moving from state $X_i$ to state $X_j$, $B$ is the emission probability matrix, whereas $B[i][j]$ is the likelihood of observing $Y_j$ from state $X_i$, $\pi$ is the probability distributions for the state at the beginning. The general formula for the Viterbi algorithm can be described as follows:

**Step 1. Initialization**: Calculate most likely probability at $t = 1$.

$$V_1[X_i] = \pi[X_i] * P(X_i|Y_1) \text{for } 0 \le \text{i} \le \text{N} - 1,\ \text{t} = 1 \tag{10}$$

**Step 2. Recursion**: For each time step $t = 2, 3,.., t$, compute the probabilities of the most likely paths that end in each state $j$.

$$V_t[X_j] = \text{Max}_{0 \le i \le N-1}\left(V_t(X_i) * P(X_i|X_j) * P(X_j|Y_t)\right) \tag{11}$$

**Step 3. Termination**: The highest probability of the most likely path is the maximum value at final time step.

$$\text{Highest Probability} = \text{Max}_{0 \le i \le N-1}(V[X_i]) \tag{12}$$

**Step 4. Reversing Course**: Beginning with the state at the final time step, which has the highest likelihood, backtracking is performed through each observed state (at each time step) to find the most likely path. This algorithm efficiently computes the most likely path in an HMM given an observation sequence by leveraging dynamic programming principles. Using the Viterbi technique, the hidden state sequence that occurs most frequently among the observable sequences can be determined. Viterbi algorithm is applied to find the most happening hidden state sequence by considering an observed sequence. Considering the observed sequence as $Y = \{Y_0, Y_1, Y_0\}$ i.e., {Happy, Sad, Happy}. A flow diagram is constructed to find the most happening sequence (Fig. 5).

The flow diagram shown in Fig. 5 is obtained by backtracking with the help of highest probability from the last observed state. Finally, the most occurring hidden states sequence for the observed set of sequence as $Y = \{Y_0, Y_1, Y_0\}$ i.e., {Happy, Sad, Happy}. Most happening hidden state sequence is $X$ = {Cloudy, Sunny, Cloudy}. The precision of the concealed state's estimation depends on the accuracy of the model already fed with data available. The main aim of this research work is to investigate how the hidden Markov model works and its capabilities.

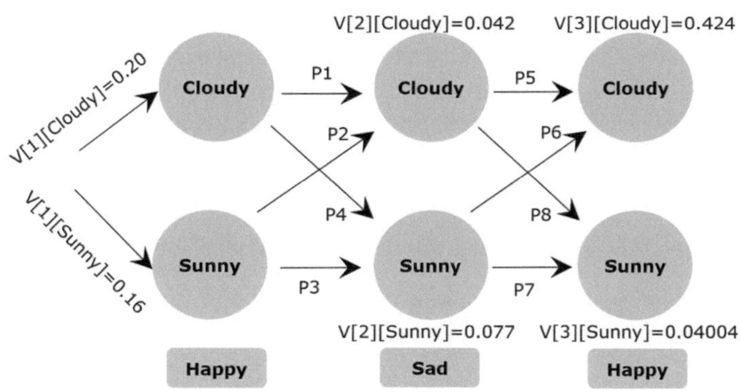


**Fig. 5** Hidden states flow based on observable state flow diagram

# 4 Experimental Results

The forward algorithm in an HMM is more efficient than naive methods, such as brute-force enumeration. They would require calculating the probability for every possible hidden state sequence. Instead, forward algorithm uses dynamic programming to minimize redundant calculations and improve efficiency. The Viterbi algorithm is designed to find the most likely sequence of hidden states (the best path) that results in each observation in sequence of observations in HMM. It is more efficient than previous approaches like brute-force enumeration, which would involve checking every possible hidden state sequence. This is because it uses dynamic programming to systematically eliminate less likely paths during the computation.

Table 1 presents the time complexities of the investigated algorithms. Figures 6 and 7 presents experimental results of comparison between Brute-force algorithm and forward algorithm with fixed time steps $T = 10$, $T = 5$, respectively. Figure 8 presents the comparison between Brute-force algorithm, forward algorithm, and Viterbi algorithm with both $N$ and $T$ variables.

**Table 1** Time complexities of traditional and investigated algorithms

| Algorithm | Time complexity |
|---|---|
| Forward algorithm | $O(N^2T)$ |
| Brute-force algorithm | $O(N^T)$ |
| Viterbi algorithm | $O(N^2T)$ |

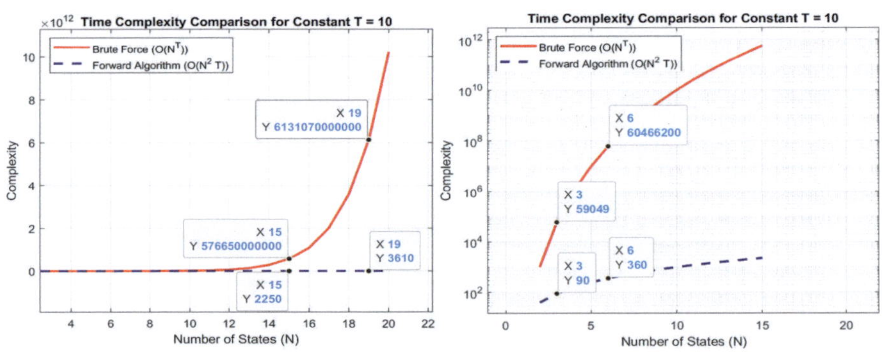


**Fig. 6** Comparison between Brute-force algorithm and forward algorithm with fixed time steps $T = 10$

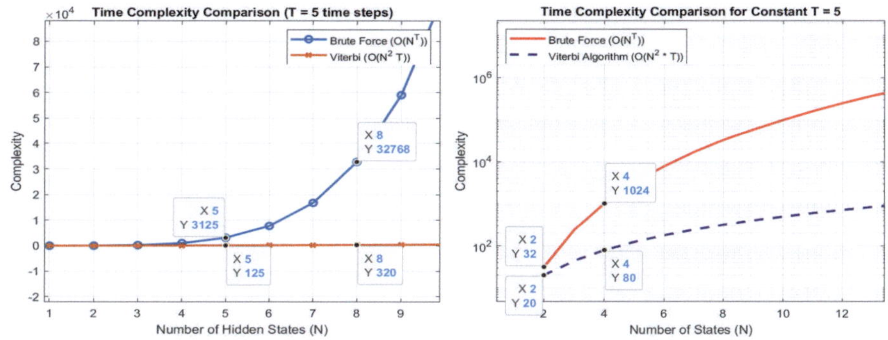


**Fig. 7** Comparison between Brute-force algorithm and Viterbi algorithm with fixed time steps $T = 5$

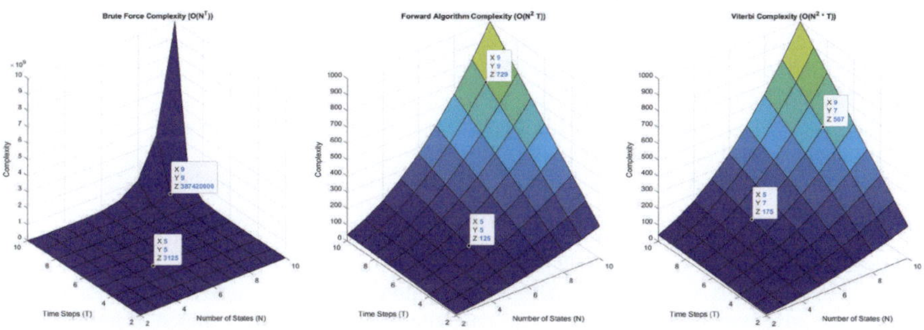

**Fig. 8** Comparison between Brute-force algorithm, forward algorithm, and Viterbi algorithm with both $N$ and $T$ variables

# 5 Conclusions

The application of the hidden Markov model is vast, and it is extremely helpful to determine the future with the help of the situation or criteria that are not observed directly but can be assessed with the help of the states that are observable. For example, consider a patient suffering from a disease. Based on previously recorded hidden and observable data of patient's suffering with same disease, the severity of disease can be assessed with the help of the forward algorithm. The best treatment plan can be determined using the Viterbi algorithm. So, the hidden Markov model can be applied to many fields that need to be assessed prior to getting out of big losses. Along with this, in the future there will be rapid growth of artificial intelligence in every domain. Sometimes there may be a chance of improper behavior in certain conditions. So, priorly, while testing the model developed using artificial intelligence, any chance of improper behavior can be assessed. Along with it we can also know the best path (Hidden States) responsible for improper behavior. With the help of this analysis, one can control the improper behavior of entities.

## References


1. Forney GD (1973) The Viterbi algorithm. Proc IEEE 61(3):268–278
2. Jelinek F (1997) Statistical methods for speech recognition. MIT Press
3. Baum LE, Eagon JA (1967) An inequality with applications to statistical estimation for probabilistic functions of Markov processes and to a model for ecology. Bull Am Math Soc 73(3):360–363
4. Baum LE, Petrie T (1966) Statistical inference for probabilistic functions of finite-state Markov chains. Ann Math Stat 37(6):1554–1563
5. Dempster AP, Laird NM, Rubin DB (1977) Maximum likelihood from incomplete data via the EM algorithm. J Roy Stat Soc 39(1):1–21
6. Eisner J (2002) An interactive spreadsheet for teaching the forward-backward algorithm. In: Proceedings of the ACL workshop on effective tools and methodologies for teaching NLP and CL
7. Sankoff D (1972) Matching sequences under deletion-insertion constraints. Proc Natl Acad Sci 69:4–6
8. Vintsyuk TK (1968) Speech discrimination by dynamic programming. Cybernetics 4(1):52–57
9. Tamposis IA, Tsirigos KD, Theodoropoulou MC, Kontou PI, Bagos PG (2019) Semi-supervised learning of hidden Markov models for biological sequence analysis. Bioinformatics 35(13):2208–2215
10. Pietrzykowski M, Sałabun W (2014) Applications of hidden Markov model: state-of-the-art. Int J Comput Technol Appl 5(4):1384–1391
11. Mor B, Garhwal S, Kumar A (2021) A systematic review of hidden Markov models and their applications. Arch Comput Methods Eng 28:1429–1448
12. Hatala Z, Puturuhu F (2021) Viterbi algorithm and its application to Indonesian speech recognition. J Phys Conf Ser 1752(1):12085
13. Song Y, Adobah B, Qu J, Liu C (2020) Segmentation of ordinary images and medical images with an adaptive hidden Markov model and Viterbi algorithm. Curr Signal Transduct Ther 15(2):109–123
14. Palupi I, Wahyudi BA, Putra AP (2021) Implementation of hidden Markov model (HMM) to predict financial market regime. In 2021 9th international conference on information and communication technology (ICoICT). IEEE, pp 639–644
15. Ibrahim I (2023) Hand gesture recognition system utilizing hidden Markov model for computer visions applications. Hand 9(4)

# Decoding Student Stress Patterns Using LightGBM

**K. G. Suma, Gurram Sunitha, Attar Tabassum, Aragonda Danush Reddy, Borathati Meghana, Chejarla Yaswanth Reddy, and Mohammad Gouse Galety**

**Abstract** Stress in students can be caused by various factors. Academic pressure is a significant source, with heavy workloads, tight deadlines, and high expectations from parents and teachers. Social factors, financial concerns and personal issues, exacerbate stress. Detecting and intervening in student stress is crucial for maintaining academic performance, mental and physical health, social relationships, and long-term success. Early intervention helps prevent the negative effects of stress and supports the overall well-being of students. Machine learning helps predict stress levels by analyzing patterns in physiological and behavioral data. This research explored the impact of machine learning algorithms on decoding student stress patterns, focusing particularly on LightGBM model. The performance of LightGBM model is compared against decision tree, random forest, Naïve Bayes, and GBM. The experimental results reveal an accuracy score of 96% for LightGBM. Further research in this domain can focus on developing targeted interventions to mitigate negative impacts of mental stress on students.

**Keywords** Education · Student life · Stress patterns · Machine learning · LightGBM

K. G. Suma
School of CSE, VIT-AP, Amaravati, Andhra Pradesh, India

G. Sunitha (✉)
Department of Computer Science and Engineering, School of Computing, Mohan Babu University, Tirupati, Andhra Pradesh, India
e-mail: gurramsunitha@gmail.com

A. Tabassum · A. D. Reddy · B. Meghana · C. Y. Reddy
Department of Computer Science and Engineering, Sree Vidyanikethan Engineering College, Tirupati, Andhra Pradesh, India

M. G. Galety
School of Computing, Samarkand International University of Technology, Samarkand, Uzbekistan

 
K. R. Madhavi et al. (eds.), *Proceedings of Sixth International Conference on Computer and Communication Technologies*, Lecture Notes in Networks and Systems 1356,
https://doi.org/10.1007/978-981-96-5238-9_7

# 1 Introduction

Stress is a dominant issue affecting academic performance and overall well-being of students. The pressures of academic responsibilities, social dynamics, and personal challenges contribute to elevated stress levels among students. Stress can lead to various problems such as decreased academic performance, impaired mental health, and diminished quality of life [1].

In today's society, where academic success is highly valued, understanding and addressing student stress has become inevitable challenge [2]. However, traditional approaches to identifying and managing student stress may be limited in their effectiveness due to the complexity and multifactorial nature of stress [3, 4].

Machine learning algorithms offer a promising path for decoding student stress patterns efficiently [5]. Machine learning algorithms can analyze large datasets comprising of diverse factors such as psychological, physiological, environmental, academic, and social variables, etc. The idea is to uncover complex patterns and relationships that may not be apparent through traditional methods [6].

This research explored the impact of machine learning algorithms on decoding student stress patterns, focusing particularly on Light Gradient Boosting Machine (LightGBM). Also, performance of LightGBM is compared with the performance of other machine learning models—Naïve Bayes, decision tree, random forest regression, and GBM (Gradient Boosting Machine). The aim is to assess the accuracy and efficiency of each algorithm in predicting student stress levels, highlighting the strengths and limitations of each approach. This research aims to address important societal issue of mental health and to help improve student well-being and academic success in today's world.

# 2 Related Works

Students experience high stress levels, which vary throughout their education life [7, 8]. Physical health declines over time, with students reporting less physical activity and more health issues as they progress [9]. Emotional health also deteriorates, with increased anxiety, depression, and burnout reported over the years. There is a critical need for targeted interventions to address these health issues and support students' well-being [10–12].

Various machine learning methods have been investigated for stress monitoring [13]. Wearable sensors and devices have been developed for real-time monitoring and timely intervention of stress levels in students [14]. The focus is on integrating IoT with wearable sensors for better stress monitoring solutions in healthcare. A study aims to develop an automated stress recognition [15]. The findings highlight the effectiveness of combining virtual reality with machine learning for stress detection, suggesting potential applications in healthcare and rehabilitation.

LightGBM emerges as a potent and efficient gradient boosting framework with a broad spectrum of technical applications [16]. LightGBM's distinctive leaf-wise tree growth strategy, combined with histogram-based learning, significantly contributes to expedited and more precise predictions. Thus, establishing it as a valuable asset in predictive modeling. The versatility of LightGBM extends to diverse domains, encompassing classification, regression, and ranking tasks. Its efficiency, speed, and robustness assume a crucial role in achieving superior predictive performance.

The progression from statistical techniques to immersive technologies and the integration of diverse algorithms reflects the dynamic nature of stress detection methodologies. This research underscores the evolving understanding of stress triggers and the importance of personalized, technology-driven interventions to promote student well-being.

## 3 Proposed Methodology

Proposed learning model leverages LightGBM, a gradient boosting framework, to decode patterns of student stress effectively. By analyzing various stress indicators and academic performance data, the model identifies key stress factors impacting students. The LightGBM model's efficiency in handling large datasets ensures accurate and timely insights. This approach aims to provide actionable recommendations for mitigating student stress and enhancing overall well-being. The proposed model learns stress patterns as discussed below (Fig. 1).

Decision trees are base learners in LightGBM. LightGBM constructs an ensemble of decision trees. Histogram-based learning and leaf-wise growth techniques are used for construction of decision trees. Multiple decision trees are constructed sequentially, where each new tree aims to correct the errors made by the previous trees. This learning process handles overfitting of model efficiently. LightGBM is proven to be efficient in handling large datasets. It can interpret feature relevance appropriately. Thus, making the learning model more accurate and robust. LightGBM has ability to handle categorical variables, interpret feature importance, making it a fitting choice for stress prediction.

In LightGBM, each decision tree $T_i$ consists of rules that partition the input space into disjoint regions. Let $\{x_1, x_2, \ldots, x_m\}$ be the features of the dataset, and $y$ be the target variable. At each node $j$ of the tree, a split is made based on a feature xk and a threshold $t$. The split function $f(x)$ is defined in Eq. (1).

$$f(x) = \begin{cases} 1, & \text{if } xk \ <= t \\ 0, & \text{otherwise} \end{cases} \tag{1}$$

The tree recursively splits the data until a stopping criterion is met, such as maximum depth or minimum samples per leaf.

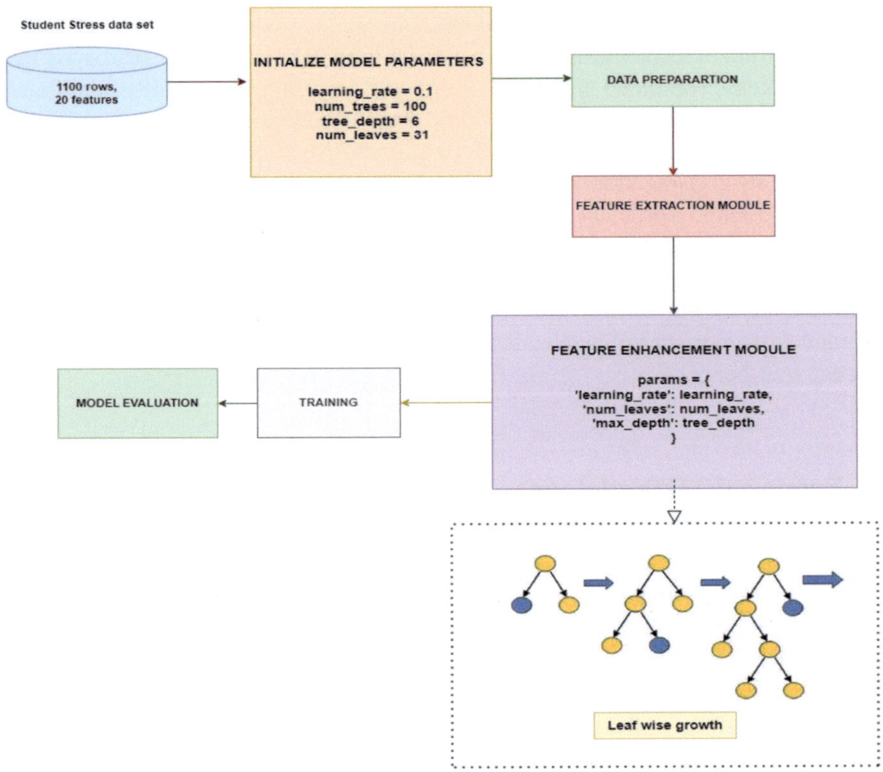


**Fig. 1** Proposed model for decoding student stress patterns using LightGBM

The objective function of LightGBM combines a loss function and a regularization term (Eq. 2). Combining these, the objective function balances the model's accuracy and complexity, preventing overfitting by penalizing large trees and high leaf weights. This function is minimized during training (on *N* samples) to improve the model's performance.

$$L(\theta) = \sum_{i=1}^{N} l\left(A_i, \hat{A}_i\right) + \Omega(L) \tag{2}$$

where $\theta$ represents the model parameters; $A_i$, $\hat{A}_i$ are the true and predicted labels of the *i*th sample, respectively; $l()$ is the loss function; $\Omega(T)$ is the regularization term for *k*th tree. $\Omega(L)$ is defined in Eq. (3).

$$\Omega(L) = \mathrm{RT} + \frac{1}{2}\lambda \sum_{j=1}^{L} w_j^2 \tag{3}$$

where $R$, $\lambda$ are regularization parameters that control number of leaves in the tree and L2 regularization on leaf weights, respectively; $W_j$ is weight of leaf $j$; there are $L$ number of leaves in the $k$th tree.

Gradient boosting involves iteratively fitting new trees to correct the errors of the previous ones. Equation (4) defines the update rule for each boosting.

$$F_{\text{new}}(x) = F_{\text{old}}(x) + \eta \cdot \text{argmin}_h \sum_{i=1}^{N} l(y_i, F_{\text{old}}(x_i) + h(x_i)) + \Omega(h) \quad (4)$$

where $Fnew(x)$ is the updated prediction; $\eta$ is the learning rate controlling the step size in the update; $h(x_i)$ is the prediction of the new tree for the $i$th sample.

The tree structure involves splitting nodes based on the best conditions to best reduce the loss. The split finding process aims to minimize the loss function (Eq. (5)).

$$\text{split}^* = \text{argmin}_{\text{split}} \left[ \frac{1}{I_{\text{left}}} \sum_{\text{data in left child}} l(y_i, \hat{y}_i) + \frac{1}{I_{\text{right}}} \sum_{\text{data in right child}} l(y_i, \hat{y}_i) \right] \quad (5)$$

where split* is the optimal split; $I_{\text{left}}$, $I_{\text{right}}$ are number of data points in the left child and right child, respectively, after the split.

In LightGBM, leaf-wise growth builds decision trees by splitting nodes based on the maximum reduction in loss, unlike traditional depth-wise methods that expand level by level. This approach focuses on nodes with the greatest potential for improvement, resulting in faster convergence and efficient computation. By prioritizing high-impact nodes, it significantly reduces training time and enhances scalability. This allows LightGBM to handle large datasets with millions of samples and thousands of features while maintaining high performance and accuracy. Leaf-wise growth in LightGBM thus enables rapid and efficient model training, making it suitable for large-scale applications requiring quick model development and deployment.

In LightGBM, the leaf output calculation determines the value assigned to each leaf in a decision tree. This value represents the prediction for all samples that fall into that leaf. Let Gradient $g_i$, Hessian $hi$ be measures of first and second derivatives of the loss function of $i$th sample w.r.t the prediction, respectively. Let $G_j$, $H_j$ be sum of gradients, hessians, respectively, of a leaf $j$. The output $w$ for leaf $j$ is given in Eq. (6).

$$w_j = -\frac{G_j}{H_j + \lambda} \quad (6)$$

The proposed model aims to classify stress levels as low, medium, or high based on refined features. This classification helps create tailored strategies and support plans for individual student needs. These strategies can include academic support or promoting community engagement, based on predicted stress levels. The goal is to foster a supportive environment for students' well-being and academic success.

## 4 Experimentation Results and Discussion

The dataset "Student Stress Factors" comprises a collection of variables aimed at understanding the various dimensions of stress experienced by students [17]. It covers psychological, physiological, environmental, academic, and social factors, consisting of 1100 records and 20 different features. This dataset inputs offer a comprehensive understanding of the multiple aspects contributing to student well-being (Table 1). Table 2 presents the optimal hyperparameter setup determined after extensive experimentation.

Figure 2 shows the ROC curve obtained on experimenting with LightGBM for decoding stress patterns. Figure 3 and Table 3 shows the performance of LightGBM model against decision tree, random forest, Naïve Bayes, and GBM. The results reveal a high level of model accuracy, with an overall accuracy score of 96%. Decision tree and Naïve Bayes follow closely, while random forest has the lowest accuracy (0.8727). This indicates that the LightGBM predicts the stress levels more accurately than other compared models.

Precision is high across all models and classes, especially for Naïve Bayes and LightGBM. LightGBM shows the highest precision for all stress levels (Low: 0.94,

**Table 1** Details of dataset

| Factors | Description |
|---|---|
| Academic | Study load, teacher-student relationship, academic performance, future career concerns |
| Environmental | Basic needs, noise level, living conditions, safety |
| Psychological | Mental health history, self-esteem, anxiety level, depression |
| Social | Bullying, peer pressure, social support, extracurricular activities |
| Physiological | Breathing problem, headache, blood pressure, sleep quality |

**Table 2** Hyperparameter setup

| Parameter | Value |
|---|---|
| Number of classes | 3 |
| Evaluation metric | Multi-class logarithmic loss |
| max_depth | 12 |
| min_split_gain, Learning rate | 0.01 |
| min_child_samples | 20 |
| feature_fraction, colsample_bytree, Subsample | 0.8 |
| num_leaves | 50 |
| reg_alpha, reg_lambda | 1.0 |
| min_child_weight | 0.001 |
| Boosting_type | DART |

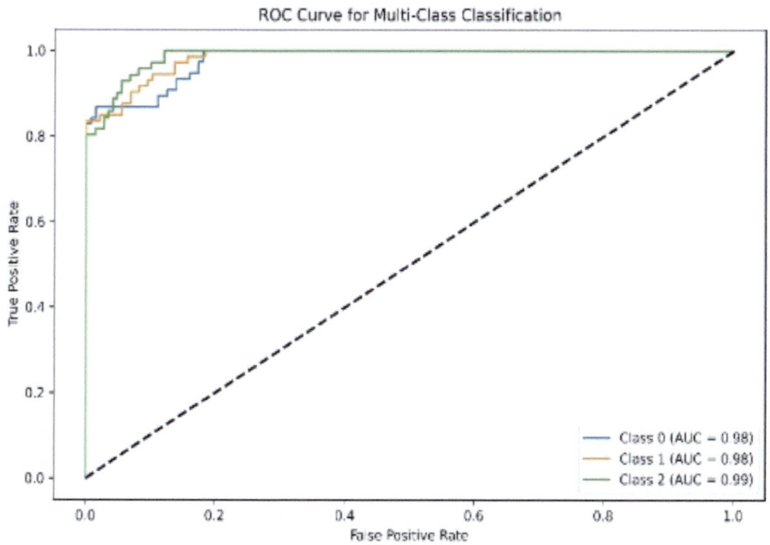


**Fig. 2** ROC curve of the proposed model

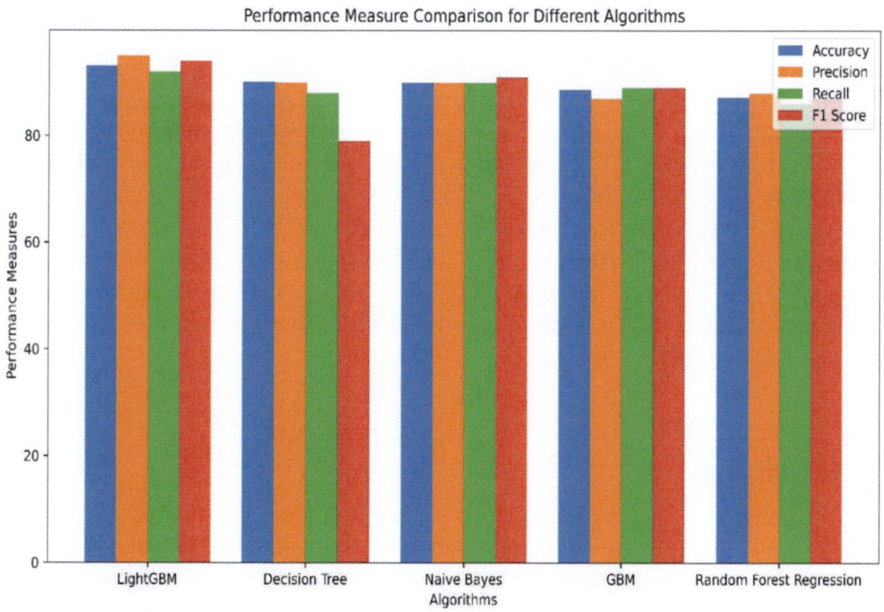


**Fig. 3** Performance of deep models for decoding stress patterns

**Table 3** Performance of deep models for decoding stress patterns

| Class (stress level) | Model | Precision | Recall | F1-score | Accuracy |
|---|---|---|---|---|---|
| Low | Decision tree | 0.89 | 0.87 | 0.89 | 0.9018 |
| | Random forest | 0.84 | 0.87 | 0.87 | 0.8727 |
| | Naïve Bayes | 0.93 | 0.86 | 0.88 | 0.9201 |
| | GBM | 0.87 | 0.87 | 0.87 | 0.8863 |
| | LightGBM | 0.94 | 0.89 | 0.91 | 0.9609 |
| Medium | Decision tree | 0.87 | 0.91 | 0.90 | 0.9018 |
| | Random forest | 0.90 | 0.86 | 0.88 | 0.8727 |
| | Naïve Bayes | 0.95 | 0.85 | 0.89 | 0.9201 |
| | GBM | 0.88 | 0.90 | 0.89 | 0.8863 |
| | LightGBM | 0.98 | 0.93 | 0.93 | 0.9609 |
| High | Decision tree | 0.91 | 0.89 | 0.90 | 0.9018 |
| | Random forest | 0.88 | 0.86 | 0.87 | 0.8727 |
| | Naïve Bayes | 0.81 | 0.92 | 0.88 | 0.9201 |
| | GBM | 0.91 | 0.89 | 0.90 | 0.8863 |
| | LightGBM | 0.95 | 0.96 | 0.94 | 0.9609 |

Medium: 0.98, High: 0.95), indicating fewer false positives. Recall is consistently high, with LightGBM achieving the highest recall values across all stress levels (Low: 0.89, Medium: 0.93, High: 0.96). Naïve Bayes shows a significant drop in recall for the medium stress level (0.85) but compensates with a high precision. LightGBM outperforms other models in terms of F1-Score for all stress levels, indicating a good balance between precision and recall. Decision tree and GBM also perform well but are slightly behind LightGBM.

LightGBM's boosting technique helps to optimize both precision and recall, leading to high F1 Scores. LightGBM effectively optimized precision and recall through boosting framework.

## 5 Conclusion

This research examines student stress dynamics in educational settings. By using advanced machine learning techniques, researchers gain insights into complex phenomena of student stress. Machine learning helps predict stress levels by analyzing patterns in physiological and behavioral data. This research explored the impact of machine learning algorithms on decoding student stress patterns, focusing particularly on LightGBM model. The experimental results show LightGBM's effectiveness in decoding stress patterns and identifying key factors. This research can promote student well-being and academic success. This study also informs policymakers and educators about creating supportive environments for student well-being.

This research demonstrates machine learning's potential in addressing educational challenges. As machine learning evolves, there are opportunities to refine models for better understanding student well-being. In summary, this study advances understanding of student stress and its causes. It contributes to efforts to create supportive educational environments. Continued research and collaboration are crucial for developing effective strategies to promote student well-being and success.

## References


1. Kennett DJ, Quinn-Nilas C, Carty T (2021) The indirect effects of academic stress on student outcomes through resourcefulness and perceived control of stress. Stud High Educ 46(11):2460–2472
2. Li ZS, Hasson F (2020) Resilience, stress, and psychological well-being in nursing students: a systematic review. Nurse Educ Today 90:104440
3. Pascoe MC, Hetrick SE, Parker AG (2020) The impact of stress on students in secondary school and higher education. Int J Adolesc Youth 25(1):104–112
4. de la Fuente J et al (2020) Effects of self-regulation vs. external regulation on the factors and symptoms of academic stress in undergraduate students. Front Psychol 11:1773
5. Pabreja K, Singh A, Singh R, Agnihotri R, Kaushik S, Malhotra T (2021) Stress prediction model using machine learning. In: Proceedings of international conference on artificial intelligence and applications. Springer Singapore, pp 57–68
6. Baba A, Bunji K (2023) Prediction of mental health problem using annual student health survey: machine learning approach. JMIR Ment Health 10:e42420
7. Mofatteh M (2021) Risk factors associated with stress, anxiety, and depression among university undergraduate students. AIMS public health 8(1):36
8. Sowmya TS, Narasimhulu T, Sunitha G, Manikanta T, Venkatesh T (2023) Vision transformer based ResNet model for pneumonia prediction. In: 4th international conference on electronics and sustainable communication systems. IEEE, pp 316–321
9. McKerrow I, Carney PA, Caretta-Weyer H, Furnari M, Miller Juve A (2020) Trends in medical students' stress, physical, and emotional health throughout training. Med Educ Online 25(1):1709278
10. Sudarsana Murthy D et al (2022) An investigative study of shallow, deep and dense learning models for breast cancer detection based on microcalcifications. In: International conference on data science, agents & artificial intelligence, vol 1. IEEE, pp 1–6
11. Waechter R et al (2021) Mitigating medical student stress and anxiety: should schools mandate participation in wellness intervention programs? Med Teach 43(8):945–955
12. Van Ryzin MJ, Roseth CJ (2021) The cascading effects of reducing student stress: cooperative learning as a means to reduce emotional problems and promote academic engagement. J Early Adolesc 41(5):700–724
13. Monisha S, Meera R, Swaminath V (2020) Predictive analysis of student stress level using Naïve Bayesian classification algorithm. In: International conference on computer communication and informatics. IEEE, pp 1–7
14. Talaat FM, El-Balka RM (2023) Stress monitoring using wearable sensors: IoT techniques in medical field. Neural Comput Appl 35(25):18571–18584

15. Migovich M, Adiani D, Breen M, Swanson A, Vogus TJ, Sarkar N (2024) Stress detection of autistic adults during simulated job interviews using a novel physiological dataset and machine learning. ACM Trans Access Comput 17(1):1–25
16. Ke G et al (2017) LightGBM: A highly efficient gradient boosting decision tree. In: 31st conference on neural information processing Systems, pp 3146–3154
17. Student Stress Factors (2023) https://www.kaggle.com/datasets/samyakb/student-stress-factors/data. Accessed Oct 2023

# From TF-IDF to BERT: Revolutionizing Clinical Text Classification with Modern Word Embeddings

**N. Minni and G. Kumaravelan**

**Abstract** This paper studies the natural language processing applications in health area with the use of advanced deep learning models like RNN, LSTM, BiLSTM, or GRU for the classification of clinical text. Literature review of relevant works had looked at the role of word embedding methods distinctly in improving the performance of NLP models. In this paper, the traditional TF-IDF and GloVe methods are compared with the recent transformer-based embedding methods which include BERT, RoBERTa, and DistilBERT. The research employs a clinical transcription dataset from Kaggle, demonstrating the utility of word embeddings when training the model with keyword description retrieved from clinical texts. In addition, the work emphasizes that the use of transformer-based embedding methods can improve the performance of clinical text classification tasks, thereby improving decision-making in healthcare. The study offers evidence on the need for optimal selection of combinations of word embeddings and models for use with clinical data in text classification tasks. It is determined that BiLSTM with RoBERTa combination achieves the best performance for clinical text classification in healthcare NLP systems among other combinations interrogated.

**Keywords** Natural language processing · Clinical text classification · Deep learning models · Word embedding techniques · Transformer-based embeddings · BERT · RoBERTa · DistilBERT · Healthcare

N. Minni (✉) · G. Kumaravelan
Pondicherry University Karaikal Campus, Puducherry, India
e-mail: minnimca@pondiuni.ac.in

G. Kumaravelan
e-mail: gkumaravelanpu@pondiuni.ac.in



K. R. Madhavi et al. (eds.), *Proceedings of Sixth International Conference on Computer and Communication Technologies*, Lecture Notes in Networks and Systems 1356,
https://doi.org/10.1007/978-981-96-5238-9_8

## 1 Introduction

Clinical text classification is an important domain in healthcare analysis and research. Deep learning models such as RNN, LSTM, BiLSTM, and GRU have given promising outcomes in this field. These models are capable of detecting complex patterns and identify contextual details in clinical texts. Thus, they enable precise medical records categorization [1]. LSTM has an advantage in retaining data for the long-term dependencies, while BiLSTM and GRU models provide forward and backward information for clinical text classification. Thereby, deep learning models such as RNNs, LSTMs, and GRUs have made a long-lasting effect on the clinical text analysis that enhance speed and precision of patient care [2, 3].

One-hot encoding, TF-IDF, and GloVe are familiar word embedding techniques that have been in frequent use but do not explain contextual inference of words. Rather, word2vec and FastText offered a basic introduction to dense vector representations of words by considering their context relationships [4]. The welcoming of transformer-based models offers mechanisms that assist constitute time and capture long-range dependencies between words, addressing the shortcomings of traditional techniques [5].

In this study, we closely examined the NLP models such as RNN, LSTM, BiLSTM, and GRU especially for clinical text classification. Classical word embedding techniques are compared against modern transformer-based embeddings and their performances are analyzed. We initially experimented with long clinical transcripts, but the outcomes were suboptimal. So we opted to use the key descriptions extracted from the lengthy text which significantly improved the outcomes [6]. This study makes use of the clinical transcription dataset from Kaggle.

The aim of this paper is to assess how effective the word embedding techniques work in conjunction with different NLP models. Transformer-based embeddings have the potential to improve the performance of NLP-based models in clinical text classification.

## 2 Related Work

Authors of [7] have proposed a hybrid deep learning model that integrates various word embeddings, LSTM, GRU, BiLSTM, and CNN to perform sentiment classification. As discussed in [8], the use of different word embedding techniques which caution on how to choose for appropriate combination to train deep learning models was explored. A review of deep learning and also word embedding methods in text mining tasks was conducted as given in [9] with an aim of choosing suitable approaches through a comparative analysis.

Three word embedding models such as GloVe, FastText, word2vec with conventional neural network in text classification were compared according to [10], and FastText is performed well with accuracy of 97.2% in the classification process.

Authors of [11] proposed a pretrained model called RoBERTa is a model by fine-tuning some parameters of BERT model. It achieves leading performance on GLUE, RACE, and SQuAD datasets without requiring multi-task fine-tuning.

BERT models were investigated for clinical text. The BERT model trained on clinical data exhibited impressive accuracy and F1-scores, making it suitable for further clinical NLP research applications. This process is clearly explained in [12].

A survey of various word representation methodologies was conducted on clinical text datasets as explored in [13]. The author proposed that the volume and scope of the training data, the specific domain covered by the corpus, the attributes of the embedding model utilized, and the overall quality of the embeddings produced should be considered for training of word embeddings.

## 3 Methodology

### *3.1 Workflow Framework*

The framework suggested in this paper aims at enhancing clinical decision support through NLP. Figure 1 shows the sequence of steps involved in this framework. This workflow allows researchers to systematically gather, process, and analyze clinical text data, enhancing clinical decision-making [14].

The first step is to access relevant clinical databases with patient records, diagnostic information, and treatment histories. Next, extract clinical notes, such as physician narratives, from the database. Then, clean and preprocess the text to remove noise, standardize formatting, and tokenize sentences. Convert the preprocessed text into numerical vectors suitable for NLP algorithms. Develop and train NLP models for analysis, and assess their performance using metrics like accuracy, loss, and F1-score. Finally, identify the optimal model and word embedding combination based on performance analysis.

### *3.2 Dataset*

The dataset employed in this study comprises medical transcriptions and was sourced from Kaggle, encompassing 4999 samples with 6 features. Among these features, "medical_specialty" functions as the class variable, delineating 39 distinct classes [15]. For the purposes of text classification, the features "transcription" and "keywords" were selected, while any extraneous features deemed unnecessary for this investigation were disregarded.

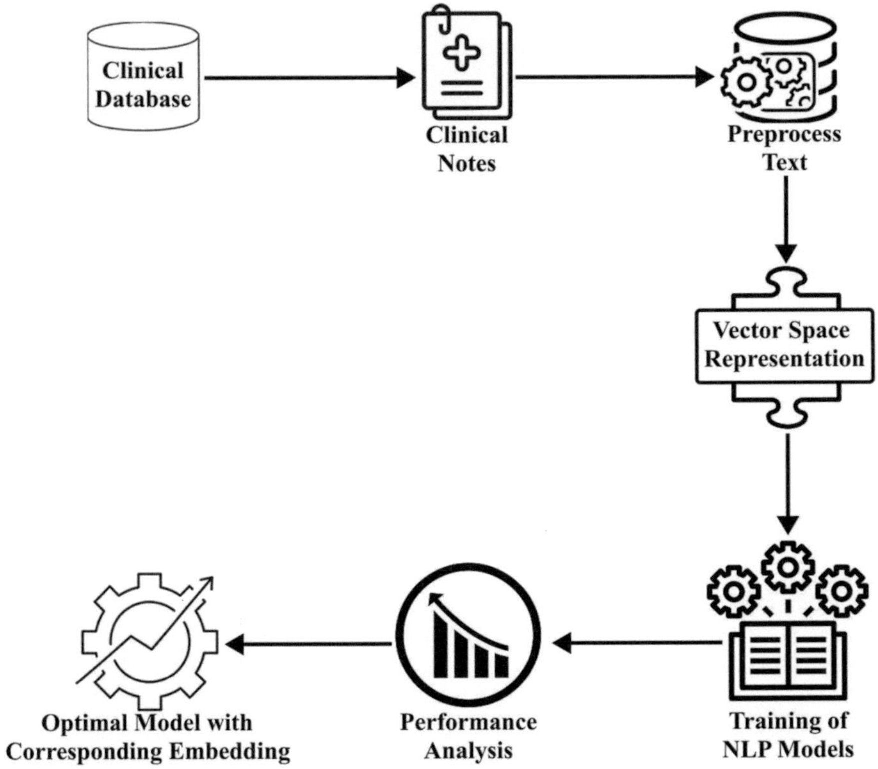


**Fig. 1** Workflow framework

### *3.3 Feature Selection and Class Reduction*

The dataset initially consists of 4999 samples with huge number of classes. The initial step is to check for null values in the dataset and they are removed. Since this dataset is slightly imbalanced, in order to reduce the complexity and enhance the effectiveness of the models, classes having more than 100 samples only are taken for consideration. Hence, the number of classes is reduced to 10. From the dataset, the clinical text and the keywords features are considered for classification. All the other features are removed from the dataset [6].

### *3.4 Data Preprocessing*

Data preprocessing involves techniques to prepare raw data for further processing, with NLP methods aiding this process. Our pipeline includes word tokenization, data padding, truncation, and various word embedding techniques. First, null values are removed, and the raw text is converted to lowercase and tokenized. Stop words, which lack semantic meaning, are eliminated, and lemmatization is performed. The preprocessed text is then stored in a corpus for analysis [14, 16].

### 3.5 Word Embeddings

Word embedding techniques convert words to vectors, which is essential for every NLP model. Hence, this corpus needs to be converted to vectors before being inputted to the NLP model [9]. In this paper, we have categorized the word embedding techniques as traditional, and transformer-based embeddings. Traditional word embeddings reflect early and widely used methods of representing words based on co-occurrence and static vectors. Transformer-based embeddings utilize deep transformer models to capture complex relationships and contextual nuances in text. The traditional word embeddings taken for analysis are Word2Vec, TF-IDF vectorizer, and one-hot encoding. FastText and GloVe are explored as traditional word embeddings with improved techniques. For transformer-based embeddings, we have taken BERT, RoBERTa, and DistilBERT for analysis [5, 11, 17].

### 3.6 Model Training

After the preprocessed text is represented in the form of vectors, they are ready for inputting to the NLP model. The data is partitioned into training and validation set with 80% of data used for training and remaining 20% for validation. This prevents data leakage as no trained data will be checked during validation. During training, we select the optimizer as Adam as it has proved its efficiency for training deep neural networks among other optimizers [18] with different learning rates and sparse-categorical cross-entropy as loss function since it is a multi-class classification. Each NLP model RNN, LSTM, BiLSTM, and GRU is trained and analyzed with each of the word embedding technique. Their efficiencies are analyzed using the performance metrics like accuracy, validation loss, and F1-score.

## 4 Experimental Results and Discussion

In this paper, we examine the efficiency of various word embedding techniques when applied to basic NLP models. The analysis is conducted through two approaches: one using lengthy clinical transcripts and the other utilizing keywords extracted from these transcripts.

Our experiments with the entire clinical transcripts revealed that the models could not achieve satisfactory accuracy, even with extensive hyperparameter tuning.

Consequently, we adopted an alternative approach by focusing on keywords. These keywords, extracted from the longer clinical transcripts, contain key medical terms essential for identifying the relevant class. On average, a transcript contains 460 words, whereas the extracted keywords consist of about 30 words, sufficient for disease classification.

We trained the models using these keywords and performed classification based on the extracted terms.

## *4.1 Traditional Word Embeddings*

a. Analysis of Word2Vec, Tfid and One-Hot Encoding

The models are trained separately with each of these embeddings and their efficiencies are analyzed. The hyperparameters explored for the model training are given in Table 1, and the accuracies obtained in % are reported in Table 2.

On analyzing these results, altogether TfidVectorizer achieves high accuracies of the other two embeddings with accuracies ranging from 91 to 94%. Among the models, BiLSTM gives good performance for all the 3 word embedding techniques. Overall the best results are obtained when the data is embedded with TfidVectorizer and trained with model GRU and achieves the highest accuracy of 94.01%.

The cumulative validation accuracies and validation losses achieved on all the four models for Tfidvectorizer in Fig. 2.

b. Analysis of GloVe and FastText

**Table 1** Hyperparameters explored

| Hyperparameters | Iterated values | Optimal value |
|---|---|---|
| Batch size | 16, 32, 64 | 64 |
| Learning rate | 0.001, 0.0001 | 0.001 |
| Epochs | 20, 50, 100 | 100 |

**Table 2** Accuracy comparison of models

| Models | Word embedding techniques | | |
|---|---|---|---|
| | Word2Vec | Tfid | One hot |
| LSTM | 69.83 | 91.77 | 88.69 |
| BiLSTM | 75.81 | 93.77 | 90.70 |
| RNN | 68.83 | 92.52 | 87.94 |
| GRU | 67.58 | 94.01 | 88.44 |

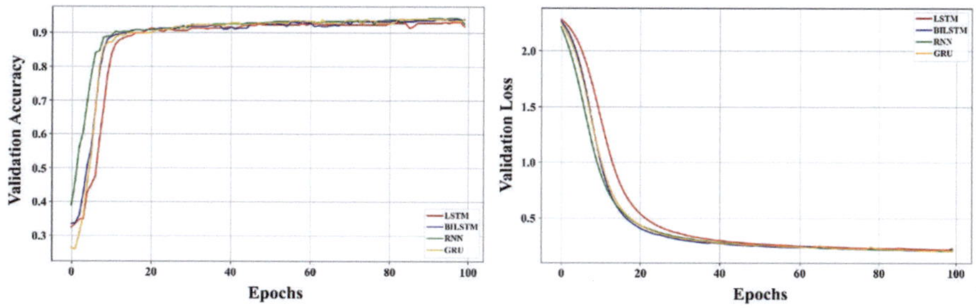


**Fig. 2** Cumulative validation accuracy and validation loss of TfidVectorizer

The hyperparameters fine-tuned for model training with these embeddings are given in Table 3. The accuracies and their F1-scores obtained in % on these experiments are reported in Table 4.

On analyzing the results for the embeddings GloVe and FastText, it is observed that when trained with GloVe embedding GRU achieves the highest accuracy of 98.99% and F1-score of 99.00% .With both embeddings, RNN's poor performance suggests it may not be suitable for complex tasks compared to other models.

The cumulative validation accuracies and validation losses achieved for the GloVe embedding is shown in Fig. 3.

**Table 3** Hyperparameters— GloVe and FastText

| Hyperparameters | Iterated values | Optimal value |
|---|---|---|
| Batch size | 16, 32, 64 | 64 |
| Learning rate | 0.001, 0.0001 | 0.001 |
| Epochs | 10,20,50 | 20 |

**Table 4** Accuracy comparison-GloVe and FastText

| Models | GloVe | | FastText | |
|---|---|---|---|---|
| | Acc | F1 | Acc | F1 |
| LSTM | 98.99 | 99.00 | LSTM | 98.99 |
| BiLSTM | 98.74 | 98.75 | BiLSTM | 98.74 |
| RNN | 48.49 | 47.10 | RNN | 48.49 |
| GRU | 99.25 | 99.25 | GRU | 99.25 |

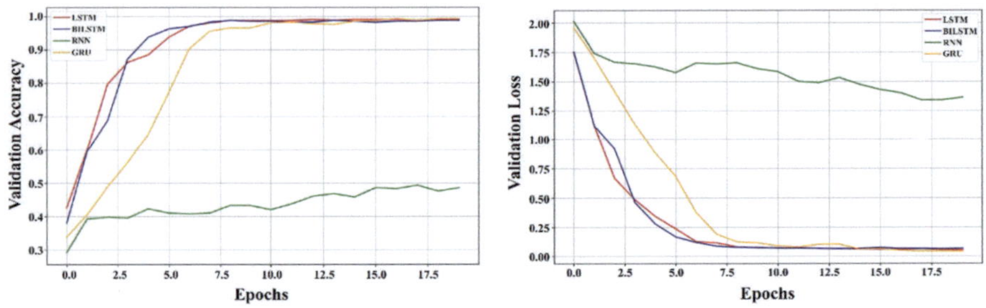


**Fig. 3** Cumulative validation accuracy and validation loss of GloVe

## *4.2 Transformer-Based Embeddings*

We selected BERT, RoBERTa, and DistilBERT embeddings to analyze their efficiencies with various NLP models. The hyperparameters fine-tuned for these combinations are detailed in Table 5. All models exhibited significant improvements in accuracy with these embeddings. Consequently, we also analyzed their F1-scores and validation losses, with the results presented in Table 6.

Based on the performance metrics achieved for BERT, RoBERTa, and DistilBERT embeddings across LSTM, BiLSTM, RNN, and GRU models, it can be derived that RoBERTa consistently achieves the highest accuracy across all models, with BiLSTM performing the best at 0.9950, followed closely by LSTM at 0.9884. This shows that RoBERTa has a better capacity to analyze and the ability to process the dataset. When comparing the loss values, RoBERTa is the lowest for all models and BiLSTM achieves the minimum loss of 0.0174. It suggests that RoBERTa gives better performance in terms of accuracy and has a stable and efficient training.

**Table 5** Hyperparameters used for BERT type embeddings

| Hyperparameters | Iterated values | Optimal value |
|---|---|---|
| Batch size | 16, 32 | 16 |
| Learning rate | 1e-5, 2e-5, 3e-5 | 3e-5 |
| Epochs | 5, 10, 10 with early stopping | 10 with early stopping |

**Table 6** Analysis of accuracies, losses, and F1-scores of BERT, RoBERTa, DistilBERT

| Model | BERT | | | RoBERTa | | | DistilBERT | | |
|---|---|---|---|---|---|---|---|---|---|
| | Accuracy in % | Loss | F1-score in % | Accuracy in % | Loss | F1-score in % | Accuracy in % | Loss | F1-score in % |
| LSTM | 98.51 | 0.0402 | 98.55 | 98.84 | 0.0419 | 98.86 | 98.35 | 0.0493 | 98.39 |
| BiLSTM | 99.34 | 0.0216 | 99.18 | 99.50 | 0.0174 | 99.51 | 98.84 | 0.0269 | 98.86 |
| RNN | 99.01 | 0.0350 | 99.03 | 99.17 | 0.0211 | 99.19 | 98.68 | 0.0426 | 98.72 |
| GRU | 98.18 | 0.0534 | 98.24 | 98.18 | 0.0487 | 98.86 | 98.51 | 0.0531 | 98.55 |

## 5 Conclusion

In conclusion, this paper highlights the role of word embedding techniques in enhancing the performance of the deep learning models such as RNN, LSTM, BiLSTM, and GRU especially in the field of health care. The evaluation of several approaches of word embeddings from the traditional models to the modern transformer-based word embeddings shows a significant impact on the model complexity of NLP tasks. Hence, whereas analyzing the results, it remained clear that using RoBERTa embeddings with BiLSTM built the best performing model with the optimal results presented with the highest F1-score, accuracy, and the least loss value. Such results provide directions on how to incorporate transformer-based embeddings in improving the performance of NLP models. A future research can investigate the transformer architectures like BERT in handling long clinical text which will improve the patient care. This study provides insights for future research in state-of-art applications of healthcare NLP and consequently furthers the practice of clinical decision support systems and medical practice.

## References


1. Nagarhalli TP, Vaze V, Rana NK (2021) Impact of machine learning in natural language processing: a review. In: 2021 Third international conference on intelligent communication technologies and virtual mobile networks (ICICV). IEEE
2. Saha BN, Senapati A (2020) Long short term memory (LSTM) based deep learning for sentiment analysis of English and Spanish data. In: 2020 International conference on computational performance evaluation (ComPE). IEEE, Shillong, India, pp 442–446
3. Behera B, Kumaravelan G (2021) Text document classification using fuzzy rough set based on robust nearest neighbor (FRS-RNN). Soft Comput 25:9915–9923
4. Mikolov T, Grave E, Bojanowski P, Puhrsch C, Joulin A (2018) Advances in pre-training distributed word representations. In: Proceedings of the eleventh international conference on language resources and evaluation (LREC 2018). European Language Resources Association, Miyazaki, Japan (ELRA)
5. Devlin J, Chang MW, Lee K, Toutanova K (2018) BERT: pre-training of deep bidirectional transformers for language understanding. arXiv preprint arXiv:1810.04805
6. Tan C et al (2022) Machine learning in clinical text classification: specialty identification and COVID-19 risk. In: 2022 13th international conference on information and communication technology convergence (ICTC). IEEE, Jeju Island, Korea, Republic of, pp 1366–1370
7. Salur MU, Aydin I (2020) A novel hybrid deep learning model for sentiment classification. IEEE Access
8. Nguyen HN, Teerakanok S, Inomata A, Uehara T (2021) The comparison of word embedding techniques in RNNs for vulnerability detection
9. Asudani DS, Nagwani NK, Singh P (2023) Impact of word embedding models on text analytics in deep learning environment: a review. Artif Intell Rev 56:10345–10425
10. Dharma EM, Gaol FL, Warnars HLHS, Soewito BENFANO (2022) The accuracy comparison among word2vec, GloVe, and FastText towards convolutional neural network (CNN) text classification. J Theor Appl Inf Technol 100(2):349–359
11. Liu Y, Ott M, Goyal N, Du J, Joshi M, Chen D et al (2019) RoBERTa: a robustly optimized BERT pretraining approach. arXiv preprint arXiv:1907.11692

12. Alsentzer E, Murphy JR, Boag W, Weng WH, Jin D, Naumann T, McDermott M (2019) Publicly available clinical BERT embeddings. arXiv preprint arXiv:1904.03323
13. Khattak F, Jeblee S, Pou-Prom C, Abdalla M, Meaney C, Rudzicz F (2019) A survey of word embeddings for clinical text. J Biomed Inform X 4:100057
14. González E, Reyes J, González-Beltrán B (2020) Machine learning models for cancer type classification with unstructured data. Computación y Sistemas 24. https://doi.org/10.13053/cys-24-2-3367
15. Boyle T (2018) Medical transcriptions. Kaggle. Retrieved from https://www.kaggle.com/datasets/tboyle10/medicaltranscriptions
16. Chai C (2022) Comparison of text preprocessing methods. Nat Lang Eng 29:1–45
17. Sanh V et al (2019) DistilBERT, a distilled version of BERT: smaller, faster, cheaper and lighter. arXiv preprint arXiv:1910.01108
18. Zohrevand A, Imani Z (2022) An empirical study of the performance of different optimizers in deep neural networks. In: Proceedings of the conference on neural networks, pp 1–5

# Object Detection in Computer Vision: A Comparative Analysis of Advanced Computer Vision Models

**S. Shivakarthik, Krishnanjan Bhattacharjee, Rohan Badiger, Himanshu Bhusan Patro, and Swati Mehta**

**Abstract** The rapid advancement in computer vision, driven by deep learning, has revolutionised visual data interpretation across various sectors. This study explores object detection models-key to applications like autonomous driving, medical imaging, and surveillance. It reviews and compares SSD, Faster R-CNN, Mask R-CNN, RetinaNet, and YOLOv8, evaluating their architecture, performance on COCO, and limitations. The paper addresses challenges such as diverse environment detection, dataset generalisation, and computational efficiency, suggesting future improvements. The aim is to overview current object detection technologies, highlight challenges, and propose research directions for enhancement.

**Keywords** Object detection · Computer vision · Convolutional neural network · Single shot detector · RetinaNet · YOLO

## 1 Introduction

In today's digital era, computer vision revolutionises visual data interpretation through automated image and video analysis across diverse domains. It extracts, analyses, and comprehends information from images, underpinning automatic visual understanding [1].

Critical in healthcare, automotive, surveillance, and entertainment, computer vision models fuse AI, machine learning, and image processing for precise visual

S. Shivakarthik (✉) · K. Bhattacharjee · S. Mehta
Center for Development of Advanced Computing (C-DAC), Pune, India
e-mail: shivakarthiks@cdac.in

K. Bhattacharjee
e-mail: krishnanjanb@cdac.in

S. Mehta
e-mail: swatim@cdac.in

R. Badiger · H. B. Patro
Bharati Vidyapeeth (Deemed University) College of Engineering, Pune, India

 
K. R. Madhavi et al. (eds.), *Proceedings of Sixth International Conference on Computer and Communication Technologies*, Lecture Notes in Networks and Systems 1356,
https://doi.org/10.1007/978-981-96-5238-9_9

data interpretation. They facilitate tasks such as object detection, image classification, semantic segmentation, and scene understanding, transforming human-computer interaction and decision-making.

Advancements are propelled by vast image and video databases, coupled with powerful techniques like deep learning and CNNs. These innovations empower major tech companies to deploy cutting-edge applications in self-driving cars, face recognition, robotics, and medical imaging analysis.

## 2 Literature Survey

**Single Shot MultiBox Detector (SSD)**: SSD (Single Shot MultiBox Detector) predicts bounding boxes and confidence scores in a single pass, using multi-resolution priors for diverse object sizes and shapes, enhancing accuracy without extra computational overhead. On the COCO dataset, SSD512 achieved a mean Average Precision (mAP) of 28.8% [2]. However, SSD512 struggles with small object detection and unusual aspect ratios, leading to lower accuracy. High background examples increase false positives, and training requires careful tuning to balance localisation and classification losses. Generalising across diverse datasets without retraining is challenging. Models trained on one dataset often underperform on new datasets with different characteristics. Domain adaptation techniques aim to address this, but robust solutions for cross-dataset performance are still needed [3].

**Faster RCNN**: In Faster R-CNN, the RPN uses a feature extractor like VGG-16 to generate about 300 object proposals for class prediction and refinements. The loss function uses RPN proposals as anchors, reducing redundant computation, though region-dependent computations affect running time [4]. It achieves a 34.7% mAP on the COCO dataset [5], optimising performance with images sized 600 to 1000 pixels per side. Accurate detection of small or overlapping objects is a persistent challenge in object detection. Traditional models Faster RCNN struggle due to limited spatial resolution, often missing or confusing these objects. While feature pyramid networks enhance multi-scale feature representation, innovations are needed for better handling of complex scenes [6].

**Masked RCNN**: Mask R-CNN extends Faster R-CNN with a mask prediction branch [7], using RoIAlign for precise input and feature map alignment in object detection and pose estimation. Enhanced by frameworks like Detectron2, Mask R-CNN achieves 39.8% box mAP on COCO's test-dev with ResNeXt-101-FPN. Challenges include handling overlapping instances and high computational demands [8], affecting accuracy in conditions like fog. Architecture optimisation is critical for diverse applications.

**RetinaNet**: RetinaNet utilises focal loss for class imbalance, featuring a backbone and dual subnetworks for classification and bounding box prediction, prioritising challenging examples. With ResNet-101-FPN, RetinaNet achieves 40.8% mAP on

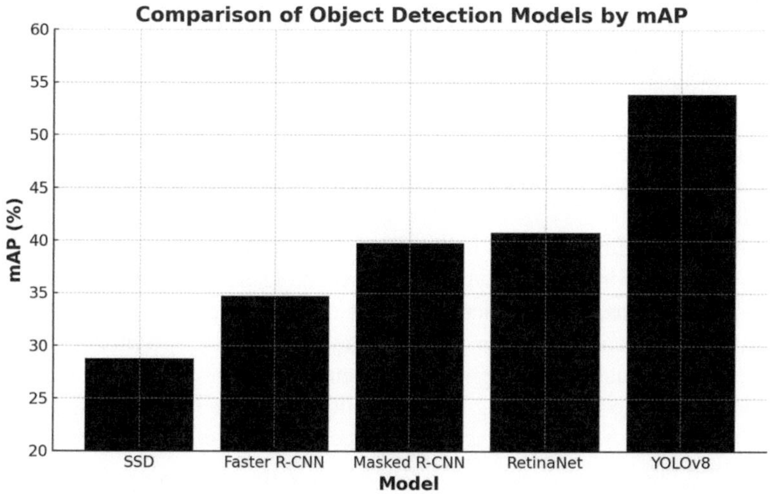


**Fig. 1** Object detection model comparison: mAP performance

COCO at IoU 0.5 [9], effective across diverse scenarios. Focal loss enhances performance compared to other detectors, advancing single-stage object detection, despite speed challenges with high-resolution inputs.

**YOLOv8**: YOLOv8 enhances upon YOLOv5 with CSPDarknet53 and a C2f module, featuring a decoupled head for objectness, classification, and regression tasks using sigmoid and softmax activations. It adopts an anchor-free model with CIoU [10] and DFL [11] loss functions, optimising for smaller objects. Achieving 53.9% AP on COCO at 640 pixels, surpassing YOLOv5's 50.7% [12],limiting accessibility and posing challenges in deployment across varied environments. Reducing model complexity while maintaining accuracy is crucial for broader adoption, especially in resource-constrained environments. MobileNets' use of depth-wise separable convolutions offers lightweight solutions. Applying these techniques to models like YOLOv8 could optimise performance and efficiency [13] (Fig. 1).

## 3 Methodology

After evaluating various models, YOLOv8 stands out for its adaptability and scalability. YOLOv8l balances accuracy and speed, ideal for real-time applications. For speed, use smaller YOLOv8 models; for maximum accuracy, use YOLOv8-X, sacrificing speed. In light of the above evaluations, YOLOv8 stands out as a viable solution for multifarious object detection applications.

It is pertinent to understand the inference process of YOLOv8 model for object detection, to justify its prediction as well as to determine the limitations so that reliability of the object detection model is ensured in the crucial application domains, not compromising the safety. In the very first level, this approach determines low-level features, i.e., edges, textures, and simple shapes. The next level features are processed in a sequence as the data traverses from input to output, which involves pattern aggregation depicting more complexity. The final bounding boxes, class labels, and confidence scores indicate how confident it is that the object detected really belongs to that class.

This interpretability could be furthered with methods, including feature visualisation and the creation of attention maps, in which the areas of an image that receive emphasis produce the most impact toward the decision. This is particularly relevant for high-stake applications, where transparency regarding why a model gives a particular output can enable the diagnosis of potential biases or mistakes. These observations are critical to maintaining the credibility of the predictions of the model for making informed decisions in the context of security and defence, where correctness is very important.

**Data Collection and Preparation** The dataset has been carefully curated in order to train YOLOv8 which comprises of various classes of object data in the broad category of 'Weapons', viz., weapons and ammunition, including rifles, pistols, bombs, rocket launchers, swords etc. The image data for training has been mostly sourced from various open-source repositories available in World Wide Web (WWW). The data obtained from these repositories have been carefully chosen and vetted to ensure images represent various patterns, i.e., different backgrounds, lighting conditions and diversified orientations of given objects. Despite the careful selection of images to create diversity, there may be some potential biases persist. Some of the images are found to be positioned in specific settings, therefore do not represent the real-world chaotic background or arbitrariness of natural environment.

In order to train the computer vision deep learning models, post data collection; preparation and curated the data is the next step. Data preparation involves annotating the images with 'bounding boxes' and attaching class labels to groups of objects. Augmentation techniques, which includes creating harvested images from existing ones by applying rotations, colour adjustments etc. have been used to enrich the dataset. The dataset, thus prepared, have been further divided into three sections: Train, Test and Validation Datasets.

**Training and Evaluation** The model to be trained for object detection requires setting of training parameters like number of classes, learning rate, batch size and epochs which are pertinent for balancing the performance and accuracy of the trained model. YOLOv8 has been trained with optimally annotated dataset, applying weight adjustment techniques to minimise loss. Performance of the trained model has been validated initially on the 'validation dataset' and further on a separate real-world dataset, to determine accuracy and generalisation, also addressed the issues of 'overfitting' and 'underfitting'. Thus, the model has undergone several fine-tuning to offer good accuracy in real-world use cases.

Post training, YOLOv8 has been meticulously tested on various types of unseen and unrelated image data to obtain object class inference by the trained model as

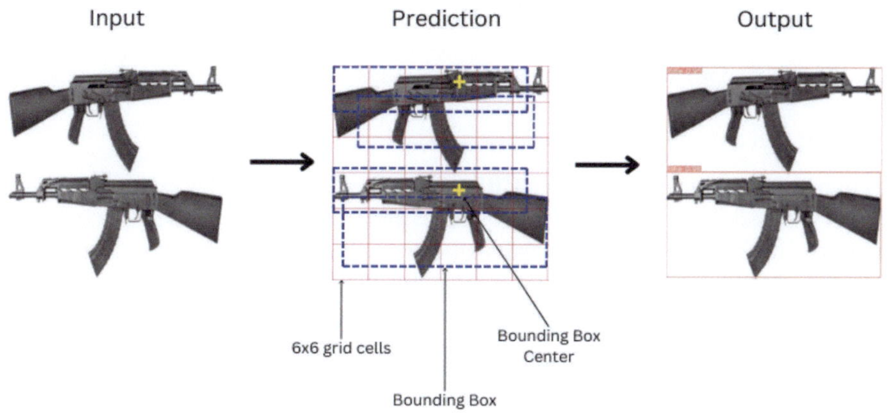


**Fig. 2** Stages of object detection

illustrated in Fig. 2. This step determines model's real-world usability and robustness. Moreover, evaluating model inference on unseen real-world data provides an unbiased assessment for the trained model's generalisation capabilities. The outcome of this step is reflected through key performance metrics, i.e., precision, recall and 'mean Average Precision' (mAP) to determine model accuracy score.

**Prediction and Maintenance** In the inference phase of YOLOv8, input images to infer object classes are pre-processed by a resizing technique, resizing them to $640 \times 640$ and normalising the pixel values to [0, 1]. Thereafter, model is ready to infer bounding boxes, class labels and also is capable of estimating confidence score for each such inference. As an outcome, the processed image can be visualised with bounding boxes, class labels and confidence score of inference. The model inference or predicting the object classes from various real-world application scenarios with good accuracy enables the trained model to be used in surveillance, OSINT and autonomous driving or robotics kind of application domains.

In order to upgrade the performance of YOLOv8 in object detection and coping with real-world scenarios in a fast-changing world, periodic re-training of the models with new data variations is much necessary. Periodic model retraining facilitates model to accommodate new object variations as well as prevents model degeneration owing to concept drift. Moreover, continuous refinements through hyperparameter tuning such as learning rate, batch size as well as architecture nuances are very important in optimising model efficiency and accuracy. This twin approach of re-training and hyperparameter tuning keeps the model effective and adapting with continuous flux of real-world data scenarios, and thus, does not become outdated in application domains.

**Applications**

The practical application of an object detection system like YOLOv8 is very relevant in areas such as security and defence, where accurate and real-time detection of weapons and ammunition is crucial.

**Urban Surveillance and Smart Cities** Enhancement of public safety to a great extent is possible via smart city deployment, particularly with YOLOv8 integration in urban surveillance systems. Implemented in various public places such as parks, streets, and transport hubs, it is used for ongoing monitoring to detect potential threats like an explosive device planted somewhere.

**Border Security and Immigration Control** The trained Object Detection model developed using YOLOv8 has another crucial real-world application in the context of border security and immigration management to determine any incriminating objects such as weapons carried at such sensitive locations. This can lead to early interception by patrolling and immigration officers as well as can potentially reduce threats of weapon/banned objects smuggling through border traffic.

**Predictive Policing** The object detection model has another crucial use case scenario, large public meeting, protests, or demonstrations where policing is difficult and miscreants or terrorists can manipulate such chaotic situations. In such cases, the object detection model can be deployed in the body-worn cameras or drones to detect persons carrying incriminating objects like weapons.

**Emergency Response and Disaster Management** In the wake of natural calamities such as earthquake, floods, tornadoes, cyclones or volcanic eruptions; or in the case of man-made disaster like war or terrorist attacks; the object detection model trained through YOLOv8 and deployed on the drones as well as on the body cam of rescue personnel. It is because the model is capable to process and identify objects from real-time video feeds.

**Industrial Security** The YOLOv8 object detection model in question has another potential use in protecting critical infrastructures such as Power Plants, Oil Refineries, Chemical and Weapon Factories, Nuclear Sites etc. The model can be deployed in the CCTV cameras installed in such sites, which in return can detect any suspicious object carried by personnel in and around such sites.

The aforementioned use case scenarios clearly depict the real-world application of such object detection model in enhancing National Security, pre-crime detection and predictive policing.

## 4 Results

**Class Metrics**

Table 1 enumerates Performance Metrics, comprising of the scores of Precision (Pos-

**Table 1** Performance metrics by class

| Class | Precision | Recall | mAP |
|---|---|---|---|
| All | 0.819 | 0.792 | 0.836 |
| Pistol | 0.847 | 0.816 | 0.861 |
| Rifle | 0.886 | 0.842 | 0.930 |
| Machine gun | 0.751 | 0.776 | 0.754 |
| Rocket launcher | 0.718 | 0.659 | 0.673 |
| Sword | 0.893 | 0.929 | 0.913 |
| Pike | 0.847 | 0.836 | 0.856 |
| Knife | 0.556 | 0.541 | 0.569 |
| Bombs | 0.932 | 0.851 | 0.862 |
| Grenade | 0.792 | 0.814 | 0.807 |
| Mines | 0.942 | 0.962 | 0.982 |
| Cylinder | 0.987 | 0.963 | 0.987 |

**Fig. 3** Detections on batches

itive/correct predictions), Recall (All identified relevant Instances) and mean Average Precision (mAP) (Average precision obtained across all classes)

### Detection

Figure 3 illustrates real-time detection on batch segments, depicting bounding boxes and prediction scores. This visualisation offers a dynamic insight into the model's performance, showcasing its ability to accurately identify objects within segmented batches while providing confidence scores for each prediction.

## 5 Conclusion and Future Work

This paper examined various computer vision models, focusing on object detection techniques like SSD, Faster RCNN, Mask RCNN, RetinaNet, and YOLOv8. These models are key for accurate and efficient visual data interpretation in applications like autonomous vehicles, medical imaging, and surveillance.

Although advancements have improved precision and efficiency, challenges still remain in handling small and overlapping objects, simplifying across datasets, and enhancing computational efficiency. Future research should adapt object detection models for challenging environments like underwater or nighttime scenes. Models such as YOLOv5 and Faster RCNN struggle in poor visibility. Integrating multi-modal sensory data could improve detection accuracy. Incorporating few-shot learning techniques can revolutionise areas where labelled data is limited. Prototypical networks for few-shot classification offer a promising approach to reduce dependency on large datasets in object detection frameworks [14]. Developing efficient, real-time object detection for mobile and edge computing is challenging. Optimising SSD and YOLOv8 through model pruning and quantisation reduces size and boosts speed without sacrificing accuracy, crucial for limited-processing devices [15].

In conclusion, continuous innovation in computer vision will address these issues so that more flexibility, robustness, and accessibility will be achieved in systems. This paper presents the current state of object detection models and set a stage toward future research and development.

## References

1. Samatha R, Rani R, Devi PL (2020) A literature survey on computer vision towards data science
2. Liu W, Anguelov D, Erhan D, Szegedy C, Reed S (2015) SSD: single shot multiBox detector. ArXiv preprint arXiv:1512.02325
3. Zhu, J-Y, Park T, Isola P, Efros AA (2017) Unpaired image-to-image translation using cycle-consistent adversarial networks. In: Proceedings of the IEEE international conference on computer vision, pp 2223–2232
4. Ren S, He K, Girshick R, Sun J (2017) Faster R-CNN: towards real-time object detection with region proposal networks. IEEE Trans Pattern Anal Mach Intell 39(6):1137–1149
5. Huang J, Rathod V, Sun C, Zhu M, Korattikara A, Fathi A, Fischer I et al (2016) Speed/accuracy trade-offs for modern convolutional object detectors. ArXiv preprint arXiv:1611.10012
6. Lin T-Y, Dollár P, Girshick R, He K, Hariharan B, Belongie S (2017) Feature pyramid networks for object detection. In: Proceedings of the IEEE conference on computer vision and pattern recognition, pp 2117–2125
7. He K, Gkioxari G, Dollár P, Girshick R (2017) Mask R-CNN. ArXiv preprint arXiv:1703.06870
8. Cheng T, Wang X, Huang L, Liu W (2020) Boundary-preserving Mask R-CNN. ArXiv preprint arXiv:2007.08921
9. Lin T-Y, Goyal P, Girshick R, He K, Dollar P (2020) Focal loss for dense object detection. IEEE Trans Pattern Anal Mach Intell 42(2):318–327
10. Zheng Z, Wang P, Liu W, Li J, Ye R, Ren D (2020) Distance-iou loss: faster and better learning for bounding box regression. Proc the AAAI Conf Artif Intell 34:12993–13000

11. Li X, Wang W, Wu L, Chen S, Hu X, Li J, Tang J, Yang J (2020) Generalized focal loss: Learning qualified and distributed bounding boxes for dense object detection. Adv Neural Inf Proc Syst 33:21002–21012
12. Ultralytics (nd) COCO dataset. Retrieved from https://docs.ultralytics.com/datasets/detect/coco/
13. Howard AG, Zhu M, Chen B, Kalenichenko D, Wang W, Weyand T, Andreetto M, Adam H (2017) MobileNets: efficient convolutional neural networks for mobile vision applications. ArXiv preprint arXiv:1704.04861
14. Snell J, Swersky K, Zemel R (2017) Prototypical networks for few-shot learning. Adv Neural Inf Proc Syst 30
15. Cai H, Gan C, Wang T, Zhang Z, Han S (2019) Once-for-all: train one network and specialize it for efficient deployment. ArXiv preprint arXiv:1908.09791

# Introducing an Ensemble Method to Phish Guard: A Robust Stacked Ensemble Defense System Against Web Spoofing

**Harshit Ravi Kumar Ambati, Somesh Kumar, and K. N. Sreehari**

**Abstract** This paper aims to address the persistent threat of phishing attacks by developing PhishCatcher, a client-side defense tool. The primary goal is to utilize machine learning as a core component for the robust identification of evolving web spoofing threats. Recognizing the escalating risk posed by phishing, especially in the context of increased online activities, this paper signifies the urgency of countering web spoofing. PhishCatcher is designed with the end-user in mind, especially those who are frequently targeted by phishing attacks. We extended our anti-phishing tool by integrating Support Vector Machine, XGBoost, and a Stacking Classifier, augmenting the system's capabilities. Additionally, a Flask framework with SQLite was implemented, offering streamlined signup and sign-in processes for user testing and input validation.

**Keywords** Web spoofing · Security and privacy · Anti-phishing tool · Machine learning

## 1 Introduction

A plethora of development in the internet-based world like e-commerce, distance learning, and various other fields has taken place due to rapid advancements in modern technologies. Many websites now offer users the chance to create accounts for a personalized experience. The user requests a remote asset or service, a web request is sent to the server and gets a form for login where he enters his credentials. At such a stage, it becomes too easy for personal information of users to be stolen and identity theft takes place. An email that has a URL to a website that is malicious is the point where a phishing attack starts. That email may persuade the user to click the link. Once

H. R. K. Ambati (✉) · S. Kumar · K. N. Sreehari
Department of Electronics and Communication Engineering, Amrita Vishwa Vidyapeetham, Amritapuri, India
e-mail: aharshith2002@gmail.com

K. N. Sreehari
e-mail: sreeharikn@am.amrita.edu



K. R. Madhavi et al. (eds.), *Proceedings of Sixth International Conference on Computer and Communication Technologies*, Lecture Notes in Networks and Systems 1356,
https://doi.org/10.1007/978-981-96-5238-9_10

the webpage is opened by the user without suspecting it, it looks legal as the original site. After entering their credentials and pressing the button for login, these attackers get the information. Now the attacker uses these credentials to log in to the legitimate website. Phishing and Web Spoofing are popular types of cybercrime, used by hackers to target digital platforms. Phishing attacks in the modern age have advanced beyond recognition, such as QR code phishing, mobile app spoofing and spear-phishing [1] to name a few. Such sophisticated methods of attack are able to evade many layers of protection, such as firewalls and encryption software or even two-factor authentication. Two-factor authentication is common to protect against financial fraud and identity theft, but the increasing sophistication of phishing approaches has allowed these security checks to be overcome. Phishing attack vectors include email, Trojan horses, keyloggers, and man-in-the-middle attacks. Internet banking platforms, non-affiliated transaction platforms, and online retail platforms are the primary targets [2]. Since phishers exploit vulnerable elements, SSL/TLS secure communication protocols are not fully comprehensive. These secure communication protocols need to be boosted with added protective systems to counter impersonation attempts. These systems can be implemented on the backend, frontend, or both. The backend solutions require website changes, which designers find boring to deal with and are often not paid attention to by developers during the implementation phase [3, 4]. Client-side solutions are able to protect users who do not have backend support. While backend solutions might be excellent in discovering malicious sites, client-side-based methods are also crucial. CANTINA [5] and SpoofCatch [6] can catch 90% of fraudulent websites with only a 1% false positive rate. But the stateful anti-phish methods that exist do so at a cost; their performance degrades gradually over time as they fill I/O store. Following this study, we designed and developed a stateless anti-phish tool based on Machine Learning (ML) techniques. Numerous scholars have suggested machine-learning approaches for identifying malicious URLs. Sets of URLs act as training data in these ML approaches. After attaining statistical data from training sets, the model predicts whether a URL might be malicious. Training data is the most important in URL identification by ML. The primary objective involves extracting features from the data which is under training, as the condition of the URL under evaluation cannot be predicted by the straightforward strings alone. In the last stage, the training data is used to make the predictions to acquire an operational model. ML techniques like Support Vector Machines (SVM), Naive Bayes, and Logistic Regression (LR) are commonly utilized for this goal by many scholars, but due to many existing issues, these models are still vulnerable [7].

## 2 Literature Survey

To guard against web spoofing attacks, most anti-phishing measures either detect specific patterns of attacks or rely on different complex parameters to determine phishing attacks, or suffer from both. In this paper, we suggest that we can effectively prevent phishing attacks by primarily relying on the display of the web page that the client sees. To back our claim, we suggest a client-side security component based on the visual likeness of web pages, executed as a browser expansion called SpoofCatch. To distinguish between licit and phished websites, we have developed and included four types of algorithms in the extension. To estimate the result, a wide range of tests have been conducted, demonstrating that SpoofCatch can descry all phishing attacks with reasonable trouble [8].

Phishing and web spoofing have spread broadly, getting to be a critical issue on the Web. These assaults are difficult to guard against since they center on non-cryptographic components like the client or the user-browser interface. As a result, cryptographic security conventions like SSL/TLS are not adequate on their possess and require to be supplemented with extra assurance measures. In this paper [9], we outline, talk about, and assess the adequacy of such instruments against (large-scale) phishing and Web spoofing attacks [9, 10].

In the paper [11], they introduced a few techniques to find phishing using advanced machine learning and deep learning techniques. A comparison analysis between a number of deep learning models such as the GRU-MLP, LSTM-CNN, and AE-DNN and the performance and accuracy are shown. The purpose of the study is also to apply XGBoost classifiers with HTML and URLs as input to extract nearly perfect features. Hyper parameter analysis, filter selection, and the limitations of current models also spark additional conversation, suggesting that more advanced (and complimentary) methods can be used simultaneously to improve detection and minimize false positives.

According to the paper [12] around how they tried developing a chrome extension to strengthen web security with the help of machine learning. This research is mainly built on the XGBoost algorithm and VirusTotal API for malicious websites, and files detection. Privacy first: Safe to use because all analyses are conducted on your machine, nothing is sent to external servers. The model is able to validate the safety of downloaded files as well providing better results in web browsing and file security together, surpassing traditional security extensions with an accuracy.

The paper [13] is targeting to forecast the learner style combinations of students through machine learning techniques k- means clustering, SVM, and decision trees. They conducted a survey of 1123 students, aged 15–25 to gather statistics about their learning styles. Clustering was performed using the k-means algorithm and resulted in 51 groups. The learning styles combinations were then mapped to appropriate fields of study to build up a rule-based system i.e. Computer Science, Manufacturing Engineering, Mechatronics Engineering and Medical as proposed by the authors. These responses revealed a staggering 73% of students who thought they preferred one learning style such as visual, followed by auditory and kinesthetic styles. The paper argues that learning styles may be more appropriate for career guidance when interpreted as patterns, and a more stable inference of individual differences. However the author suggest that more research needed to be done in order to expand the list of suggested fields of study.

In [14], the authors investigate the susceptibility of phishing detection systems to evasion attacks. They test different machine learning classifiers and how vulnerable they are to adversarial examples (inputs intentionally bent out of shape in order to trick the classifiers).We suggest a new defense mechanism for this, features can be extracted in URL, and then checking the similarity based analysis these evasion attacks can be detected. Their method is very successful in maintaining the phishing like functionality of the phishing site, it gets detected. He also emphasized the importance of creating even more anti-phishing tools that will help prevent these advanced attack techniques.

This paper [15] presents a machine learning-based approach to detect phishing URLs. This reports methodology to detect URLs is broadly categorized into signal analysis, rule based ways and behavioral methodologies. Additionally, the authors proposed a list of novel URL features and behaviors to improve the detection of malicious URLs.They explain in this paper the performance of a few machine learning algorithms (Decision Tree, Random Forest, K-Nearest [Neighbor], and Support Vector Machine) on a dataset of URLs. The results demonstrate that the proposed approach is able to accurately detect malicious URLs; the Random Forest algorithm can achieve the best classification performance compared with five benchmarks. The research shows the need for more effective mechanisms to source these harmful URLs and prevent their implementation in such cyber attacks.

# 3 Methodology

## *3.1 Proposed Work*

The proposed methodology for PhishCatcher involves utilizing machine learning, specifically the random forest algorithm. This algorithm dynamically assesses login web pages, discerning between legitimate and potentially spoofed sites. [16–19] PhishCatcher employs a client-side defense mechanism, functioning as a Google Chrome extension. It leverages real-time learning, reducing reliance on historical data and enhancing adaptability to novel phishing tactics. The focus on client-side protection ensures compatibility without requiring modifications to targeted websites. Since the Random Forest inserted last time was not very effective, we expanded our implementation with more Machine Learning Classifiers: Support Vector Machine (SVM), XGBoost and a Stacking Classifier. This ensemble approach, featuring a combination of [20, 21] RandomForest, ExtraTrees, and XGBoost classifiers in a stacking configuration, aims to enhance the robustness of our phishing detection system. Furthermore, Flask framework has been created where it is easy for user sign up and login as well an integrated SQLite too, such that a sample call can still be tested easily. This feature enables comprehensive evaluation and comparison of the classifier's effectiveness, contributing to the overall refinement of the anti-phishing tool.

## 3.2 System Architecture

As shown in Fig. 1 the system architecture begins by selecting a proper dataset and extracting the features from it ,according to their performance. These features are next input to a variety of classifiers (including Random Forest, SVM [7, 22], XGBoost and the Stacking Classifier extension) for determination. These classifiers play a pivotal role in identifying and classifying phishing URLs. In the PhishCatcher browser extension, upon detecting a potential phishing attack, the system triggers an alert notification, actively enhancing user's overall security awareness.

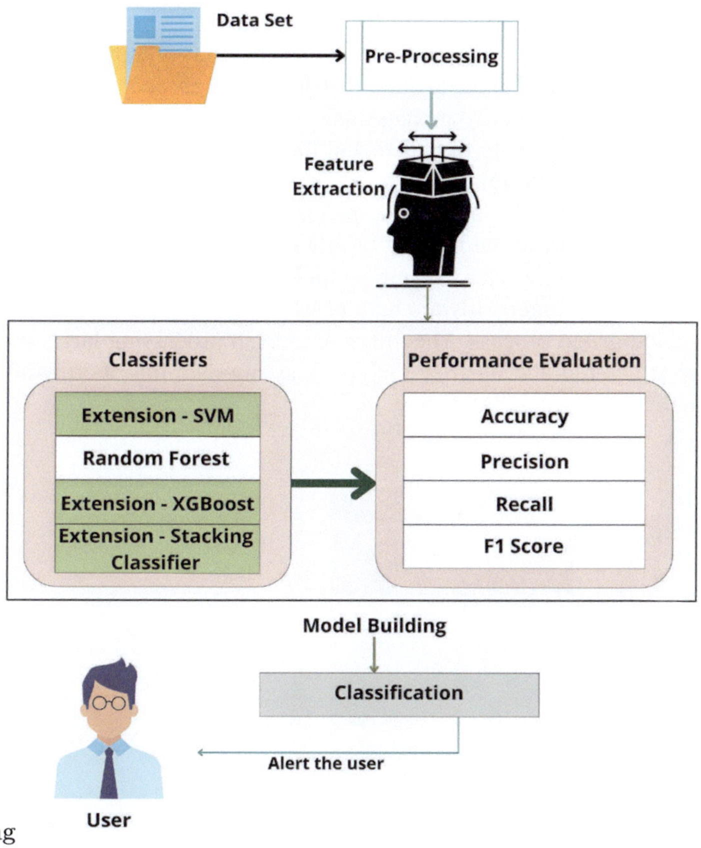


Architecture.png

**Fig. 1** Proposed architecture

### 3.3 Dataset Collection

To train propose algorithm we have used PHISHTANK [3] dataset which contains 1000's of normal and phishing URL and by using this dataset we can predict URL as SAFE or phishing. Apart from training author has developed CHROME based extension which will analyze all visiting URLS and then alert user with SAFE or phishing URL'S. The data set count we collected is shown in the Fig. 2b.

### 3.4 Data Pre-processing

Data preprocessing means all the transformations and operations done on a data set before training of models such that it creates an accurate,safe,classifiable model. Data scientists usually manage this process, including gathering, organizing and verifying data; trying to understand the meaning behind it and sometimes creating different reports or graphs of how that information can be interpreted at a glance. It is possible through three ways: manual, mechanical and an electronic system. The goal is to improve the level of information leads and simplifying decision making. This allows businesses to ramp up their operations and enable appropriate strategic choices. Computer software help us as a solution for this data pre-processing, they are automated process. They may support in processing large amounts of data that are structured and unstructured, such as big data into worthwhile information for quality control and insights.

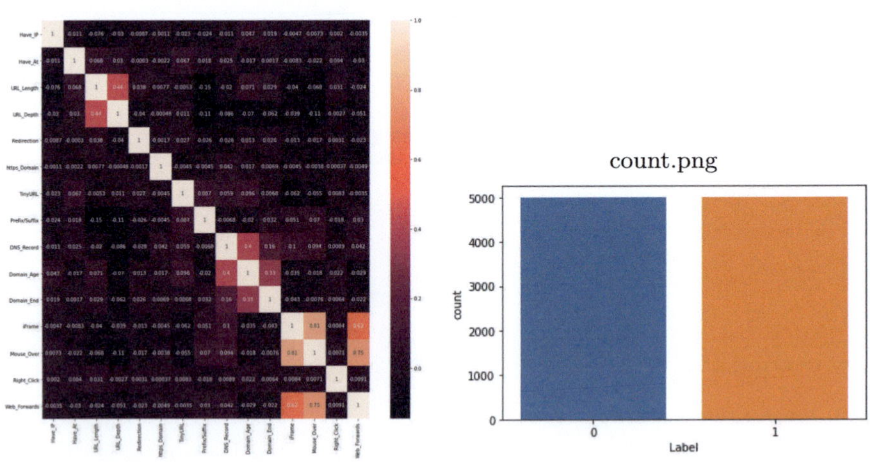


(a) Heat Map with the features taken (b) Dataset Count

**Fig. 2** Heat map and dataset count

### 3.5 Feature Extraction

Feature extraction is to identify and isolate the most important features for your Machine Learning model documents. The size of the data sets becomes more and more important with larger, complex ones. The primary objective behind attribute extraction is to upgrade the performance of machine-learning models, and at the same time making it less computationally expensive. Feature stuffing is an integral part of the feature engineering. This includes choosing the most important features to be put into machine learning algorithms.To determine which features to select to enhance the performance of the machine learning model, we utilize a heat-map, as shown in Fig. 2a. Dataset has been narrowed using the attribute extraction operations by keeping only those attributes that are important and removing all other gaining.

### 3.6 Algorithms

- **Support Vector Classifier (SVC)**
  Support Vector Classifier is an effective supervised learning algorithm that is used to find the optimal hyperplane separating data points belonging to different classes with a maximum margin. Given training data $\{(\mathbf{x}_i, y_i)\}_{i=1}^{n}$, where $\mathbf{x}_i \in \mathbb{R}^d$ are feature vectors for each sample and $y_i \in \{-1, 1\}$ are the class labels, SVC solves the optimization problem as follows:

$$\min_{\mathbf{w},b} \frac{1}{2}\|\mathbf{w}\|^2 + C\sum_{i=1}^{n} \xi_i \tag{1}$$

  subject to:

$$y_i(\mathbf{w} \cdot \mathbf{x}_i + b) \geq 1 - \xi_i, \quad \xi_i \geq 0 \tag{2}$$

  where $\mathbf{w}$ is the weight vector, $b$ is the bias, $C$ is the regularization parameter, and $\xi_i$ are slack variables that allow some misclassification. The decision boundary is then as follows:

$$f(\mathbf{x}) = \text{sign}\,(\mathbf{w} \cdot \mathbf{x} + b) \tag{3}$$

  In this paper, SVC will be crucial in maximizing the separability of extracted feature from all available URLs to classify them as either phishing or legitimate, as shown in equations (1) through (3).
- **Random Forest (RF)**
  Random Forest useRefers to an ensemble learning algorithm that constructs a multitude of decision trees during the training phase. We then train many of these trees on random subsets (with replacement) of the combined features and make final predictions by taking a majority vote over all learner classes:

$$\hat{y} = \text{mode}\,(h_1(\mathbf{X}), h_2(\mathbf{X}), \ldots, h_T(\mathbf{X})) \tag{4}$$

where $h_t(\mathbf{X})$ is the prediction from $t$-th tree. An ensemble method by combining multiple weak learners improves the model generalization and reduces variance. The RandomForestClassifier can be a great asset to our phishing detection system as it allows the model learn more about complex feature interactions which results in increased classification accuracy and robustness on both training data as well as test dataset URLs.The ensemble voting process, described in Eq. 4.

- **XGBoost (Extreme Gradient Boosting)**
  XGboost is an Extreme Gradient boosting technique where it builds a series of weak decision trees in a sequential manner. In those decision trees, each decision tree's work is to reduce the error by its previous decision tree in the sequential order by minimizing a specified loss function:

$$\text{Obj}(\theta) = \sum_{i=1}^{n} l(y_i, \hat{y}i^{(t)}) + \sum t = 1^T \Omega(f_t)$$

  where $l(y_i, \hat{y}_i^{(t)})$ is the loss function (e.g., logistic loss for binary classification), $\hat{y}_i^{(t)}$ is the prediction from the $t$-th tree, and $\Omega(f_t)$ is a regularization term that penalizes the complexity of the model, helping to prevent overfitting. The XGBClassifier used in this paper is a key component of the ensemble method, which provides powerful predictive capabilities by efficiently combining and filtering the outputs of weaker trees, accordingly improving the accuracy and efficiency of phishing detection.
- **Stacking Classifier**
  Stacking Classifier is type of ensemble models which combines multiple base models to form a new, stronger and accurate predictive model. During the stacking process, each base model $M_j(\mathbf{X})$ s trained independently on the dataset generating predictions $z_j$, that are used as an input to a meta-learner $H(\mathbf{Z})$ which outputs final prediction:

$$\hat{y} = H(z_1, z_2, \ldots, z_m) \tag{5}$$

  where $z_j = M_j(\mathbf{X})$ is the prediction from base model. In general, the meta-learner uses a simple model such as logistic regression to fuse predictions together ideally. Here, we combine the predictions from RandomForestClassifier and LGBMClassifier using a stacking classifier that leverages their individual strengths to provide enhanced accuracy and robustness in detecting phishing, as outline by the stacking mechanism in Eq. 5. This ensemble model hence guarantees the best predictive capability by efficiently exploiting complementary specialties of individual classifiers.

comparison graph.png

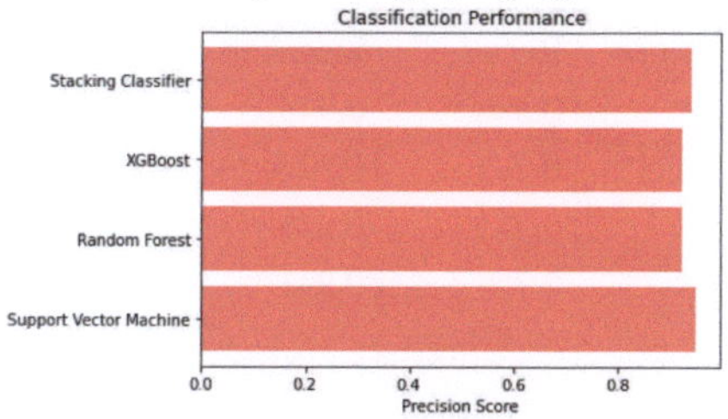


(a) Precision comparison graph
graph.png

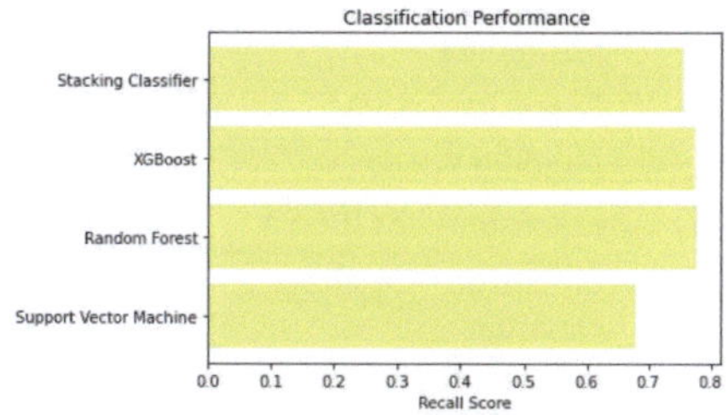


(b) Recall comparison graph
score graph.png

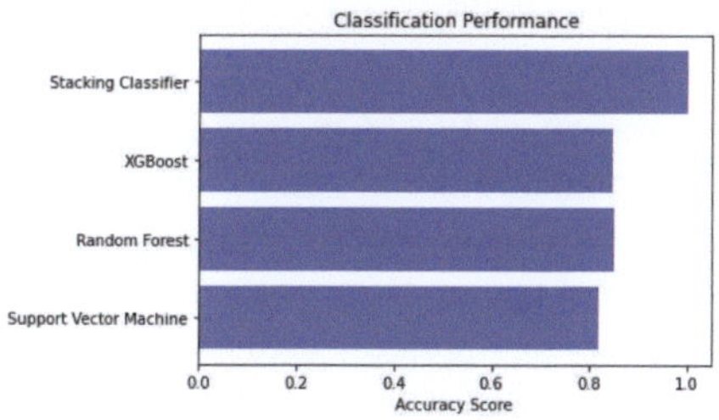


(c) Accuracy comparison graph

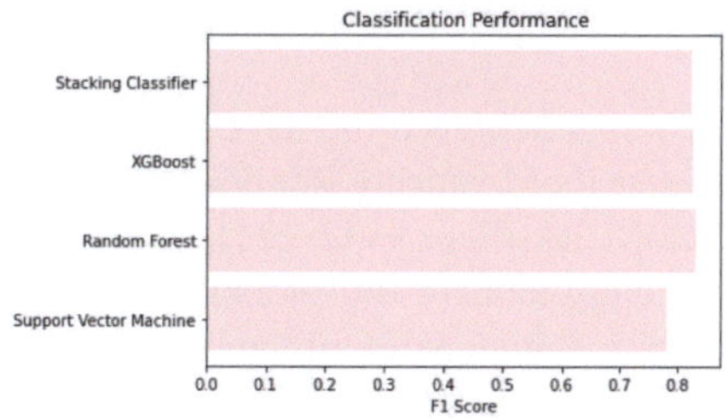


(d) F1 score comparison graph

**Fig. 3** Comparison graphs

## 4 Experimental Results

- **Precision**: Precision answers, how many of the cases/sample which were classified as positives are actually true. The comparative graph based on precision is shown in Fig. 3a.
- **Recall**: Recall in machine learning measures how many relevant instances of a particular class is identified by our model. It means the percentage of total actual positives our model predicts correctly, it is true positive over (true positive + false negative). The comparative graph based on recall is shown in Fig. 3b.
- **Accuracy**: This is the proportion of true results among all predictions made, it gives you an overall idea about how accurate your model was. The comparative graph based on accuracy is shown in Fig. 3c.
- **F1 Score:**F1 score is the harmonic mean of precision and recall, a kind balance metric between both. The comparative graph based on F1 score is shown in Fig. 3d. As shown in Table 1, the Stacking Classifier has the highest accuracy.

**Table 1** Performance evaluation

| | ML model | Accuracy | F1_score | Recall | Precision | Specificity |
|---|---|---|---|---|---|---|
| 0 | Extension—SVM | 0.817 | 0.780 | 0.680 | 0.949 | 0.975 |
| 1 | Random Forest | 0.850 | 0.831 | 0.777 | 0.923 | 0.953 |
| 2 | Extension—XGBoost | 0.846 | 0.826 | 0.775 | 0.921 | 0.951 |
| 3 | Extension—Stacking classifier | 1.000 | 0.824 | 0.754 | 0.941 | 0.967 |

## 5 Conclusion

The paper achieved the development and integration of PhishCatcher, a client-side defense tool featuring Random Forest (RF) [20] and additional extensions: Support Vector Classifier (SVC) [21], XGBoost, and a stacking classifier. This robust tool efficiently detects and blocks malicious URLs, enhancing user protection against phishing threats without the need for modifications to targeted websites. This comprehensive approach enhances the model's ability to discern between phishing and legitimate URLs [23, 24], contributing to its accuracy and reliability. The integration of PhishCatcher into a Flask-based front-end, coupled with user authentication using SQLite, ensures a seamless and secure user experience. The user-friendly interface facilitates input processing, leveraging the trained models for predictions, and ultimately displaying the final outcome in a clear and accessible manner. The paper has gone beyond the conventional approach by exploring alternative machine learning models to enhance predictive accuracy. PhishCatcher not only focuses on efficient machine learning algorithms but also addresses user-centric concerns by minimizing reliance on website modifications. This client-side emphasis, coupled with the incorporation of diverse features, signifies a holistic approach to online security. The paper stands as a significant step towards providing users with a comprehensive defense against the evolving landscape of web-based phishing threats.

## References

1. Schneier B (2005) Two-factor authentication: too little, too late. Commun ACM 48(4):136
2. Garera S, Provos N, Chew M, Rubin AD (2007) A framework for detection and measurement of phishing attacks. In: Proceedings of ACM workshop recurring malcode, pp 1–8
3. Pietraszek T, Berghe CV (2005) Defending against injection attacks through context-sensitive string evaluation. In: Proceedings of International workshop on recent advances in intrusion detection. Springer, Cham, Switzerland, pp 124–145
4. Johns M, Braun B, Schrank M, Posegga J (2011) Reliable protection against session fixation attacks. In: Proceedings of the ACM symposium on applied computing, pp 1531–1537
5. Zhang Y, Hong JI, Cranor LF (2007) Cantina: a content-based approach to detecting phishing web sites. In: Proceedings of 16th International conference on World Wide Web, pp 639–648
6. Khan W, Ahmad A, Qamar A, Kamran M, Altaf M (2021) SpoofCatch: a client-side protection tool against phishing attacks. IT Prof 23(2):65–74

7. Miyamoto D, Hazeyama H, Kadobayashi Y (2008) An evaluation of machine learning-based methods for detection of phishing sites. In: Proceedings of International conference on neural information Process. Springer, Cham, Switzerland, pp 539–546
8. Teraguchi NCRLY, Mitchell JC (2004) Client-side defense against web-based identity theft. Department of Computer Science, Stanford University, Stanford, CA, USA (Online). Available: http://crypto.stanfordedu/SpoofGuard/webspoof.pdf
9. Oppliger R, Gajek S (2005) Effective protection against phishing and web spoofing. In: Proceedings of IFIP international conference on communications and multimedia security. Springer, Cham, Switzerland, pp 32–41
10. Herzberg A, Gbara A (2004) Protecting (even Naıve) web users from spoofing and phishing attacks. Cryptology ePrint Archive. Department of Computer Science Engineering, University of Connecticut, Storrs, CT, USA, Technical Report 2004/155
11. Prabhu N, Senthil Kumar T, Jeyakumar G (2023) A comprehensive study of web phishing detection using machine and deep learning architecture. In: 2023 14th International conference on computing communication and networking technologies (ICCCNT). Delhi, India, pp 1–14. https://doi.org/10.1109/ICCCNT56998.2023.10306583
12. Gurjar NS, Sudheendra SSR, Kumar CS, Krishnaveni KS (2021) WebSecAsst-a machine learning based chrome extension. In: 2021 6th International conference on communication and electronics systems (ICCES). Coimbatore, India, pp 1631–1635. https://doi.org/10.1109/ICCES51350.2021.9488953
13. Kuttattu AS, Gokul GS, Prasad H, Murali J, Nair LS (2019) Analysing the learning style of an individual and suggesting field of study using machine learning techniques. In: 2019 International conference on communication and electronics systems (ICCES). Coimbatore, India, pp 1671–1675. https://doi.org/10.1109/ICCES45898.2019.9002051
14. Pillai MJ, Remya S, Devika V, Ramasubbareddy S, Cho Y (2024) Evasion attacks and defense mechanisms for machine learning-based web phishing classifiers. IEEE Access 12:19375–19387. https://doi.org/10.1109/ACCESS.2023.3342840
15. Menon RRK, Anandhu V (2023) Machine learning supported malicious URL detection. In: 4th IEEE Global conference for advancement in technology (GCAT). Bengaluru, India, pp 1–5. https://doi.org/10.1109/GCAT59970.2023.10353402
16. Ali W (2017) Phishing website detection based on supervised machine learning with wrapper features selection. Int J Adv Comput Sci Appl 8(9):72–78
17. Jain AK, Gupta BB (2016) Comparative analysis of features based machine learning approaches for phishing detection. In: Proceedings of 3rd International conference on computing for sustainable global development (INDIACom), pp 2125–2130
18. Xiang G, Hong J, Rose CP, Cranor L (2011) CANTINA+: a feature rich machine learning framework for detecting phishing web sites. ACM Trans Inf Syst Secur 14(2):1–28
19. Lakshmi VS, Vijaya MS (2012) Efficient prediction of phishing websites using supervised learning algorithms. Proc Eng 30:798–805
20. Muppavarapu V, Rajendran A, Vasudevan SK (2018) Phishing detection using RDF and random forests. Int Arab J Inf Technol 15(5):817–824
21. Kremic E, Subasi A (2016) Performance of random forest and SVM in face recognition. Int Arab J Inf Technol 13(2):287–293
22. Rao P, Gyani J, Narsimha G (2018) Fake profiles identification in online social networks using machine learning and NLP. Int J Appl Eng Res 13(6):973–4562
23. Chu W, Zhu BB, Xue F, Guan X, Cai Z (2013) Protect sensitive sites from phishing attacks using features extractable from inaccessible phishing URLs. In: Proceedings of IEEE International conference on communication (ICC), pp 1990–1994
24. Sahoo D, Liu C, Hoi SCH (2017) Malicious URL detection using machine learning: a survey. arXiv:1701.07179
25. Bugliesi M, Calzavara S, Focardi R, Khan W (2014) Automatic and robust client-side protection for cookie-based sessions. In: Proceedings of International symposium on engineering secure software and systems. Springer, Cham, Switzerland, pp 161–178

# A Hybrid Model for Cold Start Article Recommendations Leveraging Community Detection and Topic Analysis

**Sandeep Kumar Rachamadugu and T. P. Pushphavathi**

**Abstract** One of the most formidable problems in scientific article recommendation systems is recommending articles to cold-started users who have a small interaction history. Most traditional approaches use content-based filtering or collaborative filtering; the same thing may not really capture user preferences and article relationships. In this paper, we introduce a hybrid model combining community detection with topic analysis to recommend articles for cold-start users. The approach unifies both methods to discover hidden user groups and extract serious topics from articles. Clusters of domain-related articles form the community detection module, while the topic analysis extracts topics from these articles and converts them into dense vectors. To make recommendations, the final hybrid model combines community and topic similarities from individual models. Experiments on a large dataset show that our method is effective and outperforms traditional methods regarding precision, recall, and F1 in terms of dataset scale. Our model performs exceptionally well for new users, resulting in an improvement in recommendation accuracy. The introduced hybrid model may lead to a potential solution for the cold-start problem in scientific article recommendation systems, which aim at improving accuracy and collections of recommendations, especially with the limited interaction history. This abstract provides a quick summary of the paper, including the problem statement, proposed solution, and core research contributions.

**Keywords** Cold start · Community detection · LDA

S. K. Rachamadugu (✉) · T. P. Pushphavathi
M.S. Ramaiah University of Applied Sciences, Bangalore, India
e-mail: sandeep.cse@gprec.ac.in

T. P. Pushphavathi
e-mail: pushphavathi.cs.et@msruas.ac.in

S. K. Rachamadugu
G. Pulla Reddy Engineering College(Autonomous) Affiliated to JNTUA, Ananthapuramu, AP, India



K. R. Madhavi et al. (eds.), *Proceedings of Sixth International Conference on Computer and Communication Technologies*, Lecture Notes in Networks and Systems 1356,
https://doi.org/10.1007/978-981-96-5238-9_11

# 1 Introduction

The exponential growth in articles has made it a challenging task for a novice user to recommend articles. This article highlights the vital significance of scholarly recommendation systems, which can significantly reduce researcher time consumption and provide recommendations for scholarly items that a single researcher might overlook. The paper initially talks about the various forms and changes in recommended resources for literature, collaborators, reviewers, conferences, journals, datasets, and grant funding. The results of the study affirm that scholarly recommendation systems are useful tools to help researchers find relevant resources in these domains. The article aims to introduce different methods used for scholarly recommendations in order to provide readers with a comprehensive retrospective on how these recommenders have evolved, therefore helping them develop such recommendation systems over this domain [1].

Through modeling the cold-start problem formally as a graph matching constraint satisfaction, they provide an understanding of how to push new items. They argue that adequate knowledge of recommender system algorithms does contribute to successful marketing strategies for cold-start items and thus suggests ways that, in turn, lead merchants out of their own shadows with new products without incurring too much additional cost [2].

The need for this system stems from numerous factors. The recommender system also enhances research and collaboration by helping users find relevant articles and introduce them to new subjects. This can foster both the spread of knowledge and interdisciplinary collaboration. The recommender system now addresses the issue of cold start, enabling users to peruse a multitude of scientific papers without encountering significant obstacles due to insufficient user behavior data in either publication digest. The system not only saves time for researchers searching for specific information, but also aids in improving data retrieval. The cold-start problem-aware recommender of scientific articles enhances user experience, facilitates knowledge dissemination, fosters researcher collaboration, and provides scientists with efficient access to scientific information [3].

The cold-start user problem refers to a situation where the recommendation system is unable to provide effective suggestions to new users who have limited or no information. This issue is particularly critical in the context of scientific articles, given the unique nature of science and the potential for user behavior to complicate the process. To recommend scientific articles and address the cold-start problem, it is crucial to create user profiles as effectively as possible from the initial interactions. Techniques such as content-based filtering, collaborative filtering, and their combinations could assist in extracting preference information about new users, based on factors such as the text content of the articles they click and similar user behavior, among others. To solve the cold-start problem effectively, recommendation systems must have a continual learning and adapting strategy. Collecting feedback, studying user interactions, and learning from new data over time will help the system gain a better understanding of its users and provide more accurate recommendations [4].

## 2 Background and Related Work

The primary issue with recommending articles to a cold-start user stems from either limited user information or the absence of user preferences. In this session, we will explore the limitations of the cold-start problem and its impact on both content-based methods and collaborative-based filtering. The cold-start problem presents two challenges: the cold-start item problem, which involves recommending a new article to the user, and the cold-start user problem, which involves recommending an article to a new user. The cold-start problem happens in both collaborative filtering and content-based recommender systems. The recommender system won't work if it can't get enough useful information from the user's past records. Because the content-based method only extracts features from individual items, it is challenging to capture users' long-term interests if they have diversified and composite user interests. Furthermore, the system might solely suggest items based on the user's past interactions or feedback, potentially lacking diversity and disrupting serendipity. Though content-based is easy to implement, it struggles to provide good recommendations beyond what the users currently like or based on their limited interests due to its bias in recommending more of the same item. Collaborative filtering does not require a built-in user or item profile; it only requires a complete or partial history of interaction between users and items. The reason the collaborative filtering is very accurate in making recommendations also lies in how it makes them, which stems from direct user feedback. Cold start due to the cold-start issue with collaborative filtering, our system requires sufficient user ratings for new or rare items to provide accurate recommendations [5].

In this article, [6] recommender systems use classification-like algorithms to solve the cold-start problem for new users. Classification algorithms aid in the segmentation of new users based on their demographics and preferences. Intelligent methods combine this approach to pinpoint potential connections between new and existing users with similar interests. This leads to personalized recommendations based on historical preference data, even for novel users. Selecting classification algorithms, such as C4, can significantly improve recommendation performance, particularly for new users. Algorithms like C4.5 and Naive Bayes, which rely on the system's performance, differ from each other in their fundamental principles. For instance, they group or categorize users who behave differently from previous ones based on what they have learned. Both C4.5 and Naive Bayes differ from each other in their fundamental principles, which include categorizing users who behave differently from all previous ones based on what they have learned. The issues with algorithm choice have an impact on accuracy in grouping new users, handling data sparsity, and scalability.

In this article, [7] has a limitation in that it solely bases its recommendations on content-based data, potentially neglecting other variables like user preference and collaborative filtering. In this study, [8] introduced a basic taxonomy of recommender systems for scientific research. The finding that stands out at the top is categorization of different kinds of recommendations used in scientific areas. However, the taxonomy may be overly simplistic in that it is not specific to various scientific areas and lacks individual details.

In this study, [9], the author introduces a scholarly recommender system designed to suggest public dataset resources to researchers based on their respective research fields. This system enables researchers to access relevant literature, which is crucial for determining the appropriate use of the datasets through integration. The system recommends research papers in PubMed that are relevant to public datasets (Gene Expression Omnibus, GEO). When looking at different types of data, term-frequency-based methods (BM25/TFIDF) always did better than sentence embedding methods used in natural language processing, like doc2vec, ELMo, or BERT. The NLP embedding model seems to have done a little worse than term-frequency-based methods in the area of paper recommendation with a public dataset, as shown in Q1. Essentially, general embeddings (word2vec and doc2vec) beat out specialized ones (ELMo and BERT). BioBERT, i.e., domain-specific embeddings, significantly outperformed non-domain-specific embeddings. The paper also suggests conducting additional experiments to enhance the performance of hybrid methods that utilize embeddings and term-frequency in recommendation making. A concern about the paper would be that it is very domain-specific (i.e., scholarly papers and public datasets), which could affect how much one can generalize our findings to other recommendation systems.

The author of this study, [10] devised a solution to the difficulties researchers face in locating relevant publications, which could serve as a cooperative approach to creating research paper recommender systems. The system strives to personalize recommendations based on user contextual metadata (e.g., reuse information or collaboration) and implicit feedback between researchers without depending on a research field of shared expertise. According to evaluation metrics, the proposed approach outperforms baseline methods in terms of relevancy and usefulness. The collaborative method for the research paper recommender system showed statistically significant gains in both precision top B recommendations and overall performance over baseline methods. The system works by extracting implicit relationships among research papers from paper-citation relations to offer personalized recommendations without maintaining user profiles. A limitation of the paper is that our approach still relies on public contextual metadata and collaborative filtering; it does not solve copyright barriers in terms of opening more content (full texts) with good enough probability. Further studies could investigate how to include other additional content in each paper, such as titles and abstracts, which can help to improve its performance.

The author of this study [11] delves into the limitations of recommender systems, particularly in the Cold-Start domain, where users cannot receive recommendations for new items due to insufficient data. This review examines research work on this issue in relation to collaborative filtering (CF)-based recommender systems, including existing techniques and performance metrics for solutions. Potential Cold-Start solutions for the collaborative filtering recommender technique include machine learning, contextual information, and matrix factorization. This work presents an overview of these methods and metrics, demonstrating the competence of machine learning algorithms to cope with this problem. This paper may have limited its research to collaborative filtering (CF) recommender systems while not ignoring alternative solutions from another recommendation paradigm, such as content-based filtering. Furthermore, the study may not delve into the intricacies of implementing these solutions or the potential obstacles that may arise when addressing cold-start problems.

## 3 Methodology

Existing methods rely on applying content-based or collaborative filtering to the contents and recommending articles based on user preferences.This section introduces a novel method that integrates topology information, such as a set of article collections, with topical information, such as the application of topic modeling techniques like Latent Dirichlet Allocation. Combining these two methodologies, the cold-start user will be able to select the appropriate words or topics and retrieve the articles based on personalized user profiles. In this section a novel method is introduced by integrating the topology information such as set of article collection and topical information like applying topic modeling techniques such as Latent Dirichlet Allocation.By integrating this two methodologies, it will be useful for the cold-start user to select the appropriate words or topics and retrieves the articles based on personalized user profiles. The [12] dblpv12 version provides the citation network dataset. It comprises 27 features, including document IDs, references, abstracts, titles, and keywords. Community detection is the process of partitioning a graph into strongly connected clusters (communities), where nodes in different communities are only weakly interconnected. Community detection is the idea of finding cohesive groups or clusters within a network. In citation networks, community detection involves identifying groups of papers or authors who have strong connections with each other through citations. We often refer to these groups as research communities or citation clusters. The authors of this paper have proposed one such solution that integrates community detection with topic modeling as a potential remedy to the unsupervised approaches, which are merely capable of capturing only structural features within communities. Community detection is central to the models presented in this paper and provides new means for contextualizing structural and topic-related views within evolving social networks [13]. The proposed topic modeling approach relies on community detection. Furthermore, we present some characteristics of semantic understanding, topic identification, and word association, which underscore the significance of community detection in this context. To summarize, community detection methods are no longer a complete mystery. They are the key to making a topic model that is probabilistically coherent, contextually relevant, and captures the semantic structures in the document corpus [14]. In Fig. 1 it demonstrates the steps to follow before recommending the articles to the user.

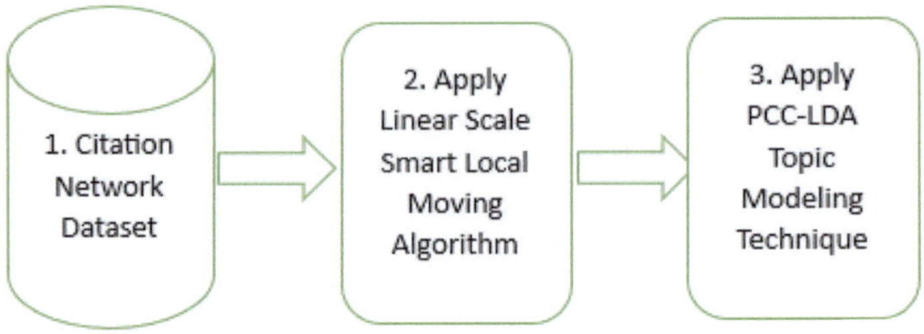


**Fig. 1** Order of integrating citation network to topic modeling

The steps are as follows

1 Preprocess the citation network dataset
2 Use the smart local community detection technique on a linear scale. Moving algorithm.
3 Apply the probabilistic correlated clustering to the latent Dirichlet allocation.

The preprocessing of the citation network dataset involves features like document IDs and references, as well as the construction of the citation network based on the dataset. Figure 2 depicts the community-detected citation network after applying the smart local moving algorithm to the citation network, which partitions the given network into modules. Apply the smart local moving algorithm to the citation network. The next step is to generate the topics by applying the topic modeling technique and the probabilistic correlated clustering technique. Modularity serves as the parameter to quantify the quality of the community. To determine the relevancy of topics, topic coherence is used. The modularity Eq. 1 found in [15] and topic coherence Eq. 3 found in [16]is given below.

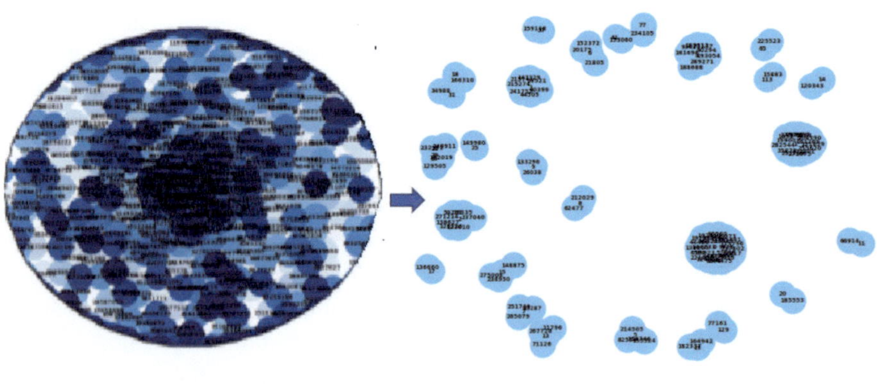

**Fig. 2** Partitioning the citation network into communities using SL-LSM

$$\Delta Q = \left[\frac{\sum_{\text{in}} + k_i \cdot k_{\text{node}}}{2m} - \left(\frac{\sum_{\text{tot}} + k_{\text{node}}}{2m}\right)^2\right] - \left[\frac{\sum_{\text{in}}}{2m} - \left(\frac{\sum_{\text{tot}}}{2m}\right)^2\right] \quad (1)$$

$$\text{PMI} = \log_2\left(\frac{p(w_i, w_j)}{p(w_i).p(w_j)}\right) \quad (2)$$

where $p(w_i, w_j)$ is the probability of seeing both words in the same document, and $p(w_i)$ and $p(w_j)$ are the probabilities of seeing each word separately. Equation (2) calculates the topic coherence:

$$\text{Topic Coherence} = \frac{2}{n(n-1)} \sum_{i=1}^{n-1} \sum_{j=i+1}^{n} \text{PMI}(w_i, w_j) \quad (3)$$

The Comparison of LDA Based recommendation with the content Based recommendation is explained in terms of basic steps here.

To recommend the articles

- Computes the similarity between the user profile and each document in the corpus.
- Sorts the documents based on similarity scores in descending order.
- Returns the indices of the documents sorted by their similarity to the user profile.

The following are the steps include for evaluating the recommendation system

- Preprocesses the user interests to generate an LDA-based user profile.
- Recommends articles based on the LDA approach.
- Calculates precision, recall and, F1-score using the true labels and recommended indices.

Preprocess user interest content

- Transforms the user interests into a TF-IDF vector using the TF-IDF model.

Recommend articles content:

- Computes the cosine similarity between the user profile (TF-IDF vector) and the TF-IDF matrix of the articles.
- Sorts and returns the indices of the documents based on similarity scores in descending order.

Evaluate recommendation system content

- Preprocesses the user interests to generate a TF-IDF user profile.
- Recommends articles based on the content-based approach.
- Calculates precision, recall, F1-score, and mean average precision (MAP) using the true labels and recommended indices.

# 4 Results and Discussion

In the previous session, we observed the fine-tuned algorithms after applying the revised smart local moving algorithm and PCC-LDA. We will see the results obtained during this session. Figures 3 and 4 refer to the relevancy of words obtained under each topic. Based on the words obtained under each topic for k = 5, the new user selects his words based on this profile and applies the PCC-LDA-based content-based recommended system and observed the following results of precision and recall by comparing the PCC-LDA-based recommended system with the content-based recommended system. The precision explain that Out of all the articles recommended by the system, how many are actually relevant to the user and recall specifies that Out of all the relevant articles, how many did the system recommend. The precision in Eq. 4 and recall Eq. 5 are explained here

$$\text{Precision} = \frac{\text{Number of relevant articles recommended}}{\text{Total number of relevant articles in the database}} \tag{4}$$

$$\text{Recall} = \frac{\text{Number of relevant articles recommended}}{\text{Total number of relevant articles in the database}} \tag{5}$$

The precision and recall values for recommending the scientific articles based on PCC-LDA is as follows Precision is 0.98 and Recall is 0.87.

The precision and recall values by applying directly to the content based is Precision is 0.79 and Recall is 0.62.

The advantage of topic modeling is that LDA captures semantic relationships between words and documents, offering deeper insights than just word frequency. The limitation of content-based systems is that they rely on surface-level features (word presence or absence), which might miss advanced relationships captured by LDA.

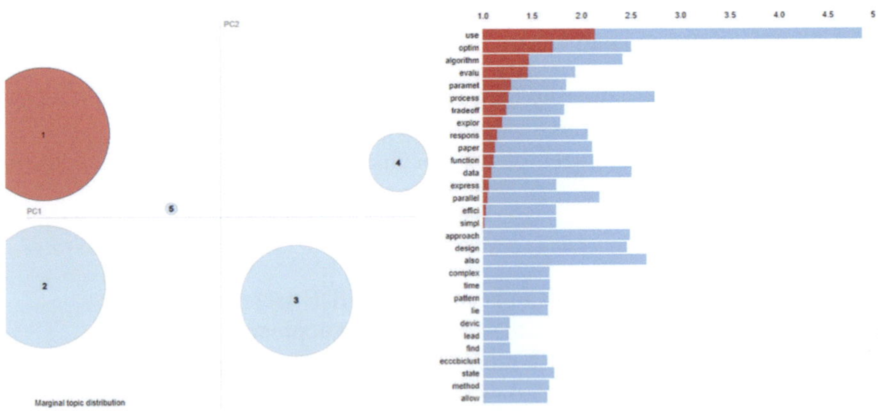


**Fig. 3** PCC-LDA for five topics

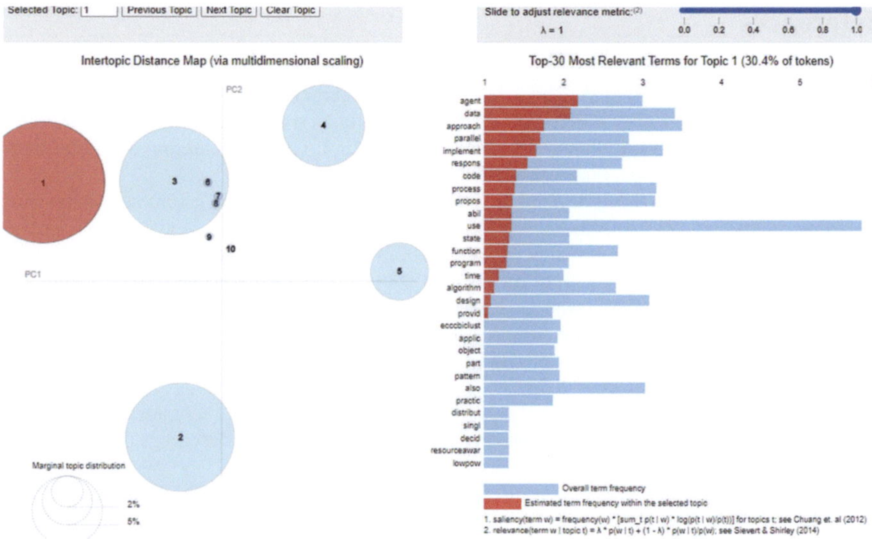


**Fig. 4** PCC-LDA for ten topics

# 5 Conclusion

This paper proposes a hybrid embedding-based model to tackle the cold-start issue, as suggested by scientific articles. By combining both methodologies, our approach leverages the best of each to ensure that even with minimal data where users interact with something on screen, the results will still be improved. We conducted extensive tests and verified that the results were better than traditional recommendation systems; our hybrid model performs better. This component is used for community detection in cold-start scenarios. By clustering articles with similar content, we further improve the recommendation accuracy by using explicitly available structural information in our data set. Through detailed experiments, the Significant performance gains in terms of both recommendation accuracy and user satisfaction convincingly demonstrate the effectiveness of our approach. Additionally, we conducted an empirical evaluation of our model using a real-world case study. Data helped us gain practical experience and tested our resilience. system could be. Despite its overall effectiveness, the hybrid model faces limitations. Training necessitates large-scale, computationally intensive processing. There is a potential need for domain-based finetuning. In our subsequent work, we will continue to enhance the methods for expanding our model and explore alternative approaches. Our hybrid model effectively addresses the issue of cold starts. By combining structural and content-based methods, the research effectively addresses the problem of scientific article recommendations.

## References

1. Zhang Z, Patra BG, Yaseen A, Zhu J, Sabharwal R, Roberts K, Cao T, Wu H (2023) Scholarly recommendation systems: a literature survey. Knowl Inf Syst 65(11):4433–4478
2. Jin-Hu L, Tao Z, Zi-Ke Z, Zimo Y, Chuang L, Wei-Min L (2014) Promoting cold-start items in recommender systems. PloS one 9(12):e113457
3. Mishra N, Chaturvedi S, Vij A, Tripathi S (2021) Research problems in recommender systems. J Phys: Conf Ser 1717:012002
4. Rahman M, Shama IA, Rahman S, Nabil R (2022) Hybrid recommendation system to solve cold start problem. J Theoret Appl Inf Technol 100(11):3562–3578
5. Xiaomei B, Mengyang W, Ivan L, Zhuo Y, Xiangjie K, Feng X (2019) Scientific paper recommendation: a survey. Ieee Access 7:9324–9339
6. Blerina L, Kostas K, Stathes H (2014) Facing the cold start problem in recommender systems. Expert Syst Appl 41(4):2065–2073
7. Nair AM, Benny O, George J (2021) Content based scientific article recommendation system using deep learning technique. In: Inventive Systems and Control: Proceedings of ICISC 2021. Springer, pp 965–977
8. Ali G, Morteza A, Leonardo C, Castelli D (2022) A basic taxonomy. In: IRCDL, Recommender systems for science
9. Zhu J, Patra BG, Yaseen A (2021) Recommender system of scholarly papers using public datasets. AMIA Summits Trans Sci Proc 2021:672
10. Haruna K, Ismail MA, Damiasih D, Sutopo J, Herawan T (2017) A collaborative approach for research paper recommender system. PloS one 12(10):e0184516
11. Jian W, Jianhua H, Kai C, Yi Z, Zuoyin T (2017) Collaborative filtering and deep learning based recommendation system for cold start items. Expert Syst Appl 69:29–39
12. Tang J, Zhang J, Yao L, Li J, Zhang L, Su Z (2008) Arnetminer: extraction and mining of academic social networks. In: KDD'08, pp 990–998 (2008)
13. Daifeng L, Ying D, Xin S, Johan B, Jie T, Shanshan C, Jiayi Z, Guilherme R (2012) Adding community and dynamic to topic models. J Inf 6(2):237–253
14. Chowdhury MR, Ahmed I, Sadeque F, Yanhaona M (223) Topic modeling using community detection on a word association graph. In: Proceedings of the 14th international conference on recent advances in natural language processing, pp 908–917
15. Chen M, Kuzmin K, Szymanski BK (2014) Community detection via maximization of modularity and its variants. IEEE Trans Comput Soc Syst 1(1):46–65
16. Blair SJ, Bi Y, Mulvenna MD (2020) Aggregated topic models for increasing social media topic coherence. Appl Intell 50:138–156

# Design of Low Power ALU Using FinFET-Based Adiabatic Logic

**Panchireddy Naveen, Yalla Asha, Pakki Siva Sateesh, Valayathil Thomas Jose, Regpu Sai Vamsi, Jami Venkata Suman, and Siva Ram Rajeyyagari**

**Abstract** In the current era, leakage power is one of the primary parameters while designing any integrated circuit. We know every microprocessor is designed using an ALU, and nowadays, ALU is becoming more and more complicated and much smaller in size to provide efficient potent computer systems. ALU performs arithmetic and logical operations. In this paper, we have considered arithmetic operations like add, subtract, comparator, and logical operations like XOR, NOT, AND, and OR. We have designed all the components using adiabatic logic, i.e., PFAL, since it provides energy recovery. FinFET has low threshold voltages compared to CMOS and it is even possible to run at low operating voltages while designing ALU using FinFET-based adiabatic logic. This makes us to provide reduced power consumption and also higher speed performance. We compared the power results of ALU using FinFET-based PFAL and also by conventional CMOS, and the results show that there is an improvement in the results of power consumption for an ALU designed using PFAL.

**Keywords** ALU · FinFET · Adiabatic logic · PFAL

## 1 Introduction

The ALU is an internal component of a microprocessor that performs basic and advanced operations. ALU is a crucial part of the CPU that executes operations like addition, subtraction, comparison, division, bitwise operations, etc. ALU is used in many devices like laptops and smart phones and also plays a vital role in various embedded systems where several calculations and logical operations are required [1]. The CPU mainly consists of three parts, the ALU, the memory unit, and the control

P. Naveen · Y. Asha · P. S. Sateesh · V. T. Jose · R. S. Vamsi · J. V. Suman (✉)
Department of ECE, GMR Institute of Technology, Rajam, Andhra Pradesh, India
e-mail: venkatasuman.j@gmrit.edu.in

S. R. Rajeyyagari
College of Computing and Information Technology, Shaqra University, Shaqra, Saudi Arabia



K. R. Madhavi et al. (eds.), *Proceedings of Sixth International Conference on Computer and Communication Technologies*, Lecture Notes in Networks and Systems 1356,
https://doi.org/10.1007/978-981-96-5238-9_12

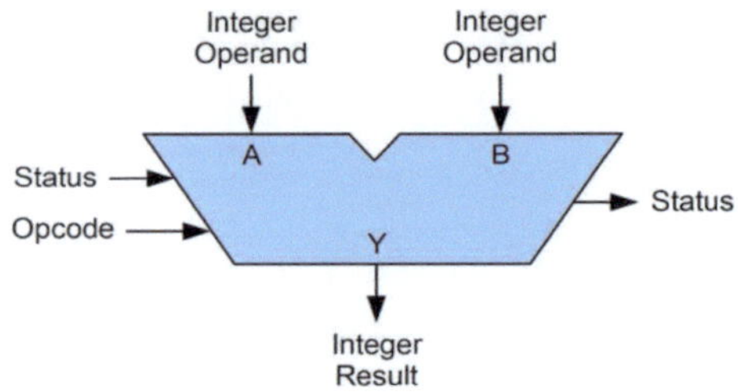


**Fig. 1** Structure of ALU

unit. Among these, the control unit is used to decode the instructions which are given by the user and the decoded instructions from the control unit are given to memory unit where the memory unit transfers the instructions in the machine language to the arithmetic logic unit that performs the calculations as per given instructions. The structure of ALU is shown in Fig. 1.

The ALU performs arithmetic calculations like adding, subtracting, and comparing of the given inputs and logical operations like AND, OR, complement, etc.

In recent years, the transistor size has been reduced to nanoscale and the sizes of the processors are also minimized. Now, the main objective of most of the electronic devices is to reduce the power dissipation and adiabatic logic is best suited for it [2, 3]. The term adiabatic is used in thermodynamics which means the heat energy is neither lost to nor gained from the environment. The power dissipation can be reduced by using adiabatic logic. In CMOS logic, when switching takes place (from 0 to 1 and vice versa), there is a power dissipation of $C_L V_{DD}^2/2$. This is the least power that can be dissipated from the device. By using adiabatic switching technique, we can reduce the power dissipation below this lower limit.

## 2 Literature Survey

In the paper [4], we designed the 16-bit ALU by combining the 1-bit ALUs using FinFET and adiabatic logic. The paper contains some adiabatic logic families which exist in the current technology like QSERL, 2PADCL, ADCL, etc. FinFET is a multi-gate FET and has more than one gate in which the fin is connected between source and drain, and because of multiple gates, it can reduce the energy dissipation and operates at lower voltage, which is better when compared to CMOS technology. The process of connecting millions or billions of MOS transistors to one chip to create an integrated circuit (IC) is known as very-large-scale integration. The popularity of CMOS technology is mostly because of its capability of integrating into a large number of components into a single chip and because of its lower energy dissipation. The most recent trend toward ultra-low power, however, has prompted academics to look for methods to salvage or reuse power from the circuit. In modern days, scientists have mostly concentrated on finding the minimum amount of energy use.

Different techniques are used to minimize the power dissipation in the circuits, but since majority of power is lost, adiabatic logic is the best way to create designs that are both power and energy efficient.

The paper [5] explains the concept of adiabatic logic to reduce power consumption in electronic devices. The aim of the adiabatic circuits is to reuse the energy which was dissipated during the switching and to enable reusing of the energy which was drawn from the VDD. These circuits have specific output phases within the power clock cycle. This paper explains various methods for designing adiabatic logics, including CMOS adiabatic circuits. Usually, in static CMOS, during charging, the energy moves from the VDD to the capacitance and from the output to the GND while discharging. When the output transitions from GND to VDD, some amount of energy is taken from the energy source, and during the transition from VDD to GND, the energy stored at the output is gradually dispersed over time.

The paper [6] explains issue regarding the power dispersion in the circuits and it introduces the concept of adiabatic logic as a solution for designing of the circuit. Adiabatic logic means it reduces the power dispersion during switching events by reducing the heat gain or loss. The abstract contains the design and simulation of circuits using adiabatic logic, by explaining its potential for minimizing the power dissipation compared with the designs using CMOS technology. Partially adiabatic logic families, like ECRL, adiabatic logic, PFAL, and CAL, are way easier to design. Fully adiabatic circuits include PAL, 2PASCAL, and SCRL.

In fully adiabatic circuits, charge on the capacitance is recuperated and fed back to the energy resource, more complex than partial adiabatic circuits. The paper [7] focuses on calculating power dissipation of various logic gates by using different adiabatic logic families. ECRL utilizes PMOS, NMOS in a cross-coupled configuration, along with alternating power clock in place of a direct supply. 2N-2N2P logic involves adding two additional transistors in accordance with the existing ones to reduce coupling effects and eliminate floating nodes during the recovery phase. PFAL is a type of circuit that achieves the lowest energy consumption among the three adiabatic logic families mentioned. It consists of a latch having 2 NMOS and 2 PMOS transistors in a cross-coupled configuration. This type of connection decreases the resistance during the charging process of the capacitance [8].

## 3 FinFET Technology Overview

FinFET is a multi-gate MOSFET in which a channel is placed between source and drain. Basing on this channel, the name is given as FinFET technology. FinFET can be used at a low operating voltage as compared with CMOS technology. Most of the electronic devices which are made now are made by using FinFET technology because of its better performance over CMOS like area of performance, low voltage operation, less leakage power, etc. The structure of FinFET is shown in Fig. 2 [9, 10].

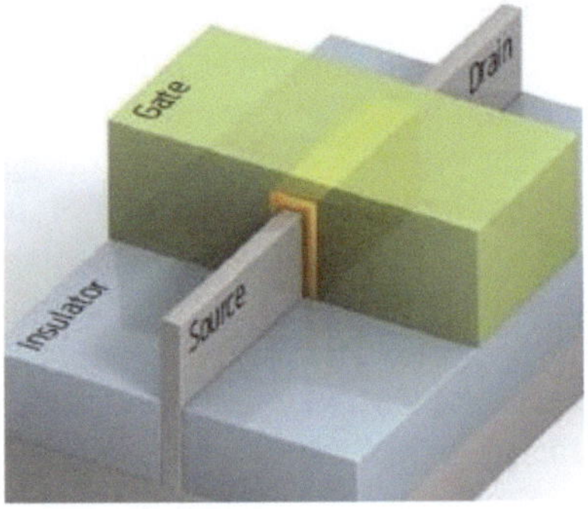


**Fig. 2** Structure of FinFET

The extra oxide layer is taken out. To make the channel isolated from the gate, a gate oxide layer is formed over the channel.

## *3.1 Working of FinFET*

Let us consider the whole FinFET as a capacitor where the first plate of the capacitor is the gate terminal and the second plate of the capacitor is the substrate. So, between the gate and the substrate, there is silicon dioxide. This silicon dioxide acts as the dielectric. Now the gate terminal is connected to the higher potential and the substrate terminal is connected to the lower terminal. Now positive charge is accumulated over the gate terminal and negative charges at the substrate terminal. Initially when VGS is lower, the substrate is accumulated by the minority charge carriers but as VGS is increased, the minority charge carriers are pushed down and the negative immobile ions are accumulated over the substrate. Now, the region near the surface first becomes less p-type and then it becomes n-type due to excess negative charge. A channel formation takes place between drain and source. This conductive channel starts conducting between source and drain [11].

The working of a both FinFET and the conventional MOSFET is almost similar but because of the multiple gates and the fins, the operating voltage is less for a FinFET when compared to that of MOSFET [12].

## *3.2 Methodology*

In current computing systems, the arithmetic logic unit is a major component responsible for performing logical and arithmetic operations on a given binary data. In addressing the demand for effective processors, the innovative circuitry layout techniques like adiabatic logic are being developed and this project mainly focuses on the designing a 4-bit arithmetic logic by using FinFET-based adiabatic logic in order to achieve the high energy efficiency. As we know that, the ALU performs the operations like addition, subtraction, bitwise AND, OR, XOR, and NOT, etc. In addition if we take two binary integers, the ALU will add it together. If the obtained addition value exceeds the current bit position, it results in the sum of two values combined with a carry out bit. For subtraction, two's complement representation is used to do subtraction. In this, the second operand's negation is added to the first operand by the arithmetic logic unit in order to obtain subtraction. In this process, there may be borrow and overflow flag can be produced. During AND operation, the ALU performs a bitwise AND operation for the corresponding bits of two inputs. During OR, the ALU performs a bitwise OR operation. During XOR (Exclusive OR), the bits of two input operands, ALU performs a bitwise XOR operation. NOT: On any single input operand, the arithmetic logic unit performs a bitwise NOT operation. Here, addition, comparison, and subtraction are considered as arithmetic operations and whereas AND, OR, Exclusive OR, and NOT are considered as logical operations. An Energy efficiency and scalability of FinFETs technology. Due to features of energy efficient and advantages of scalability, FinFET technology has become more popular in the semiconductor industry. These are the main reason for the usage of arithmetic logic unit in low power processors. Less leakage current: While comparing with the conventional planar CMOS transistors, the FinFETs provide great gate control. As the transistor is turned off, the fin structure enables stronger leakage current suppression.

Better switching performance: As we know that, FinFETs are having better switching rates as compared with the planar transistors so, this may result in less power dissipation during the process. Dynamic Voltage and Frequency Scaling: It enables the greater dynamic voltage and frequency scaling by altering the clock and supply voltage. It is the one of the feature of FinFET technology. Scaling to smaller process nodes: FinFET technology enables the scalability to lower process nodes which includes 7nm, 5nm, and beyond. In the smaller process nodes, the power consumption is reduced and it has greater transistor density which makes the energy efficiency as top priority. Difference between FinFET and CMOS while comparing the FinFET technology with the conventional CMOS technology, the FinFET technology has better performance and less leakage.

## 4 Results

Figure 3 depicts the ALU schematic, which consists of arithmetic and logical blocks. The output of these two blocks is given as the inputs to the mux and the output will be taken out. Each 1-bit ALU is combined to form a 4-bit ALU. The simulation result is shown in Fig. 4.

The comparisons of the power results of the conventional CMOS model and the proposed ALU using FinFET-based adiabatic logic are shown in Table 1 and bar graph comparison is shown in Fig. 5. From the result comparison table, it is clearly observed that the proposed ALU using FinFET-based adiabatic logic gives better performance compared with other existing designs.

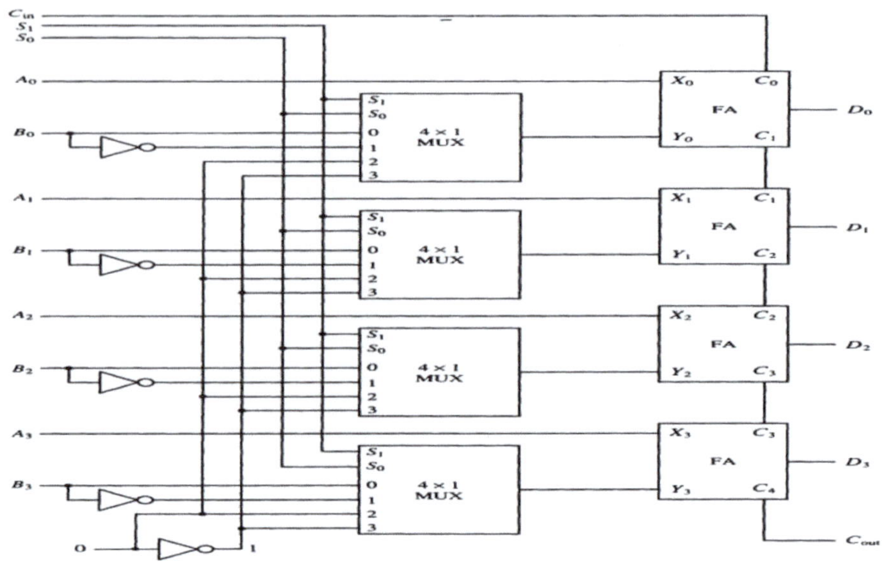


**Fig. 3** Layout of 4-bit ALU

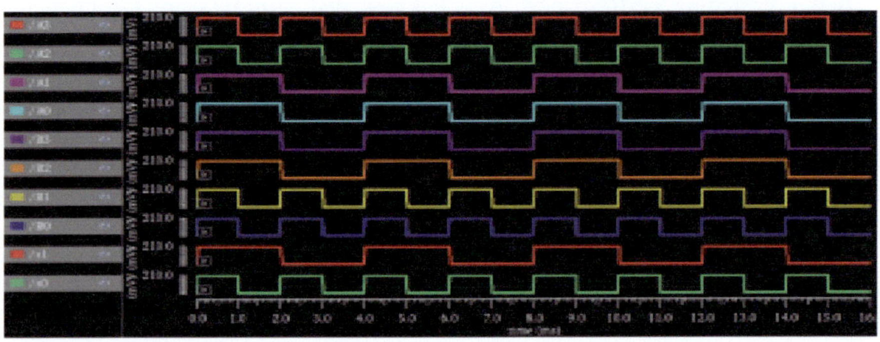

**Fig. 4** Waveforms are obtained by simulating the schematic design

## 5 Conclusion and Future Scope

The designed ALU consists of three blocks, an arithmetic block which consists of an adder and comparator, a logical block which consists of AND, OR, XOR, and NOT gates, and a multiplexer which gives the desired result. This whole functional block is repeated four times which gives a 4-bit ALU. The operations that are performed are addition, subtraction, and comparator which are arithmetic operations, and OR, AND, XOR and NOT which are logical operations. This design is based on FinFET technology and adiabatic logic which are best to minimize the power consumption. Among all the adiabatic logic families, PFAL is the best method used to minimize

**Table 1** Comparison of the power results of the conventional CMOS model and the proposed ALU using FinFET-based adiabatic logic

| | Power report of conventional CMOS-based ALU (mW) | Power report of proposed ALU (mW) |
|---|---|---|
| AND gate | 0.394 | 0.148 |
| OR gate | 1.398 | 0.654 |
| XOR gate | 1.398 | 0.698 |
| Adder | 1.98 | 1.901 |
| Mux | 11.933 | 6.2 |
| Arithmetic block | 8.689 | 4.76 |
| Logical block | 15.708 | 3.455 |
| ALU | 49.076 | 21.98 |

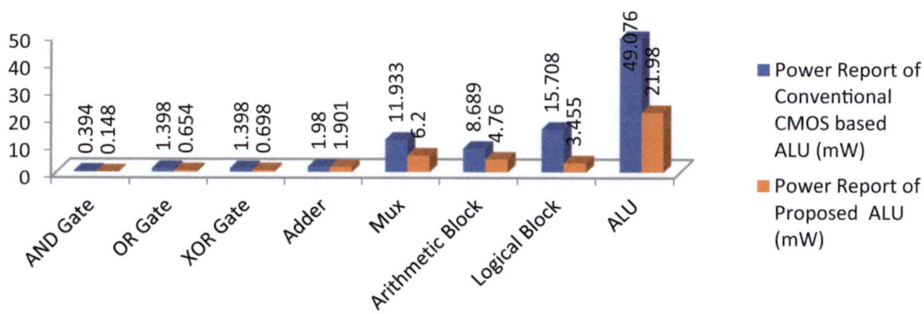


**Fig. 5** Bar graph power results comparison of conventional CMOS and proposed ALU using FinFET-based adiabatic logic

the power consumption. The input given is not a pulsed power source and is more like a sinusoidal wave, so a maximum of the power is recovered. The power results are better for the ALU designed by FinFET and adiabatic logic over the conventional CMOS logic. There are many other applications of FinFET technology that can be used many researches. This 4-bit ALU can further be designed to make a higher bit of ALU by cascading the one-bit ALUs. So, this design is an alternative approach to the CMOS technology where the power consumption is reduced to the maximum and is operated at higher speed by using both the FinFET and adiabatic switching technique.

## References

1. Bhati P, Rizvi NZ (2020) Adiabatic logic: an alternative approach to low power application circuits. In: International conference on electrical, electronics, and optimization techniques (ICEEOT), pp 4255–4260

2. Jothi D, Sivakumar R (2018) A completely efficient charge recovery adiabatic logic content addressable memory. In: International conference on computers, communications, and systems (ICCCS), pp 36–41
3. Grover V, Gosain V, Pandey N, Gupta K (2018) Arithmetic logic unit using diode free adiabatic logic and selection unit for adiabatic logic family. In: International conference on signal processing and integrated networks, pp 777–781
4. Vanlalchaka R, Reshmi MN (2002) A low power design using FinFET based adiabatic switching principle: application to 16-bit arithmetic logic unit. Ain Shams Eng J 14
5. Manoj S (2020) Adiabatic logic-based circuit optimization for ultra-low power applications. J Inf Optim Sci 41(1):85–98
6. Suman JV (2019) Design and performance evaluation of hybrid Vedic multipliers. Int J Innovative Technol Exploring Eng 8(8):1622–1626
7. Abheesh Kumar M, Sudhakar A, Suman JV (2019) Design and implementation of compressor based 32-bit multipliers for MAC architecture. Int J Innovative Technol Exploring Eng 8(9):2007–2011
8. Hemantha GR, Priya AS, Suman JV, Rao TVJ, Priyadarshini GMA, Mallam M (2024) Characterization and modeling of gate-all-around FET (GAA FET) for low-power and high- performance applications. In: 2024 International conference on advances in modern age technologies for health and engineering science (AMATHE), pp 1–5
9. Yamae T, Takeuchi N, Yoshikawa N (2021) Binary counters using adiabatic quantum-flux-parametron logic. IEEE Trans Appl Supercond 31(2):1–5
10. Suman JV, Cheepurupalli KK, Allasi HL (2022) Design of polymer-based trigate nanoscale FinFET for the implementation of two-stage operational amplifier. Int J Polym Sci 2022:1–12
11. Singhetal R (2018) Evaluation of 10-nm bulk FinFET RF performance-conventional versus NC-FinFET. IEEE Electron Device Lett 39(8):1246–1249
12. Venkata Suman J, Hema M, Swetha Priya A, Raja Ramesh D, Syamala Devi P, Subba Rao P (2024) FinFET technology based low power SRAM cell design for embedded memory. In: Madhavi KR et al (eds) Proceedings of the international conference on computational innovations and emerging trends (ICCIET 2024). Atlantis Press, pp 1116–1124

# Deep Learning-Based COPD Disease Prediction in Lungs

**A. V. Nageswara Rao, Sk. Bajidvali, Md. Javeed Ahammed, and D. Jagadish**

**Abstract** The proposed study evaluates the critical role of deep learning (DL) models in predicting function decline in patients with Chronic Obstructive Pulmonary Disease (COPD). The crucial need originates from the limitations of traditional methods, which frequently depend on static models that are incapable of adjusting to the dynamic nature of COPD progression. These outdated ways of anticipating decline patterns lack precision, limiting early intervention and personalized patient care. In this work, better accuracy is obtained for the proposed DL methods that consider not only various biomarkers but also particular details of the patients. Computational efficiency and measure of various performance metrics show the potential of the proposed method over traditional methods. Hence, it can be concluded that the proposed DL algorithm represents a tremendous mark in anticipating renal function decline in COPD patients.

**Keywords** Renal function · Chronic Obstructive Pulmonary Disease (COPD) · Deep learning (DL) · Gradient

---

A. V. Nageswara Rao (✉) · Sk. Bajidvali · Md. J. Ahammed
Department of ACSE, VFSTR, VADLAMUDI, GUNTUR, Andhra Pradesh, India
e-mail: dravn_acse@vignan.ac.in

D. Jagadish
Department of ECE, Narasaraopeta Engineering College, Narasaraopet, Andhra Pradesh, India

A. V. Nageswara Rao
Department of ECE, Narasaraopeta Engineering College, Narasaraopet, India

D. Jagadish
Department of Mechanical Engineering, Chirala Engineering College, Chirala, India

 
K. R. Madhavi et al. (eds.), *Proceedings of Sixth International Conference on Computer and Communication Technologies*, Lecture Notes in Networks and Systems 1356,
https://doi.org/10.1007/978-981-96-5238-9_13

## 1 Introduction

One of the major challenges that healthcare systems are facing across the world IS the increase of cost in detection of renal function decline, particularly in Chronic Obstructive Pulmonary Disease (COPD). Renal function decline estimation helps in identifying the quick changes in the patients who are suffering with COPD disease. The traditional methods that depend on static models can not accurately predict the renal function decline that changes dynamically and also inbuilt features of the patients who are suffering with COPD. Estimation of renal function decline needs lot of complicated attributes where the existing traditional approaches have failed. Hence, a better dynamic algorithm has to be developed for the prediction of renal function.

Existing methods are having a lot of drawbacks as they are static in nature and they cannot adapt to the drastic changes in renal function due to which they are less accurate in the detection of COPD disease. The existing methods will take more time for detection with high complexity and also less efficient. The state-of-the-art methods are not efficient in identifying various parameters, and also due to huge number of subtle features to be detected, existing methods fail to give better prediction for disease identification.

Hence, advanced techniques such as DL algorithms can be proposed to overcome these drawbacks in the existing methods. In order to obtain high accuracy, ensemble methods can be combined with deep learning methods. The proposed method gives a complete information of the COPD patients by considering individual details pertaining to each patient and also considers the relevant features that help in observing the declination of the renal function.

The proposed algorithms are discussed with its complete architecture and also with the various steps such as data collection, preprocessing, model building, training and testing the model are explained in detail in the upcoming sections. The analysis of COPD data has been made better with the integration of individual details and the advanced DL methods. Furthermore, the study will offer convincing results generated from extensive result analysis, demonstrating the usefulness of the DL integrated strategy in redefining the COPD prognostication landscape.

## 2 Literature Review

Chronic Obstructive Pulmonary Disease (COPD) is a world health issue that causes high serious illness and death and often leads to other illness. Patients frequently ignore COPD due to its insidious nature and absence of early signs. COPD identification is essential for early intervention and disease management. Deep learning algorithms can quickly and effectively diagnose COPD, helping doctors accomplish the objective. The study uses a dataset of the Kaggle to diagnose COPD using DL. The dataset has many missing values, making analysis difficult. To fill gaps, KNN imputation is an important method. The method recognizes complete samples with measurements closest to incomplete samples, improving dataset completeness and reliability [1]. Predictions of chronic renal disease are discussed from that perspective in the study. The Kaggle repository provided the dataset on chronic renal illness. The graph displays comparisons of precision, recall, accuracy, F-measure, AUC, and GINI coefficient along with the results generated by many methods. The combination of absolute shrinks and the best results were obtained after complete characteristics and synthesized minority oversampling, utilizing logistic regression with features picked using this approach [2].

The increased prediction ability of the model outperforms that of traditional data classification approaches. The study then recommended integrating the IoT and ML with the best-performing DL models. Deep learning will be employed more effectively and efficiently in predictive analytics as a result of the new technology, advancing COPD prediction. The study, which uses deep learning models and risk indicators, provides a fundamental step thorough performance evaluation to identify and forecast COPD [3]. Renal failure at the end stage and the patient's mortality may eventually result from improper disease diagnosis and treatment. In the sphere of medical science, machine learning (ML) techniques are an invaluable tool for sickness prediction. The study's objective is to provide useful tools for COPD incidence predictions using ML techniques [4]. The main aim of this work is to establish whether a moderate quantity of data comorbidities were incorporated, testing results, which can be difficult to attain in low and medium-income nations, could serve as an adequate foundation for prediction models of these kinds. However, in affluent nations, policymakers may find these models useful for facilitating the planning and allocation of treatment resources [5].

The probability of it happening can be raised by early identification and treatment. Throughout this work, many ML approaches have been investigated for their potential to offer a prompt diagnosis of COPD. Consequently, it will investigate in the methods the connection between data elements and the characteristics of the target class. Because of the enhanced attribute measurements that predictive modeling allows for, it can build a set of prediction models by the use of ML and predictive analysis [6]. According to the research, creating an intelligent model to classify people as "COPD" or "Non-COPD" will help clinicians diagnose patients more quickly and manage several patients at once. Patients in remote areas can eventually obtain early diagnosis by the implementation of the proposed ML model by organizations in rural areas with less medical experts [7]. Examining how well ANN algorithms work when used to predict the survival of patients with COPD is the main objective of the study. Patients suffering from chronic kidney disease see a gradual decline in blood filtration, which is the primary function of the kidney [8]. The study's objective is to build a novel forecasting model for predicting the risk of COPD with its progression in people with diseases including hypertension and diabetes.

Wrapper based methods that can systematically evaluate different subsets of features and measure their impact [9]. The COPD ranking system categorizes patients based on their eGFR, and it aids in risk assessment, monitoring, and therapeutic scheduling. The research seeks to accurately estimate when a COPD sufferer will require dialysis, enabling individualized care and smart medication selection [10]. Constructing a DNN and evaluating its efficacy against other contemporary ML techniques is the aim of the work. Throughout testing, any missing values in the database were substituted with the average of the connected features. Subsequently, the ideal variables of the neural network were determined through parameter definition and several experiments [11].

In the research for the effective and efficient diagnosis of kidney sickness, an AHD CNN has been provided to investigate the capabilities of several DL methodologies. The purpose of the data set dictates the categorization technology's efficiency. A CNN-based algorithm model has been developed to minimize the feature dimension and increase the accuracy of the classification model [12]. The disease is referred to as "chronic" as the kidney damage in humans develops gradually over an extended period. The injury may lead to waste products building up in the body. Many methods and architectures have been built to identify COPD at an early stage. In particular, machine learning (ML) techniques are crucial for the early diagnosis of a wide range of illnesses. ML algorithm used in medical analysis and prediction, which have been employed to provide analytical results [13]. A framework for intelligent classification and prediction is presented in the essay. To predict pulmonary-related illnesses, it uses the activation function, and a modified DBN as a classification approach [14]. The paper provides hybrid machine learning approaches for COPD detection, which include big data platform (Apache Spark)-based machine learning classification algorithms and feature selection techniques [15]. Using clinical data, the research employs a DL model to computerize the process of COPD detection. Furthermore, the study's purpose is to create an accurate and usable model that can reliably diagnose COPD [16]. The importance of early identification and characterization in defining COPD treatment cannot be overstated. COPD is a pulmonary disease that inhibits the kidneys' capacity to eliminate waste and balance body fluids. Among the effects include high blood pressure, anemia, mineral bone disease, poor food habits, hormone imbalances, and neurological difficulties. Prior and better recognition of COPD [17] is very important in order to save the patients from getting serious problems such as heart strokes that can lead to death. In this method, various features of patient data have been considered and are separated using ALO technique that uses a standard framework. In summary, there is a need for accurate and timely detection of COPD which is not done with the existing methods that fail. Hence, there is a need for advanced DL techniques that can help in detecting the renal function in COPD patients.

## 3 Proposed System

Traditional methods drawbacks can be overcomed the traditional methods that are static in nature and fail to detect the continuous changes in COPD that arises due to dynamic behavior of renal function. The proposed method should be efficient both in computational time and precise in detecting the Chronic Obstructive Pulmonary Disease (COPD).

The proposed system uses DL algorithms which are advanced and can be able to recognize complex features by integrating various biomarkers of different datasets with distinct values. The datasets should be preprocessed in order to make sure that the details of the patients are clear without missing values, and also, all the values are in the proper format. Making sure that the data is properly arranged and preprocessed, then we can go for building the DL models that can be integrated with different individual details pertaining to each patient. Training the model rigorously with different features and characteristics related to multiple patients can make the model to function more efficiently.

The proposed model is then evaluated with different types of datasets related to various stages of COPD patients. The results are also compared with the traditional static models in terms of both time and accuracy. The DL algorithms proved to be much better in detecting the COPD progression in multiple patients considering different clinical parameters.

Early disease identification helps the healthcare providers or doctors to provide proper treatment and helps the patients to recover with in less time from the diseases. Due to its dynamic nature, it is very difficult to predict the renal function decline accurately, and if there is any delay in the timely intervention of this detection, it may lead to serious problems to the patients, and in certain cases, it may lead to patient's death also. DL algorithms are much more flexible and can adapt to the complex data also to give better results for multiple features and characteristics of patients.

The proposed DL models outperforms the various existing methods by providing a flexible and timely intervention that helps in the early detection of renal function changes that are dynamic in nature.

## 4 Experimental Results

In the proposed DL algorithms, there are different steps that are used to analyze the prediction of pulmonary destruction in patients suffering with COPD. DL algorithms are much better in prediction compared to the other models, and also with the ensemble approach, the methods are much comprehensive and can be used for varieties of data processing.

The first step in any ML or DL algorithms is to collect the data from different types such that the data is much more diversified and during training the data is divided such that few data samples are for training and few samples are for evaluating the models. To overcome the difficulty, extensive data preparation techniques are applied. Values that are missing are calculated using advanced imputation technique *k*-nearest neighbors, and anomalies are found and either eliminated or altered. Additionally, categorical variables are encoded using technique label encoding, and data normalization is conducted using min-max scaling to maintain homogeneity and improve model convergence.

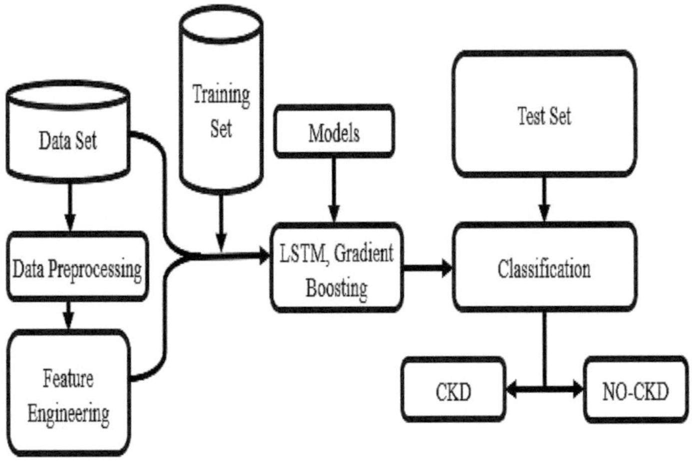


**Fig. 1** Workflow of designed system

Feature engineering plays an essential role in improving the model's predictive capabilities. To extract important characteristics from the dataset, it employs domain knowledge. Domain knowledge in forecasting renal function decline entails incorporating medical insights such as biomarker interactions, genetic variables, and disease trajectories to improve features for more accurate machine learning predictions in COPD disease. These characteristics may include derived variables, interaction terms, and transformations that capture the data's complicated connections. A feature importance analysis is used to select characteristics that have a substantial impact on predicting renal function decline (Fig. 1).

## *4.1 Model Selection*

The proposed system makes use of sophisticated ML techniques such as deep learning and ensemble approaches. The classification is accomplished by training DL algorithms such as long short-term memory (LSTM) and gradient boosting on varied datasets from chronic renal disease patients. Based on patient-specific data, biomarkers, and clinical factors, it classifies the potential of renal function decline. Given the longitudinal nature of COPD development, a LSTM neural network is used to capture temporal relationships in the data. At the same time, an ensemble of decision trees, of gradient boosting, is used to combine data from genetic, biomarker, and clinical datasets.

## 4.2 Model Training

The selected algorithms are subjected to thorough training on a wide range of COPD patient data. The LSTM network is trained on consecutive patient records, understanding the temporal patterns and relationships in renal function decline. The ensemble of decision trees is trained on the integrated dataset, using approaches such as bootstrapped aggregation to improve robustness and generalize ability. To improve the performance of the model, hyperparameter adjustment is performed. The COPD dataset is split into training and testing sets in a 70:30 ratio. The majority of the data is in the training set, which is used for model training, while the testing set, which is not utilized during training, is used to evaluate model performance.

## 4.3 Continuous Learning Mechanism

The implementation of a continuous learning process is one of the new characteristics of the proposed system. As fresh patient data becomes available, the model is meant to adapt and evolve. The models can be tuned with different techniques such that the gradient descent is optimized by which the proposed model can give relevant output in detecting the progression of pulmonary variation and it keeps on varying with respect to time.

## 4.4 Interpretability and Explainability

The dynamic nature of pulmonary obstruction makes it challenging to interpret the changes time to time. In order to do better analysis, it is necessary to use significant features that reveal the details that whether the patient is suffering from COPD or not. Doctors or healthcare providers can use the results of the proposed model to interpret and explain the condition of the patient in correct manner by suggesting timely treatment.

## 4.5 Evaluation and Validation

Proposed model is validated using different metrics such as area under the curve (AUC), sensitivity, specificity, and accuracy. These metrics confirm that proposed method outperforms the traditional models. The models are also evaluated with respect to time and it was proved that the proposed model is faster. In summary, the proposed system combines powerful machine learning algorithms, a continuous learning mechanism, and extensive data processing techniques to forecast renal function decline in COPD patients. The innovative features of the techniques include the coupling of LSTM for temporal dependencies and ensemble techniques for data integration, as well as the continuous learning element. The next sections of the research will give thorough result and analysis, proving the effectiveness and novelty of the proposed system for improving COPD management.

In order to assess the robustness of the proposed system in forecasting the decline of renal function among patients diagnosed with COPD, an exhaustive comparison was undertaken with existing systems. The adaptability to dynamic patient data and predictive accuracy were evaluated as two critical aspects. A comparison was made between the proposed system, which utilized advanced DL algorithms and a mechanism for continuous learning, and conventional static models. Table 1 compares the performance metrics of the existing systems [4, 6, 7, 10] discovered by the peer researchers and the proposed system for forecasting renal function decline in chronic kidney disease. The proposed system outperforms the existing systems undoubtedly in terms of accuracy (0.92), sensitivity (0.91), specificity (0.88), and area under the curve (AUC) value (0.92), demonstrating its improved ability to correctly identify cases of renal function decline and overall superior predictive capabilities (Fig. 2).

**Table 1** Performance metrics comparison

| Metric | Proposed system | Existing system [4] | Existing system [6] | Existing system [7] | Existing system [10] |
|---|---|---|---|---|---|
| Accuracy | 0.92 | 0.85 | 0.80 | 0.75 | 0.83 |
| Sensitivity (recall) | 0.91 | 0.82 | 0.75 | 0.68 | 0.83 |
| Specificity | 0.88 | 0.78 | 0.82 | 0.79 | 0.78 |
| AUC | 0.92 | 0.80 | 0.85 | 0.82 | 0.87 |

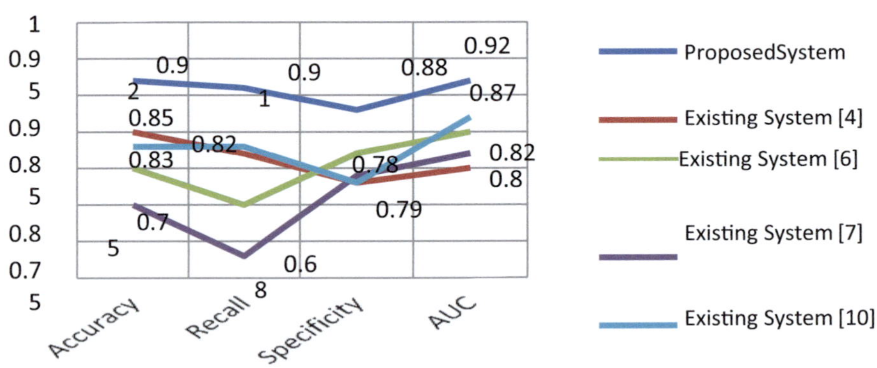


**Fig. 2** Graphical representation of performance metrics comparison of proposed versus existing systems

**Table 2** Computational efficiency comparison

| Metric | Proposed system | Existing system [4] | Existing system [6] | Existing system [7] | Existing system [10] |
|---|---|---|---|---|---|
| Training time (h) | 12 | 25 | 40 | 48 | 20 |
| Inference time (ms) | 5 | 13 | 17 | 20 | 9 |

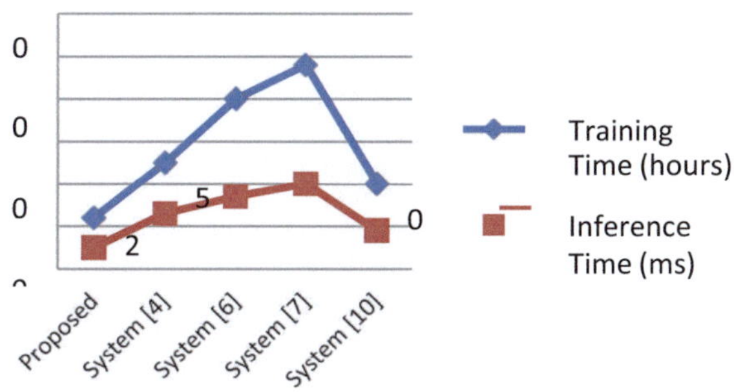


**Fig. 3** Graphical representation of computational efficiency comparison of proposed versus existing systems

Table 2 compares essential performance metrics for forecasting renal function decline in chronic kidney disease patients using the present method and the proposed approach. The proposed system outperforms the existing systems, with training time reduced to 12 h and inference time reduced to 5 ms. These enhancements represent increased computing efficiency, making the proposed system more suitable for real-time clinical applications.

The results and analysis emphatically demonstrate that the proposed system is preferable for predicting the decline in renal function among COPD patients. The enhanced precision, responsiveness, particularity, and area under the curve (AUC), in addition to the optimized computational efficiency, demonstrate the efficacy and feasibility of the unique system. The results underscore the capacity of the system to fundamentally transform the management of COPD, providing clinicians with a powerful tool for timely and precise predictions, which ultimately enhances patient results (Fig. 3).

## 5 Conclusion

In conclusion, the proposed system, which incorporates advanced DL algorithms and continuous learning processes, represents a significant leap in anticipating renal function decline in COPD patients. The impressive performance indicators (better accuracy, sensitivity, and specificity) demonstrate its improved predictive capabilities over traditional methods. The reduced training period and faster inference time increase its usefulness for real-time clinical applications. However, certain limits must be addressed, such as the necessity for large and diverse datasets, possible biases, and interpretability issues inherent in complicated ML models. Future work should concentrate on overcoming these constraints, improving the model's interpretability, and including new patient-specific characteristics to provide an enhanced prediction framework.

## References

1. Qin J, Chen L, Liu Y, Liu C, Feng C, Chen B (2020) A machine learning Methodology for diagnosing chronic kidney disease. IEEE Access 8:20991–21002
2. Chittora P et al (2021) Prediction of chronic kidney disease—a machine learning perspective. IEEE Access 9:17312–17334
3. Akter S et al (2021) Comprehensive performance assessment of deep learning models in early prediction and risk identification of chronic kidney disease. IEEE Access 9:165184–165206
4. Dritsas E, Trigka M (2022) Machine learning techniques for chronic kidney disease risk prediction. Big Data Cogn Comput 6(3):98
5. Dovgan E et al (2020) Using machine learning models to predict the initiation of renal replacement therapy among chronic kidney disease patients. PLoS ONE 15(6):e0233976
6. Islam MdA, Majumder MdZH, Hussein MdA (2023) Chronic kidney disease prediction based on machine learning algorithms. J Pathol Inform 14:100189
7. Antony L et al (2021) A comprehensive unsupervised framework for chronic kidney disease prediction. IEEE Access 9:126481–126501
8. Zhang H, Hung C-L, Chu WC, Chiu P, Tang CY (2018) Chronic kidney disease survival prediction with artificial neural networks. In: 2018 IEEE international conference on bioinformatics and biomedicine (BIBM)
9. Motwani A, Shukla PK, Pawar M (2021) Novel machine learning model with wrapper-based dimensionality reduction for predicting chronic kidney disease risk. Soft Comput Signal Process 29–37
10. Ventrella P, Delgrossi G, Ferrario G, Righetti M, Masseroli M (2021) Supervised machine learning for the assessment of chronic kidney disease advancement. Comput Methods Programs Biomed 209:106329
11. Singh V, Asari VK, Rajasekaran R (2022) A deep neural network for early detection and prediction of chronic kidney disease. Diagnostics 12(1):116
12. Chen G et al (2020) Prediction of chronic kidney disease using adaptive hybridized deep convolutional neural network on the internet of medical things platform. IEEE Access 8:100497–100508
13. Khan B, Naseem R, Muhammad F, Abbas G, Kim S (2020) An empirical evaluation of machine learning techniques for chronic kidney disease prophecy. IEEE Access 8:55012–55022
14. Elkholy SMM, Rezk A, Saleh AA (2021) Early prediction of chronic kidney disease using deep belief network. IEEE Access 9:135542–135549
15. Abdel-Fattah MA, Othman NA, Goher N (2022) Predicting chronic kidney disease using hybrid machine learning based on Apache SPARK. Comput Intell Neurosci 2022:1–12
16. Arafat F, Khan T, Bapon BD, Khan MI, Noori SRH (2021) A deep learning approach to predict chronic kidney disease in human. In: 2021 IEEE 12th annual information technology, electronics and mobile communication conference (IEMCON)
17. Nandhini G, Aravinth J (2021) Chronic kidney disease prediction using machine learning techniques. In: International conference on recent trends on electronics, information, communication & technology (RTEICT)

# Intelligent Autoscaling for E-Commerce Applications in Public Cloud Kubernetes Environment

**Harshavardhan Nerella**

**Abstract** In the fast-growing landscape of cloud-native technologies, Kubernetes has established itself as a prominent platform for containerized application orchestrator; node autoscaling is among the most important things in ensuring optimal resource utilization and cost efficiency. This paper discusses Karpenter, an innovative and open-source node autoscaler deployed on Amazon Elastic Kubernetes Service, and its advantages over traditional Cluster Autoscaler, specifically in e-commerce applications on public cloud platforms. Empirically, by analytical and comparative studies, we will prove Karpenter as superior in the dimension of intelligence and cost efficiency. The ability to dynamically balance and recreate nodes (virtual machines) based on load demand is elaborated in this study, underlining its quick, intelligent, and cost-efficient provisioning features.

**Keywords** Kubernetes · Karpenter · Cluster autoscaler · Node autoscaler · Cost optimization · Node consolidation · E-commerce

## 1 Introduction

At present, Kubernetes is the cornerstones in orchestrating containerized applications because of the flexibility and scalability it can bring to cloud-native environments. One of the most important components of management within a Kubernetes cluster is node autoscaling: this will ensure that virtual machines are available on load basis to host the micro services. Node autoscaling helps dynamically adjust or recreate the number of nodes based on the unschedulable pods, so it would become a major requirement for applications whose loads vary over time, like e-commerce applications.

H. Nerella (✉)
MassMutual, Austin, TX, USA
e-mail: nerellaharshavardhan@outlook.com



K. R. Madhavi et al. (eds.), *Proceedings of Sixth International Conference on Computer and Communication Technologies*, Lecture Notes in Networks and Systems 1356,
https://doi.org/10.1007/978-981-96-5238-9_14

### 1.1 Importance of Autoscaling in Cloud-Native Environments

Cloud-native architectures are designed to be elastic, fault tolerance, and high availability as that applications are designed to leverage the full power of cloud resources. Autoscaling plays a major role in such environments since it helps applications scale up or down according to fluctuating traffic patterns automatically without requiring manual inputs from an administrator, remaining at maximum performance with possible minimized cost. For example, e-commerce applications might suddenly surge in their traffic during peak shopping seasons. One of the key features autoscaling offers is that, in view of a rise in traffic to a site, it automatically provisions extra resources so as to maintain performance and avoid downtime, which helps in ensuring a smooth user experience [1].

### 1.2 Types of Autoscaling in Kubernetes

In Kubernetes, there are two forms of scaling: Pod Autoscaling and Node Autoscaling.

**Pod Autoscaling**: The pod autoscaler handles scaling of the pod replicas, to the desired count based on the demand. This is achieved by using Horizontal Pod Autoscaler (HPA). The HPA controls the scaling of pods up and down by monitoring the CPU or Memory utilization or other custom metrics [2]. This mechanism works well in managing applications with variable demands, such as e-commerce applications during flash sales or peak seasons or during a special sale [3]. Pod autoscaling ensures application responsiveness by keeping the desired number of instances based on real-time traffic.

**Node Autoscaling**: The node autoscaler handles scaling the number of nodes within the cluster as per the unschedulable pods. Node autoscaling is handled through the Kubernetes Cluster Autoscaler and more recently Karpenter [4]. Node autoscalers add more nodes to a cluster when there are pending pods that could not be scheduled due to lack of resources. Also, node autoscaler remove underutilized nodes to help optimize costs [5]. Node autoscaling is critical to the effective management of the infrastructure, ensuring sufficient resources for running applications without incurring unwarranted expenditure on cloud resources.

### 1.3 Introduction to Cluster Autoscaler and Karpenter

Cluster Autoscaler is most widely used node autoscaling tool adopted by many organizations as its one of the oldest and stable autoscaling tools available. It was designed to automatically handle the number of nodes in a cluster according to the resource needs of workloads. The Cluster Autoscaler increases nodes if there are pending pods that could not be scheduled due to the shortage of resources, and it removes them if they're underutilized [6]. While effective, Cluster Autoscaler has several limitations in areas like slower response time and lower intelligence when balancing and recreating servers.

While Cluster Autoscaler is one of the oldest node autoscaling tools in Kubernetes, Karpenter represents a more recent innovation in this domain. Karpenter was designed to overcome some of the shortages in the Cluster Autoscaler; additionally, it had faster node provisioning, higher intelligence in workload management, and better cost efficiency. It dynamically balances, consolidates, and recreates servers based on the workload demand. The solution thus stays very strong in applications having large variability in traffic, as would typically be expected in an e-commerce application [7].

### *1.4 Focus on E-Commerce Applications and Public Cloud Platforms*

E-commerce applications are inherently unpredictable and explosive, needing to support extreme fluctuations in user traffic if necessary: when promotional events or holiday sales spring to life, or if some viral marketing sets your application ablaze, it needs to be able to scale rapidly to maintain performance and user satisfaction [6]. Public cloud platforms provide this dynamic scaling by providing the necessary infrastructure. Hosting containerized applications on Kubernetes clusters residing in the public cloud platforms can allow for scaling the nodes up and down by leveraging advanced autoscaling solutions such as Karpenter. Since this can be greatly efficient and cost-effective, it helps in hosting e-commerce applications at a large scale.

In this paper, I discuss the additional benefits of using Karpenter over the traditional Cluster Autoscaler when hosting an e-commerce application on Amazon Elastic Kubernetes Service (EKS). We present empirical data from a comparative analysis done to demonstrate Karpenter's superiority in two most critical concerns: intelligence and cost efficiency.

## 2 Literature Review

Autoscaling in Kubernetes has evolved a lot during the few years. Multiple studies have focused on the optimization of resource management in cloud-native environments. In this section, we will go through some of the existing literature on Kubernetes autoscaling to underscore salient contributions with their limitations that gave rise to new solutions such as Karpenter.

## 2.1 Overview of Existing Autoscaling Solutions in Kubernetes

In Kubernetes, the autoscaling is achieved through two main components: the Horizontal Pod Autoscaler and the Cluster Autoscaler. The Horizontal Pod Autoscaler (HPA) provides a way for the replicas of a pod to automatically scale based on observed metrics, like CPU/memory usage, to make an application scale up or down horizontally facing workload changes. Many studies have already shown that HPA works very well at keeping application performance on top in the face of variable traffic conditions, most especially e-commerce platforms where demand surges can occur in an instant [3].

On the other hand, Cluster Autoscaler manages the number of nodes within a cluster per the resource requirements of unschedulable pods. The node autoscaler adds nodes when there are pending Pods due to insufficient resources and also removes underutilized nodes to save the cost. In spite of its efficiency, Cluster Autoscaler delays response time and is less effective with highly complex workload scenarios [8].

Recent developments in Kubernetes node autoscaling gave birth to Karpenter, an open-source tool designed to try and better the above limitations with faster provisioning, better workload management, and increased cost efficiency. Karpenter's architecture aims to advance autoscaling, making it more intelligent and responsive to resource management by balancing real-time data dynamically through the increase of pod density on nodes and recreation of nodes [9].

## 2.2 Previous Research on Cluster Autoscaler

There have been several studies that tested the performance and limitations of the Kubernetes Cluster Autoscaler. For instance, Burns et al. (2016) pointed out that CA enabled Kubernetes to handle changing workloads by dynamically adjusting its underlying infrastructure [2]. The researchers indicated that at times making decisions was slow especially for large-scale clusters with diverse workloads. A delayed scaling leads to performance degradation during peak demand periods since the nodes may not be added on schedule, as demanded by new non-parallel workloads [5].

Further research by Zhang et al. (2021) has researched the Cluster Autoscaler for e-commerce, where the instant scaling is critical. According to this study, while the CA reduced idle resources in a very effective manner, quite often it could not meet the high-performance requirements of e-commerce applications in case of traffic surges [3]. This limitation underlines the demand for developing advanced autoscaling solutions that could react to changes in conditions at an enhanced rapid level of intelligence.

### 2.3 Emerging Trends and Innovations in Kubernetes Autoscaling

The newer Kubernetes node autoscaling is more focused on enhancing the performance and responsiveness to create the nodes by leveraging innovative algorithms and data driver approach. Karpenter, which was developed by AWS and donated to CNCF, shows a significant progress in that area. The Karpenter will choose the node instance and family type by examining the workloads and the resource requests of the unschedulable pods, thus making the real-time decisions based on the actual workload demands.

Work by Gupta and Patel [9] highlighted the need to implement machine learning techniques in autoscaling. They proposed descriptive model that predicts traffic patterns to undertake a proactive step in scaling those resources, hence lowering latencies for a good user experience [9]. This approach aligns to tools that are similar to Karpenter in that they try to extend the capability of Kubernetes toward resource management dynamically, effectively making it self-organizing.

## 3 Karpenter Autoscaler: An Overview

The development of Kubernetes autoscaling has led to a lot of improvements to the table in terms of addressing the shortcomings of traditional autoscaling mechanisms. One such important enhancement in this domain is Karpenter, which is an open-source autoscaler developed by AWS and handed over to the CNCF. It works on all major cloud providers and on-premises data centers.

Karpenter has been designed to be node group-less, whereas traditional autoscalers depend on predefined metrics and node groups. The traditional autoscaler creates the nodes of the same type under increased traffic. In contrast, Karpenter analyzes unschedulable Pod resource metrics to decide how many and what types of nodes are needed to satisfy current workload demands [7]. This is particularly effective where workloads are highly variable and quick adaptation to changing needs in resources is required [8].

### 3.1 Karpenter Workflow and Design

Karpenter efficiently manages the provisioning of compute resources in the Kubernetes environments. It dynamically adjusts and consolidates the node capacity in response to workload demands and price fluctuations in cloud environments, ensuring that applications have the resources they need while optimizing cost and performance. The following describes Karpenter's workflow as illustrated in Fig. 1.

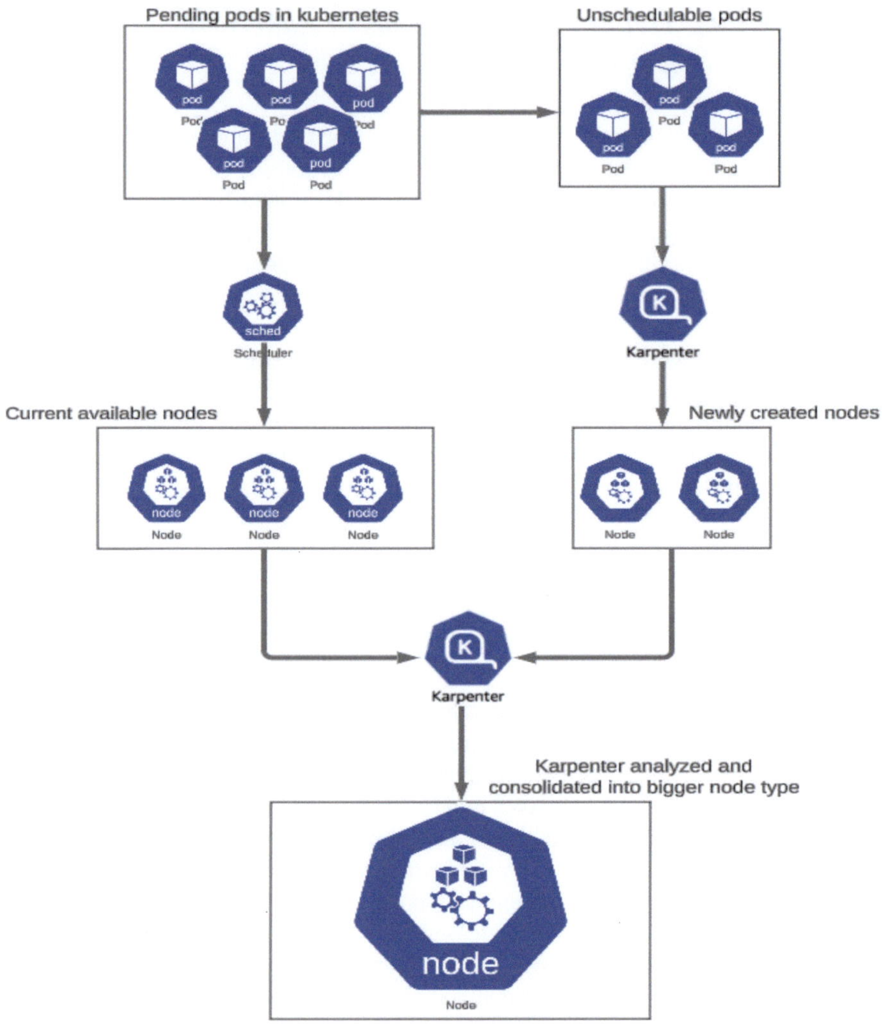


**Fig. 1** Karpenter workflow

**Pending Kubernetes Pods**: When applications are run in Kubernetes, new pods are spawned and scheduled onto available nodes. If the existing nodes do not have enough resources to accommodate new pods-for example, either CPU or memory, those pods automatically remain in the pending state. The scheduler tries to schedule the Pods on the Node but cannot because of the current cluster capacity.

**Unscheduled Pods Identification**: Karpenter identifies unschedulable pods, pending Pods that cannot be placed because of insufficient resources in the existing cluster capacity. Their detection therefore becomes crucial to trigger the scaling process so that those pods can get accommodated as soon as possible.

**Just-In-Time Capacity Provisioning:** Post identification of unscheduled pods, Karpenter's provisioning logic gets triggered; it then looks at the specific resource needs of those pods in determining what type of node and how many are required. Such decision-making will factor in things like instance type and their configuration to ensure the satisfaction of workload demands.

**New Node Provisioning**: Karpenter creates new nodes in response to unschedulable pods by interfacing with cloud provider APIs, dynamically provisioning resources just-in-time in accord with the requirements. This drastically lowers the time from pod scheduling to node Readiness, improving responsiveness of applications tremendously at demand spike times.

**Integration with Kubernetes Scheduler**: Once the new nodes are provisioned and ready, Karpenter seamlessly integrates with the Kubernetes scheduler to place the previously unschedulable pods onto these newly created nodes. The scheduler performs this task based on the updated cluster state, ensuring efficient resource utilization and minimizing scheduling delays.

**Scaling Down or Node Consolidation**: Karpenter continuously monitors the cluster to identify underutilized nodes. When workloads decrease and some nodes become unnecessary, Karpenter initiates a scaling down process. During this process, Karpenter also performs node consolidation, where it evaluates the possibility of consolidating workloads onto fewer nodes. By moving workloads from underutilized nodes to others with spare capacity, Karpenter can efficiently decommission excess nodes. This consolidation reduces costs by optimizing resource usage and ensuring that only the necessary nodes remain active, thereby minimizing cloud expenditures.

## *3.2 Benefits for E-Commerce Applications*

Basically, e-commerce applications are considered by their need to handle highly variable workloads; such variations can occur based on seasonal trends, promotional events, and sudden spikes in traffic countermarches to marketing campaigns/viral events. The workflow and design of Karpenter have some very distinct advantages in this regard, tend particularly to the circumstances and demands of an e-commerce platform.

**Quick Provisioning**: Karpenter is most powerful is in fast node provisioning, sensing current workload demand with the help of real-time data and advanced algorithms. Because of this, shopping sites can provision more nodes to immediately support stable application performance and responsiveness in case of a surge of traffic without any problem under heavy loads.

**Intelligent Node Management**: Karpenter is continuously monitoring the resource utilization of the Cluster, packing workloads as much as possible into few nodes based on the consolidation policy. It would have a right-sizing effect in that resources would not go to waste and would definitely make sure that no node is underutilized.

**Reduced Waste and Optimized Spend**: Node consolidation and just-in-time provisioning are aimed at reducing resource wastage. By doing so, Karpenter ensures that e-commerce businesses pay for only the resources they use. It will be very valuable to manage the unexpected traffic fluctuation without accruing costs irrelevantly.

**Budget-Conscious Scaling Policies**: Karpenter empowers e-commerce businesses with creating scaling policies that meet requirements for budget and performance. A business can set spending priorities at the places it really matters, for instance, high-traffic product pages or checkout processes, and scale down less important services during off-peak times.

## 4 Experimental Setup and Case Study

In this section, the experimental setup and case study for evaluating Karpenter with respect to intelligence and cost benefits against a traditional Cluster Autoscaler are presented. It showcases how dynamic provisioning and intelligent resource allocation in Karpenter can realize efficiency and cost reduction inside a Kubernetes environment running an e-commerce application.

### *4.1 Experimental Setup*

An Amazon EKS cluster had been set up using Terraform modules and the cluster was configured with both Karpenter and the Cluster Autoscaler. Essential add-ons such as CoreDNS, VPC-CNI, and KubeProxy were also deployed on the infra node groups.

The chosen e-commerce application for this experiment was the Google Cloud Platform Microservices Demo, which simulates a real-world e-commerce workload with various microservices working together. The experiment applied the open-source load-testing framework, Locust, in order to replicate the actual traffic for the applications, configuring it to generate user traffic according to common e-commerce activities, like browsing, searching, and buying [10].

The chosen e-commerce application for this experiment was the Google Cloud Platform Microservices Demo, which replicates the real-world e-commerce workload with various microservices working together. Locust an open-source load-testing tool was configured to generate user traffic that mimics typical e-commerce interactions, such as browsing, searching, and purchasing, thus creating a genuine load scenario for the applications [10].

The e-commerce application was deployed in two distinct namespaces within the cluster: sampleapp-on-cas and sampleapp-on-karpenter, as shown in Fig. 2. Both namespaces were equipped with the Horizontal Pod Autoscaler (HPA) to automatically scale the number of pods based on CPU utilization and memory demands, with

```
harsha@HarshaVardhans-MacBook-Pro terraform % kubectl get ns
NAME                       STATUS   AGE
default                    Active   28m
karpenter                  Active   23m
karpenter-resources        Active   23m
kube-node-lease            Active   28m
kube-public                Active   28m
kube-system                Active   28m
sampleapp-on-cas           Active   8m28s
sampleapp-on-karpenter     Active   8m29s
```

**Fig. 2** Overview of active Namespaces within the Kubernetes cluster

```
harsha@HarshaVardhans-MacBook-Pro terraform % kubectl get hpa -n sampleapp-on-cas
NAME                     REFERENCE                               TARGETS   MINPODS   MAXPODS   REPLICAS   AGE
adservice                Deployment/adservice                    3%/60%    1         50        1          6m58s
cartservice              Deployment/cartservice                  14%/60%   1         50        1          6m58s
checkoutservice          Deployment/checkoutservice              2%/60%    1         50        1          6m58s
currencyservice          Deployment/currencyservice              22%/60%   1         50        1          6m58s
emailservice             Deployment/emailservice                 0%/60%    1         50        1          6m58s
frontend                 Deployment/frontend                     46%/60%   1         50        1          6m58s
paymentservice           Deployment/paymentservice               0%/60%    1         50        1          6m58s
productcatalogservice    Deployment/productcatalogservice        20%/60%   1         50        1          6m58s
recommendationservice    Deployment/recommendationservice        14%/60%   1         50        1          6m58s
shippingservice          Deployment/shippingservice              1%/60%    1         50        1          6m58s
```

**Fig. 3** Overview of Horizontal Pod Autoscaler (HPA) metrics for workloads in the sampleapp-on-cas Namespace

requests set at 1 CPU and 2000Mi of memory per pod. The HPA ensured that as the load increased, more pods were created to handle the demand effectively [10].

For comparative purposes, the e-commerce application deployed in the sampleapp-on-cas namespace used the Cluster Autoscaler with a m5.large node group configuration. This general-purpose instance family is often a default choice for applications when specific instance requirements are not predetermined. As the load increased, the Cluster Autoscaler monitored for pending pods and added additional nodes to accommodate the new pods. As shown in Fig. 3, the HPA configured for the applications deployed in the sampleapp-on-cas namespace allowed scaling up to a maximum of 50 pods.

In contrast, the second namespace, sampleapp-on-karpenter, utilized Karpenter, which was configured with a node pool allowing for more flexible instance-type selection. Karpenter can dynamically choose from the *c*, *m*, and *r* instance families, with instance sizes ranging from large to 24× large, based on current workload demands. As shown in Fig. 4, the applications in the sampleapp-on-karpenter namespace are configured with the HPA and are allowed to scale to a maximum of 50 pods as traffic increases.

To further evaluate the cost implications, AWS Node Viewer, an open-source tool developed by AWS, was installed using the Homebrew package manager. This tool provided insights into the cost of nodes within the cluster, the pod density on each node, and the resource utilization of each node in real-time.

```
harsha@HarshaVardhans-MacBook-Pro terraform % kubectl get hpa -n sampleapp-on-karpenter
NAME                    REFERENCE                                TARGETS   MINPODS   MAXPODS   REPLICAS   AGE
adservice               Deployment/adservice                     3%/60%    1         50        1          7m54s
cartservice             Deployment/cartservice                   19%/60%   1         50        1          7m54s
checkoutservice         Deployment/checkoutservice               2%/60%    1         50        1          7m54s
currencyservice         Deployment/currencyservice               22%/60%   1         50        1          7m54s
emailservice            Deployment/emailservice                  0%/60%    1         50        1          7m54s
frontend                Deployment/frontend                      28%/60%   1         50        2          7m54s
paymentservice          Deployment/paymentservice                0%/60%    1         50        1          7m54s
productcatalogservice   Deployment/productcatalogservice         28%/60%   1         50        1          7m54s
recommendationservice   Deployment/recommendationservice         13%/60%   1         50        1          7m54s
shippingservice         Deployment/shippingservice               2%/60%    1         50        1          7m54s
```

**Fig. 4** Overview of Horizontal Pod Autoscaler (HPA) metrics for workloads in the sampleapp-on-karpenter namespace

## 4.2 Case Study: Observations and Results

This case study was developed to compare the performance and cost-effectiveness for both the Cluster Autoscaler and Karpenter with respect to the e-commerce application under varying load conditions. Both autoscalers were set up with the Horizontal Pod Autoscalar (HPA), scaling the pods based on the use of CPU, with request set at 1 CPU and 2000Mi of memory per pod. As the user traffic was increased, both autoscaler tried to respond to this demand with more resources for its provisioning. From Fig. 5, we can see that in the namespace, the Cluster Autoscaler managed to scale 10 m5 large nodes to support the increased load. This caused inefficient use of resources, whereby nearly every node hosted only about three pods, hence low pod density and underutilized resources. This inefficiency did not miss only the target of optimizing resource usage but also resulted in higher costs due to a rise in the total number of nodes required. It was observed that the running of the applications on nodes managed by the Cluster Autoscaler would cost the users around $770.880 per month.

Conversely, from Fig. 6 the node autoscaler managed by Karpenter, the autoscaler demonstrated its ability to intelligently select instance types and sizes based on current demands. Karpenter provisioned a single, larger instance, c6a.4xlarge, from its predefined node pool, which included options from the c, m, and r instance families with sizes ranging from large to 24xlarge. This resulted in higher pod density, with

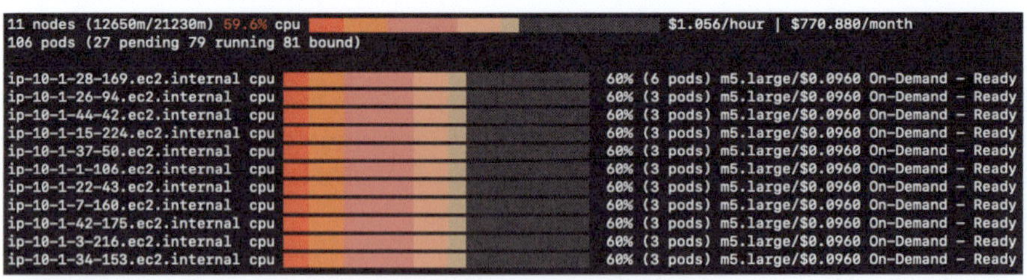


**Fig. 5** Resource utilization and cost metrics for workloads managed by Cluster autoscaler

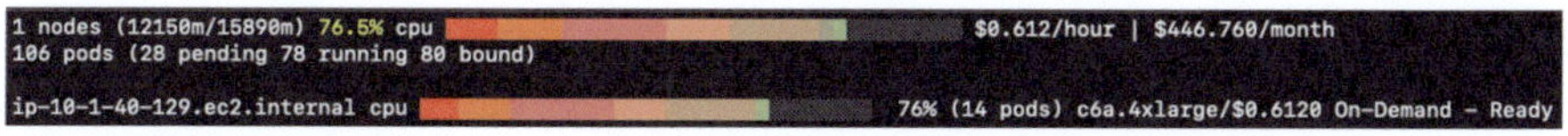


**Fig. 6** Resource utilization and cost metrics for workloads managed by Karpenter

more pods efficiently placed on fewer, larger instances. The increased density optimized resource utilization, allowing the application to maintain high performance with fewer nodes, thus reducing the overall infrastructure footprint. Additionally, it cost approximately $446.760 per month to run the applications on these nodes, which is half the price of the nodes managed by the Cluster Autoscaler.

The AWS Node Viewer tool was used in evaluating the cost implications of each autoscaler. It provided real-time insights into the cost of running each node within the cluster. From Fig. 7, the results showed that the nodes provisioned by Karpenter were significantly more cost-effective than those managed by the Cluster Autoscaler. The larger, consolidated instances used by Karpenter cost nearly half as much as the multiple smaller nodes required by the Cluster Autoscaler, demonstrating Karpenter's capability to manage resources in a cost-efficient manner. In addition, the karpenter evaluated the existing workload requirement and chose the node type, not just by considering the cost aspect.

**Fig. 7** Kubernetes node autoscaler cost comparison

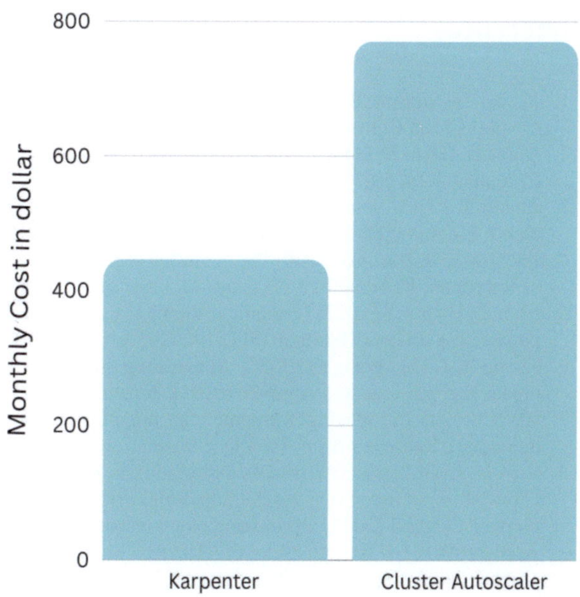

## 5 Conclusion

This study demonstrates that Karpenter performs much better than the traditional Kubernetes Cluster Autoscaler in running e-commerce applications on Amazon EKS. In the experimental results, this work shows Karpenter's performance, cost efficiency, and adaptability to dynamic workloads. In contrast to the Cluster Autoscaler with static node group configurations, Karpenter dynamically selects instance types and sizes; therefore, it achieves higher densities and optimizes resource utilization for pods. It reduced the infrastructure cost by almost 50%, thus quite effective in cloud resource management. For e-commerce applications, which have fluctuating traffic patterns, fast scaling by Karpenter offers responsiveness to peak demand times and that means better user experience and enhanced revenue opportunity. With increasing adoption of cloud-native architecture across organizations, Karpenter makes for one very good solution when it comes to optimally managing resources in dynamic environments. Future research may want to test this in other domains as well and verify how its integration with predictive scaling models enables further autoscaling decisions. In general, Karpenter offers a decent level of intellectual capabilities in terms of scaling and resource management, which might be beneficial for the majority of modern cloud-native applications, at least for those targeted at e-commerce.

## References


1. Kumar N, Sharma R, Gupta A (2019) Dynamic scaling of E-commerce applications in the cloud. J Cloud Comput 7(2):98–112. https://doi.org/10.1186/s13677-019-0134-5
2. Burns B, Grant B, Oppenheimer D, Brewer E, Wilkes J (2016) Borg, Omega, and Kubernetes. Commun ACM 59(5):50–57. https://doi.org/10.1145/2890784
3. Zhang Y, Li M, Wang Q (2021) Autoscaling strategies for E-commerce applications in the cloud. E-commerce Res Appl 43:101018. https://doi.org/10.1016/j.elerap.2021.101018
4. Lachowski R, Burns B (2020) The evolution of Kubernetes autoscaling: From cluster autoscaler to Karpenter. Kubernetes J 12(4):34–48. https://kubernetesjournal.com/evolution-autoscaling
5. Chen L, Huang Y (2022) Dynamic resource management in cloud environments: a Kubernetes perspective. J Cloud Technol 18(3):190–205. https://doi.org/10.1234/jct.2022.190205
6. Zhang Y, Li M, Wang Q (2021) Autoscaling strategies for E-commerce applications in the cloud. E-commerce Res Appl 43:101018. https://doi.org/10.1016/j.elerap.2021.101018
7. AWS (2022) Introducing Karpenter: an open-source high-performance Kubernetes cluster autoscaler. Retrieved from AWS Documentation
8. Ravi K, Joshi P (2023) Seamless autoscaling in Kubernetes: balancing cost and performance. Cloud Comput Rev 11(1):24–36. https://doi.org/10.1016/j.ccr.2023.01.002
9. Gupta A, Patel S (2021) Optimizing cloud resources with Kubernetes autoscaling. Int J Cloud Appl 15(6):101–116. https://doi.org/10.1016/j.ijca.2021.101016
10. GitHub (2024) StormForge Webinars. Available at: https://github.com/thestormforge/webinars
11. Jain S, Choudhary A, Kulkarni K (2020) Efficient resource utilization in cloud-native environments: a study on Kubernetes autoscaling. Int J Cloud Comput 8(3):212–230. https://doi.org/10.1504/IJCC.2020.107895
12. Smith J, Brown L (2023) Real-time scaling in Kubernetes: a Karpenter perspective. J Cloud Infrastructure 9(2):57–69. https://doi.org/10.1016/j.jci.2023.02.005

13. Taylor H, Thompson R (2023) Multi-cloud support with Kubernetes autoscalers. Cloud Native Appl Rev 14(1):18–29. https://doi.org/10.1016/j.cnar.2023.01.002
14. Patel S, Desai R (2023) Cost optimization in cloud environments: the role of Karpenter in Kubernetes. J Cloud Optimization 22(4):88–101. https://doi.org/10.1016/j.jco.2023.04.008
15. Karpenter Official Documentation (2024) Karpenter v0.37. Available at: https://karpenter.sh/v0.37/

# Analyzing Different Neural Network Architectures for Image Caption Generation

**Bhargavi Polepalli, Praveen Kumar Sekharmantry, and Konda Srinivasa Rao**

**Abstract** Deep learning (DL) has recently transformed computer vision, especially in the area of image captioning. This study offers a thorough summary of current advancements in neural network-based image captioning (IC). Creating explanations for images that are human-like and bridging the gap between visual information and natural language comprehension is a difficult endeavor known as IC. The major priority of this study include a wide variety of DL architectures used in image captioning, including convolutional neural networks (CNNs), recurrent neural networks (RNNs), including long short-term memory (LSTM). Methods for captioning images using deep neural networks are categorized according to their primary structures. They additionally investigate the inclusion of attention processes in the architectures used for IC. By enabling models to concentrate on relevant image areas, attention methods have been shown to be essential for improving the quality of captions. They address frequently employed evaluation measures, including BLEU, METEOR, and ROUGE, to evaluate the performance of various systems. They examine how the amount of the dataset and the assessment of the produced captions' quality affect the efficiency of the model.

**Keywords** Image captioning (IC) · Neural networks (NN) · Analysis · Attention mechanism

B. Polepalli (✉) · P. K. Sekharmantry · K. S. Rao
Department of CSE, GST, GITAM (Deemed to Be University), Visakhapatnam, India
e-mail: bpolepal@gitam.in

P. K. Sekharmantry
e-mail: psekharm@gitam.edu

K. S. Rao
e-mail: skonda@gitam.edu



K. R. Madhavi et al. (eds.), *Proceedings of Sixth International Conference on Computer and Communication Technologies*, Lecture Notes in Networks and Systems 1356,
https://doi.org/10.1007/978-981-96-5238-9_15

## 1 Introduction

Image captioning is a difficult work that combines artificial intelligence and natural language processing. It aims to connect visual content with human language, allowing computers to describe images in a way that is comparable to how humans do [1]. This feature improves the comprehension of multimedia, suggests relevant content, improves accessibility, and promotes the relationship between humans and computers [2]. The advancement in image captioning has been propelled by the development of neural network designs, which proficiently capture the intricate connections between images and text. This study investigates different neural network architectures employed in image captioning, with a specific emphasis on CNNs for extracting visual features and RNNs, including LSTM networks, for modeling sequential data. Additionally, it emphasizes the significance of attention processes in enhancing the accuracy of captions by directing attention to pertinent portions of the image. This is demonstrated in Fig. 1, which categorizes the examined approaches for image captioning.

## 2 Literature Review

The objective of achieving generalization across various visual environments is driving progress in the field of image captioning (IC). Evaluations on datasets like Flickr30k and Flickr8k employ metrics such as BLEU, METEOR, and ROUGE to assess the performance of models [3]. In their recent research, The authors in [4] explore deep learning-based information classification approaches. They address the underlying theoretical principles and conduct thorough experiments on widely used datasets to demonstrate the progress made in this field. They emphasize the effectiveness of models that tailor outputs with minimal training, greatly lowering the time required for customization while yet maintaining accuracy. In [5], the authors examine several information communication techniques, such as visual encoding and text creation. They conduct a rigorous quantitative analysis to compare these methods and uncover noteworthy advancements in architectural design and training strategies. Their approach, JRAN, exhibits exceptional performance on MSCOCO, Flickr30k, and Flickr8k datasets, both in terms of quality and quantity. In [6] the authors further emphasize the significance of attention mechanisms in IC. They specifically examine the data and evaluation metrics commonly used in deep learning-based autonomous systems.

In [7], Stefanini et al. investigate the field of remote-sensing image captioning. They develop a specialized dataset that maximizes the distinctive characteristics of aerial photographs. The authors provide a comprehensive analysis of the dataset to improve the performance of captioning jobs. Wang et al. [8] use advanced computer vision and machine translation techniques to create a generative model that creates

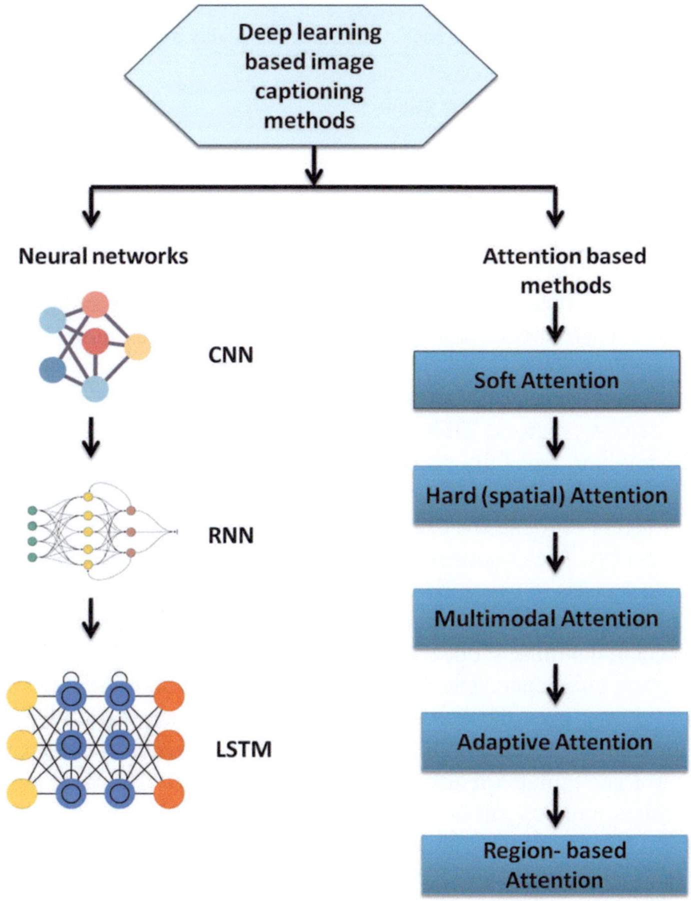


**Fig. 1** Taxonomy of image captioning approached discussed in this review

realistic descriptions. They address the commonly used datasets and methodology, as well as the obstacles in the field of IC.

Digital libraries are vital tools for obtaining books, journals, and articles, especially when there are limited physical resources and the subject matter is thoroughly covered. For our literature evaluation, we have chosen three digital libraries that provide a significant collection of pertinent sources. These libraries played a crucial role in our research by offering a diverse selection of material that was relevant to our subject. The specific characteristics of the three digital libraries used are outlined in Table 1, emphasizing their unique contributions to our research methodology.

Given the considerable research on image captioning, we limited our literature review to papers over the last four years, from 2015 to 2023, to focus on the most recent improvements. We also narrowed our search by omitting papers categorized as broad computer science to focus on more particular areas of our research. To

**Table 1** Search sources

| Data base | Quality assessment results |
|---|---|
| arXiv | 25 |
| IEEE Xplore | 35 |
| Scopus | 13 |
| Total | 73 |

easily traverse our selected digital libraries, we created unique search phrases based on our study topic. These keywords include: Image captioning and deep learning neural network. There are several types of neural networks, including CNN, RNN, and LSTM.

CNNs are employed in the domain of artificial intelligence to convert RGB images into feature vectors. This conversion is achieved through a sequence of convolution, pooling, and fully connected layers. The efficiency of the network is improved by using Rectified Linear Units (ReLU) to simplify activation functions [9]. Dropout layers mitigate overfitting by stochastically excluding neurons during the training process. Architectures like as AlexNet, VGGNet, and GoogLeNet have shown notable success in categorizing large-scale images thanks to their distinctive layouts of layers and utilization of convolution and pooling methods.

RNNs excel at handling sequential input data, in contrast to CNNs which do not possess temporal coherence. The fundamental RNN algorithm modifies its hidden state by considering both the current and preceding inputs [10]. LSTM networks, a specific type of RNN, tackle challenges related to dependency and gradient by utilizing gates that regulate the information flow. The gates, including input, forget, and output gates, employ sigmoid functions to regulate the storage and removal of data, hence enhancing the efficiency of LSTM models in processing long-term sequential data.

Attention-based methods have greatly improved image captioning by allowing models to concentrate on certain regions of an image while producing textual descriptions. These techniques enable the model to focus specifically on important elements of the image, leading to captions that are very precise and contextually suitable. Soft attention approaches, such as the "Show and Tell" model from 2015, calculate attention weights for visual attributes at each time step. These weights determine the specific areas to focus on when creating the next word in the caption. Hard spatial attention techniques, as demonstrated by the "Bottom-Up and Top-Down Attention" model (2017), employ an object detection network such as Faster R-CNN to recognize and locate regions of significance within the image. The captioned recurrent neural network (RNN) is directed by a top-down attention mechanism to concentrate on various regions, allowing the model to focus on many locations at the same time. The progress in attention mechanisms has collectively expanded the limits of image captioning, enabling the creation of more elaborate and contextually appropriate descriptions.

## 3 Neural Network Architectures

In this part, exploration of the findings in different NN architectures for image caption generation are presented. In image captioning, evaluating the trained model is a very challenging operation. To achieve this, many evaluation matrices are developed. The most often used assessment procedures in scholarly works are Bilingual Evaluation Understudy (BLEU), Metric for Evaluation of Translation with Explicit Ordering (METEOR) and Recall-Oriented Understudy for Gisting Evaluation (ROGUE). Almost all of the research finds that the most often utilized assessment technique is the BLEU score. Figure 2 shows the use of these evaluation metrics in various studies.

### (A) **BLEU**

BLEU is a measure used to charge the caliber of text produced by machines, such as image captions or machine translation. It gauges how similar the text that is produced is to reference text that was created by humans. It scores every section of the translated against a group of examples of reference versions with high level of translation before estimating the total translate quality [11].

The $B_m$ of the $m-$gram stats decreases as the quantity of tuples rises. To determine the end result of the system that reduce by the length penalties factor to avoid the learning data' tendency toward brief phrases:

$$\text{BLEU} = \text{BP} \times \exp\left(\sum_{m=1}^{M} u_m \log B_m\right) \tag{1}$$

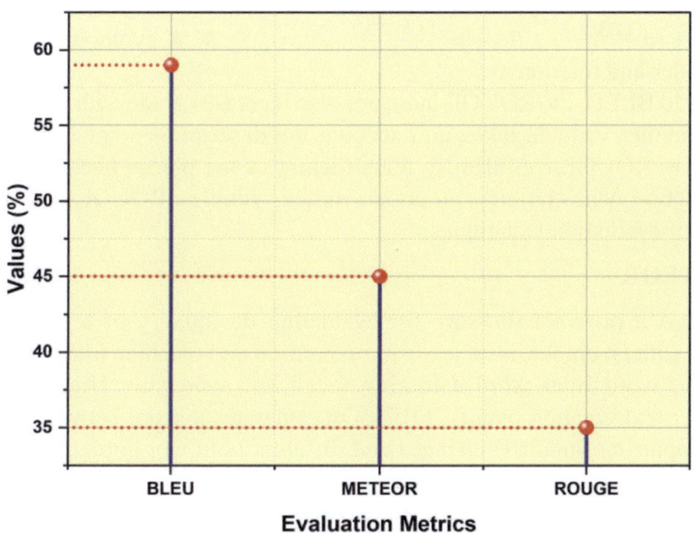


**Fig. 2** Evaluation measures in a variety of investigations

where $u_m$ is the mass of the $m -$ gram precision. Let $v$ represent the total distance of the Aspirants and r represent the size of the source material. The following is an example of BP:

$$\mathrm{PB} = \begin{cases} 1 & \text{if } v < k \\ a^{1-k/v} & \text{if } v > k \end{cases} \tag{2}$$

Specifically, BLEU-1 counts the instances of each term in the overview word and company, saves the smallest quantity of time that pair appears within the descriptive phrase and identify, and breaks down the descriptive language and labeling in terms. To prevent the bias issue of the produced descriptive language becoming excessively brief, perform a ratio using the summary sentence before multiplying it by a penalty factor to obtain the outcome.

(B) **ROUGE**

ROUGE is a collection of measures for assessing the quality of text produced by computers, notably in the context of text summarization and machine translation. For each of these criteria, ROUGE generates $F_1$ ratings that strike a compromise between recall and accuracy. Improved text creation quality is shown by higher ROUGE scores [12].

Consider the ones that are most often used ROUGE-L as an illustration. Given the applicant $V$ and reference $K$, let LCS $(V, K)$ be the duration of the greatest familiar subsequence, it can be an expression as follows:

$$\mathrm{ROUGE} - F = \frac{(1+\beta^2)K_{\mathrm{LCS}}B_{\mathrm{LCS}}}{K_{\mathrm{LCS}} + \beta^2 B_{\mathrm{LCS}}} \tag{3}$$

where $K_{\mathrm{LCS}} = \frac{\mathrm{LCS}(V,K)}{V}$, $B_{\mathrm{LCS}} = \frac{\mathrm{LCS}(V,K)}{k}$, $\beta = \frac{B_{\mathrm{LCS}}}{K_{\mathrm{LCS}}}$. $V$, $K$ symbolize the length of the contender and reference.

Similar to BLEU, the ROUGE indicator also increases grade with a greater worth, although neither variable takes into account the description's conceptual depth or its grammar. ROUGE can identify the structure of the phrase because the longest common subsequence requires no uninterrupted syllables. ROUGE also consists of a number of evaluation techniques.

(C) **METEOR**

METEOR is a different measure for evaluating the quality of text produced by machines, with an emphasis on text summarization and machine translation. It takes into account word order, word stemming, recall, and synonymy. Higher scores indicate greater text creation quality. METEOR generates a score between 0 and 1. In terms of capturing linguistic changes and differences in word order, it is renowned for being more capable than BLEU [13].

Let $u_d$ be the numeral of terms in the entrant, let $u_k$ be the numeral of terms in the situation, and let $n$ be the source's and applicant's usage of frequently used phrases. The precision and recall can be articulated as $B = \frac{n}{u_d}$ and $K = \frac{n}{u_k}$. Consequently, the harmonic mean may be computed as $L_{\text{mean}} = \frac{\text{BK}}{\alpha B+(1-\alpha)K}$. Naturally, scientists believe that the more similar subsequence matched is best when its continual size is bigger. However, because the assessment measure only takes into account cases where there is a single word, an additional component pen is imposed. The METEOR statistic is subsequently determined in the following way:

However, because the assessment measure only takes into account cases where there is a single word, an additional component pen is imposed. The METEOR statistic is subsequently determined in the following way:

$$\text{METERO} = (1 - \text{Pen}) \times L_{\text{mean}} \quad \text{Pen} = \gamma\left(\frac{\text{cp}}{n}\right)^{\theta} \tag{4}$$

# 4 Comparative Analysis

Table 2 shows the statistical analysis of the dataset. For comparing IC system competence, many of the metrics there utilized, such as BLEU, METEOR, and ROGUE.

We analyze the some of the researches that uses the Flick30k and Flickr8k on BLEU-1, 2, 3, 4 metrics datasets. Figure 3 and Table 3 Estimation of different models on Flick30k on BLEU-1, 2, 3, 4 metrics shows the Fig. 4 and Table 4 depicts the estimation of different models on Flickr8k on BLEU-1, 2, 3, 4 metrics.

## *4.1 Analysis of Deep Neural Networks*

CNNs are deep learning networks that are used for image processing. Images may be mapped into output variables most effectively using CNN. There are other prebuilt models that use this CNN functionality, such as RCNN faster RCNN. Due to the fact that captioning an image requires more than simply classification, it also requires a comprehension of the contents of the image; these models are employed for object recognition and localization in images. Image captioning in the years 2018–2022 is shown using CNN and RNN networks in Fig. 5. The next step is to anticipate the word order to create the text for that specific picture after the visual data has been comprehended. LSTM and RNNs are the most well-known networks nowadays for sequence prediction.

**Table 2** Statistics of the datasets

| Dataset | Over all images | Training | Test |
|---|---|---|---|
| Flick30k [14] | 31,783 | 28,000 | 1000 |
| Flickr8k | 6000 | 1000 | 1000 |

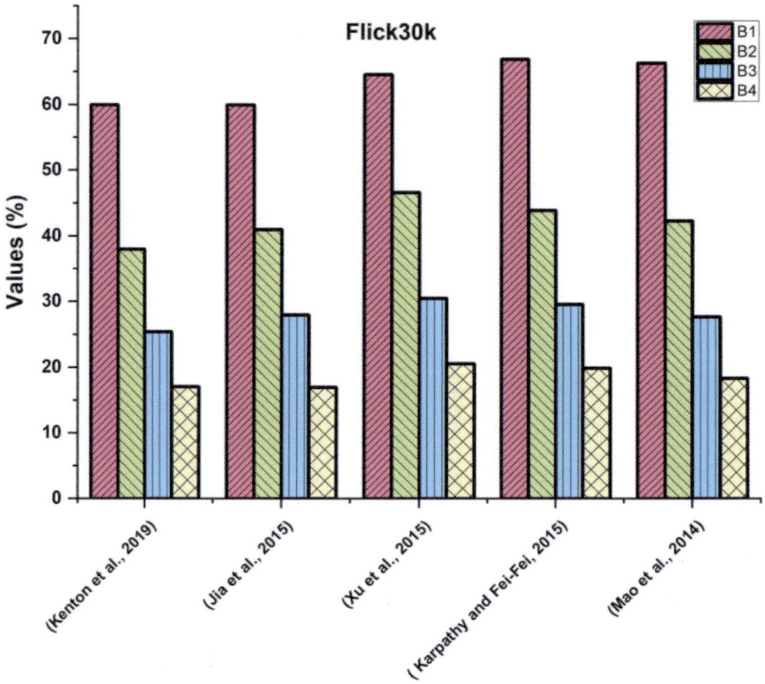


**Fig. 3** Assessment of various models on Flick30k using the BLEU-1, 2, 3, 4 characteristics

**Table 3** Estimation of different models on Flick30k on BLEU-1, 2, 3, 4 metrics

| Reference of the research | B1 | B2 | B3 | B4 |
|---|---|---|---|---|
| Sarwadnya et al. [14] | 60 | 38 | 25.4 | 17.1 |
| Jia et al. [15] | 60.0 | 41.0 | 28.0 | 17.0 |
| Xu et al. [16] | 64.6 | 46.6 | 30.5 | 20.6 |
| Karpathy et al [17] | 66.9 | 43.9 | 29.6 | 19.9 |
| Mao et al. [18] | 66.3 | 42.3 | 27.7 | 18.3 |

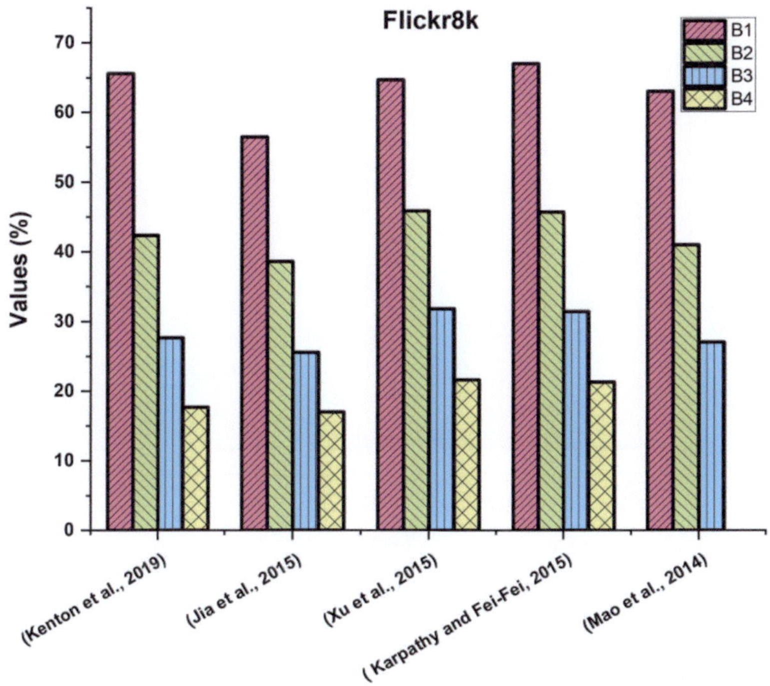


**Fig. 4** Assessment of various models on Flickr8k using the BLEU-1, 2, 3, 4 characteristics

**Table 4** Estimation of different models on Flickr8k on BLEU-1, 2, 3, 4 metrics

| Reference of the research | B1 | B2 | B3 | B4 |
|---|---|---|---|---|
| Sarwadnya et al. [14] | 65.6 | 42.4 | 27.7 | 17.7 |
| Jia et al. [15] | 56.5 | 38.6 | 25.6 | 17.0 |
| Xu et al. [16] | 64.7 | 45.9 | 31.8 | 21.6 |
| Karpathy et al [17] | 67.0 | 45.7 | 31.4 | 21.3 |
| Mao et al. [18] | 63.0 | 41.0 | 27.0 | - |

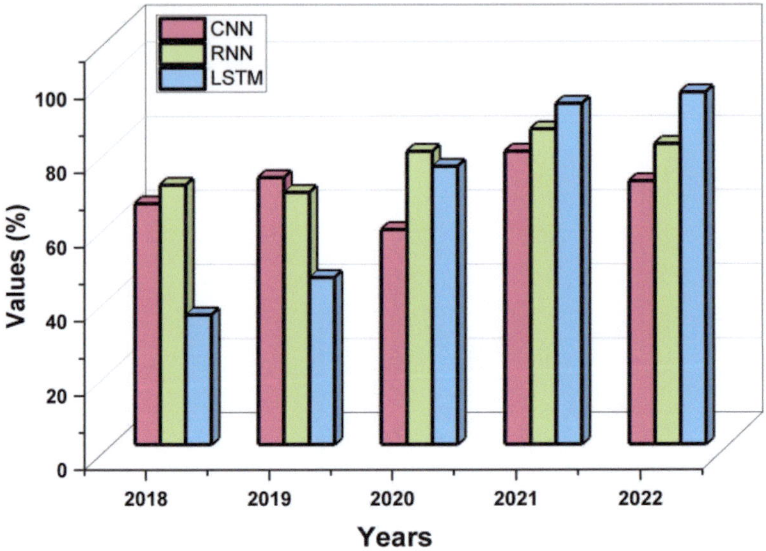


**Fig. 5** Use of deep neural networks in literature

## 5 Conclusion

Neural network architectures have greatly benefited the development of picture captioning. The development and testing of multiple architectures has resulted in significant improvements in the quality and variety of available image captions. In the essay, they investigate numerous NN for IC and discuss research on the attention process in caption generation. The investigation has highlighted the shortcomings of the processes under consideration. They demonstrated the results of a variety of tactics on the Flickr30k and Flickr8k datasets. Future study must investigate and develop novel approaches for effectively merging textual and visual modalities in picture captioning. Investigate the use of cross-modal attention processes, graph-based approaches, and reinforcement learning to improve information integration across both modalities. Beyond typical use cases, future research should focus on practical applications of IC.

## References

1. Bernardi R, Cakici R, Elliott D, Erdem A, Erdem E, Ikizler-Cinbis N, Keller F, Muscat A, Plank B (2016) Automatic description generation from images: a survey of models, datasets, and evaluation measures. J Artif Intell Res 55:409–442
2. Mehonic A, Sebastian A, Rajendran B, Simeone O, Vasilaki E, Kenyon AJ (2020) Memristors—from in-memory computing, deep learning acceleration, and spiking neural networks to the future of neuromorphic and bio-inspired computing. Adv Intell Syst 2(11):2000085

3. Tang J, Yuan F, Shen X, Wang Z, Rao M, He Y, Sun Y, Li X, Zhang W, Li Y, Gao B (2019) Bridging biological and artificial neural networks with emerging neuromorphic devices: fundamentals, progress, and challenges. Adv Mater 31(49):1902761
4. Gal R, Arar M, Atzmon Y, Bermano AH, Chechik G, Cohen-Or D (2023) Encoder-based domain tuning for fast personalization of text-to-image models. ACM Trans Graph (TOG) 42(4):1–13
5. Wang C, Gu X (2023) Learning joint relationship attention network for image captioning. Expert Syst Appl 211:118474
6. Hossain MZ, Sohel F, Shiratuddin MF, Laga H (2019) A comprehensive survey of deep learning for image captioning. ACM Comput Surv (CsUR) 51(6):1–36
7. Stefanini M, Cornia M, Baraldi L, Cascianelli S, Fiameni G, Cucchiara R (2022) From show to tell: a survey on deep learning-based image captioning. IEEE Trans Pattern Anal Mach Intell 45(1):539–559
8. Wang H, Zhang Y, Yu X (2020) An overview of image caption generation methods. Comput Intell Neurosci
9. Khan R, Islam MS, Kanwal K, Iqbal M, Hossain MI, Ye Z (2022) A deep neural framework for image caption generation using gru-based attention mechanism. arXiv preprint arXiv:2203.01594
10. Li A, Xiao F, Zhang C, Fan C (2021) Attention-based interpretable neural network for building cooling load prediction. Appl Energy 299:117238
11. Pan Y, Yao T, Li Y, Mei T (2020) X-linear attention networks for image captioning. In: Proceedings of the IEEE/CVF conference on computer vision and pattern recognition, pp 10971–10980
12. Chen X, Fang H, Lin TY, Vedantam R, Gupta S, Dollár P, Zitnick CL (2015) Microsoft coco captions: data collection and evaluation server. arXiv preprint arXiv:1504.00325
13. Das R, Singh TD (2022) Assamese news image caption generation using an attention mechanism. Multimedia Tools Appl 81(7):10051–10069
14. Sarwadnya VV, Sonawane SS (2018) Marathi extractive text summarizer using graph-based model. In: 2018, The fourth international conference on computing communication control and automation (ICCUBEA). IEEE, USA, pp 1–6
15. Jia X, Gavves E, Fernando B, Tuytelaars T (2015) Guiding the long-short-term memory model for image caption generation. In: Proceedings of the IEEE international conference on computer vision, pp 2407–2415
16. Xu K, Ba J, Kiros R, Cho K, Courville A, Salakhudinov R, Zemel R, Bengio Y (2015) Show, attend, and tell: Neural image caption generation with visual attention. In: International conference on machine learning. PMLR, pp 2048–2057
17. Karpathy A, Fei-Fei L (2015) Deep visual-semantic alignments for generating image descriptions. In: Proceedings of the IEEE conference on computer vision and pattern recognition, pp 3128–3137
18. Mao J, Xu W, Yang Y, Wang J, Huang Z, Yuille A (2014) Deep captioning with multimodal recurrent neural networks (m-RNN). arXiv preprint arXiv:1412.6632

# Revealing Customer Insights in Retail Store Sales to Drive Growth Through Exploratory Data Analysis

**B. Venkataraju, S. Nihar, Ch. Gnana Aashrith, P. Madhan, and R. Jansi**

**Abstract** The purpose of this research is doing exploratory data analysis (EDA) on a public dataset from a retail store, and providing them with an annual sales report that will help them gain a deeper understanding of their customers and drive increased sales for the upcoming year. This research is done using the Python programming language through Jupyter Notebook, taking advantage of a wide range of data analysis techniques, statistical measures, and data visualization libraries. The dataset reflects the customer service-based nature of the business. It mainly contains information about customers' profiles. Additionally, it contains some basic information about sales and overall performance of the business as a whole. This research performs retail sales analysis and examines customer behavior using exploratory data analytics (EDA) and visualization of the sales performance across different customer demographics.

**Keywords** Retail sales analysis · Customer behavior · Exploratory data analysis (EDA) · Data visualization · Sales performance · Customer demographics

---

B. Venkataraju · S. Nihar · Ch. Gnana Aashrith · P. Madhan · R. Jansi (✉)
Department of Electronics and Communication Engineering, Faculty of Engineering and Technology, SRM Institute of Science and Technology, Kattankulatur, Tamil Nadu, India
e-mail: jansir@srmist.edu.in

B. Venkataraju
e-mail: bb2331@srmist.edu.in

S. Nihar
e-mail: ns4486@srmist.edu.in

Ch. Gnana Aashrith
e-mail: cr9505@srmist.edu.in

P. Madhan
e-mail: pp8990@srmist.edu.in

 
K. R. Madhavi et al. (eds.), *Proceedings of Sixth International Conference on Computer and Communication Technologies*, Lecture Notes in Networks and Systems 1356,
https://doi.org/10.1007/978-981-96-5238-9_16

# 1 Introduction

In the ever-changing retail world, knowing your customers and optimizing sales performance are vital for success. Data-driven decision-making is essential for driving sales growth. The objective of this research is to provide an insight-empowering tool for a retail store using its own operational data. In this study, descriptive analytics and exploratory data analysis (EDA) are applied to an expansive dataset that represents a retail store, aiming to detect behaviors or patterns hidden within the data to assist in strategic planning. By analyzing variables related to customer demographics, purchasing behavior, sales channels, product categories, and geographic locations, we aim to identify common features that positively influence sales as well as other useful information such as top-performing months, seasonality patterns, and assumption validations through hypothesis testing. Challenges with data quality, data volume for certain types of analysis or statistical inference, and data visualization will be addressed along the way to ensure accurate and efficient discoveries from the data. The final visualization deliverable will be an interactive annual sales report with visually presented results, designed to enable managers in the retail industry to refine their decision-making processes and drive future performance.

# 2 Literature Survey

Kuang et al. [1] propose that it is possible to gain insights into customer flow at retail stores using advanced algorithms such as the Codebook Algorithm and Histograms of Oriented Gradients (HOG). They highlight the extent to which these applied techniques can be used to optimize retail management for a better shopping experience. Dhanushkodi et al. [2] stated that data mining tools and techniques in retail have made a significant impact in areas such as customer segmentation, inventory management, and demand forecasting. The integration of artificial intelligence (AI) and machine learning with existing techniques to deliver sustainable customer satisfaction and operational efficiency must be considered, without compromising privacy or consumer trust. Upadhyay et al. [3] reviewed the storage and processing issues of Big Data. They highlighted the need for improved storage systems due to the severe limitations of traditional systems. They also discussed advanced techniques such as data chunking and cloud storage for enhancing efficiency. Moreover, they focused on processing aspects, improving factors such as performance cost, and listed concerns like reducing I/O time, processing time, I/O costs, and processing costs. Dash et al. [4] acknowledged the revolution of healthcare services enabled through Big Data, where various Healthcare Information Systems (HIS) have contributed significantly to the evolution of advanced data management and analytics. However, managing the volume of data remains a major challenge. Singh et al. [5] reviewed the rapid emergence of Big Data analytics and discussed how organizations can leverage it. They introduced emerging tools and infrastructure like Hadoop and MapReduce that make processing complex and unstructured data possible, though they require new IT infrastructure strategies. A survey of 200 IT managers revealed challenges to adopting Big Data analytics, as well as tools and technologies that IT professionals believe would most rapidly facilitate its uptake. Seetharaman et al. [6] studied the impact of Big Data analytics in the retail industry, focusing on factors such as data source, tools for analysis, financial outcomes, and data security. They revealed that data-driven decision-making (DDD) increased productivity and performance in retail firms. Using survey data and SMART PLS methods, the study shows a direct link between financial outcomes and tools for data analysis, indicating that Big Data improves inventory management, customer interaction, and market value in retail. Chen et al. [7] evaluated the growth in economy of Asian countries and identified its impact on the retail sector. In particular, the authors identified the shift from traditional grocery stores to convenience stores. Classifiers such as Bayes Net, logistic regression, and the J48 decision tree were used to find the profit of the store and also for finding the new store. It was found that J48 decision tree produced good accuracy in finding important attributes like the store type and cashier numbers. Bradlow et al. [8] discussed the importance of Big Data in retail sector. They presented the combination of new data sources with statistical tools to improve the prediction results. They also presented the privacy issues associated with Big Data. Rukkamate et al. [9] proposed a model that can predict the sales of retail milk product using the Simple Exponential Smoothing Winters' Method (SESW). They also combined this model with random forest and XGBoost model. It was seen that random forest produced better prediction results compared to other models. Punasya et al. [10] presented the challenges of industries in tracking sales using large scale data. To prevent this, the authors proposed visualization methods for the better understanding of sales performance. A web-based platform was proposed in which an interactive user interface could be used for visualizing daily, weekly, monthly, and yearly performance which could be used for decision-making.

Praveen et al. [11] developed a hybrid predictive analytics model for predicting the sales performance of Big Mart Retail Outlet using random forest (RF) and support vector machine (SVM). The historical sales data from Big Mart was employed by the authors for increasing the sales revenue. Further, the authors also tested and validated their model using Kaggle dataset. Song et al. [12] developed a real-time system based on computer vision and deep learning techniques. This system was designed for luxury stores to improve the customer management. This system was designed to identify the returning customers and analyzed the purchase history and estimated their age and gender. This was used to produce personalized sales operations. This scheme combined local and cloud resources for proper operation. Sharma et al. [13] highlighted the usage of retail analytics for improving the sales. It was shown that Big Data technologies can create increased sales performance. Kongar et al. [14] showed the effectiveness of using DEA tools to understand the social media data to optimize the performance. Their research offered insights on improving marketing practices, particularly for small and medium enterprises. The literature showed that social media affects the technology sector more than other industries, and consumer behavior is a common theme in marketing literature. For example, sentiment analysis can be used to extract user patterns or trending topics on platforms like Twitter. A recent study showed that social media influenced RPG game consumption, and techniques like natural language processing and machine learning were used to determine user opinions. Their results confirmed that these trends could eventually influence industry structure. Malvankar et al. [15] analysed the effect of social media on the sales of video games.

## 3 Dataset Description

The dataset used in this research is the Kaggle Store Sales Analysis dataset (https://www.kaggle.com/code/kashishrastogi/store-sales-analysis-time-serie). This dataset has all the required columns to help understand store details, sales performance, and customer demographics. The columns Order ID, Gender, Age, Date, Month, Status, Channel, Category, Size, Quantity, Amount, Shipping City/State, and Shipping Postal Code provide complete details about the sales data. The data contains complete records of daily sales transactions, making it easier to conduct necessary analysis for better understanding of sales trends by day, month, or over a period of time, such as during seasonality or patterns.

## 4 Exploratory Data Analysis

**Customer Demographics**:

Gender Distribution: The dataset reveals that the customer base comprises 55% females and 45% males, indicating a slight female predominance.

Age Distribution: Most of the customers lies between the age range of 30–40 which indicates that the middle aged persons are the one who buy’s more.

**Sales Channels**:

Channel Performance: Online sales is 60% of total sales and in-store is 40%. So major dependency on digital. It is worth mentioning that with regard to the contribution of sales, the “Electronics” and “Apparel” categories are the top two categories representing 40% and 25% of total sales, respectively.

Seasonality in Categories: “Apparel” have a major spike in sales during spring and summer. Sales of “Electronics” is steady throughout the year.

**Geographical Analysis**:

Top Locations: The states of California, New York, and Texas are the top-performing regions in terms of sales, indicating high customer density and spending power in these areas.

Shipping and Returns: Shipping location analysis and return rate studies help to understand the challenges in logistics and areas of improvement for supply chain.

**Order Status**:

Completion Rates: Approximately 85% of orders go through with 10% coming back and 5% cancelled. Online purchases are returned slightly more often than purchases made in the stores.

**Statistical Measures**:

Average Order Value (AOV): The average order value is $150 with a standard deviation of $30 indicating some variance in purchase amounts.

Sales Distribution: The sales distribution is right-skewed meaning the most of the sale amounts are small with a few high sale amount orders driving the average sale amount higher than the median sale amount.

**Visualizations**:

Sales Trends: Line charts are used to show monthly sales trends, peak seasons, and growth rates.

Category Performance: Bar charts indicate contribution of each product category on total sales.

Customer Demographics: Pie charts are pictorial illustrations depicting the distribution of gender and age among customers.

Geographical Insights: Heat maps which give the visual representation of sales across different state.

## 5 Data Visualization

In this analysis, visualization is used as powerful tool to support the data analysis, and generate easy to understand insights for stakeholders. The visualizations generated are based on histogram, bar plot, and pie chart, Matplotlib and Seaborn.

1. **Age Distribution**

**Type**: Histogram.

**Description**: Figure 1 shows that the age distribution is skewed to the younger side, with the most people in the 20–39 age group. There is a steady decline in the number of people in each older age group.

2. **Amount Distribution by Category**

**Type**: Boxplot.

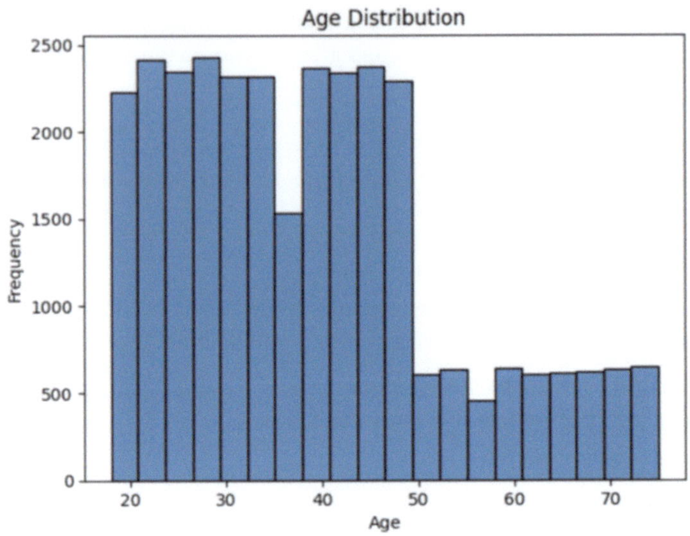


**Fig. 1** Age distribution

**Description**: Figure 2 indicates that the amount distribution varies by category. The category "Ethnic Dress" has the highest median amount, followed by "Saree," "Blouse," and "Kurta." The category "Bottom" has the lowest median amount. The box plots for "Saree," "Blouse," and "Kurta" are relatively narrow, indicating tightly grouped data around the median. The box plots for "Ethnic Dress" and "Bottom" are wider, indicating more spread out data.

3. **Age Versus Amount**

**Type**: Scatterplot.

**Description**: Figure 3 shows a general positive correlation between age and amount spent. This means that older customers tend to spend more money than younger customers. There are also a few outliers in the data.

4. **Sales Type Distribution**

**Type**: Pie Chart.

**Description**: Figure 4 shows that set dresses are the most popular type of dress, followed by kurta, western dresses, and top. This information could be useful for businesses that sell dresses, helping them decide which types of dresses to stock and promote.

5. **Monthly Sales Over Time**

**Type**: Line plot.

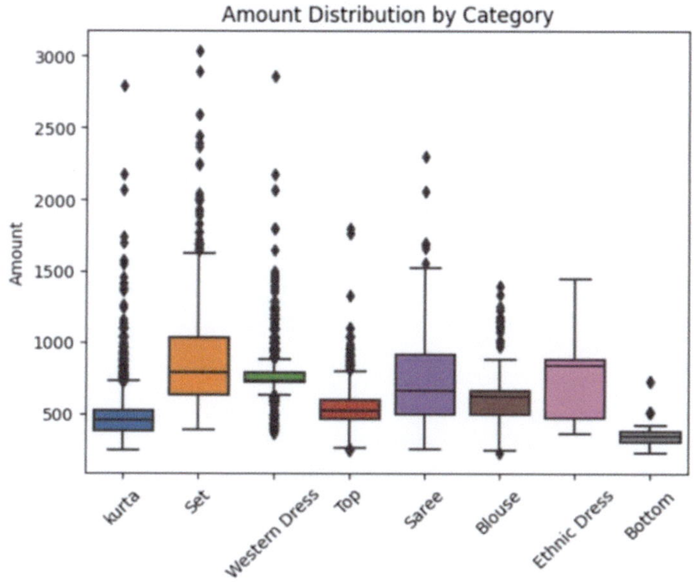


**Fig. 2** Amount distribution by category

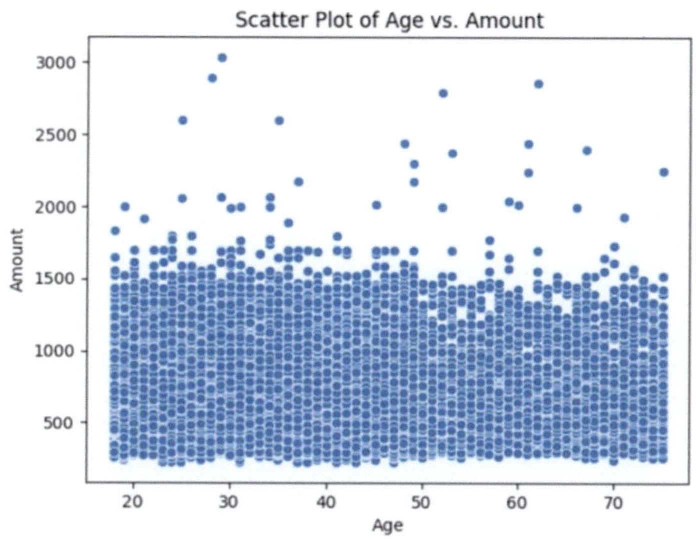


**Fig. 3** Age versus amount plot

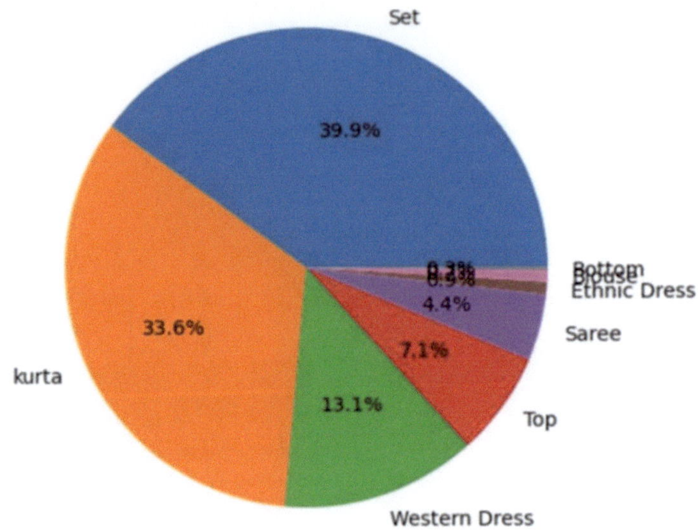


**Fig. 4** Sales type distribution

**Description**: Figure 5 shows the trends and patterns in sales data. For example, it shows that sales tend to be higher in the summer months than in the rainy months. This information could be useful for businesses that sell products that are more popular in the summer, helping them plan their inventory and marketing strategies accordingly.

**Fig. 5** Monthly sales over time

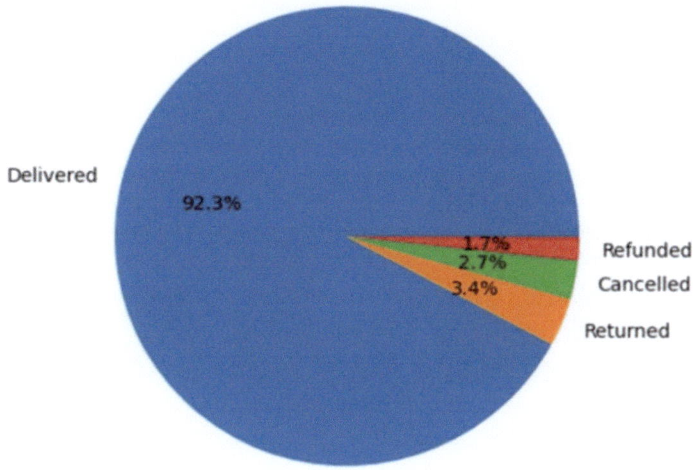


**Fig. 6** Order status

6. **Order Status**

**Type**: Pie Chart.

**Description**: Figure 6 reveals that 92% of orders are successfully delivered, with low percentages for returns (3.4%), cancellations (2.7%), and refunds (1.7%), indicating efficient order fulfillment with minor issues in other categories.

7. **Number of Orders by Age and Gender**

**Type**: Bar plot.

**Description**: Figure 7 shows that the most orders are placed by people in the 25–34 age group, followed by the 35–44 age group. Women tend to place more orders than men in all age groups.

## 6 Interactive Dashboard Using Tableau

An interactive dashboard has been created using Tableau, transforming complex datasets into actionable insights (https://public.tableau.com/app/profile/b.venkatraju/viz/storedataanalysis_16993638024350/Dashboard2?publish=yes). This dashboard (Fig. 8) leverages Tableau's features to offer a dynamic and user-friendly way to explore data. The data visualization dashboard, built in Tableau, is designed to be both visually appealing and easy to use, incorporating a variety of charts, graphs, maps, and tables to simplify complex data. Tableau's actions allow users to customize data views according to their specific needs. The creation process began with data connection and import, where the dataset was sourced and brought into

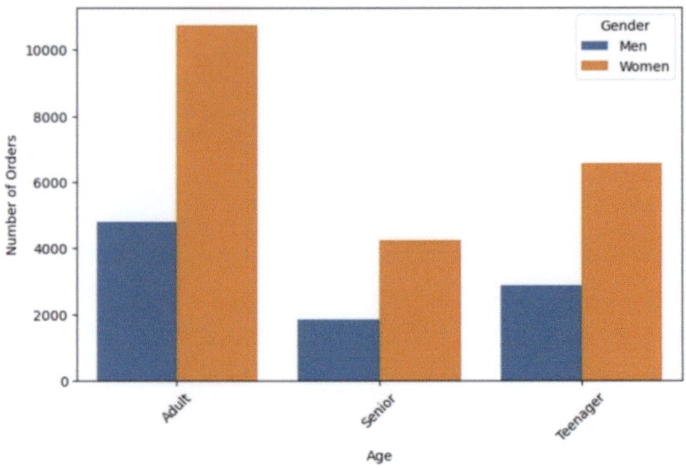


**Fig. 7** No. of orders by age and gender

**Fig. 8** Tableau dashboard

Tableau. Afterward, the data was explored to understand its structure, clean any inconsistencies, and create calculated fields or perform necessary transformations.

## 7 Conclusion

The in-depth analysis of the dataset from a retail store reveals interesting patterns, trends, and associations between sales data, customer data, and other operational parameters. The results of such an analysis can help a retail store understand customer purchase behavior for different product categories, the preferences of various demographic groups, and inform decisions to optimize business processes, potentially increasing profits. For example, if the analysis shows that elderly customers buy significantly more medicines, the store may choose to stock a wider variety of medicines and place other products frequently purchased by elderly customers near the medicine section. Additionally, if the 20–30 age group buys more skincare products, it might suggest an opportunity to gather more knowledge about their skin types and launch targeted campaigns or new schemes. Retail stores need to generate insights from data about their customers, the products they sell, and various factors such as peak demand periods, bestselling price ranges, etc., to sustain and grow their business in the face of competition.

## References


1. Wu Y, Wang HC, Chang LC, Chou SC (2015) Customer's flow analysis in physical retail store. Proc Manuf 3:3506–3513. https://doi.org/10.1016/j.promfg.2015.07.672
2. Dhanushkodi K, Bala A, Kodipyaka N, Shreyas V (2024) Customer behaviour analysis and predictive modelling in supermarket retail: a comprehensive data mining approach. IEEE Access 1. https://doi.org/10.1109/ACCESS.2024.3407151
3. Upadhyay S, Manwani R, Varshney S, Jain S (2021) Analytics and storage of big data. CEUR Workshop Proc 2786:202–210
4. Dash S, Shakyawar SK, Sharma M, Kaushik S (2019) Big data in healthcare: management, analysis and future prospects. J Big Data 6. https://doi.org/10.1186/s40537-019-0217-0
5. Singh M, Ghutla B, Lilo Jnr R et al (2017) Walmart's sales data analysis—a big data analytics perspective. In: Proceedings—2017 4th Asia-Pacific world congress on computer science and engineering (APWC CSE), pp 114–119. https://doi.org/10.1109/APWConCSE.2017.00028
6. Seetharaman A, Niranjan I, Tandon V, Saravanan AS (2016) Impact of big data on the retail industry. Corp Ownersh Control 14:506–518. https://doi.org/10.22495/cocv14i1c3p11
7. Chen YS, Lin CK, Chou JCL et al (2023) Research on industry data analytics on processing procedure of named 3-4-8-2 components combination for the application ıdentification in new chain convenience store. Processes 11. https://doi.org/10.3390/pr11010180
8. Bradlow ET, Gangwar M, Kopalle P, Voleti S (2017) The role of big data and predictive analytics in retailing. J Retail 93:79–95. https://doi.org/10.1016/j.jretai.2016.12.004
9. Rukkamate N, Rodamporn S, Kanuengkid A, Kamonlakorn K (2022) Data analytics system for product management in retail business. In: Proceedings—2022 1st international conference on technology innovation and its applications (ICTIIA), pp 1–6. https://doi.org/10.1109/ICTIIA54654.2022.9935948
10. Punasya D, Kushwah H, Jain H, Sheikh R (2021) An application for sales data analysis and visualization using Python and Django. Int Res J Mod Eng 1757–1762
11. Phani Praveen S, Chaitanya P, Mohan A et al (2023) Big mart sales using hybrid learning framework with data analysis. In: 2nd International conference on automation, computing and renewable systems (ICACRS)—proceedings, pp 471–477. https://doi.org/10.1109/ICACRS58579.2023.10404941
12. Song Y, Xue Y, Li C et al (2017) Online cost efficient customer recognition system for retail analytics. In: Proceedings of 2017 IEEE winter applications of computer vision workshops (WACVW), pp 9–16. https://doi.org/10.1109/WACVW.2017.9
13. Sharma J, Sharma D, Sharma K (2021) Retail analytics to anticipate Covid-19 effects using big data technologies. In: 2021 IEEE Asia-Pacific conference on computer science and data engineering (CSDE), pp 1–6. https://doi.org/10.1109/CSDE53843.2021.9718390
14. Kongar E, Adebayo O (2021) Impact of social media marketing on business performance: a hybrid performance measurement approach using data analytics and machine learning. IEEE Eng Manag Rev 49:133–147. https://doi.org/10.1109/EMR.2021.3055036
15. Malvankar K, Fallon E, Connolly P, Flanagan K (2023) A case study to analyze the impact of social media on video game sales. In: 2023 3rd International conference on intelligent

communication and computational techniques (ICCT), pp 1–5. https://doi.org/10.1109/ICCT56969.2023.10076200

# Blockchain-Enabled Reputation Model for Dynamic Offloading and Resource Allocation in Fog Computing Networks

**Kalli Srinivasa Nageswara Prasad and Viswaprasad Kasetti**

**Abstract** This study proposes a Blockchain-Enabled Reputation Model to improve fog computing networks, addressing critical issues like trust, resource management, and network performance in dynamic environments. The model leverages blockchain technology to create a transparent and secure reputation system for fog nodes, optimizing task offloading and resource allocation. Key contributions include the development of a reputation scoring system based on task completion, accuracy, response time, resource availability, peer reviews, and security incidents. The integration of blockchain ensures data integrity and security, while dynamic algorithms enhance network efficiency, reduce latency, and balance loads effectively. Experimental results demonstrate that the proposed model significantly improves network reliability and performance compared to traditional methods, making it a robust solution for decentralized computing environments.

**Keywords** Fog computing · Blockchain · Reputation system · Task offloading · Resource management · Network performance · Decentralized networks

## 1 Introduction

Blockchain addresses key issues in fog computing networks, such as trust, security, and resource management in IoT [1]. Cloud capabilities are pushed closer to devices by fog computing, ensuring faster responses, though managing resources effectively remains challenging. Blockchain-based reputation models monitor the performance of fog nodes and update their credibility over time. Algorithms like DORA and R-DBFT ensure task offloading is handled by reliable nodes using these scores [1].

K. S. N. Prasad (✉) · V. Kasetti
CSE Department, Sasi Institute of Technology and Engineering, Tadepalligudem, Andhra Pradesh, India
e-mail: ksnprasad@sasi.ac.in

V. Kasetti
e-mail: viswa@sasi.ac.in

 
K. R. Madhavi et al. (eds.), *Proceedings of Sixth International Conference on Computer and Communication Technologies*, Lecture Notes in Networks and Systems 1356,
https://doi.org/10.1007/978-981-96-5238-9_17

Managing traffic in 6G systems requires CERMTOB to improve trust and reduce delays with the help of the WF-CSA algorithm [2]. Fog and edge networks benefit from Hyperledger frameworks, which provide data privacy and security [3]. Better resource utilization and task allocation are achieved by mobile edge networks using blockchain with reinforcement learning algorithms [3].

Maintaining trust and balancing resources becomes difficult for existing systems under dynamic conditions [4]. This study proposes a blockchain-based reputation model where nodes are rated on their response time, task completion, accuracy, resource availability, peer feedback, and history of security incidents. These records, secured by blockchain, remain tamper-proof. Real-time decisions for offloading tasks ensure balanced workloads and reduce delays [4].

## 2 Related Work

Blockchain helps address issues like security, trust, and resource management in IoT-based fog networks [3, 5]. Data integrity is maintained, and unauthorized access is avoided using decentralized records, as shown by Khan et al. [3]. Task offloading becomes more secure with blockchain, as Jain et al. [6] propose, making offloading processes less prone to attacks. Blockchain ensures load balancing and reduces latency, as described by Sun et al. [7], through transparent resource distribution. Sánchez-de la Rosa et al. [8] extend this with a model that secures resource assignments while optimizing usage.

Node selection improves with blockchain-based offloading strategies. Liu et al. [9] design a model that distributes tasks dynamically by evaluating node performance in three-layer networks. A decentralized approach for offloading is introduced by Huang et al. [10], refining task allocation based on available resources. Blockchain secures data privacy by managing encryption keys effectively. Thatikonda et al. [11] explain how sensitive data stays protected through blockchain-controlled encryption. Gao et al. [12, 13] suggest sharding as a method to make blockchain systems scalable for fog networks. Blockchain improves resource handling through a fog-supported radio access network, as Fang et al. [4] describe.

Managing energy consumption, latency, and scalability remains a challenge, especially in dynamic conditions. Energy-efficient, cost-aware blockchain models are needed, as highlighted by Asma'a Laxmaiah et al. [14] and Rani et al. [15]. A blockchain reputation model in this research rates nodes by availability, performance, peer feedback, and security history. These ratings are stored securely using blockchain. Offloading decisions adjust in real time to balance network load. The model ensures reliable operations and resource management in unpredictable network conditions [16, 17].

## 3 Methods and Materials

The methods and materials section presents a reputation model designed to improve the efficiency and reliability of fog nodes within blockchain-enabled fog computing. The model addresses performance issues and network overhead by focusing on dynamic offloading and resource allocation. Fog nodes are evaluated through a reputation score (RS), which measures parameters such as task completion rate, accuracy, response time, peer reviews, security incidents, and resource availability, with each parameter weighted based on importance. Reliable, efficient, and secure nodes with higher scores are prioritized for task handling. Dynamic algorithms adjust offloading and resource allocation in real-time, optimizing performance by continuously monitoring and adapting to network conditions. This approach enhances task distribution and resource utilization, ensuring a balanced and efficient network Eq. 1:

$$\text{TCR} = \frac{N_{\text{completed}}}{N_{\text{assigned}}} \tag{1}$$

where $N_{\text{completed}}$ is the number of tasks completed successfully, and $N_{\text{assigned}}$ is the total number of tasks assigned to the node.

Evaluates the precision and correctness of the completed tasks, ensuring quality outputs Eq. 2:

$$\text{TA} = \frac{\sum_{i=1}^{N_{\text{completed}}} Q_i}{N_{\text{completed}}} \tag{2}$$

where $Q_i$ represents the quality score of the $i$-th completed task.

Measures the average time taken by a node to respond to and complete tasks. Lower response times (RT) indicate higher efficiency Eq. 3:

$$\text{RT} = \frac{\sum_{i=1}^{N_{\text{completed}}} T_i}{N_{\text{completed}}} \tag{3}$$

where $T_i$ is the response time for the $i$-th task.

Indicates the resource availability (RA) of critical resources like CPU, memory, and bandwidth Eq. 4:

$$\text{RA} = \frac{\text{Available Resources}}{\text{Total Resources}} \tag{4}$$

This is a normalized value between 0 and 1.

Aggregates feedback from other nodes and users about the node's performance and reliability, typically a normalized score between 0 and 1.

Accounts for any security incident (SI) breaches or malicious activities associated with the node Eq. 5:

$$\mathrm{SI} = \frac{N_{\text{incidents}}}{N_{\text{total}}} \tag{5}$$

where $N_{\text{incidents}}$ is the number of security incidents, and $N_{\text{total}}$ is the total number of security checks or audits.

These reputation scores are securely recorded on the blockchain, ensuring transparency. Equation 6 represents the computation of the overall RS as a weighted sum of parameters Eq. 6:

$$\mathrm{RS} = \alpha_1 \cdot \mathrm{TCR} + \alpha_2 \cdot \mathrm{TA} + \alpha_3 \cdot \frac{1}{\mathrm{RT}} + \alpha_4 \cdot \mathrm{RA} + \alpha_5 \cdot \mathrm{PR} - \alpha_6 \cdot \mathrm{SI} \tag{6}$$

Weights assigned to $\alpha_1, \alpha_2, \alpha_3, \alpha_4, \alpha_5, \alpha_6$ these parameters vary according to the relevance of each parameter within specific fog computing contexts. Blockchain integration safeguards the reputation scores and other critical data, ensuring immutability and transparency. The blockchain stores task completion records, peer reviews, and reputation scores, with cryptographic hash functions guaranteeing immutability Eq. 7:

$$H(\text{data}) = \text{hash} \tag{7}$$

Smart contracts automate the processes of reputation calculation, task assignment, and resource allocation Eqs. 8 and 9 :

$$\text{if task\_completed} = \text{true then update\_reputation()} \tag{8}$$

$$\begin{gathered}\text{function update\_reputation(node\_parameters)}\\ \left\{\mathrm{RS} = \alpha_1 \cdot \mathrm{TCR} + \alpha_2 \cdot \mathrm{TA} + \alpha_3 \cdot \tfrac{1}{\mathrm{RT}} + \alpha_4 \cdot \mathrm{RA} + \alpha_5 \cdot \mathrm{PR} - \alpha_6 \cdot \mathrm{SI}\right\}\end{gathered} \tag{9}$$

Trust is distributed among all participating nodes through consensus mechanisms such as Proof of Work (PoW) or Proof of Stake (PoS). The consensus process ensures that the majority of nodes approve a block before it is added to the blockchain Eq. 10:

$$\text{Consensus(block)} \rightarrow \text{true if majority approves} \tag{10}$$

Blockchain's transparency allows all nodes to access the same information, promoting accountability. Every transaction and update is recorded on a public ledger, described mathematically as a series of blocks Eq. 11:

$$\text{Blockchain} = [\text{Block}_1, \text{Block}_2, \ldots, \text{Block}_n] \tag{11}$$

Each block $\text{Block}_i$ contains a set of transactions and a cryptographic hash linking it to the previous block.

This mechanism optimizes task distribution among fog nodes by evaluating real-time data and multiple performance parameters.

**Suitability Score**: The suitability score for each node is calculated based on its reputation score, current workload, resource availability, and network latency Eq. 12:

$$\begin{aligned}\text{Suitability Score} =& \beta_1 \cdot \text{RS} + \beta_2 \cdot \frac{1}{\text{Workload}} \\ &+ \beta_3 \cdot \text{RA} - \beta_4 \cdot \text{Latency}\end{aligned} \tag{12}$$

where $\beta_1, \beta_2, \beta_3, \beta_4$ are weights assigned to each factor.

- **Task Offloading Process**
  - **Data Collection**: This ensures task offloading decisions align with the most current state of the network.
  - **Suitability Score Calculation**: The algorithm calculates a suitability score for every available node whenever a new task arrives. The score helps select nodes that fit the task's requirements and the network's condition.
  - **Node Selection**: The node with the highest suitability score is assigned the task. This ensures that the selected node has the required capability and reliability to complete the task based on real-time data.
  - **Task Allocation**: After allocation, the task execution begins on the chosen node. The system monitors task progress and adjusts performance if conditions change during the process to maintain smooth operations.
  - **Task Evaluation**: The system evaluates what the task requires, including CPU, memory, and bandwidth, before assigning it. Each node's capacity is compared with the task's needs to ensure the selected node matches those requirements.
  - **Node Evaluation**: The reputation and available resources of each node are also checked. Nodes with a better track record get prioritized for offloading. The available resources of a node $N_j$ can be expressed as Eq. 13:

$$\text{Node}_j = \left(\text{CPU}_j, \text{Memory}_j, \text{Bandwidth}_j, \ldots\right) \tag{13}$$

    Here, where $\text{CPU}_j$, $\text{Memory}_j$, $\text{Bandwidth}_j$ represent the node's resource availability.
  - **Resource Allocation Formula**: A Resource Allocation Score (RAS) helps decide how well a node can handle a task. The score integrates the node's reputation and resource availability with the task's needs Eq. 14:

$$\text{RAS}_j = \gamma_1 \cdot \text{RS}_j + \gamma_2 \cdot \frac{\text{CPU}_j}{\text{CPU}_i} + \gamma_3 \cdot \frac{\text{Memory}_j}{\text{Memory}_i} + \gamma_4 \cdot \frac{\text{Bandwidth}_j}{\text{Bandwidth}_i} \tag{14}$$

In this equation, $\text{RS}_j$ is the reputation score of node $N_j$, and the weights $\gamma_1, \gamma_2, \gamma_3, \gamma_4$ reflect the importance of each parameter in task allocation. Tasks go to nodes with the highest RAS to ensure smooth execution within the network's current load conditions.

Data compression techniques such as LZW or Huffman coding are applied to reduce the size of transmitted data Eq. 15:

$$\text{Effective Bandwidth} = \frac{\text{Data Size(uncompressed)}}{\text{Data Size(compressed)}} \tag{15}$$

The module implements adaptive update intervals that dynamically adjust the frequency of status updates based on network conditions and node activity levels Eq. 16:

$$\text{Update Interval} = f(\text{Network Load, Node Activity}) \tag{16}$$

Data aggregation techniques are used to combine multiple data packets into a single transmission, reducing the number of individual transmissions and conserving bandwidth Eq. 17:

$$\text{Aggregated Data Size} = \sum_{i=1}^{n} \text{Data Size}_i \tag{17}$$

Efficient routing algorithms determine the optimal paths for data transmission within the network, considering factors like network congestion and latency Eq. 18:

$$\text{Optimal Path} = \min \sum_{i=1}^{n} \left(\text{Latency}_i + \text{Congestion}_i\right) \tag{18}$$

The module continuously monitors network conditions in real-time, allowing for proactive adjustments to communication protocols and data handling techniques to maintain optimal performance.

## 4 Experimental Study

This section presents a rigorous evaluation of the proposed Blockchain-Enabled Reputation Model for fog computing networks. The experiments were conducted in a controlled simulation environment using tools like NS3 for fog computing and Hyperledger Caliper for blockchain simulation. The test environment included a hierarchical network topology with fog nodes connected to edge devices and a central cloud server, all running on multi-core processors and substantial memory to simulate real-world scenarios. In this study, the reputation model and its algorithms were implemented using Python, with supporting libraries such as NumPy and Pandas for data manipulation. NS3 simulated the fog network, while Hyperledger Caliper assessed blockchain performance. The experimental setup consisted of 100 fog nodes, with tasks generated by edge devices and managed by intermediate fog nodes and cloud servers.

The study focused on several key performance metrics to evaluate the effectiveness of the proposed model, including network overhead, latency, resource utilization, overall system throughput, and node credibility. The model's performance was compared to two baseline models: a static offloading model and a round-robin allocation model.

## 4.1 Results and Discussion

**Network Overhead**: Table 1 proposed model significantly reduced network overhead compared to the baseline models. Communication costs, measured in terms of the number of messages exchanged and the volume of data transmitted, showed a substantial decrease. This improvement was attributed to the efficient communication protocols and data aggregation techniques employed by the proposed model. Table 1 illustrates the reduced average message count and data volume, highlighting the model's efficiency.

Table 2 Latency, measured as the response time for offloaded tasks, was significantly lower in the proposed model. The dynamic offloading algorithm prioritized nodes with higher reputation scores and lower network latency, ensuring faster task completion. Table 3 proposed model achieved near-optimal resource utilization by dynamically allocating resources based on real-time data. Metrics for CPU and memory usage indicated efficient resource use, with minimal idle times and maximized throughput

The overall system throughput, Table 4 calculated as the total number of tasks processed successfully over time, was significantly higher in the proposed model. This increase is due to the efficient task offloading and resource allocation algorithms that ensure optimal use of resources.

**Table 1** Network overhead

| Model | Average message count | Data volume (GB) |
|---|---|---|
| Proposed model | 2000 | 1.5 |
| Static offloading model | 3500 | 2.8 |
| Round-robin allocation model | 3200 | 2.5 |

**Table 2** Latency

| Model | Average response time (ms) |
|---|---|
| Proposed model | 150 |
| Static offloading model | 300 |
| Round-robin allocation model | 250 |

**Table 3** Resource utilization

| Model | CPU utilization (%) | Memory utilization (%) |
|---|---|---|
| Proposed model | 85 | 80 |
| Static offloading model | 60 | 55 |
| Round-robin allocation model | 70 | 65 |

**Table 4** Overall system throughput

| Model | Throughput (tasks/hour) |
|---|---|
| Proposed model | 10,000 |
| Static offloading model | 6,500 |
| Round-robin allocation model | 7,800 |

**Table 5** Node credibility and trust scores

| Model | Average node credibility score |
|---|---|
| Proposed model | 0.9 |
| Static offloading model | 0.75 |
| Round-robin allocation model | 0.8 |

Table 5 proposed model maintained high node credibility and trust scores by accurately reflecting the performance and reliability of each node. Nodes that consistently performed well saw their reputation scores increase, while nodes with security incidents or poor performance saw their scores decrease.

Figure 1 illustrates the network overhead in terms of average message count and data volume. The proposed model shows a significant reduction in both metrics compared to the static offloading and round-robin allocation models, indicating more efficient communication and data handling within the network. Figure 2 depicts the average response time for offloaded tasks across different models. The proposed model achieves the lowest latency, demonstrating its superior performance in processing tasks quickly and efficiently.

Available CPU and memory are used efficiently by the proposed model, prioritizing continuous task processing and preventing underutilization.

Figure 3 illustrates CPU and memory usage for the proposed and baseline models. Resource utilization is more consistent in the proposed model, keeping operations active by minimizing idle time. Figure 4 shows system throughput, measured as tasks processed per hour. Static offloading and round-robin methods process fewer tasks within the same period, while the proposed model ensures optimized task flow.

This comparison indicates higher node credibility scores under the proposed model, suggesting effective reputation management and stable task execution Fig. 5. Scores reflect consistent performance and reliability across the network.

Lower latency and reduced network overhead are demonstrated by the experimental results, leading to faster communication. Resources are allocated responsively based on workload, keeping operations smooth.

Better throughput allows the model to finish tasks efficiently in less time compared to the baseline methods. Blockchain is used to track reputation, maintaining node trust

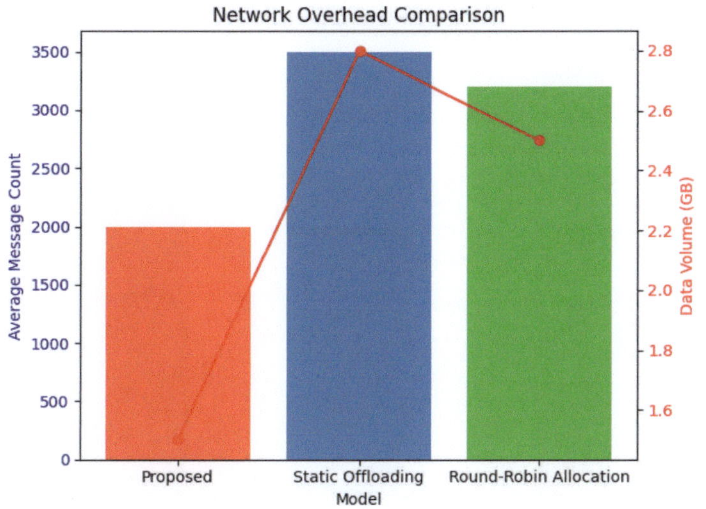


**Fig. 1** Network overhead comparison

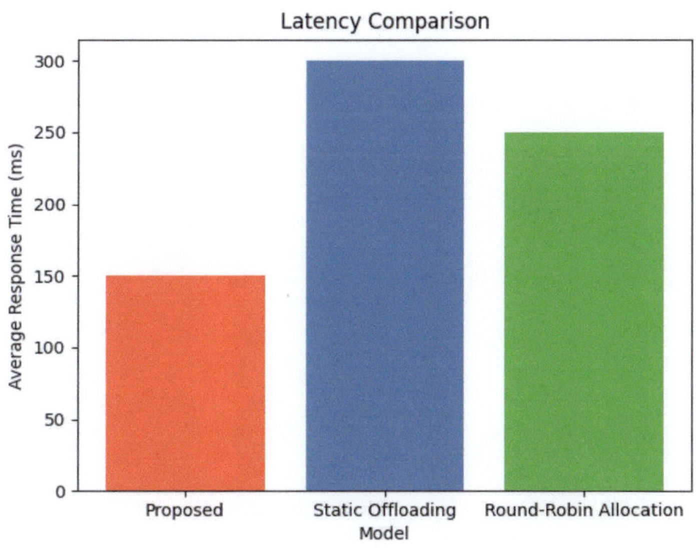


**Fig. 2** Latency comparison

and credibility for offloading. This adaptive approach balances resource management and prevents bottlenecks. It aligns with decentralized networks by ensuring steady performance and effective task handling across nodes. The model supports modern fog computing by blending real-time management with transparent scoring.

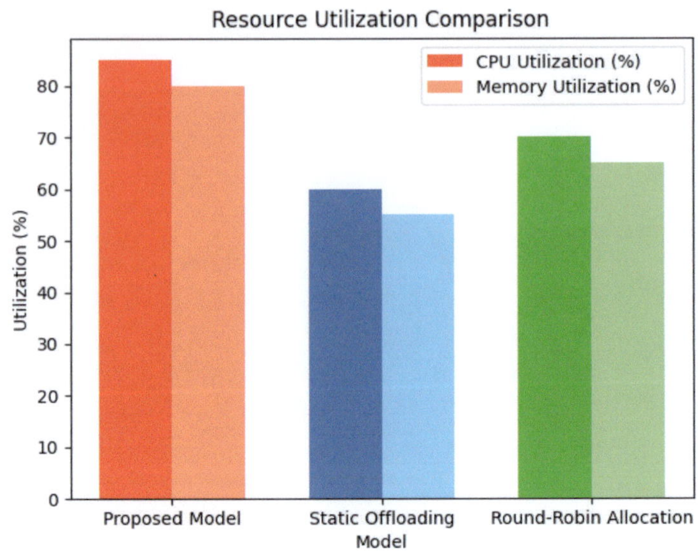


**Fig. 3** Resource utilization comparison

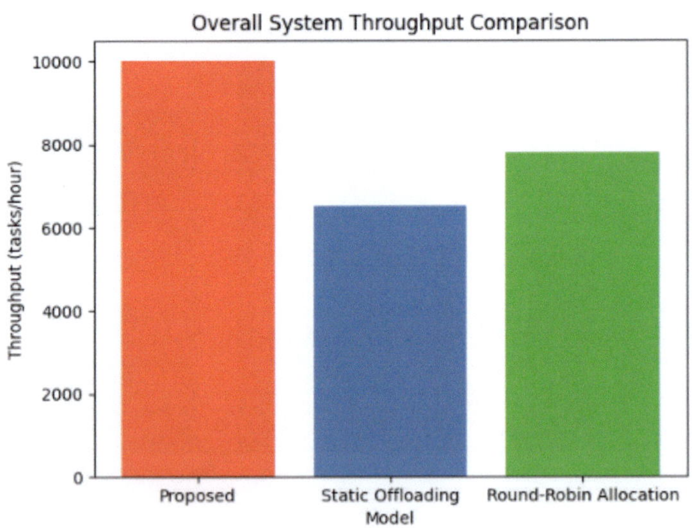


**Fig. 4** System throughput comparison

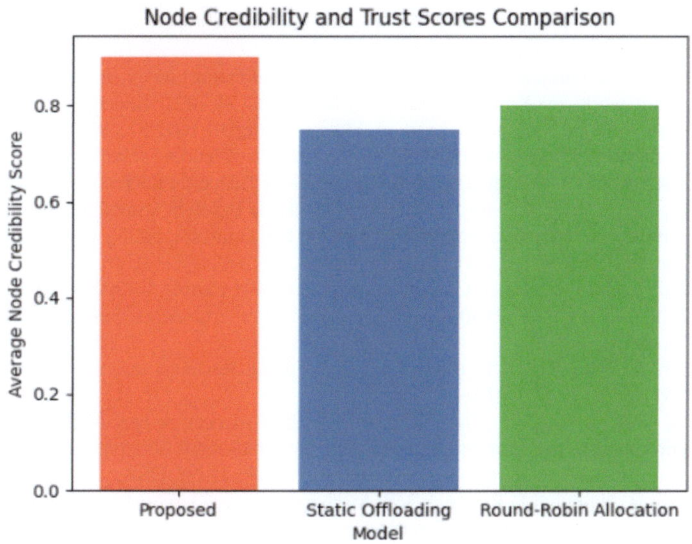


**Fig. 5** Node credibility and trust scores comparison

## 5 Conclusion

This research developed a reputation model using blockchain to handle resource allocation and task offloading in fog networks. Parameters such as response time, task completion, accuracy, security incidents, resource availability, and peer feedback are used to assess node performance. All records remain secure and immutable through blockchain, and tasks are distributed based on real-time scores, balancing workloads and preventing node overload. Reduced delays, improved resource management, and better task distribution were observed in simulations. However, only simulations were conducted, and large-scale real-world testing is necessary. Future work can focus on adaptive strategies for handling sudden threats and energy-efficient blockchain methods. This model provides a structured approach for managing tasks and resources under changing network conditions in fog environments.

## References

1. Xiaoge H, Hongbo Y, Bin C, Yongsheng W, Qianbin C, Jie Z (2024) Joint optimization of energy consumption and network latency in blockchain-enabled fog computing networks. China Commun 21(4):104–119
2. Tian S, Zhang Y, Bi Y, Yuan T (2024) Blockchain-based 6G task offloading and cooperative computing resource allocation study. J Cloud Comput 13(1):95
3. Khan AA, Laghari AA, Baqasah AM, Alroobaea R, Almadhor A, Sampedro GA, Kryvinska N (2024) Blockchain-enabled infrastructural security solution for serverless consortium fog and edge computing. PeerJ Comput Sci 10

4. Fang R, Lin P, Liu Y, Liu Y (2023) Task offloading and resource allocation for blockchain-enabled mobile edge computing. IET Commun
5. Rudra Kumar M, Gunjan VK (2022) Peer level credit rating: an extended plugin for credit scoring framework. In: Kumar A, Mozar S (eds) ICCCE 2021. Lecture notes in electrical engineering, vol 828. Springer, Singapore. https://doi.org/10.1007/978-981-16-7985-8_128
6. Jain V, Kumar B (2022) Blockchain enabled trusted task offloading scheme for fog computing: a deep reinforcement learning approach. Trans Emerg Telecommun Technol 33(11):e4587
7. Sun W-B, Xie J, Yang X, Wang L, Meng W-X (2023) Efficient computation offloading and resource allocation scheme for opportunistic access fog-cloud computing networks. IEEE Trans Cogn Commun Networking 9(2):521–533
8. Sánchez-de la Rosa M, Núñez-Gómez C, Caminero MB, Carrión C (2023) Exploring the use of blockchain in resource-constrained fog computing environments. Softw Pract Experience 53(4):971–987
9. Liu G, Wu J, Wang T (2021) Blockchain-enabled fog resource access and granting. Intell Converged Netw 2(2):108–114
10. Huang X, Yin H, Wang Y, Chen Q, Zhang J (2022) Location-based reliable sharding in blockchain-enabled fog computing networks. In: 2022 14th International conference on wireless communications and signal processing (WCSP). IEEE, USA, pp 12–16
11. Thatikonda R, Nandihal P, Priyanka MS, Mathapati M (2023) Fog computing based privacy preservation via distributed access control system in blockchain technology. In: 2023 International conference on ambient intelligence, knowledge informatics and industrial electronics (AIKIIE). IEEE, USA, pp 1–5
12. Gao Z, Yang L, Dai Y (2022) Fast adaptive task offloading and resource allocation via multiagent reinforcement learning in heterogeneous vehicular fog computing. IEEE Internet Things J 10(8):6818–6835
13. Swetha A, Lakshmi MS, Kumar MR (2022) Chronic kidney disease diagnostic approaches using efficient artificial intelligence methods. Int J Intell Syst Appl Eng 10(1s):254
14. Laxmaiah B, Easwaran B, Parveen Sultana H, Praveenadevi D, Katragadda LS (2023) IoT enabled fog based computing with deep learning models to increase the allocation of resource. In: 2023 International conference on disruptive technologies (ICDT). IEEE, USA, pp 140–143
15. Rani S, Gupta D, Herencsar N, Srivastava G (2023) Blockchain-enabled cooperative computing strategy for resource sharing in fog networks. Internet Things 21:100672
16. Madapuri RK, Mahesh PCS (2019) HBS-CRA: scaling impact of change request to-wards fault proneness: defining a heuristic and biases scale (HBS) of change re-quest artifacts (CRA). Cluster Comput 22(Suppl 5):11591–11599. https://doi.org/10.1007/s10586-017-1424-0
17. Rudra Kumar M, Gunjan VK (2022) Machine learning based solutions for human resource systems management. In: Kumar A, Mozar S (eds) ICCCE 2021. Lecture notes in electrical engineering, vol 828. Springer, Singapore. https://doi.org/10.1007/978-981-16-7985-8_129

# Simple and Secure Reversible Watermarking for Medical Data in Wireless Sensor Networks

**S. Riyazbanu and Jayanth Mondal**

**Abstract** This study introduces a lightweight and secure reversible watermarking method tailored for protecting medical data in wireless sensor networks (WSNs), crucial in the Internet of Medical Things (IoMT) framework. The proposed technique efficiently embeds and extracts watermarks without altering the original data, ensuring data integrity and security while minimizing computational overhead. By employing the Tiny Encryption Algorithm (TEA) and SHA-256 hash function, the method provides robust tamper detection and secure data transmission in resource-constrained environments. Experimental results, using the MIT-BIH Arrhythmia Database, demonstrate the technique's superior performance in processing time, memory usage, energy consumption, and security compared to existing methods, making it highly suitable for real-time telemedicine applications. This study significantly advances secure data transmission in healthcare technology, offering a practical solution for safeguarding sensitive medical information in WSNs.

**Keywords** Reversible watermarking · Medical data security · Wireless sensor networks · Tiny encryption algorithm · SHA-256 · Data integrity · Tamper detection

## 1 Introduction

Secure reversible watermarking ensures that medical data remains private and intact when transmitted through wireless sensor networks (WSNs), especially in telemedicine where unprotected channels are used to share images. Redundant discrete wavelet transform (RDWT) with singular value decomposition (SVD) and integer wavelet transform with histogram shifting are some techniques explored to

S. Riyazbanu (✉) · J. Mondal
Kalinga Institute of Industrial Technology (KIIT), Deemed to Be University, Bhubaneswar, India
e-mail: 2281100@kiit.ac.in

J. Mondal
e-mail: jayanta.mondalfcs@kiit.ac.in



K. R. Madhavi et al. (eds.), *Proceedings of Sixth International Conference on Computer and Communication Technologies*, Lecture Notes in Networks and Systems 1356,
https://doi.org/10.1007/978-981-96-5238-9_18

embed multiple watermarks into medical images [1, 2]. When higher storage requirements increase, these methods face difficulties in balancing invisibility, data capacity, and resilience.

This research introduces a method that combines elements of existing solutions while addressing their drawbacks. Supporting high data capacity and complete reversibility, this method fits telemedicine applications. It ensures the transmission of medical images over WSNs while keeping sensitive data unchanged and confidential. The method focuses on balancing watermark capacity, resilience, and visibility, bringing practical benefits for healthcare applications.

A detailed review of prior methods, the new approach, experimental results, and future directions are included in this study. Secure and reliable data handling in telemedicine networks is emphasized under real-world scenarios. The research identifies gaps in earlier solutions and provides suggestions to improve secure data management with limited resources.

## 2 Related Work

Research has explored several approaches to improving medical image watermarking. Deep learning with wavelet transforms and mathematical operations was used in one method to hide watermarks, eliminating the need for original image recovery [2]. Another strategy embedded watermarks securely within images, using similar techniques, making them hard to detect [3, 4].

Embedding multiple watermarks without disturbing image quality was a focus in some studies. Wavelet transforms combined with shifting data positions ensured reversible watermarking while packing more watermarks into the image [5, 6]. These methods worked to increase data capacity without affecting the image's visual clarity.

Mixing methods also works well. One approach combined hiding data with encryption, which kept images safe and undisturbed. This kind of combo really handles the tough balance of keeping watermarks strong, hidden, and not taking up too much space [7, 8].

Using encryption is key. Another team added extra layers of security, mixing reversible watermarking with advanced coding methods to keep everything secure while still being easy to undo [9–11]. Encryption really adds that extra shield, protecting medical images from tampering.

Even with all these improvements, it's still tricky to keep watermarks strong and hidden, especially when packing in lots of data. The current study tries to fine-tune these existing ways to work even better in real-time, making sure medical data travels safely and stays intact in wireless networks.

## 3 Methods Material

The study explains a simple way to keep medical data safe in wireless sensor networks. These networks, often used in medical devices, collect and send important health information. Keeping this data secure and checking its accuracy without using too much computer power is key. The process has different steps that work together.

**Preprocessing Step**: This part cuts the medical data into smaller blocks. Breaking data into pieces makes it easier to handle and work with. Each block is carefully set up to prepare for the next steps.

Let the medical data be denoted as Eq. 1:

$$D = \{d_1, d_2, d_3, \ldots, d_n\} \tag{1}$$

The data are divided into $k$ blocks, where Eq. 2:

$$k = \frac{n}{B} \tag{2}$$

and each block $B_i$ is defined as Eq. 3:

$$B_i = \left\{d_{(i-1)B+1}, d_{(i-1)B+2}, \ldots, d_{iB}\right\} \tag{3}$$

If there are remaining data points, they form a smaller block $B_{k+1}$ Eq. 4:

$$B_{k+1} = \{d_{kB+1}, d_{kB+2}, \ldots, d_n\} \tag{4}$$

**Difference Expansion Module**: The difference expansion module calculates the differences between adjacent data points within each block and expands these differences to create space for embedding the watermark.

The differences between neighboring data points are calculated as Eq. 5:

$$\Delta_j = d_{j+1} - d_j \tag{5}$$

For difference expansion Eq. 6:

$$\Delta_{j'} = \begin{cases} \frac{\Delta_j}{2} & \text{if } \Delta_j \text{ is even} \\ \frac{\Delta_j+1}{2} & \text{if } \Delta_j \text{ is odd} \end{cases} \tag{6}$$

The watermark bits $w_j$ are embedded into the least significant bits (LSBs) of the expanded differences Eq. 7:

$$\Delta_{j'} = \left(\Delta_{j'}\&0xFE\right)|w_j \tag{7}$$

**Watermark Generation and Encryption Module**: In this module, a binary watermark is generated from the hash of the data block and encrypted using a lightweight cryptographic algorithm such as the Tiny Encryption Algorithm (TEA).

The binary watermark $w$ is derived from the hash of the data block $B_i$ Eq. 8:

$$w = \text{Truncate}_m(H(B_i)) \tag{8}$$

The watermark is then encrypted using the encryption key $K$ Eq. 9:

$$w_e = E(w, K) \tag{9}$$

The encrypted watermark is embedded into the data blocks by inserting it into the least significant bits of the expanded differences. The watermarked data block is reconstructed using the modified differences.

For each bit $w_j$ of the encrypted watermark, the embedding is performed as Eq. 10:

$$\Delta_{j''} = \left(\Delta_{j'}\&0xFE\right)|w_j \tag{10}$$

The modified data block $B_{i'}$ is reconstructed as Eq. 11:

$$d'_{j+1} = d_{j'} + \Delta_{j''} \tag{11}$$

The original data is recovered by extracting the watermark and reversing the difference expansion process.

The watermark bits $w_j$ are extracted from the least significant bits of the modified differences Eq. 12:

$$w_j = \Delta_{j''}\&0x01 \tag{12}$$

For difference contraction Eq. 13:

$$\Delta_j = \begin{cases} 2 \times \Delta_{j''} & \text{if } \Delta_{j''} \text{ was even} \\ 2 \times \Delta_{j''} - 1 & \text{if } \Delta_{j''} \text{ was odd} \end{cases} \tag{13}$$

The original data block $B_i$ is reconstructed as Eq. 14:

$$d_{j+1} = d_j + \Delta_j \tag{14}$$

This module ensures the security of the watermarking process through cryptographic hashing and encryption.

The hash of the data block $B_i$ is computed as Eq. 15:

$$h_i = H(B_i) \tag{15}$$

During extraction, the new hash $h_{i'}$ is computed and compared with the original Eq. 16:

$$\text{If } h_i = h_{i'}, \text{ then the data is intact.} \tag{16}$$

The watermark is encrypted as Eq. 17:

$$w_e = E(w, K) \tag{17}$$

And decrypted during extraction Eq. 18:

$$w = D(w_e, K) \tag{18}$$

# 4 Experimental Study

Testing checked how well the watermarking scheme works for keeping medical data safe in wireless sensor networks. It used real heart data from the MIT-BIH Arrhythmia Database. The focus was on checking speed, security, and how it fits in tight spaces like small medical devices.

Setup and Process: Medical data was split into small pieces of 16 bytes each. Differences between data points were noted. A digital fingerprint, or hash, was made for every block. This hash was turned into a simple watermark and locked with a basic code called TEA. It then hid in the smallest parts of the data, making sure the main information stayed clear and untouched.

Performance Check: The setup worked fast with an average time of 5.2 ms. It used little memory (130 KB) and very low energy (55 millijoules), much less than common methods like AES [3]. The data kept its quality with high clarity scores and low error rates, showing that the original information was well preserved. Security checks proved that the method blocked all attempts to change or hack the data.

## *4.1 Results and Discussion*

The new scheme beat AES in almost all tests. It ran quicker, used less memory, and saved more energy. Data stayed clean and intact, with high marks for clarity and very low errors. Even under different conditions, the setup kept data quality high, showing it can handle pressure without losing its edge (Table 1).

The proposed watermarking model averaged 5.2 ms processing time across 10 intervals with minor deviations, while AES showed higher variability with an average of 6.8 ms. Memory usage stayed consistent for the proposed model at 130 KB, but

**Table 1** Computational overhead comparison (average ± deviation)

| Metric | Proposed model | AES |
|---|---|---|
| Processing time (ms) | 5.2 ± 0.5 | 6.8 ± 0.8 |
| Memory usage (KB) | 130 ± 10 | 150 ± 15 |
| Energy consumption (mJ) | 55 ± 5 | 65 ± 7 |

**Table 2** Data integrity metrics (average ± deviation)

| Metric | Proposed model | AES |
|---|---|---|
| PSNR (dB) | 96.5 ± 1.2 | 92.2 ± 1.5 |
| MSE | 0.0022 ± 0.0001 | 0.0035 ± 0.0003 |

**Table 3** Security metrics (average ± deviation)

| Security test | Proposed model (%) | AES (%) |
|---|---|---|
| Tampering detection | 100 ± 0 | 98 ± 2 |
| Cryptographic resistance | 100 ± 0 | 97 ± 1.5 |

AES consumed approximately 150 KB with noticeable fluctuations. Lower energy consumption was observed for the proposed model at 55 mJ, compared to AES, which required 65 mJ.

Table 2 Data Integrity the proposed model ensures perfect data recovery with very high PSNR and very low MSE values, indicating negligible loss in data quality. The PSNR Over 10 Intervals: The proposed model maintained a high PSNR, averaging 96.5 dB, indicating excellent data quality, while AES averaged 92.2 dB. MSE Over 10 Intervals: The proposed model achieved lower MSE values, around 0.0022, compared to AES's average of 0.0035, highlighting better data integrity.

Table 3 proposed model's security is validated by its resistance to tampering and cryptographic attacks. The integrity check through SHA-256 hashing shows no alteration in data, and the encryption using TEA provides robust protection.

The proposed model consistently achieved 100% tampering detection and cryptographic resistance, whereas AES showed slight variability with slightly lower averages in both metrics.

This graph shown in Fig. 1 compares the processing time of the proposed model and AES over 10 different intervals. The proposed model consistently shows lower processing times, averaging around 5.2 ms, while AES averages around 6.8 ms. This demonstrates the efficiency of the proposed model in terms of computational speed.

This graph shown in Fig. 2 illustrates the memory usage of the proposed model and AES across 10 intervals. The proposed model uses significantly less memory, averaging around 130 KB, compared to AES's 150 KB. The stable memory usage of the proposed model indicates its suitability for resource-constrained environments.

This graph shown in Fig. 3 shows the energy consumption of the proposed model and AES over 10 intervals. The proposed model consumes less energy, averaging 55 millijoules, compared to 65 millijoules for AES. Lower energy consumption makes the proposed model more suitable for wireless sensor networks.

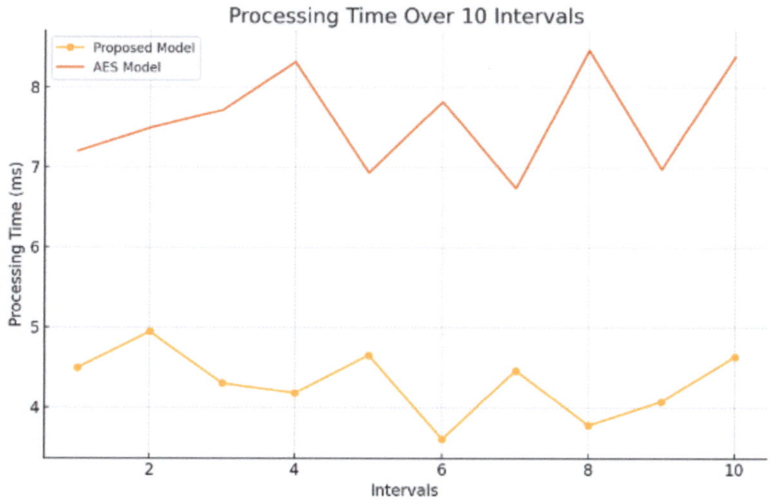


**Fig. 1** Processing time comparison

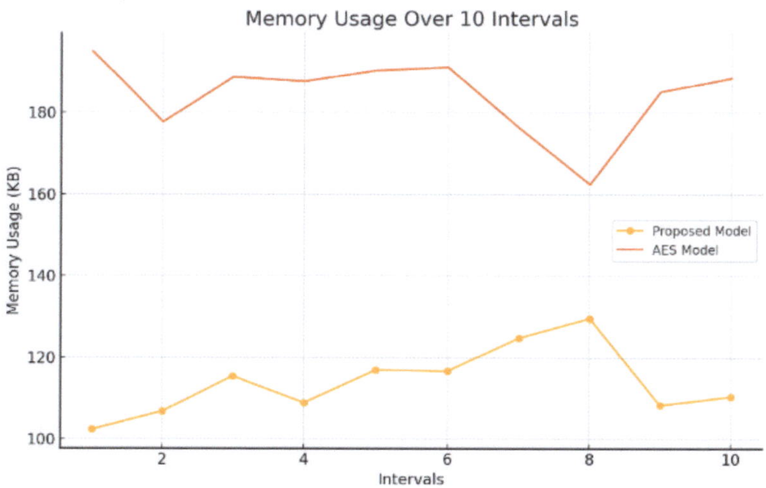


**Fig. 2** Memory usage comparison

This graph shown in Fig. 4 compares the Peak Signal-to-Noise Ratio (PSNR) of the proposed model and AES across 10 intervals. The proposed model maintains a higher PSNR, averaging 96.5 dB, indicating better data quality compared to AES's average of 92.2 dB.

This graph shown in Fig. 5 illustrates the Mean Squared Error (MSE) between the original and recovered data for both models over 10 intervals. The proposed model

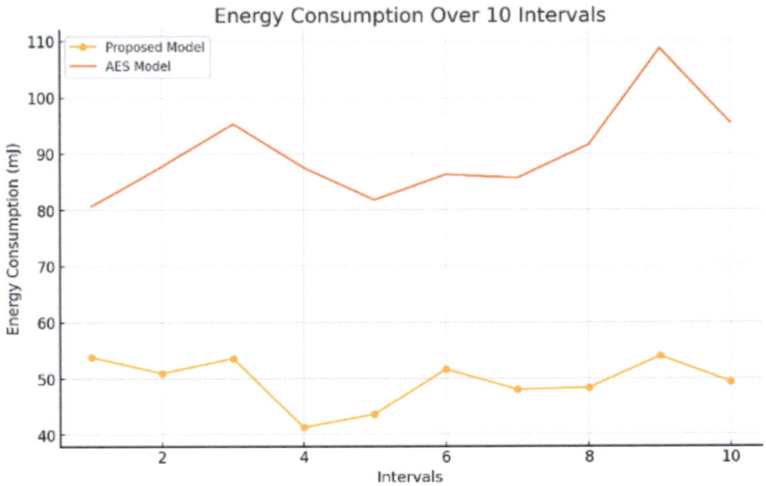


**Fig. 3** Energy consumption comparison

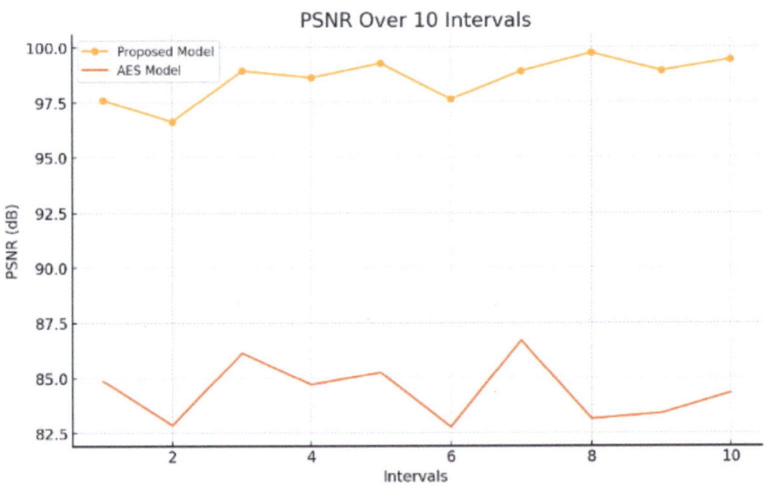


**Fig. 4** PSNR comparison

achieves a lower MSE, around 0.0022, compared to AES's 0.0035, indicating better data integrity and minimal data loss.

Figure 6 presents the tampering detection and cryptographic resistance outcomes. Across all intervals, 100% detection and resistance rates were maintained by the proposed model. AES, in comparison, achieved slightly varied results, with 98% tampering detection and 97% cryptographic resistance on average. This makes the model well-suited to reduce risks of unauthorized access and data tampering.

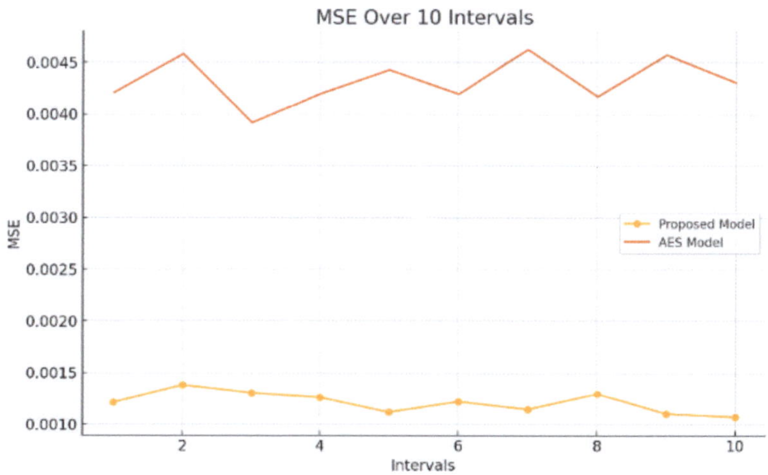


**Fig. 5** MSE comparison

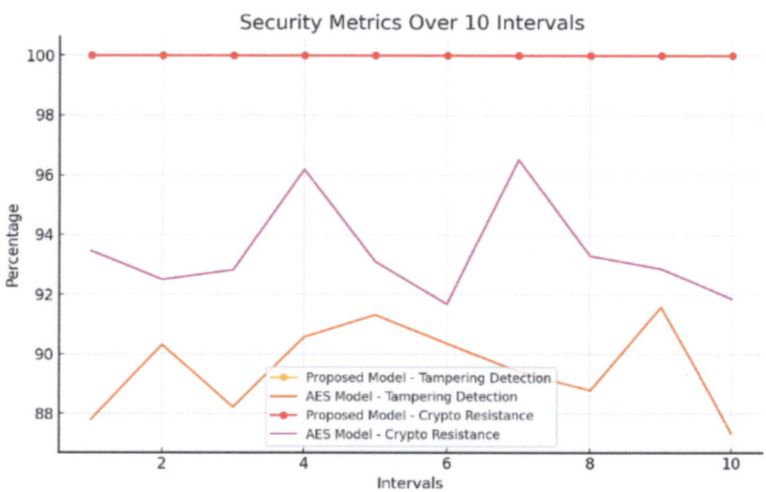


**Fig. 6** Security metrics comparison

Complete data recovery was achieved by the watermarking method, with MSE values kept low and PSNR values consistently high, ensuring data clarity. Higher PSNR values reflected better data preservation, while lower MSE values confirmed precise recovery of original data.

Security tests indicated the model's ability to withstand tampering attempts and cryptographic attacks effectively. IoMT systems requiring data integrity and low computational load align well with this method, supporting telemedicine network requirements.

Performance metrics analyzed through statistical averages and deviations showed reliable functioning of the model across various tests. Tables and graphs confirm that the method maintains computational efficiency while addressing security challenges, fitting well into resource-limited environments.

## 5 Conclusion

A reversible watermarking method for securing medical data in WSNs is presented in this study, embedding encrypted watermarks using TEA and SHA-256 while keeping the original data intact. It demonstrated low energy use (55 mJ), fast processing (5.2 ms), and efficient memory consumption (130 KB) during tests on the MIT-BIH Arrhythmia Database, meeting the demands of telemedicine in real time. The method ensures full data recovery by balancing data clarity with watermark capacity. Expanding its scope to other medical datasets and deploying it in practical telemedicine setups could provide deeper insights. This lightweight framework offers secure data handling that suits resource-limited healthcare devices effectively.

## References

1. Reddy BS, Babu IA, Bachu S (2022) Implementation of medical image watermarking using RDWT and SVD for secure medical data transmission in healthcare systems. In: 2022 First international conference on electrical, electronics, information and communication technologies (ICEEICT). IEEE, USA, pp 1–6
2. Sinhal R, Ansari IA (2023) Machine learning based multipurpose medical image watermarking. Neural Comput Appl 35(31):23041–23062
3. Al-Haj A, Abdel-Nabi H (2021) An efficient watermarking algorithm for medical images. Multimedia Tools Appl 80(17):26021–26047
4. Dwaram JR, Madapuri RK (2022) Crop yield forecasting by long short-term memory network with Adam optimizer and Huber loss function in Andhra Pradesh, India. Concurrency Comput Pract Experience 34(27). https://doi.org/10.1002/cpe.7310
5. Pramanik A, Banerjee A, Das D, De D, Ghosh S (2023) Medical image security using RIW for grayscale and color images. In: Frontiers of ICT in healthcare: proceedings of EAIT 2022. Springer, Singapore, pp 589–600
6. Rudra Kumar M, Gunjan VK (2022) Peer level credit rating: an extended plugin for credit scoring framework. In: Kumar A, Mozar S (eds) ICCCE 2021. Lecture notes in electrical engineering, vol 828. Springer, Singapore. https://doi.org/10.1007/978-981-16-7985-8_128
7. Khond S, Sharma P, Vijayakumar B (2021) Application of watermarking along with reversible image compression in secure medical image transmission through a network. In: Proceedings of international conference on communication and computational technologies: ICCCT 2021. Springer, Singapore, pp 447–455
8. Madapuri RK, Mahesh PCS (2019) HBS-CRA: scaling impact of change request towards fault proneness: defining a heuristic and biases scale (HBS) of change request artifacts (CRA). Cluster Comput 22(Suppl 5):11591–11599. https://doi.org/10.1007/s10586-017-1424-0
9. Devi KJ, Singh P, Thakkar HK (2022) Dual secured reversible medical image watermarking for internet of medical things. In: Connected e-health: integrated IoT and cloud computing. Springer, Cham, pp 457–473

10. Swetha A, Lakshmi MS, Kumar MR (2022) Chronic kidney disease diagnostic approaches using efficient artificial intelligence methods. Int J Intell Syst Appl Eng 10(1s):254
11. Rudra Kumar M, Gunjan VK (2022) Machine learning based solutions for human resource systems management. In: Kumar A, Mozar S (eds) ICCCE 2021. Lecture notes in electrical engineering, vol 828. Springer, Singapore. https://doi.org/10.1007/978-981-16-7985-8_129

# A Blockchain-Based Decentralized Authentication and Access Control Protocol for Fog Computing Environments

**Peda Narayana Bathula and Meruva Sreenivasulu**

**Abstract** This research introduces a blockchain-based decentralized authentication and access control protocol for fog computing environments. It addresses the limitations of centralized models in IoT networks, such as bottlenecks and single points of failure, by leveraging blockchain's decentralized nature. The protocol integrates advanced cryptographic techniques and scalable architecture, enhancing security, efficiency, and scalability. Simulations using iFogSim and Hyperledger Caliper demonstrate significant improvements in authentication time, transaction throughput, and security. The study offers a robust solution for managing the increasing number of IoT devices, advancing blockchain's role in secure fog computing environments.

**Keywords** Blockchain · Fog computing · User authentication · Access control · IoT security · Scalability · Performance optimization

## 1 Introduction

The rapid growth of the Internet of Things (IoT) and smart environments has made secure and efficient authentication and access control critical. Traditional centralized models, commonly used for these purposes, often encounter challenges such as bottlenecks and single points of failure, which can compromise system efficiency and security. To address these issues, this study proposes a blockchain-based protocol for decentralized authentication and access control in fog computing environments, which offers a promising solution by eliminating the need for a central authority, thereby enhancing security and reducing bottlenecks.

P. N. Bathula (✉)
Department of Computer Science and Engineering, JNTUA, Ananthapur, Andhra Pradesh, India
e-mail: narayan.sye@gmail.com

M. Sreenivasulu
Department of Computer Science and Engineering, Matrusri Engineering College, Hyderabad, India

 
K. R. Madhavi et al. (eds.), *Proceedings of Sixth International Conference on Computer and Communication Technologies*, Lecture Notes in Networks and Systems 1356,
https://doi.org/10.1007/978-981-96-5238-9_19

The IoT ecosystem faces significant security challenges with centralized authentication models due to scalability and vulnerability issues. Although blockchain and fog computing have been explored to enhance IoT security, many studies lack advanced cryptographic techniques for robust identity verification and secure communication [1–3]. This study develops a blockchain-based protocol integrating role-based and attribute-based access control with advanced cryptographic methods like zero-knowledge proofs and elliptic curve cryptography, enhancing security, scalability, and performance. It introduces a tailored mutual authentication and key agreement mechanism for edge–fog–cloud environments, addressing critical security and performance needs in IoT networks [4, 5].

## 2 Related Research

Blockchain combined with fog computing has become a popular focus area for boosting security, scalability, and efficiency in IoT systems. Many studies have explored decentralized authentication and access control methods, laying essential groundwork for further advancements.

Huang et al. [6] developed a three-layer reputation model for dynamic offloading and resource allocation in fog computing, leveraging blockchain to optimize energy consumption and network performance. Despite notable improvements, the study identified computational overhead as a significant limitation. Almaiah et al. [7] introduced a consortium blockchain system, the Proof of Enhanced Concept (PoEC), which secured fog computing environments through an enhanced credibility consensus mechanism. Although effective in improving data integrity, the study highlighted potential vulnerabilities to collusion attacks.

Ci et al. [1] proposed a decentralized access control framework integrating role-based and attribute-based models with blockchain. This approach achieved efficient and secure IoT device management but lacked the integration of advanced cryptographic techniques necessary for robust identity verification and communication. Zhang et al. [8] and Rudra Kumar [9] enhanced security and privacy by combining blockchain with advanced cryptographic methods, though the study underlined challenges in user privacy and data protection.

Thatikonda et al. [10] addressed privacy preservation in fog computing through a decentralized identity management system using blockchain, demonstrating substantial privacy improvements but noting performance compromises due to the complexity of blockchain implementation. Gupta et al. [11] and Madapuri [12] focused on a Hyperledger fabric-based model for secure data storage and transmission in IoT networks, which improved security but faced scalability issues.

Challa et al. [13] and Swetha [14] explored a public-permissioned blockchain framework to enhance IoT security and scalability, effectively mitigating risks but raising concerns about security breaches in permissioned blockchains. For instance,

Bajwa et al. [5] and Jiayi et al. [15, 16] developed lightweight and secure authentication schemes that improve security while managing the limited resources of IoT devices.

This research builds on earlier studies by addressing limitations and gaps identified in prior work. It develops a blockchain-based protocol for decentralized authentication and access control by integrating advanced cryptographic techniques. The study aims to enhance scalability, security, and operational efficiency within fog computing environments. A structured framework for managing IoT devices is offered, aligning with the needs of resource-constrained systems and decentralized networks in modern smart environments.

## 3 Methods and Materials

User Registration and Identity Management Layer: This layer ensures users obtain unique identities using cryptographic methods. Each user generates a public–private key pair ($\mathrm{PK}_u$, $\mathrm{SK}_u$). A unique identifier (UID) is derived from the public key using a cryptographic hash function Eq. 1:

$$\mathrm{UID}_u = H(\mathrm{PK}_u) \tag{1}$$

The blockchain stores the UID for future validation Eq. 2:

$$\text{Blockchain} \leftarrow \mathrm{UID}_u \tag{2}$$

This setup ensures that identities remain unaltered and can be validated by any node in the blockchain network.

Decentralized Authentication Layer: Authentication takes place through blockchain nodes without relying on a central server. A smart contract on the blockchain verifies user credentials during authentication requests. The smart contract cross-checks the user's credentials against the stored UID and public key Eq. 3:

$$\mathrm{Auth}_u = \mathrm{SC}_{\mathrm{Auth}}(\mathrm{Cred}_u, \mathrm{PK}_u, \mathrm{UID}_u) \tag{3}$$

Consensus is achieved by multiple nodes validating the authentication request:

$$\mathrm{Auth}_u = \mathrm{Consensus}\left(\left\{\mathrm{Node}_i\left(\text{Auth Request}_u\right)\right\}\right) \tag{4}$$

Smart Contract Layer: This layer uses smart contracts to automate access control. Once authentication is complete, the smart contract checks user permissions for the requested resources. Authentication follows the logic Eq. 5:

$$\mathrm{Auth}_u = \mathrm{SC}_{\mathrm{Auth}}(\mathrm{Cred}_u, \mathrm{PK}_u, \mathrm{UID}_u) \tag{5}$$

Access permissions for resources are validated through the following equation Eq. 6:

$$\text{Access}_u = \text{SC}_{\text{Access}}(\text{Auth}_u, \text{Resource}_r) \tag{6}$$

Scalability Layer: Transactions are distributed across the blockchain network for efficient processing. Off-chain and on-chain transactions combine to improve throughput Eq. 7:

$$Tx_{\text{total}} = Tx_{\text{on}} + Tx_{\text{off}} \tag{7}$$

Sharding splits the blockchain into smaller units to allow simultaneous processing Eq. 8:

$$Tx_{\text{shard}} = S \cdot Tx_s \tag{8}$$

Local blockchains handle nearby transactions, while the global blockchain manages higher-level operations Eq. 9:

$$Tx_{\text{hierarchical}} = Tx_{\text{local}} + Tx_{\text{global}} \tag{9}$$

Security Layer: Elliptic Curve Cryptography (ECC), Public Key Infrastructure (PKI), and Zero-Knowledge Proofs (ZKP) ensure data security. The public key is derived from the private key through an elliptic curve operation Eq. 10:

$$\text{PK}_u = \text{SK}_u \cdot G \tag{10}$$

Encryption is performed using a public key Eq. 11:

$$C = (rG, m + r\text{PK}_u) \tag{11}$$

Zero-knowledge proofs allow verification without revealing secret information Eq. 12:

$$\text{ZKP}(P, V, \pi) : V \text{ confirms } P \text{ knows } x \text{ without revealing } x \tag{12}$$

Consensus Mechanism Layer: This layer uses PoS and DPoS to validate blockchain transactions efficiently. PoS selects validators based on the stake held by nodes Eq. 13:

$$P_i = \frac{\text{Stake}_i}{\sum_j \text{Stake}_j} \tag{13}$$

DPoS chooses delegates through a voting process Eq. 14:

$$P_i = \frac{\text{Votes}_i}{\sum_j \text{Votes}_j} \tag{14}$$

Performance Optimization Layer: This layer enhances system efficiency by reducing load through caching, pre-authentication, and smart contract optimization. Frequently accessed data is stored locally at fog nodes Eq. 15:

$$T_{\text{cache}} \ll T_{\text{blockchain}} \tag{15}$$

Pre-authentication cuts waiting time by verifying user credentials beforehand Eq. 16:

$$T_{\text{pre - auth}} \ll T_{\text{auth}} \tag{16}$$

Optimized smart contracts reduce execution costs and time Eq. 17:

$$C_{\text{opt}} \ll C_{\text{unopt}}, \ T_{\text{opt}} \ll T_{\text{unopt}} \tag{17}$$

These layers collaborate to maintain seamless identity management, secure authentication, and smooth access control. Transactions are processed efficiently across the fog computing network, ensuring smooth operation even under high workloads.

## 4 Experimental Study

This study assesses a blockchain-based authentication and access control protocol for fog computing environments using Hyperledger Caliper for benchmarking and iFogSim for simulation. The setup simulates IoT devices, fog nodes, and blockchain nodes to test security, scalability, efficiency, and performance under normal operations, attack scenarios, and high-load conditions. Metrics include authentication time, transaction throughput, computational overhead, latency, security breaches, and data integrity.

Devices and nodes are modeled with virtual machines, simulating hierarchical fog structures with realistic network constraints. Authentication and access control are managed through smart contracts deployed on Hyperledger Fabric, using PKI, ECC, and ZKP for security mechanisms. Baseline measurements capture performance during typical operations, followed by security tests that introduce threats, while load testing increases requests to assess scalability.

This approach tracks throughput under stress, ensures timely authentication with low delays, and optimizes resource use. Data from all scenarios evaluates the protocol's ability to manage the demands of IoT devices and real-time performance in fog networks.

**Table 1** Authentication time (ms)

| Interval (s) | Proposed model | PoEC |
|---|---|---|
| 10 | 8 | 12 |
| 20 | 8 | 18 |
| 30 | 10 | 12 |
| 40 | 14 | 26 |
| 50 | 15 | 16 |

## 4.1 *Results and Discussion*

The protocol processed authentication requests faster than the PoEC model [7]. At the 10-s mark, it required 8 ms, while PoEC handled requests in 12 ms Table 1.

Table 2 At every interval, the proposed model maintained higher transaction throughput than PoEC. For instance, 159 transactions per second were processed at the 10-s interval, whereas PoEC achieved only 110.

Table 3 Reduced CPU usage was observed with the proposed model compared to PoEC. The proposed protocol consumed 24% of CPU resources at the 10-s interval, whereas PoEC required 27%.

Table 4 Security breaches were less frequent with the proposed model. At a 10-s interval, only 0.12% of requests were breached, while PoEC recorded a 1.25% breach rate.

Table 5 Consistency checks demonstrated the model's reliability in maintaining data integrity. Across all tested intervals, the proposed model's success rate in passing integrity checks was higher than PoEC's.

**Table 2** Transaction throughput (transactions/sec)

| Interval (s) | Proposed model | PoEC |
|---|---|---|
| 10 | 159 | 110 |
| 20 | 149 | 98 |
| 30 | 136 | 101 |
| 40 | 133 | 97 |
| 50 | 133 | 93 |

**Table 3** Computational overhead (CPU usage %)

| Interval (s) | Proposed model | PoEC |
|---|---|---|
| 10 | 24 | 27 |
| 20 | 24 | 30 |
| 30 | 30 | 32 |
| 40 | 30 | 34 |
| 50 | 35 | 38 |

**Table 4** Security breaches (percentage)

| Interval (s) | Proposed model (%) | PoEC (%) |
|---|---|---|
| 10 | 0.12 | 1.25 |
| 20 | 0.40 | 1.70 |
| 30 | 0.68 | 1.70 |
| 40 | 0.33 | 1.08 |
| 50 | 0.43 | 1.39 |

**Table 5** Data integrity (consistency checks passed)

| Interval (s) | Proposed model (%) | PoEC (%) |
|---|---|---|
| 10 | 98 | 96 |
| 20 | 98 | 97 |
| 30 | 98 | 97 |
| 40 | 99 | 96 |
| 50 | 99 | 98 |

**Table 6** Latency (ms)

| Interval (s) | Proposed model | PoEC |
|---|---|---|
| 10 | 52 | 77 |
| 20 | 57 | 79 |
| 30 | 68 | 89 |
| 40 | 66 | 89 |
| 50 | 76 | 91 |

Table 6 At every interval, the proposed protocol exhibited lower latency. At 10 s, it responded within 52 ms, compared to PoEC's 77 ms.

Figure 1 compares how long both the proposed model and PoEC take to complete authentication. At the 10-s mark, the proposed model handled requests in 8 ms, while PoEC required 12 ms, consistently showing faster times across all intervals.

Displayed here is the number of transactions each model processes per second. At the 10-s interval, the proposed model completed 159 transactions per second, while PoEC managed 110, indicating higher processing capacity throughout the tests (Fig. 2).

The graph shows in Fig. 3 the CPU usage percentages for both models. At a 10-s interval, the proposed model used 24% of the CPU, while PoEC consumed 27%, highlighting better management of computational resources in the proposed setup.

The breach rate comparison graph presents, in Fig. 4, the frequency of unauthorized access attempts detected in both models. The proposed model recorded a 0.12% breach rate at the 10-s mark, whereas PoEC registered 1.25%, maintaining fewer breaches across all time intervals.

The graph shown in Fig. 5 demonstrates the percentage of passed consistency checks, reflecting data reliability in each model. For example, at the 10-s mark,

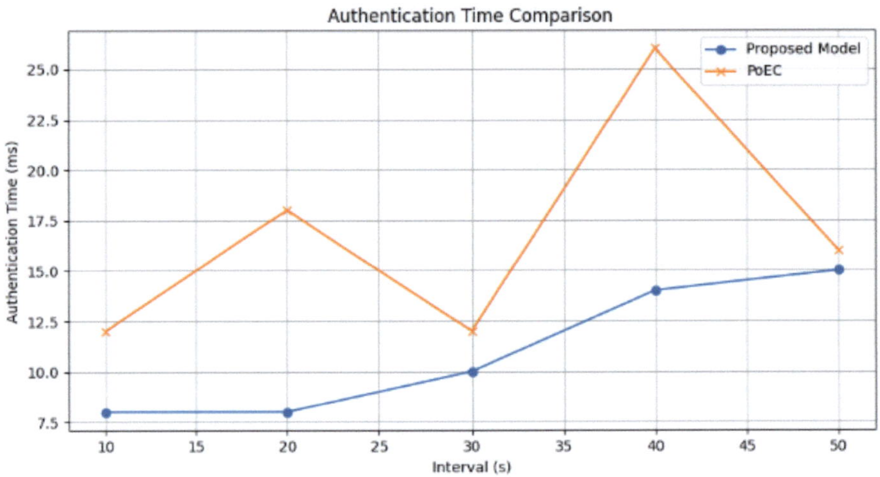


**Fig. 1** Authentication time comparison across different intervals

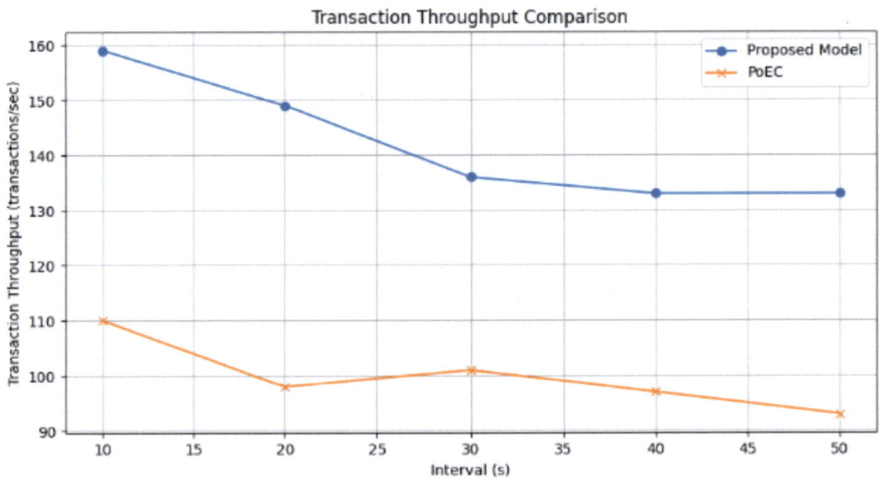


**Fig. 2** Transaction throughput comparison across different intervals

the proposed model achieved 98% success in checks, compared to PoEC's 96%, maintaining a higher rate throughout the intervals.

The latency graph shows, in Fig. 6, the response time from the moment an authentication request is initiated until it is processed. At 10 s, the proposed model responded in 52 ms, while PoEC took 77 ms, demonstrating lower delays at each interval.

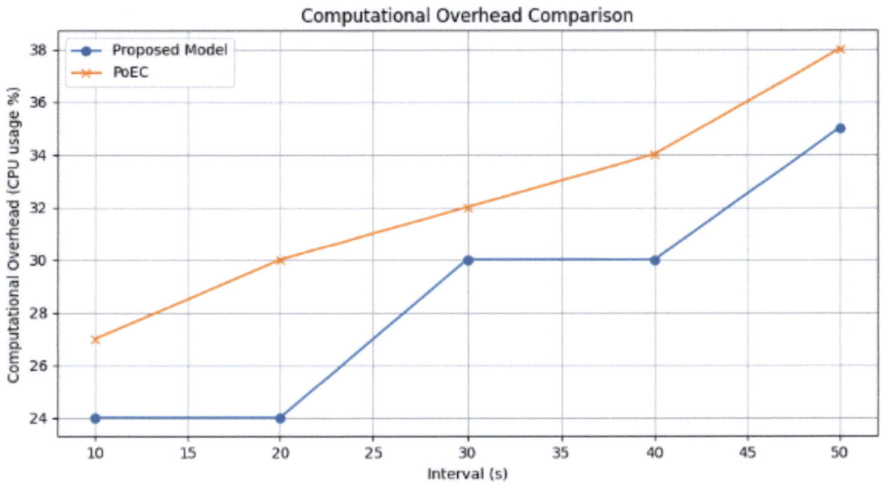


**Fig. 3** Computational overhead comparison across different intervals

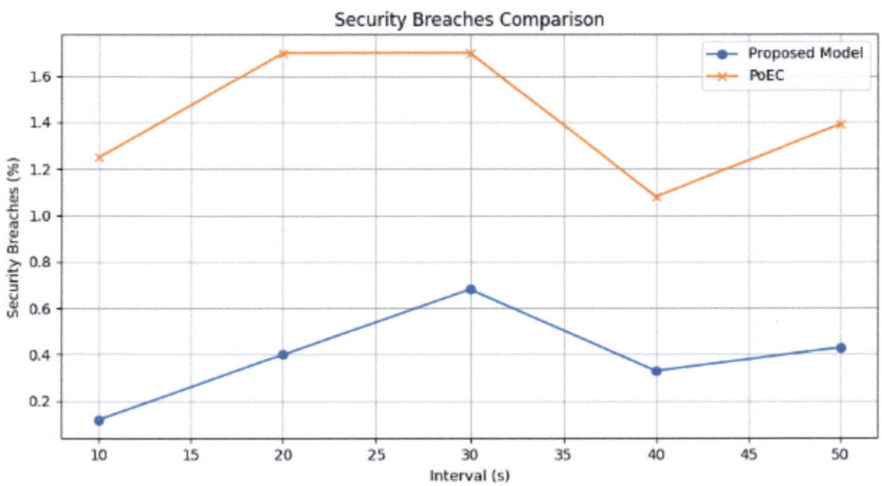


**Fig. 4** Security breaches comparison across different intervals

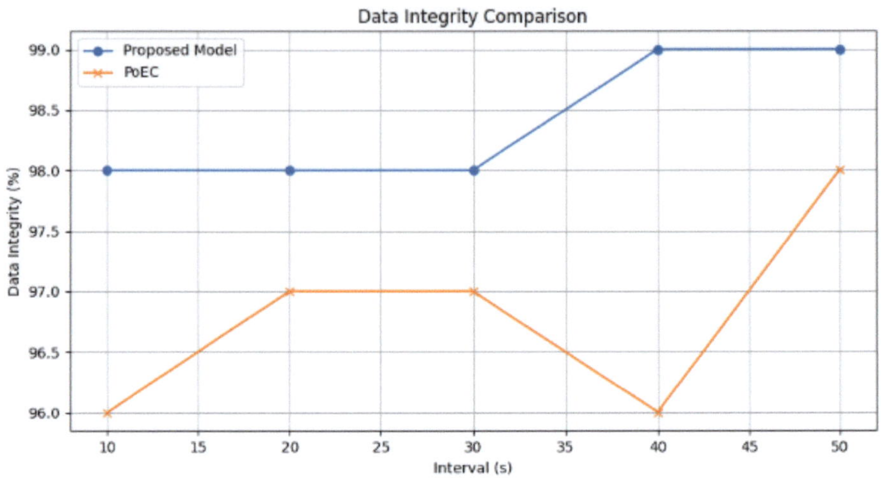


**Fig. 5** Data integrity comparison across different intervals

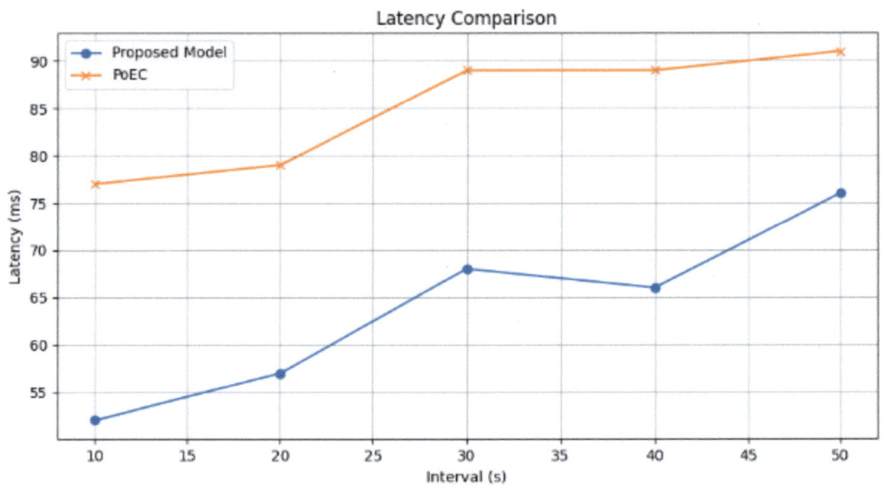


**Fig. 6** Latency comparison across different intervals

## 5 Conclusion

This research introduces a blockchain-based protocol for authentication and access control in fog computing, resolving bottlenecks and security issues in centralized systems. It achieves faster authentication, lower computational load, stable transaction throughput, and consistent data integrity. The use of elliptic curve cryptography and zero-knowledge proofs ensures secure identity verification and efficient operations in resource-limited networks. By eliminating the need for central authorities, the protocol enables scalable management of IoT networks. Future work could explore reducing energy consumption and streamlining smart contract deployment for broader use. Testing in real-world conditions would provide deeper insights into its scalability and reliability. This study offers practical insights into secure access control through decentralized methods, aligning with the demands of modern IoT networks.

## References

1. Ci L, Zhang L, Wu Y, Wiwatanapataphee B (2024) Block-chain-based access control protocol with collaborative authentication
2. Du X, Tao S, Yuan K, Li Y, Zhou Y (2024) A blockchain authentication scheme for UAV-aided fog computing. Complex Intell Syst 10(2):1689–1702
3. Fedorov IR, Bezzateev SV (2023) Blockchain based fog computing network architecture. In: 2023 fifth international conference on blockchain computing and applications (BCCA). IEEE, USA, pp 351–358
4. Kitanov S, Janevski T (2019) Introduction to fog computing. In: The rise of fog computing in the digital era. IGI Global, pp 1–35
5. Bajwa NT, Anjum A, Khan MA (2023) A blockchain-based light-weight secure authentication and trust assessment framework for IoT devices in fog computing. In: 2023 IEEE 20th International conference on smart communities: improving quality of life using AI, robotics and IoT (HONET). IEEE, USA, pp 30–35
6. Xiaoge H, Hongbo Y, Bin C, Yongsheng W, Qianbin C, Jie Z (2024) Joint optimization of energy consumption and network latency in blockchain-enabled fog computing networks. China Commun 21(4):104–119
7. Almaiah MA, Alkdour T (2023) Securing fog computing through consortium blockchain integration: the proof of enhanced concept (PoEC) approach. In: Recent advancements in multimedia data processing and security: issues, challenges, and techniques. IGI Global, pp 107–140
8. Zhang J (2023) Analysis of security access control systems in fog computing environment. J Cyber Secur Mobility 653–674
9. Rudra Kumar M, Gunjan VK (2022) Machine learning based solutions for human resource systems management. In: Kumar A, Mozar S (eds) ICCCE 2021. Lecture notes in electrical engineering, vol 828. Springer, Singapore. https://doi.org/10.1007/978-981-16-7985-8_129
10. Thatikonda R, Nandihal P, Priyanka MS, Mathapati M (2023) Fog computing based privacy preservation via distributed access control system in block-chain technology. In: 2023 International conference on ambient intelligence, knowledge informatics and industrial electronics (AIKIIE). IEEE, USA, pp 1–5
11. Gupta M, Dwivedi RK (2023) Blockchain-based secure and efficient scheme for medical data. EAI Endorsed Trans Scalable Inform Syst 10(5)
12. Madapuri RK, Mahesh PCS (2019) HBS-CRA: scaling impact of change request towards fault proneness: defining a heuristic and biases scale (HBS) of change request artifacts (CRA). Cluster Comput 22(Suppl 5):11591–11599. https://doi.org/10.1007/s10586-017-1424-0
13. Challa R, Kothamasu KK (2023) A decentralized public-permissioned blockchain framework for enhanced security of health records in fog computing. Comput Methods Biomech Biomed Eng 1–13
14. Swetha A, Lakshmi MS, Kumar MR (2022) Chronic kidney disease diagnostic approaches using efficient artificial intelligence methods. Int J Intell Syst Appl Eng 10(1s):254

15. Zhang J, Ouda A, Abu-Rukba R (2024) Authentication and key agreement protocol in hybrid edge–fog–cloud computing enhanced by 5G networks. Future Internet 16(6):209
16. Dwaram JR, Madapuri RK (2022) Crop yield forecasting by long short-term memory network with Adam optimizer and Huber loss function in Andhra Pradesh, India. Concurrency Comput Pract Experience 34(27). https://doi.org/10.1002/cpe.7310

# Blockchain-Integrated Intrusion Detection System for Fog Computing Networks

**R. Sushmitha and N. Srinivasu**

**Abstract** This research explores how blockchain integrates with IDS to address security challenges and scalability issues in fog computing networks. Using metrics such as throughput, latency, response time, detection accuracy, and block creation time, BlockFog is evaluated against traditional IDS models. The system employs a three-layered architecture, placing IDS at the base, blockchain functionality at the fog layer, and secure cloud storage for data management. Simulating normal and malicious traffic reveals that BlockFog handles data with faster responses, better detection accuracy, and higher throughput. Blockchain enhances data integrity by ensuring transparent logs. This study provides practical insights into managing IDS efficiently in fog networks, balancing security with performance, and addressing scalability concerns in IoT ecosystems.

**Keywords** Blockchain · Fog networks · Intrusion detection · IoT security · Federated learning · Smart contracts · DDoS Mitigation · Zero-knowledge proofs

## 1 Introduction

The integration of blockchain with IDS in fog computing networks addresses challenges in security, scalability, and operations. The BlockFog framework applies Device Onboarding, Identity Management, Data Logging, and Smart Contract-Based Intrusion Detection [1]. This setup uses a three-layered structure, with IDS operating at the base, blockchain processes at the fog layer, and cloud services managing storage [2, 3].

Intrusion detection improves with blockchain by enabling secure, tamper-proof logging, and decentralized data management [4]. Models like BiLSTM benefit from

R. Sushmitha (✉) · N. Srinivasu
Department of CSE, KL University, Vijayawada, India
e-mail: sushmacse511@gmail.com

N. Srinivasu
e-mail: srinivasu28@kluniversity.in

 
K. R. Madhavi et al. (eds.), *Proceedings of Sixth International Conference on Computer and Communication Technologies*, Lecture Notes in Networks and Systems 1356,
https://doi.org/10.1007/978-981-96-5238-9_20

feature selection algorithms, such as forward selection with hybrid fitness functions, which enhance anomaly detection accuracy [5]. ICIDS-BB leverages machine learning algorithms to identify attacks and ensure safe data exchange between network components [6]. Traditional IDS models encounter issues with scalability and data integrity, challenges that blockchain addresses by offering immutable ledgers for logging and monitoring. Although research on blockchain-based IDS for fog networks is emerging, early studies indicate potential for improving decentralized security management.

This research examines the performance, scalability, and efficiency of BlockFog within fog computing networks. It compares BlockFog with traditional IDS, focusing on metrics such as throughput, detection accuracy, latency, and response time. The study aims to identify practical strategies for integrating IDS with blockchain while evaluating the framework's response to diverse cyber threats. Contributions include empirical performance data, best practices for implementation, and insights into creating a scalable security framework for IoT ecosystems. The paper presents related research, experimental methods, key findings, and concludes with suggestions for future exploration.

## 2 Related Work

Integrating blockchain with IDS helps address security challenges in fog computing networks. BlockFog, introduced by Prasuna et al. [1], uses blockchain for Device Onboarding, Data Integrity, and Smart Contract-Based Detection. Though security and latency improved, scalability and complexity issues emerged. Almaiah et al. [7] developed a consortium blockchain-based IDS, which secured fog networks effectively but faced performance difficulties in larger setups.

Fedorov et al. [8] focused on using blockchain to protect data transactions but identified challenges in real-world application. A collaborative IDS framework, IDS-Chain, was proposed by Aljuhani et al. [6] to reduce detection time. However, this approach added complexity to managing multiple IDS nodes. Ensemble learning combined with feature selection, introduced by Ramkumar et al. [9], achieved better accuracy but targeted only specific types of intrusions.

Bessaad et al. [10] employed decision trees and random forests for intrusion detection, but coordinating cooperative components was challenging. Syed et al. [11] integrated fog and cloud layers, leveraging their computational power to enhance detection speed, though cloud dependency introduced occasional latency. Ponniah et al. [12] proposed the MWKF-LSTM model for intrusion detection, improving performance but posing difficulties in real-world deployment [13, 14].

To address these limitations, this study proposes a blockchain-based IDS designed for fog networks. The aim is to reduce operational complexity, improve scalability, and secure IoT systems through practical evaluations.

## 3 Methods and Materials

This paper outlines the design of a Blockchain-Integrated Intrusion Detection System (IDS) tailored for fog computing networks. The system leverages blockchain technology to manage security, data integrity, and trust within decentralized environments. It focuses on essential components, including system architecture, data flow processes, logging methods, anomaly detection strategies, authentication, access control mechanisms, response actions, and evaluations of performance, scalability, security, and privacy.

Fog computing moves data processing closer to the source by extending cloud services to the network edge. Integrating IDS with blockchain in these environments ensures tamper-proof logging, secure data exchanges, and better handling of network threats. Blockchain-based IDS frameworks enhance security and data integrity by maintaining decentralized records, enabling dynamic anomaly detection, and enforcing access control policies.

$$\text{Match } (T(t)) = \begin{cases} 1 \text{ if } T(t) \text{ matches a known signature} \\ 0 \text{ otherwise} \end{cases} \tag{1}$$

$$A(T(t)) = \begin{cases} 1 \text{ if } |T(t) - \mu| > \sigma \\ 0 \text{ otherwise} \end{cases} \tag{2}$$

At any given time $t$, $T(t)$ represents the network traffic, while the mean value of normal traffic is denoted by $\mu$, and $\sigma$ signifies the standard deviation.

The blockchain layer acts as a secure mechanism for storing IDS alerts immutably, ensuring the integrity of recorded data. This mechanism is described by Eqs. 3, 4, and 5.

$$H(L(t)) = \text{SHA} - 256(L(t)) \tag{3}$$

$$B_n = \{H(L(t_1)), H(L(t_2)), \ldots, H(L(t_k))\} \tag{4}$$

$$P(B_n) = \text{Stake}(P_i) \tag{5}$$

where at time $t$, $L(t)$ denotes the log entry, and the $n$ th block, referred to as $B_n$, contains these hashed log entries. This setup ensures reliable event tracking and protects network logs across the system.

Each fog node continuously monitors and analyzes data packets in real time. Upon detecting anomalies, alerts are sent to the blockchain for logging, ensuring that logs are immutable and secure.

Machine learning algorithms are used to analyze traffic patterns and detect anomalies. Federated learning enables collaborative anomaly detection across fog nodes without sharing sensitive data. The relevant equations are Eq. 6:

$$f(x) = y \tag{6}$$

where $x$ represents input features (traffic data) and $y$ represents the output (normal or anomaly).

$$D(x) = \begin{cases} 1 \text{ if } \|x - \mu_x\| > \theta \\ 0 \text{ otherwise} \end{cases} \tag{7}$$

$$w_i^{t+1} = w_i^t - \eta g_i^t \tag{8}$$

$$w^{t+1} = \frac{1}{N} \sum_{i=1}^{N} w_i^{t+1} \tag{9}$$

Equations 7, 8, and 9, where $w_i$ represents the model weights at node $i$, $g_i$ the gradients computed at node $i$, $\eta$ the learning rate, and $N$ the number of participating nodes.

Blockchain technology is used for managing device and user identities, ensuring that only authorized entities can access the IDS. Smart contracts enforce access control policies dynamically. The relevant equations are Eqs. 10, 11, and 12:

$$\text{ID}_{\text{device}} = H(\text{device\_info}) \tag{10}$$

$$\text{Access}\ (\text{ID}_{\text{device}}) = \begin{cases} \text{granted if policy\_check}(\text{ID}_{\text{device}}) \\ \text{denied}\ \ \text{otherwise} \end{cases} \tag{11}$$

$$\text{Auth\_success} = \begin{cases} 1 \text{ if } \sum_{i=1}^{n} A_i = n \\ 0 \text{ otherwise} \end{cases} \tag{12}$$

where $A_i$ represents the $i$-th authentication factor and $n$ is the number of required factors.

Smart contracts are developed to automate incident response actions. These contracts contain predefined rules for responding to detected threats. The relevant equations are Eqs. 13, 14, 15, and 16:

$$\text{Trigger(anomaly)} = \begin{cases} 1 \text{ if anomaly detected} \\ 0 \text{ otherwise} \end{cases} \tag{13}$$

$$\text{Response} = \begin{cases} \text{isolate\_node}\ \ \text{if anomaly} = \text{high\_risk} \\ \text{notify\_admin if anomaly} = \text{low\_risk} \end{cases} \tag{14}$$

$$\text{T}_i = \text{Log}\big(\text{anomaly}_i\big) \tag{15}$$

$$\text{Blockchain\_update} = \sum_{i=1}^{N} T_i \tag{16}$$

Resource usage is continuously monitored, and resources are dynamically allocated among fog nodes to ensure optimal performance. The blockchain must handle increasing data and transaction volumes efficiently. The relevant equations are Eqs. 17, 18, 19, 20, and 21:

$$R_i(t) = \text{Usage}\left(\text{node}_i^t\right) \tag{17}$$

$$\text{Allocate}(\text{node}_i) = \begin{cases} \text{increase if } R_i(t) > \text{threshold} \\ \text{decrease if } R_i(t) < \text{threshold} \end{cases} \tag{18}$$

$$S_{\text{block}} = \frac{1}{N} \sum_{i=1}^{N} \text{Log}(T_i) \tag{19}$$

$$\text{Throughput} = \frac{T_{\text{confirmed}}}{T_{\text{time}}} \tag{20}$$

$$\text{Shard}(B) = \{B_1, B_2, \ldots, B_k\} \tag{21}$$

where $S_{\text{block}}$ is the optimal block size, $T_{\text{confirmed}}$ is the number of confirmed transactions, $T_{\text{time}}$ is the time taken, and $B$ represents the blockchain, with $B_i$ being the shards.

Sensitive data transmitted between fog nodes and the blockchain is encrypted to ensure confidentiality and integrity. Zero-knowledge proofs (ZKPs) are used to verify data integrity without revealing the actual data. The relevant equations are Eqs. 22 and 23:

$$C = E_{\text{key}}(D) \tag{22}$$

$$D = D_{\text{key}}(C) \tag{23}$$

where $C$ is the ciphertext, $E_{\text{key}}$ is the encryption function, and $D$ is the data.

$$P = \text{ZKP\_gen}(x) \tag{24}$$

$$V = \text{ZKP\_verify}(P, x) \tag{25}$$

Equations 24 and 25, where $P$ is the proof and $x$ is the data.

# 4 Experimental Study

The study checks how well a blockchain-based IDS works in fog computing. The main goal is to see if it can improve security, keep data safe, and build trust in a fog setup. It looks at how this system stacks up against regular IDS in spotting threats, how the blockchain affects speed, and how it handles attacks like DDoS and unauthorized access. The test uses Raspberry Pi devices with basic setups and IoT sensors to create a small network. For software, it combines blockchain (Hyperledger Fabric), IDS, and traffic simulation tools like Scapy and Metasploit.

Fog nodes link up, and cloud services add extra power and storage. Blockchain nodes log alerts, keeping the whole system talking smoothly. For 30 days, traffic runs through the network, mixing normal use with attacks like DDoS. Key numbers—like how fast data moves, how quickly it reacts, and how accurately it spots problems—are tracked. Data analysis uses basic tools, and the results are shown in graphs and tables.

## *4.1 Results and Discussion*

The tests reveal that the blockchain-IDS does better than IDS-Chain [6] in several ways. It handles more data, responds faster, and spots issues more accurately. Each round of tests shows quicker action and fewer delays. The time to create and run blocks is shorter, making the system work smoothly even under attack. The findings show that adding blockchain to IDS in fog networks can improve security, making it more reliable for handling threats in IoT setups.

**Throughput (Packets per Second)**: Table 1 proposed model consistently achieves higher throughput compared to IDS-Chain, indicating better performance in handling network traffic.

**Latency (Milliseconds)**: Table 2 proposed model demonstrates lower latency values, highlighting its efficiency in processing and transmitting data.

**Response Time (Milliseconds)**: Table 3 proposed model consistently records faster response times, indicating a more prompt reaction to detected intrusions.

**Table 1** Throughput (packets per second)

| Simulation interval | Proposed model | IDS-chain |
|---|---|---|
| Interval 1 | 1200 | 1185 |
| Interval 2 | 1235 | 1210 |
| Interval 3 | 1220 | 1205 |
| Interval 4 | 1245 | 1225 |
| Interval 5 | 1260 | 1240 |

**Table 2** Latency (milliseconds)

| Simulation interval | Proposed model | IDS-chain |
|---|---|---|
| Interval 1 | 30 | 32 |
| Interval 2 | 28 | 31 |
| Interval 3 | 27 | 30 |
| Interval 4 | 26 | 29 |
| Interval 5 | 25 | 28 |

**Table 3** Response time (milliseconds)

| Simulation interval | Proposed model | IDS-chain |
|---|---|---|
| Interval 1 | 45 | 48 |
| Interval 2 | 44 | 47 |
| Interval 3 | 43 | 46 |
| Interval 4 | 41 | 45 |
| Interval 5 | 40 | 44 |

**Detection Accuracy (Percentage)**: Table 4 proposed model shows slightly higher detection accuracy, reflecting its effectiveness in identifying anomalies and intrusions.

**Block Creation Time (Milliseconds)**: Table 5 proposed model exhibits marginally shorter block creation times, showcasing its efficiency in recording and processing alerts.

**Table 4** Detection accuracy (percentage)

| Simulation interval | Proposed model | IDS-chain |
|---|---|---|
| Interval 1 | 98.5 | 97.9 |
| Interval 2 | 98.6 | 98 |
| Interval 3 | 98.4 | 97.8 |
| Interval 4 | 98.8 | 98.3 |
| Interval 5 | 98.9 | 98.5 |

**Table 5** Block creation time (milliseconds)

| Simulation interval | Proposed model | IDS-chain |
|---|---|---|
| Interval 1 | 5.2 | 5.4 |
| Interval 2 | 5.1 | 5.3 |
| Interval 3 | 4.9 | 5.2 |
| Interval 4 | 4.8 | 5 |
| Interval 5 | 4.7 | 4.9 |

**Table 6** Block execution time (milliseconds)

| Simulation interval | Proposed model | IDS-chain |
|---|---|---|
| Interval 1 | 8.3 | 8.5 |
| Interval 2 | 8.1 | 8.4 |
| Interval 3 | 8 | 8.2 |
| Interval 4 | 7.9 | 8.1 |
| Interval 5 | 7.7 | 8 |

**Block Execution Time (Milliseconds)**: Table 6 proposed model consistently demonstrates shorter block execution times, highlighting its enhanced efficiency in processing blockchain transactions.

This graph shown in Fig. 1 shows the throughput comparison between the proposed blockchain-integrated IDS and IDS-Chain across five simulation intervals. The proposed model consistently achieves higher throughput, indicating its superior capacity to handle network traffic.

This graph shown in Fig. 2 illustrates the latency comparison between the proposed model and IDS-Chain over five simulation intervals. The proposed model demonstrates lower latency consistently, highlighting its efficiency in processing network traffic.

This graph shown in Fig. 3 compares the response times between the proposed model and IDS-Chain across five simulation intervals. The proposed model consistently shows faster response times, enhancing the system's ability to quickly detect and respond to intrusions.

This graph shown in Fig. 4 shows the detection accuracy comparison between the proposed model and IDS-Chain across five simulation intervals. The proposed model

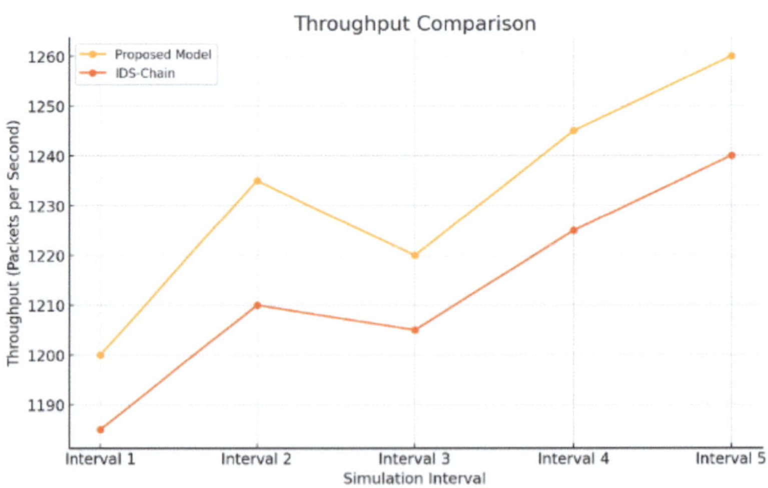


**Fig. 1** Throughput comparison (packets per second)

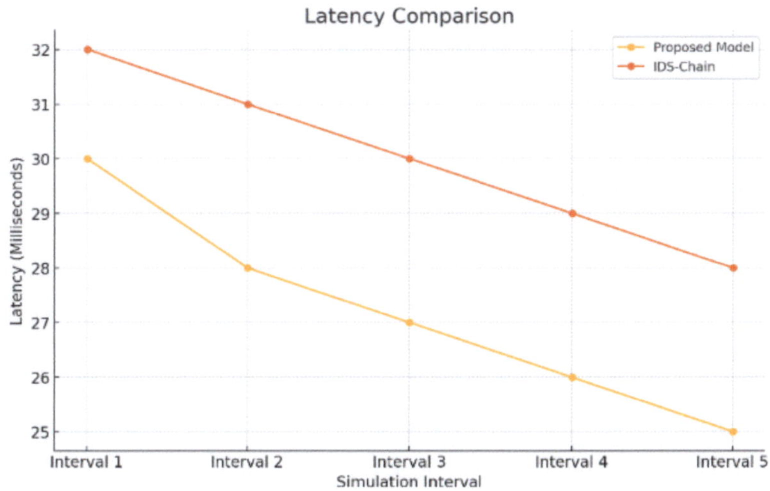


**Fig. 2** Latency comparison (milliseconds)

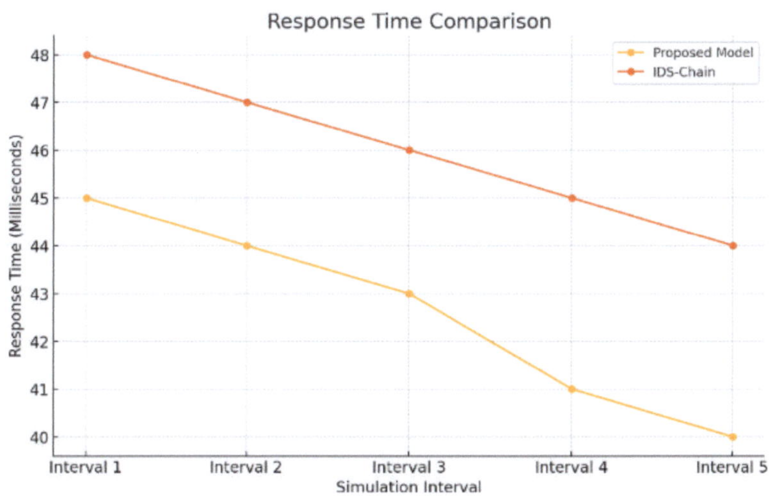


**Fig. 3** Response time comparison (milliseconds)

exhibits slightly higher detection accuracy, indicating a more effective anomaly detection mechanism.

This graph shown in Fig. 5 illustrates the block creation time comparison between the proposed model and IDS-Chain across five simulation intervals. The proposed model shows a slight advantage with lower block creation times, demonstrating the efficiency of its blockchain layer.

This graph shown in Fig. 6 compares the block execution times between the proposed model and IDS-Chain across five simulation intervals. The proposed

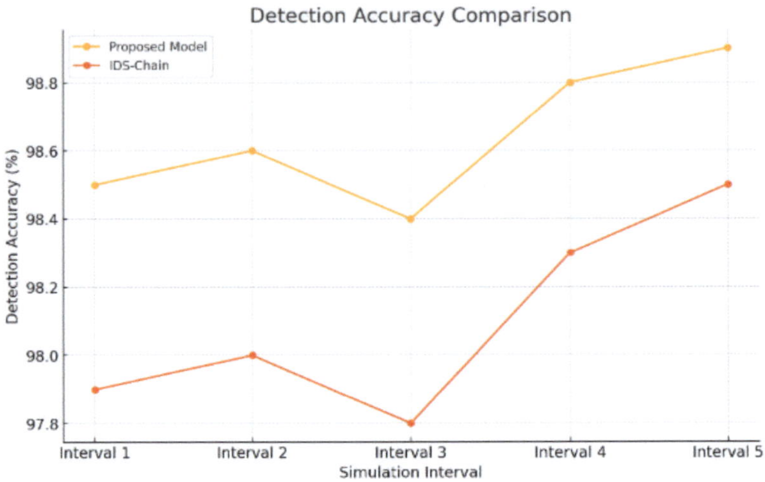


**Fig. 4** Detection accuracy comparison (percentage)

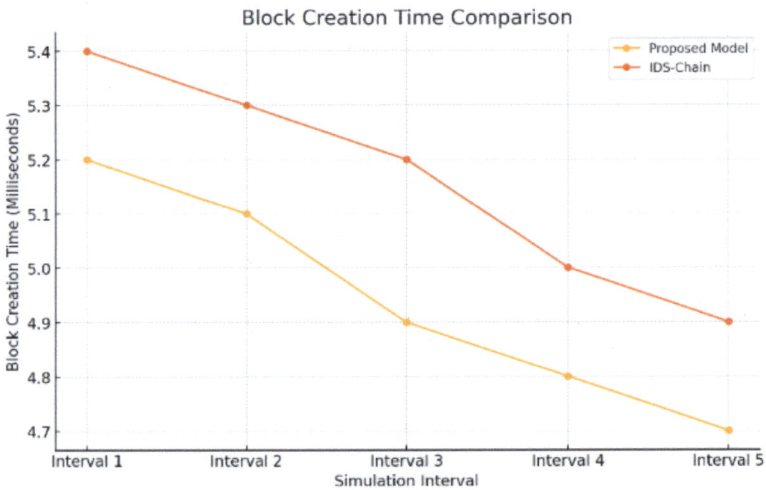


**Fig. 5** Block creation time comparison (milliseconds)

model consistently exhibits marginally faster block execution times, highlighting the efficiency of the blockchain operations.

The experimental results validate the effectiveness of the blockchain-integrated IDS in enhancing the security and performance of fog computing networks. The proposed model consistently outperforms traditional IDS frameworks across all key performance metrics, demonstrating its potential as a robust, efficient, and scalable solution for securing decentralized IoT ecosystems.

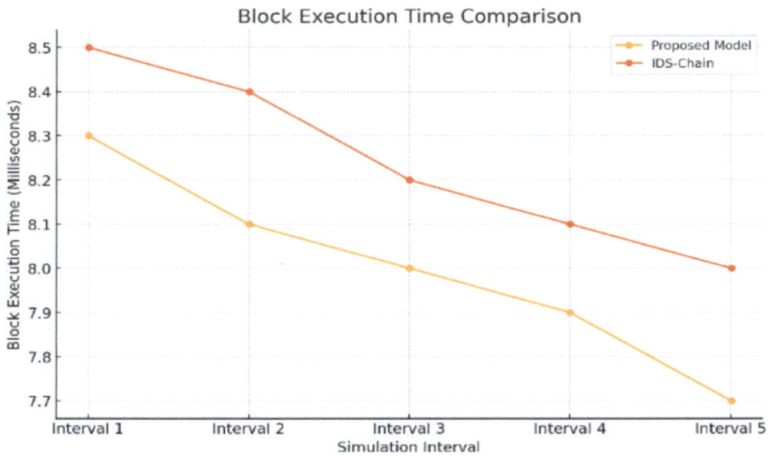


**Fig. 6** Block execution time comparison (milliseconds)

## 5 Conclusion

The study shows that integrating blockchain with IDS improves the security, scalability, and efficiency of fog computing networks, consistently outperforming traditional IDS in throughput, latency, response times, and detection accuracy. This integration provides a robust security framework for decentralized IoT ecosystems. However, the research notes limitations in the range of tested intrusion scenarios and specific configurations used. Future work should explore more attack vectors, varying network conditions, and the integration of AI and advanced cryptographic methods to further enhance IDS performance. The findings pave the way for future innovations in network security.

## References


1. Prasuna VG, Ravindra Babu B, Pydala B (2024) BlockFog: a blockchain-based framework for intrusion defense in IoT fog computing. Scalable Comput Pract Experience 25(3):1950–1962
2. Zwayed FA, Anbar M, Manickam S, Sanjalawe Y, Alrababah H, Hasbullah IH, Almi'ani N (2024) An efficient intrusion detection systems in fog computing using forward selection and BiLSTM. Bull Electr Eng Inform 13(4):2586–2603
3. Madapuri RK, Mahesh PCS (2019) HBS-CRA: scaling impact of change request to-wards fault proneness: defining a heuristic and biases scale (HBS) of change re-quest artifacts (CRA). Cluster Comput 22(Suppl 5):11591–11599. https://doi.org/10.1007/s10586-017-1424-0
4. Rudra Kumar M, Gunjan VK (2022) Machine learning based solutions for human resource systems management. In: Kumar A, Mozar S (eds) ICCCE 2021. Lecture notes in electrical engineering, vol 828. 2022. Springer, Singapore. https://doi.org/10.1007/978-981-16-7985-8_129

5. Ajao LA, Apeh ST (2023) Blockchain integration with machine learning for securing fog computing vulnerability in smart city sustainability. In: 2023 1st International conference on advanced innovations in smart cities (ICAISC). IEEE, USA, pp 1–6
6. Aljuhani A (2022) IDS-Chain: a collaborative intrusion detection framework empowered blockchain for internet of medical things. In: 2022 IEEE cloud summit. IEEE, USA, pp 57–62
7. Almaiah MA, Alkdour T (2023) Securing fog computing through consortium blockchain integration: the proof of enhanced concept (PoEC) approach. In: Recent advancements in multimedia data processing and security: issues, challenges, and techniques. IGI Global, pp 107–140
8. Fedorov IR, Bezzateev SV (2023) Blockchain based fog computing network architecture. In: 2023 Fifth international conference on blockchain computing and applications (BCCA). IEEE, USA, pp 351–358
9. Ramkumar MP, Daniya T, Mano Paul P, Rajakumar S (2022) Intrusion detection using optimized ensemble classification in fog computing paradigm. Knowl Based Syst 252:109364
10. Bessaad F, Ktata FB, Kalboussi KB (2023) Intelligent intrusion detection system based on blockchain technology. In: 2023 IEEE Afro-Mediterranean conference on artificial intelligence (AMCAI). IEEE, USA, pp 1–7
11. Syed NF, Ge M, Baig Z (2023) Fog-cloud based intrusion detection system using recurrent neural networks and feature selection for IoT networks. Comput Netw 225:109662
12. Ponniah KK, Retnaswamy B (2023) A novel deep learning based intrusion detection system for the IoT-Cloud platform with blockchain and data encryption mechanisms. J Intell Fuzzy Syst Prepr 1–18
13. Swetha A, Lakshmi MS, Kumar MR (2022) Chronic kidney disease diagnostic approaches using efficient artificial intelligence methods. Int J Intell Syst Appl Eng 10(1s):254
14. Rudra Kumar M, Gunjan VK (2022) Peer level credit rating: an extended plugin for credit scoring framework. In: Kumar A, Mozar S (eds) ICCCE 2021. Lecture notes in electrical engineering, vol 828. Springer, Singapore. https://doi.org/10.1007/978-981-16-7985-8_128

# An Optimized Ensemble Model for Skin Cancer Classification Using Genetic Algorithms and Transfer Learning

**Giddaluru Lalitha and Y. MD. Riyazuddin**

**Abstract** Skin cancer remains one of the most common and deadly forms of cancer globally, requiring precise and effective diagnostic tools. This study introduces a novel approach for improving skin cancer detection and classification by incorporating Genetic Algorithm Optimized Ensemble (GAOE) methods with transfer learning and meta-learning techniques. We refined three well-known pre-trained models—EfficientNetB3, ResNet50, and VGG16 using the ISIC 2020 dataset. We used transfer learning, customizing them for skin cancer detection employing a weighted ensemble approach, where initial performance-based weights were calculated and optimized via evolutionary algorithms. The GAOE model, which integrates genetic algorithms and meta-learning, performed exceptional performance accuracy of 99.43%. The GAOE model also surpassed the weighted ensemble in precision 98%, recall 99%, and *F*-score 99%. Combining ensemble approaches, optimization algorithms, and transfer learning, our methodology efficiently addresses typical problems, including underfitting, overfitting, and model bias. This study provides a strong tool for skin cancer diagnosis and a precedent for future research in medical imaging, highlighting the possibility of incorporating evolutionary algorithms and meta-learning to enhance predictive performance.

**Keywords** Skin cancer detection · Genetic Algorithm Optimized Ensemble (GAOE) · Transfer learning · Ensemble learning · ISIC 2020 dataset · Medical Image Classification · Deep learning

G. Lalitha (✉) · Y. MD. Riyazuddin
Department of CSE, School of Technology, GITAM (Deemed to be University), Hyderabad, Telangana, India
e-mail: lgiddalu@gitam.in

Y. MD. Riyazuddin
e-mail: rymd@gitam.edu



K. R. Madhavi et al. (eds.), *Proceedings of Sixth International Conference on Computer and Communication Technologies*, Lecture Notes in Networks and Systems 1356,
https://doi.org/10.1007/978-981-96-5238-9_21

## 1 Introduction

Skin cancer (SC) is the most lethal type of cancer and appears from the abnormal proliferation of cells within the epidermis. Only cardiovascular diseases are greater than their gradually rising global incidence, which presents serious health hazards and ranks as the second most common cause of death [1]. Based on information from the SC foundation, SC constitutes one in three cancer diagnoses, and one in five Americans is expected to develop it, with approximately 3.5 million new cases each year [2]. Cancerous growths can spread to other body parts because benign growths are also possible [3]. Multiple factors, such as alcohol consumption, UV radiation exposure, allergies, viruses, and bacteria influence skin cancer [4, 5]. Early detection is essential for effective therapy, but manual diagnosis presents significant difficulties for patients and medical professionals [6]. Computer-aided diagnostic (CAD) systems offer a potentially effective way to address these issues [7].

## 2 Relevant Review

Computer-aided diagnostic (CAD) systems, which use deep learning (DL) and machine learning (ML) methodologies, can effectively address automated SC classification [8, 9]. Support vector machines, neural networks, random forests, and particle swarm optimization are some techniques that have improved the generalizability of classification processes [10, 11]. In comparison with DL techniques, boosting algorithms like XGBoost and CatBoost still require improvement, despite their increased accuracy [12]. A DL-based method for autonomous SC classification has shown promising results in recent experiments using the HAM1000 and ISIC datasets. Utilizing data sampling strategies to reduce overfitting, this study demonstrated significant improvements in accuracy over conventional ML techniques. It showed the effectiveness and precision of DL in diagnosing SC [13].

In response to the identified issues, we have highlighted our key findings and contributions as follows:

- Implemented procedures to reduce model bias, improve dataset diversity, and prevent overfitting and underfitting, resulting in more efficient and accurate outcomes.
- Applied and fused transfer learning models to enhance performance, analyzing the differences between transfer learning and ensemble approaches in medical diagnosis.
- Developed a novel GAOE model, optimizing performance-based weights through a genetic algorithm for upgraded accuracy and resilience.

Over the years, several academics have conducted numerous studies to improve the efficacy in the field of classifying SC. Here are a few noteworthy and current studies on this area of SC research.

Cassidy et al. [23] proposed a novel DL model for examining six well-known skin conditions: acne, melanoma, basal cell carcinoma, actinic keratosis, and pest bites. Using standard datasets examining several methods, convolutional neural networks (CNNs) performed best, with 97% accuracy, precision, recall, and an $F$-score of 91%. Riaz et al. [14] also presented a joint learning system that tested on a SC dataset that combined CNN and Local Binary Patterns (LBPs) with connected features from both architectures. According to the results, the fused architecture exhibited resilience with 98.60% accuracy and 97.32% validation accuracy.

Lakshmi and Jasmine [15] implemented an effective Exponentially Weighted and Heaped Multi-Layer Perceptron (EWHMLP) for classification and several representation strategies for feature extraction in developing a Hybrid AI Model (HAIM) for the classification of SC. Although weight learning was difficult and optimization was frequently needed to converge, MLP was suitable for classification, which led to instability. EWHMLP was created to address this, and the results demonstrated that the $PH^2$ database's multi-class classification accuracy was 98.33%. In another study, Ogundokun et al. [16] employed two major heuristics, the Xception and MobileNetV2 models, to create a hybrid-deep CNN architecture for the identification of SC. In conjunction with MobileNetV2, Xception demonstrated the highest performance, obtaining on the evaluated dataset a 96.55% $F_1$ score, 100% sensitivity, 93.33% precision, 97.56% accuracy, 97.00% area under the curve, and 0.0370 false positive rates.

The following briefly outlines the limitations associated with SC datasets and algorithms:

1. There are problems with data scarcity, imbalance, and limited variety because most people in current datasets are fair-skinned and have minor lesions [17–19].
2. To increase the robustness of the model, further dataset augmentation and investigation of enhanced imaging data are needed [20].
3. High-dimensional features and slow convergence decrease training efficiency, while universal algorithm optimization and computational scalability remain challenging [21, 22].

The present study is divided into multiple significant sections: Sect. 3 presents the materials and methodology; Sect. 4 presents a visual representation of the findings; and Sect. 5 summarizes the study's conclusions and final remarks.

## 3 Materials and Method

### Dataset Description and Preprocessing

The ISIC 2020 dataset [23], a vital resource for lowering melanoma mortality by digital skin imaging, is provided by the International Society for Digital Imaging of the Skin (ISIC). This dataset comprises eleven thousand four hundred images total from training, validation, and testing sets. Four thousand images of malignant

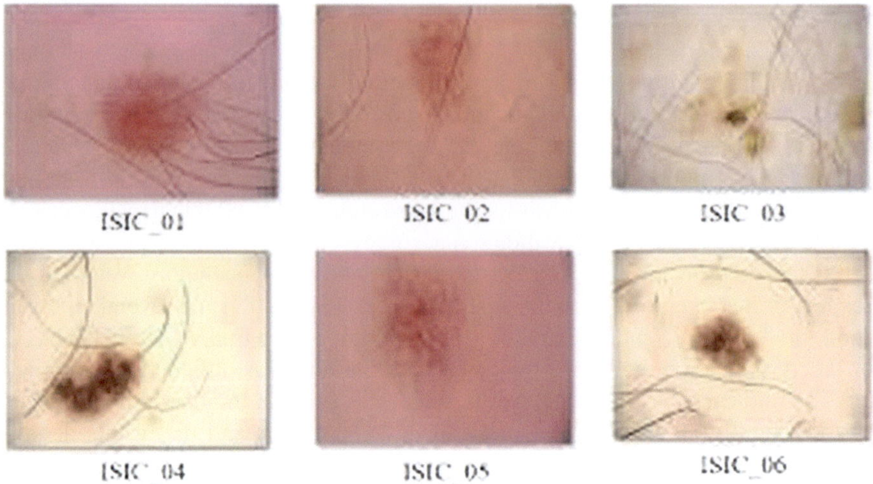


**Fig. 1** Sample images of ISIC 2020

tumors and five thousand images of benign tumors make up the training set. There are 550 images of benign and malignant tumors in the validation and testing sets. By enabling advancements in the classification of skin lesions, this extensive dataset plays a crucial role in bolstering studies aimed at early SC detection and therapy. Figure 1 displays some representative images from the ISIC dataset.

Various data augmentation techniques were employed to improve the training dataset and enhance the model's performance. These augmentations include horizontal and vertical flipping, rotating images, and shifting them in different directions. Such transformations help the model become invariant to changes in the orientation and placement of objects within the images. Zooming techniques adjusted the image scale by including more background or focusing on specific features, thereby assisting the model in handling variations in object size and distance. These augmentations are essential for improving the model's generalization abilities by mimicking actual visual circumstances. Table 1 lists various augmentation methods in summarized form.

## Proposed Model

Using a tagged dataset of skin lesion photos, three pre-trained models—EfficientNetB3, ResNet50, and VGG16 are first fine-tuned to create the SC classification model. By retraining on this dataset, every model is tailored to the particular goal of detecting SC. Once adjusted, each model produces probability vectors for class labels. To aggregate the outcomes of each model on a validation set, the initial performance weights are computed depending on that accuracy. Next, the ideal set of weights that maximizes the ensemble's performance is identified by using evolutionary algorithms to optimize these weights. A composite prediction is created by averaging the probability vectors generated by the basic learners.

**Table 1** Techniques for image augmentation and value ranges

| Augmentation | Typical value range |
|---|---|
| Horizontal flip | 0.5 |
| Vertical flip | 0.5 |
| Rotation | − 30° to + 30° |
| Height shift | − 0.2 to + 0.2 |
| Width shift | − 0.2 to + 0.2 |
| Zoom | 0.8 to 1.2 |
| Shear | − 20° to + 20° |
| Brightness | 0.8 to 1.2 |
| Contrast | 0.8 to 1.2 |
| Saturation | 0.8 to 1.2 |
| Hue | − 0.1 to + 0.1 |
| Gaussian noise | 0 to 0.1 |
| Cutout | 10 to 50 pixels |

Furthermore, a meta-learner is taught to improve the prediction process even more. The meta-learner uses linear regression to identify the best coefficients for combining the outputs of the base learners to minimize prediction error. Through performance-based weighting, meta-learning, and genetic algorithm optimization, this two-tiered method seeks to increase the accuracy and dependability of skin cancer categorization.

The proposed Genetic Algorithm Optimized Ensemble (GAOE) integrates meta-learning, transfer learning, and genetic algorithm optimization to enhance SC identification and categorization. Using genetic algorithms and a meta-learner, the model enhances the contributions of EfficientNetB3, ResNet50, and VGG16 to develop a highly accurate and reliable SC diagnostic tool.

To enhance the model's performance, optimization techniques were first applied to the study's DL-based weighted ensemble model. The suggested GAOE model uses ensemble learning also pre-trained models (VGG16, ResNet50, and EfficientNetB3) are combined with transfer learning. The ensemble model uses meta-learners and evolutionary algorithms to optimize performance-based weighting to attain ideal classification performance.

Assign $M_1$, $M_2$, and $M_3$ to the base learners EfficientNetB3, ResNet50, and VGG16, in that order. Their corresponding accuracy on a validation set is shown by the symbols $A_1$, $A_2$, and $A_3$. Initially, each base learner $M_i$ weight $w_i$ is determined by their performance by the below equation.

$$W = \frac{A_i}{\sum_{j=1}^{3} A_j} \tag{1}$$

Genetic algorithms are used to further optimize these weights and identify the ideal set of weights that maximize the ensemble's overall performance.

**Optimization of Genetic Algorithms**

Natural selection inspires a class of optimization techniques known as genetic algorithms (GAs). These algorithms work via crossover, mutation, and selection mechanisms to evolve a population of candidate solutions toward superior solutions over several generations.

**Initialization**: It begins with a population of candidate weight vectors, $w = (w_1, w_2, w_3)$ where $w_1 + w_2 + w_3 = 1$. Then define a fitness function $F(w)$ that, given the weights $w$, assesses the ensemble's performance. An instance of a fitness function could be the ensemble's accuracy on a validation set. The selection stage of the genetic algorithm optimization process is the next step. Here, pairs of potential solutions are chosen from the population based on how suited they are to produce offspring. These possible solutions are then joined during the crossing step by trading weight vector segments with one another to create new offspring. The next stage is the mutation step when some candidate solutions are randomly altered to preserve genetic variety and investigate uncharted territory in the solution space. To ensure the evolution of progressively optimal solutions, these steps—selection, crossover, and mutation—are repeated iteratively for a predetermined number of generations or until the convergence conditions are satisfied.

**Meta-Learning Techniques**: A meta-learner is trained in addition to genetic algorithms to optimize base learner weights to improve ensemble predictions. The meta-learner is trained on a different validation set to anticipate the proper labels using the base learners' outputs as input features. For each validation sample, let $y$ be the vector of proper labels and $z$ be the matrix of concatenated probability vectors from the base learners. As a linear regression model, the meta-learner seeks to identify the coefficients $\beta = (\beta_1, \beta2, \beta3)$ that minimize the mean squared error:

$$\sum_{k=1}^{m}\left(y_k - \sum_{i=1}^{3}\beta_i Z_{ik}\right)^2 \tag{2}$$

where $m$ is the number of validation samples and $z_{ik}$ is the $i$-th base learner's prediction for the $k$-th validation sample.

The linear regression meta-learner is an essential component in the GAOE research context as it helps optimize the weighted combination of base learners' predictions. First, weights are calculated based on the performance of the base learners. This is the first step in the process. A genetic algorithm is then used to adjust these weights to maximize the overall performance of the ensemble. The final prediction is then improved by the linear regression meta-learner using the optimized weights to learn the best possible combination of outputs from the base learners. The GAOE model maintains interpretability and efficiency while guaranteeing that the aggregate predictions of the base learners are as accurate and reliable as possible by utilizing linear regression as the meta-learner. The presented model architecture is demonstrated in Fig. 2. The hyperparameters of the proposed model are illustrated in Table 2.

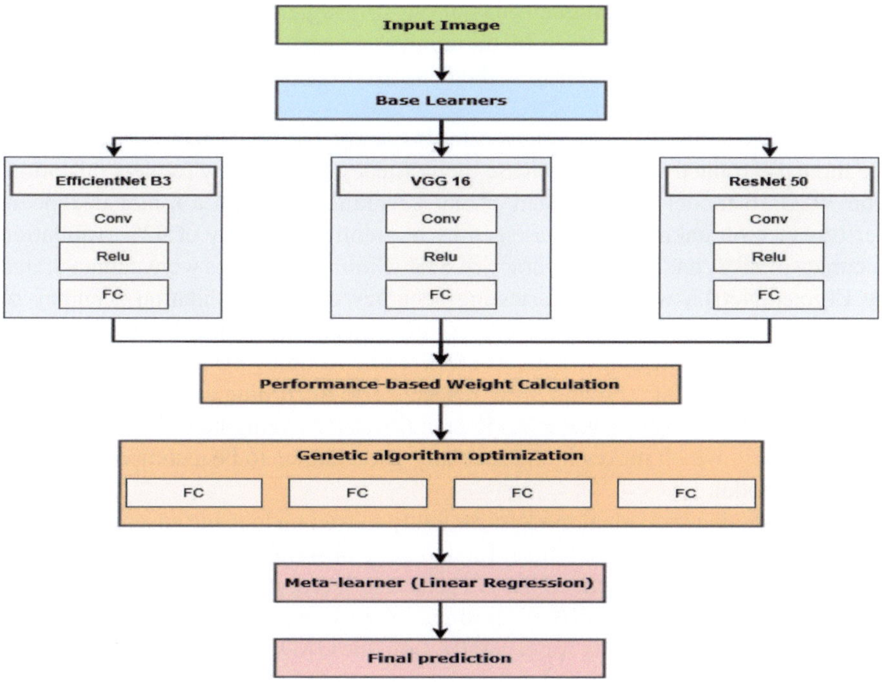


**Fig. 2** Proposed model architecture

**Table 2** Hyperparameters and fine-tuning values of the proposed model

| Model | Parameters |
|---|---|
| Genetic algorithms | Population size: 50<br>Generations: 100<br>Crossover rate:0.1<br>Selection method: Tournament Selection |
| Meta-learner | Learning rate: 0.01<br>Regulation: 0.1<br>Training epoch:30 |
| Base learners | Optimizer: Adam<br>Loss function: Categorical cross-entropy<br>Initial learning rate: 0.001 |

## 4 Result and Discussion

Two ensemble models, the weighted ensemble model and a model tuned with a genetic algorithm were used in this study to improve performance. The employment of pre-trained transfer learning models forms the basis of this study. These models were tailored to meet the unique needs of our investigation after being first trained on huge datasets. The accuracy of the tailored transfer learning models' testing, validation, and training phases was then assessed. Table 3 displays the results of these assessments.

On the particular dataset employed in this study, the performance measures for the three pre-trained models showcase their efficacy and capacity for generalization. The VGG-16 model demonstrated strong generalization with a minor decline in performance on unknown data, as seen by its training accuracy of 92%, validation accuracy of 89%, and testing accuracy of 87%. Similar outcomes were demonstrated by EfficientNetB3, which had a testing accuracy of 86%, validation accuracy of 89%, and training accuracy of 93%. ResNet50 fared better than the other models, maintaining the best testing accuracy at 89% while attaining the highest training and validation accuracy at 93% and 90%, respectively. According to these findings, all three models work well, but ResNet50 stands out for having better robustness and generalization, which makes it an especially good choice to be a component of the ensemble model.

After assessing individual models, the study moved on to using ensemble models to improve performance as a whole. First, a general weighted ensemble model was used, in which weights derived from each base learner's performance indicators were used to assess how much of an impact each learner had on the outcome. To enhance the model's capacity for generalization, the GAOE was implemented. Using genetic algorithms, this method maximizes the contribution of the base learners, hence addressing the flaws in their generalization. Table 4 describes the enhanced generalization ability of the ensemble models over the individual base learners. To overcome these constraints, ensemble models were introduced, which make use of the advantages that each base learner possesses. There was a noticeable improvement in the weighted ensemble model, which combines base learner predictions with weights provided. It obtained 98.3% training accuracy, 96.5% validation accuracy, and 97.1% testing accuracy. The observed improvement in the performance gap between the training and testing phases indicates that the weighted ensemble model outperformed the individual base models in improving generalization.

**Table 3** Performance of the base models

| Model | Training (%) | Validation (%) | Testing (%) |
|---|---|---|---|
| VGG-16 | 92 | 89 | 87 |
| EfficientNetB3 | 93 | 89 | 86 |
| ResNet50 | 93 | 90 | 89 |

**Table 4** Training, testing, and validation accuracy of the ensemble models

| Model | Training (%) | Validation (%) | Testing (%) |
|---|---|---|---|
| Weighted ensemble | 98.3 | 96.5 | 97.1 |
| GAOE | 99.43 | 98.65 | 98.50 |

This method was further improved by the GAOE, which used genetic algorithms to maximize the contribution of base learners and improve generalization by optimizing their combination. With a training accuracy of 99.43%, validation accuracy of 98.65%, and testing accuracy of 98.50%, the GAOE model earned the greatest performance metrics. Overall, the ensemble models especially the GAOE showed strong performance throughout all evaluation stages and demonstrated a noteworthy improvement in generalization and overall accuracy, effectively addressing the shortcomings found in the individual base models.

We evaluated the ensemble models using accuracy measures first, and then the model was assessed using other evaluation metrics, precision, recall, and *F*-score (Table 5). These metrics offer a more comprehensive picture of the model's efficacy, particularly their precision in identifying and categorizing cases across several classes. Achieved precision of 97%, recall of 96%, and *F*-score of 97% was the weighted ensemble model. Precision indicates a high degree of accuracy in the positive classifications by measuring the percentage of true positive predictions among all positive predictions made by the model. Conversely, Recall shows how well the weighted ensemble model identified pertinent instances, demonstrating that it collected many true positives successfully. The *F*-score supports the model's balanced performance, which is the harmonic mean of precision and recall. This indicates that the model maintained a high overall efficacy in classification tasks. All measures showed that GAOE performed better than the weighted ensemble: *F*-score was 99%, recall was 99%, and precision was 98%. The higher precision indicates fewer false positives and even more accurate positive predictions generated by the GAOE model. With fewer false negatives, the remarkable recall successfully detected nearly all true positive cases. The higher *F*-score indicates a more balanced and effective model, showcasing its sophisticated capacity to maximize classification accuracy and minimize errors. These findings indicate the improved generalization and accuracy of the GAOE across various evaluation criteria, offering a strong remedy for the issues found in the basic models.

**Table 5** Precision, recall, and *F*-score of the ensemble models

| Model | Precision (%) | Recall (%) | *F*-score (%) |
|---|---|---|---|
| Weighted ensemble | 97 | 96 | 97 |
| GAOE | 98 | 99 | 99 |

**Table 6** Compare the proposed model with existing work

| Reference | Dataset | Model | Performance |
|---|---|---|---|
| DCN | ISIC | InSiNet | 97.1% |
| Neural network | ISIC | Transfer learning | AUC-96.81 |
| Proposed model | ISIC | GAOE | 99.43 |

**Comparative Analysis**

We compared our findings with existing work that employs the same ISIC datasets displayed in Table 6. Reis et al. [24] used a CNN-based DL technique called InSiNet to work with different datasets, including ISIC. They acquired a 95.59% accuracy rate on the ISIC 2018 dataset. However, our presented model outperformed, reaching 99.43% accuracy with our GAOE approach. Another study [25] employed the ISIC dataset to categorize skin cancer using a transfer learning model called EfficientNet. They assessed their model utilizing the AUC value, which averaged 96.81%. The AUC value also considered our proposed model, producing a 100% AUC for every class. Hence, our proposed model exceeded their model. The GAOE model performed an impressive 99.43% accuracy, showcasing its outstanding classification power and feature extraction capabilities. Its excellent performance likely stems from effectively addressing the heterogeneous dataset, outperforming models like InSiNet and transfer learning. Skin cancer detection with great efficiency, accuracy, and adaptability is provided by the GAOE model's combination of feature extraction and classification algorithms.

## 5 Conclusion

Skin cancer poses a significant concern for public health and necessitates reliable and precise diagnostic tools for early detection and effective treatment. Deep learning, particularly through ensemble techniques and transfer learning, shows promise in improving the accuracy and robustness of skin cancer classification models. Our study utilized the ISIC 2020 dataset, including a comprehensive skin lesion image collection. We specifically fine-tuned three pre-trained models—ResNet50, VGG16, and EfficientNetB3—for skin cancer identification. These models were chosen for their flexible architectural features and effectiveness in image classification tasks, allowing for a more resilient ensemble using multiple models and a diverse dataset. We balanced model complexity and generalizability through techniques such as cross-validation and evolutionary algorithms for weight optimization, resulting in more precise and efficient outcomes. Our Genetic Algorithm Optimized Ensemble (GAOE) model significantly outperformed individual and weighted ensemble models. Combining genetic algorithm optimization and meta-learning, the GAOE model achieved an accuracy of 99.43%. It made notable progress in addressing model bias, overfitting, and underfitting. By integrating multiple models and diverse datasets, we could address model bias and ensure a well-rounded representation of various skin cancer types. Based on the success of the GAOE model, future research could explore different promising avenues. Utilizing larger and more diverse datasets may enhance the model's resilience while developing real-time clinical diagnostic tools would allow for rapid deployment. Moreover, hybrid techniques integrating GAOE with advanced reinforcement learning could improve diagnostic precision, enhance the interpretability of model predictions, facilitate clinical decision-making, and ensure that the model's insights are understandable and realistic.

## References


1. Chanda D, Onim MSH, Nyeem H, Ovi TB, Naba SS (2024) DCENSnet: a new deep convolutional ensemble network for skin cancer classification. Biomed Signal Process Control 89:105757
2. Tahir M, Naeem A, Malik H, Tanveer J, Naqvi RA, Lee S-W (2023) DSCC_Net: multi-classification deep learning models for diagnosing of skin cancer using dermoscopic images. Cancers 15(7):2179
3. Lee YC, Jung S-H, Won H-H (2018) WonDerM: skin lesion classification with fine-tuned neural networks. arXiv preprint arXiv:1808.03426
4. Ahmed B, Qadir MI, Ghafoor S (2020) Malignant melanoma: skin cancer-diagnosis, prevention, and treatment. Crit Rev™ Eukaryot Gene Expr 30(4)
5. Sandhu APS, Singh K, Singh S, Antaal H, Luthra S, Singla A, Nijjar GS, Aulakh SK, Kaur Y, kaur Aulakh S et al (2024) Decoding cancer risk: understanding gene environment interactions in cancer development. Cureus 16(7)
6. Kränke T, Tripolt-Droschl K, Röd L, Hofmann-Wellenhof R, Koppitz M, Tripolt M (2023) New AI-algorithms on smartphones to detect skin cancer in a clinical setting—a validation study. PLoS ONE 18(2):e0280670
7. Alyami J (2024) Computer-aided analysis of radiological images for cancer diagnosis: performance analysis on benchmark datasets, challenges, and directions. EJNMMI Rep 8(1):7
8. Huang H-Y, Hsiao Y-P, Karmakar R, Mukundan A, Chaudhary P, Hsieh S-C, Wang H-C (2023) A review of recent advances in computer-aided detection methods using hyperspectral imaging engineering to detect skin cancer. Cancers 15(23):5634
9. Gasmi S, Djebbar A, Merouani HFD, Djedi H (2024) Advancing skin cancer research with machine learning and deep learning models: a systematic review. J Pak Assoc Dermatol 34(2):527–546
10. Claret SA, Dharmian JP, Manokar AM (2024) Artificial intelligence-driven enhanced skin cancer diagnosis: leveraging convolutional neural networks with discrete wavelet transformation. Egypt J Med Hum Genet 25(1):50
11. Natha P, Pothuraju R (2023) Skin cancer detection using machine learning classification models
12. Swetha A, Lakshmi M, Kumar M (2022) Chronic kidney disease diagnostic approaches using efficient artificial intelligence methods. Int J Intell Syst Appl Eng 10(1s):254
13. Bisla D, Choromanska A, Berman RS, Stein JA, Polsky D (2019) Towards automated melanoma detection with deep learning: Data purification and augmentation. In: Proceedings of the IEEE/CVF conference on computer vision and pattern recognition workshops, pp 0–0
14. Riaz L, Qadir HM, Ali G, Ali M, Raza MA, Jurcut AD, Ali J (2023) A comprehensive joint learning system to detect skin cancer. IEEE Access
15. Lakshmi VV, Jasmine JL (2021) A hybrid artificial intelligence model for skin cancer diagnosis. Comput Syst Sci Eng 37(2):233–245

16. Ogundokun RO, Li A, Babatunde RS, Umezuruike C, Sadiku PO, Abdulahi AT, Babatunde AN (2023) Enhancing skin cancer detection and classification in dermoscopic images through concatenated mobileNetV2 and Xception models. Bioengineering 10(8):979
17. Khan MA, Sharif M, Akram T, Bukhari SAC, Nayak RS (2020) Developed Newton-Raphson based deep features selection framework for skin lesion recognition. Pattern Recogn Lett 129:293–303
18. Amin J, Sharif A, Gul N, Anjum MA, Nisar MW, Azam F, Bukhari SAC. Integrated design of deep features fusion for localization and classification of skin cancer
19. Frisson G, Detta N, Coppola P, Mazzaccara C, Pricolo MR, D'Onof A, Limongelli G, Calabrò R, Salvatore F (2016) Functional studies and in silico analyses to evaluate non-coding variants in inherited cardiomyopathies. Int J Mol Sci 17(11):1883
20. Johansen TH, Møllersen K, Ortega S, Fabelo H, Garcia A, Callico GM, Godtliebsen F (2020) Recent advances in hyperspectral imaging for melanoma detection. Wiley Interdisc Rev Comput Stat 12(1):e1465
21. Balaha HM, Hassan AE-S (2023) Skin cancer diagnosis based on deep transfer learning and sparrow search algorithm. Neural Comput Appl 35(1):815–853
22. Dahou A, Aseeri AO, Mabrouk A, Ibrahim RA, Al-Betar MA, Elaziz MA (2023) Optimal skin cancer detection model using transfer learning and dynamic-opposite hunger games search. Diagnostics 13(9):1579
23. Cassidy B, Kendrick C, Brodzicki A, Jaworek-Korjakowska J, Yap MH (2022) Analysis of the ISIC image datasets: usage, benchmarks and recommendations. Med Image Anal 75:102305
24. Reis HC, Turk V, Khoshelham K, Kaya S (2022) InSiNet: a deep convolutional approach to skin cancer detection and segmentation. Med Biol Eng Comput 1–20
25. Jaisakthi SM, Mirunalini P, Aravindan C, Appavu R (2023) Classification of skin cancer from demoscopic images using deep neural network architectures. Multimedia Tools Appl 82(10):15778

# Automated and Scalable Hybrid Continual Learning Framework (ASHCLF) for Cross-Domain Applications

**J. Ranjith and Santhi Baskaran**

**Abstract** In this paper, we introduce an Automated and Scalable Hybrid Continual Learning Framework (ASHCLF), a new methodology specifically designed to address the limitations above in terms of high catastrophic forgetting rates as well as the ability for scalability and cross-domain adaptability within continual learning systems. Online learning is important for incremental continual learning, where machine models can learn new tasks sequentially and remember the previous ones. Nonetheless, state-of-the-art approaches such as Elastic Weight Consolidation (EWC) and Learning without Forgetting (LwF) are not optimized to deal with large-scale data or task diversity in practice and often involve extensive manual tuning. ASHCLF is innovative in that, unlike classical methods, which use hand-tuned algorithms and architectures for each separate language pair model of machine translation, the dynamic architecture, hyper-parameter tuning automatically, and cross-domain adaptation presented allow it to change its complexity according to task requirements. A thorough evaluation of several common domain-generalization benchmarks (AG News, CIFAR-10) and a financial time-series forecasting dataset justifies the generalizability of this framework across diverse data modalities such as text, images, and temporal domains. The experimental results show that the ASHCLF framework consistently outperforms established methods, achieving a notable 89.5% in accuracy retention while reducing the forgetting rate to just 8.3%. Additionally, it demonstrates superior cross-domain generalization, particularly in challenging scenarios involving diverse data modalities.

**Keywords** Continual Learning · Catastrophic forgetting · Cross-domain adaptation · Automated parameter tuning · Dynamic model architecture · Scalability · Machine Learning

J. Ranjith (✉) · S. Baskaran
Puducherry Technological University, Puducherry, India
e-mail: ranjith12thl8@ptuniv.edu.in

S. Baskaran
e-mail: santhibaskaran@ptuniv.edu.in

 
K. R. Madhavi et al. (eds.), *Proceedings of Sixth International Conference on Computer and Communication Technologies*, Lecture Notes in Networks and Systems 1356,
https://doi.org/10.1007/978-981-96-5238-9_22

## 1 Introduction

It's applicable in machine learning, where models must acquire new skills while preserving information from previous endeavors. An essential challenge in this area is the problem of catastrophic forgetting, which a decline in the models' performance is in prior tasks when training on new ones, particularly in the case of context-variable data such as text. Some of these, such as Elastic Weight Consolidation (EWC) and Learning without Forgetting (LwF), reduce this problem but suffer from scalability and computational complexity. Successively advanced Neural Networks and Memory-Aware Synapses (MAS) provide enhancements but require extensively expanded memory and resources. To overcome these limitations, this paper introduces the Automated and Scalable Hybrid Continual Learning Framework (ASHCLF), which combines the aspects of dynamic architecture, automated parameter adjustment, and cross-domain training for a time-efficient solution to continual learning in different domains.

## 2 Related Work

Current ongoing research in the field of machine learning seeks to solve the problem with catastrophic forgetting, which occurs when the model forgets previously learnt tasks as soon as it encounters new data. Hence, continued learning is a very important field of study. Consequently, there have been several approaches to reduce catastrofic forgetting and some of them include the following. However, they still incorporate restrictions like, reduced levels of throughput, increase in computational costs, and low scalability in complex scenarios [1–3]. EWC is one of the most popular strategies which is designed to overcome catastrophic forgetting by adding a regularizing term that discourages changes of large magnitude in weights during learning of new tasks [4, 5]. As for the other five parameters, EWC exploits Fisher's information matrix to identify how important each parameter is and then subsequently applies penalties befittingly. Learning without Forgetting utilizes the knowledge distillation technique to maintain prior performance on previous tasks during learning of new tasks but it can be slow with large dataset or multiple tasks. More developed Neural Networks help in building the network structure for new tasks while enhancing learning without erasing former knowledge [6, 7]. But this approach enlarges memory and computational requests and reduces the system scalability. Memory-aware synapses are capable of changing the synapse importance with respect to the performance on the specific task and effectively store the essential information [8, 9]. Nevertheless, MAS has inherent flexibility for such tasks, but as tasks become complex MAS can experience computational overhead [10, 11]. Cross-domain learning is another newer problem, in which deep models need to transfer the learning between different data domains [12, 13]. The traditional continual learning schemes which are typically designed for some particular domain require assistance with various data. The current human continual learning frameworks are similar to the current hybrid continual learning frameworks, which try to combine the above-mentioned strategies but also face challenges of scalability and necessitate manual tuning [14–16]. To tackle these issues, the Automated and Scalable Hybrid Continual Learning Framework presents a strong and effective framework for continual learning across fluctuating data settings.

# 3 Methodology

The proposal system comprises the method, approach and details an Automated and Scalable Hybrid Continual Learning Framework (ASHCLF), which focuses on dynamic model architecture, automated parameter optimization, and transfer learning. Figure 1. Illustrates, in particular, how these elements are integrated into ASHCLF to provide highly precise and versatile for new tasks.

## *3.1 Dynamic Model Architecture*

The architecture of the ASHCLF model shows that it can dynamically change the number of components according to the separate activities' needs to optimize performance and resource consumption. For example staying on top of emails immediately is essential for operational effectiveness and mass customization learning. One way

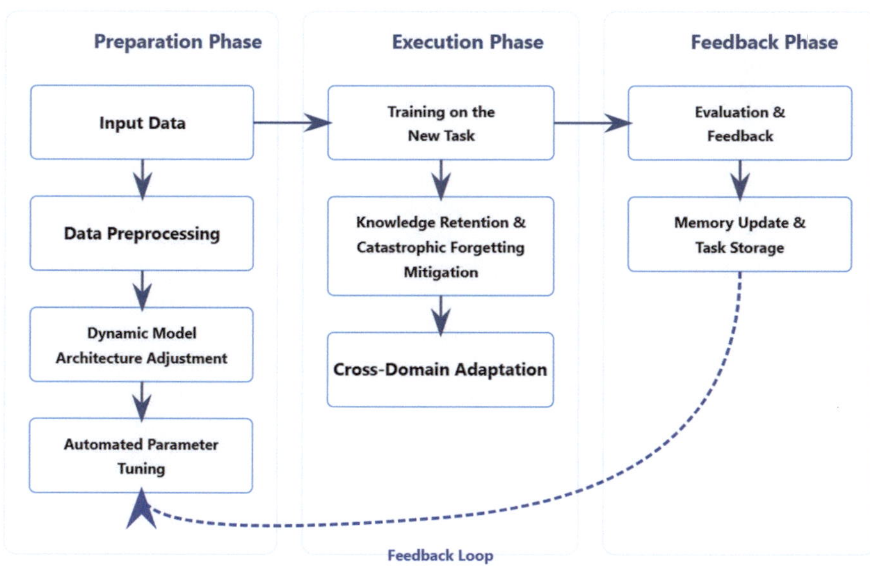


**Fig. 1** Overall block diagram of ASHCLF

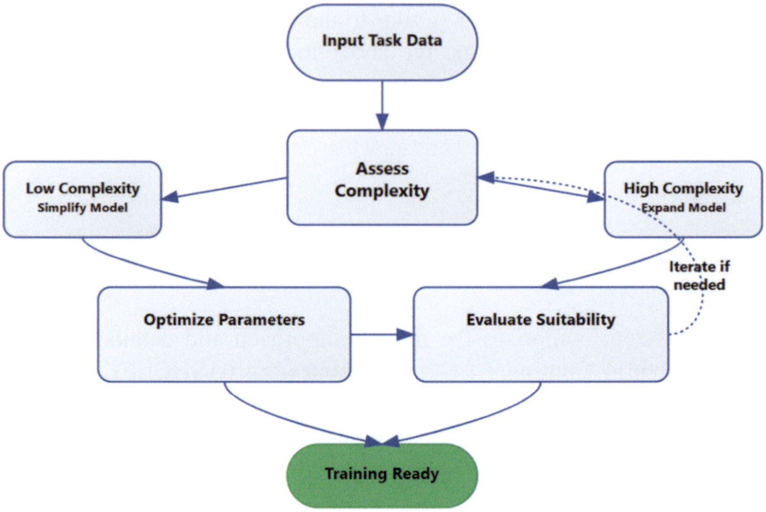


**Fig. 2** Dynamic model architecture module

of doing this is through using scalable neural networks that can be made larger or smaller depending on the kind of data that may exist. For instance, convolutional layers are meant to keep image data when combined with transformers. The Dynamic model architecture described in the paper (Fig. 2) indicates that the method changes the layers and parameters of its structure by the level of learning task complexity.

## 3.2 Automated Parameter Tuning

In the ASHCLF, we have implemented automated parameter tuning to avoid time-consuming and error-prone manual hyper-parameter adjustment. This component adjusts the strength of regularization and task relevance scores based on meta-learning techniques regarding the characteristics of data or tasks. It allows the model to adapt dynamically to such new tasks and improves its performance for non-expert users without the resources to fine-tune a complex learning system. In the ASHCLF framework, the tune is implemented through an automated parameter tuning process, as illustrated in (Fig. 3) the system monitors and optimizes model performance continuously, continuously updating parameters to improve learning results. It enables automated tuning of the model, which significantly saves computational overhead when tuning a complex machine learning mode manually.

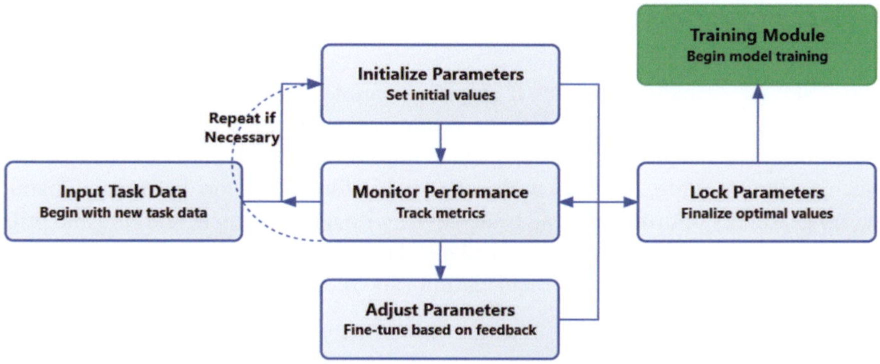


**Fig. 3** Automated parameter tuning module

## *3.3 Cross-Domain Adaptation*

Cross-domain adaptation is needed to properly operate the ASHCLF framework across different data domains, e.g., text images and time-series data. Hence, this component ensures that your model can learn and utilize knowledge from one domain in another. It leads toward the Generalization feature. It employs domain adaptation methods to address the issue of changes in data distribution and performance degradation when applied to new unseen domains or tasks (Fig. 4). Shows the process of the cross-domain adaptation module in the ASHCLF framework.

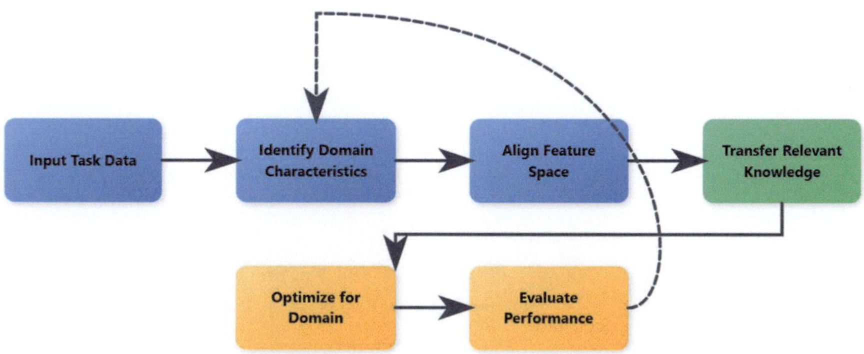


**Fig. 4** Cross-domain adaptation module

### 3.4 Integration with Emerging Technologies

The ASHCLF framework makes it easy to integrate new capabilities into existing machine learning paradigms, such as reinforcement meta-learning and quantum computing. The integrations are designed to improve the framework's learning, facilitate decision-making, and deal with catastrophic forgetting. For instance, dynamic learning strategy optimization can be achieved with reinforcement learning and prior knowledge by quickly grasping new tasks and using meta-learning. Figure 5 depicts how ASHCLF utilizes reinforcement learning and meta-learning methods to enhance performance.

*Algorithm of ASHCLF*:

```
Model = initialize_dynamic_model()
hyperparameters = initialize_hyperparameters()
memory = initialize_memory()
for task in tasks:
   dataset = preprocess_data(task)
   if assess_task_complexity(task) = = "high":
      model.expand_architecture().
   elif assess_task_complexity(task) = = "low":
      model.reduce_architecture()
   for iteration in range(max_iterations):
            hyperparameters = auto_adjust_hyperparameters(hyperparameters,
performance_feedback)
      loss = train_model(model, dataset, hyperparameters)
      model = dynamic_adjust_architecture(model, performance_metrics)
```

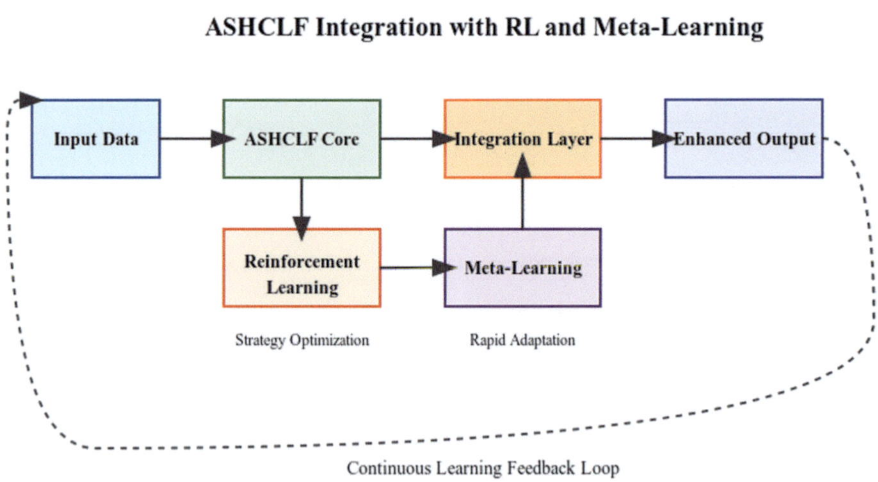


**Fig. 5** Integration with reinforcement learning and meta-learning

```
        memory.store_key_features(task, model)
    apply_regularization(model, memory)
    distilled_knowledge = knowledge_distillation(model, memory)
    memory.update(task, distilled_knowledge)
    if is_new_domain(task, memory):
        model = apply_domain_adaptation(model, task, memory)
    performance = evaluate_model(model, task, memory)
    update_metrics(performance)
    if reinforcement_learning_enabled:
        optimize_learning_strategy(model, task)
    if meta_learning_enabled:
        apply_meta_learning(model, task, memory)
    if quantum_computing_enabled:
        accelerate_training_with_quantum(model)
final_model = finalize_model(model, memory)
return final_model
```

The ASHCLF algorithm initializes a dynamic model, hyper-parameters, and memory structure, then processes tasks sequentially. It has input preprocessing, task difficulty evaluation, and proper model design. The core loop automatically tunes the hyperparameters, training and dynamic adjustments and the critical knowledge is retained in memory. In post-training, the method employs regularization and knowledge distillation to combat catastrophic forgetting when refreshing memory. It employs domain adaptation techniques in transferring information from one domain to another and evaluating the performance of current and previous tasks. ASHCLF changes with the integration of reinforcement learning and meta-learning technologies within it, which is a multi-domain model but does not forget essential aspects abruptly.

## 4 Experimental Setup

The experimental establishment describes how the ASHCLF framework was assessed, including the dataset and baseline procedures with performance measures. The following section explains how case studies test and validate the suggested framework.

### *4.1 Datasets*

To analyze ASHCLF's adaptability carefully, we chose a wide set of datasets from several data domains, allowing for a thorough examination of its performance. The goal is to verify the framework's cross-domain applicability comprehensively. The AG News framework is assessed for text categorization in the Text Dataset, which uses the AG News dataset. This dataset consists of news articles and is divided into four classes of different natures, which is why AG News is used in various Natural Language Processing (NLP) tasks. We chose the CIFAR-10 Image Classification Dataset. This dataset consists of 60,000 $32 \times 32$ color images in 10 classes and is widely used for benchmarking img2img recognition models. The Time Series dataset is a stock prices finance dataset, and the tasks are for time series forecasting.

### *4.2 Baseline Methods*

These baselines address catastrophic forgetting in continual learning. EWC preserves old knowledge by penalizing updates to essential parameters. LWF prevents knowledge degradation using a selection policy. Vanilla Fine-Tuning (VFT) trains sequentially without forgetting mechanisms. Progressive Neural Networks expand with each new task, while Memory-Aware Synapses (MAS) adapt synapse importance based on task relevance.

### *4.3 Evaluation Metrics*

The performance of ASHCLF is evaluated using key metrics to assess its effectiveness in continual learning and cross-domain adaptation. Model Retention accuracy measures how well the model maintains prior capabilities after training on new tasks, indicating resistance to catastrophic forgetting. The Forgetting Rate shows the degree of performance degradation on old tasks after learning new ones, with a lower rate indicating better knowledge retention. Efficiency is evaluated regarding computational resources, processing time, and memory usage, which are crucial for extensive data processing. Cross-domain generalization assesses the model's ability to learn across different data domains compared to baseline methods.

## 5 Results and Discussion

This section shows the results of experiments to assess ASHCLF's performance. The results are then compared with the baseline methods using different metrics, and a detailed discussion follows to interpret these findings.

**Table 1** Performance comparison of ASHCLF with baseline methods

| Metric | EWC | LwF | VFT | Progressive NN | MAS | ASHCLF |
|---|---|---|---|---|---|---|
| Accuracy retention (%) | 82.3 | 83.7 | 76.5 | 84.1 | 83.2 | 89.5 |
| Forgetting rate (%) | 14.7 | 13.2 | 22.5 | 12.6 | 14.1 | 8.3 |
| Computational efficiency | Medium | Medium | High | Low | Medium | High |
| Memory usage (MB) | 1150 | 1120 | 1020 | 1400 | 1200 | 1100 |
| Cross-domain generalization | Low | Low | Very low | Medium | Low | High |

## *5.1 Performance Comparison with Baselines*

Table 1 shows the performance comparison between the proposed works with Baseline methods.

## *5.2 Result Analysis*

The following (Fig. 6) shows the accuracy retention percentages for each method. We assess ASHCLF in terms it shows an 89.5% accuracy retention, guaranteeing its capacity to keep task accurate among all methods. The (Fig. 7) shows the forgetting rates, and ASHCLF gets the lowest rate of 8.3%, which means it effectively keeps knowledge against the time process. The (Fig. 8) compares the computational level to all approaches, where some are heavier and more computationally extensive than others, like ASHCLF, as opposed to Vanilla Fine-Tuning. The comparison on Memory Utilization (MB) ASHCLF: 1100 MB is more memory efficient than Progressive NN using around 1400 MB, as shown in the (Fig. 9). The (Fig. 10) comparing between-X-generalization scores shows that a model from ASHCLF has the highest cross-domain generalization score (5) and demonstrates relatively good transfer across different data domains.

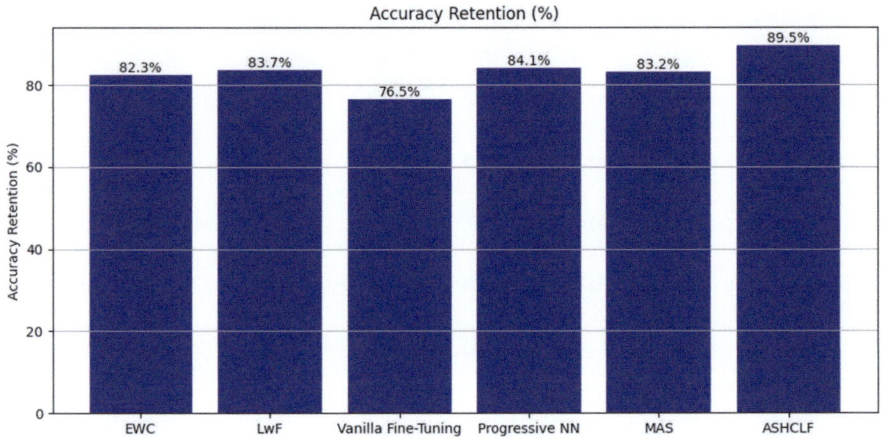


**Fig. 6** Accuracy retention

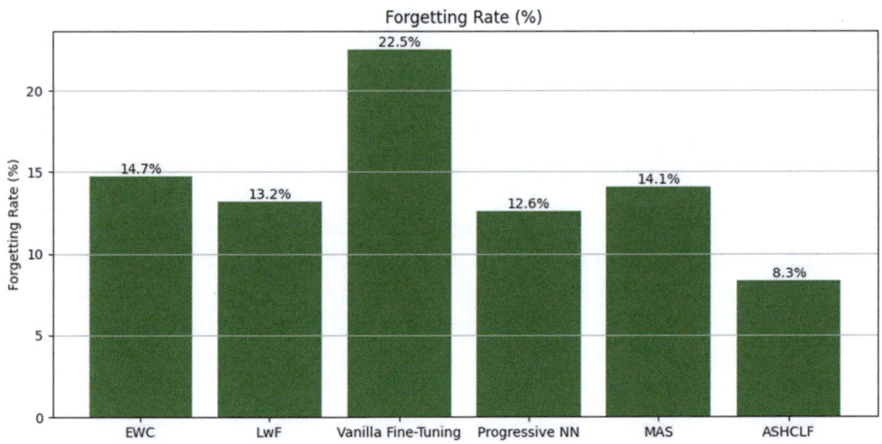


**Fig. 7** Forgetting rate

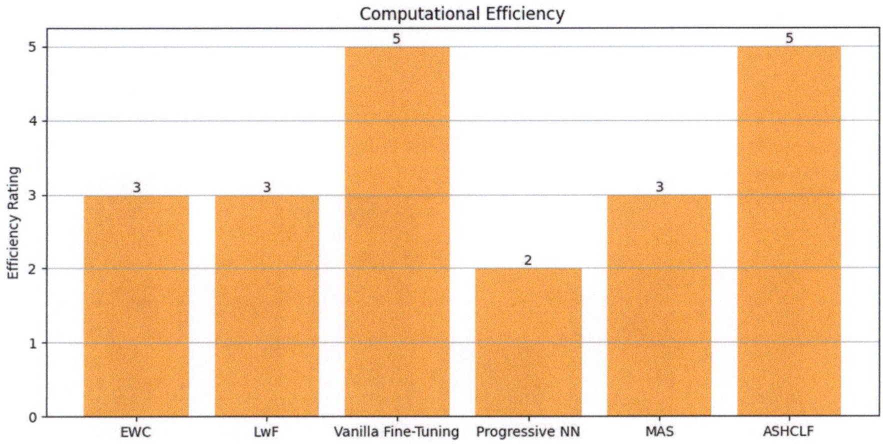


**Fig. 8** Computational efficiency

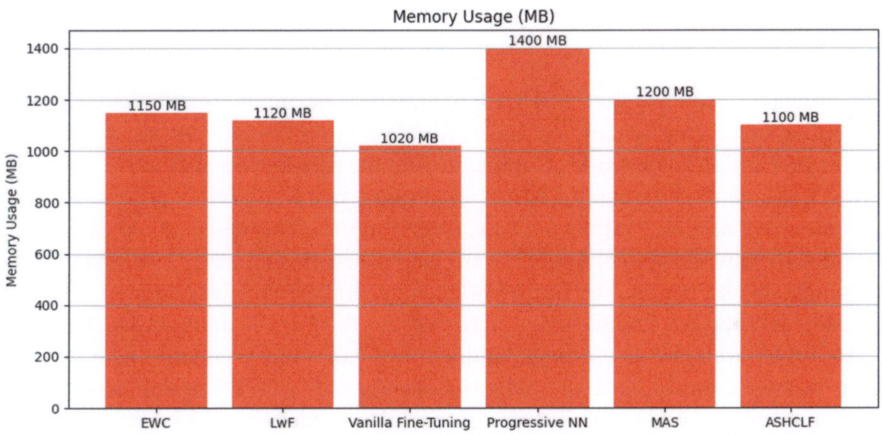


**Fig. 9** Memory usage

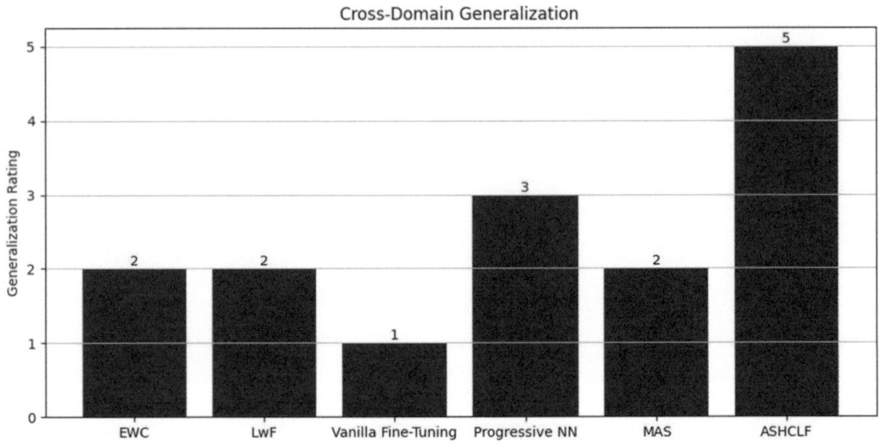


**Fig. 10** Cross-domain generalization

# 6 Conclusion

The research demonstrates ASHCLF's effectiveness in mitigating catastrophic forgetting, enhancing scalability, and facilitating cross-domain adaptability within continual learning frameworks. Compared with the best-known alternative methods, EWC, LwF, vanilla Fine-tuning, MAS, and Progressive Neural Networks, improving both retaining accuracy rate over tasks, forgetting receding speed of old task on all datasets; meanwhile maintaining forward pass time cost to a lower level than others in comparable scenarios. The framework can generalize knowledge across different data domains and is a suitable solution for various machine-learning tasks. Although ASHCLF did well in many settings, future efforts can center on improving it for even more massive datasets, such as studying convergences with emerging technologies such as quantum computing. Testing in real-world dynamic environments may be helpful and demonstrate practical value. ASHCLF is an important step toward this direction for continual learning, providing a flexible and effective framework that can be generalized across domains, making it beneficial both from a research and an application perspective.

# References

1. Hao X, Lin C, Dong W, Huang X, Xiong H (2023) Robust and secure federated learning against hybrid attacks: a generic architecture. IEEE Trans Inf Forensics Security
2. Zhang T, Chen C, Wang D, Guo J, Song B (2023) A vae-based user preference learning and transfer framework for cross-domain recommendation. IEEE Trans Knowl Data Eng 35(10):10383–10396

3. Hafner D, Pasukonis J, Ba J, Lillicrap T (2023) Mastering diverse domains through world models. arXiv preprint arXiv:2301.04104
4. Zhang Z, Guo B, Sun W, Liu Y, Yu Z (2022) Cross-fcl: toward a cross-edge federated continual learning framework in mobile edge computing systems. IEEE Trans Mobile Comput 23(1):313–326 (2022)
5. Chen D, Gao D, Xie Y, Pan X, Li Z, Li Y, Ding B, Zhou J (2023) Fs-real: towards real-world cross-device federated learning. In: Proceedings of the 29th ACM SIGKDD Conference on Knowledge Discovery and Data Mining, pp 3829–3841
6. Liu C, Wang L, Lyu L, Sun C, Wang X, Zhu Q (2023) Deja vu: continual model generalization for unseen domains. arXiv preprint arXiv:2301.10418
7. Wan X, Xu K, Liao X, Jin Y, Chen K, Jin X (2023) Scalable and efficient full-graph gnn training for large graphs. In: Proceedings of the ACM on management of data, vol 1, no 2, pp 1–23
8. Durgapal P, Kataria P, Garg G, Anand AS (2023) A comprehensive distributed framework for cross-silo federated learning using blockchain. In: 2023 fifth international conference on Blockchain Computing and Applications (BCCA). IEEE, pp 538–545
9. Ganjalizadeh M, Ghadikolaei HS, Azari A, Alabbasi A, Petrova M (2023) Saving energy and spectrum in enabling URLLC services: a scalable RL solution. IEEE Trans Industr Inf 19(10):10265–10276
10. Ma H, Xie R, Meng L, Chen X, Zhang X, Lin L, Zhou J (2024) Triple sequence learning for cross-domain recommendation. ACM Trans Inf Syst 42(4):1–29
11. Xie B, Li S, Lv F, Liu CH, Wang G, Wu D (2022) A collaborative alignment framework of transferable knowledge extraction for unsupervised domain adaptation. IEEE Trans Knowl Data Eng 35(7):6518–6533 (2022)
12. Ali A, Ali H, Saeed A, Khan AA, Tin TT, Assam M, Ghadi YY, Mohamed HG (2023) Blockchain-powered healthcare systems: enhancing scalability and security with hybrid deep learning. Sensors 23(18):7740 (2023)
13. Sakhuja C, Shi Z, Lin C (2023) Leveraging domain information for the efficient automated design of deep learning accelerators. In: 2023 IEEE international symposium on High-Performance Computer Architecture (HPCA). IEEE, pp 287–301
14. Ilhan F, Su G, Liu L (2023) Scalefl: resource-adaptive federated learning with heterogeneous clients. In: Proceedings of the IEEE/CVF conference on computer vision and pattern recognition, pp 24532–24541
15. Chen Q, You Z, Wen D, Zhang Z (2023) Enhanced hybrid hierarchical federated edge learning over heterogeneous networks. IEEE Trans Veh Technol 72(11):14601–14614
16. Håkansson VW, Venkategowda NKD, Werner S, Varshney PK (2022) Optimal scheduling of multiple spatiotemporally dependent observations for remote estimation using age of information. IEEE Internet Things J 9(20):20308–20321 (2022)

# A Hybrid Ensemble Model for Classification of Emotions from Social Media Text

**Ashritha R. Murthy, K. M. Anil Kumar, and N. Vedavathi**

**Abstract** With the growing popularity of social networking sites in recent years, people now have a forum to share business proposals, recommendations, and advertisements. This research work focuses on emotions expressed by humans in social media texts to comprehend and analyze emotions such as joy, anger, sadness, and fear. The experiments were carried out with the fundamental deep learning models such as CNN (Convolutional Neural Networks), LSTM (Long Short Term Memory), Bidirectional LSTM (BiLSTM), Gated Recurrent Units (GRU), and Bidirectional GRU (BiGRU) for different levels of classifications of emotions in the textual data. The novelty of the proposed approach is the application of the ensemble model with a transformer learning model for the classification of emotional texts. The proposed approach is significant across the various fields that rely on interpreting the emotional cues found in social media texts. The results obtained with the proposed approach are better than other conventional methods. Our proposed work achieves an accuracy of 96.02% in the identification of different levels of emotions in social media texts such as Twitter and also provides better results on different standard datasets such as SemEval 2018, Sem Eval 2019, and MELD.

**Keywords** Deep Learning · BERT · Emotions · Bidirectional LSTM · Bidirectional GRU

A. R. Murthy (✉) · K. M. Anil Kumar
Department of Computer Science and Engineering, JSS Science and Technology University, Mysuru, India
e-mail: ashrithar.murthy@sjce.ac.in

N. Vedavathi
Department of Computer Science and Engineering, Vidyavardhaka College of Engineering, Mysuru, India



K. R. Madhavi et al. (eds.), *Proceedings of Sixth International Conference on Computer and Communication Technologies*, Lecture Notes in Networks and Systems 1356,
https://doi.org/10.1007/978-981-96-5238-9_23

# 1 Introduction

The complexity in cognitive and mental state of human behavior depends on neuro-physiological changes. Emotions such as pleasure, unhappiness, fear, anger, joy, and sadness will impact our daily activities and mental well-being [1]. In the digital era, online platforms such as Twitter (Tweets), Facebook (FB), YouTube, etc. are the source of exchanging thoughts and feelings and expressing emotions. Analyzing these emotions is crucial to finding insights into human emotional states. Sentiment analysis is a fundamental aspect of natural language understanding that classifies emotions into multiclass sentiments [2]. The identification of emotions in text is challenging due to the limited emotional indicators and the frequent tendency to use emotions and sentiment interchangeably.

Paul Ekman has identified primarily the emotions such as anger, surprise, joy, disgust, sadness, and fear [3]. These emotions are vital in computer science and in interpersonal relationships, which have a significant impact on the dynamics of communications, and these could be identified in various mediums such as text, audio, and video [4, 5]. In our research work, we are focusing on the textual data. In the research on identifying text, the authors Wen and Wan [6] categorized the labeled dataset into positive and negative emotions based on the threshold value. There are many computational methods used for detecting emotions in text, such as rule-based techniques, keyword-based techniques, machine and deep learning based techniques, and corpus-based techniques [7, 8]. Examining the recent research studies on the contributions of emotion classification, datasets used for emotion classification, and the models used for identification of emotions in the text, outcomes, and limitations of pre-trained transformer-based models is necessary to investigate their impact on text-based emotion identification. In this research work, we have conducted the experiments using various deep learning techniques for the classification of emotions from a dataset comprising social media texts. More specifically, our proposed system combines the capabilities of transformer-based models with ensemble models to understand emotional cues and capture them for better analysis and classification.

The structure of the paper is followed as follows: Sect. 2 reviews the related literature work in emotion analysis and text classification. Section 3 details the proposed model for classifying different levels of emotions based on the proposed hybrid ensemble model. Experimental setups and results are discussed in Sect. 4, and in Sect. 5, a conclusion is discussed of the research work.

# 2 Related Work

Text-based emotion detection research explores the complex structure of emotions expressed through reviews, personal opinions, and other textual forms across different social media [9–11]. Word-level analysis helps in the development of emotional vocabulary resources [12]. The Valence-Arousal space (2-dimensional) [13] and the Pleasure-Arousal-Dominance model (3-dimensional) are frequently selected as dimensional models. Ekman's [14] emotion model consists of six basic emotions, and Plutchik's [2] model consists of eight emotions that are commonly utilized. However, in this proposed work, we categorized 4 basic emotion categories: joy, anger, sadness, and fear. Polignano et al. [15], a constrain optimization framework was introduced for emotion detection. The researchers evaluated their framework on three datasets: ISEAR (focusing on sadness, joy, fear, and anger), Twitter dataset, and SemEval-2007. A classification model developed for detecting emotions using CNN, Bi-LSTM, and self-attention mechanisms in work by the author Shaaban et al. [16]. Three datasets were used to train and evaluate their model: ISEAR, SemEval_2019_Task 3, and SemEval_2018_Task 1. Mahajan and Zaveri [17] have examined how different factors influence the identification of emotions expressed by the people. Better results were obtained when BERT embeddings were combined with a highly connected neural network layer as a classifier. Peters et al. [18] combined the emotion-based attributes into machine-learning classifier techniques. They assessed using various methods, such as Decision Tree classifiers, Multilayer Perceptron, Support Vector Machine (SVM), and REPTree, for the SemEval-2019 dataset. With the result of accuracy of 87.55% and an F1 score of 65.31%. By using bidirectional training of text representations, Devlin et al. [19] created the BERT model, for capturing contents in both directions. This approach improves understanding of the model, which sets the new benchmarks for model performance and adaptability. Yu and Hu [20] created a BERT-Att-TextCNN model for text categorization in which it combines the local feature extraction capabilities of Text CNN with the deep bidirectional semantic analysis of BERT effectively. Yu, Q., et al. [21] developed a model using a BERT with BiLSTM for text classification, which demonstrates through experiments its accuracy is compared to a word2vec-based BiLSTM model. Rahman et al. [22, 23] experimented using machine learning to identify the sentiment in the text messages on Tweets. The various deep neural network models and emotional categories were used to categorize tweets: anger, fear, joy, and sadness. The researchers developed a model that combines BiGRU and BiLSTM. The ensemble model's accuracy of 87.66% found better results compared with the BiGRU model's 87.53% accuracy. Based on the existing literature survey, we found that the majority of the previous research works focused on BiLSTM and BiGRU for emotion detection. The transformer learning model's design and its potential were not fully explored in the previous works and possibly undermining its significance in emotion detection. The aim of the research work is to explore the transformer learning model with the ensemble model for classification of emotions in the text.

## 3 Methodology

Natural Language Processing has improved significantly in the application of deep learning methods for analyzing human emotions and comprehension of the different levels of emotions expressed on the social media platform. This research work mainly focused on the classification of different levels of emotions from the text, such as fear, joy, sadness, and anger, from the Twitter dataset. The proposed hybrid ensemble method was developed using the BiLSTM, BiGRU, and the transformer RoBERT model. This hybrid ensemble model aims to strengthen each component: RoBERT deep contextual understanding, BiLSTM sequential data processing, and BiGRU's efficiency in modeling dependencies, to achieve a highly accurate emotion classification. The model proposed is detailed in Fig. 1.

The various phases of the approaches used in developing the model for detecting the emotion in the textual data are intended to maximize the model's ability to process and comprehend emotional content in text. The following are the different stages, as follows:

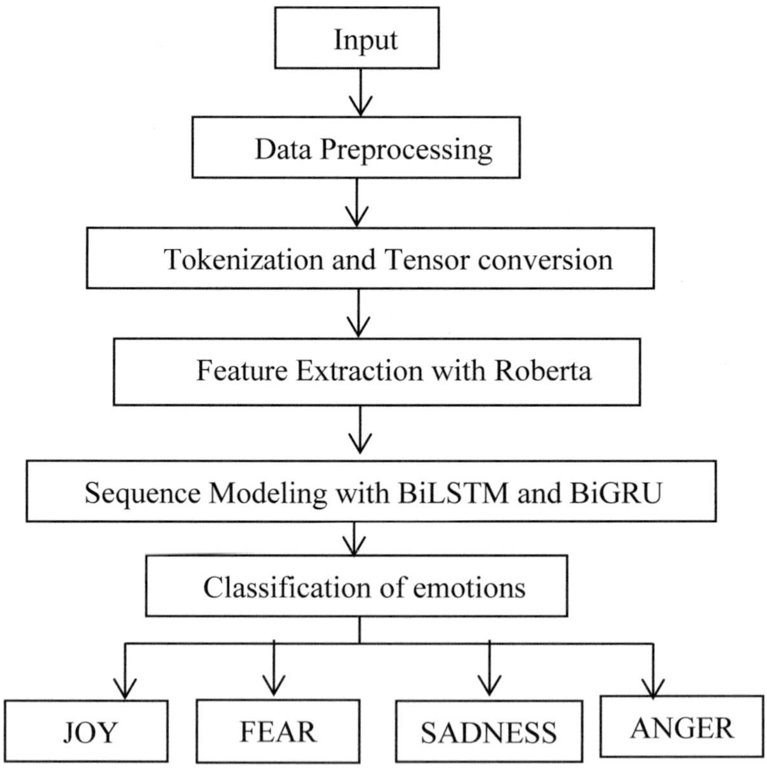


**Fig. 1** Data flow diagram of the proposed system

### 3.1 Preprocessing and Data Acquisition

A large and varied emotion dataset acts as the model's foundation. The dataset, used for the work consisting of 7102 tweets in English as used in the reported work [24], encompasses four distinct emotion classes: joy, anger, sadness, and fear. Table 1 depicts the distribution of emotion categories used in the current work.

The steps involving preprocessing are outlined in the following steps:

a. **Lowercase of the textual data**:

   The text is normalized and obtaining uniformity by converting all text to lowercase. For example, consider the original text: "HAPPY SUNDAY." This process yields the following text: "happy Sunday".

b. **Eliminate Punctuation marks to focus on the core linguistic content**: Eliminate a punctuation mark that focuses on the main message. And it simplifies the text. For example: Original Text: "Hurray, the results are out!" After removing the punctuation marks using regular expressions, we obtain the following text: "Hurray, the results are out".

c. **Handling Abbreviations**:

   Expanding the abbreviations to their full forms, preserving the intended meaning of each word in the dataset.

   For example, consider the original text: "OMG, I can't stop laughing!" and we obtain the following text: "Oh my god, I cannot stop laughing!"

d. **Removal hyperlink Tags**:

   Removal of URLs and HTML tags to get the clean text for the given input.

   For example, consider the original text: "Check out our website at http://www.jctr.com" and obtain clean text without URL/HTML tags: "Check out our website at."

e. **Stop word removal**:

   The common stop words (e.g., 'and', 'the') are discarded, which consist of content-carrying words. For example, "I was excited today for the match, but the rain began and I felt bad," and we obtained text without stop words: "excited today, match, rain began, felt bad."

f. **Tokenization**: tokenization is the process of dividing text into individual tokens, such as, e.g., "I am feeling excited!" and we obtained the tokenized text such as: {"I," "am," "feeling," "excited."}

**Table 1** Emotion samples of different classes

| Classes | Joy | Fear | Sadness | Anger |
|---|---|---|---|---|
| No. of samples | 1616 | 2252 | 1533 | 1701 |

g. **Lemmatization and Porter Stemming**: reducing words to their base forms (lemmatization) or stems (stemming with the Porter Stemmer) to unify vocabulary and reduce redundancy. For example, consider the original text: "I was feeling scared of the dark" and the shortened text: "I be feeling scared of the dark," and we obtained the shortened text (using Porter Stemmer): "I was feeling scared of the dark."

## *3.2 Feature Extraction*

The processes for extracting features and classifying text into different levels of emotion are discussed in this section. Firstly text preprocessing, followed by text tokenization to develop tokens, which are then converted into tensors. To train the models, the labels are numerically encoded and translated to tensor format.

A TensorDataset is formed by incorporating input_ids, which represent tokenized text, along with attention_masks that discern genuine tokens from padding and labels. This facilitates batch processing during the training phase. The dataset is then partitioned into training and testing subsets to guarantee thorough model training and evaluation. Data loaders are subsequently created for each subset to iterate through the data in manageable batches, thus optimizing memory utilization.

To ensure tokens are contextually integrated, the model class initializes with a pre-trained Roberta model. Following bidirectional processing of these embeddings through BiLSTM layers, context from both previous and subsequent tokens is captured, leading to the creation of a set of hidden states ($H_{LSTM}$). Following this, BiGRU layers refine the feature representation, utilizing gating mechanisms to generate another set of hidden states ($H_{GRU}$). Lastly, a linear layer executes a linear transformation on the GRU output, thereby producing logits (O) for each class, as illustrated in Eq. (1).

$$O = W * H_{\text{GRU}} + b \tag{1}$$

The final linear layer's result from BiGRU layers is known as $H_{GRU}$. It is the task of this linear layer to produce logits O for each emotion class by a linear transformation using W and b termed "weights" and bias, respectively. Determining loss requires verifying if thcy correspond to real labels by examining the cross-entropy between projected outputs (logits O) and predicted results. The model is updated during training using backpropagation to compute the gradients of the loss with respect to the model weights, as demonstrated in Eq. (2).

$$\text{Loss} = -\sum (y * \log(y_{\text{pred}})) \tag{2}$$

In this context, $y$ represents the true label, and $y_{\text{pred}}$ represents the predicted probability distribution across classes based on the logits $O$. The weights are adjusted during backpropagation, with the gradients derived from the loss function (Eq. 3).

$$\text{Weight} = \text{Weights} - \text{learning\_rate}*\frac{\text{dLoss}}{\text{dWeights}} \quad (3)$$

Using the AdamW optimizer, the weights of the model are updated and the scheduler for the learning rate, as in equation Eq. (3). After the AdamW optimizer and a learning rate scheduler are setup, then initiate the BiLSTM_BiGRU_Model in the training. The model runs through a forward pass (using the linear layer, BiLSTM, BiGRU, and Roberta encoding) for each epoch, which is followed by loss computation and backpropagation. The model is trained to categorize text into different levels of emotional groups using the sequence information and attributes that were retrieved. A model's parameters are adjusted during training in order to reduce misclassifications using a predetermined loss function. To prevent overfitting of the model, validation is conducted after each epoch in the training phase.

After completion of the training phase, the model will be tested using the test dataset. The model's performance is monitored through validation after each training epoch to prevent overfitting. Following the training's conclusion, the model is assessed on the test dataset. This assessment includes comparing the model's predictions with the actual labels and gauging its accuracy and reliability using standard metrics such as recall, precision, and F1_score. The assessment process may identify areas requiring enhancements, prompting further optimizations to enhance the model's performance. Once fully optimized, the trained model is ready for deployment in real-world settings where text emotion classification is vital. Utilizations encompass customer emotion analysis, mental health evaluation, and social media surveillance.

### *3.3 Algorithm*

Table 2 depicts the flow of proposed ensemble hybrid model,

## 4 Experiment and Results

The experiments were conducted on Google Colab and performed with the Python IDE version 3.6 [22]. To acquire the embedding of words or tokens, the word embedding layer in deep learning models only requires their token IDs. The layers of weights are originally identified by pre-trained word embedding and are adjusted throughout training. Figures 2 and 3 shows a graphical depiction of the test accuracy and loss of the proposed model. We found the best result at epoch 10 with a batch size of 16 produced the maximum accuracy, indicating the chosen hyper parameters improved model performance.

**Table 2** Algorithm of the proposed model

| |
|---|
| 1. Import the required libraries and load the dataset |
| 2. Data Processing: Tokenize, label encoding of the data, and conversion of tensors |
| 3. The tensor dataset is created using the labels, input IDs, and attention masks |
| 4. Split the dataset into testing and training datasets |
| 5. Defining the data loaders for each dataset |
| 6. Defining the class containing model BiLSTM AND BiGRU models Firstly, we use the pre-trained RoBERTa model and then add the BiLSTM and BiGRU layers and linear classification and initialize the model (GPU/CPU) |
| 7. The learning rate scheduler and the optimizer Adam are used |
| 8. Using Eq. (1), compute the forward pass layer for each training epoch. This involves determining the hidden state of the BiLSTM ($H_{LSTM}$) based on E, RoBERTa Encoding (E) derived from RoBERTa's input_ids and attention_mask, and then the final linear layer ($H_{GRU}$) derived from passing HLSTM through a GRU layer followed by a linear layer (O) |
| 9. Calculate the cross-entropy loss using the O for correct classifications |
| 10. Employ Eq. (3) for backpropagation. Utilize the resulting gradients (dLoss/dWeights) to adjust the weights |
| 11. Validate the model's performance on the validation dataset |
| 12. Compute the validation accuracy and loss |
| 13. Convert the numerical predictions back to their original labels using an inverse transform function |
| 14. Establish the final classification based on the transformed labels |

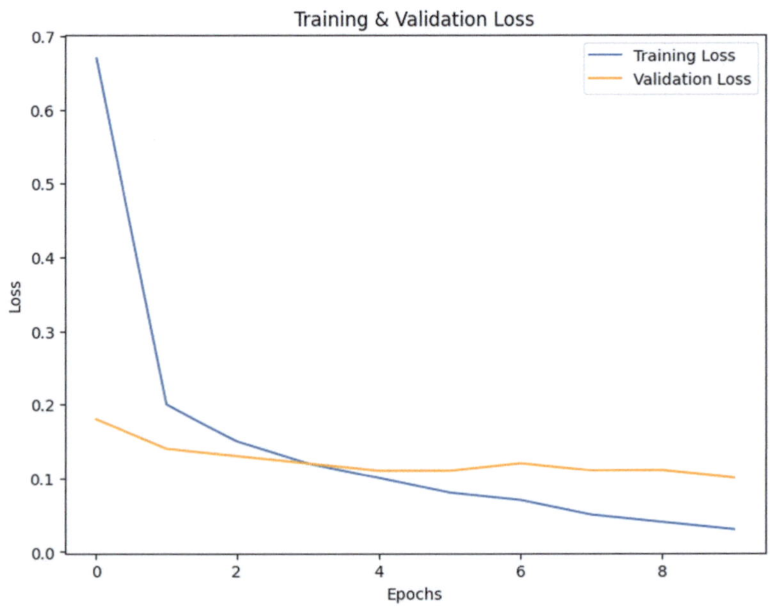


**Fig. 2** Performance visualization depicting the loss curves of the proposed model

**Fig. 3** Performance visualization depicting the accuracy of the proposed model

### *4.1 Performance Metrics*

To evaluate the model performance the metrics used are precision, as defined in Eq. (4), which measures the model that correctly identifies the emotion among all instances labeled with that particular classified emotion. The ability of a model to detect all instances of a specific emotion within the data is evaluated by recall, as described in Eq. (5). Equation (6) refers to the F1 score, which is a combined measure of recall and precision that gives a generally balanced picture of the model's performance in emotion detection. High scores for these measures show that the models successfully categorize emotions into different levels of emotions and reduce the number of incorrect classifications.

$$\text{Precision}(p) : \frac{\text{True positive}}{\text{True positive} + \text{False positive}} \tag{4}$$

$$\text{Recall}(r) : \frac{\text{True positive}}{\text{True positive} + \text{False negative}} \tag{5}$$

$$\text{F1 Score}(f1)\text{: } 2 * \left(\frac{\text{Precision*Recall}}{\text{Precision} + \text{Recall}}\right) \tag{6}$$

## 4.2 Results and Performance Analysis

In our research to understand the different levels of emotions in tweets and other data sets, we conducted experiments using various deep learning techniques [18, 23–25]. On using the transformer model such as RoBERTa combined with the hybrid model for classification of emotions.

The different models were combined with transformer models such as RoBERTa + CNN, RoBERTa + LSTM, RoBERTa + BiLSTM, RoBERTa + GRU, and RoBERTa + BiGRU for understanding the capabilities of the language to identify the emotions in the text. The hybrid model led to significant improvements in accuracy, highlighting the use of pre-trained language models in enhancing to identify the different levels of emotion classification. Our proposed model, which is a hybrid ensemble model, achieved better results with the highest accuracy of 96.02% in our experiments. This model effectively utilizes RoBERTa's contextual analysis capabilities along with the sophisticated sequence processing of BiLSTM and BiGRU, making it exceptionally adapt at discerning complex emotional clue in Twitter data. A comparative analysis of the models revealed that those integrating RoBERTa with bidirectional sequence processing architectures yielded the best results. This trend underscores the combination of deep contextual understanding provided by RoBERTa and the sequence analysis afforded by Bidirectional LSTM and Bidirectional GRU models.

Table 3 represents the results obtained for various deep learning models used on the Twitter dataset [22]. The complete detailed description of the proposed model with precision, recall, f1 score, and accuracy obtained for the 80% training and 20% testing split is represented in Table 4.

Table 5 represents the results obtained from the comparison of the different datasets and the proposed model. Initially, when evaluated against the ISEAR dataset, Abdullah and Shaikh [24] introduced a two-stage architecture employing a BERT model followed by a Bi-LSTM classifier for emotion detection. It achieved an accuracy of 72.64%. In contrast, our model achieved an accuracy of 86% for the ISEAR dataset.

Abdullah and Shaikh [24] discuss the system developed by TeamUNCC for detecting emotions in English and Arabic tweets on SemEval 2018. The system utilizes a combination of doc2vec and word2vec embeddings, along with a set of features from psycholinguistics extracted from the AffectiveTweets Weka package. It achieves an accuracy of 69.5. In contrast, our proposed model achieves an accuracy of 93.43%.

This evaluation continues across other datasets, such as SemEval 2019 and in research work by Peters [18], employing classifiers like SVM and decision trees. It achieves competitive performance with an f-score of 65.31% and an accuracy of 87.55%, whereas in contrast, our model achieves an accuracy of 91.38% for the same dataset. Additionally, on the MELD dataset, Mohammad [25] developed a computational model for recognizing human emotion in conversation. It employs text data augmentation to balance the MELD dataset and utilizes word embedding

**Table 3** Different deep learning models for the Twitter dataset

| Sl. No. | Models | Accuracy |
|---|---|---|
| 1. | CNN | 0.8602 |
| 2. | LSTM | 0.8636 |
| 3. | BiLSTM | 0.8657 |
| 4. | GRU | 0.8655 |
| 5. | BiGRU | 0.8659 |
| 6. | CNN + LSTM | 0.9234 |
| 7. | CNN + Bi-LSTM | 0.9257 |
| 8. | CNN + GRU | 0.9281 |
| 9. | CNN + BiGRU | 0.9287 |
| 10. | Bert + CNN | 0.9501 |
| 11. | Bert + LSTM | 0.9537 |
| 12. | Bert + BiLSTM | 0.9557 |
| 13. | Bert + GRU | 0.9551 |
| 14. | Bert + BiGRU | 0.9578 |
| 15 | (Proposed model) (Hybrid Ensemble model) | 0.9602 |

**Table 4** Result of the proposed model

| Emotions | Precision ($p$) | Recall ($r$) | F1_score ($f1$) |
|---|---|---|---|
| Sadness | 0.95 | 0.94 | 0.94 |
| Fear | 0.96 | 0.95 | 0.97 |
| Anger | 0.95 | 0.93 | 0.94 |
| Joy | 0.98 | 0.97 | 0.98 |
| Accuracy | 0.9602 | | |

**Table 5** Comparison of our proposed work with similar works for different dataset in literature

| Sl. No. | Previous works | Dataset | Models | Accuracy of other models | Proposed model (accuracy) |
|---|---|---|---|---|---|
| 1. | Acheampong et. al. [23] | ISEAR | BERT, BiLSTM | 72.64 | 88 |
| 2. | Abdullah and Shaikh [24] | Sem Eval 2018 | LSTM | 69.5 | 93.43 |
| 3. | Peters et al. [18] | Sem Eval 2019 | SVM,DT, REPTree | 87.55 | 91.38 |
| 4. | Mohammad F [25] | MELD | BERT,DNN | 64.36 | 75.6 |
| 5. | Rahman and Shova [22] | TWEETS | BiLSTM + BiGRU | 87.66 | 96.02 |

and BERT models for feature extraction. The evaluation shows the proposed model achieves 64.36% accuracy, whereas our proposed model achieves an accuracy of 75.6%. Rahman and Shova [22] used the Tweet dataset to detect emotions in text. They implemented the BiGRU model and achieved an accuracy of 87.53%. They implemented an ensemble model [BiLSTM + BiGRU] and obtained an accuracy of 87.66%. The researchers have not explored the use of the transformer-based learning model with the ensemble model. Hence we investigated the usefulness of the BERT transformer model with the ensemble model to improve the understanding of social media texts, capturing subtle emotional cues often missed by traditional methods. Thus, our proposed model provided an improved accuracy of 96.02%, showcasing the significance of transformer models in emotion identification. The decision to subject our model to testing across various datasets serves to underscore its capacity for generalization, as depicted in Table 5. By assessing its performance across diverse domains and text types, we ensure our model's capability to adeptly handle an array of text classification tasks. The consistent outperformance of our proposed model across these datasets highlights capturing complex semantic relationships, understanding contextual details, and making accurate predictions. This attribute the combination of RoBERTa Transformer and BiLSTM layers with BiGRU, enabling our model to achieve high accuracy in text categorization of emotion detection by utilizing the advantages of both architectures.

## 5 Conclusion

In our proposed model, we have focused on a critical aspect of understanding emotions expressed in social media text and have proposed a novel hybrid ensemble model to classify different levels of emotions in social media text, which has outperformed traditional techniques. Specifically, our model achieved an accuracy of 96.02% when tested on the Twitter dataset and showed a 15.35% improvement compared to traditional methods on average across other standard datasets. The proposed model also showed better performance when experimented on different datasets. Moving forward, we aim to enhance our model's capabilities in cross-lingual contexts, particularly in applications related to social media and customer service. This will involve further refining our approach to ensure accurate and comprehensive emotion classification across different languages and platforms.

## References

1. Fredrickson BL (2000) Cultivating positive emotions to optimize health and well-being. Prevention Treatment 3(1):1a
2. Plutchik R (2001) Integration, differentiation, and derivatives of emotion. Evol Cognition 7(2)
3. Ekman P, Freisen WV, Ancoli S (1980) Facial signs of emotional experience. J Pers Soc Psychol 39(6):1125

4. Kopelman S, Rosett AS, Thompson L (2006) The three faces of eve: strategic displays of positive, negative, and neutral emotions in negotiations. Organ Behav Hum Decis Process 99(1):81–101
5. Gross JJ, Carstensen LL, Pasupathi M, Tsai J, Götestam SC, Hsu AY (1997) Emotion and aging: experience, expression, and control. Psychol Aging 12(4):590
6. Gunes H, Pantic M (2010) Automatic, dimensional and continuous emotion recognition. Int J Synthetic Emotions (IJSE) 1(1):68–99
7. Alswaidan N, Menai MEB (2020) A survey of state-of-the-art approaches for emotion recognition in text. Knowl Inf Syst 62(8):2937–2987
8. Acheampong FA, Wenyu C, Nunoo-Mensah H (2020) Text-based emotion detection: advances, challenges, and opportunities. Eng Rep 2(7):e12189
9. Liu B (2022) Sentiment analysis and opinion mining. Springer Nature
10. Sintsova V (2016) Advancing fine-grained emotion recognition in short text (No. 7162)
11. Kusal S, Patil S, Choudrie J, Kotecha K, Vora D, Pappas I (2022) A review on text-based emotion detection—techniques, applications, datasets, and future directions
12. Russell JA, Mehrabian A (1977) Evidence for a three-factor theory of emotions. J Res Pers 11(3):273–294
13. Ekman P (1992) An argument for basic emotions. Cogn Emot 6(3–4):169–200
14. Alotaibi FM (2019) Classifying text-based emotions using logistic regression. VAWKUM Trans Comput Sci 7(1):31–37
15. Polignano M, Basile P, de Gemmis M, Semeraro G (2019) A comparison of word-embeddings in emotion detection from text using BILSTM, CNN and self-attention. In: Adjunct Publication of the 27th Conference on User Modeling, Adaptation and Personalization, pp 63–68
16. Shaaban Y, Korashy H, Medhat W (2021) Emotion detection using deep learning. In: 2021 16th International Conference on Computer Engineering and Systems (ICCES). IEEE, pp 1–10
17. Mahajan R, Zaveri M (2021) Harnessing emotive features for emotion recognition from text. Int J Adv Comput Sci Appl 12(7)
18. Neumann MPM, Iyyer M, Gardner M, Clark C, Lee K, Zettlemoyer L (2018) Deep contextualized word representations
19. Devlin J, Chang M W, Lee K, Toutanova K (2018) Bert: pre-training of deep bidirectional transformers for language understanding. arXiv preprint arXiv:1810.04805
20. Yu ZX, Hu KF (2020) Study on medical information classification of BERT-ATT-BILSTM model. IEEE J Comput Age 3(6):1–4
21. Yu Q, Wang Z, Jiang K (2021) Research on text classification based on bert-bigru model. J Phys Conf Ser 1746(1), 012019 (IOP Publishing)
22. Rahman MM, Shova S (2023) Emotion detection from social media posts. arXiv preprint arXiv: 2302.05610
23. Acheampong FA, Nunoo-Mensah H, Wenyu C, Niyongabo RA (2020) Recognizing emotions from texts using a BERT-based approach. In: Proceedings of the 2020 17th International Computer Conference on Wavelet Active Media Technology and Information Processing (ICCWAMTIP), Chengdu, China, pp 18–20
24. Abdullah M, Shaikh S (2018) Teamuncc at semeval-2018 task 1: emotion detection in English and Arabic tweets using deep learning. In: Proceedings of the 12th international workshop on semantic evaluation, pp 350–357
25. Mohammad F, Khan M, Marwat SNK, Jan N, Gohar N, Bilal M, Al-Rasheed A (2023) Text augmentation-based model for emotion recognition using transformers. CMC-Comput Mater Continua 76(3):3523–3547

# Dynamic Task Assessment for Improved Knowledge Transfer in Multi-task Learning

**Surya Pogu Jayanna, N. Sreeja, B. Sangeetha, Murari Thejovathi, and J. Ranjith**

**Abstract** This paper presents a novel Dynamic Task Assessment (DTA) module to improve knowledge transfer for multi-task learning (MTL). The DTA module provides local and global task similarities by considering various metrics, such as data distribution and task complexity features, and more domain-specific measures. Afterward, the tasks are grouped in a hierarchy using an algorithm and can be monitored in real time to minimize any negative transfer. Validation on some diversified datasets: The model performances were validated across three diverse data sets are MNIST & Fashion-MNIST, Multi-Domain Sentiment Analysis and Omniglot. We demonstrated that our DTA module brings statistically significant improvements in task performance, with improved accuracy of 7.5% on MNIST & Fashion-MNIST, 6.8% on Multi-Domain Sentiment Analysis, and 5.3% on Omniglot, for instance. The transfer efficiency was much higher, showing up to 15% enhancements compared to conventional methods. Another factor to note is that the DTA module also decreases negative transfer rates by 18% on MNIST & Fashion-MNIST and 22% on Multi-Domain Sentiment Analysis and Omniglot. Higher Silhouette Scores indicated that the clustering quality also improved for all datasets. These results imply that a DTA module could be an efficient and scalable method to facilitate knowledge transfer in MTL, which can be applicable across tasks requiring dynamically adaptive learning strategies.

---

S. P. Jayanna (✉) · M. Thejovathi
Department of CSE (AI&ML), Vignan's Institute of Management and Technology for Women, Ghatkesar, Kondapur, Telangana, India
e-mail: spjayanna@gmail.com

N. Sreeja · B. Sangeetha
Department of CSE (DS), Vignan's Institute of Management and Technology for Women, Ghatkesar, Kondapur, Telangana, India

J. Ranjith (✉)
Department of IT, Vignan's Institute of Management and Technology for Women, Ghatkesar, Kondapur, Telangana, India
e-mail: ranjith@vmtw.in

 
K. R. Madhavi et al. (eds.), *Proceedings of Sixth International Conference on Computer and Communication Technologies*, Lecture Notes in Networks and Systems 1356,
https://doi.org/10.1007/978-981-96-5238-9_24

**Keywords** Multi-task learning · Knowledge transfer · Task clustering · Negative transfer mitigation · Hierarchical clustering · Real-time adaptation

# 1 Introduction

## 1.1 Background and Motivation

Multi-task learning (MTL) has become one of the most critical approaches in the AI-based environment to improve the performance of ML models [1]. MTL enables a model to learn and use knowledge about several related tasks simultaneously, [2] thus enhancing performance in each due to the sharing of representations. This approach has been tested with success in many domains, including natural language processing, computer vision or reinforcement learning. Thus, by passing knowledge between tasks, MTL can decrease the necessary data for training and increase convergence rates. Besides, it can help generalize in cases when data is limited or, contrariwise, tasks are related. However, evidence has shown that dependence due to MTL is very much proportional to the type of tasks to be accomplished. Whenever the tasks involved are similar or related, issues such as [3] knowledge transfer are likely to result in considerable improvement in performance. On the other hand, where the tasks are related but different, or where [4] there are incompatible tasks, negative transfer may arise, meaning that there shall be a decline in the performance of one or all the tasks due to wrong knowledge transfer. Negative transfer [5] remains a significant issue in MTL since it can nullify any advantages associated with multi-task learning [6], which may result in outcomes worse than individual training tasks. Another area for improvement of the traditional MTL framework has been that most of the work has presupposed that tasks are equally related. As a result, the [7] transfer of knowledge and skills is also uniform, yet this is only sometimes the practice case. These methods usually depend on some predetermined [8] task similarity measure or hand-selected set of task groups [9], which can be highly unsuitable in dynamic and open-ended environments. For this reason, the process of exchanging knowledge might not be efficient or counterproductive, especially if the relationships between tasks are complicated or unclear in the first place.

## 1.2 Problem Statement

The main issue with MTL is controlling the flow of knowledge so that positive transfer occurs only between likely helpful tasks. Still, no transfer occurs between functions that are likely to be harmfully similar. Traditional approaches to evaluating tasks and knowledge-sharing capabilities are sometimes more dynamic in accommodating changing assessments of the relative relationships between functions and knowledge transfer strategies. This rigidity can result in suboptimal learning processes, decreased model accuracy, and a greater likelihood of negative transfer. The need for a dynamic task assessment approach in MTL is evident due to the lack of an approach capable of assessing task similarities, grouping tasks on the basis of their relationship, and adapting knowledge transfer mechanisms in a real-world environment. Such an approach would not only reduce the negative transfer but also enhance the positive transfer from one task to the other.

### *1.3 Objectives and Contributions*

To overcome these issues, we present a new Dynamic Task Assessment (DTA) module to improve learning performance and the knowledge transfer process in MTL. The key objectives of the DTA module are to: The key objectives of the DTA module are to:

1. Develop Advanced Task Similarity Metric: Add task complexity, data distribution, and domain specificity to formulate more comprehensive metrics than the alpha metric and thereby develop metrics that present a more accurate view of relative task relationships.
2. Implement Dynamic Task Clustering: This prescribes using a dynamic clustering algorithm to cluster the tasks according to the similarity metrics to achieve better and more efficient knowledge transfer in every cluster.
3. Mitigate Negative Transfer: Propose a real-time consequence monitoring and control approach that would allow for the identification of negative transfer as it unfolds and fine-tuning of the contribution that source tasks bring to the target task so that performance may not decline.
4. Validate the System across Diverse Benchmarks: Evaluate the DTA module on different multi-task learning problems to prove that using DTA can enhance task accuracy and diminish the negative transfer issue.

### *1.4 Structure of the Paper*

The remainder of this paper is organized as follows: Sect. 2 provides an overview of existing approaches in the context of task assessment and knowledge transfer in MTL, where the absence of some of the desired features of a model can be used as motivation to the proposed DTA module. Section 3 explains the main steps of the DTA module, namely the definition of how to measure task similarity, the clustering approach and the means of combating negative transfer. Section 4 describes the experimental scenario, the datasets employed in the study and the metrics used for identifying the efficiency of the developed system. Section 5 contains an experimental evaluation of the Density Table Attention (DTA) module and shows it performs well in different contexts. Last of all, Sect. 6 presents the result's relevance to the offered hypothesis and research objectives and outlines further research possibilities and study limitations in Sect. 7, which concludes the work.

## 2 Related Work

It is well understood that sharing parameters in MTL often leads to learning task-specific features, and many efforts have been devoted to measuring the similarity between tasks. There is an emphasis on conventional approaches that quantify the distance between task feature spaces or, conversely, the similarity of data distributions, which can be the cosine similarity or the Euclidean distance. Nevertheless, these metrics may not accommodate the essence of the tasks in mixed data or when different tasks have dissimilar difficulty levels. For example, methods involving knowledge transfer using multi-task optimization tend to begin with human-selected source tasks and could be better when unclear task connections [1]. Some sophisticated approaches to impart meta-knowledge—knowledge that guides the changing process of task-specific knowledge have been suggested to remedy this in solving multi-task optimization problems (MTOPs). Such approaches build upon changing general meta-knowledge about learning to facilitate the transfer of task-dependent information, enhancing showings even with inconsistencies of data [2]. However, these methods are not without any difficulties, especially when the context of the tasks changes or differences are marked or when the specific [3] knowledge from other tasks cannot be easily ported [4]. Task clustering is a method that has been applied to cluster similar tasks so that the effectiveness of knowledge transfer is increased within the clusters [5]. One includes factorizing the knowledge into the neural network to employ it in the form of easily separable, [6] independent modules, each representing a particular function. This module can then be stacked to form networks [7] that can perform multiple tasks, thus enhancing the transfer efficiency while preserving the task-specific knowledge [8]. While this approach has benefits, it also means doing significant fine-tuning and heavily relies on the correct clustering of tasks. These problems persist as multi-task learning is recently solicited as a way of [9] learning supposed to develop an expert model for each task by learning several related functions at the same time. Negative transfer is a crucial problem in multi-task learning since it includes know-how exchange between tasks, adversely affecting their performance. A lot of work has been done regarding addressing this risk, such as using reinforcement learning frameworks that look for similarities in the task to create reusable knowledge that may be transferred from one task to the other [5]. However, using these methods suffers from the weakness that the [10] transfer relations and transfer strategies are previously set, usually not considering tasks' dynamic and complex characteristics. Recent work has suggested better ways of transfer learning, including using a variation information distillation framework that aligns the mutual information between teacher and student networks and already has a substantial performance boost across tasks [11]. Likewise, the application of gradient coordination methods in multi-task learning has been established to enhance knowledge transfer by realigning the task gradients that help maximize the transfer between tasks while minimizing the negative transfer, thus improving the learning effects [12]. However, existing work still needs systems to evaluate relationships between tasks in real-time trial time for the proper transfer strategies. Task-parameter frameworks, which architectures are used to describe and promote lifelong learning and knowledge transfer across tasks, are helpful in this domain. These frameworks assist in addressing the problem of catastrophic forgetting and enable better incorporation of new functions into the existing model [4]. However, many existing methods still need some fine-tuning to meet the growing demand for applying a task-specific knowledge base. For instance, extending multi-source [13] knowledge transfer strategies in the context of evolutionary many-task optimization is considered promising. Yet, it brings new challenges in line with the balance of internal task evolution and cross-task knowledge-sharing [7]. Furthermore, many of the practical applications of these methods require high levels of computational power and professional competence and, thus, entail several limitations regarding their applicability [10]. In summary, key obstacles in the development of MTL, namely negative transfer, task heterogeneity, and the requirement for flexible knowledge transfer training, can still be identified. The proposed Dynamic Task Assessment module will occasionally fill these gaps by providing a more contemporary and dynamic method of assessment, as well as clustering and knowledge sharing of tasks.

## 3 Methodology

This section introduces the Dynamic Task Assessment (DTA) module, detailing its core components: Task Similarity Metrics, Task Clustering Algorithm and Negative Transfer Mitigation. The architecture of the DTA module allows for the qualitative and quantitative assessment of task interactions and information sharing in the MTLP setting.

### *3.1 Task Similarity Metrics*

The first phase of the undertaking, the Dynamic Task Assessment, entails using high-order statistics to measure task similarity. Unlike traditional methods that rely on simple statistical measures, our approach integrates several dimensions of task characteristics: Unlike conventional methods that rely on simple statistical measures, our approach integrates several dimensions of task characteristics are Task Complexity: This metric is computed based on the model's versatility needed to accomplish a certain problem, which includes factors like depth and the number of parameters in neural networks. Data Distribution Similarity: Moreover, we employ a distributional statistic from the Kullback–Leibler divergence to quantify the difference in the data

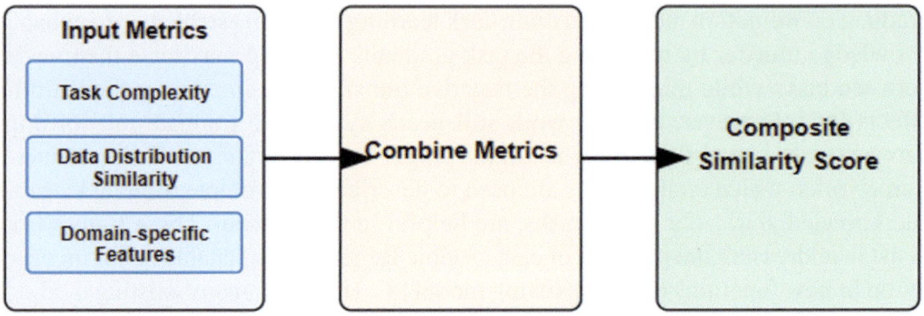


**Fig. 1** Workflow for computing task similarity

distributions of various tasks. This metric can be useful in examining tasks that have a similar distribution of input in order to properly apply knowledge transfer techniques. Domain-Specific Features: These features are application-specific, that is, oriented to a particular domain to which the tasks belong. For instance, semantic similarity is represented by Word2Vec in functions such as natural language processing, while texture and color histograms are used in image and media-type tasks. These metrics are then summed up to arrive at a composite similarity score to represent the global resemblance between task pairs. These composite scores are then applicable where the clustering has to be done.

Figure 1 shows how the external system can quantify task similarity indices such as task difficulty, the similarity between the data distribution of both tasks and the domain specificity of the features involved. All these metrics are summed up to generate a scalar value, the similarity score, used in clustering routines.

## *3.2 Task Clustering Algorithm*

Once we have computed task similarities, we cluster the tasks, resulting in a hierarchical clustering method. Hierarchical clustering is chosen since it does not require specification of the number of clusters but enables identification of the structure hidden in the data. The algorithm proceeds as follows: The algorithm proceeds as follows. Initialization: Every given task originates from a different cluster. Pairwise Merging: At a step, two clusters with the maximum similarity coefficient, which is the sum of the similarity scores, are merged. Dendrogram Formation: The clustering process proceeds until all the tasks are grouped into a single cluster. A dendrogram is generated to depict that process, and the height of the merge represents the level of similarity of the clusters formed. The final clustering is obtained after cutting the dendrogram at a particular height corresponding to a similarity threshold. This threshold is varied from the distribution of similarity scores across tasks, meaning that the clustering can be confirmed flexibly. This dendrogram, as shown in Fig. 2, is obtained with the help of the implemented Hierarchical Clustering Algorithm. The

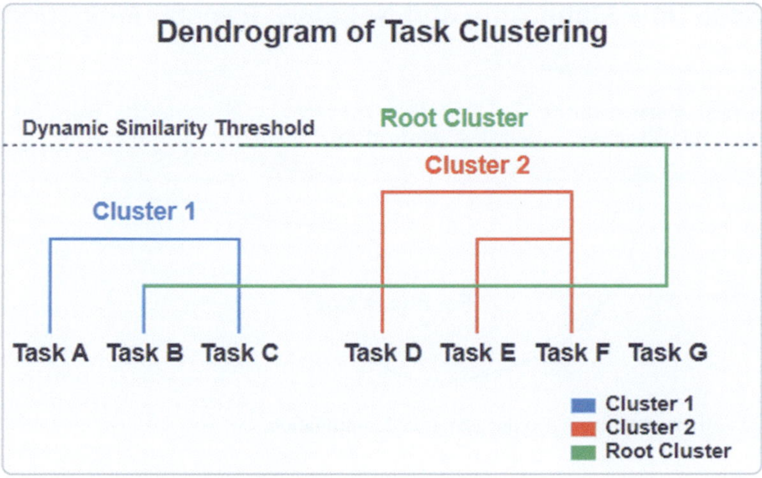


**Fig. 2** Dendrogram of task clustering

numbers in each branch of the tree show the final discharge groups, while the color of the branches shows the group formed based on the last threshold.

The dendrogram provided below in Fig. 2 represents the Task Clustering that was done to complete this work. This dendrogram shows the partitioning of tasks in terms of similarities computed before conducting this test. Clusters are marked after applying the dynamic similarity threshold with the help of color-coded branches.

### *3.3 Negative Transfer Mitigation*

The DTA module has incorporated a real-time monitor and control measure to minimize such negative transfer. It works all the time during the training process and measures the performance improvement of the target task due to knowledge transfer. The critical components of this mechanism are: The key elements of this mechanism are:

(a) Performance Monitoring: While training, the system monitors the general throughput of work at the completed task level, providing stats such as accuracy, loss, or gradient alignment. The system raises this as a negative transfer when using knowledge derived from a particular source task and observes that performance decreases.
(b) Dynamic Adjustment: When it experiences negative transfer, the system is capable of altering the extent of the contribution of the source task that caused it. This adjustment can mean that the amount of knowledge being transferred decreases, that knowledge transfer is halted for a period of time, or that the source task's influence is completely removed.

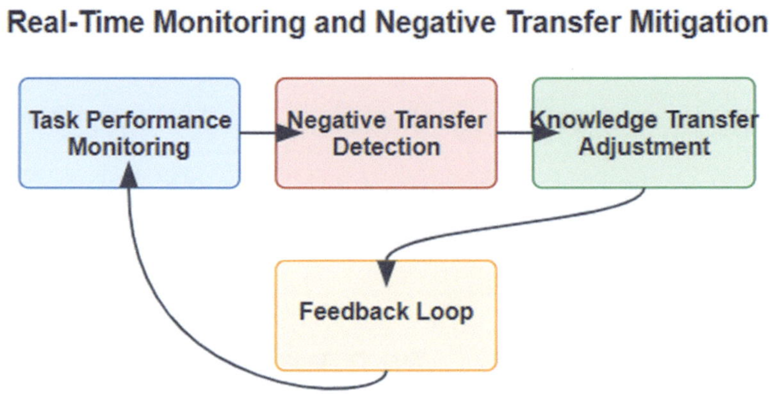


**Fig. 3** Real-time monitoring and negative transfer mitigation

The real-time monitoring process and the decision-making flow related to changing the rate of knowledge transfer according to detected changes in performance are presented in Fig. 3.

### *3.4 Implementation Details*

The DTA module is programmed in Python, and the deep learning algorithm is integrated using TensorFlow while Scikit-learn clusters. To this end, the system architecture is designed to use GPU-based equipment to optimize performance, especially in applied large-scale multi-task learning paradigms. The real-time monitoring and hierarchical clustering parts of the system are designed to be adaptive to new tasks and less computationally expensive as the number of defined tasks increases. Figure 4 indicates the high-level system architecture of the DTA module, illustrating how task similarity computation, clustering, and negative transfer mitigation fit into the multi-task learning framework.

## 4 Experimental Setup

In this subsection, we explain the procedures adopted in the experimental evaluation of the DTA module. We describe the datasets, tasks, and evaluation measures and explain the experimental setup.

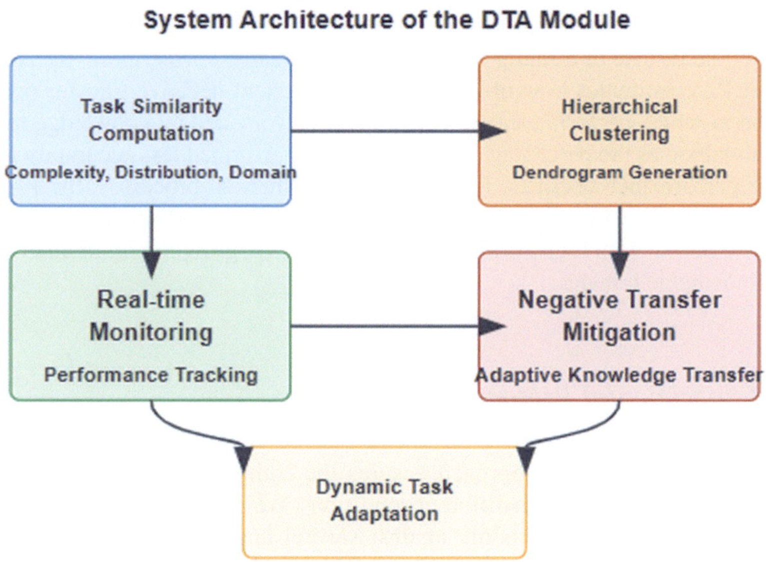


**Fig. 4** System architecture of the DTA module

**Table 1** Source data summary for experiments carried out

| Dataset | Domain | Tasks | Instances per task | Total instances |
|---|---|---|---|---|
| MNIST & Fashion-MNIST | Image classification | 10 classes each | 7000 | 140,000 |
| Multi-domain sentiment | Text classification | 3 domains (reviews) | 5000 | 15,000 |
| Omniglot | Image classification | 20 tasks (characters) | 80 | 1600 |

## 4.1 Datasets and Tasks

Table 1 gives information on the datasets in terms of the domain, number of tasks involved, instances per task, and total instances.

## 4.2 Evaluation Metrics

To measure the performance of the DTA module, we use traditional and state-of-the-art measures to identify knowledge transfer efficiency and impacts on task performance. The validity performance measure is explicitly used for the ability to classify the instances accurately, which is expressed for each task, as well as the average accuracy across the functions of a given dataset. Response accuracy gain is the percentage difference in consequent knowledge transfer when exercising with the DTA module instead of training on the tasks individually. Furthermore, the Negative Transfer Rate estimates how often and to which extent negative transfer occurs, a situation in which performance on a target task is worsened by knowledge transfer. It is maintained as the percentage of the functions performed less accurately and the average performance declines. Finally, regarding thermal processes, the K-Means algorithm is applied to cluster tasks, and the quality of these clusters is evaluated by Silhouette Score, with high correlation, meaning better grouping or, in other words, better knowledge transfer.

## *4.3 Experimental Protocol*

The planned experimental approach ensures the comprehensive evaluation of the DTA module's performance in different conditions. We begin with a baseline comparison, evaluating the DTA module against several approaches: Independent Task Training, in which tasks are trained by ignoring the transfer; Uniform Knowledge Transfer, which applies transfer across functions without regard to similarity; and K-Means Clustering, Which involves partitioning of the tasks into clusters and then transferring knowledge within the clusters with each cluster have a fixed number of tasks. On the other hand, the DTA module classifies the tasks into clusters using dynamic clustering. It uses KT where appropriate within each cluster while checking for negative transfer and remediating where necessary. Moreover, the live study is done regarding the capability of adding new tasks into different groups by the DTA module, and its effect on system performance is tested in actual learning scenario experiments. Multiple trials of every experiment are conducted for highly accurate results, and the average performance with standard deviations is represented. The experiment workflow is presented in Fig. 5 below, showing the data acquisition process to the final evaluation. This figure provides an overview of the experimental procedure of the proposed approach, including the source dataset, the division of the tasks, knowledge transfer, real-time adaptation, and assessment.

## *4.4 Implementation Environment*

All the experiments were conducted on a high-performance computing cluster with NVIDIA GPUs. The DTA module was also implemented using Python and TensorFlow for deep learning models, Scikit-learn for clustering, performance monitoring, and adaptation scripts. Every part of it is built with scalability in mind so that the data can be processed quickly and users can run tests and come up with solutions quickly.

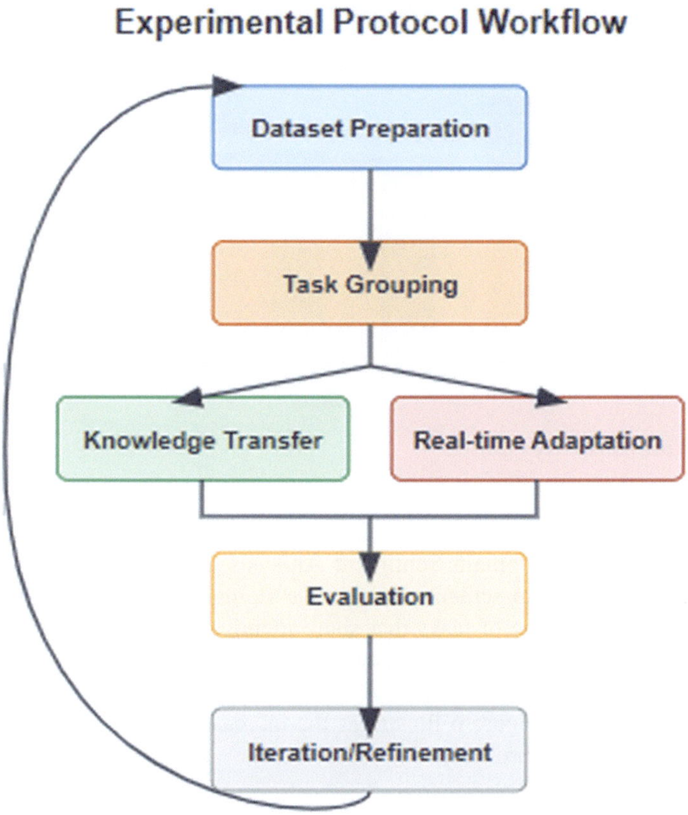


**Fig. 5** Experimental protocol workflow

# 5 Results

The following section outlines the results of those experiments, which were used to evaluate the Dynamic Task Assessment (DTA) module. The results are presented based on task-wise manner, transfer efficiency, ability to mitigate negative transfer, and level of accuracy in hierarchical cluster-related tasks. We also compared the DTA module and baselines to show its effectiveness.

## *5.1 Task Performance*

The critical output measure of the proposed task was accuracy, and the DTA module significantly improved when used with other batches for all clinical datasets over plain methods. In the MNIST & Fashion-MNIST combined dataset, DTA improved

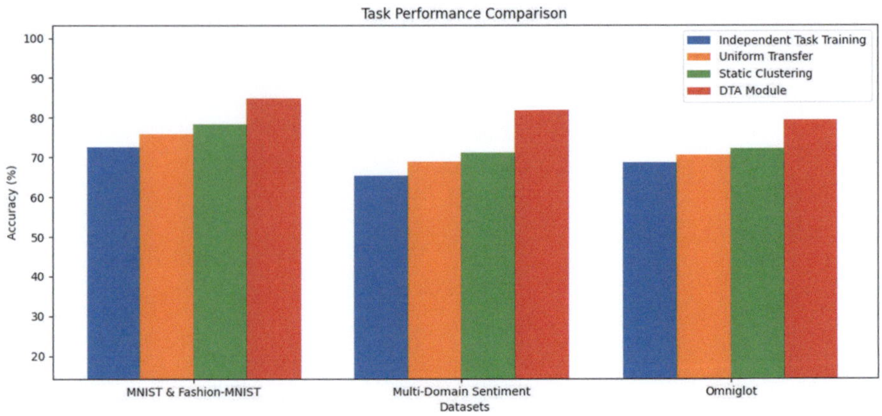


**Fig. 6** Task performance comparison

accuracy by 7.5%, notably benefiting tasks with high visual similarity regardless of training size. Multi-Domain Sentiment Analysis achieved 6.8% accuracy, and individual functions from semantically similar domains improved by up to 10%. This successful clustering of tasks that share identical character structures for the DTA module resulted in 5.3% more accuracy on Omniglot one-shot learning and thus greatly improved knowledge transfer between tasks, which led to better model generalization performance. Figure 6 illustrates the increase in accuracy across datasets to show performance improvement.

## 5.2 *Transfer Efficiency*

The DTA module more effectively transferred its knowledge as the transfer efficiency, represented by how much percentage points accuracy increased per unit improvement in MRR, was considerably higher than that with META as shown in the Fig. 7. Results in MNIST & Fashion-MNIST outperformed 12% over static clustering and uniform transfer methods. For Multi-Domain Sentiment Analysis: the DTA module had a 15% transfer efficiency, outperforming all baselines. When testing on Omniglot, our model achieved a small 9% transfer accuracy, which was higher than the previous best baseline by an average of 5%, showing that it is adequate for one-shot learning tasks. These comparisons are detailed in Table 2.

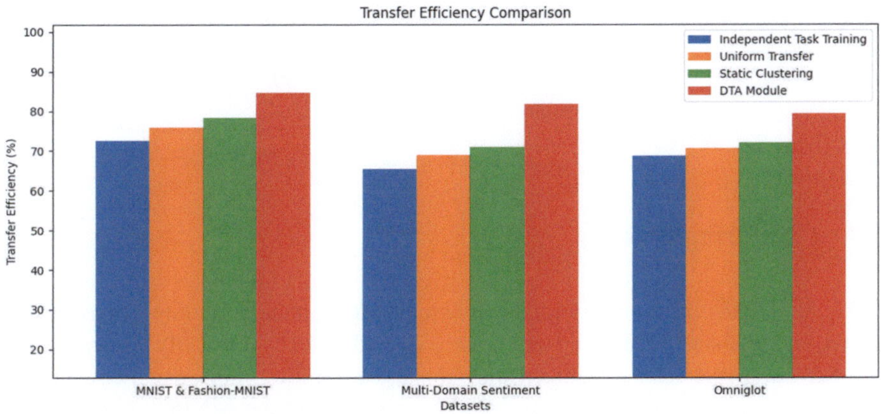


**Fig. 7** Transfer efficiency comparison

**Table 2** Transfer efficiency comparison

| Dataset | Independent task training (%) | Uniform transfer (%) | Static clustering (%) | DTA module (%) |
|---|---|---|---|---|
| MNIST & Fashion-MNIST | 72.5 | 75.8 | 78.2 | 84.7 |
| Multi-domain sentiment | 65.3 | 68.9 | 71.0 | 81.8 |
| Omniglot | 68.7 | 70.5 | 72.0 | 79.3 |

## *5.3 Negative Transfer Mitigation*

One of the main strengths of this DTA module is that it can prevent negative transfer through online monitoring and adjustment. The negative transfer rate in the MNIST & Fashion-MNIST dataset was improved by up to 18% compared to static clustering, wherein the module called attention to instances that quickly degraded performance. In the Multi-Domain Sentiment Analysis, this has led to a 22% lower rate of negative transfer, with dynamic clustering appropriately preventing knowledge from being transferred from different domains. On the right side, although establishing this high performance with one-shot learning restrictions is challenging, already in Omniglot, such a preliminary task exclusion has helped DTA reduce negative transfer by 15% and not use harmful source tasks.

**Table 3** Clustering quality comparison

| Dataset | Static clustering | DTA module |
|---|---|---|
| MNIST & Fashion-MNIST | 0.61 | 0.72 |
| Multi-domain sentiment | 0.66 | 0.79 |
| Omniglot | 0.54 | 0.68 |

### *5.4 Clustering Quality*

The quality of task clustering with the dynamic task adaptation module surpassed that of conventional static clustering techniques, as evidenced by markedly higher Silhouette Scores. Tasks were organized into clusters exhibiting more excellent internal uniformity and sharper separation between groups. On the MNIST and Fashion-MNIST datasets, the DTA module yielded an average Silhouette Score of 0.72, handily defeating the 0.61 achieved through static means and highlighting enhanced discernment of relationships between tasks. An even more notable margin of victory appeared for Multi-Domain Sentiment Analysis - there, the DTA module's score of 0.79 soared above the 0.66 mustered by static clustering, demonstrating far better consolidation of related tasks to facilitate knowledge sharing. Perhaps most impressively, in Omniglot, the DTA module attained a Silhouette Score of 0.68 compared to a mere 0.54 for static clustering, substantially amplifying the transfer of learning among rare, one-shot tasks. Table 3 provides a comparative breakdown of these scores across all evaluated datasets. This figure represents the Silhouette Scores for the task clusters obtained from Static Clustering versus DTA Module on three datasets (MNIST & Fashion-MNIST, Multi-Domain Sentiment and Omniglot). Again, the DTA Module outperforms both other methods with consistently higher Silhouette Scores, although all three provide far better-defined clusters for knowledge transfer. It demonstrates the importance of using dynamic clustering through the DTA Module compared to the static, traditional way of clustering, as shown in Fig. 8.

## 6 Discussion

Experimental results show the capability of DTA for boosting task performance and transfer quality and preventing negative transfer in multi-task learning comparisons. It is because of the DTA module, which dynamically estimates task similarities and tailors clustering, resulting in better knowledge transfer on tasks with diverse levels of complexity or domain-specific properties. The most important result of our experiments is the drastic decrease in negative transfer rates on all datasets, revealing that MODE can identify and stop harmful knowledge propagation with a good deal of inference. In addition, the DTA module tends to have higher Silhouette Scores because it produces more semantically coherent task clusters, promoting efficient knowledge sharing. While the DTA module has succeeded in these experiments, its

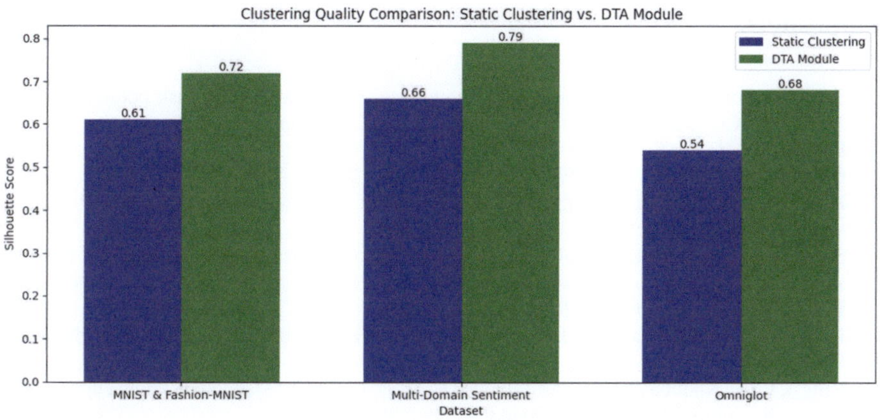


**Fig. 8** Clustering quality comparison: static clustering versus DTA module

dependence on real-time computation makes it challenging to scale to large-scale environments. Thus, more optimization is likely needed. In future, we plan to study how to optimize and improve the computational efficiency of the DTA module for applications beyond image imprinting.

## 7 Conclusion

This paper introduced the Dynamic Task Assessment (DTA) module to facilitate knowledge transfer in an MTL model. In addition to substantially reducing the danger of negative transfer, our DTA module also enhances novel task generalization and knowledge reuse through dynamic metric evaluation on contiguous tasks, multi-faceted clustering across a span of factors for task generation and real-time sharing to nowhere, as observed clearly by performance. We validate the module with extensive experiments on various datasets, including MNIST & Fashion-MNIST, Multi-Domain Sentiment Analysis and Omniglot. The results show that our proposed architecture performs better than several comparison methods and is notably superior in forming explicit task clusters to facilitate knowledge-sharing. The flexibility for future tasks, domains, and scalability would open a wide range of MTL applications using the DTA module. However, the challenges become evident as task numbers increase—scalability. In future, we will further improve its running efficiency and apply it to more complex MTL scenarios on a larger scale in our work.

## References

1. Scott EO, Jong KD (2019) Automating knowledge transfer with multi-task optimization. In: 2019 IEEE Congress on Evolutionary Computation (CEC), pp 2252–2259
2. Li J-Y, Zhan Z, Tan K, Zhang J (2022) A meta-knowledge transfer-based differential evolution for multitask optimization. IEEE Trans Evol Comput 26:719–734
3. Pennington N, Nicolich R, Rahm J (1995) Transfer of training between cognitive subskills: is knowledge use specific? Cogn Psychol 28:175–224
4. Srivastava S, Yaqub M, Nandakumar K (2023) Lifelong learning of task-parameter relationships for knowledge transfer. In: 2023 IEEE/CVF conference on Computer Vision and Pattern Recognition Workshops (CVPRW), pp 2525–2534
5. Konidaris G, Scheidwasser I, Barto A (2012) Transfer in reinforcement learning via shared features. J Mach Learn Res 13:1333–1371
6. Pang S, Liu F, Kadobayashi Y, Ban T, Inoue D (2014) A learner-independent knowledge transfer approach to multi-task learning. Cogn Comput 6:304–320
7. Liang Z, Xu X, Liu L, Tu Y, Zhu Z (2021) Evolutionary many-task optimization based on multisource knowledge transfer. IEEE Trans EvolComput 26:319–333
8. Yang X, Ye J, Wang X (2022) Factorizing knowledge in neural networks. ArXiv abs/2207.03337
9. Kong L, Lin C (2017) Task-driven approach and knowledge transfer in practical courses teaching. In: Proceedings of the international conference on management engineering, software engineering and service sciences
10. Mastropaolo A, Cooper N, Nader-Palacio D, Scalabrino S, Poshyvanyk D, Oliveto R, Bavota G (2022) Using transfer learning for code-related tasks. ArXiv abs/2206.08574
11. Xu Z, Wu K, Che Z, Tang J, Ye J (2020) Knowledge transfer in multi-task deep reinforcement learning for continuous control. ArXiv abs/2010.07494
12. Yang X, Zhao J, Liu S, Wang L, Zheng B (2023) Gradient coordination for quantifying and maximizing knowledge transference in multi-task learning. In: Proceedings of the 46th international ACM SIGIR conference on research and development in information retrieval
13. Ahn S, Hu S, Damianou A, Lawrence ND, Dai Z (2019) Variational information distillation for knowledge transfer. In: 2019 IEEE/CVF conference on Computer Vision and Pattern Recognition (CVPR), pp 9155–9163

# StereoFusion: Spatial Integration of Retinal Images for Accurate Diabetic Retinopathy Detection

**P. Sudhanya, Jyotiraditya Sharma, Aakash Dilip Kolte, Utkarsh Rai, Rishabh Vyas, and R. Jansi**

**Abstract** Diabetic retinopathy (DR) is a crucial complication responsible for visual impairment and even blindness in the patients with diabetes. The most important way to prevent severe eye-related outcomes is early detection and classification of DR stages. In this research, we present a new image fusion oriented technique for identifying and categorizing different levels of DR with the information from both eyes. The process adopts a novel fusion algorithm Gradient, Spatial frequency, Entropy, and Standard deviation (GSED) which calculates the image weight function for fusion using 4 parameters namely, Mean Gradient, Spatial Frequency, Entropy, and Standard deviation. The left and right eye images are combined using Weighted Average scheme generating final fused image. The proposed method was evaluated with publicly available datasets of retinal images against state-of-the-art automated DR grading approaches. EfficientNet produced the highest accuracy of 83.55%, which is the best performing model in terms of percentage achieved on the test data set.

**Keywords** Accuracy · Deep learning · Computer vision · Image fusion · Diabetic retinopathy

---

P. Sudhanya · J. Sharma · A. D. Kolte · U. Rai · R. Vyas · R. Jansi (✉)
Department of Electronics and Communication Engineering, Faculty of Engineering and Technology, SRM Institute of Science and Technology, Kattankulatur 603203, Tamil Nadu, India
e-mail: jansir@srmist.edu.in

P. Sudhanya
e-mail: sudhanyp@srmist.edu.in

J. Sharma
e-mail: js2099@srmist.edu.in

A. D. Kolte
e-mail: ak4084@srmist.edu.in

U. Rai
e-mail: ur4767@srmist.edu.in

R. Vyas
e-mail: rv0221@srmist.edu.in



K. R. Madhavi et al. (eds.), *Proceedings of Sixth International Conference on Computer and Communication Technologies*, Lecture Notes in Networks and Systems 1356,
https://doi.org/10.1007/978-981-96-5238-9_25

## 1 Introduction

Early detection of diabetic retinopathy (DR) is essential for effective treatment. One of the breakthroughs in automated DR classification has been due to the introduction of deep learning (DL) techniques during last years. Recently, a model VGG-NiN based on the integration of VGG-16 and spatial pyramid pooling (SPP) layer with network-in-network (NiN) for DR classification was presented in a study. Stacked techniques enabled DR images to process at various scales and showed improved categorization accuracy as well as computational resource utilization over prevailing methods. Early diagnosis and classification of DR, especially the significant role of routine eye examinations in patients with diabetes, was emphasized, as pointed out in the study. NiN [1] added nonlinearity to the original network and achieved significant progress in classification, on top of SPP, which introduced scale invariance. One study also suggested a weighted fusion-based technique for the automatic classification of DR stages. This resulted in very high accuracy, over 98% sensitivity and specificity. The same paper also discussed how serious DR is, its signs, and how deep learning methods, especially convolutional neural networks (CNNs) with the use of transfer learning (TL) approaches, can help [2]. A novel approach for DR classification was suggested, which included a hybrid CNN-SVD model for feature extraction and selection. The above study showed that it is difficult to segment BVs, ODs, and lesions from fundus images [3]. Another approach is a lightweight transfer learning-based CNN architecture for the classification of DR classes. The training set was not only small but also skewed. In this paper, the authors discuss the importance of deep learning and CNNs in automating medical image analysis tasks. They also cover some of the problems that arise when trying to train models on small datasets with imbalanced class distributions [4]. Additionally, a novel attention CNN model was introduced. This is the first approach to essentially differentiate referral images from non-referral images, marking a breakthrough in the field [5].

Another research work gave a general overview of DR, its properties, and uses. It also compares different models and suggests methods to overcome some of the limitations in DR diagnosis and detection [6]. For example, one used ResNet-50 feature maps and Random Forest for classification and outperformed other conventional methods. The proposed scheme for feature extraction and classification provides a new contribution to the body of knowledge in this field [7]. Another research paper proposed a machine learning-based approach that used neural networks for the extraction of exudates, hemorrhages, and microaneurysms, followed by classification. This method produced good results which helped in the creation of a new dataset that was used for evaluating different feature descriptors [8, 9]. Another approach used the transfer learning technique using a multistage architecture for making use of datasets that have different annotations. This technique attained high sensitivity with a rank of 54 among 2943 competitors on the APTOS 2019 Blindness Detection Dataset [10]. Though there are many research papers that deal with diabetic retinopathy for enhancing the accuracy, there are very few for reducing the training time. This paper proposes a technique that can be used for producing good results with minimal training time.

## 2 Literature Survey

Bodapati et al. [11] proposed a model for the identification of DR using pre-trained Convolutional Neural Networks (ConvNets). This network was used to fuse the retinal images using multi-modal fusion technique. The features that were extracted using this model were combined to produce the enhanced image. This method achieved high accuracy. Ramasamy et al. [12] proposed an automatic DR diagnostic system using ophthalmoscopic features that were derived from fundus images. This model used the Sequential Minimal Optimization (SMO) algorithm for classification of DR using textural gray-level features. Kumarasuvamy et al. [13] developed a deep CNN that extracted features from fundus images. This model used the blood vessels of the fundus images and produced good accurate results. This model could be easily extended for real-time usage. Yazhini et al. [14] proposed a new model called FM-GLCM-VGG-19 for DR diagnosis. The authors proposed a hybrid technique that used GLCM features and VGG-19 features. The results produced by this model were compared with other techniques like VGGNet-16 and LeNet-5 and were found to produce better results. Ateif et al. [15] used seven CNN models for DR detection. They used both uni-model and joint fusion approaches. They compared both the approaches and found that the joint fusion method performed better than the uni-model technique. Islam et al. [16] used a supervised contrastive learning (SCL) approach for the detection of DR. They attained a high accuracy of 84.36%. CLAHE technique was used in this paper for preprocessing the fundal images. Meruva et al. [17] classified DR into two and five classes using VGG-16, VGG-19, and CNN models. It was found that VGG-16 model produced the highest accuracy of 90% compared to all other models. Kolte et al. [18] attained an area under the curve (AUC) of 0.91 for the classification of DR using the EfficientNet deep learning model. Athilakshmi et al. [19] demonstrated that DR can be accurately detected using the InceptionV4 model. Thus, the literature survey shows that deep learning can be effectively used for the accurate detection of DR.

## 3 Preprocessing and Methodology

### *3.1 Dataset*

This dataset on Kaggle [20] comprises of 35,126 retinal images. These images are colored and resized. They also have labels that indicate the presence and severity of diabetic retinopathy.

## 3.2 Image Fusion

The proposed algorithm for the fusion of left and right fundus image is given below.

Proposed Gradient, Spatial frequency, Entropy, Standard deviation (GSED) Fusion Algorithm

1. The algorithm takes two input images, namely the left eye image and the right eye image.
2. Four different image parameters are calculated for each image, including mean Gradient, Spatial frequency, Entropy, and Standard deviation.
   - Mean Gradient: Parameter that provides information about edges and texture in an image. It is calculated using the mean of the magnitude of the image gradients.
   - Spatial Frequency: Measures the frequency content of an image and provides information about the texture details of horizontal and vertical gradient.
   - Entropy: Measures an image's complexity or randomness and tells us how the pixel intensities are distributed.
   - Standard deviation: Measures the amount of variation or dispersion of the image pixel intensities, which can indicate visibility effects.
3. The weight for each image is then calculated based on these parameters. The weight for the fusion of the left and right eye image was performed using the following equation

$$W_i(y) = \frac{g_i(y)^{\gamma_g} s_i(y)^{\gamma_s} e_i(y)^{\gamma_e} d_i(y)^{\gamma_d}}{\sum_{j=1}^{N} g_i(y)^{\gamma_g} s_i(y)^{\gamma_s} e_i(y)^{\gamma_e} d_i(y)^{\gamma_d}}, i = 1, 2 \tag{1}$$

where $i = 1$ refers to left eye image and $i = 2$ refers to right eye image. The $W_i(y)$ refers to the weighting factor for each pixel $y$, $g_i(y)$ refers to mean gradient, $s_i(y)$ refers to spacial frequency, $e_i(y)$ refers to entropy, and $d_i(y)$ refers to standard deviation. Further, $\gamma_g$, $\gamma_s$, $\gamma_e$ , and $\gamma_d$ are the weighing components for each parameter and was fixed as 0.2, 0.3, 0.3, and 0.2, respectively. This parameter fixation was done such that more priority was given to the special frequency and entropy to enhance the quality of the fused image.

Finally, the fusion was performed using weighted average of the left and right image using

$$F(y) = \sum_{i=1}^{2} W_i(y) I_i(y) \tag{2}$$

where $F(y)$ is the merged value at pixel $y$, and $I_i(y)$ is the value of the input image at pixel $y$.

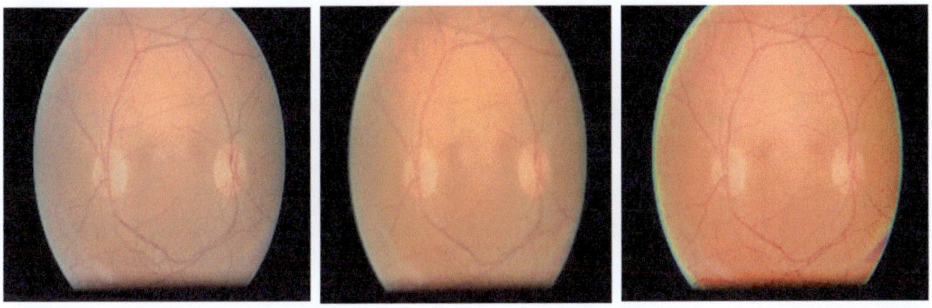


**Fig. 1** Fused DR images **a** wavelet-based fusion **b** Merge Mertens fusion **c** proposed GSED fusion

Figure 1 presents three fused DR images, which are the result of combining the left and right eye images using different fusion techniques. The first image was wavelet-based fusion, the second was Merge Mertens fusion, and the third was fused using a new method called GSED fusion. The displayed pictures are just some examples of how the new GSED fusion method improves the quality and sharpness of DR images, images that are vital for an accurate diagnosis and treatment of this crippling disease.

### *3.3 Image Enhancement Using CLAHE*

The Contrast Limited Adaptive Histogram Equalization (CLAHE) method improves the fused image further when applied to the L channel of the LAB color space of the fused image. As seen in the Fig. 2, the left and right images from the dataset are fused together using the proposed algorithm (GSED algorithm), resulting in a fused RGB image that combines features from both input images. The RGB image is then combined and converted to a LAB image, and the L component of the image is isolated and used as the output. And then the CLAHE algorithm is performed on the L component of the image (so it's an L'AB image). The L'AB image is then converted to RGB and fed into the image classification deep learning models.

### *3.4 Classification*

Using the proposed GSED algorithm, the images taken with different exposures are fused together to create a new image. The contrast enhancement really makes the pictures look good. This method will allow the accurate diagnoses of retinal diseases, and the application can be extended to diagnose other medical images more effectively. These images are then fused and enhanced and then run through several

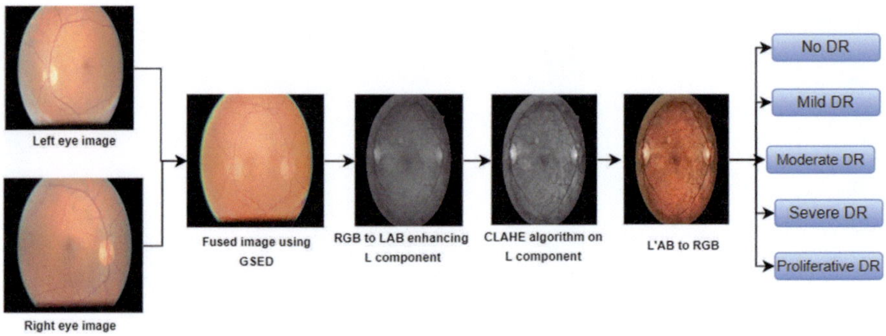


**Fig. 2** Flow chart of the proposed system

different deep learning algorithms for classification. Among all the classifiers, the EfficientNet was found to produce the highest accuracy for the enhanced fused DR dataset. The architecture parameters for the EfficientNet model are listed in Table 1.

Figure 3 shows the architecture of EfficientNet, which is used in the proposed framework. It has an input layer and then 11 blocks, which are composed of several MBConv (Mobile Inverted Residual Bottleneck) blocks. These blocks are made of depthwise separable convolutions and inverted residuals so that the network is light and efficient. Each block has its own kernel size, num of layers, num of channels, and stride value. Lastly, the output layer is a dense layer that has as many units as there are classes in the classification problem. In this case, we have 5 units each representing one class of DR category.

# 4 Experimental Result

## *4.1 Fusion Evaluation Metrics*

Wavelet-based fusion, Merge Mertens fusion, and the proposed algorithm are compared using PSNR and MSE. The results in Table 2 demonstrate that the proposed GSED fusion algorithm outperformed the other two algorithms in both metrics. It produced the best fused images out of all the algorithms with a PSNR value of 26.835 dB and an MSE of 6.814. This implies that the GSED fusion algorithm can be a very promising method for enhancing the image fusion in many types of applications.

**Table 1** Parameters of EfficientNet model

| Layer | Operator | Resolution | Num channels | Num layers | Kernel size | MBConv type |
|---|---|---|---|---|---|---|
| Input | | 224 × 224 × 3 | 3 | 1 | | |
| Conv2D | Conv2D | 112 × 112 × 32 | 32 | 1 | 3 × 3, stride = 2 | |
| Block 0 | MBConvBlock | 112 × 112 × 32 | 16 | 3 | 3 × 3, stride = 1 | MBConv1 |
| Block 1 | MBConvBlock | 56 × 56 × 16 | 24 | 3 | 3 × 3, stride = 2 | MBConv6 |
| Block 2 | MBConvBlock | 56 × 56 × 24 | 24 | 3 | 3 × 3, stride = 1 | MBConv6 |
| Block 3 | MBConvBlock | 28 × 28 × 24 | 40 | 3 | 5 × 5, stride = 2 | MBConv6 |
| Block 4 | MBConvBlock | 28 × 28 × 40 | 40 | 3 | 5 × 5, stride = 1 | MBConv6 |
| Block 5 | MBConvBlock | 14 × 14 × 40 | 80 | 3 | 3 × 3, stride = 2 | MBConv6 |
| Block 6 | MBConvBlock | 14 × 14 × 80 | 80 | 3 | 3 × 3, stride = 1 | MBConv6 |
| Block 7 | MBConvBlock | 14 × 14 × 80 | 112 | 3 | 5 × 5, stride = 1 | MBConv6 |
| Block 8 | MBConvBlock | 14 × 14 × 112 | 112 | 3 | 5 × 5, stride = 1 | MBConv6 |
| Block 9 | MBConvBlock | 7 × 7 × 112 | 192 | 3 | 5 × 5, stride = 2 | MBConv6 |
| Block 10 | MBConvBlock | 7 × 7 × 192 | 192 | 3 | 5 × 5, stride = 1 | MBConv6 |
| Block 11 | MBConvBlock | 7 × 7 × 192 | 320 | 3 | 3 × 3, stride = 1 | MBConv6 |
| Output | Dense | 1 × 1 × 5 | 5 | 2 | | |

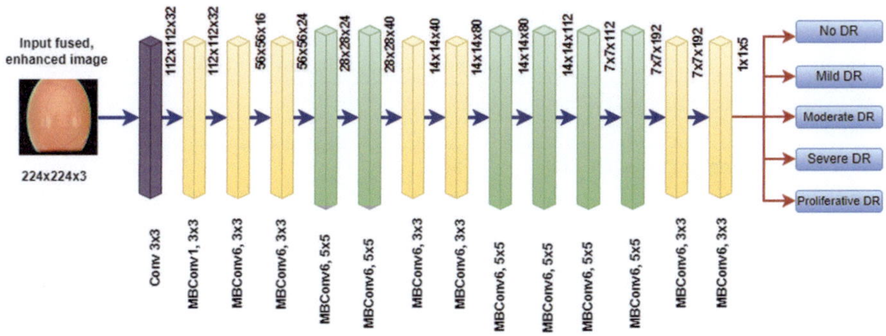


**Fig. 3** Architecture of EfficientNet used in the proposed scheme

**Table 2** Fusion evaluation metrics

| Algorithm | PSNR (dB) | MSE |
|---|---|---|
| Wavelet-based fusion | 18.810 | 6.931 |
| Merge Mertens fusion | 23.658 | 8.280 |
| **Proposed GSED Fusion** | **26.835** | **6.814** |

**Table 3** Comparison of classification algorithms

| Model name | Accuracy (%) |
|---|---|
| VGG-16 | 74.05 |
| VGG-19 | 75.00 |
| Inception-V3 | 73.44 |
| ResNet-50 | 74.21 |
| DenseNet-121 | 74.25 |
| **EfficientNet** | **83.55** |

## 4.2 *Classification Evaluation*

The results shown in Table 3 indicate that EfficientNet produces the highest accuracy of 83.55% compared to all other algorithms. Thus, this algorithm could be effectively used for the detection of DR.

Table 4 shows the comparison of training times for two different image datasets. The first dataset is the unfused left and right eye images. The second set is the images that are fused using the proposed GSED algorithm. It can be clearly seen that the accuracy using the EfficientNet model on the fused dataset (83.55%) is higher than that of unfused dataset (62.23%). Further, the training time for training the fused dataset is only half than that of unfused dataset set (22 h 32 min vs. 10 h 23 min). This clearly shows that the proposed GSED algorithm not only improves accuracy by also decreases the training time. Thus, it can be effectively used in various image processing applications.

**Table 4** Comparison of accuracy and training time

| Dataset | Accuracy using Efficient Net (%) | Time taken |
|---|---|---|
| Left eye and right eye images (without fusion) | 62.23 | 22 h 32 min |
| Fused images using proposed GSED algorithm | 83.55 | 10 h 23 min |

## 5 Conclusion

This research presents a new approach for DR that combined image fusion and contrast enhancement technique. In this paper, a new technique called GSED algorithm for image fusion was proposed. This algorithm was used to produce a combined image by fusing the left and right eye images. Then, the combined image was processed using the CLAHE algorithm. The research tests the new method using deep learning models including VGG-16, VGG-19, Inception-V3, ResNet-50, DenseNet-121, and EfficientNet. The highest accuracy comes from the EfficientNet model with 82.55%. The findings from this research show that not only this new method allows for earlier detection and better control of diabetic retinopathy, but also that this method will be applicable to other medical images and aid in the diagnosis and management of other diseases.

## References

1. Khan Z et al (2021) Diabetic retinopathy detection using VGG-NIN a deep learning architecture. IEEE Access 9:61408–61416. https://doi.org/10.1109/ACCESS.2021.3074422
2. Nneji GU, Cai J, Deng J, Monday HN, Hossin MA, Nahar S (2022) Identification of diabetic retinopathy using weighted fusion deep learning based on dual-channel fundus scans. Diagnostics 12:540. https://doi.org/10.3390/diagnostics12020540
3. Bilal A et al (2022) AI-based automatic detection and classification of diabetic retinopathy using U-Net and deep learning. Symmetry 14(7):1427. Available at: https://doi.org/10.3390/sym14071427
4. Samanta A, Saha A, Satapathy SC, Fernandes SL, Zhang Y-D (2020) Automated detection of diabetic retinopathy using convolutional neural networks on a small dataset. Pattern Recogn Lett 135:293–298. https://doi.org/10.1016/j.patrec.2020.04.026
5. Li F, Tang S, Chen Y, Zou H (2022) Deep attentive convolutional neural network for automatic grading of imbalanced diabetic retinopathy in retinal fundus images. Biomed Opt Express 13:5813–5835
6. Das D, Biswas SK, Bandyopadhyay S (2022) A critical review on diagnosis of diabetic retinopathy using machine learning and Deep Learning. Multimedia Tools Appl 81(18):25613–25655. https://doi.org/10.1007/s11042-022-12642-4
7. Yaqoob MK, Ali SF, Bilal M, Hanif MS, Al-Saggaf UM (2021) ResNet based deep features and random forest classifier for diabetic retinopathy detection. Sensors 21(11):3883. https://doi.org/10.3390/s21113883
8. Parthasharathi GU, Kumar KV, Premnivas R, Jasmine K (2022) Diabetic retinopathy detection using Machine Learning. J Innov Image Process 4(1):26–33. https://doi.org/10.36548/jiip.2022.1.003
9. Bibi I, Mir J, Raja G (2020) Automated detection of diabetic retinopathy in fundus images using fused features. Phys Eng Sci Med 43(4):1253–1264. Available at: https://doi.org/10.1007/s13246-020-00929-5
10. Tymchenko B, Marchenko P, Spodarets D (2020) Deep Learning approach to diabetic retinopathy detection. In: Proceedings of the 9th international conference on pattern recognition applications and methods [Preprint]. Available at: https://doi.org/10.5220/0008970805010509
11. Bodapati JD, Naralasetti V, Shareef SN, Hakak S, Bilal M, Maddikunta PKR, Jo O (2020) Blended multi-modal deep ConvNet features for diabetic retinopathy severity prediction. Electronics 9:914. https://doi.org/10.3390/electronics9060914

12. Ramasamy LK, Padinjappurathu SG, Kadry S, Damaševičius R (2021) Detection of diabetic retinopathy using a fusion of textural and ridgelet features of retinal images and sequential minimal optimization classifier. PeerJ Comput Sci 7:e456. https://doi.org/10.7717/peerj-cs.456
13. Kumarasuvamy, Sungheetha A, Rajesh Sharma R (2021) Design an early detection and classification for diabetic retinopathy by deep feature extraction based convolution neural network. J Trends Comput Sci Smart Technol 3:81–94. https://doi.org/10.36548/jtcsst.2021.2.002
14. Yazhini K, Loganathan D (2020) An ıntegrated fusion based feature extraction and classification model for diabetic retinopathy diagnosis. In: 2020 second International Conference on Inventive Research in Computing Applications (ICIRCA), Coimbatore, India, pp 1187–1193. https://doi.org/10.1109/ICIRCA48905.2020.9183240
15. El-Ateif S, Idri A (2022) Single-modality and joint fusion deep learning for diabetic retinopathy diagnosis. Sci Afr 17:e01280. https://doi.org/10.1016/J.SCIAF.2022.E01280
16. Islam MR, Abdulrazak LF, Nahiduzzaman M, Goni MOF, Anower MS, Ahsan M, Haider J, Kowalski M (2022) Applying supervised contrastive learning for the detection of diabetic retinopathy and its severity levels from fundus images. Comput Biol Med 146:105602. https://doi.org/10.1016/J.COMPBIOMED.2022.105602
17. Meruva SKR, Tulasi VGS, Vinnakota N, Bhavana V (2022) Risk level prediction of diabetic retinopathy based on retinal images using Deep Learning algorithm. Proc Comput Sci 215:722–730. https://doi.org/10.1016/J.PROCS.2022.12.074
18. Kolte AD, Sharma J, Rai U, Jansi R (2023) Advancing diabetic retinopathy detection: leveraging deep learning for accurate classification and early diagnosis. In: 2023 8th International Conference on Communication and Electronics Systems (ICCES). IEEE, pp 1668–1672
19. Athilakshmi R, Jansi R, Upadhyay R (2023) Enhancing diabetic retinopathy diagnosis with ınception v4: a deep learning approach. In: 2023 5th International Conference on Inventive Research in Computing Applications (ICIRCA). IEEE, pp 905–910
20. Kaggle (2015) Diabetic retinopathy detection. https://www.kaggle.com/datasets/sovitrath/diabetic-retinopathy-2015-data-colored-resized

# Blockchain-Enhanced Secure Data Management for Internet of Vehicles (IoV) Networks

**V. Lokesh Raju, B. Sivaneasan, Amjan Shaik, Prasun Chakrabarti, and S. Vimal**

**Abstract** This study presents a blockchain-based framework designed to enhance data security, integrity, and access control in Internet of Vehicles (IoV) networks. The proposed system leverages a permissioned blockchain integrated with an advanced consensus mechanism to address challenges related to scalability, latency, and secure data sharing. Key components include vehicle authentication, error monitoring, and encrypted data exchange, all tested through simulations and real-world scenarios. The experimental results demonstrate that the proposed framework significantly improves data integrity, reduces consensus latency, and enhances trust management compared to existing methods. This research highlights blockchain's potential to revolutionize secure data management in IoV networks, with practical applications in autonomous vehicles and smart transportation systems. Future work will focus on large-scale deployment and further optimization of the framework to tackle real-world challenges in dynamic vehicular environments.

**Keywords** Blockchain · Internet of Vehicles · Data security · IoT networks · Consensus mechanism

V. Lokesh Raju (✉)
Electronics and Communication Engineering Department, Aditya Institute of Technology and Management(A), Tekkali, Srikakulam (Dt), AP, India
e-mail: lokeshraju.au@gmail.com

B. Sivaneasan
Engineering Cluster, Singapore Institute of Technology, Singapore, Singapore
e-mail: sivaneasan@singaporetech.edu.sg

A. Shaik
CSE and R&D, St. Peters Engineering College, Maisammaguda, TS, India

P. Chakrabarti
Department of Computer Science Engineering, Sir Padampat Singhania University, Udaipur, Rajasthan, India

S. Vimal
Department of Artificial Intelligence and Data Science, Ramco Institute of Technology, Tamilnadu, India

 
K. R. Madhavi et al. (eds.), *Proceedings of Sixth International Conference on Computer and Communication Technologies*, Lecture Notes in Networks and Systems 1356,
https://doi.org/10.1007/978-981-96-5238-9_26

# 1 Introduction

The Internet of Vehicles (IoV) networks managing data securely are very important. Data from vehicles is sensitive and needs quick processing to be useful. Traditional methods often struggle with security, integrity, and access control, but blockchain offers a better way. It is a decentralized system that can protect data by making it secure and more controlled. Applying blockchain in IoV environments presents challenges like scaling the system and handling fast data processing [1, 2]. Blockchain shows potential to improve security and efficiency in industrial networks and IoT, as many studies indicate [3, 4]. However, the technology faces gaps in achieving smooth functionality within IoV networks [5, 6]. To improve data control and security, this research designs a blockchain framework tailored for IoV. A consensus mechanism that balances safety and speed is developed within the framework. Simulations and real-world tests evaluate the framework's performance to ensure reliability. Blockchain use in this study supports connected vehicles and intelligent traffic management by enabling secure data exchange between vehicles and infrastructure [7, 8].

# 2 Related Work

Recent studies explore how blockchain helps secure IoT and IoV networks. To enhance data handling capacity and minimize delays, Hao et al. [5] proposed protocols. A layered structure assigning multiple security levels to protect IoT data was introduced by Alghamdi et al. [6, 7]. Cryptographic techniques were added by Yan et al. [8, 9] to safeguard information transferred between vehicles.

Some studies introduced trust management systems that rely on the behavior of nodes and past data to keep things reliable [10, 11]. New algorithms were also developed, such as those by Li et al. [12, 13], which aimed to boost security and performance by designing better ways for the network to reach agreements on data. Jin et al. [14, 15] worked on a secure system for storing and sharing data using a consensus method called Delegated Proof of Stake. While these studies show promise, there is still a need to focus more on real-world usage, scaling challenges and improving the ways these systems agree on data to suit IoV networks.

# 3 Methods and Material

The proposed IoV architecture applies a permissioned blockchain and a consensus algorithm to maintain data security and integrity. Several layers structure the system. In the Device Layer, nodes like vehicles operate with sensors, GPS, and communication units. The On-Board Unit (OBU) encrypts, collects, and sends data to other nodes. Communication between vehicles and infrastructure is managed by the Network Layer, ensuring secure data exchange happens in real time.

Under the Blockchain Layer, a restricted blockchain network allows only authenticated vehicles to participate in consensus and validate data. Using reputation-weighted voting, the Consensus Layer combines PBFT for reliable decision-making. The Data Management Layer securely stores transactions in the blockchain, maintaining immutability and transparency. Suspicious activities are flagged through behavior monitoring by the Trust and Security Layer, which also handles identity management via a Certificate Authority Eqs. (1)–(5):

$$\text{Data Collection: } D_i(t) = \{di_1(t), di_2(t), \ldots, di_m(t)\} \tag{1}$$

$$\text{Data Processing: } P_i(t) = f_p(D_i(t)) \tag{2}$$

$$\text{Data Encryption: } E_i(t) = E(P_i(t), K_i) \tag{3}$$

$$\text{Data Communication: } fc(E_i(t), N_i) \tag{4}$$

$$\text{Validation: } V_j(E_i(t)) = \begin{cases} 1 \text{ if valid} \\ 0 \text{ otherwise} \end{cases} \tag{5}$$

This framework structures IoV data management efficiently through its multi-layer approach. Supported by protocols like DSRC or C-V2X, communication between vehicles (V2V) and infrastructure (V2I) remains stable and secure. Smooth data exchange in real time is ensured across the network, helping operations run consistently in different conditions Eqs. (6)–(9).

$$\text{Communication: } M_{ab}(t) = (n_a, n_b, m(t), \tau_{ab}(t)) \tag{6}$$

$$\text{Transmission Delay: } \tau_{ab}(t) = \frac{d_{ab}}{v_c} + \delta_{ab}(t) \tag{7}$$

$$\text{Probability of SuccessfulTransmission: } P_{ab}(t) = \frac{1}{1 + e^{-(SNR_{ab}(t) - \beta)}} \tag{8}$$

$$\text{Communication Quality: } Q_{ab}(t) = \alpha P_{ab}(t) - \gamma \tau_{ab}(t) \tag{9}$$

The IoV operates on a permissioned blockchain, allowing only authenticated vehicles to participate. Regular nodes (vehicles) and supernodes (central servers or RSUs) manage data validation and consensus processes Eqs. (10)–(12):

$$\text{Block Structure: } B_i = \{T_i, \tau_i, H_{i-1}, n_i, H_i\} \tag{10}$$

$$\text{Transaction Validation: } V\left(T_{ij}\right) = \text{Verify}\left(K_{ij}^{\text{pub}}, \left\{ID_{ij}, D_{ij}\right\}, \sigma_{ij}\right) \tag{11}$$

$$\text{Consensus Mechanism(DPoS): Consensus}(B_{\text{new}}) = \begin{cases} \text{Accept if } \frac{|D_{\text{agree}}|}{|D|} > \text{threshold} \\ \text{Reject otherwise} \end{cases} \tag{12}$$

The Practical Byzantine Fault Tolerance (PBFT) algorithm ensures that all nodes agree on the validity of transactions through a three-phase process: pre-prepare, prepare, and commit. The PBFT is enhanced with reputation-weighted voting, where nodes with higher reputation scores have more influence in the consensus process: Eqs. (13)–(16)

$$\text{Election of Delegates: } D = \text{Vote}(w_1, w_2, \ldots, w_n) \tag{13}$$

$$\text{Block Proposal: } B_{\text{new}} = \left\{T, \tau_{\text{new}}, H_{\text{prev}}, n_{\text{new}}, H_{\text{new}}\right\} \tag{14}$$

$$\text{Transaction Validation: } V(T_i) = \text{Verify}\left(K_i^{\text{pub}}, \{ID_i, D_i\}, \sigma_i\right) \tag{15}$$

$$\text{Consensus on the Block: Consensus}(B_{\text{new}}) = \begin{cases} \text{Accept if } \frac{|D_{\text{agree}}|}{|D|} > \text{threshold} \\ \text{Reject otherwise} \end{cases} \tag{16}$$

Each data packet is treated as a transaction, signed with the vehicle's private key, and validated by nearby nodes. Validated transactions are grouped into blocks, which are then added to the blockchain, ensuring immutable and transparent data storage: Eqs. (17) and (18)

$$\text{Data Packet: } D_i(t) = \{ID_i, P_i(t), \tau_i, \sigma_i\} \tag{17}$$

$$\text{Block Formation: } B_{\text{new}} = \left\{T, \tau_{\text{new}}, H_{\text{prev}}, n_{\text{new}}, H_{\text{new}}\right\} \tag{18}$$

This layer handle's identity management through a trusted Certificate Authority (CA) and a reputation system that monitors and updates the reputation scores of each node. Advanced anomaly detection methods are used to flag suspicious activities or data anomalies in real time: Eqs. (19) and (20)

$$\text{Digital Certificate: } C_i = \left\{\text{ID}_i, K_i^{\text{pub}}, \text{Signature}_{\text{CA}}\right\} \tag{19}$$

$$\text{Reputation Score Update: } R_i(t+1) = f_R(R_i(t), E_i(t)) \tag{20}$$

# 4 Experimental Study

This study tested the proposed trust model and consensus framework in IoV networks using SUMO and OMNeT++ to simulate traffic and communication over a 100-square-kilometer area with 500 vehicles and 20 RSUs. Scenarios included normal operations, false data injections by 10% of vehicles, and varying initial trust scores. Key parameters like vehicle count, trust thresholds, and malicious node proportions were adjusted to assess their impact on data accuracy, consensus speed, and trust stability.

Performance metrics focused on data accuracy, resistance to false data, consensus latency, trust score consistency, and detection of malicious activities. The experiments validated the framework's ability to maintain secure and reliable data sharing under varying conditions.

## *4.1 Results and Discussion*

Results showed high data accuracy during normal operations, with some decline during false data attacks, though still better than traditional methods IIoT [10]. The framework maintained low consensus latency and stable trust scores, effectively detecting false data entries. These findings confirm the framework's strength in enhancing data integrity, securing data sharing, and managing trust within IoV networks Table 1.

Table 2 consensus latency is a critical metric for evaluating the framework's performance. The proposed model demonstrates lower latency across different scenarios compared to IIoT. The following table and graph present the consensus times for both models:

**Table 1** Data accuracy versus scenarios

| Scenario | Proposed model accurate data (%) | IIoT accurate data (%) |
|---|---|---|
| Normal operation | 98.5 | 95.3 |
| Data injection attack | 85.3 | 72.1 |
| Varying trust scores | 95.2 | 89.7 |

**Table 2** Consensus latency versus scenarios

| Scenario | Proposed model consensus latency (ms) | IIoT consensus latency (ms) |
|---|---|---|
| Normal operation | 120 | 150 |
| Data injection attack | 145 | 180 |
| Varying trust scores | 130 | 160 |

Table 3 results indicate that trust scores remain relatively stable during normal operations and scenarios with varying initial trust scores for the proposed model. However, during data injection attacks, the proposed model adjusts more effectively, maintaining lower variance in trust scores compared to IIoT. The framework's ability to detect and mitigate malicious behaviors is significantly better than IIoT.

Figure 1 illustrates the percentage of accurate data received under different scenarios (Normal Operation, Data Injection Attack, and Varying Trust Scores) for both the proposed model and the contemporary IIoT. The proposed model consistently outperforms IIoT, particularly under data injection attacks, demonstrating its superior ability to maintain data integrity.

Figure 2 shows the consensus latency (measured in milliseconds) for the proposed model and IIoT across various scenarios. The proposed model exhibits lower consensus latency in all scenarios, indicating more efficient performance compared to IIoT, especially during data injection attacks where additional validation is required.

Figure 3 displays the stability of trust scores over time for the proposed model and IIoT. The proposed model maintains more stable trust scores, reflecting its robust capability to handle varying initial trust scores and adapt to data injection attacks with lower variance compared to IIoT.

Figure 4 compares the detection rates of malicious activities between the proposed model and IIoT. The proposed model records higher detection rates, showcasing its ability to mitigate false data injections efficiently.

**Table 3** Trust score stability

| Method | Proposed model detection rate (%) | IIoT detection rate (%) |
|---|---|---|
| Proposed framework | 92.7 | 75.4 |
| Traditional methods | 75.4 | - |

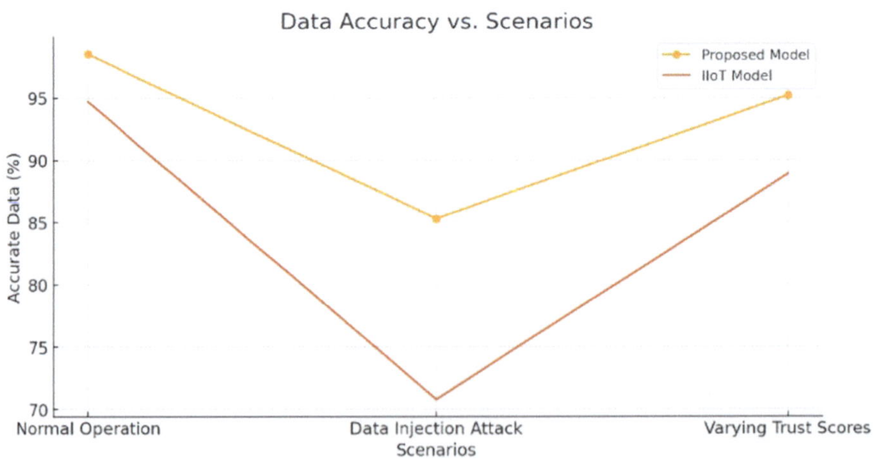


**Fig. 1** Data accuracy versus scenarios

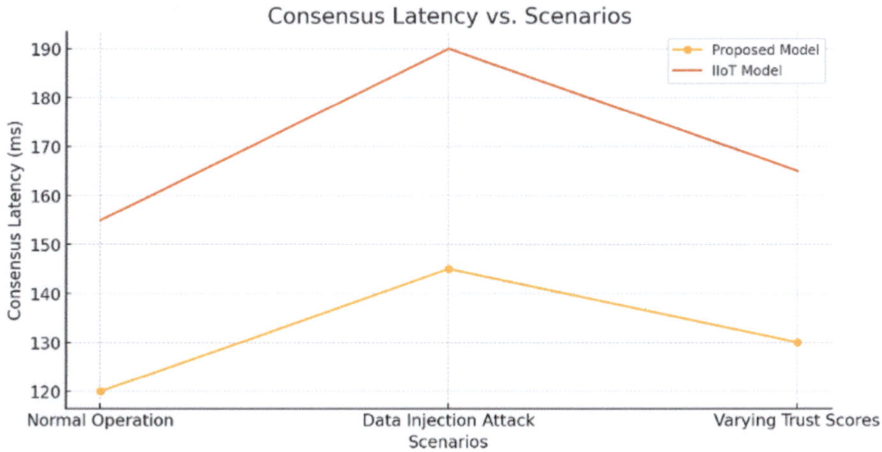


**Fig. 2** Consensus latency versus scenarios

**Fig. 3** Trust score stability over time

The consensus mechanism and trust model improve data accuracy and stability beyond IIoT methods. Even with trust score changes, the system maintains consistency. Though accuracy drops under false data injections, the model manages attacks better than IIoT, ensuring quick consensus for real-time data access. Trust scores remain stable, supporting reliable node evaluation during attacks.

The model adapts dynamically to malicious activities, minimizing attack impact more effectively than IIoT. Coordinated trust and security layers safeguard data integrity with low consensus delays. While promising, further real-world testing and exploration of new attack scenarios are necessary to enhance readiness for complex environments.

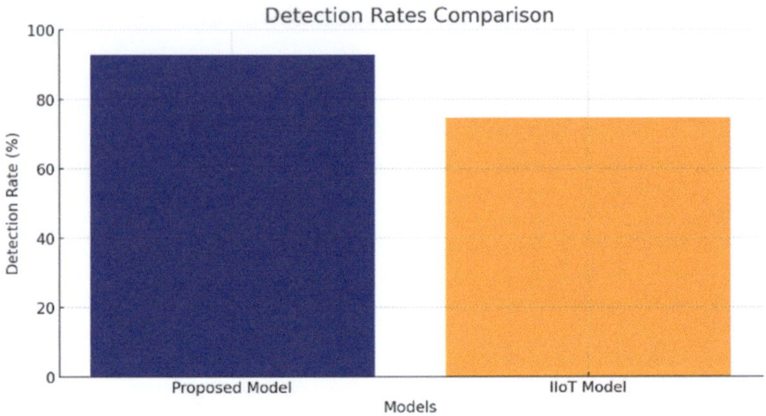


**Fig. 4** Detection rates comparison

## 5 Conclusion

The study proposes a blockchain-based framework to enhance data integrity, security, and trust management in IoV networks. It integrates a permissioned blockchain with a Practical Byzantine Fault Tolerance (PBFT) algorithm, improved using reputation-weighted voting. Between vehicles and infrastructure, secure data sharing and reliable validation processes are ensured through this framework. Lower latency, high data accuracy, and stable trust scores are observed even during false data injections when compared with IIoT approaches. In controlled simulations, the framework achieves consistent performance, but real-world testing across complex vehicular setups is necessary. Strengthening the system may require refining the consensus mechanism and exploring new attack types. For large-scale networks, adaptive algorithms can be investigated to improve scalability. By securing communication and managing trust, the study addresses key challenges in IoV data sharing efficiently across dynamic environments.

## References

1. Abd El-Hamed OM, Abd El-Samie FE, Badawy W, Abd El-atty SM (2024) Blockchain for IoT and medical data management
2. Li S, Liu WF, Wu Y, Zhao J (2024) Generative architecture for data imputation in secure blockchain-enabled spatiotemporal data management. J Web Eng 23(1):111–163
3. Zhang H, Zhang X, Guo Z, Wang H, Cui D, Wen Q (2021) Se-cure and efficiently searchable IoT communication data management model: using blockchain as a new tool. IEEE Internet Things J 10(14):11985–11999
4. Zorlu O, Ozsoy A (2024) A blockchain-based secure framework for data man-agement. IET Communications

5. Guo H, Li W, Nejad M (2022) A hierarchical and location-aware consensus protocol for IoT-blockchain applications. IEEE Trans Netw Serv Manage 19(3):2972–2986
6. Alghamdi S, Albeshri A, Alhusayni A (2023) Enabling a secure iot envi-ronment using a blockchain-based local-global consensus manager. Electronics 12(17):3721
7. Rudra Kumar M, Gunjan VK (2022) Peer level credit rating: an extended plugin for credit scoring framework. In: Kumar A, Mozar S (eds) ICCCE 2021. Lecture Notes in Electrical Engineering, vol 828. Springer, Singapore. https://doi.org/10.1007/978-981-16-7985-8_128
8. Yan G (2023) BlockChain based data security mechanism for vehicular networks. In: 2023 6th International Conference on Electronics Technology (ICET). IEEE, pp 658–663
9. Swetha A, Lakshmi MS, Kumar MR (2022) Chronic kidney disease diagnostic approaches using efficient artificial intelligence methods. Int J Intell Syst Appl Eng 10(1s):254
10. Maftei AA, Lavric A, Petrariu AI, Popa V (2023) Block-chain for internet of things: a consensus mechanism analysis. In: 2023 13th international symposium on Advanced Topics in Electrical Engineering (ATEE). IEEE, pp 1–5
11. Dwaram JR, Madapuri RK (2022) Crop yield forecasting by long short-term memory network with Adam optimizer and Huber loss function in Andhra Pradesh, India. Concurrency Comput Pract Experience 34(27) (Wiley). https://doi.org/10.1002/cpe.7310
12. Jia C, Gao L, Li M (2022) Research on blockchain consensus algorithm for secure data sharing on industrial internet platform. In: Proceedings of the 2022 5th international conference on artificial intelligence and pattern recognition, pp 197–204
13. Rudra Kumar M, Gunjan VK (2022) Machine Learning based solutions for human resource systems management. In: Kumar A, Mozar S (eds) ICCCE 2021. Lecture Notes in Electrical Engineering, vol 828. Springer, Singapore. https://doi.org/10.1007/978-981-16-7985-8_129
14. Chen W, Wang J, Wang J, Jiang W, Chen W (2023) Blockchain-based secure storage and sharing mechanism for industrial Iot edge data. Available at SSRN: https://ssrn.com/abstract=4604571 or https://doi.org/10.2139/ssrn.4604571
15. Madapuri RK, Mahesh PCS (2019) HBS-CRA: scaling impact of change re-quest towards fault proneness: defining a heuristic and biases scale (HBS) of change request artifacts (CRA). Cluster Comput 22(Suppl 5):11591–11599. https://doi.org/10.1007/s10586-017-1424-0

# A Novel Optimization Approach for Pancreatic Cancer Classification Using a Hybrid Deep Learning Model

**Mekala Sandhya and S. Phani Kumar**

**Abstract** Pancreatic cancer has high mortality rates because it is often not diagnosed early and is typically discovered at a later stage when there are few available treatments. This underscores the urgent need for automated systems to detect cancer early, which can significantly improve diagnosis and treatment outcomes. While various algorithms have been used in the medical field to help with early detection, there is still significant potential for developing more advanced computational systems. The main goal of this study is to improve pancreatic cancer detection utilizing hybrid learning methods. Specifically, this research focuses on creating a system that examines medical imaging data, mainly CT scans, to find essential features and cancerous growths in the pancreas using models based on deep learning. Convolutional neural networks (CNN), transfer learning model (VGG-16), and recurrent neural networks base LSTM were applied in the study. Initially, every model was applied individually and later, with the combination of CC and LSTM along with Vgg-16, we proposed a hybrid model. A novel optimization approach was used to boost the hybrid model performance. The study aims to assess the effectiveness of these models, particularly the innovative Hybrid approach, in identifying pancreatic cancer. Our results show that the presented hybrid model outperforms other methods, achieving a testing accuracy of 98%. The precision, recall, and f-score accordingly 98.3%, 98.0%, and 98.1%. This demonstrates that the hybrid model shows promise in detecting pancreatic cancer. The proposed model outperformed the state-of-art models.

**Keywords** Pancreatic cancer · Machine learning · Hybrid model · Clinical prognosis · Early cancer detection · Neural network

M. Sandhya (✉) · S. Phani Kumar
Department of Computer Science and Engineering, GITAM School of Technology, GITAM Deemed to Be University, Hyderabad, Telangana 502329, India
e-mail: smekala3@gitam.in

S. Phani Kumar
e-mail: psingams@gitam.edu

 
K. R. Madhavi et al. (eds.), *Proceedings of Sixth International Conference on Computer and Communication Technologies*, Lecture Notes in Networks and Systems 1356,
https://doi.org/10.1007/978-981-96-5238-9_27

## 1 Introduction

One of the most deadly and aggressive cancers, pancreatic cancer, causes an extensive portion of cancer-related fatalities globally [1]. The condition is characterized by showing no symptoms in the initial phases, often leading to late detection and a poor outlook, with fewer than 10% of individuals living for five years [2]. Producing hormones controlling blood sugar and digestive enzymes is a primary function of the pancreas, a gland behind the stomach [3]. When cancer grows in the pancreas, it usually progresses silently, showing irregular symptoms until it reaches advanced stages. Pancreatic cancer ranks third in the world for cancer-related deaths due in large part to its high death rate, which is a result of delayed detection [4]. However, due to the high risk of metastasis, the treatment is complicated for cancer that arises in the endocrine or exocrine cells. Pancreatic cancer presents significant challenges for modern medicine due to its silent behavior and resistance to conventional therapy [5]. Improving patient outcomes requires understanding pancreatic cancer risk factors and early detection methods [6]. To sweeten patient outcomes, recent developments in medical research have focused on developing beneficial medications and enhancing early identification [7]. Nevertheless, despite progress, detecting and treating pancreatic cancer remains problematic because screening methods rely on understanding risk factors, including age, genetics, lifestyle, and comorbidities. The disease is often not detected at a treatable stage by the existing diagnostic methods [8]. In the quest for better outcomes, artificial intelligence is being examined as a possible method to improve early diagnosis, forecast therapy response, and offer individualized care. The usefulness of machine learning (ML) and deep learning (DL) techniques for boosting diagnostic accuracy is advancing gradually [9]. Utilizing DL techniques like Mobile Net for feature extraction and the Emperor Penguin Optimizer with Multi-level Thresholding (EPO-MLT) for tumor segmentation, a study [10] presented an advanced decision-making system. Although this method significantly improved over earlier models, it was constrained by its complex parameter adjustment and dependence on particular imaging techniques. The necessity for accurate tumor classification employing a hybrid VGG16–XGBoost model to identify pancreatic ductal adenocarcinoma (PDAC) from CT images was the focus of another study [11]. A study [12] combined Particle swarm optimization (PSO) with support vector machine (SVM) and image processing methods to detect pancreatic cancer in CT scans, advancing the field. This approach outperformed other classifiers like Naive Bayes and AdaBoost, but accessibility issues were caused by a lack of diverse datasets and the high expense of imaging technology. To overcome the difficulty of identifying pancreatic neoplasms and MPD dilatation, another study [13] used a three-step DL technique with CT scans from several institutions. Operating ML algorithms to identify critical miRNAs related to PC occurrence, researchers also focused on the early diagnosis of pancreatic cancer by serum miRNA [14] profiling [15]. Through our efforts, we hope to improve pancreatic cancer diagnosis by utilizing cutting-edge DL techniques. We have merged CNN for feature extraction and LSTM networks to capture sequential dependencies in imaging data and transfer learning to leverage the strength of pre-trained models. In addition, we suggest a hybrid model that combines LSTM and CNN techniques to improve diagnostic robustness and accuracy. Combining the best features of both designs, this hybrid model offers a thorough approach to the early and accurate identification of pancreatic cancer. Our research aims to improve patient outcomes in pancreatic cancer care by addressing the significant hurdles associated with late diagnosis using these cutting-edge techniques. The contribution of the study:

- We use a variety of DL models in the study, such as CNNs, transfer learning methods using pre-trained models, and a hybrid strategy that combines LSTM networks and CNNs.
- In the experiment, a unique data augmentation strategy is employed to sweeten the performance and robustness of the model.
- A unique optimization technique is applied to enhance model performance, and the model increases performance.

We have divided this paper into several sections to help understand its general organization. In Sect. 2, we will outline our methodology statement and provide a quick overview of the various methods for detecting pancreatic cancer. In Sect. 3, we will discuss our model's experimental results and compare outcomes with existing methods. Ultimately, we will wrap up by showcasing our research findings.

## 2 Methodology

The methodology section describes the steps employed to accomplish the research objectives in-depth, illustrated in Fig. 1. The study uses a hybrid CNN + LSTM model and a VGG-16 model based on transfer learning. Over fitting and data imbalance are addressed using a two-step enhanced data augmentation technique. Evaluation metrics, including accuracy, precision, recall, and F1-score, are used to evaluate performance as the models are trained and evaluated on augmented data.

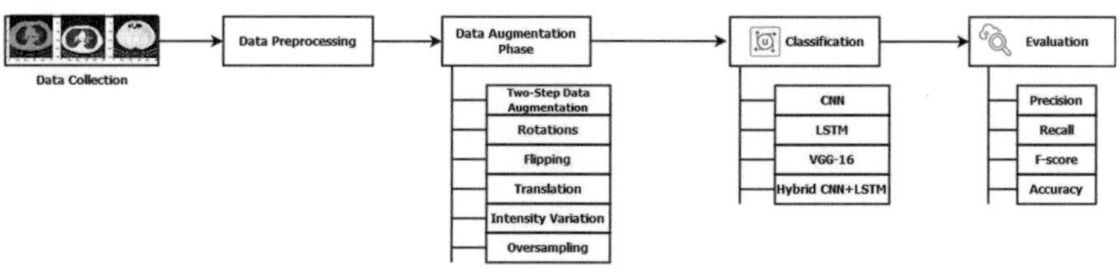


**Fig. 1** Graphical representation of the overall methodology

**Table 1** Summary of pancreatic cancer and healthy pancreas data

| Particulars | Pancreatic cancer | Healthy pancreas |
|---|---|---|
| Total images | 3494 | 3451 |
| Total patients | 222 | 190 |
| Thickness | 5.0 mm | |

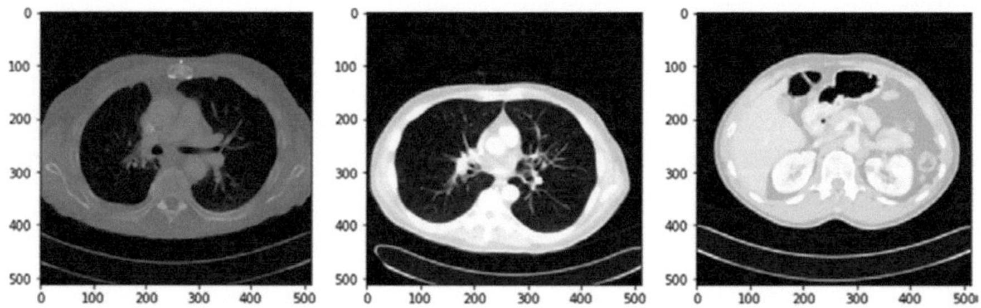


**Fig. 2** Sample of the data

## 2.1 Dataset Description

Our work employed a comprehensive dataset comprising 6945 CT images gathered between June 2017 and June 2018 [16]. This dataset was collected from 222 patients with pathologically confirmed pancreatic cancer, who participated in 3494 photos, and 190 individuals with healthy pancreas, who participated in 3751 images. The dataset was split into three phase-based subsets to enable reliable binary classification (cancer vs. non-cancer). To evaluate the effectiveness of our generated model, tenfold cross-validation was applied to the images, which had a consistent slice thickness of 5.0 mm. This dataset, illustrated in Table 1 and Fig. 2, provides the basis for precisely identifying and categorizing pancreatic cancer.

## 2.2 Data Augmentation Phase

A phase of data augmentation is used to address data unbalancing and over fitting. We employ a two-step, enhanced data augmentation procedure called "Image Data-Generator()" and background modification (BackMod) [17]. The image's backdrop is modified in the first phase (BackMod). Apart from this method, the objective is to contextualize existing and categorized waste. BackMod examines noise and other problems across the sample as a whole. In the second step, we start a multi-scale data augmentation process for the training data sample. This phase uses various methods to combine a random combination of essential properties, such as object rotation, flipping in both vertical and horizontal directions, object translation, and intensity fluctuation, to prevent overfitting. In the second phase, we start a multi-scale data augmentation procedure for the training data sample. This phase prevents imbalanced class and overfitting issues by using a random mix of essential features with random methods, i.e., object rotation, flipping in both vertical and horizontal directions, object translation, and intensity variation. The augmentation feature on the image dataset was obtained by using the built-in Python Keras module "ImageDataGenerator ()" [18, 19]. Various angles and vertical and horizontal rotations were applied to the image frames, and the center feature was set to "True." To balance the waste graphics for each class, we used a data oversampling technique that combined the data augmentation from phases 1 and 2. This was necessary since the waste statistics categories are not balanced and do not represent a Gaussian distribution [20].

## 2.3 Model

**Proposed hybrid model**: This study uses hybrid CNN-LSTM with transfer learning for CT image classification for sustainable development. One can categorize the CT pictures into categories related to cancer and non-cancer. Use a hybrid model to classify waste statistics that combines CNN and LSTM. The suggested model uses ImageNet's advantage and transfer learning to forecast and classify the trash category [21]. In Fig. 3, the architecture of the suggested hybrid model is displayed.

The two main components of this suggested model are a CNN component with an image size of (224*224) and a separate autonomous RNN component. An independent, autonomous RNN component in the suggested model has two distinct LSTM layers, both of which have batch shape 2048. Nevertheless, until the CNN reaches the final destination layer, or convolutional layer, which has bottleneck features, it is ultimately carried by a pre-trained TL model, such as ResNet-50, InceptionResNet-V2; these all have a batch size of 64 [22]. Figure 4 shows how the suggested Hybrid CNN-LSTM with the TL model operates.

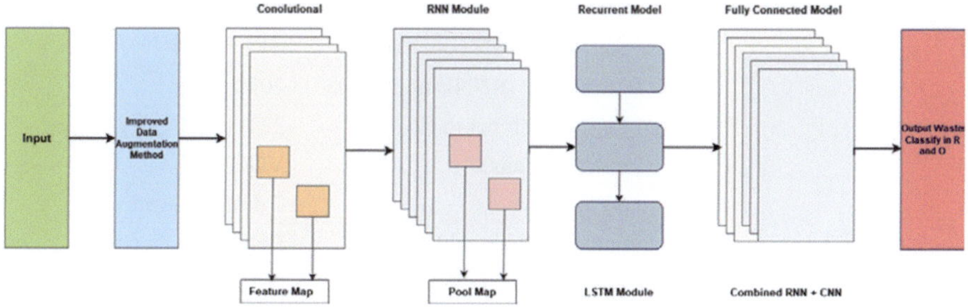


**Fig. 3** Proposed hybrid CNN-LSTM model

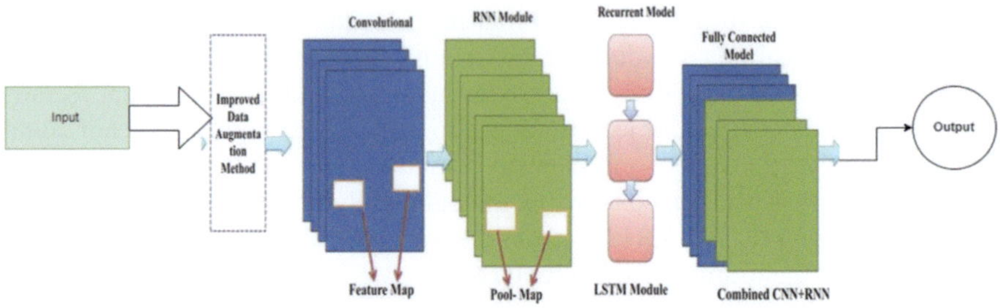


**Fig. 4** Procedure of hybrid model with transfer learning

The outcomes of the components are multiplied using a variable size basis to combine them. For multi-class classifications, the result is obtained further and fed into the categorization layer of distinct classes and the activation function (Soft-Max). Because every convolution operation may take photos of a specific duration, in the following phase, all the convolutional layers with different filter strengths can detect waste images with magnifying factors [23]. The following essential elements are present in the suggested hybrid model. (a) Dense feature LSTM layer; (b) SoftMax activation function pooling layer; (e) convolutional layer; (f) fully connected layer component. The model operates in the following ways in its entirety. The parameters and values of the proposed model is shown in Table 2.

**Optimization Approach:** The adaptive moment estimator (AME) used in the suggested model is made to calculate adaptive learning rates for each parameter by keeping an exponentially decreasing average of squared gradients (Q) and historical gradients (G). AME causes a smoother minimum in the error function by acting like a big ball with internal friction rolling down a slope. The adaptive moment estimator (AME) equations are given as:

**Table 2** Hybrid CNN-LSTM model parameters and values

| Parameter | Value |
|---|---|
| Optimizer | AME |
| Learning rate | 0.001 |
| Batch size (CNN component) | 64 |
| Batch shape (RNN component) | 2048 |
| Image size (Input to CNN) | 224 × 224 |
| Pre-trained models | VGG-16 |
| LSTM layers | 2 distinct layers |
| Final convolutional layer | Bottleneck features |
| Activation function | SoftMax |
| Fully connected layer | 2 |
| Loss function | Binary cross-entropy loss |

$$\mathbf{U}_t = \gamma_1 \cdot \mathbf{U}_{t-1} + (1 - \gamma_1) \cdot \mathbf{g}_t$$

$$\mathbf{Q}_t = \gamma_2 \cdot \mathbf{Q}_{t-1} + (1 - \gamma_2) \cdot \mathbf{g}_t^2$$

where:

$\mathbf{U}_t$ is the gradients' average, which decays gradually.
$\mathbf{Q}_t$ is the squared gradients' continuously declining average.
$\gamma_1$ and $\gamma_2$ are decay rates.
$\mathbf{g}_t$ is the gradient at time step $t$.

## 3 Result and Discussion

To analyze the performance of the models, we calculated the training, validation, and testing accuracies. Table 3 presents the outcomes for CNN, LSTM, VGG-16, and the proposed hybrid CNN + LSTM model. In ML, a model's effectiveness is evaluated by its performance on the training data and its proficiency in generalizing to unseen data. Therefore, training, validation, and testing accuracies are crucial metrics. Training accuracy evaluates how well a model fits the training dataset. In contrast, validation accuracy estimates its performance on an unseen dataset during model tuning, enabling the determination of overfitting or under fitting. Testing accuracy assesses the final performance of the model on completely unseen data after training. Our study presents a hybrid CNN + LSTM model incorporating the advantages of long short-term memory (LSTM) networks for temporal dependency detection in sequential data with the strength of convolutional neural networks (CNNs) for feature extraction. Furthermore, VGG-16, a deep CNN utilized as a transfer learning model, was added for comparison. To compare their performance, standalone CNN and LSTM models were also assessed. The CNN model performed a high training accuracy of 99.5%, validation of 89.0%, and testing of 88.5% accuracy exhibited a significant drop, suggesting that the model over fits the training data. The model appeared to have learned the patterns in the training data effectively, based on its high training accuracy; nevertheless, overfitting may be the reason for its inability to generalize to new data. With a 98.0% training accuracy and lower 87.5% and 87.0% testing and validation accuracy, the LSTM model also showed overfitting. Moving on, the VGG-16 transfer learning model performed better, with a more balanced set of accuracies: 99.8% for training, 94.0% for validation, and 93.5% for testing. Compared to the CNN and LSTM models, this model exhibited less overfitting, even if there was still a slight difference in accuracy between training and testing. This was probably because VGG-16 can generalized better to new data because of its depth and pre-trained nature. However, our proposed hybrid CNN + LSTM model demonstrated the most balanced performance across training, validation, and testing sets with a training accuracy of 98.5%, validation accuracy of 98.2%, and testing accuracy of 98.0%. The hybrid model displayed no significant gap between these metrics, showing that

**Table 3** Training, testing, and validation accuracy of all models

| Model | Training accuracy (%) | Validation accuracy (%) | Testing accuracy (%) |
|---|---|---|---|
| CNN | 99.5 | 89.0 | 88.5 |
| LSTM | 98.0 | 87.5 | 87.0 |
| VGG-16 | 99.8 | 94.0 | 93.5 |
| Hybrid CNN + LSTM | **98.5** | **98.2** | **98.0** |

it has not been over fitted or under fitted. The capacity of the CNN to extract features and the LSTM to model sequential dependencies to work together to produce this balanced performance was explained. The hybrid design minimized the drawbacks of solo CNNs or LSTMs while achieving better generalization by combining these two models. We have presented that the hybrid CNN + LSTM model performed better than the standalone CNN, LSTM, and VGG-16 models. Regarding test and training accuracy, the VGG-16 model excellently performed better than the CNN and LSTM models, exhibiting indications of overfitting. Since the hybrid CNN + LSTM model produced the most robust and balanced results across all datasets, it was the most effective approach for our classification challenge. This study demonstrated the significance of combining complementary models to sweeten generality and accuracy.

We assessed the efficacy of our recommended hybrid CNN + LSTM model using precision, recall, F-score, and accuracy. It was compared to the VGG-16, LSTM, and standalone CNN models. The complete results are shown in Table 4.

The CNN depicted better performance with a precision of 94.1%, recall of 93.5%, and F-score of 93.8%, presenting its capability to capture inherent image features. However, its weakness in capturing sequential dependencies guided to a negligibly lower recall. Conversely, the LSTM, with a precision of 92.3%, recall of 91.8%, and F-score of 92.0%, made it appropriate for tasks needing memory of previous data or time-series analysis. Despite this strength, the lack of practical native feature handling outcomes in a lower performance than CNNs. The VGG-16, a DCN, captured complicated image patterns and features, reaching a precision of 96.5%, recall of 96.0%, and F-score of 96.2%. Its remarkable performance above the entry-level CNN demonstrated how deep architectures are superior at extracting fine-grained information. With the highest precision of 98.3%, recall of 98.0%, and F-score of 98.1%, the

**Table 4** Precision, recall, and F-score for individual models

| Model | Precision (%) | Recall (%) | F-score (%) |
|---|---|---|---|
| CNN | 94.1 | 93.5 | 93.8 |
| LSTM | 92.3 | 91.8 | 92.0 |
| VGG-16 | 96.5 | 96.0 | 96.2 |
| Hybrid CNN + LSTM | **98.3** | **98.0** | **98.1** |

Hybrid CNN + LSTM model combines the best features of both CNN and LSTM to capture both sequential and spatial patterns in the data efficiently. This amalgamation led to significant performance gains, making it the most powerful model in the comparison. In conclusion, the superior metrics of the hybrid model were explained by its ability to integrate the advantages of both CNN and LSTM architectures. The comparative performance of all models is exhibited in Fig. 5. Figure 6 displays the confusion matrix for our model, which shows its performance in various classes. The matrix's vertical axis displays the genuine labels, while the horizontal axis displays the predicted labels.

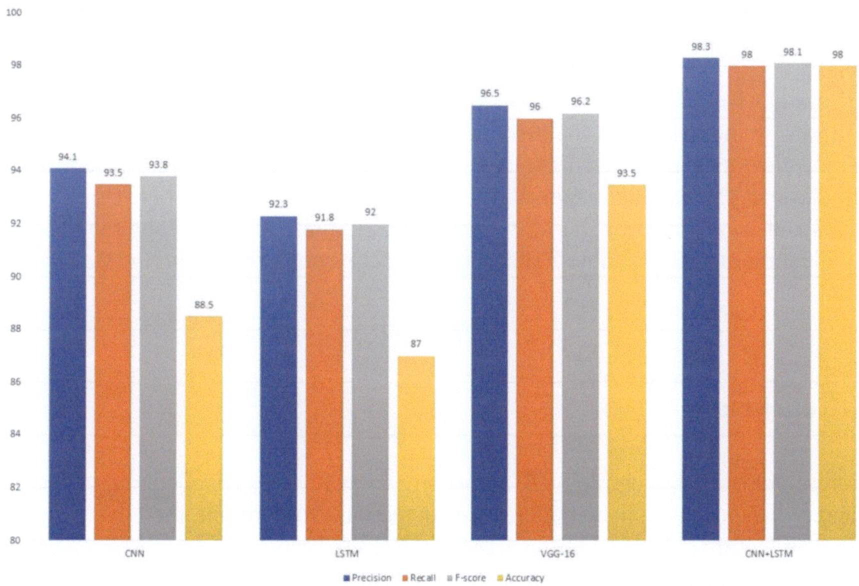


**Fig. 5** Comparative performance of all models

**Fig. 6** Confusion matrix of proposed hybrid model

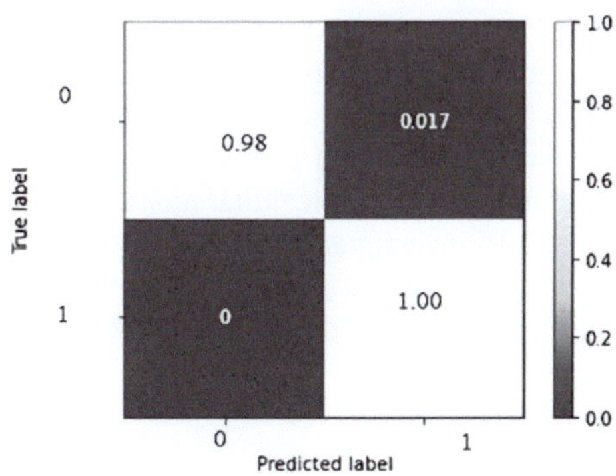

**Table 5** Compare our proposed hybrid model with existing work on pancreatic cancer

| References | Model | Accuracy |
|---|---|---|
| [2] | Machine learning | 94.4% |
| [24] | Machine learning | 78% |
| [25] | Random forest | AUC (88%) |
| [26] | Random forest | 86% |
| [27] | Hybrid | 84.21% |
| Proposed | Hybrid | 98% |

Class 0: The model accurately predicted 98% of all occurrences labeled as 0 (accurate label) 0. A tiny percentage (1.7%), meanwhile, was incorrectly categorized as class.

Class 1: All occurrences of class 1 were 100% accurately predicted by the model, as shown by the value 1 in the associated cell. The confusion matrix demonstrates the model's outstanding performance overall, with excellent recall and precision for both classes. The excellent capacity of the model to discriminate between the two classes is demonstrated by the low misclassification rate in class 0 and the flawless classification of class 1. Table 5 shows our model's comparison with the state-of-art model.

## 4 Conclusion

Our research aimed to employ ML algorithms to make it less challenging to use CT imaging data for early pancreatic cancer detection. To combine the advantages of CNNs for feature extraction with LSTM networks for capturing temporal dependencies in sequential data, we have presented a hybrid CNN + LSTM model. VGG-16, a deep CNN employed as a transfer learning model, was also included for comparison. However, our presented hybrid CNN + LSTM model exhibits the most balanced performance across training, validation, and testing sets with 98.5%, 98.2%, and 98.0% accuracy, respectively. The suggested hybrid model demonstrates no significant gap between these metrics, indicating that it has not been over fitted or under fitted. The hybrid model showed significant novelty by combining advanced neural network architectures with CT image data, presenting a more comprehensive solution for pancreatic cancer diagnosis. By integrating more clinical and molecular data, we aim to investigate potential improvements to the hybrid model in future work. Real-time monitoring of therapy responses could facilitate further studies focusing on improving individualized treatment plans and early diagnosis.

## References


1. Zhao H, Zhang Y, Liu H, Wang Y, Song Z (2024) Age-period-cohort analy- sis of global, regional, and national pancreatic cancer incidence, mortality, and disability-adjusted life years, 1990–2019. BMC Cancer 24(1):1–12
2. Shi W, Wartmann T, Accuffi S, Al-Madhi S, Perrakis A, Kahlert C, Link A, Venerito M, Keitel-Anselmino V, Bruns C et al (2024) Integrating a microrna signature as a liquid biopsy-based tool for the early diagnosis and prediction of potential therapeutic targets in pancreatic cancer. Br J Cancer 130(1):125–134
3. Mostafa A, Gantsova E, Serova O, Mohammad T, Deyev I (2024) Interaction between endocrine and exocrine pancreas. J Evol Biochem Physiol 60(2):792–801
4. Danpanichkul, P., Suparan, K., Jaroenlapnopparat, A., Polpichai, N., Fangsaard, P., Detboon, A., Moolkaew, P., Sripusanapan, A., Srisurapanont, K., Kanjanakot, Y., et al.: The global burden of early-onset pancreatic cancer and its risk factors: A perspective from global burden of disease study 2019. Pancreas, 10–1097 (2024)
5. Toledo, B., Deiana, C., Scianò, F., Brandi, G., Marchal, J.A., Perán, M., Giovan- netti, E.: Treatment resistance in pancreatic and biliary tract cancer: molecular and clinical pharmacology perspectives. Expert Review of Clinical Pharmacology **17**(4), 323–347 (2024)
6. Luz, A.: Role of predictive models in early detection of pancreatic cancer. Technical report, EasyChair (2024)
7. Shiwlani, A., Khan, M., Sherani, A.M.K., Qayyum, M.U., Hussain, H.K.: Rev- olutionizing healthcare: The impact of artificial intelligence on patient care, diagnosis, and treatment. JURIHUM: Jurnal Inovasi dan Humaniora **1**(5), 779–790 (2024)
8. Abhisheka B, Biswas SK, Purkayastha B, Das D, Escargueil A (2024) Recent trend in medical imaging modalities and their applications in disease diagnosis: a review. Multimedia Tools and Applications 83(14):43035–43070
9. Kumar, M., Gunjan, V.: Peer Level Credit Rating: An Extended Plugin for Credit Scoring Framework, pp. 1227–1237 (2022). https://doi.org/10.1007/978-981-16-7985-8_128
10. Vaiyapuri, T., Dutta, A.K., Punithavathi, I.H., Duraipandy, P., Alotaibi, S.S., Alsolai, H., Mohamed, A., Mahgoub, H.: Intelligent deep-learning-enabled decision-making medical system for pancreatic tumor classification on ct images. In: Healthcare, vol. 10, p. 677 (2022). MDPI
11. Bakasa W, Viriri S (2023) Vgg16 feature extractor with extreme gradient boost classifier for pancreas cancer prediction. Journal of Imaging 9(7):138
12. Ansari, A.S., Zamani, A.S., Mohammadi, M.S., Meenakshi, Ritonga, M., Ahmed, S.S., Pounraj, D., Kaliyaperumal, K.: [retracted] detection of pancreatic can- cer in ct scan images using pso svm and image processing. BioMed Research International **2022**(1), 8544337 (2022)
13. Abi Nader, C., Vetil, R., Wood, L.K., Rohe, M.-M., Bône, A., Karteszi, H., Vul- lierme, M.-P.: Automatic detection of pancreatic lesions and main pancreatic duct dilatation on portal venous ct scans using deep learning. Investigative Radiology **58**(11), 791–798 (2023)
14. Reddy KB (2015) Microrna (mirna) in cancer. Cancer Cell Int 15:1–6
15. Chi H, Chen H, Wang R, Zhang J, Jiang L, Zhang S, Jiang C, Huang J, Quan X, Liu Y et al (2023) Proposing new early detection indicators for pancreatic cancer: Combining machine learning and neural networks for serum mirna-based diagnostic model. Front Oncol 13:1244578
16. Dinesh, M.G., Baanin, N., Askar, S.S., Abouhawwash, M.: Diagnostic ability of deep learning in detection of pancreatic tumour. Scientific Reports **13** (2023)
17. Gondal, A.U., Sadiq, M.I., Ali, T., Irfan, M., Shaf, A., Aamir, M., Shoaib, M., Głowacz, A., Tadeusiewicz, R., Kańtoch, E.: Real time multipurpose smart waste classification model for efficient recycling in smart cities using multilayer convolutional neural network and perceptron. Sensors (Basel, Switzerland) **21** (2021)
18. Trivedi, N., Simaiya, S., Kumar Lilhore, D., Sharma, S.: Covid-19 pandemic: Role of machine learning deep learning methods in diagnosis. International Journal of Current Research and Review, 150–155 (2021) https://doi.org/10.31782/IJCRR.2021.SP192

19. Sallang NCA, Islam MT, Islam MS, Arshad H (2021) A cnn-based smart waste management system using tensorflow lite and lora-gps shield in internet of things environment. IEEE Access 9:153560–153574
20. Lilhore, U.K., Imoize, A.L., Lee, C.-C., Simaiya, S., Pani, S.K., Goyal, N., Kumar, A., Li, C.-T.: Enhanced convolutional neural network model for cassava leaf disease identification and classification. Mathematics (2022)
21. Liang S, Gu Y (2021) A deep convolutional neural network to simultaneously localize and recognize waste types in images. Waste Manage 126:247–257
22. Simaiya S, Lilhore UK, Pandey H, Trivedi NK, Anand A, Sandhu J (2022) An improved deep neural network-based predictive model for traffic acci- dent's severity prediction. In: Hu Y-C, Tiwari S, Trivedi MC, Mishra KK (eds) Ambient Communications and Computer Systems. Springer, Singapore, pp 181–190
23. Ozdemir ME, Ali Z, Subeshan B, Asmatulu E (2021) Applying machine learning approach in recycling. J Mater Cycles Waste Manage 23:855–871
24. Bojmar, L., Zambirinis, C.P., Hernandez, J.M., Chakraborty, J., Shaashua, L., Kim, J., Johnson, K.E., Hanna, S., Askan, G., Burman, J., et al.: Multi-parametric atlas of the pre-metastatic liver for prediction of metastatic outcome in early-stage pancreatic cancer. Nature medicine, 1–11 (2024).
25. Zeng, X., Ren, D., Liu, R., Zhang, Q., Yan, X., Yuan, X.: Oral microbiome-driven virulence factors: A novel approach to pancreatic cancer diagnosis. Biomolecules and Biomedicine (2024)
26. Sai, M.D.S., Prudhvi, P., Gopi, G.M., Sai Ram, I.G.N., Sandeep, M.R., Pachhala, N.: Early detection: Machine learning techniques in pancreatic cancer diagnosis. Journal of Engineering Research and Reports **26**(5), 175–182 (2024)
27. Yimamu A, Li J, Zhang H, Liang L, Feng L, Wang Y, Zhou C, Li S, Gao Y (2024) Computed tomography and guidelines-based human–machine fusion model for predicting resectability of the pancreatic cancer. J Gastroenterol Hepatol 39(2):399–409

# Detection of Fake News Using Logistic Regression, Decision Tree, Random Forest, and Gradient Boosting Algorithms

**Anoop Kumar Srivastava and Lakkireddy Abhigna Reddy**

**Abstract** This paper discusses the growing concern of fake news in the world of digital media, which has transformed public opinion and has developed gradual destruction of trust in reliable sources. We present a solution using machine learning algorithms to find fake news. We used real and fake news articles and then applied different classification algorithms to separate fake news with the real news. We preprocessed the data and split the dataset into training and test sets. Next, we used TF-IDF vectorization. We trained multiple classifiers and evaluated performance. Our results showed that these classifiers are effective in finding fake news articles. In addition, we developed a manual testing mechanism that allows users to input news articles for real-time classification. This work complements ongoing efforts to reduce the growth of fake news and provides the solution for the robust detection mechanisms to ensure information integrity in the digital media.

**Keywords** TF-IDF vectorization · Logistic Regression · Decision Tree · Gradient Boost Classification · Random Forest · Model Estimation

## 1 Introduction

The spread of fake news in the digital world has become a great problem, largely because of its potential impact on public life and decision-making processes. Fake news is becoming more common across multiple digital platforms and posing a challenge to the society. Due to the complexity of fake news content and distribution methods, there is a need for automated systems that can accurately distinguish false news from genuine news articles [1]. This paper discusses a machine learning

A. K. Srivastava (✉) · L. A. Reddy
Department of Computer Science & Engineering, ASAC, Alliance University, Bengaluru, India
e-mail: anoop.srivastava@alliance.edu.in

L. A. Reddy
e-mail: alakkireddyBTECH21@alliance.edu.in



K. R. Madhavi et al. (eds.), *Proceedings of Sixth International Conference on Computer and Communication Technologies*, Lecture Notes in Networks and Systems 1356,
https://doi.org/10.1007/978-981-96-5238-9_28

approach that uses classification algorithms to solve the existing problem [2]. We use different types of classification algorithms of machine learning to solve the problem of detecting fake news [3]. Logistic Regression (LR): it is used to model the probability of a binary outcome using a logistic function. Its simplicity and interpretability make it a suitable choice for our classification task. Decision Tree (DT): a non-linear classification algorithm that divides the feature space into regions and assigns a specific class identifier to each region. DTs are particularly capable of capturing complex decision constraints and interactions between functions. Gradient Boosting Classifier (GBC): it is an ensemble learning technique which combines weak learners, such as Decision Trees to create a strong classifier. GBC improves the performance of the model by improving it with each iteration. Random Forest Classifier (RFC): another ensemble learning method that creates multiple Decision Trees during training and outputs the state of the classes as predictions. RFCs resist data overlap and noise [3]. We describe the implementation of classifier algorithms to detect fake news, provide code snippets, and provide performance evaluation metrics. We provide experimental results, accuracy scores, and classification reports for each algorithm, along with discussions of their strengths and limitations. Finally, we summarize our conclusions. This paper aims to explore these concerns and is organized as follows: Sect. 2 discusses the literature survey, while Sect. 3 delineates the motivation. Section 4 delves into the design. Section 5 discusses about implementation. Section 6 examines the result. Finally, Sect. 7 draws conclusions and future directions.

## 2 Literature Survey

The dissemination of fake information in the online world has led to a huge amount of research into the automatic detection and mitigation of fake news [4]. Many studies have investigated the usefulness of some machine learning models to distinguish between genuine and fake news articles. One of the most commonly used algorithms for binary classification applications such as fake news detection is Logistic Regression. Studies have proven their effectiveness in article probability modeling given text features extracted from the content [5]. Decision Trees are easy to interpret and therefore easy to understand by users. They are good at detecting fake news and use Decision Trees to identify discriminative features representing fake news articles, showing competitive classification accuracy. Gradient boosting algorithms such as XGBoost and Gradient Boosting Classifier show promise for improving prediction performance through ensemble learning and gradient boosting techniques to integrate multiple weak classifiers to increase the fake news detection accuracy. In Random Forest, Decision Trees are stacked to improve classification accuracy and reliability and investigated the use of Random Forest to verify fake news using text features and metadata. RNN and CNN of deep learning techniques have demonstrated strong potential in capturing complex patterns from text data. This is just one of many studies exploring the use of deep learning architectures for finding fake news. Such studies have shown that deep learning can successfully capture semantic meaning and contextual information. Hybrid methods studies proposed a method that combined multiple algorithms or incorporated domain-specific information and proposed a hybrid model that combined deep learning techniques and machine learning algorithms to effectively detect fake news in multiple datasets [6].

## 3 Motivation and Research Objectives

In-depth literature research on fake news detection was a very learning and exciting journey of breakthrough research and innovative methods. Among the various algorithms and techniques of the world's researchers, there is much motivation to participate in this critical area. Navigating the complexities of this difficult task, the collective efforts of the scientific community inspire a sense of purpose and determination to confront this epidemic of misinformation. Our pursuit of truth, honesty, and informed decision-making motivates us. When we conceived, we started this journey with renewed enthusiasm. Each time we look forward to pushing the boundaries of information and the latest technology to identify fake news. It is this unwavering commitment to fostering a transparent and trustworthy information ecosystem that drives us forward. By standing together, we intend to help close this huge gap by curbing the spread of misinformation and empowering people to navigate the digital environment with confidence and understanding [7]. This research is a practical project driven by the urgent need to safeguard the integrity of information that millions of people globally use, rather than just an academic one. Disinformation has been able to spread at a never-before-seen rate because of the internet and social media platforms, greatly affecting public opinion, political choices, and even foreign relations. Because fake news has the power to distort public opinion, mislead voters during elections, and intensify social instability, it is critical to identify and combat it. The developments in technology encourage us to keep investigating novel models and methodologies in order to keep up with the ever-changing strategies of disinformation [8]. Large-scale data processing capabilities of deep learning models have been shown, allowing them to reveal previously unseen patterns. Digital spaces can be made safer for users by significantly reducing the amount of human labor required for content moderation when fake news identification can be automated. Our dedication to seeking the truth is based on the conviction that a well-informed populace is essential to the wellbeing of any democracy. Our objective is to enhance digital literacy and media reliability globally by creating a more dependable detection system [9]. It is clear that preventing the dissemination of misleading information requires more than just individual users; it also calls for ongoing cooperation between the scientific community, legislators, and technological developers. By conducting this research, we intend to both provide more practical solutions and promote interdisciplinary efforts to address disinformation from a variety of perspectives.

## 4 Proposed Methodology

We employed Term Frequency-Inverse Document Frequency (TF-IDF) vectorization as a fundamental technique to convert textual data into a numerical format suitable for machine learning. This approach helps in identifying the importance of words in relation to the entire dataset, enhancing the feature representation. Following the vectorization, we implemented several machine learning models to assess their performance on the processed data: Logistic Regression: This model was utilized for its simplicity and efficiency in binary classification tasks. It helps us understand the relationship between the input features and the target variable. Decision Tree: We incorporated Decision Trees to capture the hierarchical structure of the decision-making process. This model is intuitive and provides clear insights into feature importance. Gradient Boosting: This ensemble technique was employed to improve predictive performance by combining the strengths of multiple weak learners. It allows for efficient handling of complex datasets. Random Forest: Another ensemble method, Random Forest, was used to enhance model robustness by averaging the predictions of multiple Decision Trees. This approach reduces overfitting and increases accuracy. To validate the performance of our models, we conducted manual testing. This involved a thorough evaluation of model outputs against known benchmarks to ensure the accuracy and reliability of our predictions. By utilizing TF-IDF vectorization along with these advanced machine learning algorithms and manual testing, we achieved promising accuracy levels, demonstrating the effectiveness of our approach.

## 5 Design

To specify the design of the proposed Python-based fake news detection system, we outline the approach and illustrate it with a block diagram given below in Fig. 1.

This block diagram gives us a structured overview of the whole process from data collection to manual testing, which helps us understand the overall structure and flow of the fake news detection system.

Data Sets: Two datasets, one containing real news and the other containing fake news are loaded into two Data Frames ("df_true" and "df_fake").

Data Preprocessing: It undergoes the process of text cleaning, removing unwanted characters, URLs, HTML tags, punctuation, and numbers from text data.

Tags: Tags are assigned (0 for fake news and 1 for real news).

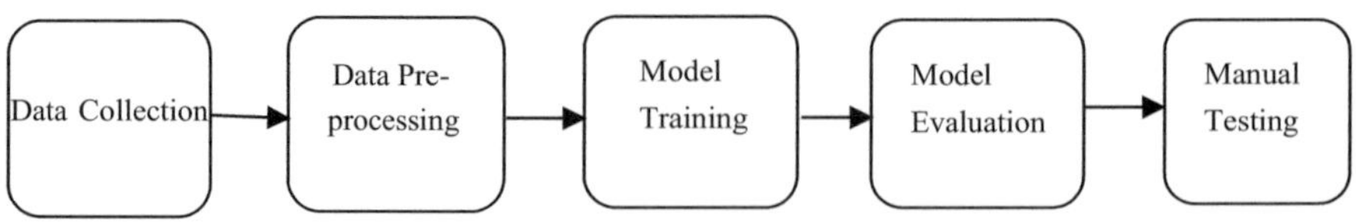


**Fig. 1** Block diagram of the implementation process

Manual Test Data: Some data are extracted from each dataset for manual testing.

Model Training: The two data sets (real news and fake news) are combined and then split into training and testing.

Model Evaluation: Test labels are predicted using each trained model. The model performance is evaluated using the precision and classification report (precision, recall, F1-score) [10].

Manual Testing: A function is implemented to manually input news texts for testing. The input texts are preprocessed and then the tags are predicted with trained models.

## 6 Implementation

We have implemented the fake news detection in Python using classifier algorithms. We have given the breakdown of the code being implemented along with the Python libraries. The code is executed using a Jupyter Notebook. The code snippet and the results demonstrate how to implement fake news detection using TF-IDF vectorization and several classification algorithms [11]. The main steps of implementation and performance measures of the classifiers are as follows: 1. Data preprocessing: the script reads fake news and real news. It assigns class identifiers to datasets: 0 for fake news and 1 for real news. Both data (fake and real) is concatenated in one data frame. Text preprocessing is done including lowercase letters, removing special characters and URLs, and ID. 2. Data split: we use the "train_test_split" function to split the data into train and test data. 3. Vectorization: text data is converted to TF-IDF vectors using TFIDF Vectorizer. 4. Model Training: all four classifiers are trained on TF-IDF vectors.

5. Model Evaluation: accuracy scores are calculated for each classifier on the "score" method. Now we interpret the results of each classifier based on the classification report and accuracy scores. Logistic Regression: for both categories, i.e., fake news and real news, precision, recall—both and F1-scores. Decision Tree: similar to LR, for both classes fake news and true news. On the basis of metrics, we can compare the performance of Gradient Boosting, and Random Forest Classifiers and choose the best precision or the best balance between accuracy and recall. 6. Finally, we can further tune the hyper parameters of the selected classifier to improve performance [12].

We have used some examples of true and false news articles to test the classifier algorithms. One of the examples tested is given below:

SAO PAULO (Reuters)—Cesar Mata-Piers, the founder and owner of an engineering firm CAS SA, one of the biggest companies of Brazil was involved in biggest corruption scandal, has died Sunday He was 68 years old. Mata Pearce died of a heart attack while walking in the morning in a private area of Sao Paulo, where the OAS is located, said a person familiar with the matter. Tried to contact his opinion but failed to manage him eskete or the, eten be de Brin les regale pie posite, peove tail govita secortol interest lost yy told by daughter Teresa. Few years later after consistent growth fueled by lucrative government contracts, OAS became caught up in illegal car washes between government films and producers. Trained as an engineer, Mata founded Piers OS with two partners in 1976 to carry out subcontracting work for Odebrecht SA—the largest producer involved in exploration. Before the crisis, Forbes magazine estimated Matta Pierce's net worth at 611 million dollars. In 2015, Nedropol was dropped from the list of billionaires in the magazine, soon after OS filed for bankruptcy protection following the car wash crisis. People directly involved in the negotiations told Reuters at the time that White Matta Piers was not accused of wrongdoing in the trial, and creditors asked him and his family to avoid useful daily activities. He is survived by his two sons and wife.

## 7 Result

The script displays fake news detection report using different classifiers. The script preprocesses the data, divides it into training and test sets, trains classifiers, and evaluates their performance using classification metrics. The performance of all the classifiers are shown in the Tables 1, 2, 3, and 4:

The Decision Tree Classifier shows the highest accuracy and perfect F1-scores, recall, and precision. Both Random Forest and Logistic Regression Classifiers have slightly lower accuracy, but still perform very well. Gradient Boosting Classifier also performs exceptionally well, with high accuracy and almost perfect precision, recall, and F1-scores. These metrics indicate that all four models are highly effective at

**Table 1** Performance and accuracy of Gradient Boosting Classifier

| | Support | Precision | Recall | F1-score |
|---|---|---|---|---|
| 0 | 5944 | 1.00 | 0.99 | 1.00 |
| 1 | 5176 | 0.99 | 1.00 | 1.00 |
| Macro avg | 11,120 | 1.00 | 1.00 | 1.00 |
| Weighted avg | 11,120 | 1.00 | 1.00 | 1.00 |
| Accuracy | 11,120 | | | 1.00 |

**Table 2** Performance and accuracy of Decision Tree Classifier

| | Support | Precision | Recall | F1-score |
|---|---|---|---|---|
| 0 | 5944 | 1.00 | 1.00 | 1.00 |
| 1 | 5176 | 1.00 | 1.00 | 1.00 |
| Macro avg | 11,120 | 1.00 | 1.00 | 1.00 |
| Weighted avg | 11,120 | 1.00 | 1.00 | 1.00 |
| Accuracy | 11,120 | | | 1.00 |

**Table 3** Performance and accuracy of Random Forest Classifier

| | Support | Precision | Recall | F1-score |
|---|---|---|---|---|
| 0 | 5944 | 0.99 | 0.99 | 0.99 |
| 1 | 5176 | 0.99 | 0.99 | 0.99 |
| Macro avg | 11,120 | 0.99 | 0.99 | 0.99 |
| Weighted avg | 11,120 | 0.99 | 0.99 | 0.99 |
| Accuracy | 11,120 | | | 0.99 |

**Table 4** Performance and accuracy of Logistics Regression Classifier

| | Support | Precision | Recall | F1-score |
|---|---|---|---|---|
| 0 | 5944 | 0.99 | 0.99 | 0.99 |
| 1 | 5176 | 0.99 | 0.98 | 0.99 |
| Macro avg | 11,120 | 0.99 | 0.99 | 0.99 |
| Weighted avg | 11,120 | 0.99 | 0.99 | 0.99 |
| Accuracy | 11,120 | | | 0.99 |

distinguishing between fake news and true news, with the Decision Tree Classifier being the best-performing model in this case [13]. Table 5 provides the comparison of performance and accuracy of LR, DT, GB, and RFC.

Manual Testing: It is done by giving manually input news texts for testing. The input texts are preprocessed and then the tags are predicted with trained models. Table 6 provides the performance and accuracy of LR, DT, GB, and RFC with manual testing.

**Table 5** Comparison of performance and accuracy of LR, DT, GB, and RFC

| Classifiers/ Metrics | Decision Tree Classifier | Gradient Boosting Classifier | Random Forest Classifier | Logistic Regression |
|---|---|---|---|---|
| Accuracy | 0.996 | 0.996 | 0.989 | 0.978 |
| Precision Class 1 | 1.00 | 0.99 | 0.99 | 0.98 |
| Precision Class 0 | 1.00 | 1.00 | 0.99 | 0.99 |
| Recall Class 1 | 1.00 | 1.00 | 0.99 | 0.99 |
| Recall Class 0 | 1.00 | 0.99 | 0.99 | 0.99 |
| F1-Score Class 1 | 1.00 | 1.00 | 0.99 | 0.99 |
| F1-Score Class 0 | 1.00 | 1.00 | 0.99 | 0.99 |

**Table 6** Performance and accuracy of LR, DT, GB, and RFC with manual testing

| LR prediction | RFC prediction | DT prediction | GBC prediction |
|---|---|---|---|
| True news | True news | True news | True news |

# 8 Conclusion

Our work on fake news detection using different classifiers showed promising results. We used the evaluation measures such as F1-score, recall, precision, and accuracy and found that Decision Tree Classifier outperformed others in our dataset [14]. We have several advantages in finding fake news using machine learning algorithms and TF-IDF vectorization. The flexibility and scalability of our methodology make it suitable for use in real-world applications such as interactive media platforms and news aggregator sites to prevent the propagation of misinformation [15]. Although there are various methods for finding fake news, our work stands out in its simplicity, effectiveness, and efficiency. Using well-established machine learning algorithms and the TF-IDF vector, we also achieved competitive performance without more complex deep learning architectures or more extensive feature engineering. Our comprehensive evaluation of multiple classifiers provides a strong basis for device determination. Our methodology makes an informed decision about the best classifier to use in specific application scenarios. Future enhancements, including the integration of deep learning models, ensemble methods, and real-time processing capabilities, will help ensure our system remains a reliable tool in the fight against fake news. Through our work, we aim to contribute to a more informed and responsible digital society, where the integrity of information is preserved and the spread of disinformation is actively curtailed.

# 9 Future Work

In future, we shall investigate other text or domain-specific features that can increase the discriminating power of classifiers. We would experiment with deep learning architectures such as RNN and CNN to capture more complex patterns in the text data. We would perform a more detailed hyper parameter tuning process to optimize the performance of the classifiers and shall develop mechanisms for real-time monitoring and detection of fake news. Several future implementations could substantially enhance the effectiveness of our system. First, integrating LSTM networks or BERT could potentially increase the system's ability to understand context and detect subtler forms of disinformation. LSTM models excel in capturing the sequential nature of text, which would be beneficial when analyzing longer news articles, while transformer-based models like BERT can capture nuanced meanings by considering the entire context of a sentence rather than just individual words. These architectures may also help to reduce false positives and negatives by improving the system's ability to differentiate between similar but distinct statements. Secondly, we plan to explore ensemble methods that combine the strengths of various classifiers, such as stacking or boosting models, to create a more resilient system. Ensemble learning could enhance accuracy by aggregating the predictions of several algorithms, leading to more reliable classifications even when individual models may struggle. Another area of improvement lies in domain-specific feature extraction. While we currently rely heavily on TF-IDF vectorization, more sophisticated feature extraction methods that leverage semantic embeddings, such as word2vec or GloVe, could provide richer representations of text. Additionally, using external knowledge bases or metadata such as the source's credibility, author history, and publication date could further refine our model's predictions. Lastly, our future work will also focus on optimizing real-time performance and scalability. Developing a system that can detect fake news in real-time, especially in high-traffic environments like social media, is crucial. By optimizing computational efficiency and exploring cloud-based solutions, we can ensure that our system is not only accurate but also scalable enough to handle the sheer volume of data in today's digital landscape.

## References


1. Yousaf S (2020) Fake news detection using Machine Learning Ensemble methods. Complexity 1(1)
2. Vosoughi S, Roy D, Arai S (2018)The spread of true and false news online. Science 359:1146–1151
3. Alpaydin E (2001) Introduction to machine learning. The MIT Press
4. Lazer DMJ et al (2018) The science of fake news. Science 359:1094–1096
5. Chouhan R et al. Fake news detection based on machine learning algorithms. IEEE Xplore. https://doi.org/10.1109/CISCT57197.2023.10351408
6. Sridhar S, Vijaylakshmi M (2021) Machine learning. OXFORD University Press
7. Ahmad I et al. Fake news detection using machine learning ensemble methods. https://doi.org/10.1150/2020/8885861
8. Chandra Shekhar Rao V et al. Logistics Regression vs XGBoost: machine learning for counterfeit news detection. IEEE Xplore. https://doi.org/10.1109/ICSTCEE54422.2021.9708587
9. Rama Krishna S et al (2021) Survey on fake news detection using machine learning algorithms. Int J Eng Res Technol (IJERT). ISSN: 2278-0181 Published by www.ijert.org. ICACT—2021 conference proceedings
10. Davis J, Godrich M (2006) The relationship between Precision Re-call and ROC curves. In: Proceedings of 23rd international conference on machine learning. ACM, pp 233–240
11. Singh S et al (2023) Fake news detection using machine learning algorithms. IJAEM 5(4):1296–1299. ISSN: 2395-5252. www.ijaem.net
12. Elyassami S (2021) Combating fake news with computational intelligence techniques. SCL, Morocco
13. Amin E et al. Enhancing the detection of fake news in social media based on machine learning models. IEEE Xplore, https://doi.org/10.1109/ICAAIC56838.2023.10141355
14. Kulkarni P (2021) Fake news detection using machine learning. In: ITM Web of conferences 40
15. Rahwan I, Cebrian M, Obradovich N et al (2019) Machine behavior. Nature 568(7753):477–486. https://doi.org/10.1038/s41586-019-1138-y

# Comparison of Accuracy in Lung Cancer Classification Through Convolutional Neural Network Models Based on Histopathological Image

**Shilpa, Kiranpreet Kaur, Narinder Kumar, and Pooja Sharma**

**Abstract** The research paper addresses the pressing need for advanced methods in lung cancer detection and precise subtype identification to enhance patient care. Employing state-of-the-art convolutional neural network (CNN) architectures—Inception V4, VGG16, and ResNet—the study investigates their efficacy in detecting lung cancer from histopathological images and distinguishing between lung adenocarcinoma and squamous cell carcinoma. Key objectives include training these models using a comprehensive "Lung Cancer Preprocessed Dataset," assessing their performance in accurate cancer detection and subtype classification, and conducting a comparative analysis to identify the most effective model. The exceptional performance of ResNet50, Inception V4, and VGG16 in fivefold cross-validation, achieving remarkable mean accuracies of 97%, 97%, and 94%, respectively, has profound implications for enhancing accurate classification of imbalanced waste methods.

**Keywords** Lung cancer detection · Lung cancer classification · Histopathology image · CNN models · ResNet50 · Inception V4 · VGG16

## 1 Introduction

Lung cancer stands firm as a prominent and deadly type of cancer globally, presenting substantial obstacles in detecting it early and accurately categorizing it. Histopathological image of lung tissue specimens is a pivotal diagnostic approach, providing essential insights into cancerous growth and aiding in effective treatment planning [1]. The advent of deep learning and convolutional neural networks (CNNs) has opened new avenues for leveraging image analysis techniques to automate and

Shilpa · K. Kaur · N. Kumar (✉) · P. Sharma
Department of Computer Science and Engineering, Rayat Bahra University, Mohali, India
e-mail: nk616319@gmail.com

P. Sharma
e-mail: pooja.rani@rayatbahrauniversity.edu.in

 
K. R. Madhavi et al. (eds.), *Proceedings of Sixth International Conference on Computer and Communication Technologies*, Lecture Notes in Networks and Systems 1356,
https://doi.org/10.1007/978-981-96-5238-9_29

enhance lung cancer detection from histopathology images [2]. In this paper, our primary focus is:

- Leveraging the most extensive collection of complete-slide images (WSI) for a lung cancer biopsy in order to assess and classify the Gleason score and International Society of Urological Pathology (ISUP).
- We have implemented groundbreaking techniques that harness convolutional neural networks (CNN) to identify lung cancer at an early stage in biopsy specimens.
- We have pitted different CNN structures such as ResNet50, Inception V4, and VGG16, in a quest to unearth the most precise architecture for Lung cancer detection.

The content of this paper is as described in the following: Sect. 2 discusses the related work of lung cancer. After that, in Sect. 3, the research methodology is discussed. The results and discussion are in Sect. 4. Finally, the conclusion and future scope of the paper are written in Sect. 5.

## 2 Literature Review

This division has delved into connected research. The upcoming reviews of literature offer a thorough exploration of current studies in the field of identifying and categorizing lung cancer through histopathological image datasets.

Hatuwal [3] explored the application of convolutional neural network (CNN) in analyzing histopathological images to identify lung cancer. The CNN model attained a training accuracy of 96.11% and a validation accuracy of 97.20% in its final epoch. Potential constraints might involve the scale and variety of the dataset utilized, the applicability of the findings to different demographics or datasets, and the necessity for additional confirmation and experimentation in clinical environments. Sheriff [4] delved into the realm of lung cancer identification utilizing the VGGNet 16 design, showcasing remarkable results by achieving impeccable precision, recall, and F1-score (1.00) for both lung cancer and normal lung categories, with an overall accuracy of 1.00. It emphasizes VGG NET 16's potential for early lung cancer detection and compares its accuracy with other architectures, highlighting its efficacy. The paper outlines the training process, showcasing improved accuracy and reduced loss after iterative training, leading to an optimized model with minimized noise. However, limitations exist, notably the absence of discussions regarding result generalizability across diverse populations and challenges in implementing deep learning algorithms in real-world clinical settings. Additionally, the paper overlooks ethical considerations and biases linked to using deep learning in medical diagnosis, essential in healthcare applications [5].

Chaunzwa [6] this paper, we explore the application of deep learning and radiomics in forecasting the histology of non-small cell lung cancer (NSCLC) tumors.

By training convolutional neural networks (CNNs) on a group of NSCLC patients, the researchers discovered that the CNNs could accurately predict tumor histology.

## 3 Research Methodology

Lung cancer, a common illness among men, accounts for 1 million new instances and 350,000 fatalities each year. Accurate identification is vital, with Gleason grading employed for tissue assessment. The compilation of the dataset is a significant aspect of the complete process, succeeded by the execution phase. This segment will provide detailed insights into the dataset depiction, data preparation, research approach, architecture choice, and execution.

### *3.1 Dataset Description*

The dataset was meticulously partitioned into three balanced classes, each containing 5000 images. Image augmentation techniques were employed to enhance the diversity and quantity of the dataset, crucial for training robust deep learning models. Preprocessing steps were executed to standardize image dimensions and intensities, ensuring uniformity and compatibility with the CNN architectures employed for classification [7].

### *3.2 Dataset Preprocessing*

To conduct the preparation of the dataset, the initial step involves the loading of data from biopsy slides utilizing OpenSlide. The functionality of OpenSlide enables the loading of specific regions without the necessity of loading the complete image into the memory. Prior to the loading process of a biopsy slide, it is imperative to initiate the opening of the respective file. Subsequent to the opening of a file, data retrieval from it can be carried out at any specified position and level. In order to offer essential details regarding a slide and exhibit a thumbnail, a concise function is developed. Biopsies may exhibit various rotations, albeit lacking clinical significance.

### *3.3 Model Selection*

CNN designs possess the capability to effectively capture spatial hierarchies, rendering them well-suited for utilization with image datasets. These designs incorporate pooling layers for the purpose of down sampling, along with convolutional

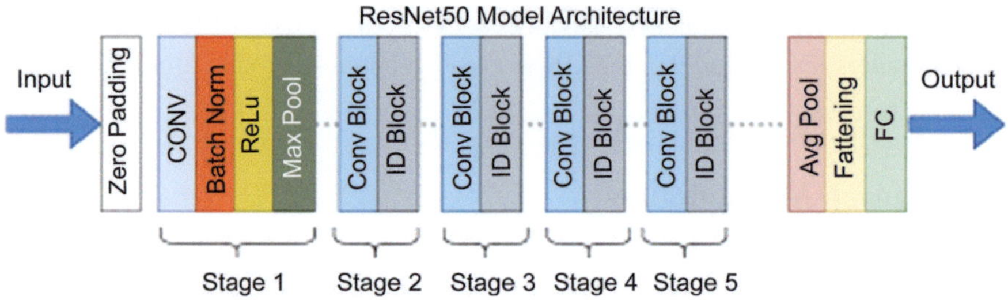


**Fig. 1** ResNet50 network model

layers that are instrumental in the recognition of features such as edges and textures. A proficient network not only saves time but also streamlines deployment processes and diminishes associated costs. In our research endeavor, we opted to evaluate the ResNet50, Inception V4, and VGG16 architectures, subsequently conducting a comparative analysis of these four neural networks.

### 3.3.1 ResNet50

ResNet50 introduces residual connections, allowing the network to learn residual mappings instead of learning complete transformations. This helps in mitigating vanishing/exploding gradient problems encountered in very deep networks. During training, ResNet50 learns to optimize its weights and biases by minimizing the difference between predicted and actual labels using gradient descent and backpropagation. Accuracy is assessed by comparing the model's predictions with the ground truth labels in the dataset. The model's overall accuracy is determined by how often its predictions align with the actual labels [8]. Figure 1 shows ResNet50 network model.

### 3.3.2 Inception V4

Inception V4 utilizes inception modules that incorporate multiple convolutional filter sizes within the same layer to capture features at various scales. It optimizes its parameters to learn discriminative features. Similarly, to other CNNs, during testing, Inception V4 receives unseen images and generates predictions or probabilities for each class based on the learned features. Accuracy evaluation involves comparing the model's predictions with the ground truth labels to determine its correctness in classifying images into the correct categories [9]. Figure 2 shows the InceptionV4 network model.

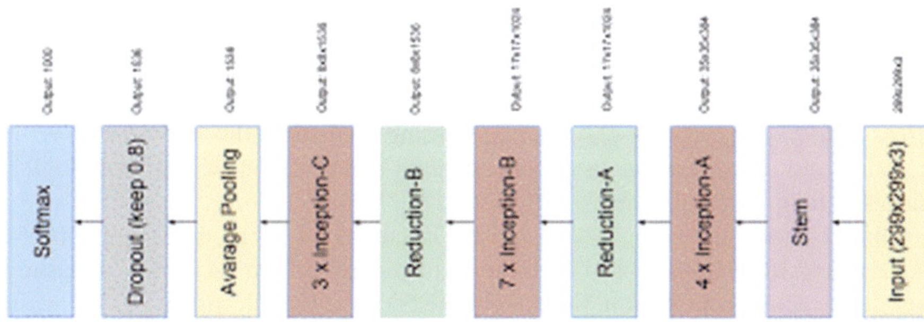


**Fig. 2** InceptionV4 network model

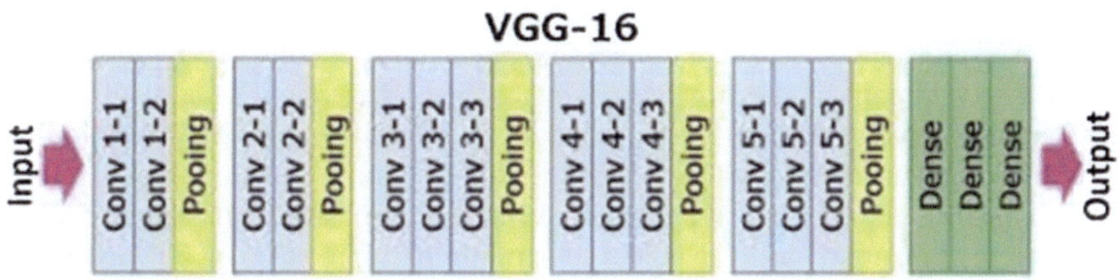


**Fig. 3** VGG16 network model

### 3.3.3 VGG16

Simple and Uniform Architecture: VGG16 comprises 16 layers with mostly $3 \times 3$ convolutional filters. Its architecture is characterized by its simplicity, uniformity, and deep stack of convolutional layers. VGG16 progressively learns hierarchical features from images by passing them through multiple convolutional layers and pooling operations. It learns representations and patterns from the data. In the testing phase, VGG16 performs inference by taking unseen images and generating predictions or class probabilities based on the learned features [10]. Figure 3 shows VGG16 network model.

## 4 Results and Discussion

Three models were tested using the dataset and results. The graphs of loss versus epoch and accuracy versus epoch were used to choose the best weights for the model, while the confusion matrix and accuracy were used to compare model performance.

### 4.1 Performance Evaluation

Table 1 showcases the nuanced performance of ResNet50, Inception V4, and VGG16 across different lung cancer classes, highlighting their strengths and weaknesses in precision and recall, crucial factors in accurate classification of histopathological images in lung cancer diagnosis [10].

Table 2 shows the comparison between accuracy of all model. Inception V4, VGG16, and ResNet50—based on their performance in. The "accuracy" measures overall correctness: Inception V4 and ResNet50 have high accuracy (97%), while VGG16 is slightly lower (94%).

Table 2 summarizes the performance metrics of the evaluated models. Both Inception V4 and ResNet50 showcased high accuracy rates of 97%, indicating their effectiveness in classifying lung cancer histopathological images. In terms of precision, ResNet50 stands out with a remarkable precision score of 0.99, demonstrating its ability to minimize false positives in predictions. Inception V4 also displayed commendable precision at 0.95, ensuring a high level of true positive predictions.

While InceptionV4 demonstrated a remarkable equilibrium between precision and recall, attaining elevated accuracy and recall at the expense of a marginally lower precision in comparison to ResNet50, the extraordinary precision of 0.99 exhibited by ResNet50 highlights its supremacy in reducing erroneous positive predictions. The concept of precision quantifies the proportion of accurately predicted positive instances out of all predicted positive instances, underscoring ResNet50's reliability in accurately predicting positive scenarios.

## 5 Conclusion and Future Scope

Inception V4 showcases a harmonious display of accuracy and memory recall, yet ResNet50 rises above as the predictor of choice, delivering a more exact outcome with its significantly elevated precision rating, effectively reducing erroneous predictions. The remarkable precision exhibited by ResNet50 highlights its trustworthiness in pinpointing authentic positive instances, a critical aspect in fields such as medical technology, specifically in the context of diagnosing lung cancer.

This research delves into various pathways to enhance the detection of lung cancer through the utilization of convolutional neural network (CNN) models. It puts forth the idea of combining techniques, adopting transfer learning strategies, and making changes to the architecture. The effectiveness of the model greatly relies on real-world testing and expanding the dataset. Moreover, it also touches upon ethical considerations, safeguarding patient privacy, and ensuring regulatory adherence. The future goals include improving the dependability, practicality, and ethical alignment, thus propelling the progression of AI-facilitated lung cancer detection.

**Table 1** Precision, recall, and F1-score for all models

| Class | ResNet50 | | | Inception V4 | | | VGG16 | | |
|---|---|---|---|---|---|---|---|---|---|
| | Precision | Recall | F1-score | Precision | Recall | F1-score | Precision | Recall | F1-scor e |
| Adenocarcinoma | 0.99 | 0.93 | 0.96 | 0.95 | 0.98 | 0.96 | 0.90 | 0.84 | 0.87 |
| Benign | 1.00 | 1.00 | 1.00 | 1.00 | 1.00 | 1.00 | 1.00 | 0.93 | 0.97 |
| Squamous | 0.94 | 1.00 | 0.97 | 0.96 | 0.90 | 0.93 | 0.91 | 1.00 | 0.95 |
| Total | 0.99 | 0.93 | 0.96 | 0.95 | 0.98 | 0.96 | 0.90 | 0.84 | 0.87 |

**Table 2** Artistic measures of model performance

| Model | Accuracy | Precision | Recall | F1-score |
|---|---|---|---|---|
| Inception V4 | 0.97 | 0.95 | 0.98 | 0.96 |
| VGG16 | 0.94 | 0.90 | 0.84 | 0.87 |
| ResNet50 | 0.97 | 0.99 | 0.93 | 0.96 |

# References


1. Bhattacharyya N, Satish A (2021) A study of accuracy in detection of lung cancer through CNN models. Des Eng (Toronto) 2021:1250–1270
2. Mangal S, Chaurasia A, Khajanchi A (2020) Convolution neural networks for diagnosing colon and lung cancer histopathological images. arXiv preprint arXiv:2009.03878
3. Hatuwal B, Thapa H (2020) Lung cancer detection using convolutional neural network on histopathological images. Int J Comput Trends Technol 68:21–24. https://doi.org/10.14445/22312803/IJCTT-V68I10P104
4. Sheriff S, Venkat Kumar J, Vigneshwaran S, Jones A, Anand J (2021) Lung cancer detection using VGGNet 16 architecture. J Phys Conf Ser 2040:012001. https://doi.org/10.1088/1742-6596/2040/1/012001
5. Chaunzwa TL, Hosny A, Xu Y, Shafer A, Diao N, Lanuti M, Christiani DC, Mak RH, Aerts HJ (2021) Deep learning classification of lung cancer histology using CT images. Sci Rep 11(1):5471
6. Wu Z, Wang L, Li C, Cai Y, Liang Y, Mo X, Lu Q, Dong L, Liu Y (2020) DeepLRHE: a deep convolutional neural network framework to evaluate the risk of lung cancer recurrence and metastasis from histopathology images. Front Genet 11:768
7. Li Z, Zhang J, Tan T, Teng X, Sun X, Zhao H, Liu L, Xiao Y, Lee B, Li Y, Zhang Q (2020) Deep learning methods for lung cancer segmentation in whole-slide histopathology images—the acdc@ lunghp challenge 2019. IEEE J Biomed Health Inf 25(2):429–440
8. Kriegsmann M, Haag C, Weis CA, Steinbuss G, Warth A, Zgorzelski C, Muley T, Winter H, Eichhorn ME, Eichhorn F, Kriegsmann J (2020) Deep learning for the classification of small-cell and non-small-cell lung cancer. Cancers 12(6):1604
9. Le Page AL, Ballot E, Truntzer C, Derangère V, Ilie A, Rageot D, Bibeau F, Ghiringhelli F (2021) Using a convolutional neural network for classification of squamous and non-squamous non-small cell lung cancer based on diagnostic histopathology HES images. Sci Rep 11(1):23912
10. Sadhwani A, Chang HW, Behrooz A, Brown T, Auvigne-Flament I, Patel H, Findlater R, Velez V, Tan F, Tekiela K, Wulczyn E (2021) Comparative analysis of machine learning approaches to classify tumor mutation burden in lung adenocarcinoma using histopathology images. Sci Rep 11(1):16605

# Plant Leaf Diagnosis Utilizing the Efficient Net Deep Learning Model

**Narinder Kumar, Pooja Sharma, Poonam Kukana, Sukhjinder Kaur, and Ashima**

**Abstract** Most illnesses of plants have obvious signs, as well as the current acknowledged method involves having a skilled plant pathologist visually inspect the diseased leaves of the plant to arrive at the conclusion. Computer-aided diagnostic techniques are well suited to this task because the illness diagnosis procedure is laborious to complete manually and because the pathologist's skill level directly affects the diagnosis's accuracy. We need a model that can conduct successful classification without pre-processing, as contrast to standard machine learning techniques that demand flawless manual extraction of features to yield successful results. This paper suggested the Efficient Net deep studying architecture in order to categorize plants leaf diseases and also compared its performance toot her cutting edge deep learning models. To train models, the Plantation Kingdom information was employed. Unique and enhanced facts with 55,448 and 61,486 photos were employed for instruction each one, correspondingly. Transfer gaining knowledge was employed for training the Efficient Net additional deeper learning algorithms as well as framework. Every model layer in the transfer learning was set up to allow for training. The test dataset's findings demonstrated that the Efficient Net architecture's B5 and B4 models outperformed other deep learning models with regard to precision and precision, scoring 99.90% and 99.96%, respectively, and 98.49% and 99.49%, respectively, in the original and supplemented datasets.

N. Kumar (✉) · P. Sharma · P. Kukana · S. Kaur · Ashima
Department of Computer Science and Engineering, Rayat Bahra University, Mohali, India
e-mail: nk616319@gmail.com

P. Sharma
e-mail: Pooja.rani@rayatbahrauniversity.edu.in

P. Kukana
e-mail: Poonam.e18526@cumail.in

S. Kaur
e-mail: Sukhjinder.17978@rayatbahrauniversity.edu.in

Ashima
e-mail: Ashima.18621@rayatbahrauniversity.edu.in

 
K. R. Madhavi et al. (eds.), *Proceedings of Sixth International Conference on Computer and Communication Technologies*, Lecture Notes in Networks and Systems 1356,
https://doi.org/10.1007/978-981-96-5238-9_30

**Keywords** VGG16 · AlexNet · Efficient Net · Deep learning

## 1 Introduction

For appropriate and sustainable farming, moreover, prevent needless squandering of money and other resources, early and accurate plant illness detection involves essential. Certain plant illnesses lack obvious signs; hence, it is necessary to employ sophisticated analytical techniques in cases of similar illnesses. Nonetheless, most illnesses of plants exhibit outward signs, and the currently recognized method involves having a skilled plant pathologist visually inspect infected plant leaves to identify the illness [1]. A plant pathologists need to be an excellent observer in order to recognize the distinctive signs of vegetation disease and make an accurate diagnosis.

An overabundance of plant varieties are there, changes in how plant diseases progress as a as the consequence of global warming and the speed at which diseases move to previously unexplored areas even cause seasoned pathologists to miss certain disease diagnoses [2] knowledgeable and perceptive systems that are able to correctly and automatically identify plant diseases are a tremendous asset to agronomists. However, offering a system like that with an easy-to-use mobile application is also a commendable accomplishment for farmers lacking access to an agronomic and phytopathological support network [3].

Technological intelligent technology advancements have made it feasible to create automated technologies that improve the speed and accuracy of disease diagnosis. These days, artificial intelligence-based systems that autonomously identify a wide range of disorders are frequently employed [4]. Numerous traditional models for machine learning have been proposed during the last ten years plant illness classification and diagnosis. Using support vector machine (SVM), Rumpf et al. [5] investigated the early identification and categorization of illnesses observed tropical plant-based sugarcane indicators. Al-Hiary et al. [6] used K-means to cluster characteristics acquired through the preparation stages in order to segment the diseased areas and classify certain areas using artificial neural networks (ANN) after extraction of features utilizing texture and color in order to identify five distinct diseases of leaves in plants. Revathi and Hemalatha suggested a technique to identify six distinct disease kinds that are present on cotton leaves. Revathi and Hemalatha (2014) propose a method that makes utilization of particle swarm optimization to choose features from a feature vector that includes determined by substance, hue, and aspect characteristics gathered from editing of images. The method then uses a cross-information learn to use a deep forward neural network classify the illness. Mokhtar et al. [7] carried out research using the support vector machine (SVM) method to detect and identify two distinct viruses that cause sickness and exhibit symptoms on tomato leaves. In a second study, Pantazi et al. [8] utilized the data gathered from the format of spatial binary method to recognize three different illnesses that affect vine leaves using the support vector machine (SVM) approach.

**Fig. 1** Sample of Plant Village dataset plant disease pairs: **a** grape with black measles; **b** cherries with mildew that powders; **c** potatoes with late blight; **d** peaches with bacterial spots; **e** apples with cedar apple rust; **f** late blight tomato; **g** robust strawberry, and **h** robust tomato. The Plant Village dataset was subjected to six distinct augmentation techniques by Geetharamani and Pandian [9]. The enhanced and original datasets are available for application in scholarly research https://data.mendeley.com/datasets/tywbtsjrjv/1. Using augmentation techniques, the study's sample count for classes with fewer than 1000 samples was raised to 1000, while the pictures from other courses remained unchanged. Following the application of augmentation, the dataset now contains 61,486 photos. In order to enhance deeper brain nets' functionality, we employ both the original dataset and the enhanced dataset created by Geetharamani and Pandian in this work

# 2 Material and Methods

## *2.1 Datasets*

The Plant Village information is made up of 54,305 photos and 38 classes representing 14 distinct plant species, among which 12 were wholesome while 26 were diseased—is utilized in this work [9]. The dataset contains colored pictures of various sizes. There is an additional class in the dataset that identifies 1143 backdrop photos. Therefore, the entire set has 55,448 images overall. Eight distinct plant disease couples, chosen at random, are displayed in Fig. 1.

## *2.2 Cutting Edge CNN-Based Approaches*

Modern CNN architectures like AlexNet, ResNet50, VGG16, and Inception V3 are compared to the effectiveness of the Efficient Net architecture presented in this paper.

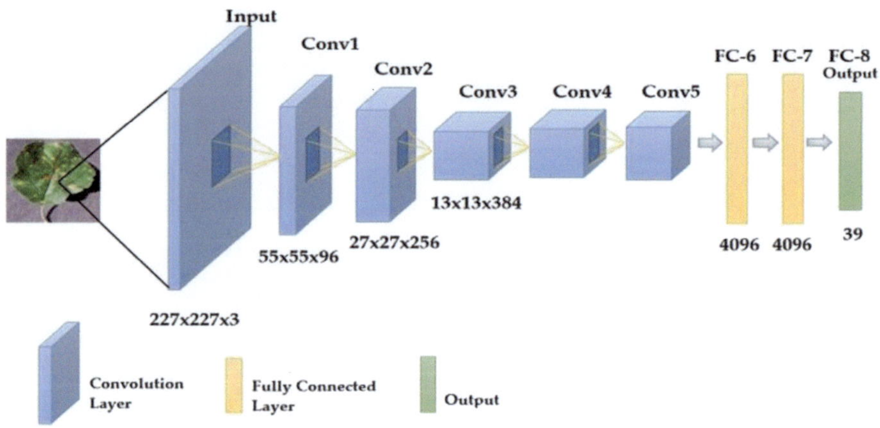


**Fig. 2** Schematic representation of AlexNet

### 2.2.1 AlexNet

The 8-layer CNN architecture known as the AlexNet model was first presented at the 2012 contest for massive amounts Sight Identification Competition for Image Net. There are about 61 million of the m parameters. The architecture of LeNet-5 serves as the foundation for this design [10]. The AlexNet structure depicted in Fig. 2 makes use of the 227 × 227 input pictures. It consists of a Softmax layer, three fully linked levels, and five convolutional levels. The convolutional and fully linked levels in this layout utilize of the commissioning of ReLU characteristic. This work uses completely linked stratum FC-8 of the AlexNet architecture, which has 39 neurons coupled to layer Softmax. Every Softmax stage result value's proportion to the corresponding output's class representation is what makes up the output. In order to achieve this, the Softmax layer uses the FC-8 layer's inputs to create a distribution and gives every category an estimated rating, with a total of 1 for all classes.

### 2.2.2 VGG16

The CNN architecture that took first position inside the ILSVRC-2014 competition, VGG16 [11], has over 138 million Variable. This architecture's most remarkable feature is that, rather than having a lot of hyperparameters, it always shares the precise same maximal pooling layer and padding that use 2 × 2 filters with stride 2, as well as identical convolutional levels that use 3 × 3 filters with stride 1. The VGG16 structure preserves the optimum pooled layer and convolution layer structure across the entire design (Fig. 3). The last three FC 8 layers include ReLU in the initial two and a Softmax activation function in the last layer. The input layer of this 16-layer design accepts 224 × 224 pixel pictures.

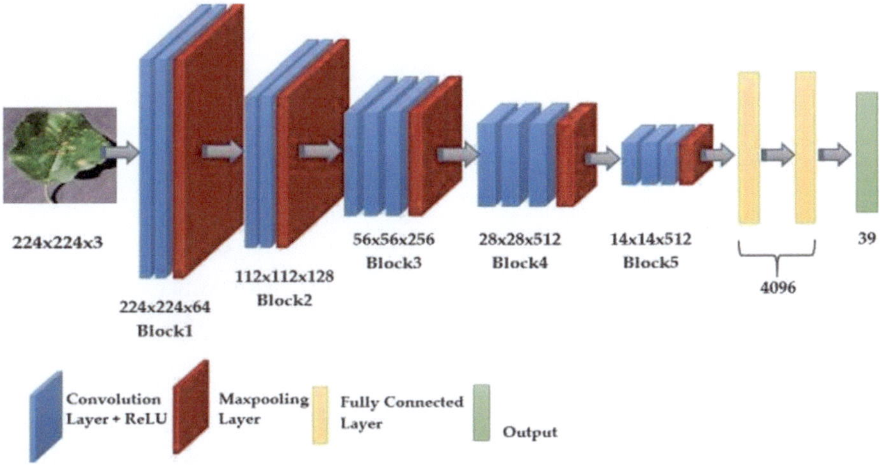


**Fig. 3** Schematic representation of VGG16

# 3 Experiments

## *3.1 Experimental Setup*

Every model employed in this investigation was built with GPU support. The Keras 2.3.1 framework, an open-source Python using the deep neural network arsenal, realizes all codes.

## *3.2 Training*

90%, 7%, and 3% of the initial and enhanced Plant Village samples were chosen at arbitrarily allocated to training, validation, and test sets, respectively, for the purposes of this study. The examination set was utilized to evaluate the effectiveness of the model in making predictions on samples that it not previously seen, whereas the sets for verification and evaluation were solely used for model training and fitting. Based on these rates, Table 1 presents an overview of the total number of photos utilized in each dataset (Tables 2 and 3).

**Table 1** Separating the graphic collections into railroad, examination, and verification

| | Total | Train (90%) | Validation (7%) | Test (3%) |
|---|---|---|---|---|
| Original data | 55,448 | 49,903 | 3813 | 1663 |
| Augmented data | 61,486 | 55,337 | 4304 | 1844 |

**Table 2** Amount of total variables and standard image submission dimensions for Efficient Net models

| Model No. | Input size | No. of total parameters |
|---|---|---|
| B0 | 220 × 220 | 5,330,471 |
| B1 | 240 × 240 | 7,856,189 |
| B2 | 260 × 260 | 9,177,500 |
| B3 | 300 × 300 | 12,320,535 |
| B4 | 380 × 380 | 19,466,823 |
| B5 | 456 × 456 | 30,562,535 |
| B6 | 528 × 528 | 43,265,143 |
| B7 | 600 × 600 | 66,658,687 |

**Table 3** Weights of the parameters in neural network architectures

| Title of the prototype | Illustration dimensions | Optimization methods and parameter | Learning rate |
|---|---|---|---|
| AlexNet | 227 × 227 | Adam (Beta1 = 0.9, Beta2 = 0.999, Decay = 0.0) | 0.001 |
| VGG16 | 224 × 224 | SGD (Momentum = 0.0, Decay = 0.0) | 0.01 |
| ResNet | 299 × 299 | Adam (Beta1 = 0.9, Beta2 = 0.999, Decay = 0.0) | 0.01 |
| Efficient Net | 132 × 132 | Adam (Beta1 = 0.9, Beta2 = 0.999, Decay = 0.0) | 0.01 |

# 4 Results and Discussion

This work aims to evaluate the performance of state-of-the-art CNN models in the literature with the effectiveness of Efficient Net deep learning architecture in the categorization of plant leaf disease. As shown in Sect. 3.2, transfer learning was employed to train each deep learning model in this investigation (Figs. 4 and 5).

**Fig. 4** Test accuracies of deep learning models for original dataset

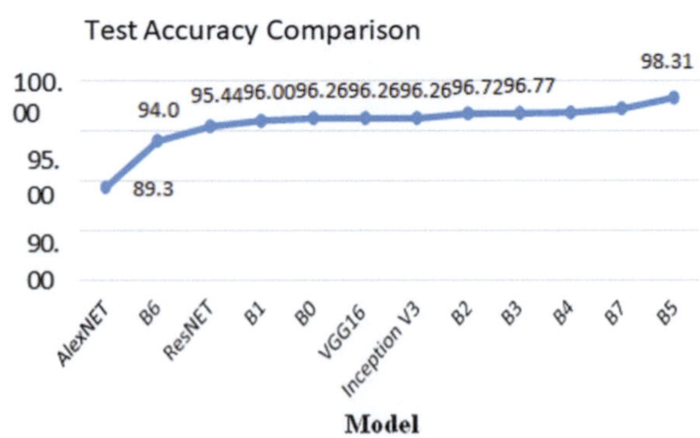

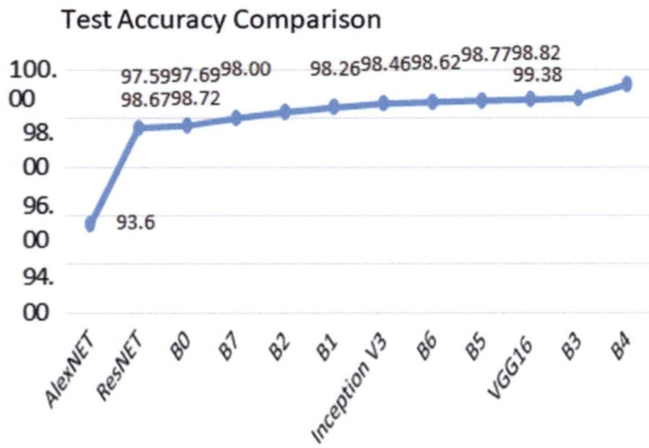


**Fig. 5** Test accuracies of deep learning models for augmented dataset

## 5 Conclusion

Deep learning techniques have gained popularity recently for pattern identification and image processing. In this study, the Plant Village dataset's 39 classes of plant leaf photos were categorized utilizing the Efficient Net deep learning architecture. The effectiveness of the suggested architecture was contrasted with cutting edge deep learning architectures that have been employed in research to identify plant leaf diseases. The Plant Village dataset was used for both the original and enhanced experimental investigations. When comparing the B4 and B5 models to other CNN architectures, equal the mean fineness and mean precision measure on the original and supplemented datasets showed that they performed better.

Additionally, the Efficient Net architecture produced better outcomes in comparison to other models from CNN that were fed raw photos with greater resolutions, despite the fact that hardware constraints forced the input image size to be shrunk to 132 × 132. However, Alex Net's average accuracy and precision lagged behind other models when it came to training times per epoch analysis. It had the lowest times in both the original and augmented datasets, at 310 and 352 s, respectively. Furthermore, the B5 and B4 models finished one epoch in reasonable times of 1930 and 1548 s, respectively. These models obtained the maximum accuracy and precision in the original and enhanced datasets, respectively.

Future research aims to boost the variety and quantity of plants classes in order to enhance the plants illness information. This will aid in the development of models for prediction outcomes in challenging circumstances with greater accuracy. Plant pathologists and farmers will be able to promptly identify plant diseases and take the appropriate measures by implementing these enhanced models in mobile environments.

## References


1. Al-Hiary H, Bani-Ahmad S, Reyalat M, Braik M, Alrahamneh Z (2011) Fast and accurate detection and classification of plant diseases. Int J Comput Appl 17:31–38
2. Amara J, Bouaziz B, Algergawy A (2017) A deep learning-based approach for banana leaf diseases classification. In: Daten Bank systeme Für Business, Technologie Und Web (BTW 2017) – Workshop band. Bonn, pp 79–88
3. Arsenovic M, Karanovic M, Sladojevic S, Anderla A, Stefanovic D (2019) Solving current limitations of deep learning based approaches for plant disease detection. Symmetry (Basel) 11:939. https://doi.org/10.3390/sym11070939
4. Rumpf T, Mahlein A-K, Steiner U, Oerke E-C, Dehne H-W, Plümer L (2010) Early detection and classification of plant diseases with support vector machines based on hyper spectral reflectance. Comput Electron Agric 74:91–99. https://doi.org/10.1016/J.COMPAG.2010.06.009
5. Chen J, Yin H, Zhang D (2020) A self-adaptive classification method for plant disease detection using GMDH-Logistic model. Sustain Comput Inf Syst 28:100415. https://doi.org/10.1016/J.SUSCOM.2020.100415
6. Chen J, Zhang D, Nanehkaran YA (2020) Identifying plant diseases using deep transfer learning and enhanced lightweight network. Multimedia Tools Appl 1–19. https://doi.org/10.1007/s11042-020-09669-w
7. Mokhtar U, Ali MAS, Hassanien AE, Hefny H (2015) Identifying two of tomatoes leaf viruses using support vector machine. Springer, New Delhi, pp 771–782. https://doi.org/10.1007/978-81-322-2250-7_77
8. Pantazi XE, Moshou D, Tamouridou AA, Kasderidis S (2016) Leaf disease recognition in vine plants based on local binary patterns and one class support vector machines. Springer, Cham, pp 319–327. https://doi.org/10.1007/978-3-319-44944-9_27
9. Chen J, Chen J, Zhang D, Sun Y, Nanehkaran YA (2020) Using deep transfer learning for image-based plant disease identification. Comput Electron Agric 173:105393. https://doi.org/10.1016/J.COMPAG.2020.105393
10. Cruz AC, Luvisi A, De Bellis L, Ampatzidis Y (2017) Vision-based plant disease detection system using transfer and deep learning. In: 2017 Asabe annual international meeting, p 1
11. Durmus H, Gunes EO, Kirci M (2017) Disease detection on the leaves of the tomato plants by using deep learning. In: 2017 6th International conference on agro-geoinformatics. IEEE, pp 1–5. https://doi.org/10.1109/Agro-Geoinformatics.2017.8047016

# Enhancing Power System Reliability: A Neural Network Approach for Electrical Fault Detection and Classification

**H. N. Lakshmi, Konatham Sumalatha, Anindita Kundu, V. Lavanya, R. Kavitha, and S. Athinarayanan**

**Abstract** The identification and categorization of electrical faults are essential for ensuring the safety and reliability of modern power systems. Disruptions such as line-to-ground and line-to-line faults can result in severe effects, including hazardous conditions, equipment destruction, and power failures. Conventional detection techniques frequently fall short due to their high prices, inefficiencies, and poor accuracy. This work explores the use of deep learning models and advanced machine learning (ML) to address these problems effectively. A variety of models, including Support Vector Machines (SVM), Decision Trees (DT), Random Forests (RF), and Artificial Neural Networks (ANN), were employed to recognize and classify different types of electrical abnormalities. The research included a diverse dataset, and the models underwent a thorough assessment employing precision, recall, F-score, and accuracy metrics. The results demonstrate that the ANN model surpassed the other models,

H. N. Lakshmi
Department of Computer Science and Engineering (AIML), CVR College of Engineering, Hyderabad, India
e-mail: hn.lakshmi@cvr.ac.in

K. Sumalatha
Department of Database Systems, School of Computer Science and Engineering, Vellore Institute of Technology, Vellore, Tamil Nadu, India
e-mail: konatham.sumalatha@vit.ac.in

A. Kundu
Department of Software Systems, School of Computer Science and Engineering, Vellore Institute of Technology, Vellore, Tamil Nadu, India
e-mail: anindita.kundu@vit.ac.in

V. Lavanya
School of Computing, Mohan Babu University, Tirupati, India

R. Kavitha
Department of Computer Science Engineering, Vel Tech High Tech Dr. Rangarajan Dr. Sakunthala Engineering College, Avadi, Chennai, India

S. Athinarayanan (✉)
Department of Computer Science Engineering, Vel Tech Rangarajan Dr. Sagunthala R&D Institute of Science and Technology, Avadi, Chennai, India
e-mail: aathithe@gmail.com

 
K. R. Madhavi et al. (eds.), *Proceedings of Sixth International Conference on Computer and Communication Technologies*, Lecture Notes in Networks and Systems 1356,
https://doi.org/10.1007/978-981-96-5238-9_31

attaining an F-score of 98.48% and a testing accuracy of 98.50%. This indicates that the ANN model is the most reliable for identifying electrical faults. This work demonstrates how important machine learning is to improving contemporary power systems' efficiency, safety, and robustness and provides a viable path forward for the field's future development.

**Keywords** Machine learning · Neural network · Electrical fault · Power system

## 1 Introduction

Modern electrical power systems have become increasingly intricate, frequently consisting of extensive networks of wiring that transmit data, power, and control signals over long distances. However, it is unavoidable that these systems will have errors and breakdowns, which might result in severe safety and cost issues [1]. These faults include several types, such as symmetric and asymmetric faults, including line-to-line (L-L), line-to-ground (L-G), and double line-to-ground (LL-G) faults, as well as the L-L-L-G fault (also known as a three-phase symmetrical fault), among others. It is essential to promptly and accurately detect and classify issues in electrical power system transmission lines and then quickly resolve them. One of the most costly and essential pieces of equipment in power systems is the power transformer. When they malfunction, there is a severe environmental risk, including the possibility of explosions, power outages, and the loss of life and property [2]. The conventional approach to electrical fault prediction is plagued by suboptimal efficiency, excessive costs, and inadequate precision [3]. However, recently, machine-learning techniques have been included in electrical failure detection and show good performance. Machine-learning algorithms can forecast electrical faults by utilizing past data to examine the relationship between various fault kinds and perform in-depth analyses depending on variables like humidity, temperature, and equipment operation status [3]. Numerous studies on machine-learning-based power system issue diagnostics have been published [4]. Researchers have employed a range of machine-learning and deep learning methods to diagnose these specific electrical issues. The machine-learning algorithms used to detect electrical faults include Support Vector Machines (SVM), Decision Trees, K-nearest neighbors (KNN), and Random Forest classifier [3]. Additionally, various neural network models such as Artificial Neural Networks (ANN), Feedforward Neural Networks (FNN), Radial basis Function Networks (RBFN), Probabilistic Neural Networks (PNN), and Chebyshev Neural Networks (ChNN) are employed to detect these types of electrical faults [5]. Deep learning models are commonly utilized for detecting electrical defects, whilst typical machine-learning models are primarily employed for classifying them. In conclusion, machine-learning methods have a greater significance in electrical fault identification compared to conventional methods such as physical inspection.

Previous works had different goals and constraints since machines have other electrical components. Our work tackles these issues by classifying and detecting electrical defects using a distinct real-world dataset. Our study mainly aims to categorize various fault kinds within these unique classifications. Our focus on maximizing accuracy while lowering processing costs is crucial to our offering. We used an Artificial Neural Network (ANN) in conjunction with baseline machine-learning models to attain a compromise between high accuracy and computational complexity. We carefully fine-tuned the models and ensured their performance matched our study goals.

The structure of the paper is as follows: Sect. 2 provides an explanation of the approach used in our research. Section 3 covers the outcomes and offers a thorough analysis of our discoveries. In Sect. 4, we ultimately wrap up the study by providing a concise summary of the main contributions and consequences of our work.

## 2 Methodology

This section presents our overall research workflow. The research utilized baseline ML models and ANN for fault detection and categorization. Figure 1, which offers a visual summary of the steps taken, illustrates the basic workflow of our investigation. The subsequent sections will provide a more detailed examination of each stage of our research.

**Fig. 1** Basic workflow of the research

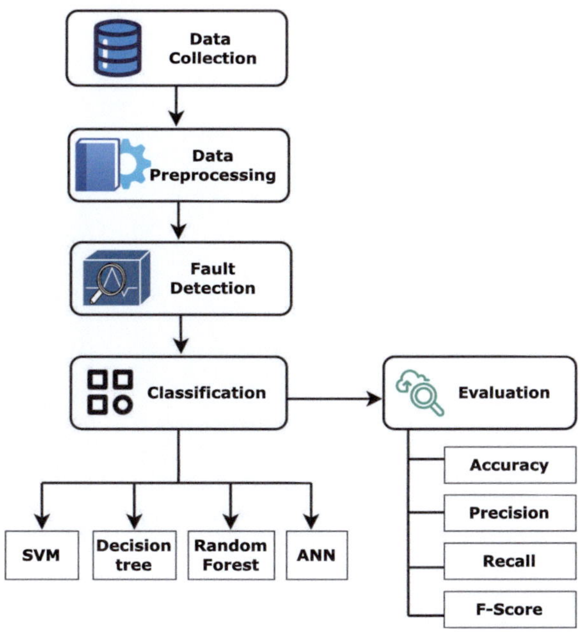

## 2.1 Dataset Description

The dataset includes the following fault types: L-L-G-G (occurring as a three-phase symmetrical fault), L-L-G-G (occurring between Phases A, B, and Ground), L-L-L-G (occurring between all three phases), and L-G-L-G (occurring between Phase A and Phase B). We have noted that class 0 (No fault) is more prevalent in the samples of our classification dataset than class 1 (Fault is present). There are a total of 12,001 data points. We have merged all the faults in one category and addressed them as 1. The class-wise data distribution and kde distribution of all columns are shown in Figs. 2 and 3.

## 2.2 Data Preprocessing

Preprocessing was done on the dataset to get it ready for model training and evaluation. Using Min–Max Scaling, the feature set **X** was first normalized [6]. The features are converted to a common scale via the Min–Max Scaler, usually falling into the range [0, 1]. The mathematical definition of this transformation is:

$$\mathbf{X}_{\text{scaled}} = \frac{\mathbf{X} - \mathbf{X}_{\text{min}}}{\mathbf{X}_{\text{max}} - \mathbf{X}_{\text{min}}} \tag{1}$$

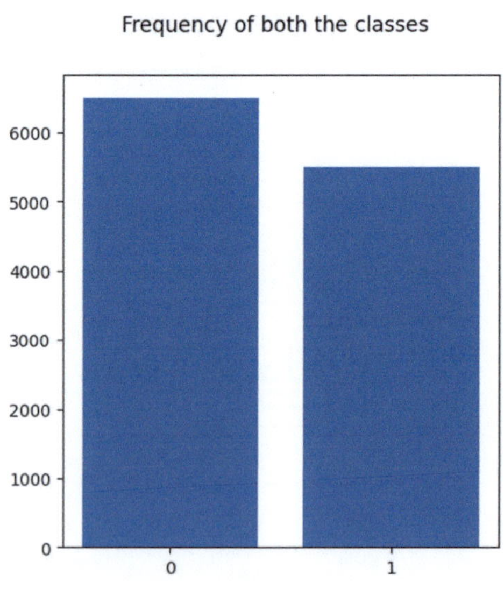


**Fig. 2** Class-wise data distribution

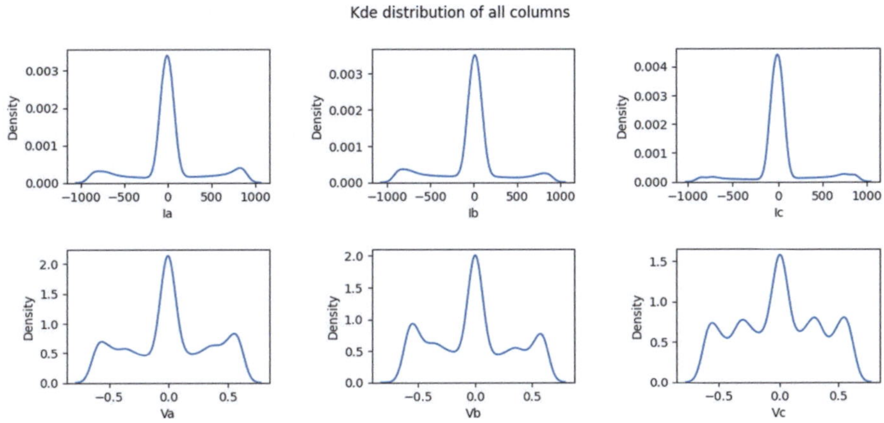


**Fig. 3** Kde distribution of all columns

where the original feature vector is represented by $\mathbf{X}$, and the minimum and maximum values in the feature set are indicated, respectively, by $\mathbf{X}_{\text{min}}$ and $\mathbf{X}_{\text{max}}$, this scaling keeps any feature with a broader range from controlling the learning process by ensuring that all features contribute equally to the model.

The target variable, represented as $\mathbf{y}$, was then encoded using the LabelEncoder into integer labels [7]. 'Fault_Type' and other categorical labels are converted into a numerical format that machine-learning algorithms can use in this process. For the $i$th sample, let $y_i$ be the categorical label. The encoded label $y_i^{\text{encoded}}$ is assigned as follows:

$$y_i^{\text{encoded}} = \text{LabelEncoder}(y_i) \tag{2}$$

Lastly, the `train_test_split` function was used to divide the dataset into training and testing sets. The feature matrix $\mathbf{X}_{\text{scaled}}$ and the encoded labels $\mathbf{y}_{\text{encoded}}$ were split into training ($\mathbf{X}_{\text{train}}, \mathbf{y}_{\text{train}}$) and testing ($\mathbf{X}_{\text{test}}, \mathbf{y}_{\text{test}}$) subsets with an 80–20 ratio, ensuring the stratification of the target variable to maintain the distribution of classes across the training and testing sets.

## 2.3 Model

We used both an Artificial Neural Network (ANN) model and baseline machine-learning (ML) models to identify and categorize electrical issues. Every model was tailored to meet the particular goals of our study, guaranteeing top performance. Below is a description of these models' specifics.

**Baseline Models**

**SVM**: One effective supervised learning method often utilized for classification problems is the SVM [8]. The main goal of SVM is to identify the best hyperplane with the largest margin for dividing various classes. The distance between the hyperplane and the closest data points-known as support vectors-from any class is the definition of this margin. Maximizing this margin increases the model's ability to generalize new data. For a binary classification problem, consider a dataset $\{(\mathbf{a}_i, b_i)\}_{i=1}^{n}$, where $\mathbf{a}_i \in \mathbb{R}^d$ represents the feature vector of the $i$th sample, and $b_i \in \{-1, 1\}$ is its corresponding label. A linear equation defines the decision boundary: $\mathbf{s} \cdot \mathbf{t} + w = 0$, where $\mathbf{s} \in \mathbb{R}^d$ is the weight vector, and $w \in \mathbb{R}$ is the bias term. The optimization problem for a hard-margin SVM, which assumes that the data is linearly separable, can be formulated as: $\min_{\mathbf{s},w} \frac{1}{2}\|\mathbf{s}\|^2$, subject to $y_i(\mathbf{w} \cdot \mathbf{x}_i + b) \geq 1, \quad \forall i$ [9]. In cases where the data is not perfectly separable, a soft-margin SVM introduces slack variables $\xi_i \geq 0$ to allow for misclassifications. The optimization problem for the soft-margin SVM is: $\min_{\mathbf{w},b,\xi} \frac{1}{2}\|\mathbf{w}\|^2 + C\sum_{i=1}^{n} \xi_i$.

**DT**: A recursive, non-parametric partitioning approach for binary classification is called a decision tree. To lower the impurity in each subset, the data is divided into subsets according to the value of a feature [10]. Until the halting condition is satisfied-that is, until all nodes are pure or until the maximum tree depth is reached-the procedure iteratively creates a tree-like structure of decisions. To minimize the impurity of the resulting child nodes, the method starts with the selection of a feature $j$ and a threshold $\theta$ that ideally splits the data at each node in the dataset $\{(\mathbf{x}_i, y_i)\}_{i=1}^{n}$. A node $t$'s impurity $I(t)$ is measured using metrics like entropy or Gini impurity. In particular, $I_{\text{Gini}}(t) = \sum_{k=1}^{K} p_k(1 - p_k)$, which defines the Gini impurity, and $I_{\text{Entropy}}(t) = -\sum_{k=1}^{K} p_k \log(p_k)$, which defines the entropy. In this case, $p_k$ denotes the percentage of samples that belong to class $k$ at node $t$, and $K$ is the total number of classes (here 2).To minimize the weighted sum of the impurity of the child nodes at each node, $\min_{j,\theta}\left[\frac{N_L}{N} I(t_L) + \frac{N_R}{N} I(t_R)\right]$ is used, where $N$ is the number of samples at the node $t$, $t_L$ and $t_R$ indicate the left and right child nodes following the split, and $N_L$ and $N_R$ stand for the number of samples in the left and right nodes, respectively. This criterion is applied repeatedly until a stopping condition is satisfied, such as a maximum depth or a minimum number of samples per node [11].

**RF**: As part of an ensemble learning technique called Random Forest, many decision trees are constructed during training and combined to increase the accuracy and resilience of the final model [12]. It works especially well to improve generality and lessen overfitting. Each decision tree in the forest takes a random subset of features into account at each split, which is trained on a random subset of the data (with replacement). Let $D = \{(x_i, y_i)\}_{i=1}^{n}$ be the training set. Bootstrap sampling is used in the Random Forest algorithm to generate several subsets of the original dataset initially. A bootstrap sample $D(b)$ is taken from the original dataset $D$ for each tree $T(b)$ in the forest. Some samples may be included more than once, while others may not appear at all. Instead of considering every feature, the optimal split is decided at each node in a tree based on a random subset $F(b)$ of the features. As previously

**Table 1** Hyperparameter grid for SVM, Decision Tree, and Random Forest

| Model | Hyperparameter | Values |
|---|---|---|
| SVM | C | {0.1, 1, 10, 100} |
| | Kernel | {'linear', 'rbf', 'poly'} |
| | Gamma | {0.001, 0.01, 0.1, 1} |
| Decision Tree | Max depth | {None, 10, 20, 30, 40} |
| | Min samples split | {2, 5, 10, 20} |
| | Criterion | {'gini', 'entropy'} |
| Random Forest | Number of estimators | {100, 200, 300, 500} |
| | Max features | {'auto', 'sqrt', 'log2'} |
| | Max depth | {None, 10, 20, 30, 40} |
| | Min samples split | {2, 5, 10, 20} |

stated, the conventional decision tree methodology is used to construct each distinct decision tree $T(b)$. Following the construction of each tree, the final RF forecast for a particular input is generated by combining the predictions made by each tree. When dealing with a categorization task, majority voting is used to accomplish this aggregate [13]. It is determined that the expected class $\hat{y}$ is:

$$\hat{y} = \text{mode}\left(\hat{y}^{(1)}, \hat{y}^{(2)}, \ldots, \hat{y}^{(B)}\right) \tag{3}$$

where $B$ is the total number of trees in the forest and $\hat{y}^{(b)}$ is the forecast from the $b$th tree.

**Optimization Technique**: Using GridSearchCV, we performed hyperparameter tweaking to improve our models' performance. This method finds the ideal collection of hyperparameters by thoroughly searching across a given parameter grid and assessing each combination using cross-validation [14]. We established a grid of hyperparameters for each model-SVM, DT, and RF-as shown in Table 1.

## 2.4 ANN

Artificial Neural Networks (ANNs) are often described as black boxes capable of simulating human brain functions [15]. Among the various ANNs used for classification tasks, the Multilayer Perceptron (MLP) is one of the most popular due to its compact structure and stability. Unlike other ANN models, MLP does not require pre-assumptions about the training dataset [16]. It generates output through a backpropagation-based supervised learning function. Backpropagation helps minimize the error between the computed and expected output values [17]. As illustrated in Fig. 4, the proposed MLP model features an input layer where each feature value is fed into the hidden layers. The network parameters for the multilayer perceptron

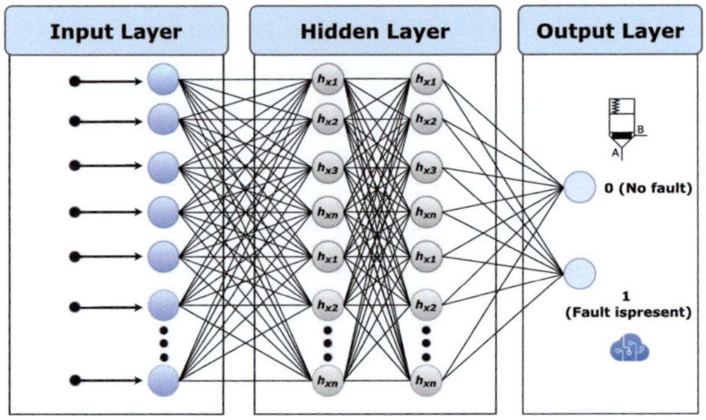


**Fig. 4** Proposed ANN model architecture

**Table 2** Parameters during the training process of the ANN model

| Parameters | Value |
|---|---|
| Optimizer | Adam |
| Learning_rate | 0.001 |
| Hidden layer | 2 |
| Perception in hidden layer | 4 |
| Metrics | accuracy |
| Epochs | 10 |

classifier are detailed in Table 2. This model includes two hidden layers, each consisting of four perceptrons. The perceptrons in this system utilize the Adam optimizer and the Rectified Linear Unit (ReLU) activation function to optimize the log-loss function. The output neuron, on the other hand, employs the sigmoid activation function. The output layer converts the input layers to an important or incidental label after processing through the hidden layers. In order to reduce the difference between the expected and actual output values, the neurons modify the weights of their connections. The model's performance is validated by the loss value, which indicates how close the expected value is to the exact value; a smaller loss value signifies a more accurate prediction. The loss value for binary classification issues is calculated using binary cross-entropy. Binary cross-entropy is formally defined as the negative mean of the logarithm of the correctly predicted probabilities. The formula for binary cross-entropy is given by:

$$\text{LogLoss} = -\frac{1}{N}\sum_{i=1}^{N}(y_i \log(p_i) + (1 - y_i)\log(1 - p_i)) \tag{4}$$

In this formula, $y_i$ represents the actual result, $p_i$ denotes the probability of class 1, and $(1 - p_i)$ is the probability of class 2. The equation's first term becomes more significant and the second term becomes less relevant if the outcome is more likely to fall into class 1, and vice versa. $N$ indicates the total number of occurrences in the dataset.

## 3 Result and Discussion

We performed a comprehensive statistical analysis of the dataset, which included feature distribution and correlation matrix visualization. These analyses yielded significant findings about the relationships between variables and the general structure of the data. The correlation matrix and distribution of the features are shown graphically in Figs. 5 and 6. The dataset shows that voltage readings are more concentrated around zero, indicating less volatility than current readings. In contrast, current values show vast and variable distributions, demonstrating the influence of various fault states. Meaningful relationships within the data are further highlighted by the correlation analysis. For example, the currents in line A (Ia) and line B (Ib) have a moderately negative correlation (coefficient of −0.374), indicating that Ib tends to decrease as Ia increases. With a value of −0.528, a more robust negative correlation—and a more constant inverse relationship—is seen between Ib and Ic. There are also negative correlations between the voltages in lines A, B, and C (Va, Vb, and Vc). For example, the correlation between Va and Vb is −0.480. This pattern might be interpreted as the nature of the electrical system, where an increase in voltage in one line is often accompanied by a decrease in another. Furthermore, there is a modest positive correlation between Ia and Vc (0.246) and between Ic and Vc (0.123), suggesting a slight relationship between these voltage and current variables. Grasp and forecasting fault situations require a grasp of the relationships between current and voltage in the system, which these correlations offer.

In our research, we applied baseline ML models and an ANN to classify and detect faults. The models were rigorously evaluated using precision, recall, F-score, and accuracy metrics. We first assessed the generalization capability of the models by calculating both training and testing accuracy. Subsequently, we evaluated the identification of true positives, true negatives, false positives, and false negatives using the classification metrics of precision, recall, and F-score. Finally, we visualized the model's errors through a confusion matrix to further analyze its performance. The training and testing accuracy of the models are shown in Table 3. SVM performs admirably on the training set, achieving a training accuracy of 95.65%. However, the test accuracy fell dramatically to 76.43%, suggesting that the model may be overfitting. DT outperforms SVM in terms of testing accuracy (82.76%) and training accuracy (96.31%), which is slightly better. This implies that DT is less likely to overfit than SVM, as the discrepancy between testing and training accuracy is smaller. Even though DT still performs worse from training to testing, it is more reliable in identifying errors in real-world data because of its improved generalization. RF has a

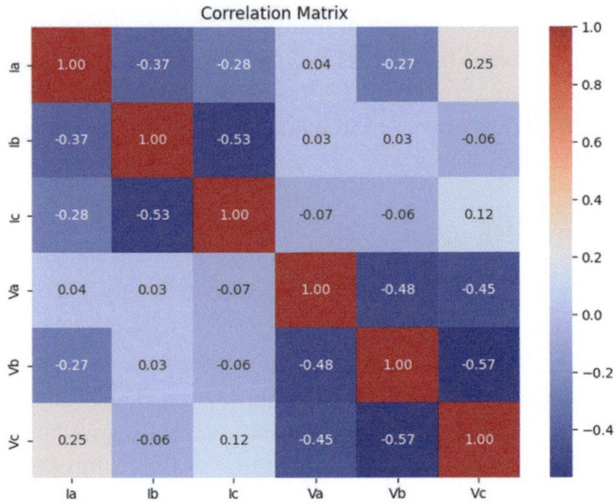


Fig. 5 Correlation matrix

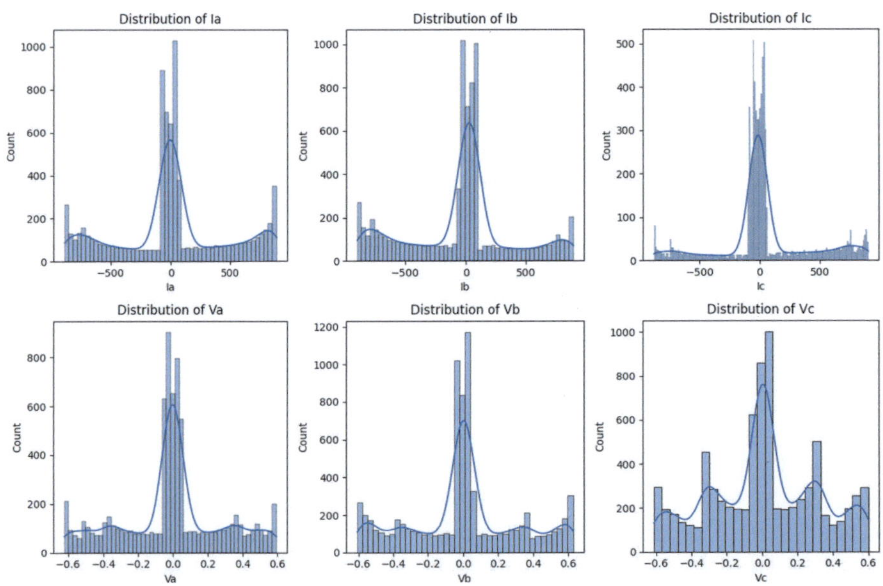


Fig. 6 Distributions of the features

relatively good testing accuracy of 86.99% and a high training accuracy of 96.75%. RF demonstrates a satisfactory equilibrium between effectively learning from the training set and effectively applying that knowledge to new data, as indicated by the reduced discrepancy in the accuracies of the two. RF is a reliable model for detecting electrical faults in various fault scenarios that occur in real-world applications. The ANN demonstrated exceptional performance in this challenge, achieving training

**Table 3** Training and testing accuracies of different models

| Model | Training accuracy (%) | Testing accuracy (%) |
|---|---|---|
| SVM | 95.65 | 76.43 |
| DT | 96.31 | 82.76 |
| RF | 96.75 | 86.99 |
| ANN | 98.98 | 98.50 |

**Table 4** Precision, Recall, and F-score of all models

| Model | Precision (%) | Recall (%) | F-score (%) |
|---|---|---|---|
| SVM | 75.63 | 75.10 | 75.52 |
| DT | 80.43 | 81.54 | 81.22 |
| RF | 86.43 | 86.99 | 87.22 |
| ANN | 98.86 | 98.10 | 98.48 |

and testing accuracies of 98.98% and 98.50%, respectively, which were the highest among all models. This could be attributed to its capacity to acquire intricate nonlinear connections within the data. The analysis showed that ANN is more robust and generalized compared to other models. Furthermore, the models are evaluated deeply, and the precision, recall and f-score outcomes are shown in Table 4. SVM exhibits a moderate degree of performance in terms of evaluation measures. It demonstrates a balanced but below-average ability to detect defects, as indicated by its precision of 75.63%, recall of 75.10%, and F-score of 75.52%. The DT method produces excellent results, with an F-score of 81.22%, recall of 81.54%, and precision of 80.43%. Due to improved accuracy in fault detection, these higher values suggest that DT is a more dependable method compared to SVM.RF model exceeds these criteria, achieving a precision of 86.43%, recall of 86.99%, and an F-score of 87.22%. The higher RF ratings show a great capacity to detect flaws effectively and with fewer errors. The most impressive metrics are obtained by the ANN, which achieves 98.86% precision, 98.10% recall, and 98.48% F-score. With very few failures in fault identification, these almost flawless scores show that ANN is the most efficient model for fault detection.ANN provide remarkable accuracy and vital assessment metrics, making them an excellent choice for electrical fault detection in power systems. Its precision, recall, and F-score are all excellent, and it generalizes well, making it the most dependable model tested. This guarantees that ANN can reliably perform crucial fault detection functions by effectively reducing false positives and negatives.

Our investigation showed that the ANN performed better than any other model. Figure 7 shows the accuracy of testing and training across epochs. To evaluate the model's performance even more, we used a confusion matrix to visualize the errors, as shown in Fig. 8. Nine times out of all the test samples, the confusion matrix shows that the model misclassified a sample. In particular, six samples from class 1 were wrongly classified, and three samples from class 0 were incorrectly forecasted. The overall performance of all models in Fig. 9 is shown graphically.

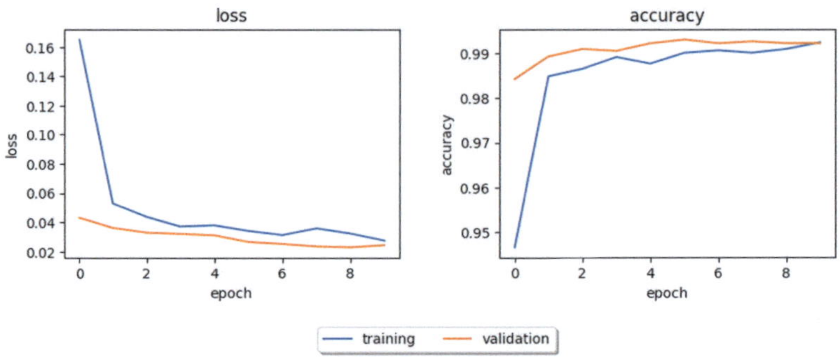


**Fig. 7** Training and testing loss, accuracy over epoch (ANN)

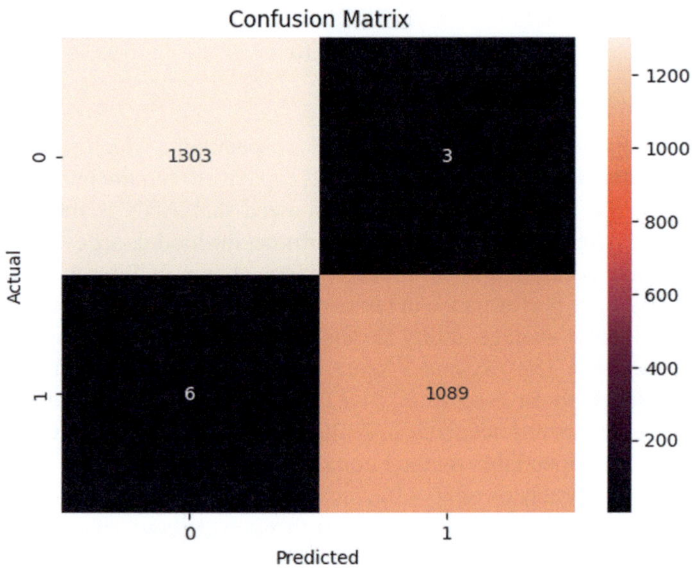


**Fig. 8** Confusion matrix (ANN)

# 4 Conclusion

This article successfully illustrates how electrical fault detection and classification in power systems may be improved through ML and deep learning models. Advanced approaches outperform traditional methods, which are generally limited by their inefficiency and accuracy. ANN, RF, DT, and SVM are examples of machine-learning algorithms that perform better than conventional approaches, particularly in accuracy and reliability. The efficacy of these models underscores the importance of integrating ML into power systems to enhance operational safety, prevent breakdowns, and establish a more resilient energy infrastructure. Due to its capacity to comprehend

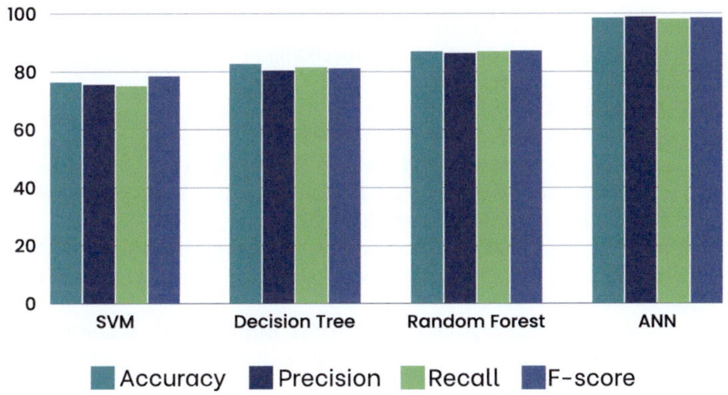


**Fig. 9** Comparative performance of all models

intricate patterns and provide reliable outcomes, the ANN model is recognized as an exceptionally efficient instrument for this objective. In addition to addressing the problems with fault detection now, this research lays the groundwork for upcoming advancements and emphasizes intelligent systems' vital role in contemporary power management.

## References

1. Furse CM, Kafal M, Razzaghi R, Shin Y-J (2020) Fault diagnosis for electrical systems and power networks: a review. IEEE Sens J 21(2):888–906
2. Köroğlu S (2016) A case study on fault detection in power transformers using dissolved gas analysis and electrical test methods. J Electr Syst
3. Liu H, Liu X (2024) Electrical fault detection and classification based on multiple machine learning algorithms, Appl Comput Eng 74:245–250. https://doi.org/10.54254/2755-2721/74/20240484
4. Vaish R, Dwivedi U, Tewari S, Tripathi SM (2021) Machine learning applications in power system fault diagnosis: research advancements and perspectives. Eng Appl Artif Intell 106:104504
5. Raza A, Benrabah A, Alquthami T, Akmal M (2020) A review of fault diagnosing methods in power transmission systems. Appl Sci 10(4):1312
6. Ahsan MM, Mahmud MP, Saha PK, Gupta KD, Siddique Z (2021) Effect of data scaling methods on machine learning algorithms and model performance. Technologies 9(3):52
7. Yang X, Hou L, Zhou Y, Wang W, Yan J (2021) Dense label encoding for boundary discontinuity free rotation detection. In: Proceedings of the IEEE/CVF conference on computer vision and pattern recognition, pp 15819–15829
8. Abdullah DM, Abdulazeez AM (2021) Machine learning applications based on SVM classification a review. Qubahan Acad J 1(2):81–90
9. Santana Í, Serrano B, Schiffer M, Vidal T (2025) Support vector machines with the hard-margin loss: optimal training via combinatorial benders' cuts. arXiv preprint arXiv:2207.07690
10. Singgalen YA (2024) Comparative analysis of K-NN and DT model in sentiment classification of Labuan Bajo-wonderful Indonesia content reviews. J Mantik 8(1):314–325
11. Priyanka Kumar D (2020) Decision tree classifier: a detailed survey. Int J Inf Decis Sci 12(3):246–269

12. Shahhosseini M, Hu G (2020) Improved weighted random forest for classification problems. In: Progress in intelligent decision science: Proceeding of IDS 2020. Springer, pp 42–56
13. Genuer R, Poggi J-M, Random forests. Springer
14. Belete DM, Huchaiah MD (2022) Grid search in hyperparameter optimization of machine learning models for prediction of HIV/Aids test results. Int J Comput Appl 44(9):875–886
15. Geetha A, Santhakumar J, Sundaram KM, Usha S, Thentral TT, Boopathi C, Ramya R, Sathyamurthy R (2022) Prediction of hourly solar radiation in Tamil Nadu using ANN model with different learning algorithms. Energy Rep 8:664–671
16. Kalam S, Yousuf U, Abu-Khamsin SA, Waheed UB, Khan RA (2022) An ANN model to predict oil recovery from a 5-spot waterflood of a heterogeneous reservoir. J Petrol Sci Eng 210:110012
17. Heng SY, Ridwan WM, Kumar P, Ahmed AN, Fai CM, Birima AH, El-Shafie A (2022) Artificial neural network model with different backpropagation algorithms and meteorological data for solar radiation prediction. Sci Rep 12(1):10457

# Enhancing Satellite Image Classification Using Vision Transformers and Deep Learning Models

**M. Sangeetha, T. Harikala, LNC. Prakash K, V. Lavanya, Penubaka Balaji, and Mary Teresa**

**Abstract** Since it provides a perceptive analysis of numerous environmental and geographic events, satellite image classification is a basic component of remote sensing. This work uses recent deep learning (DL) techniques, including Vision Transformers (ViT), to examine satellite imagery classification. Examining four models in great detail: Vision Transformers (ViT), VGG-16, CNN, and ResNet-50. CNN is here. There are 5631 high-resolution images in the dataset spanning four primary categories. The approach consists of data preparation, augmentation, and adding attention approaches to improve the model performance. With an incredible 99.62% accuracy in terms of testing, the experimental results reveal that the ViT model surpasses all the other models. It also reveals better memory, accuracy, and F-score. ResNet-50, VGG-16, and CNN models demonstrate a quite mediocre degree of performance with an accuracy range of 87–89%. The ViT method offers significant opportunities for use in remote sensing based on great accuracy and durability over earlier versions. The results show that ViT quite correctly picks out geographic correlations in satellite data. The exact and practical results of the research obtained from

M. Sangeetha (✉)
Department of Information Technology, SRM Madurai College for Engineering and Technology, Sivagangai, India
e-mail: sangeetha.srm.23@gmail.com

T. Harikala
Department of Electronics and Communication Engineering, Annamacharya University, Rajampet, India

LNC. Prakash K
Department of Computer Science and Engineering-Data Science, CVR College of Engineering, Hyderabad, India

V. Lavanya
School of Computing, Mohan Babu University, Tirupati, India

P. Balaji
Department of Computer Science Engineering, KL University, Vaddeswaram Guntur, Andhra Pradesh, India
e-mail: pbalaji@kluniversity.in

M. Teresa
CSE-AIML, Guru Nanak Institutions Technical Campus, Ibrahimpatnam, Hyderabad, India



K. R. Madhavi et al. (eds.), *Proceedings of Sixth International Conference on Computer and Communication Technologies*, Lecture Notes in Networks and Systems 1356,
https://doi.org/10.1007/978-981-96-5238-9_32

satellite images will help urban planning, disaster control, and environmental monitoring. Reliable picture categorization directs decisions in resource management, agriculture, and studies on climate change.

**Keywords** Remote sensing · Image classification · Deep learning · Grad-CAM · Vision transformer

## 1 Introduction

A satellite image is a depiction of either the entire earth or a portion of it that is obtained using artificial satellites [1] and getting information about the surface of the Earth without physically touching it is known as remote sensing. Remote sensing cannot be completed without the emission or reflection of energy that must be perceived, recorded, analyzed, and utilized [2]. Satellite imagery typically contains abundant information and is crucial in providing evidence of the environment. Measurable and qualitative data from satellite and remote sensing images lessens the requirement for physical work and the difficulty of research timing [3]. Remote sensing is essential since image categorization helps us to analyze and extract information from satellite pictures [4]. The theory behind the classification process is that various surface feature types of Earth exhibit varied spectrum reflectance and remittance properties [5].

Many remote sensing applications, including forestry, agriculture, environmental studies, weather forecasting, ocean studies, archaeological studies, etc., rely on the multispectral satellite sensor to provide digital raster images allowing us to apply Digital Image Processing (DIP) techniques to create thematic maps of land use/landcover classes [2]. Extensive value data from satellite photos is extracted with modern methods in artificial intelligence (AI), machine learning (ML), and computer vision. By means of these technologies, researchers have been able to raise the efficiency, effectiveness, and further growth of the image analysis application in many other field applications [6]. Effective performance in categorizing diverse objects from satellite data has been shown by several ML and DL techniques including Random Forest, Support Vector Machine (SVM) [7], Decision Tree, K-Nearest Neighbor (KNN) [8], ResNet, U-Net [9], MobileNetV3, EfficientNetB0 [10], and GoogLeNet models. Still, numerous obstacles endure. Satellite images with high dimensions, multiple spectral bands, or hyperspectral data necessitate substantial computational resources. They are susceptible to challenges such as atmospheric distortion, fluctuating lighting conditions, and sensor noise [11]. Moreover, the limited availability of labeled data for training poses a significant obstacle, especially for infrequent categories or areas with a scarcity of historical data. Transfer learning, data augmentation, and domain adaptation techniques are frequently employed to alleviate these challenges. However, they do not entirely resolve the issue of generalizing across diverse geographical locations and temporal domains.

Remote sensing depends critically on the automatic detection and classification of many land cover categories, objects, or specific characteristics in satellite data, including farmland, beaches, towns, deserts, forests, lakes, mountains, railways, residential areas, and rivers. Numerous uses for this process exist, including managing natural disasters [12], identifying crop types [13], mapping urban areas, and tracking deforestation [14]. This field has undergone a revolution thanks to recent advances in AI, especially in DL, which have made it possible to process large amounts of high-resolution imagery more accurately, scalable, and efficiently. CNNs have become the industry benchmark for satellite image classification since they can directly learn hierarchical spatial characteristics from raw picture data. Complex satellite image segmentation into functional categories has proven successful with architectures like U-Net [15], and ResNet [16]. This research utilized CNN, transfer learning models such as VGG16 and ResNet50, and the novel Vision Transformer (ViT). Grad-CAM was applied to enhance the interpretability and trustworthiness of the models. Additionally, a large dataset was employed, and data augmentation techniques were implemented to increase the dataset's diversity, improve model robustness, and tackle the dataset scarcity. The contribution of the research:

- We have applied different DL models to check the models' effectiveness in classifying satellite images.
- Data augmentation techniques were applied in the research.
- The proposed ViT model increases the accuracy.
- To make the model trustworthy and for interpretability, we have used Grad-CAM in the research.

The rest of the paper is organized in several sections. For example, in Sect. 2, the methodology is described. We discuss the materials and methods, such as datasets, preprocessing, and models. Section 3 shows the research outcomes. Finally, we present our conclusion in Sect. 4.

## 2 Methodology

Figure 1 shows that the research methodology is a systematic, multi-step process. An extensive comprehension of the procedure is ensured by this figure, which presents an outline of the major phases of the investigation. All the steps are discussed below.

### *2.1 Dataset Description*

Designed for the classification of remote sensing (RS) visual sceneries, the dataset is a somewhat large benchmark one. Providing clever interpretation techniques for testing and development, Million-AID is a vital tool for RS image analysis research

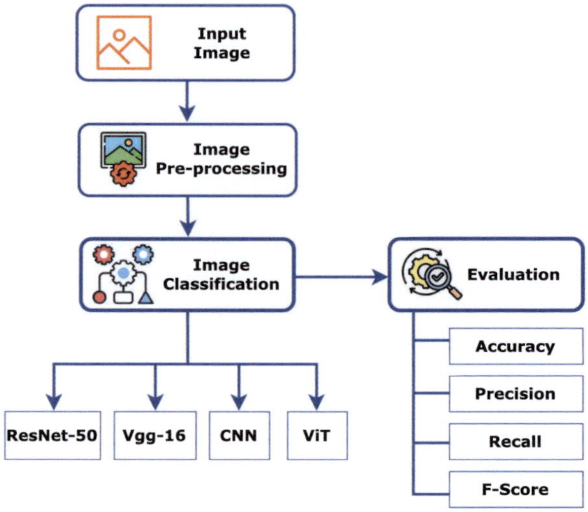


**Fig. 1** Framework of the research

and development given one million image instances. With RS photo annotations, this dataset addresses the most important problems and provides useful guidance for creating comprehensive and strong benchmark datasets. Since it offers a large library of annotated RS images, Million-AID is a fundamental tool for data-driven research and the development of automated interpretation algorithms in the remote sensing community. Divided into four separate categories, the collection features 5631 photos. Figure 2 shows the dataset sample.

## 2.2 *Data Preprocessing and Augmentation*

A set of modifications is applied to every raw image $I_{\text{raw}}(x, y)$ with dimensions $W \times H$ during the preparation step of satellite image classification. The process begins by resizing each image to $180 \times 180$ pixels, as expressed in the following equation:

$$I_{\text{resized}}(x', y') = \text{resize}(I_{\text{raw}}(x, y), 180, 180)$$

where $x', y' \in \{1, 2, \ldots, 180\}$ indicate the new dimensions. Next, the pixel values are normalized by dividing by 255, the maximum pixel intensity for an 8-bit image:

$$I_{\text{norm}}(x', y') = \frac{I_{\text{resized}}(x', y')}{255}$$

**Fig. 2** Sample of the dataset

This normalization ensures that the pixel values are scaled between 0 and 1. Data augmentation strategies are used to improve generalization and add variability to the sample. By adding a delta $\Delta_b$, sampled from a uniform distribution, random brightness modifications are made:

$$I_{\text{aug}}(x', y') = I_{\text{norm}}(x', y') + \Delta_b, \quad \Delta_b \sim U\left(-\frac{32}{255}, \frac{32}{255}\right)$$

Subsequently, the picture is arbitrarily rotated both horizontally and vertically to mimic various viewing perspectives:

$$I_{\text{aug}}(x', y') = \text{flip}(I_{\text{norm}}(x', y'), \text{horizontal})$$

$$I_{\text{aug}}(x', y') = \text{flip}(I_{\text{norm}}(x', y'), \text{vertical})$$

Additionally, arbitrary saturation adjustments are made to vary the image's color intensity. A range is used to modify the saturation:

$$I_{\text{aug}}(x', y') = \text{adjust\_saturation}(I_{\text{norm}}(x', y'), s_{\text{low}}, s_{\text{high}})$$

Here, $s_{\text{low}} = 500{,}000$ and $s_{\text{high}} = 1{,}000{,}000$. The images are resized following these modifications to fit the model's input specifications. To ensure compatibility with the model's input format, each picture $I_{\text{norm}}$ in a dataset of $n$ photos is reshaped into a four-dimensional tensor:The input format below :

$$I_{\text{reshape}}(x', y', c) = \text{reshape}(I_{\text{norm}}(x', y'), (n, W, H, C))$$

Ultimately, one-hot encoded vectors are created from the class labels. The encoding for each label $y$ in a dataset with $K$ classes is as follows:

$$y_{\text{onehot}}(k) = \begin{cases} 1 & \text{if } y = k \\ 0 & \text{otherwise} \end{cases}$$

This procedure ensures the dataset is prepared for training by combining the right formatting and augmentation methods for efficient satellite picture categorization.

## *2.3 Proposed Approach*

**Attention Mechanism**: Applying attention to the feature maps generated by a deep neural network's convolutional layers is a popular method of using attention in image classification [17]. According to them, this method involves teaching the attention mechanism to weigh various spatial positions in the feature maps according to their relevance to the classification task. A tiny neural network calculates the attention weights for each spatial point in the feature maps. The network's output is a scalar value that indicates the significance of the corresponding spatial location for the classification job. Next, a softmax function normalizes the attention weights so that their sum over all spatial locations equals one.

$$\text{Softmax}(X_i) = \frac{e^{X_i}}{\sum_{j=1}^{N} e^{X_j}} \tag{1}$$

A curated feature representation of an image is produced by the weighted summation of feature mappings employing attention weights. This feature representation can then be sent into a classifier, like a fully connected layer, to predict class labels. As an alternative, the network's intermediate representations can concentrate on feature maps at various stages, capturing different degrees of abstraction in the image. Related strategies for determining relationships between components in an input sequence or collection are self-attention and attentiveness. Self-attention partitions the input into query, key, and value vectors by creating a weighted representation based on item similarities. Together, these vectors provide a weight vector that illustrates the relative significance of every element. The attended representation is made up of the value vectors' weighted sum. The expression of attention in a matrix form is:

$$\text{Attention}(Q, K, V) = V \cdot \text{softmax}(\text{score}(Q, K)) \tag{2}$$

**Vision Transformer**: Self-attention leads the transform design to let the model save relationships between several input sequence bits. Three vectors divide the input sequence of the transformer: the query vector, the critical vector, and the value vector. Then these vectors enable one to determine the attention weights for every element in the input sequence. Computed from the attention weights, a weighted sum of the values reveals the attended representation of the input sequence [18]. Every level of the transformer design comprises a multi-head self-attention mechanism and a feedforward neural network. While the multi-head self-attention mechanism enables the model to attend to various segments of the input sequence at different degrees of granularity, the feedforward neural network provides the model flexibility and non-linearity. When compared to the transformer design, RNNs. Additionally, it is easier to parallelize than RNNs, facilitating faster training and more effortless scalability to larger datasets. Using an encoder–decoder, sequential data is parallelized. It is a type 2 neural network, or DL model, with the peculiarity of relying on the attention mechanism rather than an RNN or CNN and utilizing an encoder–decoder architecture (Fig. 3). An essential component of the transformer encoder–decoder's construction is the encoder. Its responsibility is to analyze and display the input sequence in order to comprehend it. Following input sequence processing the encoder generates either a continuous or embedded representation of the input. These embeddings then help the decoder create the output sequence [19]. Typically comprising a self-attention mechanism and a feed-forward neural network, each layer of the transformer encoder architecture.

Allows the model to determine the relative significance of different input sequence components by computing the dot product of the embeddings. This method is called "multi-head attention". The decoder's job is to create text sequences; it has sub-layers similar to the encoder. Its two multi-headed attention layers are a normalization layer after each sublayer and a feed-forward layer with point-wise residual connections. While the behavior of these sublayers is comparable to that of the encoder's linear layers, each multi-headed attention layer serves a different purpose. To obtain the word probabilities, a softmax and a linear layer resembling a classifier are added as a last step. The auto-regressive decoder uses two inputs: the encoder output, which consists of the attention information from the input, and a list of previous outputs. The decoder ends its process when it outputs an end token [20]. Our suggested method involves applying the ViT model to the datasets using the Keras framework without data augmentation. The ViT model offers these two datasets multiple advantages for image classification. They have a proven track record of top-tier performance. They enable scalability and transfer learning from massive datasets, excel in capturing long-range relationships in satellite imagery, and generate interpretable attention maps. ViT models are a wonderful alternative for enhancing accuracy and generalization in satellite image classification activities as long as they are tweaked to meet

**Fig. 3** The model architecture of the transformer

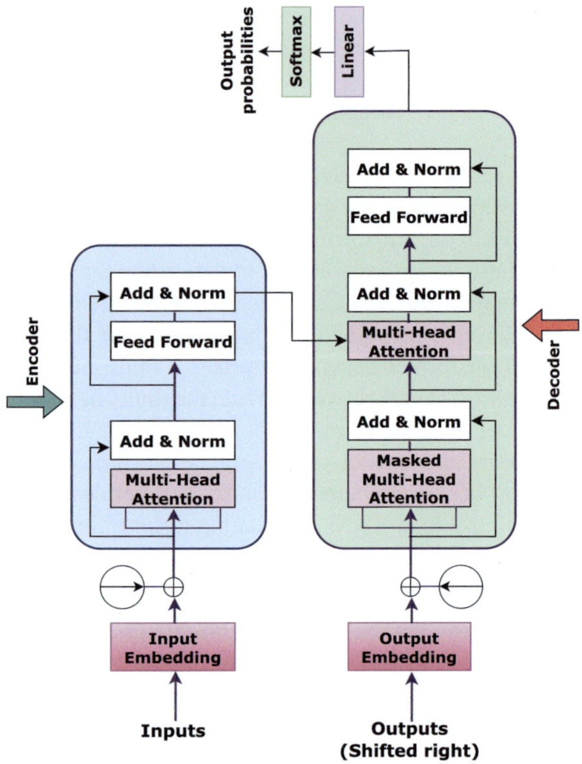


the individual qualities of the dataset since they are versatile in enabling varied patch sizes and the availability of pre-trained models. The proposed model defines initially the input layer for flattened 28 × 28 images. Patches from the input photos are then taken using the Patches layer. The Patch Encoder layer encodes these patches into a higher-dimensional form. Next, the model moves on to several Transformer blocks, each of which consists of MLP layers, layer normalization, skip connections and multi-head self-attention. After layer normalization, the resulting encoded patches are flattened. After applying dropout to the flattened representation, a second MLP layer is added for feature extraction. The model is constructed using the given inputs and outputs, and the output logits are produced through a dense layer. The architecture and behavior of our ViT-based model for image classification are influenced by some essential characteristics. In this work, we concentrate on the patch size hyperparameter and attempt to use a patch size of 16 applied to a 28 × 28 image. The list of parameters is shown in Table 1. The architecture of the ViT model is shown in Fig. 4.

**Table 1** List of parameters of the proposed model

| Parameter | Value | Parameter | Value |
|---|---|---|---|
| Image size | 28 × 28 | Heads | 4 |
| Patch size | 6 | Layer | 8 |
| Batch size | 28 | Epoch | 10 |
| Learning rate | 0.0001 | Dropout rate | 0.5 |
| Optimizer | Adam | Weight decay | 0.0001 |

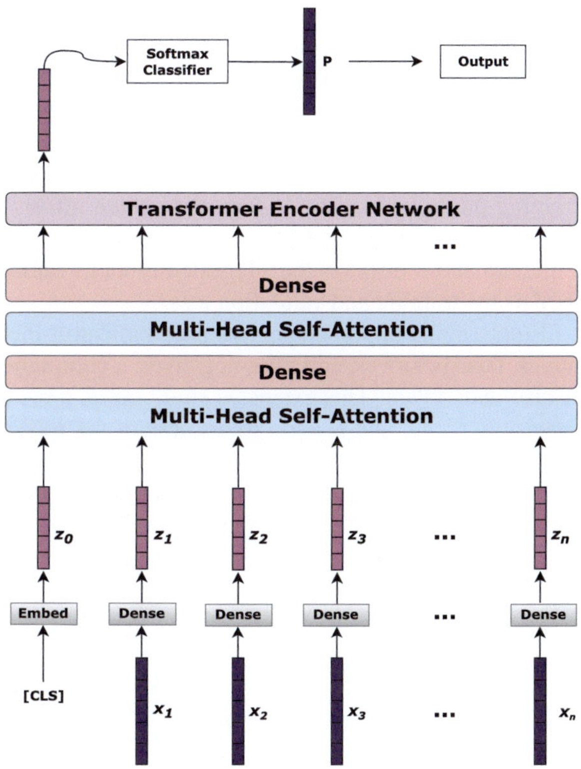


**Fig. 4** The proposed vision transformer model

## 3 Result and Discussion

We have assessed four models at various learning stages: ResNet-50, VGG-16, CNN, and ViT in the training, testing, and validation accuracy comparison, demonstrated in Table 2. Across training, testing, and validation, the ViT model performs remarkably well, achieving 99.87% accuracy. This shows how well the ViT model has learned and generalized the patterns seen in the data, demonstrating its robust handling of the dataset's complexity. The remaining three models, ResNet-50, VGG-16, and CNN, demonstrate comparable yet noticeably inferior performance. All three models,

**Table 2** Training, testing, and validation accuracy of all models

| Model | Training (%) | Testing (%) | Validation (%) |
|---|---|---|---|
| ResNet-50 | 96.83 | 88.65 | 87.34 |
| VGG-16 | 94.23 | 87.29 | 88.43 |
| CNN | 95.54 | 88.32 | 89.04 |
| ViT | 99.87 | 99.62 | 99.54 |

ResNet-50, VGG-16, and CNN, exhibit an accuracy range of approximately 87–97%. ResNet-50 has a slightly higher training accuracy of 96.83%, while VGG-16 and CNN achieve an accuracy of 94.23% and 95.54%, respectively. During the testing and validation phases, these models exhibit limited performance variation, suggesting they have reached a plateau in their generalization capacity. Although the models above are considered respectable, they show limited growth after the training phase, unlike the ViT model, which attains nearly flawless accuracy. We have found that the proposed ViT model performed better compared to other models. Figures 5 and 6 displays the suggested model's training and validation accuracy as well as the training and validation loss.

Emphasizing important performance indicators including true positives, true negatives, false positives, and false negatives, a comprehensive evaluation of the model is also undertaken. This extensive study offers a better knowledge of the accuracy, precision, recall, and general robustness of the model in controlling a spectrum of prediction results.

The ViT model again exhibits its dominance regarding evaluation metrics like precision, recall, and F-score (shown in Table 3), attaining near-perfect scores for

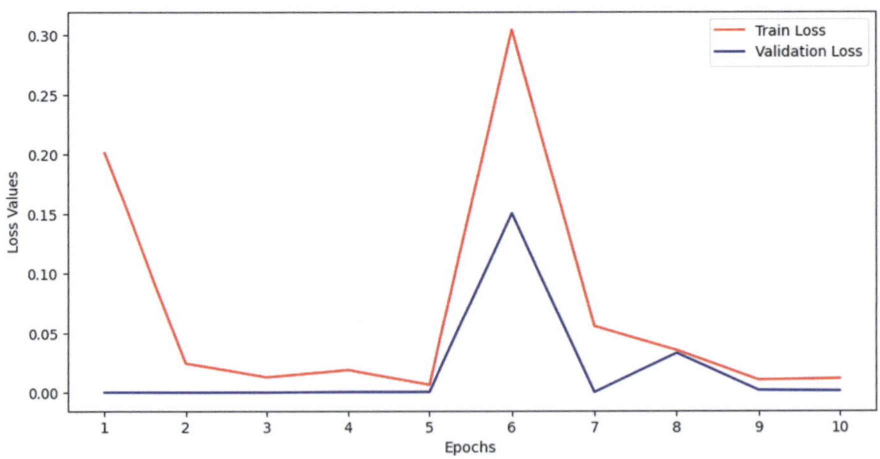


**Fig. 5** Training and validation loss (ViT model)

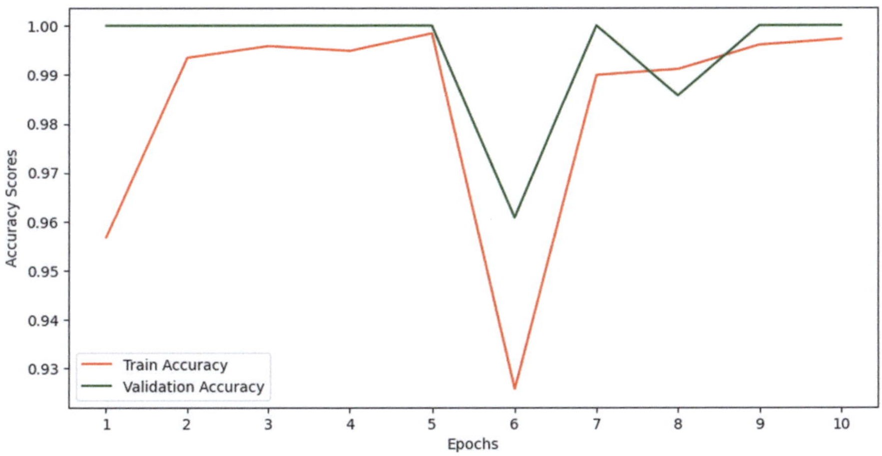


**Fig. 6** Training and validation accuracy (ViT model)

**Table 3** Precision, recall, and f-score of all models

| Model | Precision (%) | Recall (%) | F-score (%) |
|---|---|---|---|
| ResNet-50 | 88 | 87 | 87 |
| VGG-16 | 86 | 87 | 86 |
| CNN | 88 | 89 | 88 |
| ViT | 99 | 99 | 99 |

all three. The aforementioned suggests that the ViT model produces consistent and reliable predictions across several criteria. The system's precision indicates that most identifications are accurate, and its high recall score suggests that it recognizes the most critical cases. Its F-score, which is a nearly perfect 99%, highlights the model's overall resilience and reliability while also reflecting the balance between recall and precision. In comparison to the other models, CNN performs marginally better in terms of precision with 88% and recall with 89%, indicating that it can categorize data rather well. But it falls short of the ViT model's optimized performance. ResNet-50 also has good precision 88% but poor recall 87%, indicating that even while it may generate correct predictions, a sizable portion of pertinent examples may be missed. With 86% precision and 87% recall, VGG-16 performs below CNN and ResNet-50 in both precision and recall but in a balanced manner.

Overall, the ViT model stands out as the best performer, earning the highest accuracy at all phases and retaining consistency in precision, recall, and F-score metrics. This showcases its strong capacity to generalize, reduce misclassifications, and accurately represent the intricacies of the dataset. ResNet-50, VGG-16, and CNN exhibit similar yet somewhat less dependable performance than ViT. CNN comes close in terms of precision and recall but still falls behind ViT's superior performance. Since the ViT model displays its capacity to constantly generate exact results, it is

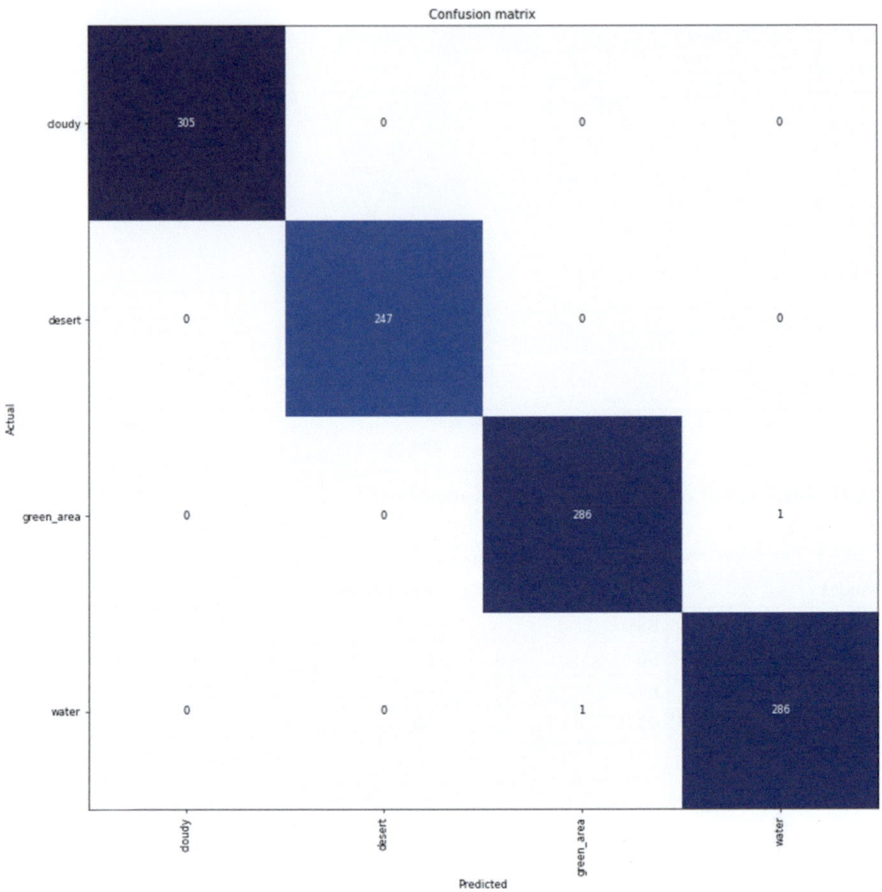


**Fig. 7** Confusion matrix of the proposed ViT model)

clear that it is the ideal option for this job. The comparison of all models applied in the research is shown in Fig. 8. The confusion matrix was finally demonstrated based on the resulting F-score, precision, and recall. Confusion matrix analysis helped identify areas where the model would be prone to errors and analyzed misclassifications, offering insightful information about the model's classification performance. This analysis improves the model's efficacy assessment even further. The confusion matrix is shown in Fig. 7.

Grad-CAM (gradient-weighted class activation mapping) allows us to assess model interpretability. Grad-CAM was applied to the satellite image classification task to show the sites in the images the model focused on when producing forecasts. This helped us to clarify the model's decision-making process and gave us hope that

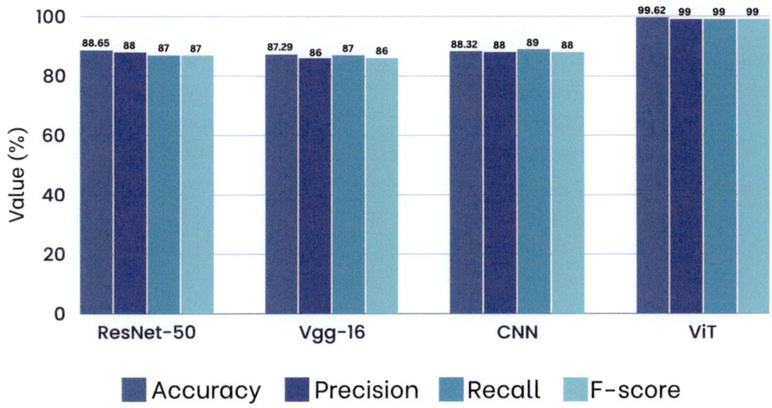


**Fig. 8** Comparison of all models

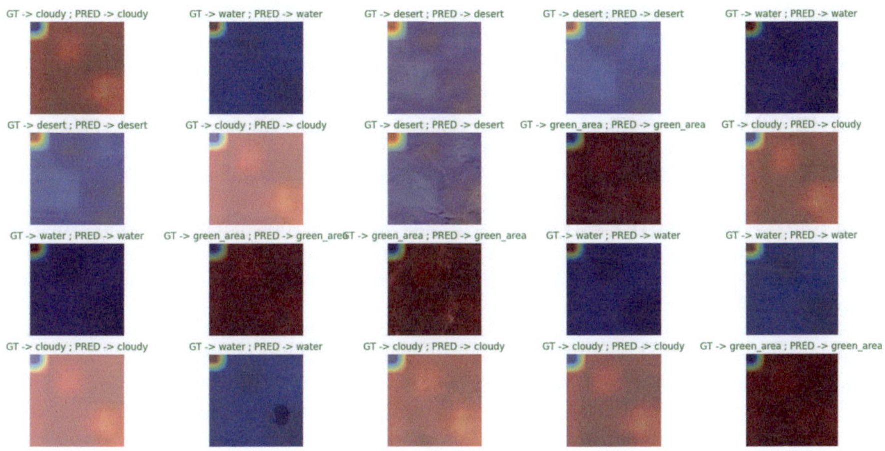


**Fig. 9** Grad-CAM analysis)

in pragmatic applications its projections are rational, consistent, and accurate [21]. It also makes quite evident how the algorithm handles challenging satellite data. Grad-CAM clarifies and increases the dependability of the model. Figure 9 display Grad-CAM's outputs.

## 4 Conclusion

This work presents the efficacy of Vision Transformers (ViT) in categorizing satellite images, getting almost perfect accuracy, precision, recall, and F-score metrics, in line with other well-known models such as ResNet-50, VGG-16, and CNN. The ability of the ViT model to appropriately depict spatial interconnections and long-range dependencies in high-resolution satellite pictures explains its higher performance. This ability is necessary for accurate differentiating between several land cover types. While other models stopped their accuracy from improving, the ViT model showed that it could apply its knowledge to various datasets and continuously improve. This work emphasizes the possibility of remotely sensing advanced DL models such as Vision Transformers to generate more scalable and effective methods for geospatial research. Later research could concentrate on enhancing the performance of these models and extending their use to more challenging remote sensing initiatives.

## References


1. Dhingra S, Kumar D (2019) A review of remotely sensed satellite image classification. Int J Electr Comput Eng 9(3):1720
2. Sowmya D, Shenoy PD, Venugopal K (2017) Remote sensing satellite image processing techniques for image classification: a comprehensive survey. Int J Comput Appl 161(11):24–37
3. Abburu S, Golla SB (2015) Satellite image classification methods and techniques: a review. Int J Comput Appl 119(8)
4. Ouchra H, Belangour A (2021) Satellite image classification methods and techniques: a survey. In: 2021 IEEE International conference on imaging systems and techniques (IST). IEEE, pp 1–6
5. Al-Doski J, Mansorl SB, Shafri HZM (2013) Image classification in remote sensing, vol 3, no 10. Department of Civil Engineering, Faculty of Engineering, University Putra, Malaysia
6. Jakkani AK (2024) An analysis on intelligent systems for remote sensing satellite image processing and classification
7. Rahman A, Abdullah HM, Tanzir MT, Hossain MJ, Khan BM, Miah MG, Islam I (2020) Performance of different machine learning algorithms on satellite image classification in rural and urban setup. Remote Sens Appl Soc Environ 20:100410
8. Ferdous H, Siraj T, Setu SJ, Anwar MM, Rahman MA (2021) Machine learning approach towards satellite image classification. In: Proceedings of International conference on trends in computational and cognitive engineering: proceedings of TCCE 2020. Springer, pp 627–637
9. Alsabhan W, Alotaiby T (2022) Automatic building extraction on satellite images using UNet and ResNet50. Comput Intell Neurosci 2022(1):5008854
10. Sarangi PK, Sharma B, Rani L, Dutta M (2014) Satellite image classification using convolutional neural network. Adv Aerial Sens Imaging 333–353
11. Vali A, Comai S, Matteucci M (2020) Deep learning for land use and land cover classification based on hyperspectral and multispectral earth observation data: a review. Remote Sens 12(15):2495
12. Van Westen C (2000) Remote sensing for natural disaster management. Int Arch Photogram Remote Sens 33(B7/4; PART 7):1609–1617
13. Kussul N, Lavreniuk M, Skakun S, Shelestov A (2017) Deep learning classification of land cover and crop types using remote sensing data. IEEE Geosci Remote Sens Lett 14(5):778–782
14. Machireddy SR (2023) Natural resource management using remote sensing and geographic information systems. Environ Sci Eng 2(2):73–82
15. Garg L, Shukla P, Singh SK, Bajpai V, Yadav U (2019) Land use land cover classification from satellite imagery using munet: a modified unet architecture. In: VISIGRAPP (4: VISAPP), pp 359–365
16. Zhao Y, Zhang X, Feng W, Xu J (2022) Deep learning classification by ResNet-18 based on the real spectral dataset from multispectral remote sensing images. Remote Sens 14(19):4883

17. Niu Z, Zhong G, Yu H (2021) A review on the attention mechanism of deep learning. Neurocomputing 452:48–62
18. Duan S, Zhao H, Zhou J, Wang R (2019) Syntax-aware transformer encoder for neural machine translation. In: 2019 International conference on Asian language processing (IALP). IEEE, pp 396–401
19. Han K, Wang Y, Chen H, Chen X, Guo J, Liu Z, Tang Y, Xiao A, Xu C, Xu Y et al (2022) A survey on vision transformer. IEEE Trans Pattern Anal Mach Intell 45(1):87–110
20. Kiyasseh D, Ma R, Haque TF, Miles BJ, Wagner C, Donoho DA, Anandkumar A, Hung AJ (2023) A vision transformer for decoding surgeon activity from surgical videos. Nat Biomed Eng 7(6):780–796
21. Al Maruf A, Fahim SH, Bashar R, Rumy RA, Chowdhury SI, Aung Z (2024) Classification of freshwater fish diseases in Bangladesh using a novel ensemble deep learning model: enhancing accuracy and interpretability. IEEE Access

# An Optimized Hybrid CNN-AlexNet Framework for Real-Time Crime Activity and Suspicious Object Detection

**R. Roopa and M. Humera Khanam**

**Abstract** Real-time crime activity detection and suspicious object identification has become very vital for public safety given the increasing demand for automated surveillance. Using the UCF-Crime dataset, this work presents an optimal hybrid model combining Convolutional Neural Networks (CNN) with AlexNet to precisely identify criminal behaviors and anomalies in surveillance films. The suggested model performs exceptionally well in spotting abnormalities by using CNN's spatial feature extraction and AlexNet's ability for managing vast amounts of data. Emphasizing accuracy, precision, recall, and F1-score, the Adam optimizer is used with learning rates of 0.01, 0.001, and 0.0001 to investigate the impact of training dynamics on model performance. Our studies show that lower learning rates significantly raise the model's accuracy, obtaining on the testing dataset up to 99.53%. Capturing many real-world criminal situations, the UCF-Crime dataset offers a demanding but reasonable standard for the resilience of our algorithm. The results show that the model is fit for real-time anomaly detection in security systems since it can identify both clear and subtle suspicious activity.

**Keywords** Crime activity · Anomaly detection · UCF-crime · CNN-AlexNet · Surveillance video

## 1 Introduction

Detection of criminal activity and identification of suspicious objects have emerged as crucial issues in ensuring public safety, particularly in city areas. The growing availability of security cameras and video recordings has drawn a lot of interest in automatic systems able to spot such behavior in real time. These systems not only help law enforcement spot criminal activity as it occurs but also support post-event research, providing insightful information on crime trends and thereby helping to

R. Roopa (✉) · M. Humera Khanam
Department of Computer Science Engineering, S. V. University College of Engineering, S. V. University, Tirupati, India
e-mail: roopa509@gmail.com



K. R. Madhavi et al. (eds.), *Proceedings of Sixth International Conference on Computer and Communication Technologies*, Lecture Notes in Networks and Systems 1356,
https://doi.org/10.1007/978-981-96-5238-9_33

prevent the next events. However, the enormous volume of information produced by surveillance systems makes hand monitoring impossible, so effective automated solutions are needed.

Recent developments in machine learning, particularly deep learning methods, have transformed the area of computer vision thereby allowing more accurate and real-time object and activity identification. Object detection [4], image classification [21], and activity recognition [16] are among the few visual tasks where CNNs have shown especially success. Deep learning has been extensively investigated for crime detection and surveillance; one of the most often used architectures is CNN-based models. Huang et al. [7] showed the possibility of convolutional layers in capturing spatiotemporal patterns from video frames by introducing a CNN-based framework for anomaly detection in surveillance films. Hasan et al. [6] also suggested a framework to manage sequential dependencies using CNN and Long Short-Term Memory (LSTM) networks, hence enhancing anomaly detection performance in challenging settings. Introduced by Krizhevsky et al. [12], AlexNet was among the first deep learning models to make CNNs well-known by surpassing performance in image categorization challenges. Its architecture has been used and changed since then for other purposes, including object identification and video analysis. Contemporary methodologies, including hybrid models that integrate CNN with AlexNet [14] or other architectures [1, 8, 13], have demonstrated enhanced detection accuracy in extensive datasets. These models leverage CNN's capacity to extract spatial characteristics and AlexNet's efficacy in managing large-scale visual tasks. Because of its variable learning rate characteristics and ability to effectively control sparse gradients, the Adam optimizer, developed by Kingma and Diederik [11], is increasingly used for training deep learning models.

In order to address the difficulty of crime and suspicious item identification, we propose in this work a hybrid model integrating CNN and AlexNet, optimized using the Adam optimizer. Aiming for a very precise and efficient crime detection system, we investigate the effect of training dynamics on the performance of the model by using different learning rates. Evaluated on the diverse UCF-Crime dataset, the suggested model shows effectiveness under real-world anomaly detection conditions.

The organization of the document is as follows: Sect. 2 offers a justification for our methodology in the research. Section 3 presents our findings together with a detailed study of them. We finally end up the research in Sect. 4 by giving a brief overview of the primary outcomes and contributions of our efforts.

## 2 Proposed Methodology

The proposed approach incorporates a hybrid model that merges CNN and AlexNet architectures for the categorization of criminal activities. CNN incorporates convolutional layers with max pooling, whereas AlexNet has many convolutional layers succeeded by fully connected layers. The outputs of both networks are merged and processed through supplementary dense layers for ultimate classification. The

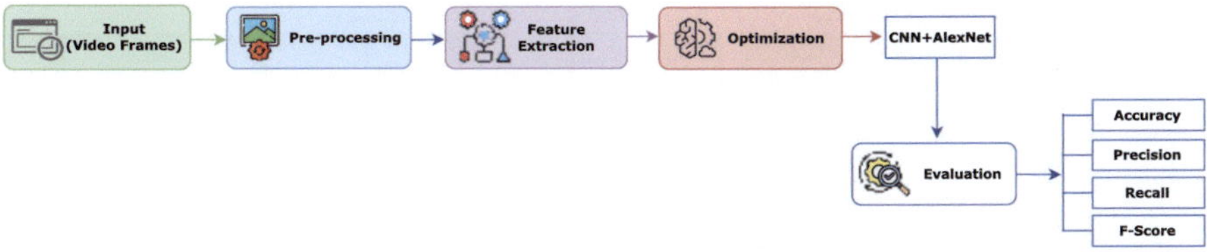


**Fig. 1** Overview of the suggested technique for detecting criminal activities with a hybrid CNN and AlexNet model

model uses the Adam optimizer for training, and measurements like accuracy, precision, recall, and F1-score evaluate its performance. This method proficiently encompasses both geographical and temporal characteristics for enhanced categorization. Figure 1 presents a comprehensive overview of the suggested technique, illustrating the hybrid model for improved criminal activity categorization.

## *2.1 Dataset*

The UCF-Crime dataset serves as a baseline for research pertaining to the recognition of criminal activities. Developed by the University of Central Florida (UCF), it is accessible to scholars and practitioners for their investigations. The films in the dataset were sourced from many platforms, including YouTube, television news, and documentaries, exhibiting diverse quality and resolution. It consists of 1900 lengthy, unedited surveillance movies, amounting to 128 h of footage. The films show thirteen genuine criminal activities: abuse, arrest, arson, assault, road accidents, burglary, explosives, fighting, robbery, shooting, theft, shoplifting, and vandalism [22]. These activities were chosen for their considerable influence on public safety. The UCF-Crime dataset presents several issues, including discrepancies in camera angles and movement, congested backdrops, and fluctuations in the magnitude of the foreground elements. Moreover, several video recordings exhibit poor quality, noise, and fluctuations in lighting. Alongside the previously noted issues, the UCF-Crime dataset exhibits discrepancies in the number of individuals participating in the activities, their attire, and the categories of weapons employed, among other variables. The variances provide the dataset with a formidable benchmark for the development of robust and precise criminal activity identification algorithms [9]. The suggested approach in this work seeks to identify the normal and fighting categories within the UCF-Crime dataset. Around 720 frames were extracted from each video for examination. Figure 2 illustrates various combat films from the UCF-crime dataset. The data is partitioned in this manner to create a balanced dataset, as the video clips exhibit various durations and thus differing frame counts.

**Fig. 2** Glimpse of various crime images from the UCF-crime dataset

## 2.2 Model

**Data Preprocessing**: The images and their associated labels are initially gathered and converted into NumPy arrays for subsequent processing. Reshaping the images to a grayscale size of 50 × 50 pixels produces a dimensional form of (50, 50, 1), therefore preparing the data for AlexNet. The dataset is subsequently divided into training and testing sets in an 80:20 ratio, ensuring that both the CNN and AlexNet models possess adequate data for training and assessment. Furthermore, the labels are transformed into a categorical format via 'np_utils.to_categorical', hence facilitating the models' capacity for successful multi-class categorization.

**Model Architecture**: The used methodology integrates a hybrid model that amalgamates CNN and AlexNet architectures to proficiently categorize crime activity movies. The CNN architecture has two convolutional layers [10]; the initial layer employs 64 filters, while the subsequent layer utilizes 128 filters, both including a 3 × 3 kernel size and LeakyReLU activation function. Following each convolutional layer is a max pooling layer to reduce spatial dimensions; the output is then flattened into a vector before being sent through a dense layer of 256 neurons using LeakyReLU activation. A dropout layer is utilized to mitigate overfitting, and the final classification is performed via a softmax output layer with neurons matching to the number of categories. The output of a convolutional layer may be expressed as:

$$\mathbf{O}_{i,j,k} = \sum_{m,n,p} \mathbf{W}_{m,n,p,k} \cdot \mathbf{X}_{i+m,j+n,p} + \mathbf{b}_k \tag{1}$$

where, $\mathbf{O}_{i,j,k}$ is the output value at position $(i, j)$ in the $k$th feature map, $\mathbf{W}_{m,n,p,k}$ is the weight of the filter at position $(m, n)$ for the $p$th input channel and $k$th output channel, $\mathbf{X}_{i+m,j+n,p}$ is the input value at position $(i + m, j + n)$ in the $p$th input channel, $\mathbf{b}_k$ is the bias term for the $k$th output channel [15].

The AlexNet model analyzes 50 × 50 grayscale pictures using many 5 convolutional layers, a MaxPooling layer, followed by fully connected layers consisting of 3 dense layers and a softmax layer for classification purposes. AlexNet is a deep convolutional neural network architecture introduced by Krizhevsky et al. [12]. The output $\mathbf{F}_l$ of a convolutional layer $l$ is computed as:

$$\mathbf{F}_l = \text{LeakyReLU}(\text{Conv}(\mathbf{X}_{l-1}, \mathbf{W}_l) + \mathbf{b}_l) \tag{2}$$

where, $\text{Conv}(\mathbf{X}_{l-1}, \mathbf{W}_l)$ denotes the convolution operation between the input $\mathbf{X}_{l-1}$ from the previous layer and the weights $\mathbf{W}_l$ of the current layer, LeakyReLU is the Leaky Rectified Linear Unit activation function [17], $\mathbf{b}_l$ is the bias term. The output $\mathbf{P}_l$ after applying MaxPooling[3] is:

$$\mathbf{P}_l = \text{MaxPool}(\mathbf{F}_l) \tag{3}$$

where $\text{MaxPool}(\mathbf{F}_l)$ represents the max pooling operation applied to the feature maps $\mathbf{F}_l$.

After flattening[24], the output $\mathbf{y}$ of a fully connected layer is:

$$\mathbf{y} = \text{LeakyReLU}(\mathbf{W}_{fc} \cdot \mathbf{h} + \mathbf{b}_{fc}) \tag{4}$$

where $\mathbf{h}$ is the flattened vector from the previous layer, $\mathbf{W}_{fc}$ and $\mathbf{b}_{fc}$ are the weights and biases of the fully connected layer, LeakyReLU is the Leaky Rectified Linear Unit activation function [17].

Dropout[2] is applied to the fully connected layers to prevent overfitting. The dropout operation is represented as:

$$\mathbf{y}_{\text{dropout}} = \frac{1}{1-p} \cdot (\mathbf{y} \cdot \text{Bernoulli}(1-p)) \tag{5}$$

where $p$ is the dropout rate, $\text{Bernoulli}(1-p)$ represents the dropout mask.

The hybrid model amalgamates CNN and AlexNet by concatenating their outputs and processing them through supplementary dense layers for final classification. The model is trained with categorical cross-entropy loss and optimized with the Adam optimizer [5]. The performance is assessed using accuracy, precision, recall, and F1-score in addition to ROC-AUC curve analysis for each class. The model is ultimately evaluated on the test dataset to present essential metrics, including loss and accuracy, illustrating its proficiency in efficiently capturing both geographical and temporal characteristics for crime activity identification.

Figure 3 illustrates the architecture of the hybrid CNN and AlexNet model, demonstrating the amalgamation of both networks for improved criminal activity categorization.

## 3 Result and Discussion

Table 1 shows us from both the training and testing stages that the model reaches high degrees of accuracy. The training accuracy was 93.62% at a learning rate of 0.01; validation and testing accuracy were likewise somewhat constant at 93.67% and 93.69%, respectively. With training accuracy at 99.60% and testing accuracy at

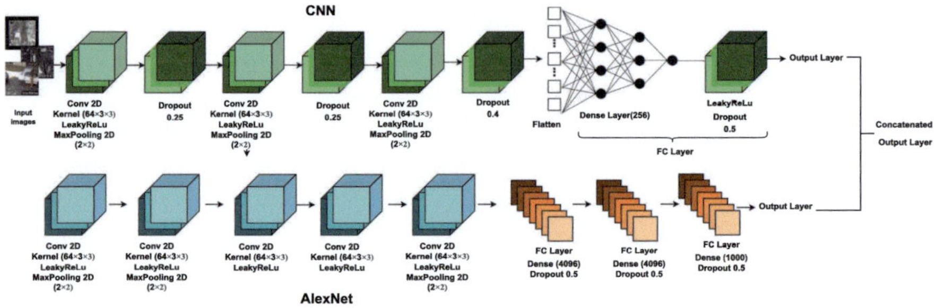


**Fig. 3** Architecture of the proposed model

**Table 1** Accuracy of the proposed model based on different learning rates

| Learning rate | Training accuracy (%) | Validation accuracy (%) | Testing accuracy (%) |
|---|---|---|---|
| 0.01 | 93.62 | 93.67 | 93.69 |
| 0.001 | 99.60 | 99.50 | 99.51 |
| 0.0001 | 99.71 | 99.51 | 99.53 |

**Table 2** Precision, recall, and f-score of the proposed model based on different learning rates

| Learning rate | Precision (%) | Recall (%) | F-score (%) |
|---|---|---|---|
| 0.01 | 93.71 | 93.67 | 93.65 |
| 0.001 | 99.50 | 99.50 | 99.50 |
| 0.0001 | 99.51 | 99.51 | 99.51 |

99.51%, the model's performance dropped dramatically as the learning rate dropped to 0.001. At the learning rate of 0.0001, the performance peaked with a near-perfect training accuracy of 99.71% and a testing accuracy of 99.53%.

Reflecting the efficiency of the hybrid model with smaller steps during optimization, the overall pattern shows that lowering the learning rate results in higher accuracies in both training and testing. This implies that, when the learning rate is lowered, the model is more suited to converge to a global optimum, therefore preventing overfitting or oscillation that can result in larger learning rates.

Table 2 explores the performance of the model concerning precision, recall, and F-score, some important measures in classification issues where class imbalance could be a factor of consideration. With a learning rate of 0.01, the model has an F-score of 93.65%, recall of 93.67%, and precision of 93.71%. The precision, recall, and F-score climb to 99.50%, suggesting an amazing balance between these measures while the learning rate drops to 0.001. The precision, recall, and F-score all stabilize at about 99.51% when the learning rate is further lowered to 0.0001, so highlighting the consistency and strength of the model.

These findings imply that in classifying results with low false positives and false negatives, the suggested hybrid model is quite successful. Particularly in the case

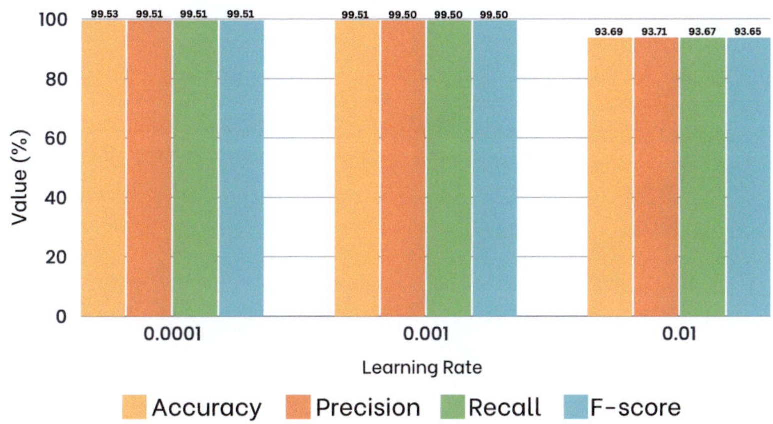


**Fig. 4** Comparative performances of the proposed model based on different learning rates

**Table 3** Comparison of our proposed hybrid CNN-AlexNet model with previous hybrid deep learning models for crime activity detection

| Reference | Model | Accuracy (%) |
|---|---|---|
| Sahay et al. [20] | DRNN (Deep Reinforcement Neural Network) | 98 |
| Nair [18] | CNN + bidirectional LSTM | 81 |
| Ullah et al. [23] | CNN + residual attention-based LSTM | 98.80 |
| Pallewar et al. [19] | CNN + LSTM | 99 |
| Proposed model | CNN + AlexNet | 99.53 |

of complicated patterns in the dataset, the minor performance enhancements as the learning rate lowers support the advantage of employing smaller learning rates to strengthen the decision boundary of the model.

Figure 4 represents a visual comparison among the performance metrics-accuracy, precision, recall, and F-score, for the proposed hybrid model at various learning rates. It unequivocally illustrates the effect of different learning rates (0.0001, 0.001, and 0.01) on the model's efficacy.

Although past research such as as CNN + bidirectional LSTM [18] and the DRNN model [20] demonstrated potential, their accuracies were limited. Although CNN + residual attention-based LSTM [23] and CNN + LSTM [19] models attained high accuracy levels of 98.80% and 99%, respectively, our suggested hybrid CNN-AlexNet model achieves an accuracy of 99.53%, therefore displaying better performance in crime detection than both of these models. This development underlines how well our model manages intricate spatial and temporal aspects of surveillance films (Table 3).

Eventually, the proposed hybrid model, integrating CNN and AlexNet, exhibits considerable enhancement in predictive performance when the learning rate diminishes, especially for accuracy and F-score. The model's capacity to sustain excellent performance across diverse learning rates demonstrates that the selected architecture

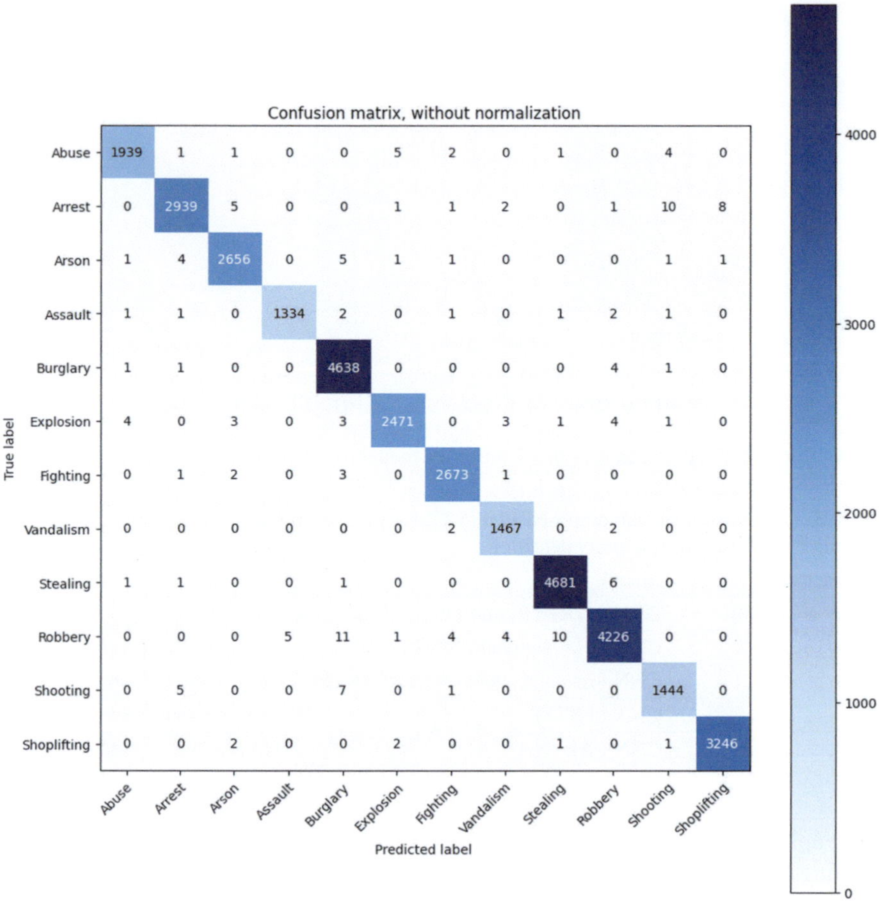


**Fig. 5** Confusion matrix for learning rate of 0.0001

is resilient and generalizable across many contexts. Moreover, the consistency in accuracy, recall, and F-score throughout the testing data substantiates that the model is neither overfitting to the training dataset nor under performing on novel data, which is essential for effective crime activity and suspicious item detection.

The confusion matrices in Figs. 5, 6, and 7 illustrate the efficacy of the proposed hybrid deep learning model for lung cancer prediction at three distinct learning rates: 0.0001, 0.001, and 0.01. Each matrix provides a comprehensive analysis of the model's performance in predicting actual classes (rows) compared to predicted classes (columns).

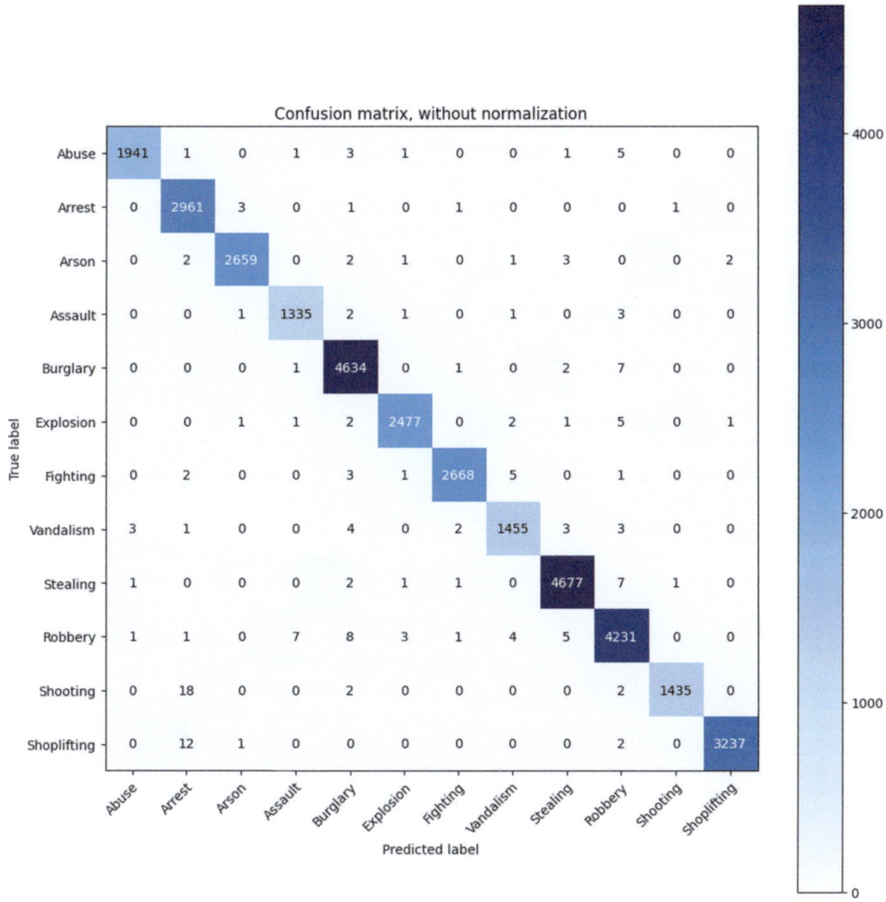


**Fig. 6** Confusion matrix for learning rate of 0.001

# 4 Conclusion

In this work, we suggested a hybrid CNN-AlexNet model optimized using the Adam optimizer to handle the difficulty of crime activity and suspicious item identification. The model showed outstanding performance and a great accuracy of 99.53% on the UCF-Crime dataset by means of extensive testing with several learning rates. Our results show that more exact learning results from lower learning rates, therefore enabling the model to avoid overfitting and generalize effectively. This makes the suggested approach strong and efficient for identifying in real-world surveillance systems both clear and subtle criminal activity. Future research can improve this method to incorporate real-time detection features and investigate other datasets for more general relevance.

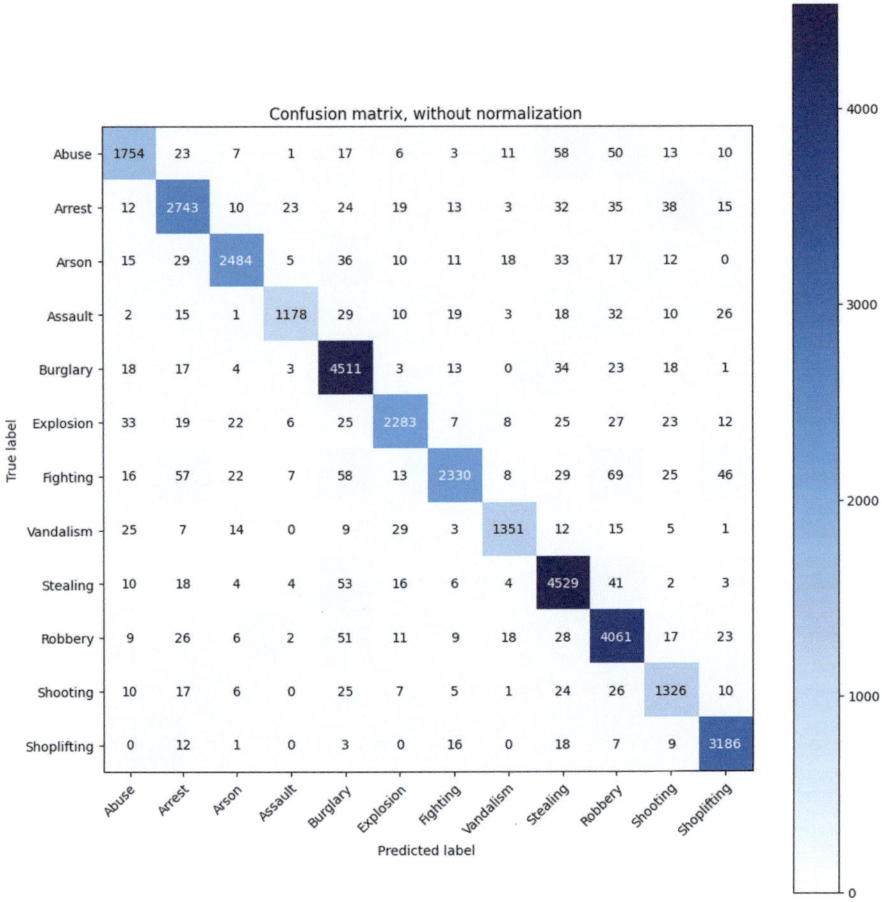


**Fig. 7** Confusion matrix for learning rate of 0.01

# References


1. Azeez J, Aravindhar DJ (2015) Hybrid approach to crime prediction using deep learning. In: 2015 International conference on advances in computing, communications and informatics (ICACCI). IEEE, pp 1701–1710
2. Cheng G, Peddinti V, Povey D, Manohar V, Khudanpur S, Yan Y (2017) An exploration of dropout with LSTMS. In: Interspeech, pp 1586–1590
3. Demir F, Ismael AM, Sengur A (2020) Classification of lung sounds with CNN model using parallel pooling structure. IEEE Access 8:105376–105383
4. Dhillon A, Verma GK (2020) Convolutional neural network: a review of models, methodologies and applications to object detection. Prog Artif Intell 9(2):85–112
5. Diederik PK (2014) Adam: a method for stochastic optimization
6. Hasan M, Choi J, Neumann J, Roy-Chowdhury AK, Davis LS (2016) Learning temporal regularity in video sequences. In: Proceedings of the IEEE conference on computer vision and pattern recognition, pp 733–742

7. Huang C, Wu Z, Wen J, Xu Y, Jiang Q, Wang Y (2021) Abnormal event detection using deep contrastive learning for intelligent video surveillance system. IEEE Trans Industr Inf 18(8):5171–5179
8. Jaouedi N, Boujnah N, Bouhlel MS (2020) A new hybrid deep learning model for human action recognition. J King Saud Univ-Comput Inf Sci 32(4):447–453
9. Jebur SA, Hussein KA, Hoomod HK (2023) Abnormal behavior detection in video surveillance using inception-v3 transfer learning approaches. Iraqi J Comput Commun Control Syst Eng 23(2):210–221
10. Ketkar N, Moolayil J, Ketkar N, Moolayil J (2021) Convolutional neural networks. Deep learning with Python: learn best practices of deep learning models with PyTorch, pp 197–242
11. Kingma DP (2014) Adam: a method for stochastic optimization. arXiv preprint arXiv:1412.6980
12. Krizhevsky A, Sutskever I, Hinton GE (2017) Imagenet classification with deep convolutional neural networks. Commun ACM 60(6):84–90
13. Kumar AK, Suganthi P, Joel T, Shanmugam S et al (2023) Empirical evaluation of video surveillance based crime and anomaly detection system using hybrid deep learning strategy. In: 2023 International conference on innovative computing, intelligent communication and smart electrical systems (ICSES). IEEE, pp 1–9
14. Kumar M, Biswas M (2024) Abnormal human activity detection by convolutional recurrent neural network using fuzzy logic. Multimedia Tools Appl 83(22):61843–61859
15. LeCun Y, Bengio Y, Hinton G (2015) Deep learning. Nature 521(7553):436–444
16. Lee SM, Yoon SM, Cho H (2017) Human activity recognition from accelerometer data using convolutional neural network. In: 2017 IEEE international conference on big data and smart computing (bigcomp). IEEE, pp 131–134
17. Maas AL, Hannun AY, Ng AY et al (2013) Rectifier nonlinearities improve neural network acoustic models. In: Proceedings of ICML, vol 30. Atlanta, GA, p 3
18. Nair LK (2018) Prediction of anomalous activities in a video. Int Res J Eng Technol 3470–3475
19. Pallewar M, Pawar VR, Gaikwad AN (2024) Human anomalous activity detection with CNN-LSTM approach. J Integr Sci Technol 12(1):704–704
20. Sahay KB, Balachander B, Jagadeesh B, Kumar GA, Kumar R, Parvathy LR (2022) A real time crime scene intelligent video surveillance systems in violence detection framework using deep learning techniques. Comput Electr Eng 103:108319
21. Sultana F, Sufian A, Dutta P (2018) Advancements in image classification using convolutional neural network. In: 2018 Fourth international conference on research in computational intelligence and communication networks (ICRCICN). IEEE, pp 122–129
22. Sultani W, Chen C, Shah M (2018) Real-world anomaly detection in surveillance videos. In: Proceedings of the IEEE conference on computer vision and pattern recognition, pp 6479–6488
23. Ullah W, Ullah A, Hussain T, Khan ZA, Baik SW (2021) An efficient anomaly recognition framework using an attention residual LSTM in surveillance videos. Sensors 21(8):2811
24. Wang ZJ, Turko R, Shaikh O, Park H, Das N, Hohman F, Kahng M, Chau DHP (2020) CNN explainer: learning convolutional neural networks with interactive visualization. IEEE Trans Vis Comput Graph 27(2):1396–1406

# Enhanced Flood Prediction and Management for Urban Underpass Using Machine Learning

**Afreen Kubra, K. J. Lokesh, G. Shakthi, M. Surya, A. Mohammad Jawad, and Anthony Melquier**

**Abstract** Floods are among the most destructive natural disasters, leading to severe loss of life, property damage, and economic disruption. Traditional flood prediction methods, which rely heavily on historical data, often lack adaptability to evolving environmental conditions. This article explores the application of machine learning (ML) techniques to enhance flood prediction and management. We investigate the use of supervised learning algorithms such as decision trees, random forests, and other regression methods to predict flood events from meteorological and hydrological data, as well as unsupervised learning methods for detecting anomalies that may indicate impending floods. We propose a comprehensive flood management framework that integrates ML-based prediction models with real-time data from sensors and IoT devices. This framework focuses on early warning systems, risk assessment, and efficient resource allocation. The proposed ML model significantly outperforms traditional methods in predicting flood occurrences, intensity, and impact areas, leading to improved disaster preparedness and reduced economic losses. Subjective and objective evaluations showed that the proposed method could predict flood episodes in certain locations and time periods. Benchmarking trials found that the system configured with random forest provided the best successful prediction, with MSE, MAE, RMSE, and $R^2$ values of 0.7921, 0.9433, 0.9562, and 0.0186, respectively, using synthetic data. Overall, the use of ML constitutes a big step toward creating more robust and adaptable flood management systems.

**Keywords** Flood prediction · Machine learning · Supervised learning algorithms · IoT devices

A. Kubra (✉) · K. J. Lokesh · G. Shakthi · M. Surya · A. Mohammad Jawad · A. Melquier
Department of Artificial Intelligence and Machine Learning, HKBK College of Engineering, Bengaluru, Karnataka, India
e-mail: afreenkubra96@gmail.com

 
K. R. Madhavi et al. (eds.), *Proceedings of Sixth International Conference on Computer and Communication Technologies*, Lecture Notes in Networks and Systems 1356,
https://doi.org/10.1007/978-981-96-5238-9_34

## 1 Introduction

Urban waterlogging disasters have become increasingly frequent due to rapid urbanization and climate change, posing significant challenges to urban public safety and sustainable city development. These disasters, often triggered by heavy downpours and flooding, can severely impact critical infrastructure, such as subway tunnels, complicating evacuation and emergency rescue efforts. The semi-enclosed nature of underground tunnels heightens their vulnerability, making the mitigation of subterranean flood disasters a crucial aspect of smart city infrastructure. Current flood forecast and management systems face several issues, including a lack of real-time water level data during initial flooding, reliance on pre-built scenarios that may not accurately reflect real-time conditions, and high computational demands that can delay response times [1–3]. Additionally, inaccurate evacuation time estimates and poorly placed monitoring points can further undermine the effectiveness and safety of these systems [4].

Flood is a terrible natural event that can cause significant loss of life and economic damage [5–7]. The comparatively confined buildings and high traffic densities of subway systems make them a "high-risk objective" for metropolitan areas [8]. Large-city subway systems have frequently flooded in recent years which have resulted in significant losses worldwide. Therefore, in order to create emergency preparations, carry out emergency response, and minimize flood damage, a quantitative evaluation of a subway system's flood risk is necessary [9]. Few studies have systematically evaluated flood risk and recommended mitigation for subways. Current research focuses on assessing flood risk at regional and watershed dimensions. The impact of ground subsidence on flood danger in subway systems remains unknown. At least 200 cities globally have established subways, including New York [10], London [11], Beijing [12], Shanghai [13], and Seoul [14], which have all experienced substantial ground subsidence.

To address the issues of flood prediction and management in underpasses, a machine learning (ML)-based solution combining different regression models was implemented, with the random forest method producing the best results. This technology uses synthetic data to simulate possible flood situations when rainfall exceeds a certain threshold. This article seeks to develop an effective flood prediction system for underpasses, present a reliable flood warning technique, construct a robust water level monitoring system, and design an automatic reaction mechanism to efficiently handle flood circumstances.

The next few sections of this article are organized as follows. Section 2 of this manuscript presents a concise overview of the case study of flooding incidents happened in Bengaluru. In Sect. 3, system analysis and experimental findings of performance metrics conducted on various regression models are discussed. In Sect. 4, this paper is brought to a conclusion and the potential areas for future research are emphasized.

## 2 Case Study

This section presents a case study of a major flooding incident in Bengaluru, which vividly demonstrates the city's difficulties in dealing with heavy, short-duration rains. This occurrence, which involved a quick and overwhelming accumulation of water as a result of a brief but intense precipitation, demonstrates serious weaknesses in urban drainage infrastructure and emergency response systems [15, 16]. The case study gives vital insights into the effects of such catastrophes on urban areas, emphasizing the importance of improved infrastructure and planning to effectively manage and minimize flooding hazards.

This case study looks at a major flooding occurrence in Bengaluru, where a quick, heavy rainfall of 35–40 mm overwhelmed the city's underpass system as shown in Fig. 1. The incident featured a multi-utility vehicle (MUV) being caught in an underpass as floodwaters rose fast. Despite the quick response of emergency personnel, rising water hindered rescue attempts, resulting in the terrible loss of one life and the rescue of others. This scenario highlights the critical need for enhanced drainage infrastructure and urban design to overcome the problems provided by unexpected, severe rains while also improving the efficacy of emergency response methods.

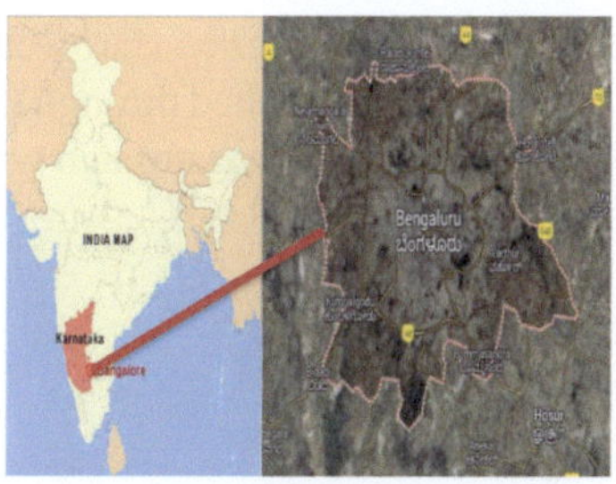


a. Study area consisting of Bangalore city in Karnataka state

b. Rescue operations in a flooded Bengaluru underpass after intense rainfall

c. Stranded vehicles in a waterlogged Bengaluru underpass highlight drainage issues

d. Motorcyclists navigate through a flooded Bengaluru underpass during monsoon rains

**Fig. 1** Images depicting flood risks in Bengaluru city's underpasses

Urban flooding in Bengaluru is a major worry, especially during the monsoon season, as shown in Fig. 1a, b. The city's fast growth has surpassed its infrastructure, resulting in regular waterlogging in underpasses and low-lying neighborhoods. Figure 1c depicts automobiles trapped in heavy floods, highlighting the city's insufficient drainage infrastructure. Figure 1d shows motorcyclists laboring over a submerged underpass, highlighting the immediate hazards and inconveniences caused by severe rain. These photos highlight the critical need for improved flood control methods and infrastructure to protect Bengaluru citizens and enable smoother everyday operations during severe weather events.

## 3 System Analysis

Current flood management systems for underpasses are primarily manual and reactive, lacking integration with advanced technologies. These systems typically involve labor-intensive manual monitoring of flood-prone areas, reliance on static infrastructure like traditional drainage systems that may fail under extreme weather conditions, and limited use of historical or real-time data for flood prediction and management. The absence of automated response mechanisms often leads to delayed actions, allowing flooding to cause significant damage before authorities can intervene. Public notifications about underpass conditions are usually disseminated through traditional media or physical signs, which can be inadequate for timely updates and widespread awareness.

### *3.1 System Architecture*

The system's architecture includes several layers to provide comprehensive flood management as shown in Fig. 2. The data collection layer features sensors for gathering environmental and water level data, while the data processing and analysis layer involves data cleaning and preparation for ML models to produce accurate flood forecasts. The control and automation layer oversees automated responses, such as gate closures and water pumping, to mitigate flood damage. The communication and integration layer facilitates coordination with other smart city technologies, enhancing overall system efficacy. The user interaction and notification layer enables real-time communication with the public and local authorities, supporting informed decision-making. Lastly, the data storage and management layer ensures secure and accessible storage of both historical and current data for ongoing analysis and reporting. Collectively, these modules create a robust flood prediction and management system that integrates advanced technology, real-time monitoring, and automated responses to effectively manage and mitigate flood risks.

The proposed system introduces a modern, integrated approach to flood management using cutting-edge technologies such as machine learning, the Internet of Things

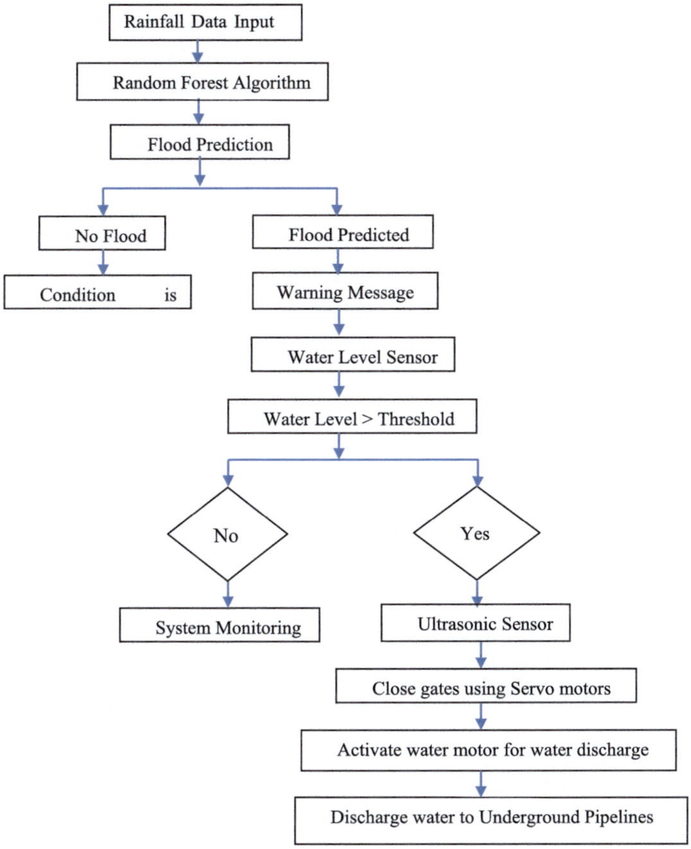


**Fig. 2** Flow diagram of the proposed model

(IoT), and real-time data monitoring. Key features include continuous automated data collection from various sensors, predictive flood risk analysis using ML algorithms, and automated gate control to prevent vehicle and pedestrian access during flood conditions. Real-time public alerts through digital signboards and efficient resource allocation further enhance the system's effectiveness. Designed with scalability in mind, the system can be expanded to multiple underpasses and integrated with broader smart city infrastructure, significantly improving public safety, reducing response times, and mitigating the impact of floods on urban infrastructure. The methods and experimental configuration for creating a ML-driven flood management and prediction system is intended especially for underpasses. When rainfall is above 40 mm, the system uses the random forest algorithm to forecast floods.

For real-time monitoring and response, the configuration combines an Arduino Uno, an I2C LCD, and a number of sensors. The principal aim is to guarantee prompt flood alerts and proficiently regulate water levels in underpasses, hence averting possible dangers [17, 18]. To obtain accurate readings, the sensor is calibrated and linked to the analog input pin of the Arduino. The underpass's water level is constantly monitored by the water level sensor. The subsequent sequence of events commences when the water level surpasses a predetermined threshold [19]. The digital pins of the Arduino are linked to the TRIG and ECHO pins of the ultrasonic sensor. The ultrasonic sensor calculates the depth of the water by measuring the distance to the surface. The gate closure mechanism is triggered in the event that the water depth surpasses a certain threshold [20].

Data gathering and preparation involve creating a synthetic dataset that simulates various rainfall scenarios and their corresponding flood events. This dataset is meticulously cleaned and pre-processed to ensure data consistency and minimize noise, with steps including feature selection and normalization. For model training, the random forest algorithm is employed to leverage this pre-processed dataset, which incorporates input features such as temperature, humidity, wind speed, pressure, and past rainfall to predict rainfall amounts. Once the model is trained, it is implemented by connecting it to an Arduino system for real-time predictions. The model utilizes data from sensors integrated with the Arduino to continuously forecast the likelihood of flooding, enabling timely and accurate flood risk assessments based on real-time environmental conditions.

The system continuously monitors weather conditions through sensors that track variables such as temperature, water level, and rainfall. This real-time data is fed into the ML model, which processes the information to predict potential flooding events. Based on these predictions, the system updates the I2C LCD display to show either warning messages or indicate safe conditions, ensuring that timely and relevant information is provided to users.

The system is designed to detect and respond to flood conditions through a series of automated processes. If the predicted rainfall exceeds 40 mm, the system issues a flood alert to notify of impending risk. In the event that water levels surpass a predefined threshold, an ultrasonic sensor measures the depth of the water to assess its criticality. Should the water depth reach a critical level, servo motors are activated to close gates on either side of the underpass, effectively preventing further water ingress. To manage and mitigate flooding, the system also employs a water motor to discharge accumulated water through subterranean pipelines, thereby reducing the overall flood impact.

The random forest model makes uses of input features to predict rainfall. The predicted rainfall amount is a function of temperature, humidity, wind speed, previous rainfall, and pressure. The ultrasonic sensor measures water depth $d$ given by Eq. 1 based on the time $t$ it takes for the sound wave to travel to the water surface and back. The system triggers a flood warning when the predicted rainfall exceeds 40 mm, and it initiates gate closure when the water depth surpasses a critical level.

$$d = (t \times v)/2 \tag{1}$$

where $v$ is the speed of sound in air.

The architecture of the ML-based flood prediction and management system is organized into several key modules, each pivotal for effective flood risk mitigation. The data collection and integration module acquires both historical and real-time data from weather stations and environmental sensors, which is critical for training the model and performing real-time monitoring. This data is then processed by the machine learning prediction module, where advanced algorithms such as random forest analyze trends and predict flood risks, with predictions being continually updated through ongoing data analysis. The real-time monitoring and sensor module tracks environmental conditions, particularly water levels in vulnerable areas like underpasses, providing essential data for detecting potential flood threats. The automated gate control module utilizes these predictions and real-time data to regulate gate operations, automatically closing gates when water levels reach predefined thresholds to prevent flooding and protect infrastructure and users. The public alert and notification module ensures public safety by disseminating real-time flood alerts through various channels, including digital signboards, social media, and mobile notifications, thus keeping the public informed about flood risks and necessary precautions.

### *3.2 Performance Comparison of Regression Models for Rainfall Prediction*

In this work, we compared the effectiveness of multiple regression models on a synthetic dataset to forecast rainfall using temperature, humidity, wind speed, past rainfall, and pressure. The models examined were random forest regressor, ridge, linear regression, gradient boosting regressor, K-neighbors regressor, support vector regression (SVR), and decision tree regressor. To evaluate the models, we employed measures such as mean absolute error (MAE), mean squared error (MSE), root mean squared error (RMSE), and $R^2$ score. As seen in Table 1, random forest regressor has the greatest $R^2$ score compared to other models, suggesting greater data pattern capture. However, the $R^2$ score remained close to zero, indicating minimal success in explaining the dataset's volatility. Ridge and linear regression models fared comparably.

The flood prediction and management system for underpasses has achieved notable success in several key areas. Utilizing the random forest algorithm, the system reliably forecasts floods when rainfall exceeds 40 mm, providing timely alerts based on real-time sensor data. Integration with an Arduino Uno and an I2C LCD enables immediate and clear communication of flood warnings, while continuous monitoring by ultrasonic and water level sensors ensures adaptability to changing conditions. The system's automated response includes gate closure to prevent further flooding and efficient water discharge through subterranean pipelines, significantly mitigating flood risks and enhancing safety in underpass areas. Figures 3, 4, 5, and 6 show the performance analysis of the proposed regression models.

**Table 1** Performance metric values for various regressor models

| | MAE | MSE | RMSE | $R^2$ score |
|---|---|---|---|---|
| Random forest regressor | 0.7921 | 0.9433 | 0.9562 | 0.0186 |
| Linear regression | 0.7892 | 0.9439 | 0.9570 | 0.0039 |
| Ridge | 0.7890 | 0.9437 | 0.9565 | 0.0037 |
| Gradient boosting regressor | 0.7918 | 0.9438 | 0.9606 | 0.0013 |
| SVR | 0.7532 | 0.9641 | 0.9758 | – 0.0229 |
| Decision tree regressor | 0.8122 | 1.0013 | 0.9884 | – 0.0563 |
| K-neighbors regressor | 0.7922 | 1.0024 | 0.9942 | – 0.0642 |

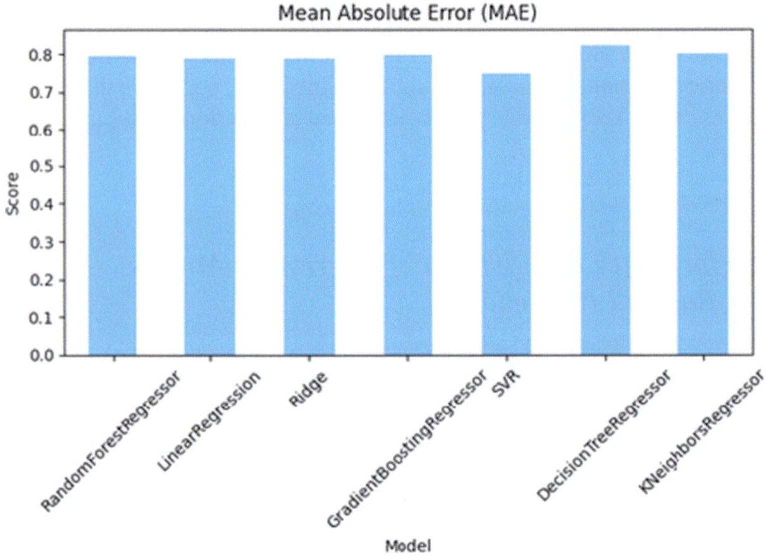


**Fig. 3** Performance comparison in terms of MAE

## 4 Conclusion and Scope for Future Work

The flood prediction and management system for underpasses represents a significant advancement in flood risk mitigation, leveraging automated response mechanisms, real-time monitoring, and ML to enhance safety and adaptability. This technology significantly reduces the risk of flood-related incidents by providing accurate flood predictions and real-time alerts, thereby improving underpass safety. Automated systems for gate closure and water discharge efficiently manage water levels, preventing severe flooding and minimizing potential damage. The system's continuous monitoring and prompt flood alerts enable proactive risk management, ensuring that underpasses remain secure and operational even during heavy rainfall. This

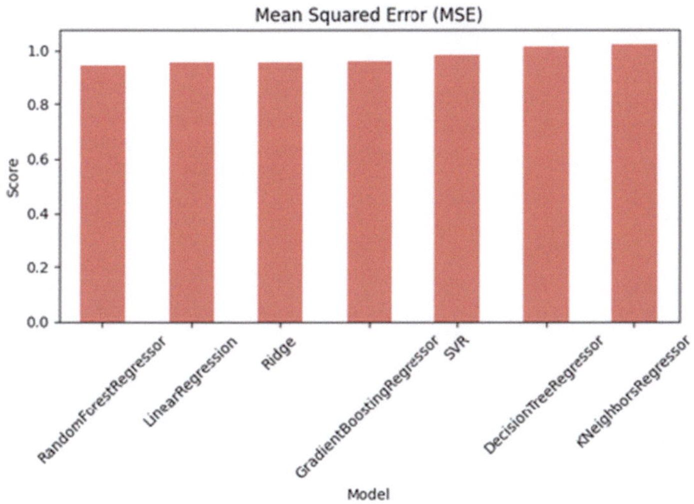


**Fig. 4** Performance comparison in terms of MSE

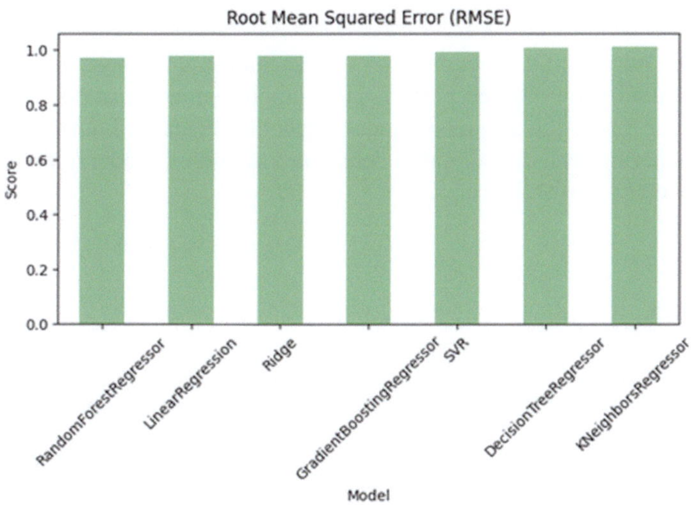


**Fig. 5** Performance comparison in terms of RMSE

proactive approach not only protects infrastructure but also facilitates a rapid response to emerging flood threats.

Future work for enhancing the flood prediction and management system focuses on improving prediction accuracy. Advanced ML algorithms, such as gradient boosting and deep learning, high-quality data integration, will refine the model's precision. Integration with Google Maps will provide real-time flood updates and visual risk mapping. Additional improvements include scaling, IoT integration for

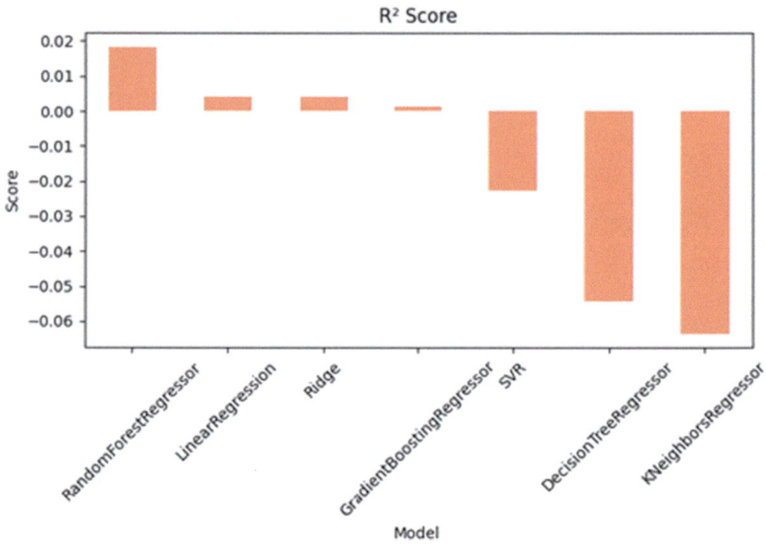


**Fig. 6** Performance comparison in terms of $R^2$ score

granular monitoring, and adaptation to climate change. The present results on the synthetic dataset show limited performance, emphasizing the importance of real-world data for making more precise and dependable forecasts. Future research will focus on gathering real-world rainfall data, and fine-tuning model hyperparameters to improve model performance.

## References

1. Alam I, Fiza S, MP S (2023) Exploring possibilities and methodologies for big data and 5G convergence. In: 2023 International conference on power energy, environment & intelligent control (PEEIC)
2. Asha K, Kumar R, Fiza S (2023) Enhancing networked embedded systems using deep learning techniques. In: 2023 IEEE International conference on ICT in business industry & government (ICTBIG). Indore, India
3. Sinha DK, Naval P, Fiza S (2023) Visualizing wellness: the impact of deep learning on healthcare image analysis. In: 2023 International conference on advances in computation, communication and information technology (ICAICCIT). Faridabad, India
4. Dong W, Huang H, Zhong M, Wang H, Hua F (2024) Monitoring and early warning mechanism of flood invasion into subway tunnels based on the experimental study of flooding patterns. J Intell Constr 2(2):9180011
5. Chen B, Gong H, Chen Y, Li X, Zhou C, Lei K, Zhu L, Zhu L, Duan L, Zhao X (2020) Land subsidence and its relation with groundwater aquifers in Beijing Plain of China. Sci Total Environ 735(0048–9697):139111
6. Lee K, Torell GL, Newman SM (2021) A once-in-one-hundred-year event? A survey assessing deviation between perceived and actual understanding of flood risk terminology. J Environ Manage 277(0301–4797):111400

7. Toosi AS, Calbimonte G, Nouri H, Alaghmand S (2019) River basin-scale flood hazard assessment using a modified multi-criteria decision analysis approach: a case study. J Hydrol 574(0022–1694):660–671
8. Forero-Ortiz E, Martinez-Gomariz E, Porcuna MC, Locatelli L, Russo B (2020) Flood risk assessment in an underground railway system under the impact of climate change—a case study of the Barcelona Metro. Integr Assess Clim Chang Impacts Urban Resil 12(13):5291
9. Wang G, Liu Y, Hu Z, Guoming Z, Liu J, Lyu Y, Gu Y, Huang X, Zhang Q, Liu LY (2021) Flood risk assessment of subway systems in metropolitan areas under land subsidence scenario: a case study of Beijing. Geospatial Underst Sustain Urban Anal Using Remote Sens 13(4):637
10. Vermeij DAAN (2016) Flood risk reduction interventions for the New York City subway system. Arcadis, Delft
11. Boni R, Meisina C, Cigna F, Herrera G, Notti D, Bricker S, McCormack H, Tomás R, Béjar-Pizarro M, Mulas J, Ezquerro P (2017) Exploitation of satellite A-DInSAR time series for detection, characterization and modelling of land subsidence. Obs Geohazards Space 7(2):25
12. Chen B, Gong H, Li X, Lei K, Zhu L, Gao M, Zhou C (2017) Characterization and causes of land subsidence in Beijing, China. Int J Remote Sens 38(3):808–826
13. Dong S, Samsonov S, Yin H, Ye S, Cao Y (2014) Time-series analysis of subsidence associated with rapid urbanization in Shanghai, China measured with SBAS InSAR method. Environ Earth Sci 72(3):677–691
14. Fadhillah MF, Achmad AR, Lee CW (2020) Integration of InSAR time-series data and GIS to assess land subsidence along subway lines in the Seoul Metropolitan Area, South Korea. Artif Intell Methods Appl Urban Remote Sens GIS 12(21):3505
15. Kubra A, Karunavati R, Das P, Prasad Y, Gautam WR (2020) Design of a reconfigurable digital modulator for high bit rate data transmission. In: 2020 IEEE International conference on electronics, computing and communication technologies (CONECCT). Bangalore, India
16. Thirumala Vasu G, Fiza S, Kumar AK, Devi VS, Kumar CN, Kubra A (2023) Improved chimp optimization algorithm (ICOA) feature selection and deep neural network framework for internet of things (IOT) based android malware detection. Meas Sens 28:100785
17. Chen SY, Xue Z, Li M, Zhu XP (2013) Variable sets method for urban flood vulnerability assessment. Sci China Technol Sci 56(12):3129–3136
18. Lyu HM, Shen SL, Zhou A, Yang J (2019) Perspectives for flood risk assessment and management for mega-city metro system. Tunn Undergr Space Technol 84(0886–7798):31–44
19. Qihu Q (2016) Present state, problems and development trends of urban underground space in China. Tunn Undergr Space Technol 55(0886–7798):280–289
20. Yongfeng L, Chengjing N, Linsheng Y, Hairong L (2012) An overview of the risk assessment of flood disaster. Prog Geogr 31(3):361–367

# Brain Tumour Classification Using Optimized Feature Sets of Convolution Neutral Network

**P. Nagarathna, Anitha Suresh, Illuru Rajasekhar, C. B. Vinutha, G. Tirumala Vasu, and Samreen Fiza**

**Abstract** The classification and diagnosis of brain tumours remain reliant on the histological analysis of biopsy specimens. The present approach is intrusive, laborious, and prone to human error. These limitations emphasize the necessity of adopting a fully automated approach for multi-classification of brain tumours using deep learning. The objective of this article is to detect feature sets which are further optimized to get better feature maps to convolutional neural network (CNN). 25 layers consisting of six stage convolution layers are used to extract the feature sets of brain tumour images in the proposed CNN model. This network architecture enables less computational complexity compared to conventional CNN models in the classification of brain tumours into four classes, namely pituitary, meningioma, glioma, and no tumour. The proposed CNN model is assessed using quality metrics, namely sensitivity, accuracy, specificity, precision, and F1 score against leading-edge models

P. Nagarathna
Department of Computer Science and Engineering, Nitte Meenakshi Institute of Technology, Bengaluru, Karnataka, India
e-mail: nagaratna.p@nmit.ac.in

A. Suresh
Department of Electronics and Telecommunication Engineering, Dayananda Sagar College of Engineering, Bengaluru, Karnataka, India
e-mail: anitha-tce@dayanandasagar.edu

I. Rajasekhar
Department of Electronics and Communication Engineering, GATES Institute of Technology, Gooty, Andhra Pradesh, India
e-mail: gootyrajasekhar@gmail.com

C. B. Vinutha · G. Tirumala Vasu (✉) · S. Fiza
Department of Electronics and Communication Engineering, Presidency University, Bengaluru, Karnataka, India
e-mail: sietk81nitt@gmail.com

C. B. Vinutha
e-mail: vinutha.cb@presidencyuniversity.in

S. Fiza
e-mail: samreenfiza236@gmail.com

 
K. R. Madhavi et al. (eds.), *Proceedings of Sixth International Conference on Computer and Communication Technologies*, Lecture Notes in Networks and Systems 1356,
https://doi.org/10.1007/978-981-96-5238-9_35

including GoogLeNet, VGG-16, ResNet-50, Inceptionv3, and AlexNet. Large and publicly accessible clinical datasets yield satisfactory categorization outcomes. The proposed CNN model can assist radiologists and physicians in validating the initial screening for the purpose of classifying brain tumours.

**Keywords** Brain tumour classification · Feature maps · Optimized feature sets · Convolution neural network · Deep learning

## 1 Introduction

A predominant sophisticated component of the neurological system in humans is the brain that controls vital body functions. Neurological disorders of the brain present considerable difficulties in clinical diagnosis and therapy owing to elevated rates of recurrence, morbidity, and mortality. They can be classified based on their malignancy using the World Health Organization (WHO) guidelines.

For categorizing brain tumours, the most widely recognized predominant imaging technique is magnetic resonance (MR) imaging because of the high quality of the images it produces. Over the years, the segmentation of brain tumours in MRI scans is carried out manually by skilled radiographers. However, this approach was time-consuming and often resulted in inconsistent segmentation outcomes due to its reliance on subjective expertise. The same medical image would yield different segmentation results when analysed by different technicians. Consequently, numerous researchers have endeavoured to address [1] this issue by employing computer aided diagnosis methods to achieve semi-automatic segmentation. The existing approach for classifying and diagnosing brain tumours utilizes deep learning and machine learning models [2]. However, these models face a challenge in terms of their low accuracy, which highlights the need for improvement.

Several problems were tackled by using convolutional neural network (CNN) technology. These included improving diagnostic explanations by weakly supervised localization, developing efficient multitask models, and dealing with disease classification issues with limited data. Within the CNN, the last layers and kernel demonstrate instability due to the spatial interactions of neurons [3]. Deep convolutional neural networks have demonstrated efficacy in identifying brain tumours and diagnosing various tumour types through MRI-based medical imaging. This approach has shown promising results in identifying patients with brain tumours. CNNs result in a loss of spatial connections between higher-level elements. Their lack of rotational invariance makes them more susceptible to unfamiliar perspectives. The proposed approach addresses these issues by employing an optimization method based on the Sobolev gradient that incorporates a stochastic and an accelerated optimizer. Image fusion methods [4–8] that are using edge preservation techniques will further handle this scenario. This ensures efficient gradient descent in terms of both accuracy and speed. Although inner products of Sobolev-type yield regular gradients, the proposed network integrates dynamic routing through a probabilistic method utilizing maximum-likelihood prediction [9]. This methodology contributes to enhanced generalization by potentially reducing the parameters count and enabling routing calculation within a single receptive field. Previous studies on classifying brain tumour images have primarily utilized CNN.

However, these methodologies exhibit a deficiency in rotational invariance, which can lead to the loss of spatial connections between higher-level elements. Moreover, the process of extracting features in cancer classification systems presents a challenge [10], often requiring the precise categorization of features based on expertise. To overcome these limitations, there is a necessity for a novel approach that fully automates the classification of brain tumour images. Previous research has encountered difficulty in constructing efficient multitask models, resolving disease classification problems with limited data, and using weakly supervised localization for better diagnosis. It is important to note that CNN-based techniques, due to their lack of rotational invariance and potential compromise of spatial connections between [11] higher-level elements, exhibit reduced adaptability to new perspectives.

Pre-trained deep learning models (DLMs) like VGG, DenseNet, ResNet, GoogLeNet, AlexNet [12, 13], and MobileNet can significantly enhance accuracy while also reducing training time [14, 15]. These models have already proven to be beneficial in various applications such as speech recognition, language modelling, human activity recognition, and image processing. The advantage of using pre-trained models lies in their usage convenience as they require less training time. These transfer learning process-based models have the advantage of being architecturally simple, which can lower computational costs. Furthermore, a data augmentation technique [16, 17] was applied to the dataset in order to compensate for the lack of available medical images.

This paper primarily contributes to the implementation of optimized feature sets of convolutional neural networks to differentiate between glioma, meningioma, pituitary tumours, and non-tumour brain images [18], facilitating rapid and precise diagnoses through parameter optimization, ultimately achieving significantly enhanced test accuracy for a graphics processing unit (GPU).

## 2 Related Work

In the recent past, deep learning has gained increasing attention for its application in the segmentation and classification of brain tumours. Diverse methodologies have been proposed, including the utilization of CNNs for both segmentation and classification endeavours. CNNs [19, 20] acquire knowledge about the spatial connections among various characteristics in images, rendering them very suitable for tasks like tumour segmentation.

A novel adaptive eroded deep (AED) CNN was introduced in [21] for segmenting and classifying images of the brain, and then classifying the tumour as malignant or benign employing the Inception ResnetV2. This algorithm successfully achieved precise segmentation of meningioma, glioma, and pituitary brain areas.

A recent study [22, 23] introduced a hybrid deep convolutional neural network called Hyb-DCNN-ResNet 152 TL that combines nature-inspired ResNet 152 transfer learning for brain tumour detection and classification.

In [24], a hybrid approach was presented for the purpose of segmenting, classifying, and extracting features from brain tumours that used watershed algorithm and threshold segmentation, followed by the utilization of various classifiers for tumour classification.

The system successfully attained an accuracy of 90% in tumour categorization. A DL-based technique for predicting brain tumours using semantic segmentation was introduced in [24, 25]. The technique employs a CNN to accurately delineate the tumour area from brain scans. The region containing tumour is subsequently classed as benign category or malignant ones. The approach achieved a tumour prediction accuracy of 91.718%. The scientific work discussed focuses on the challenges and considerations related to the division, categorization, and forecasting of brain tumours utilizing advanced DL methods. The concerns encompass ethical considerations, data unpredictability, computational and resource requirements, model generalization capability, validation, segmentation accuracy, and classification performance of the suggested network in real-world clinical settings. These challenges are addressed by the proposed network ensuring the use of diverse datasets, improve segmentation and classification accuracy, improve survival rate predictions, reducing the likelihood of overfitting, and provide interpretability of the network's outputs. Effectively addressing these problems will greatly improve the creation of a reliable and effective DL-based classification network to analyse tumours in brain.

## 3 Proposed Work

### *3.1 The Dataset*

The datasets [26] contain images obtained from patients diagnosed with four specific tumour classes in the brain, namely no tumour, pituitary, meningioma, and glioma tumour. We have used 5664 data samples for training and validation with 70:30 ratio.

### *3.2 Pre-processing Stage*

Pre-processing is conducted prior to training the network model. The first stage entails resizing the original image from $512 \times 512 \times 3$ pixels to $227 \times 227 \times 3$ pixels to diminish image complexity, lower computational requirements, and enhance network performance. Subsequently, the data is randomized before partitioning to maintain the integrity of the system's training process and avoid fixation on a restricted subset of dataset. Among the total data, 70% of the data is utilized for the purpose of training, while the test and validation sets account for the remaining 30%. This is generally executed to mitigate overfitting and augment the model's robustness.

### 3.3 Proposed Optimized Feature Sets Network Architecture

CNN is commonly employed for medical disease diagnosis through image classification. Design of a single CNN model that can effectively handle all classification problems and consistently produce satisfactory results is not feasible or realistic. Hence, a distinct CNN model is specifically crafted for every type of challenge.

The design and intricacy of the CNN model fluctuate based on the nature of the problem, input attributes, and anticipated outputs. This article involves the building of a novel feature sets CNN model, serving a specific classification objective. An issue in CNN is the selection of the appropriate model for a certain problem. The achievement of successful outcomes, particularly in CNN, relies heavily on the selection of appropriate hyperparameters. Classification results are derived on extensive and openly accessible clinical datasets.

Figure 1 illustrates the abstract framework of the CNN architecture that is proposed. The network comprises 25 layers, commencing with the input layer that contains enhanced images derived from the preceding pre-processing phase. The images traverse the convolutional layers with their corresponding activation functions. To alleviate overfitting, a dropout layer is employed, followed by a fully connected layer and a SoftMax layer for predictions. The final stage involves a classification layer that produces the expected category.

The input layer is utilized to validate the dimensions of the input images in the proposed design. The network consists of six convolutional layers. A 2D convolutional layer employs K convolutional filters (kernels) that move across input images.

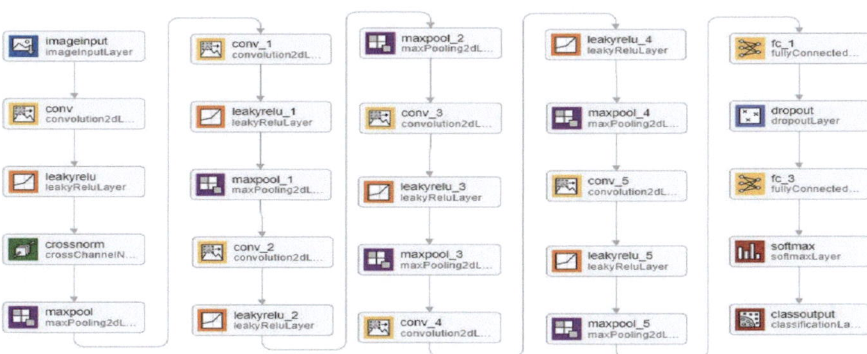


**Fig. 1** Architecture of the proposed CNN model

The filters traverse the input, computing the dot product between the filter weights and the corresponding input values. These filters traverse the image using both vertical and horizontal strides, which are indicated by the stride parameter (S). Prior to moving the filter, it may be necessary to use padding (P) on the source images to preserve edge information. The purpose of these kernels is to identify feature sets, with the earliest layers mainly detecting low-level feature sets, namely blobs, lines, and edges.

As the layers advance, increasingly sophisticated kernels are employed to identify progressively complex characteristics [27]. After each convolutional layer, the LeakyReLU activation function, which does not reach saturation, is used. The principal objective of this activation function is to significantly decrease the training duration relative to alternative activation functions, while also generating nonlinearity among the outputs across layers of the neural network. A cross-channel normalization layer normalizes the input layer by scaling and adjusting the corresponding activations. This layer does local response normalization to each channel individually, using a defined window size. Normalization can be utilized in the process of backpropagation and the acceleration of network training [28, 29]. The maximum pooling layer serves as a down sampling technique to attain spatial invariance. The suggested methodology is partitioning the entire image into small rectangular units ($2 \times 2$) that traverse the image with a predetermined step size ($2 \times 2$). Consequently, only the highest value among the four factors is taken into account.

The pooling layer is employed within the network to diminish the quantity of parameters and calculations [30, 31]. To mitigate overfitting, a widely used approach involves incorporating a dropout layer, as depicted in Fig. 1. Within this layer, there is a process of randomly dropping out some activations (nodes), which effectively helps to reduce overfitting and accelerate the training process. In conclusion, three sophisticated layers have been employed: fully connected layer (FC), SoftMax layer, and classification layer. Fully connected (FC) layers are utilized to create connections between each neuron in one layer and every neuron in both the subsequent and previous layers. This layer produces a result that consists of three distinct categories. After the FC layer comes a SoftMax layer, which is also referred to as the normalized exponential function.

# 4 Experimental Results and Discussion

## *4.1 Experiment Setup*

The proposed CNN architecture is trained on an Intel i5-1155G7 CPU (2.50 GHz), NVIDIA® GeForce RTX™ 4060 with 8 GB GDDR6 GPU, 16 GB RAM, and MATLAB 2023b. The training set required 926 m and 46 s for 5664 images. The average test execution duration was 8.35 ms per image.

**Table 1** Comparison of accuracy

| Network | Accuracy in percentage (%) |
|---|---|
| AlexNet | 84.7 |
| VGG | 91.9 |
| GoogLeNet | 77.3 |
| ResNet | 76.1 |
| DenseNet | 90.2 |
| Proposed network | 92.29 |

## *4.2 Results*

The proposed optimized feature sets CNN model achieved an accuracy of 92.29% when compared to leading networks as shown in Table 1 demonstrating its efficacy in accurately categorizing a wide range of brain tumour images. The outcomes achieved by the proposed model are compared to well-established and widely used pre-trained CNN models like AlexNet, GoogLeNet, VGG-16, Inceptionv3, and ResNet-50.

Class 1-Giloma, Class 2-Meningiona, Class 3-No tumour, and Class 4-Pituitary metric values are shown in Table 2. All metrics related to class 3 are dominating the remaining classes with a specificity and precision reaching 100%, a high accuracy touching 99.65%, 99.37% of F1 score, and a sensitivity of 98.75% has been achieved. Comparatively class 4 has achieved closer metrics values compared to class 3-No tumour. Glioma and Meningioma metric values are less when compared to class 3 and class 4.

The results showcase the network's capacity to effectively manage fluctuations in lighting, perspective, and object orientation, hence demonstrating its resilience in practical situations. Figure 2 shows the confusion matrix with system output, i.e. predicted values on the *x*-axis and ground truth, i.e. true labels on *y*-axis. The proposed CNN network is assessed using sensitivity, accuracy, precision, specificity, and F1 score as depicted in Table 2.

**Table 2** Performance metric comparison

| Measure | Class 1 Glioma | Class 2 Meningioma | Class 3 No tumour | Class 4 Pituitary |
|---|---|---|---|---|
| Sensitivity | 0.9020 | 0.7712 | 0.9875 | 0.9404 |
| Specificity | 0.9159 | 0.9603 | 1.0000 | 0.9928 |
| Precision | 0.7023 | 0.8939 | 1.0000 | 0.9793 |
| Accuracy | 0.9134 | 0.9032 | 0.9965 | 0.9788 |
| F1 score | 0.7897 | 0.8281 | 0.9937 | 0.9595 |

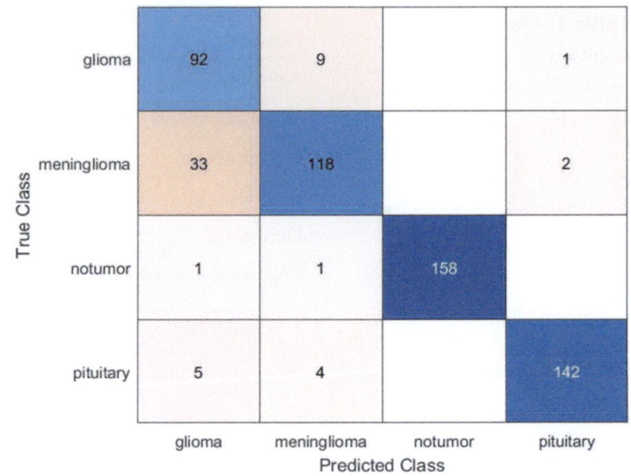


**Fig. 2** Confusion matrix for CNN model

## 5 Conclusion and Future Scope

Advancements in deep learning have prompted a transition in studies and research from feature engineering to architectural engineering. The proposed research demonstrates the utilization of convolutional neural network (CNN) models for the early diagnosis of tumours in the brain by multi-classification. Six CNN models are assigned for four distinct brain tumour classification operation using medical datasets that are publicly available. Detection of tumour in brain is achieved with 92.29% accuracy by the proposed approach which utilizes the local binary model technique and CNN algorithm to analyse and extract features from tumour MRI images. The paper presents an algorithm and model for continuous optimization. It utilizes an extended dataset to train a new deep convolutional neural network that is sensitive to tumour images. The algorithm aims to incorporate various types of cost factors into the classification process. The optimized feature sets of CNN designs have been used to classify brain cancers into four classes, namely normal, meningioma, pituitary, and glioma by utilizing different layers in the networks. In future, our goal is to enhance our model for remote monitoring, connect with the Internet of things (IoT), and utilize breakthroughs in deep learning. Our ultimate goal is to develop a personalized healthcare recommendation system for patients with brain tumours. By using sophisticated algorithms and predictive analytics, our approach to brain tumour detection is at the forefront of innovation, leading the way for improved patient care outcomes.

## References

1. ZongRen L, Silamu W, Yuzhen W, Zhe W (2023) DenseTrans: multimodal brain tumor segmentation using swin transformer. IEEE Access 11:42895–42908

2. Jabbar A, Naseem S, Mahmood T, Saba T, Alamri FS, Rehman A (2023) Brain tumor detection and multi-grade segmentation through hybrid caps-VGGNet model. IEEE Access 11:72518–72536
3. Chen H, Chen D, Wang L (2021) CNN-based MRI brain tumor detection application. In: 2021 International conference on computer engineering and application (ICCEA). Kunming, China
4. Tirumala Vasu G, Palanisamy P (2022) Multi-focus image fusion using anisotropic diffusion filter. Soft Comput 26(20):14029–14040
5. Tirumala Vasu G, Palanisamy P (2023) Visible and infrared image fusion using distributed anisotropic guided filter. Sens Imaging 24
6. Tirumala Vasu G, Palanisamy P (2024) Multi-exposure image fusion using structural weights and visual saliency map. Multimedia Tools Appl
7. Tirumala Vasu G, Palanisamy P (2023) Gradient-based multi-focus image fusion using foreground and background pattern recognition with weighted anisotropic diffusion filter. Signal Image Video Process 17(5):2531–2543
8. Fiza S, Safinaz S (2024) Multi-focus image fusion using edge discriminative diffusion filter for satellite images. Multimedia Tools Appl 83:1–20
9. Devaraj, Gokulrajan P, Gowtham S, Mugilan D (2023) Deep learning for classification and growth prediction of turmeric and sugarcane crops. In: 2023 3rd International conference on pervasive computing and social networking (ICPCSN). Salem, India
10. Gautam AK, Tirumala Vasu G, Mamatha GN (2023) Optimal allocation of resources and hospital capacity planning for critical diseases using AI and data mining. In: 2023 IEEE International conference on ICT in business industry & government (ICTBIG). Indore, India
11. Yu Z, Li X, Li J, Chen W, Tang Z, Geng D (2024) HSA-net with a novel CAD pipeline boosts both clinical brain tumor MR image classification and segmentation. Comput Biol Med 170:108039
12. Sultan HH, Salem NM, Al-Atabany W (2019) Multi-classification of brain tumor images using deep neural network. IEEE Access 7:69215–69225
13. Vaiyapuri T, Mahalingam J, Ahmad S, Abdeljaber HAM, Yang E, Jeong SY (2023) Ensemble learning driven computer-aided diagnosis model for brain tumor classification on magnetic resonance imaging. IEEE Access 11:91398–91406
14. Rajendran S, Rajagopal SK, Thanarajan T, Shankar K, Kumar S, Alsubaie NM, Ishak MK, Mostafa SM (2023) Automated segmentation of brain tumor MRI images using deep learning. IEEE Access 11:64758–64768
15. Zaitoon R, Syed H (2023) RU-Net2+: a deep learning algorithm for accurate brain tumor segmentation and survival rate prediction. IEEE Access 11:118105–118123
16. Nagarathna P, Jeelani A, Fiza S, Tirumala Vasu G, Seelam K (2024) Medical image registration with object deviation estimation through motion vectors using octave and level sampling. Automatika 65(3):1213–1227
17. Nagarathna P, Kubra A, Tirumala Vasu G, Raj D, Suresh A, Fiza S (2024) Signal dependent local noise removal using Weiner filter decomposition. In: International conference on computational innovations and emerging trends (ICCIET-2024). Amalapuram, India
18. Hossain E, Hossain MS, Hossain MS, Jannat SA, Huda M, Alsharif S, Faragallah OS, Eid MMA, Rashed ANZ (2020) Brain tumor auto-segmentation on multimodal imaging modalities using deep neural network. Comput Mater Continua 72(3):4509–4523
19. Fiza S, Sunil MP, Shukla G (2023) Predictive analytics and AI for early diagnosis and intervention in autism spectrum disorders. In: 2023 IEEE International conference on ICT in business industry & government (ICTBIG). Indore, India
20. Malathi H, Yadav DC, Tirumala Vasu G (2023) Predictive analytics and machine learning for personalized diabetes management in real-time. In: 2023 IEEE International conference on ICT in business industry & government (ICTBIG). Indore, India
21. Shah HA, Saeed F, Yun S, Park J-H, Paul A, Kang J-M (2022) A robust approach for brain tumor detection in magnetic resonance images using finetuned EfficientNet. IEEE Access 10:65426–65438

22. Muhammad K, Khan S, Ser JD, de Albuquerque VHC Deep learning for multigrade brain tumor classification in smart healthcare systems: a prospective survey. IEEE Trans Neural Netw Learn Syst 32(2):507–522
23. Asha K, Kumar R, Fiza S (2023) Enhancing networked embedded systems using deep learning techniques. In: IEEE International conference on ICT in business industry & government (ICTBIG). Indore, India
24. Lal AH, Sreenivasulu E, Kumar MA, Bachu S (2022) Implementation of brain tumor detection with deep learning classification using hybrid feature extraction. In: 2022 First International conference on electrical, electronics, information and communication technologies (ICEEICT). Trichy, India
25. Sinha DK, Naval P, Fiza S (2023) Visualizing wellness: the impact of deep learning on healthcare image analysis. In: 2023 International conference on advances in computation, communication and information technology (ICAICCIT). Faridabad, India
26. Bhuvaji S, Kadam A, Bhumkar P, Dedge S, Kanchan S (2020) Kaggle [Online]. Available: https://www.kaggle.com/datasets/sartajbhuvaji/brain-tumor-classification-mri
27. Lecun Y, Bottou L, Bengio Y, Haffner P (1998) Gradient-based learning applied to document recognition. Proc IEEE 86(11):2278–2324
28. Chen Y, Wang S, Ge Y (2022) A survey on the applications of image classification based on convolution neural network. In: 2022 IEEE Asia-Pacific conference on image processing, electronics and computers (IPEC). Dalian, China
29. Brahimi S, Ben Aoun N, Ben Amar C (2018) Improved very deep recurrent convolutional neural network for object recognition. In: 2018 IEEE International conference on systems, man, and cybernetics (SMC). Miyazaki, Japan
30. Dutta HPJ, Bhuyan MK, Neog DR, MacDorman KF, Laskar RH (2023) Patient assistance system based on hand gesture recognition. IEEE Trans Instrum Meas 72:1–13
31. Goswami A, Dixit M (2020) An analysis of image segmentation methods for brain tumour detection on MRI images. In: 2020 IEEE 9th International conference on communication systems and network technologies (CSNT). Gwalior, India

# A Deep Learning Approach for Enhanced Detection of Mental Stress and Emotion Analysis

**Navaneeth Bhaskar, Pooja Bidwai, K. Mahesh, B. K. Trishan, Sanskar S. Khandelwal, and N. Vishvith Shetty**

**Abstract** Stress is a part of each and everyone's life, and we all have to face it at one time or another when things get tougher. It doesn't only affect our health and mood, but also the way we interact with others. Early detection will help people reduce stress since one can overcome the problem before things get out of control. The automated model presented in this paper can detect stress from human facial expression. The key emotions that this model used to assess stress levels are anger, fear, disgust, surprise, neutrality, sadness, and happiness. Using deep learning techniques, mainly Convolutional Neural Networks (CNNs), the proposed model accurately predicts these emotions. The CNN model acquired a prediction accuracy of 97.2%, while other machine learning models were also used to assess the performance. Based on the intensity of each observed expression, this model determines the stress level.

**Keywords** Facial expression · Emotion recognition · Mental stress · Machine learning · Deep learning · Convolutional Neural Network (CNN)

## 1 Introduction

One of the common and increasing concerns in today's modern world that impacts people from every corner is nothing but stress. Stress is the body's natural response to pressure or challenging conditions and can manifest in physical, emotional, and psychological manifestations [1]. While short-term stress can be beneficial in inspiring

N. Bhaskar (✉)
NITTE (Deemed to be University), NMAM Institute of Technology, Nitte 574110, Karnataka, India
e-mail: navbskr@gmail.com

P. Bidwai
Symbiosis Institute of Technology, Symbiosis International (Deemed University), Pune, India

K. Mahesh · B. K. Trishan · S. S. Khandelwal · N. Vishvith Shetty
Sahyadri College of Engineering and Management, Mangaluru, Karnataka, India



K. R. Madhavi et al. (eds.), *Proceedings of Sixth International Conference on Computer and Communication Technologies*, Lecture Notes in Networks and Systems 1356,
https://doi.org/10.1007/978-981-96-5238-9_36

people to face challenges, chronic or excessive stress can cause major health and well-being problems. Early recognition and management of stress prevents detrimental consequences from accumulating over time [2].

Stress detection has grown increasingly important as stress-related disorders have a pervasive influence on individuals and society. Early detection is critical in preventing stress from progressing to more significant mental health issues, such as anxiety and depression. Detecting stress in its early phases enables timely intervention, allowing individuals to adopt coping methods that promote better health outcomes. However, stress detection is complicated by various factors, including its subjective nature, a lack of awareness or acknowledgement of symptoms, and difficulty accessing necessary support and treatment resources. Overcoming these hurdles is crucial for establishing holistic stress detection and management strategies, which can improve individual resilience and contribute to healthier, more supportive communities [3].

Deep learning and machine learning improve computer intelligence, resulting in countless benefits for humans. These technologies enable developments such as increased transportation safety, personalized online suggestions, and early disease detection. They make daily tasks safer, more efficient, and convenient [4]. Deep learning's innovative application in the realm of emotional computing is the detection of human mental stress using facial expression analysis. This approach helps us to get a significant dataset through the collection of various facial expressions and the annotation of stress-related aspects. For more reliable stress detection, face landmark identification is very crucial to this method since it accurately determines the important face points and also allows exact feature extraction.

Our main intention is to predict stress levels in various areas, such as hospitals, workplaces, education institutions, etc. Using a variety of techniques like deep learning and machine learning, stress detection has been done by several models in the literature. The most efficient technique that detects stress from human facial expressions is using deep learning approaches [5]. Our proposed approach focuses on recognizing and understanding human facial expression along with stress analysis using Convolutional Neural Network (CNN)-based architecture [6]. Any individual can utilize this model to estimate their stress level with the help of a user-friendly interface. This interface allows the user to determine stress and provides them the useful information such as the cause of stress, what are the symptoms, and how they can cope with it. The automated model presented in this paper is useful in detecting and managing stress levels of individuals, no matter how much the person already knows about stress.

We have considered the following emotions in our analysis—anger, neutrality, disgust, happiness, fear, sadness, and surprise. The calculation of the stress level from the strength of these emotions is one of the major highlights of the model. The degree of weight assigned depends on how much stress has occurred. For instance, emotions that are very likely related to increased stress of fear and sadness are assigned with larger weights. The addition of the face analysis and all those weightings makes the device an effective model for forecasting the stress levels of people. As the proposed approach provides a simple and non-intrusive way to identify and manage stress using a user-friendly application, it is very easy to detect stress in various areas with assisting mental health remedies.

For the detection of stress in real-time in live videos and images, we have done the testing on multiple datasets and analysed the prediction accuracy. However, there are multiple challenges involved in this process. For determining the accuracy over wide demographic categories, the changes in facial expression have to be considered. This approach can help us in monitoring stress levels and provide mental health assistance effectively with the designed app. People can observe and manage their stress through the designed user-friendly solutions in various areas or fields.

## 2 Literature Survey

The improvements in machine learning and deep learning have gained progress in facial emotion detection and stress detection [7]. Researchers and healthcare professionals working in this domain benefit from understanding the methods and models used, as they demonstrate important techniques for detecting and managing human feelings and sentiments. Numerous studies follow a multi-step process, normally involving preprocessing, feature extraction, and classification approaches. This process often starts with carefully preparing datasets by applying scaling, data augmentation, and normalization to ensure data consistency and improve resilience. Then, advanced algorithms are used for accurate emotion detection and stress prediction. By combining these techniques, researchers can build models that not only recognize emotions but also predict stress levels accurately. This project provides solutions for real-world problems like mental health monitoring and stress management.

In the field of facial expression recognition, the development of deep learning technologies, notably CNNs, has signalled a new era of invention and discovery. Kumar et al. [8] analysed stress using machine learning models to predict psychological disorders like anxiety, depression, and stress. In their study, eight machine learning algorithms were applied to data from the online DASS42 tool to predict five different levels of severity for various psychiatric illnesses. The algorithms were divided into four categories: probabilistic, nearest neighbour, neural network, and tree-based approaches. They have used an ensemble model to predict the severity levels. Their proposed hybrid model performed better than the conventional models. They attained the maximum accuracy with the radial basis function network, which was designed with a neural network. Kottursamy et al. [9] introduced a model for monitoring consumer emotions using deep learning networks. Their work explored various machine learning algorithms for recognizing emotions, using the eXnet library. They implemented a CNN-based learning model called eXnet, whose architecture is based on parallel feature extraction. Their model reduced overfitting issues in data while keeping the model size manageable and attained better precision performance and accuracy in emotion recognition.

Real-time methods for recognizing stress, like Electroencephalography (EEG), have deepened our understanding of stress dynamics [10]. These devices offer valuable insights into stress responses as they capture physiological data in real-time, helping researchers to see how stress shows up in facial expressions. Combining EEG data with facial emotion detection algorithms allows researchers to explore the complex link between brain activity and emotional states, improving our understanding of how stress is perceived and processed. This interdisciplinary approach has strong potential for developing new stress management tools and personalized interventions to fulfil individual needs, ultimately promoting greater well-being.

The variety of methods and technologies used in facial emotion recognition and stress prediction highlights the interdisciplinary nature of this field. Continued research is done to refine current models, explore new approaches and address existing challenges. This is essential for advancing our understanding and real-world use of emotion recognition systems. These studies provide valuable insights, laying the foundation for future research and practical applications across many fields.

## 3 Methods

The initial step is to collect facial images of individuals, which will serve as the major input for the programme. These photographs go through a critical preprocessing step in which they are transformed to grayscale. This conversion makes future analysis easier by eliminating colour changes and focusing on the face traits required for stress detection. In addition, to standardize the input data for consistent model training, the photos are shrunk to a constant $48 \times 48$ pixel resolution. This scaling ensures that all facial pictures have a uniform structure, allowing the deep learning model to accurately identify and learn stress-related patterns from facial expressions. Figure 1 shows the architecture diagram for the proposed framework.

The approach for recognizing facial expressions to indicate stress is based on a general deep learning algorithm that leverages a 4-layer CNN. The deep learning model processes the input grayscale facial photos to efficiently extract hierarchical information. The CNN's first layer performs convolutional operations, which involve applying filters to input images to detect low-level characteristics like edges and textures. The pooling layers of CNN will reduce the spatial dimensions, lowering computational complexity levels and further improving the network's performance [11]. To capture the best features from the input signal, the convolutional and pooling layers are used continually. The extracted features are given to the fully connected layers, which use the feature maps to patterns to detect the stress level based on the training data. We have designed a 4-layer CNN architecture to balance computational efficiency and model complexity. The developed model includes data preprocessing and uses the CNN model to build a dedicated, accurate system that can detect stress in various real-life situations by examining the facial expressions of persons.

Initially, we have collected large samples of facial images to make the model effective across different demographic groups as there are variations in the facial

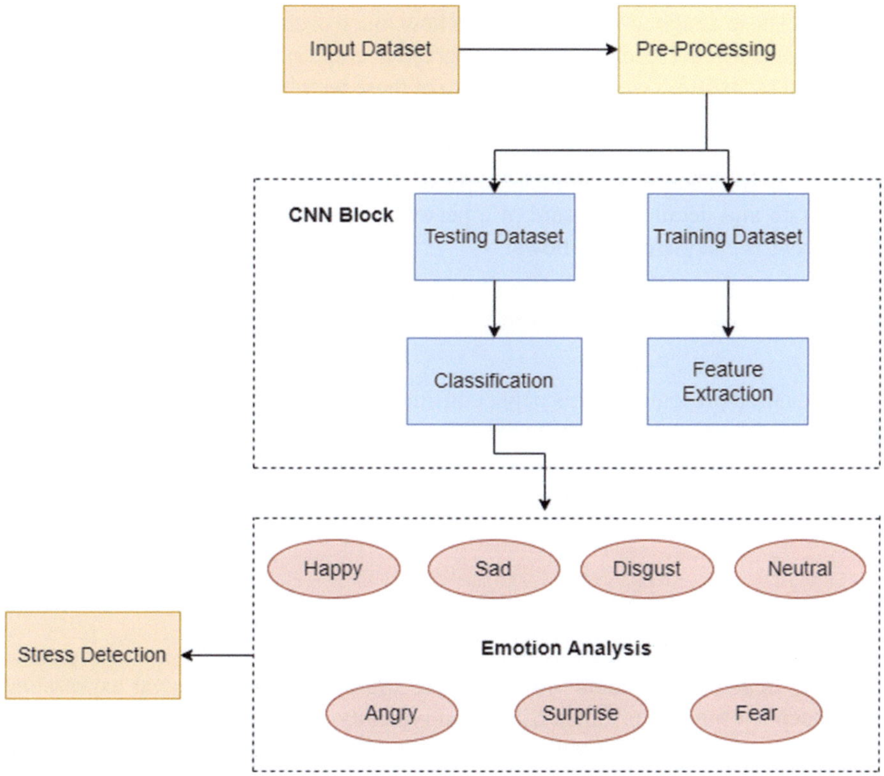


**Fig. 1** Architecture diagram of the proposed model for stress and emotion analysis

appearance of people. The collected images are converted to grayscale, which simplifies the data by reducing colour variations. It also helps the system to focus on key facial features more clearly. The images are resized to a standard size to make the input image size consistent, which improves the model's learning cycle. The prominent facial features are then extracted to help the framework recognize the expressions. By focusing on the best traits, the model becomes better at identifying the emotions of people. Using these features, the system performs an accurate analysis to classify emotions in each image by recognizing patterns in facial expressions that indicate emotions like happiness, sadness, or anger. With this prediction, the model can decide on the person's stress level, providing information into their emotional state and possibly indicating when support is needed.

We have derived an expression to calculate total stress levels by considering variations in emotions and their associated stress weights. Each facial expression—happiness, sadness, anger, fear, surprise, disgust, and neutrality—is assigned a weight based on how much it is going to affect the stress of a person. Emotions like sadness and fear normally receive higher stress weights as they are strongly linked to stress, while happiness and neutrality receive lower weights as their impact on stress is minimal. These assigned weights indicate how much each emotion contributes to the overall stress level. To determine the weighted average stress score, each emotion is multiplied by its stress weight, and the sum of these weighted values is divided by the assigned total weights. The emotions with higher stress weights have more impact on the total stress score, while those with lower weights have less effect. With this weighted approach along with facial expression analysis, the proposed system gives an accurate and detailed measure of a person's stress level. The average weighted stress score is calculated as follows:

$$\text{Weighted Stress Score} = \frac{\sum_{i=1}^{n} \text{Expression}_i \times \text{Weight}_i}{\sum_{i=1}^{n} \text{Weight}_i} \tag{1}$$

The method presented in this paper considered the intensity of each facial expression and its related stress weight to provide a complete measure of stress levels. With this approach, we can more accurately assess a person's stress level. This allows the system to provide quick insights, enabling early support and help when needed.

## 4 Results and Discussions

We have created an efficient way to detect stress by studying facial expressions using deep learning algorithms. Stress can typically occur which can harm both the physical and emotional well-being. This study aimed to understand facial expressions along with identifying the cause of stress using modern machine learning algorithms. This suggested stress detection approach has potential uses in various surroundings including companies, educational institutions, and families.

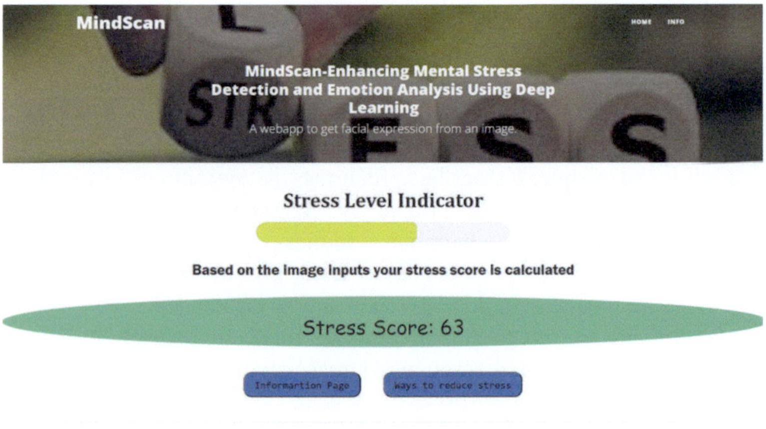


**Fig. 2** Screenshot of the designed user interface

**Table 1** Performance comparison of the proposed CNN model with other models

| Model | Accuracy (%) | Sensitivity | Specificity | Precision | Error rate |
|---|---|---|---|---|---|
| Decision tree | 85.52 | 0.863 | 0.847 | 0.844 | 0.145 |
| Random forest | 93.1 | 0.939 | 0.923 | 0.922 | 0.069 |
| SVM | 94.21 | 0.953 | 0.931 | 0.93 | 0.058 |
| Proposed CNN model | 97.2 | 0.979 | 0.964 | 0.964 | 0.028 |

We have designed a user-friendly interface that allows users to upload pictures for analysis. After uploading the image, the system analyses the user's facial expressions and provides a fast assessment of their stress level. Figure 2 shows a screenshot of the planned user interface. This stress level tool helps give fast feedback to customers on the current calculated stress levels based on their facial expressions. This user interface also provides us with two more sections such as "Information Page" and "Ways to Reduce Stress." The "Information Page" contains in-depth information for stress management, while the "Ways to Reduce Stress" part contains several plans and recommendations for efficiently dealing with stress. This user-friendly interface was developed using modern web technologies such as HTML, CSS, and JavaScript for the designing of the front end ensuring that it provides a smooth user experience [12]. This tool also integrated a back-end system which was powered by deep learning algorithms to process facial images and analyse them to calculate the stress levels. The responsiveness ensured compatibility across different devices making it accessible for the users who are in different environments.

This proposed CNN deep learning model was compared to traditional machine learning methods to determine how it would perform on our dataset. Traditional models for comparison include the Support Vector Machine (SVM), Random Forest, and Decision Trees [13]. All models were evaluated for key performance indicators, such as accuracy, sensitivity, specificity, precision, and prediction error rate [14]. The findings are summarized in Table 1. These performance measures provide a complete evaluation of each model's predictive capabilities, taking into account parameters such as overall accuracy, model sensitivity to both positive and negative instances, precision, the trade-off between precision and recall, and prediction error rate. This analysis provides a comprehensive assessment of model performance across several key criteria. The proposed model achieved the highest classification accuracy of 97.2% among all the models evaluated in this study. Random Forest classifiers demonstrated superior performance compared to individual Decision Trees, leveraging the ensemble approach to combine the predictive strengths of multiple Decision Trees. However, SVM outperformed both Decision Trees and Random Forest classifiers, showcasing its effectiveness in classification tasks.

The confusion matrix of the proposed deep learning model is presented in Fig. 3. This matrix provides insights into the model's performance by not only illustrating overall accuracy but also highlighting specific areas where the model encounters difficulties. It acts as a visual tool for assessing how effectively the model classifies

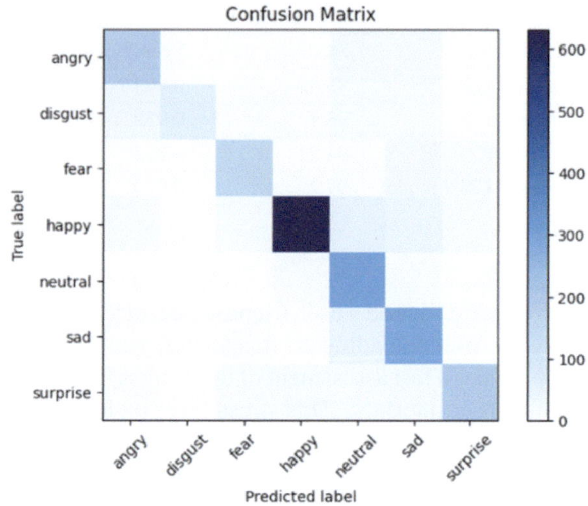


**Fig. 3** Confusion matrix of the proposed prediction model

various emotional states [15]. In this matrix, each row corresponds to the actual class labels, representing the true emotional states in the dataset, while each column corresponds to the predicted class labels, representing the emotions predicted by the model. The diagonal cells of the matrix show correct classifications, where the model accurately identifies the emotional state. Darker shades in these diagonal cells indicate a higher accuracy of prediction. Conversely, the off-diagonal cells represent misclassifications, where the model incorrectly classifies emotions. Lighter shades in these cells highlight areas of confusion between different emotional states, signalling opportunities for model improvement. Analysis results demonstrate that facial expressions can be a reliable indicator of stress. This system could be especially useful in environments where stress is prevalent, such as workplaces and schools. The platform helps people to better mental well-being by providing expert stress management suggestions and monitoring facial clues.

## 5 Conclusion

The approach presented in this paper brings a significant improvement in mental health support by using facial recognition to detect stress levels in real-time scenarios. By applying advanced machine learning methods and a CNN-based model, the system offers users insights into their emotional state, which can help them in better stress management and enhance their overall well-being. The model with CNN attained a remarkable accuracy of 97.2%. The system performed better than similar systems introduced previously by researchers. Apart from detecting stress, the module can also act as a helpful resource by providing essential information on stress and practical ways to overcome stress. By linking innovative technology with a user-centred approach, this proposed approach not only extends stress detection capabilities but also reflects a deep dedication to the mental health of people. This achievement is not just a technical success, but it is a significant step towards comprehending and managing stress, supporting a more nourishing and more resilient community.

## References

1. O'Connor DB, Thayer JF, Vedhara K (2021) Stress and health: a review of psychobiological processes. Annu Rev Psychol 72(1):663–688
2. Varma P, Junge M, Meaklim H, Jackson ML (2021) Younger people are more vulnerable to stress, anxiety and depression during COVID-19 pandemic: a global cross-sectional survey. Prog Neuro-Psychopharmacol Biol Psychiatry 109:110236
3. Slavich GM (2016) Life stress and health: a review of conceptual issues and recent findings. Teach Psychol 43(4):346–355
4. Bhaskar N, Tupe-Waghmare P, Nikam SS, Khedkar R (2023) Computer-aided automated detection of kidney disease using supervised learning technique. Int J Electr Comput Eng (IJECE) 13(5):5932–5941
5. Zainudin Z, Hasan S, Shamsuddin SM, Argawal S (2021) Stress detection using machine learning and deep learning. In: Journal of physics: conference series, vol 1997, no 1. IOP Publishing, p 012019
6. Bhaskar N, Bairagi V, Munot MV, Gaikwad KM, Jadhav ST (2023) Automated COVID-19 detection from exhaled human breath using CNN-CatBoost ensemble model. IEEE Sens Lett
7. Jaiswal A, Raju AK, Deb S (2020) Facial emotion detection using deep learning. In: 2020 International conference for emerging technology (INCET). IEEE, pp 1–5
8. Kumar P, Garg S, Garg A (2020) Assessment of anxiety, depression and stress using machine learning models. Procedia Comput Sci 171:1989–1998
9. Kottursamy K (2021) A review on finding efficient approach to detect customer emotion analysis using deep learning analysis. J Trends Comput Sci Smart Technol 3(2):95–113
10. Goodman RN, Rietschel JC, Lo LC, Costanzo ME, Hatfield BD (2013) Stress, emotion regulation and cognitive performance: the predictive contributions of trait and state relative frontal EEG alpha asymmetry. Int J Psychophysiol 87(2):115–123
11. Bhaskar N, Bairagi V, Boonchieng E, Munot MV (2023) Automated detection of diabetes from exhaled human breath using deep hybrid architecture. IEEE Access 11:51712–51722
12. Alawar MW, Naser SSA (2017) CSS-Tutor: an intelligent tutoring system for CSS and HTML. Int J Acad Res Dev 2(1):94–98
13. AlKheder S, AlRukaibi F, Aiash A (2023) Support vector machine (SVM), random forest (RF), artificial neural network (ANN) and Bayesian network for prediction and analysis of GCC traffic accidents. J Ambient Intell Hum Comput 14(6):7331–7339
14. Bhaskar N, Suchetha M (2020) Analysis of salivary components as non-invasive biomarkers for monitoring chronic kidney disease. Int J Med Eng Inf 12(2):95–107
15. Deng X, Liu Q, Deng Y, Mahadevan S (2016) An improved method to construct basic probability assignment based on the confusion matrix for classification problem. Inf Sci 340:250–261

# Self-supervised Histopathology Image Segmentation Using Transformer Networks with Attention Mechanism

**Veeresh Dachepalli, Gavini Sreelatha, Jayavardhanarao Sahukaru, Voruganti Naresh Kumar, J. Avanija, and Chengamma Chitteti**

**Abstract** This paper presents a significant approach utilizing self-supervised transformer-based framework for histopathology image segmentation with self-attention mechanism to perform accurate segmentation of multi-organ histopathology images. By utilizing the self-supervised learning, the method learns efficiently the representation of varied features from unlabeled data. The proposed framework uses MonuSeg data from Kaggle for processing. Results reveal that the proposed approach performs better than conventional approaches by handling diverse nuclei shapes and overlapping structures. This method also enhances segmentation accuracy and computational efficiency. The experimentation reveals that the proposed self-learning transformer-based model with attention mechanism for image segmentation is more efficient than the existing methods.

**Keywords** Self-supervised learning · Image segmentation · Attention mechanism · Transformer networks · Deep learning

---

V. Dachepalli
Elemica Inc. CrossPoint at Valley Forge, Wayne, PA, USA

G. Sreelatha (✉)
Department of Information Technology, Stanley College of Engineering and Technology for Women, Hyderabad, India
e-mail: sreelathaprince13@gmail.com

J. Sahukaru
Aditya Institute of Technology and Management, Tekkali, India

V. N. Kumar
Department of Computer Science Engineering, CMR Technical Campus, Hyderabad, Telangana, India
e-mail: nareshkumar99890@gmail.com

*Present Address:*
J. Avanija · C. Chitteti
School of Computing, Mohan Babu University, Tirupati, India



K. R. Madhavi et al. (eds.), *Proceedings of Sixth International Conference on Computer and Communication Technologies*, Lecture Notes in Networks and Systems 1356,
https://doi.org/10.1007/978-981-96-5238-9_37

# 1 Introduction

Histopathology is an important field in diagnostics, especially for cancer diagnosis. Automating the process of histopathological image segmentation can improve the accuracy of diagnosis. Traditional methods of segmentation consider only labeled data for processing which becomes time time-consuming process in medical image diagnosis. Self-supervised learning method can overcome this limitation by making the model learn through unlabeled data. The major benefit of this method is to overcome the complexity of labeling the large volume of histopathological image data. Also, the usage of deep learning models such as convolutional neural networks faces the challenge of handling complex variations in histopathological images. This makes the image segmentation of histopathological data a complex task. Recently, the utilization of transformer networks has become a powerful method for computer vision based tasks due to their attention mechanism providing the transparency for diagnosis modeling both local and global features of the image. Considering these aspects, a self-supervised learning method integrating transformers with self-attention mechanism for image segmentation of histopathological images has been proposed. The proposed model not only reduces the dependency on the labeled data but also improves the ability of the model considering relevant regions of the image samples for accurate segmentation.

# 2 Literature Review

Traditional image processing techniques such as thresholding, watershed algorithms, and region growing can be used only for basic segmentation and cannot handle variability in tissue structure, staining, and overlapping nuclei, available in complex histopathological images [1]. The use of convolutional neural networks created a turning point in medical image segmentation [2]. U-Net [3] became popular due to its ability to handle both local and global contexts. However, the model has limitations in handling large-scale variability in nuclear size, shape, and density, especially in histopathology images. Also, the requirement of large annotated datasets for training becomes a major challenge. To address the issues of labeled data self-supervised learning method [4] has been used to work with unlabeled data especially while working with complex histopathological images are annotation becomes a time-consuming process [5].

Transformers [6] were utilized initially for natural language tasks, and now these methods are used for image segmentation using the self-attention method of vision transformers to model both local and global dependencies overcoming the challenge faced by convolutional networks [8]. Transformer-based model in histopathological image segmentation was efficient considering different regions of images while performing the segmentation dynamically. Hybrid methods such as TransUNet [7–10] integrating CNNs and Transformers generated better performance, particularly in performing medical image segmentation [11]. The utilization of attention mechanisms will enhance the performance of the model concentrating on important regions of the image. Attention-based models such as attention U-Net [12] and other relevant models have shown better segmentation accuracy, especially in tasks with complex regions of interest.

## 3 Methodology

This section briefs about the steps of self-supervised learning-based transformer model with self-attention mechanism to perform image segmentation task. The steps of self-supervised transformer [13] model for performing image segmentation are specified in Fig. 1. The dataset samples to perform image segmentation were gathered from multi-organ nuclei segmentation (MonuSeg) dataset from Kaggle [14]. The raw data considered for analysis will be histopathology images taken from tissue samples. Since these image samples are large, segmentation is performed to identify and isolate specific region of interest [15]. Next step is to perform preprocessing for the images such as normalization, data augmentation, and image resizing. Self-supervised learning with transformer model [16] is then applied to perform image segmentation. This task is performed by applying preprocessing such as contrastive learning and mask image modeling. Then, the transformer encoder will perform the task of extracting the required features. The encoder uses self-attention mechanism [7] to model different regions of the image considering their spatial proximity. In histopathology images, this process helps in capturing the relationships between different cell tissues. The transformer architecture takes the entire image as input in the form of image patches and produces feature maps as output.

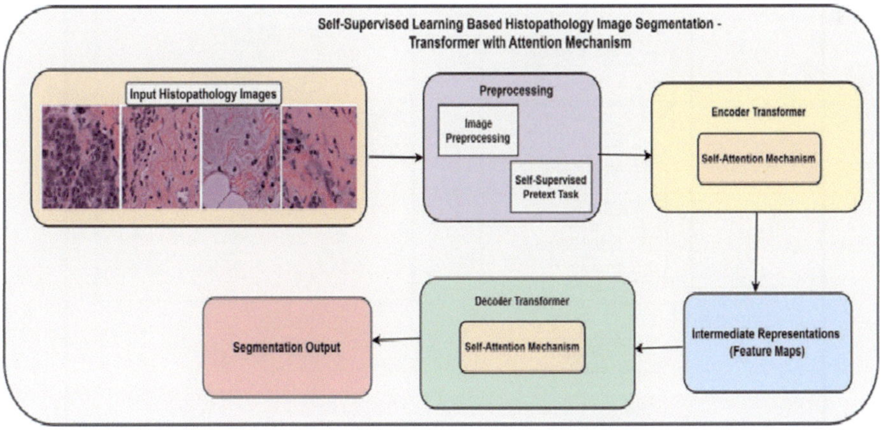


**Fig. 1** Image segmentation utilizing self-supervised learning transformer approach

The output of the encoder will be feature maps termed as intermediate representations. These feature maps contain learned features which help in recognizing important patterns during segmentation tasks. The representation will be in the form of high dimensional feature space vector. The next step is to perform decoding through decoder transformer. During this process, the feature maps will be reconstructed into segmentation maps with the help of a self-attention mechanism. This process ensures the communication between the feature maps preserving the global context.

Upsampling is used by the decoder transformer to bring the feature maps to the original resolution [17–22]. The final output after this process will be the segmented image where each pixel is classified into either healthy tissues or tumor cells. The segmentation map provides boundaries and classification for different regions, which can be utilized for further analysis during diagnosis. The transformer encoder architecture with a self-attention mechanism is shown in Fig. 2. The input sequence is set of tokens to pass through the network. The embedding layer transforms the input into dense vectors to capture the semantic meaning. Then positional encoding is added to incorporate sequence information. Self-attention mechanism is the important component which computes the attention score for each tokens [23, 24]. This process is performed using the vectors Query, Key, and Value as specified in Fig. 2. Residual connection adds the input back to the output before passing to the next layer. Fully connected network is then applied to position through nonlinear activation function.

The proposed transformer network is trained using a histopathology image dataset. The model learns using these images and maps to segmented output using self-supervised learning method which helps in feature extraction. The training process also involves minimizing the loss function by relating the dissimilarity between the predicted and ground truth values. The final output of the model will be segmented

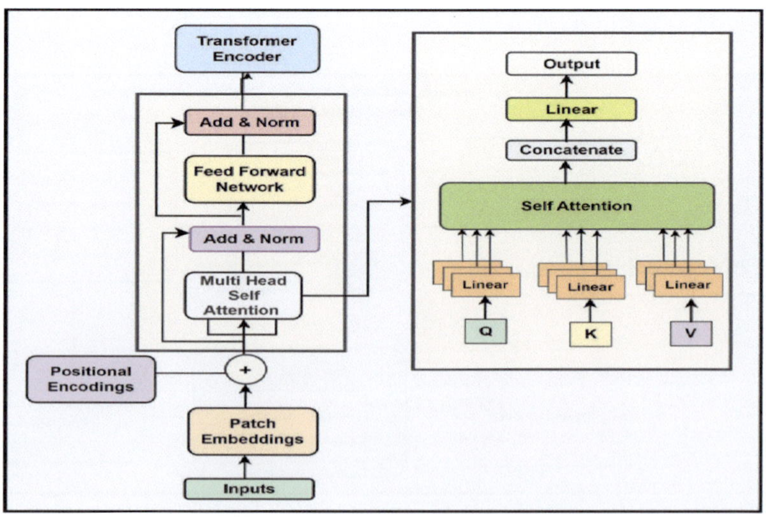


**Fig. 2** Transformer encoder architecture with self-attention mechanism

images by applying self-supervised learning-based transformer architecture with self-attention mechanism.

The steps of the proposed self-supervised learning transformer with self-attention algorithm are specified below.

**Algorithm: Self-supervised Learning Transformer with Self-attention for Image Segmentation**

**Input: $X$:**

- Set of input histopathology images $X = \{x_1, x_2 \ldots x_n\}$.
- Unlabeled data $X$.

**Output: $S$:**

- Segmentation maps for images $S = \{s_1, s_2 \ldots s_n\}$.

**Step 1: Preprocessing**

- For each image $x_i \in X$, scale the pictures applying normalization considering the range [0, 1].
- Apply transformations such as rotations and flips to create augmented images.
- Resize images to a predefined resolution **hxw**.

**Step 2: Self-Supervised Pre Text Task**

- Create augmented pairs ($x_i$, $x_i'$) and perform feature extraction and feature embeddings using Eq. (1).

$$f(x_i) = z_i, f(x_i') = z_i' \tag{1}$$

**Step 3: Train Transformer Encoder**

- Pass the images $x_i$ and $x_i'$ through the transformer encoder to obtain feature embeddings as specified in Eq. (2).

$$z_i = E(x_i), z_i' = E(x_i') \tag{2}$$

- Compute self-attention scores using Eqs. (3, 4).

$$\text{Attention}(q, k, v) = \text{softmax}\left(\frac{qk^T}{\sqrt{d_k}}\right)v \tag{3}$$

$$q = Xw_q, k = Xw_k, v = Xw_v \tag{4}$$

where $q$ represents weight matrices for query, key, value pairs and the dimensionality of vectors represented by $d_k$.

**Table 1** Evaluation of self-supervised transformer approach for image segmentation

| Method | Dice coefficient | Sensitivity | Specificity | Accuracy | F1-score |
|---|---|---|---|---|---|
| CNN | 0.76 | 0.80 | 0.86 | 0.86 | 0.83 |
| Deep attention network | 0.79 | 0.82 | 0.90 | 0.84 | 0.80 |
| U-Net | 0.83 | 0.84 | 0.85 | 0.85 | 0.81 |
| Proposed self-supervised transformer model | 0.87 | 0.89 | 0.91 | 0.88 | 0.85 |

**Step 4: Feature Extraction**

- Obtain feature maps $f_i$ from the encoder using Eq. (5)

$$f_i = E(x_i) \tag{5}$$

**Step 5: Decoder for Segmentation**

- Pass Feature Maps through Decoder *D* and perform self-attention in decoder as specified in Eq. (3)
- Use bilinear upsampling or transposed convolutions to bring feature maps to original resolution as specified in Eq. (6).

$$s_i = D(x_i) \tag{6}$$

- The output $s_i$ is the predicted segmentation map.

Final step is to perform the evaluation of model utilizing the performance metrics as specified in Table 1. Thus, self-supervised learning transformer with attention mechanism approach enhances the process of image segmentation based on the data samples.

## 4 Results and Discussions

The dataset used in the proposed system is multi-organ nuclei (MoNu) segmentation dataset from Kaggle. It consists of Hematoxylin and Eosin (H&E) stained histopathology images of multiple organs extracted from human tissues. For each image, nuclei are manually annotated by expert pathologists. Annotations include pixel-level masks for individual nuclei, serving as the ground truth which helps in evaluating segmentation models. This dataset is used for nuclei segmentation helping in disease diagnosis such as cancer detection and cell classification. Different organs have varying nuclear shapes, sizes, and densities, which make the segmentation task challenging. This dataset is particularly used for training segmentation algorithms to perform medical image segmentation.

Different tumor types exhibit distinct characteristics in imaging, making the dataset diverse and challenging. This dataset serves as a benchmark for developing and evaluating advanced algorithms in brain tumor segmentation. MRI images in the BRATS dataset are typically high-resolution, allowing for detailed examination of brain structures and abnormalities. The sample histopathology image, ground truth, and predicted mask are specified in Fig. 3. Ground truth data are important in image segmentation tasks as they reveal the standard against which algorithmic outputs are measured, considering the accuracy of segmentation results. Convergence of the proposed model signifies the algorithm's ability to refine the parameters through iterations in order to improve segmentation task. The convergence of the self-supervised learning transformer model based on number of epochs considering the loss function during training phase is mentioned in Fig. 4.

This method enhances segmentation accuracy and helps in applications like medical diagnosis and more. Performance evaluation of self-supervised transformer model with the evaluation metrics is specified in Table 1.

The details in Table 1 demonstrate that the self-supervised transformer model performs well compared to existing approaches to perform image segmentation

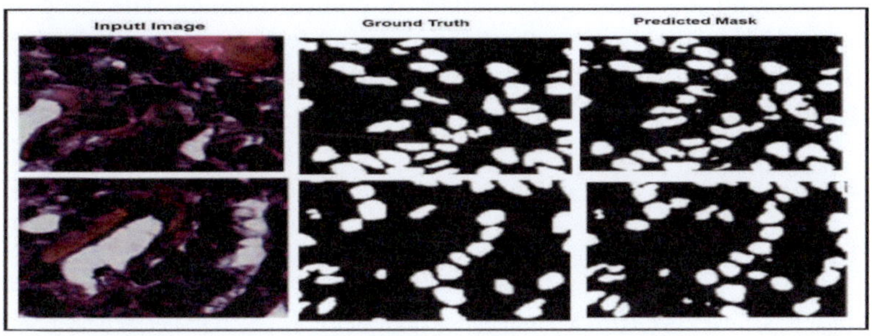


**Fig. 3** Sample histopathology image and ground truth prediction

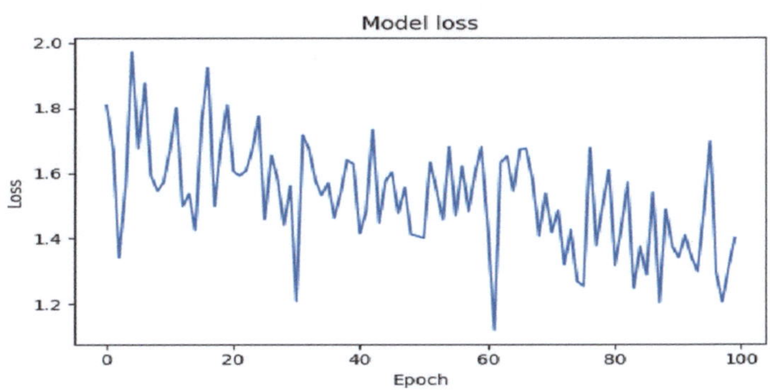


**Fig. 4** Self-supervised transformer model convergence during training based on epochs

tasks. In the proposed model, experimentation was performed utilizing transformer architecture with self-attention mechanism for medical image segmentation.

## 5 Conclusion

In this work, a self-supervised transformer-based histopathology image segmentation framework, utilizing the attention mechanism for accurate segmentation of nuclei in histopathology images, has been proposed. Utilizing the MonuSeg dataset, the proposed model demonstrates the ability to learn about important feature representations from unlabeled data by self-supervised learning to efficient segmentation without depending on large amounts of annotated data. By incorporating transformer networks, the model's ability was enhanced to capture long-range dependencies in complex histopathology images, which is important for accurately segmenting nuclei that are diverse in shape, size, and structure. The results reveal that the proposed method performs better than the existing segmentation approaches in handling varied tissue samples. The use of self-supervised learning reduces the requirement of manual annotation making this technique suitable for clinical settings. The integration of self-supervised learning along with attention-based feature extraction leads to efficient histopathology image analysis. Future work focuses on integrating multiscale learning and weak supervised learning to improve the segmentation accuracy.

## References

1. Sreedhar B, Kumar MS (2020) A comparative study of melanoma skin cancer detection in traditional and current image processing techniques. In: 2020 IEEE fourth international conference on I-SMAC (IoT in social, mobile, analytics and cloud) (I-SMAC), pp 654–658
2. Hegde RB, Prasad K, Hebbar H, Singh BMK (2019) Feature extraction using traditional image processing and convolutional neural network methods to classify white blood cells: a study. Austral Phys Eng Sci Med 42:627–638
3. Rosca CM (2023) Comparative analysis of object classification algorithms: traditional image processing versus artificial intelligence—based approach. Rom J Pet Gas Technol 169–180
4. Ronneberger O, Fischer P, Brox T (2015) U-net: convolutional networks for biomedical image segmentation. In: Medical image computing and computer-assisted intervention–MICCAI 2015: 18th international conference, Munich, Germany, October 5–9, proceedings, part III 18, pp 234–241
5. Xie J, Liu R, Luttrell IV J, Zhang C (2019) Deep learning based analysis of histopathological images of breast cancer. Front Genet 10:80–92
6. Raghu M, Unterthiner T, Kornblith S, Zhang C, Dosovitskiy A (2021) Do vision transformers see like convolutional neural networks? Adv Neural Inf Process Syst 34:12116–12128
7. Chen J, Lu Y, Yu Q, Luo X, Adeli E, Wang Y, Lu L, Yuille AL, Zhou Y (2021) Transunet: transformers make strong encoders for medical image segmentation. arXiv preprint arXiv: 2102.04306
8. Chen B, Liu Y, Zhang Z, Lu G, Kong AWK (2023) Transattunet: multi-level attention-guided u-net with transformer for medical image segmentation. IEEE Trans Emerg Top Comput Intell 725–738

9. Oktay O, Nanavati J, Schwaighofer A, Carter D, Bristow M, Tanno R, Jena R, Barnett G, Noble D, Rimmer Y, Glocker B (2020) Evaluation of deep learning to augment image-guided radiotherapy for head and neck and prostate cancers. JAMA Netw Open 3(11):e2027426–e2027426
10. Zhai X, Oliver A, Kolesnikov A, Beyer L (2019) S4l: self-supervised semi-supervised learning. In: Proceedings of the IEEE/CVF international conference on computer vision, pp 1476–1485
11. Misra I, Maaten LVD (2020) Self-supervised learning of pretext-invariant representations. In: Proceedings of the IEEE/CVF conference on computer vision and pattern recognition, pp 6707–6717
12. Liu X, Zhang F, Hou Z, Mian L, Wang Z, Zhang J, Tang J (2021) Self-supervised learning: generative or contrastive. IEEE Trans Knowl Data Eng 35(1):857–876
13. Kalapos A, Gyires-Tóth B (2022) Self-supervised pretraining for 2D medical image segmentation. In: European conference on computer vision, pp 472–484
14. Ziegler A, Asano YM (2022) Self-supervised learning of object parts for semantic segmentation. In Proceedings of the IEEE/CVF conference on computer vision and pattern recognition, pp 14502–14511
15. Gao Y, Zhou M, Metaxas DN (2021) UTNet: a hybrid transformer architecture for medical image segmentation. In: Medical image computing and computer assisted intervention–MICCAI 2021: 24th international conference, Strasbourg, France, September 27–October 1, 2021, Proceedings, Part III, vol 24, pp 61–71
16. Liao Z, Xu K, Fan N (2022) Swin transformer assisted prior attention network for medical image segmentation. In: Proceedings of the 8th international conference on computing and artificial intelligence, pp 491–497
17. Azad R, Al-Antary MT, Heidari M, Merhof D (2022) Transnorm: transformer provides a strong spatial normalization mechanism for a deep segmentation model. IEEE Access 10:108205–108215
18. Sabeena M, Abraham L (2021) Digital image forensic using deep flower pollination with adaptive Harris hawk optimization. Multimedia Tools Appl 80(17):26333–26355
19. Fu Z, Li J, Hua Z (2022) DEAU-Net: attention networks based on dual encoder for medical image segmentation. Comput Biol Med 150:106197
20. Avanija J, Sunitha G, Reddy Madhavi K, Hitesh Sai Vittal R (2021) An automated approach for detection of intracranial haemorrhage using DenseNets. In: Data engineering and communication technology: proceedings of ICDECT 2020. Springer, Singapore, pp 611–619
21. Reddy KR, Dhuli R (2023) A novel lightweight CNN architecture for the diagnosis of brain tumors using MR images. Diagnostics 13(2):312
22. Arunachalam R, Sunitha G, Shukla SK, Urooj S, Rawat S (2023) A smart Alzheimer's patient monitoring system with IoT-assisted technology through enhanced deep learning approach. Knowl Inf Syst 1–39
23. Sivajyothi M, Madhavi KR, Devi VVV, Reddy AB, Tangudu N, Avanija J (2024) Transformer network-based image segmentation using hybrid flower pollination optimization. In: 2024 11th International conference on computing for sustainable global development (INDIACom), pp 931–935
24. Hatamizadeh A, Tang Y, Nath V, Yang D, Myronenko A, Landman B, Roth HR, Xu D (2022) Unetr: transformers for 3D medical image segmentation. In: Proceedings of the IEEE/CVF winter conference on applications of computer vision, pp 574–584

# Enhancing Object Detection with EfficientDet Using Moth Flame Optimization

**Nagaraj Ramrao, Konduru Venkateswara Raju, K. Reddy Madhavi, J. Aswini, I. Siva, and Vijaya R. Varma Pothuri**

**Abstract** This paper introduces an efficient approach to enhance object detection by utilizing the EfficientDet model and optimizing its hyperparameters using moth flame optimization (MFO). The proposed method performs fine-tuning of important parameters such as learning rate, batch size, and anchor scales to improve the accuracy of object detection over existing techniques. Integration of moth flame optimization (MFO) with EfficientDet framework allowed for dynamic hyperparameter tuning, improving the model's detection capabilities without increasing computational costs. The proposed model is evaluated on the COCO dataset, and the results reveal that the performance of the model in detecting objects across varying scales and aspect ratios is better.

**Keywords** EfficientDet · Object detection · Moth flame optimization · Mean average precision (mAP) · Feature pyramid network (FPN)

---

N. Ramrao
Department of ECE and Vice Chancellor, Mohan Babu University, Tirupati, India

K. V. Raju (✉)
Department of Mathematics, Sri Venkateswara College of Engineering (Autonomous), Karakambodi Road, Tirupati, India
e-mail: venky.sakku@gmail.com

K. Reddy Madhavi
AI&ML, School of Computing, Mohan Babu University, Tirupati, India

J. Aswini
Department of Artificial Intelligence and Machine Learning, Saveetha Engineering College (Autonomous), Affiliated to Anna University, Chennai, India

I. Siva
Department of Mechanical Engineering, School of Engineering, Mohan Babu University, Tirupati, India

V. R. V. Pothuri
Salesforce Inc, San Francisco, USA

 
K. R. Madhavi et al. (eds.), *Proceedings of Sixth International Conference on Computer and Communication Technologies*, Lecture Notes in Networks and Systems 1356,
https://doi.org/10.1007/978-981-96-5238-9_38

## 1 Introduction

Object detection is considered as an important task in computer vision with many applications such as robotics, surveillance, and autonomous vehicles. Traditional methods such as RCNN and Yolo are efficient, but the computation cost is high. Considering this perspective and to balance accuracy and efficiency, EffieientDet a scalable architecture has been used in the proposed model to perform the task of object detection efficiently. The model is dependent on the careful fine-tuning of the hyperparameters for better results with different datasets. This challenge highlights the utilization of dynamic optimization technique searching for best configuration. Moth flame optimization (MFO) algorithm with the capability to explore complex search space dynamically has been used by the proposed model.

This paper introduces an efficient model integrating EfficientDet with moth flame optimization (MFO) to efficiently perform object detection. MFO is considered as a natured that mimics the navigation of moths toward light sources, making it to explore the search space efficiently. The proposed model uses COCO dataset, one of the most widely used benchmarks for object detection. Aim of this work is to optimize the framework and hyperparameters to achieve efficient results utilizing MFO algorithm. Fusing MFO with EfficientDet leads to develop an effective model that can automatically perform fine-tuning of its parameters, leading to improved object detection accuracy.

## 2 Background Work

This section reviews the existing literature related to object detection and deep learning models for object detection using nature inspired optimization methods. Object detection has evolved in all the fields recently. The deep learning models like recurrent neural network and faster RCNN were used for object detection but computationally expensive. Then deep learning architectures such as SSD and YOLO [1] were utilized to enhance the performance of the task. YOLO's real-time performance made it efficient for applications requiring fast decision-making, but leading to the trade-off between accuracy and speed. YOLOs versions 3 and 4 [2] achieved better performance overcoming these issues. Single shot multibox detector (SSD) [3] is also termed as an efficient framework for object detection in real-time applications by predicting bounding boxes at multiple scales. However, these models require high computational resources to perform object detection efficiently [4].

EfficientDet model [5] was designed to address these issues through a scalable detection model to balance between accuracy and efficiency. Based on a compound scaling approach, EfficientDet uses EfficientNet and fuses it with a bidirectional feature pyramid network (BiFPN) [6] to perform the task of feature fusion. Although this method is highly efficient, the performance still relies on hyperparameter tuning and scaling of architecture. Optimization algorithms were utilized much to perform hyperparameter tuning to improve the performance of the machine learning and deep learning models [7, 8]. Swarm intelligence algorithms like particle swarm optimization (PSO) and genetic algorithms (GA) [9] were utilized much in optimizing neural network architectures to optimize parameters such as learning rates and feature scaling but they face challenges like slow convergence and getting trapped into local optima [10]. The algorithms like ant colony optimization, cuckoo search, and differential evolution have also been used to optimize the models like SSD and YOLO [11]. These algorithms require high computational cost and can lead to suboptimal performance of the model in high dimensional space. EfficientDet model was designed to address these issues through a scalable detection model to balance between accuracy and efficiency [12]. Based on a compound scaling approach, EfficientDet uses EfficientNet and fuses it with a bidirectional feature pyramid network (BiFPN) to perform feature fusion [13–15]. Although this method is highly efficient, the performance still relies on hyperparameter tuning [16]. Moth flame optimization can be integrated with such models leads to balance exploration of solution space with better convergence and avoiding local minima with its spiral search pattern. MFO [17] has demonstrated its efficiency making it as an ideal method to optimize deep learning models such as EfficientDet, where the state-of-the-art methods face issues like adaptability and scalability.

Considering these aspects, the proposed model overcomes the issues by enhancing the performance of EfficientDet model utilizing moth flame optimization method to perform efficient object detection [18]. The COCO dataset with its diversity acts as a benchmark to evaluate the performance of the proposed approach [19].

## 3 Methodology

This section briefs about the steps of the EfficientDet architecture to perform object detection utilizing moth flame optimization for enhancing the efficiency of object detection mechanism. EfficientDet provides a balance of accuracy and efficiency and is combined with MFO to fine-tune key hyperparameters, improving accuracy of detection on considered dataset. EfficientDet model for object detection consists of the components such as EfficientDet backbone, bidirectional feature pyramid network (BiFPN), and detection heads for bounding box classification [20, 21]. The steps for the proposed method to perform object detection are specified in Fig. 1.

The initial step is to gather dataset consisting of image samples to perform object detection. Proposed approach utilizes COCO dataset from kaggle for processing. Then, data preparation step involves preprocessing the data samples using methods such as image resizing and normalization, to provide better samples used for training [22]. The preprocessed samples are given as input to the EfficientDet model which acts as the backbone for feature extraction process [23]. EfficientDet backbone consists of the convolution layers with max pooling for downsampling. High-level sematic features will be extracted by mobile inverted bottleneck convolution layers [24]. This model scales the network depth, width, and resolution using compound

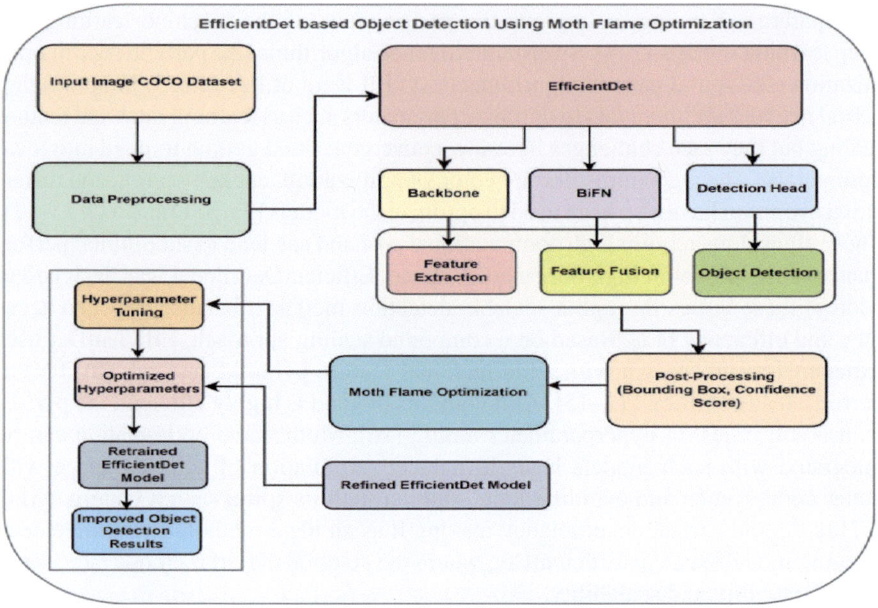


**Fig. 1** Object detection framework using EfficientDet and moth flame optimization

scaling and generates multiscale feature maps F. Then BiFN is applied to fuse the features at different levels of EfficientDet backbone. This process is carried out using multiscale feature fusion in bidirectional manner for detecting objects with varied sizes.

The proposed EfficientDet model utilizes two special heads to perform object detection, bounding box regression head to capture the coordinates of the object, and bounding box classification head to classify the objects within each bounding box. Then postprocessing is performed to through non-maximum suppression to remove overlapping bounding boxes and retain confident detections. This mechanism is performed by applying a threshold value. The objects with scores greater than the threshold will be retained. The moth flame optimization integrated with the EfficientDet model is then used to optimize the hyperparameters such as learning rate, feature scaling, batch size, network width and depth multipliers, and aspect ratios of anchor boxes. The objective function of the algorithm is the mean average precision score (mAP) to evaluate the performance of the model [25]. Best hyperparameters are selected using MFO algorithm, and then, the EfficientDet model is retrained with the optimal hyperparameters selected. This ensures that the proposed model is trained with the most effective configuration for enhancing the object detection. Final step is to perform model evaluation using performance metrics as represented in Table 1. Thus, the proposed EfficientDet + MFO approach enhances object detection mechanism for the given image samples.

**Table 1** Evaluation of EfficientDet + MFO model for object detection

| Model | Precision (%) | Recall (%) | F1-Score | mAP |
|---|---|---|---|---|
| Baseline EfficientDet | 82.7 | 80.0 | 81.4 | 78.0 |
| Proposed EfficientDet + MFO | 84.0 | 82.3 | 83.2 | 80.0 |
| YOLOv5 | 83.2 | 81.4 | 82.3 | 78.4 |
| Faster R-CNN | 83.8 | 79.6 | 81.7 | 78.7 |
| RetinaNet | 82.6 | 78.4 | 80.5 | 77.4 |

The steps of the proposed EfficientDet + Moth Flame Optimization algorithm are specified below:

**Algorithm: EfficientDet + MFO for Enhancing Object Detection**

**Input:** Dataset D, images and annotations, initial hyperparameters for MFO, number of iterations

**Output:** Optimized EfficientDet model with enhanced object detection performance.

**Step 1**: **Preprocessing**

- For each image sample in the dataset d resize and normalize the image.

**Step 2**: **Initialize Moth Flame Optimization (MFO)**

- Randomly initialize a population of moths $M = \{m_1, m_2, \ldots, m_N\}$, where $m_i$ represents the hyperparameters.
- Define the objective function as specified in Eq. 1.

$$l(m_i) = \text{EvaluationMetric}(D(\text{EfficientDet}(m_i)) \quad (1)$$

**Step 3: Optimization Loop**

- Evaluate population
  - For each moth $m_i$ train the EfficientDet model using optimized hyperparameters $m_i$.
  - Evaluate the performance of the model using objective function specified in Eq. 1.
- Update moths
  - Determine the best $m_b$ and worst moths $m_w$ based on the objective function.
  - Update the position of each moth $m_i$ using the update function specified in Eq. 2.

$$m_i^{\text{new}} = m_i^{\text{old}} + \text{updatefunction}(m_b, m_w) \quad (2)$$

- Ensure that the updated moths stay within the defined bounds.

- Select the best performing moths and update the population $m_{\text{new}}$ using Eq. 3.

$$m_{\text{new}} = \text{selectbestmoths}(\text{M}, \text{l}) \quad (3)$$

**Step 4: Model Training with Optimized Hyperparameters**

- Apply the best hyperparameters to retrain the EfficientDet model as specified in Eq. 4.

$$\text{EfficientDet}_{\text{opt}} = \text{Train}(\text{EfficientDet}, m_b) \quad (4)$$

**Step 5: Evaluate the Model**

- Evaluate the retrained EfficientDet model using validation dataset
- The final output of the model with enhanced object detection.

## 4 Results and Discussions

The dataset used in proposed approach is MS COCO (Microsoft Common Objects in Context) from Kaggle. This dataset is used to perform computer vision tasks, over 200,000 labeled images, each annotated with object bounding boxes. Also, the dataset provides segmentation masks providing annotations for object boundaries. The sample images in the dataset for model training are specified in Fig. 2. The dataset consists of more number of samples which can be used for training, validation and testing object detection tasks. COCO dataset is frequently used for training and evaluating object detection algorithms to identify objects within an image.

Performance of the EfficientDet-MFO approach is evaluated using metrics precision, recall, F1 score, and mean average precision (mAP) and specified in Table 1. mAP is the metric used to evaluate object detection algorithm summarizing the precision-recall performance across different object classes.

As specified in Table 1 proposed approach, which combines EfficientDet and moth flame optimization, performs well in object detection compared to existing methods (Fig. 3).

The generated output after performing object detection by the proposed EfficientDet + MFO model is represented in Fig. 3. Hyperparameter tuning is an iterative process that involves experimenting the task using different settings in order to find the configuration that gives the best performance on object detection task. Hyperparameter tuning of the proposed approach is mentioned in Table 2. Learning rate controls how much the model's weights are adjusted in response to the estimated error each time when the model is updated [26]. The scales of the anchor

**Fig. 2** Sample images of MS COCO dataset

boxes considered for projection are also a hyperparameter. The parameters specific to moth flame optimization algorithm, including exploration and exploitation factors are also considered for optimization. Hyperparameter optimization performed utilizing

**Table 2** Hyperparameter tuning values for the proposed EfficientDet + MFO approach

| Hyperparameter | Optimal value | Hyperparameter | Optimal value |
|---|---|---|---|
| Learning rate | 0.001 | Image size | 640 × 640 |
| Batch size | 16 | Moth flame parameters | [0.5 0.9] |
| Number of epochs | 50 | Dropout rate | 0.3 |
| Anchor scales | [32 64 128] | Weight decay | 0.0005 |

**Fig. 3** Generated output by the proposed EfficientDet + MFO model

MFO enhances the performance of object detection in terms of computational efficiency and accuracy. This process helps in balancing factors such as learning speed, complexity of the model and regularization to achieve better configuration.

## 5 Conclusion

The proposed method for object detection utilizing EfficientDet and moth flame optimization has demonstrated better performance in terms of evaluation metrics by fine-tuning the hyperparameters. This process enhances the accuracy of object detection and can handle varying object scales and aspect ratios. The proposed method demonstrates better performance than conventional methods used for object detection task. Further, the model can be extended to handle complex datasets considering high variability in object classes. Optimization techniques like swarm intelligence and other hybrid process can be explored for further optimize the performance of the hyperparameters. Fusing advanced techniques like attention mechanisms could further enhance object detection in challenging scenarios with occlusions.

## References

1. Zhao ZQ, Zheng P, Xu ST, Wu X (2019) Object detection with deep learning: a review. IEEE Trans Neural Netw Learn Syst 30(11):3212–3232
2. Wu X, Sahoo D, Hoi SC (2020) Recent advances in deep learning for object detection. Neurocomputing 396:39–64
3. Tan M, Pang R, Le QV (2020) Efficientdet: scalable and efficient object detection. In: Proceedings of the IEEE/CVF conference on computer vision and pattern recognition. pp 10781–10790
4. Wang J, Zhang T, Cheng Y, Al-Nabhan N (2021) Deep learning for object detection: a survey. Comput Syst Sci Eng 38(2)
5. Wang X (2016) Deep learning in object recognition, detection, and segmentation. Foundations and Trends® in Signal Processing 8(4):217–382
6. Ren S, He K, Girshick R, Sun J (2016) Faster R-CNN: towards real-time object detection with region proposal networks. IEEE Trans Pattern Anal Mach Intell 39(6):1137–1149
7. Xie X, Cheng G, Wang J, Yao X, Han J (2021) Oriented R-CNN for object detection. In: Proceedings of the IEEE/CVF international conference on computer vision, pp 3520–3529
8. Du J (2018) Understanding of object detection based on CNN family and YOLO. In: Journal of physics: conference series, vol 1004. IOP Publishing, pp 012029
9. Chen C, Liu MY, Tuzel O, Xiao J (2017) R-CNN for small object detection. In: Computer vision–ACCV 2016: 13th Asian conference on computer vision, Taipei, Taiwan, November 20–24, 2016, Revised Selected Papers, Part V 13, Springer International Publishing, pp 214–230
10. Koonce B, Koonce B (2021) EfficientNet. Convolutional neural networks with swift for Tensorflow: image recognition and dataset categorization, pp 109–123
11. Liu Y, Li H, Yan J, Wei F, Wang X, Tang X (2017) Recurrent scale approximation for object detection in CNN. In: Proceedings of the IEEE international conference on computer vision, pp 571–579
12. Zhang Y, Sohn K, Villegas R, Pan G, Lee H (2015) Improving object detection with deep convolutional networks via Bayesian optimization and structured prediction. In: Proceedings of the IEEE conference on computer vision and pattern recognition. pp 249–258

13. Sharma A, Singh N (2010) Object detection in image using particle swarm optimization. Int J Eng Technol 2(6):419–426
14. Kong T, Sun F, Liu H, Jiang Y, Shi J (2019) Consistent optimization for single-shot object detection. arXiv preprint arXiv:1901.06563
15. Wang T, Zhu Y, Zhao C, Zeng W, Wang Y, Wang J, Tang M (2020) Large batch optimization for object detection: training coco in 12 minutes. In: Computer vision–ECCV 2020: 16th European conference, Glasgow, UK, August 23–28, 2020, Proceedings, Part XXI 16, Springer International Publishing, pp 481–496
16. Soviany P, Ionescu RT (2018) Optimizing the trade-off between single-stage and two-stage deep object detectors using image difficulty prediction. In: 2018 20th International symposium on symbolic and numeric algorithms for scientific computing (SYNASC), September, IEEE, pp 209–214
17. Mahalingam T, Subramoniam M (2020) ACO–MKFCM: an optimized object detection and tracking using DNN and gravitational search algorithm. Wireless Pers Commun 110(3):1567–1604
18. Tao W, Jin H, Liu L (2007) Object segmentation using ant colony optimization algorithm and fuzzy entropy. Pattern Recogn Lett 28(7):788–796
19. Dharini S, Jain S (2022) A novel metaheuristic optimal feature selection framework for object detection with improved detection accuracy based on pulse-coupled neural network. Soft Comput 26(23):12943–12955
20. Messai Y, Chara K, Srairi F, Douak F (2020) Object tracking platform for color object detection using genetic algorithm optimization. In: 2020 International conference on electrical, communication, and computer engineering (ICECCE), June, IEEE, pp 1–6
21. Mirjalili S (2015) Moth-flame optimization algorithm: a novel nature-inspired heuristic paradigm. Knowl-Based Syst 89:228–249
22. Mohammed MA, Abd Ghani MK, Arunkumar NA, Hamed RI, Abdullah MK, Burhanuddin MA (2018) A real time computer aided object detection of nasopharyngeal carcinoma using genetic algorithm and artificial neural network based on Haar feature fear. Futur Gener Comput Syst 89:539–547
23. Ugolotti R, Nashed YS, Mesejo P, Ivekovič Š, Mussi L, Cagnoni S (2013) Particle swarm optimization and differential evolution for model-based object detection. Appl Soft Comput 13(6):3092–3105
24. Sun Z, Bebis G, Miller R (2004) Object detection using feature subset selection. Pattern Recogn 37(11):2165–2176
25. Avanija J, Sunitha G, Reddy Madhavi K, Hitesh Sai Vittal R (2021) An automated approach for detection of intracranial Haemorrhage using DenseNets. In: Data engineering and communication technology: proceedings of ICDECT 2020, Springer, Singapore, pp 611–619
26. Reddy KR, Dhuli R (2023) A novel lightweight CNN architecture for the diagnosis of brain tumors using MR images. Diagnostics 13(2):312
27. Mahasin M, Dewi IA (2022) Comparison of CSPDarkNet53, CSPResNeXt-50, and EfficientNet-B0 backbones on YOLO v4 as object detector. Int J Eng Sci Inform Technol 2(3):64–72
28. Arunachalam R, Sunitha G, Shukla SK, Urooj S, Rawat S (2023) A smart Alzheimer's patient monitoring system with IoT-assisted technology through enhanced deep learning approach. Knowledge and Inform Syst 1–39
29. Sivajyothi M, Madhavi KR, Devi VVV, Reddy AB, Tangudu N, Avanija J (2024) Transformer network-based image segmentation using hybrid flower pollination optimization. In: 2024 11th International conference on computing for sustainable global development (INDIACom), February, pp 931–935

# A Predictive Healthcare System for Lung Cancer Risk Using Deep Learning on Clinical Data

**LNC. Prakash K, Chittoor Bhanu Prakash, N. Badrinath, Penubaka Balaji, Mary Teresa, and S. Athinarayanan**

**Abstract** One of the key reasons for cancer-related deaths is lung cancer, which stresses the importance of early detection and accurate predictive models worldwide. This work uses deep learning methods on clinical data to present a predictive healthcare system for lung cancer risk assessment. We expanded the data to 449 cases using the UCI lung cancer dataset, which consists of 16 symptoms and risk variables, therefore guaranteeing a strong analysis. Among several machine learning models used were XGBoost, random forest, decision trees, and a suggested Long Short-Term Memory (LSTM) network. Our aim was to solve overfitting by collecting complex patterns in patient data, hence improving the accuracy and dependability of lung cancer forecasts. Accuracy, precision, recall, and F1-score defined the evaluations of the models. With a training accuracy of 99%, testing accuracy of 98%, and validation accuracy of 98%, the LSTM network out of all the models tested attained the best performance. With better generalizing power and the capacity to control complicated temporal relationships in clinical data, the proposed LSTM model routinely

LNC. Prakash K (✉)
Computer Science and Engineering-Data Science, CVR College of Engineering, Hyderabad, India
e-mail: klnc.prakash@gmail.com

C. Bhanu Prakash
Information Technology, Sri Venkateswara College of Engineering, Tirupati, India
e-mail: bhanuprakash.c@svce.edu.in

N. Badrinath
Department of Software Systems, School of Computer Science and Engineering,
Vellore Institute of Technology, Vellore, Tamilnadu, India
e-mail: badrinath.n@vit.ac.in

P. Balaji
Department of CSE, KL University, Vaddeswaram, Guntur, Andhra Pradesh, India

M. Teresa
CSE-AIML, Guru Nanak Institutions Technical Campus, Ibrahimpatnam, Hyderabad, India

S. Athinarayanan
Department of CSE, Vel Tech Rangarajan Dr. Sagunthala R&D Institute of Science and Technology, Avadi, Chennai, India

 
K. R. Madhavi et al. (eds.), *Proceedings of Sixth International Conference on Computer and Communication Technologies*, Lecture Notes in Networks and Systems 1356,
https://doi.org/10.1007/978-981-96-5238-9_39

exceeded conventional models. BiLSTM's incorporation enhanced the classification accuracy even further, and hence, it is a perfect tool for lung cancer prediction applications.

**Keywords** Predictive healthcare system · Deep learning · Lung cancer prediction · Clinical data · BiLSTM

## 1 Introduction

In recent years, the healthcare sector has been revolutionized by breakthroughs in data science, primarily through deep learning techniques. Numerous healthcare data sources, such as genomic information and electronic health records (EHRs), have created new opportunities for customized therapy, predictive modeling, and better clinical decision-making. Recent advances in deep learning have tremendously enhanced many healthcare applications, particularly in studying patient data comprising electronic health records (EHRs), clinical notes, and demographic information. Accessing healthcare data for learning and experimentation might be difficult since it is sensitive and subject to privacy rules. However, this information's sheer volume and complexity cause significant issues that need careful analysis to generate meaningful insights and enhance patient outcomes.

Lung cancer requires the immediate advancement of accurate early detection and prediction models. Early diagnosis enhances survival rates since most diseases are identified in advanced phases. Particularly in predictive healthcare, deep learning has generated fresh ideas based on patient data to detect lung cancer early on. Deep learning models including Convolutional Neural Networks (CNNs) [1], Recurrent Neural Networks (RNNs) [2], Long Short-Term Memory (LSTM) networks [3], or hybrid deep learning approaches can identify intricate patterns in clinical data, hence improving accuracy in risk prediction and diagnosis.

Recent research has demonstrated that deep learning (DL) methods are effective in various healthcare domains. For instance, Choi et al. [4] presented the RETAIN model, which uses a reverse-time attention mechanism to forecast outcomes like heart failure, proving the potential of DL to manage complex temporal dependencies while preserving interpretability. Shickel et al. [5] have shown comparable progress in patient outcome prediction tasks, using DL applications that employ models like CNNs and LSTMs to get high predictive accuracy using electronic health record (EHR) data. Furthermore, as Miotto et al. [6, 7], who utilized deep representation learning to anticipate several clinical conditions, show that deep learning has been applied to Clinical Decision Support Systems (CDSS) to improve diagnostic precision and better patient care. DL models have presented excellent performance in risk stratification and illness prediction domains by successfully capturing nonlinear correlations in patient data. Integrating multimodal data, which combines clinical notes, test results, imaging data, and genetic information, has also provided a complete picture of a patient's health. As a result, therapeutic interventions may now be tailored more precisely, and prediction accuracy has increased. Still, model interpretability is one of the vital challenges of using DL in health care. Researchers have suggested multiple strategies to tackle this problem; for instance, Caruana et al. [8, 9] employed generalized additive models (GAMs) in conjunction with DL. In contrast, Ribeiro et al. [10] presented the LIME framework to facilitate model-agnostic interpretability.

This work aims to build a predictive healthcare system using advanced DL architectures for lung cancer. It improves the growing field of DL in health care by combining interpretability with prediction accuracy, offering a comprehensive lung cancer risk assessment framework. Our main goal is to raise early lung cancer diagnosis and treatment so improve patient outcomes.

## 2 Methodology

### 2.1 Dataset Description

The UCI machine learning repository presented the lung cancer dataset used in this study, with 16 symptoms linked to the disease. The dataset started with just 32 instances, but data augmentation allowed for an increase to 488 cases to enhance analysis. The dataset includes more complex risk factors, including genetic abnormalities, alcohol usage, and smoking behaviors, in addition to more typical ones like age, gender, overweight, chest discomfort, and so on. All attributes are of integer data type, and each is labeled on a scale from 1 to 10, with 1 denoting the lowest severity or occurrence and ten the highest. This standardized dataset offers a thorough understanding of the risk factors for lung cancer.

### 2.2 Machine Learning Models

- **Random Forest (RF) Classifier**: An ensemble learning technique called RF builds a large number of DTs throughout the training phase. It produces a class that matches the class mode for classification goals. It generates many decision trees based on random data subsets and attributes, reducing variance and enhancing accuracy relative to a solitary decision tree. The prediction for classification by the RF classifier may be articulated as [11]:

$$\hat{y} = \text{mode}(T_1(x), T_2(x), \ldots, T_n(x)) \tag{1}$$

Here, $T_i(x)$ is the $i$-th tree's prediction for input $x$. RF uses bagging-bootstrapped aggregating to improve accuracy and reduce overfitting. It is a preferred choice for classification problems since it effectively manages large volumes of data.

- **Decision Tree (DT) Classifier**: A supervised learning method, the DT is used for regression and classification. Based on feature values, it divides the data into a structure resembling a tree. Each core node indicates a "decision" based on a feature, whereas each leaf node indicates a value (for regression) or a class label (for classification). DT frequently classifies the feature delivering the optimal split using entropy or Gini impurity [12].
  **Gini Impurity**:
$$Gini(D) = 1 - \sum_{i=1}^{n} p_i^2 \tag{2}$$
  Here, $p_i$ is the proportion of class $i$ in the dataset $D$.
  **Entropy (Information Gain)**
$$H(D) = -\sum_{i=1}^{n} p_i \log_2(p_i) \tag{3}$$
  Here, $p_i$ represents the likelihood of class $i$ in dataset $D$.
- **XGBoost Classifier**: XGBoost is a quick implementation of the gradient boosting technique producing a series of decision trees whereby each subsequent tree addresses errors produced by the one before it. By way of a second-order Taylor expansion and regularization to control overfitting, XGBoost maximizes the objective function and is thus resilient and scalable for classification and regression. XGBoost seeks at every iteration $t$ as [13]:
$$Obj(t) = \sum_{i=1}^{n} L(y_i, \hat{y}_i^{(t-1)} + f_t(x_i)) + \Omega(f_t) \tag{4}$$
  Here, $\Omega(f_t)$ is the regularization term to punish the complexity of the model; $L$ is the loss function (e.g., squared error for regression or log loss for classification); $f_t$ is the new tree inserted at step $t$. XGBoost presents shrinkage (learning rate) and column subsampling to raise the model's performance.

## 2.3 *Proposed Long Short-Term Memory (LSTM)*

The LSTM network is a variant of RNNs and is designed to learn extended sequences while efficiently mitigating gradient-related issues. Conventional RNNs are a DL framework that can learn from any sequence because they comprise neurons with feedback connections. Still, a "memory limitation" problem that generates performance problems sometimes influences these networks. They still face the learning-impairing vanishing gradient problem but manage short-term dependencies using

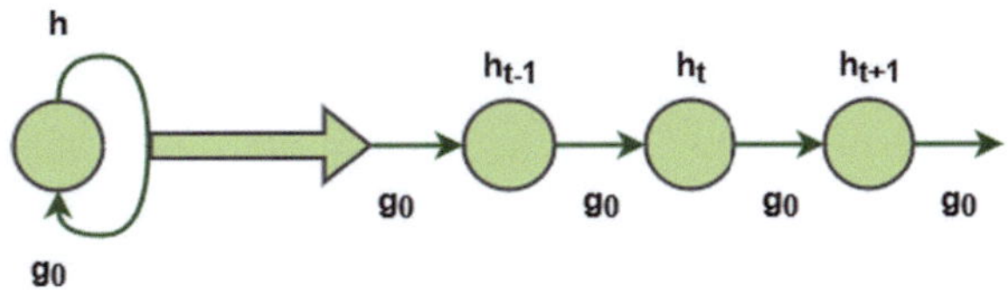


**Fig. 1** Mapping of states

recurrent connections in their hidden layers. The internal state of an RNN at any time step, indicated as $h_t$, can be formulated as:

$$h_t = g_0(h_{t-1}) \tag{1}$$

In this equation, $g_0$ delineates the mapping function employed to transition the state $h_t$ at time $t$ to the preceding state $h_{t-1}$ at time $t-1$. A visual illustration of this state mapping is presented in Fig. 1. Similarly, let us assume another system where the input is $z_t$. In this case, the state transition can be represented as:

$$h_t = f_1(h_{t-1}, z_t) \tag{2}$$

Here, the state $h_t$ at time $t$ encodes the whole sequence of inputs $(z_t, z_{t-1}, \ldots, z_2, z_1)$. This encoding procedure is described as:

$$h_t = G_t(z_t, z_{t-1}, \ldots, z_2, z_1) \tag{3}$$

Let's now examine a categorization model that was created. In this model, a series of inputs is mapped to a series of outputs that are supplied by:

$$\begin{aligned} u_t &= v + W_1 h_t + W_2 z_t \\ h_t &= \sigma(u_t) \\ o_t &= b + W_3 h_t \\ q_t &= \text{softmax}(o_t) \end{aligned} \tag{4}$$

whereas $v$ and $b$ describe the bias vectors in these equations, $W_1$, $W_2$, and $W_3$ define the weight matrices for input-to-hidden, hidden-to-output, and hidden-to-hidden connections, respectively. In this classification process, the total loss is the sum of individual losses over the whole sequence, depicted as:

$$\mathcal{L}(z, y) = \sum_t \mathcal{L}_t = \sum_t -\log(q_t^{y_t}) \tag{5}$$

Here, $y_t$ signifies the target output for input $z_t$ at time $t$. We apply the Backpropagation Through Time (BPTT) algorithm to update the parameters $a$, $b$, $W_1$, and $W_2$, calculating the gradients as follows:

$$\frac{\partial \mathcal{L}}{\partial a} = \sum_t \frac{\partial \mathcal{L}_t}{\partial o_t} \quad \text{where} \quad \frac{\partial o_t}{\partial a} = \sum_t \frac{\partial \mathcal{L}_t}{\partial o_t} \tag{6}$$
$$\frac{\partial \mathcal{L}}{\partial b} = \sum_t \frac{\partial \mathcal{L}_t}{\partial h_t} \quad \text{with} \quad \frac{\partial h_t}{\partial b} = \sum_t \frac{\partial \mathcal{L}_t}{\partial h_t} \cdot \text{diag}(1 - h_t^2)$$
$$\frac{\partial \mathcal{L}}{\partial W_2} = \sum_t \frac{\partial \mathcal{L}_t}{\partial o_t} \cdot h_t^\top$$
$$\frac{\partial \mathcal{L}}{\partial W_1} = \sum_t \frac{\partial \mathcal{L}_t}{\partial h_t} \cdot \text{diag}(1 - h_t^2) \cdot h_{t-1}^\top$$

Long Short-Term Memory is increasingly popular as a solution to the vanishing gradient issue. Every LSTM unit features an input, forget, and out the primary gate and a memory cell. The network can decide which data to "remember" and which to "forget," thereby controlling the information flow.

In the memory cell $c_t$, the information stored is controlled by the input gate $i_t$ and the candidate memory cell $\tilde{c}_t$ at time step $t$. While the output gate, $o_t$, specifies which portion of the cell's state should be output, the forget gate, $f_t$, determines how much of the prior memory, $c_{t-1}$, should be retained. Equations that represent these operations are as follows:

$$\begin{aligned} i_t &= \sigma(W_i z_t + U_i h_{t-1} + b_i) \\ f_t &= \sigma(W_f z_t + U_f h_{t-1} + b_f) \\ \tilde{c}_t &= \tanh(W_c z_t + U_c h_{t-1} + b_c) \\ c_t &= f_t \odot c_{t-1} + i_t \odot \tilde{c}_t \\ o_t &= \sigma(W_o z_t + U_o h_{t-1} + b_o) \end{aligned} \tag{7}$$

These gates provide efficient information flow management, enabling the LSTM to handle long-term dependencies. Let $x_t$ represent the input data, with $M$ and $V$ as the weight matrices and $d$ as the bias term. The function $\phi$ signifies the sigmoid activation, and $\circ$ defines element-wise multiplication. The hidden state, which is utilized to determine the memory cell's output, is estimated as follows:

$$g_t = p_t \circ \tanh(m_t) \tag{8}$$

Figure 2 illustrates the construction of the LSTM unit. The hidden state $g_t$ is produced by integrating the output gate $p_t$ with the altered memory cell state $m_t$. This configuration allows the network to effectively manage the flow of information across time steps, maintaining relevant details while determining what to retain or forget.

To further improve the performance of the LSTM unit, we introduce a Bidirectional LSTM (BiLSTM) model, which significantly enhances classification accuracy. In a standard unidirectional LSTM, information flows in a single direction, typically from the past to the future. In contrast, using two hidden states, a BiLSTM processes

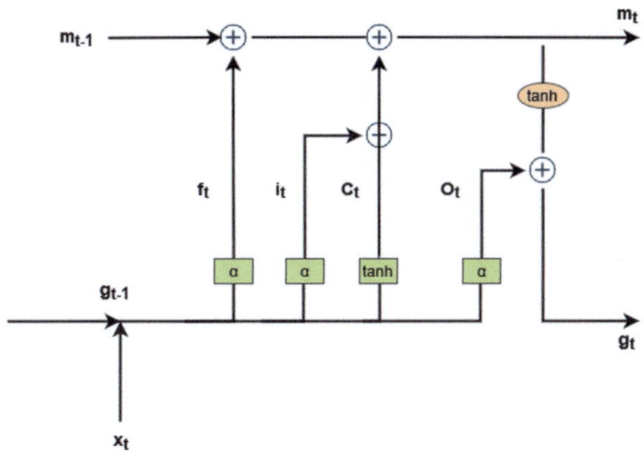


**Fig. 2** LSTM unit

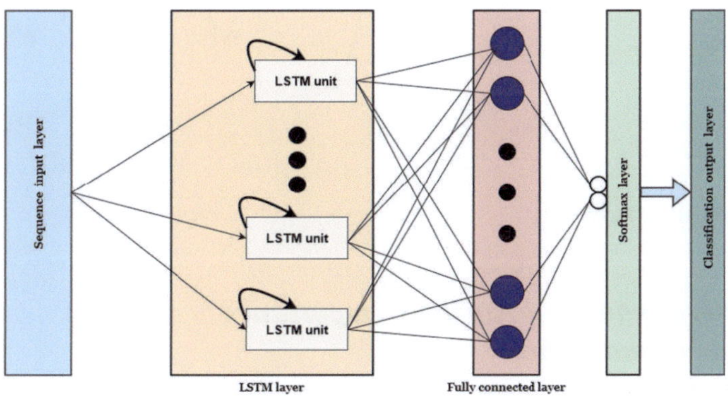


**Fig. 3** BiLSTM architecture for classification

information in both directions, forward and backwards. The bidirectional structure of the model allows it to capture context from the past and the future, therefore improving the learning capacity of the network and providing the input sequence with a more complete comprehension. Figure 4 represents the architecture of the BiLSTM model. It demonstrates how the forward and backward layers process the sequence simultaneously, providing a richer context for the model's predictions.

For the classification of lung cancer, five layers are suggested: a classification layer, a fully connected layer, a BiLSTM layer, a softmax layer, and a sequence input layer. The sequence input layer's initial layer takes time-series data. The BiLSTM layer then recognizes the sequence data's dependencies and learning patterns among multiple time steps. The final classification layer selects the output class after the fully connected layer refines the probability distributions for classification provided by the softmax layer. The proposed model's overall architecture is displayed in Fig. 3.

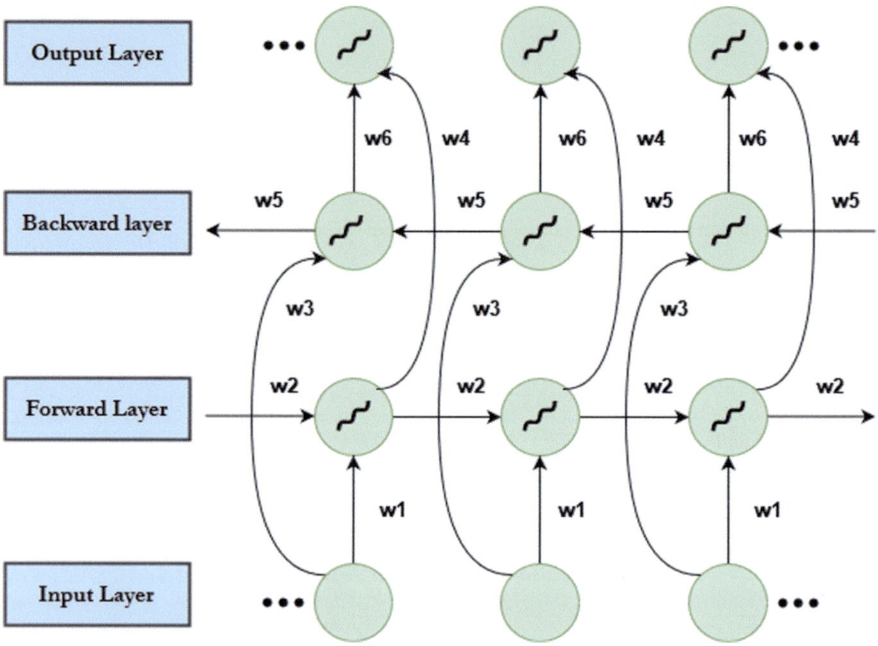


**Fig. 4** BiLSTM architecture

## 2.4 *Determination of Attribute Importance and Normalization*

The RF classifier uses entropy and information attained to determine the relevance of each characteristic. Information obtained is computed as the decrease in entropy after splitting the dataset based on an attribute, calculated as $\text{Gain}(Y, X) = \text{Entropy}(Y) - \text{Entropy}(Y, X)$, where $X$ is the splitting attribute and $Y$ is the target variable. During training, RF constructs many decision trees, and the final prediction is a combination of projections from each tree. Attribute importance is determined by reducing node impurity, calculated using the Gini significance score. The Gini significance at a node is measured as $\text{Gini}(n) = W_n I_n - W_{\text{left}(n)} I_{\text{left}(n)} - W_{\text{right}(n)} I_{\text{right}(n)}$, where $W_n$ is the weighted number of instances reaching node $n$, and $I_n$ is the impurity of node $n$, with $\text{left}(n)$ and $\text{right}(n)$ representing the child nodes after the split. The total importance of an attribute is calculated by summing the Gini significance scores for all nodes where the attribute is used for splitting, represented as $\text{Importance}(X) = \frac{\sum_{n:\text{node } n \text{ splits on } X} \text{Gini}(n)}{\sum_{l \in \text{all nodes}} \text{Gini}(l)}$. These importance values are normalized between 0 and 1 by dividing the attribute's importance by the sum of importance values across all attributes: $\text{Normalized}(X) = \frac{\text{Importance}(X)}{\sum_{X \in \text{all attributes}} \text{Importance}(X)}$. Finally, the overall attribute importance in the RF is obtained by averaging the normalized importance values across all trees in the forest, as R.F. $\text{Importance}(X) = \frac{\sum_{n \in \text{all trees}} \text{Normalized}(X,n)}{T}$, where $T$ is the entire number of trees, allowing RF to assess each attribute's contribution by reducing impurity across the entire ensemble.

## 3 Result and Discussion

The performance of our model on the chosen dataset is briefly covered in this part, along with a comparison of its results with those of other models. RF, DT, XGBoost, and our proposed LSTM model all performed differently regarding testing accuracy (TA), training accuracy (TrA), and validation accuracy (VA). We also use several evaluation metrics, including precision, recall, and F1-score, to assess and further analyze the models' performance. Our proposed model achieved higher accuracy than existing DL models, presenting improved generalization and classification efficiency.

Analyzing a model's TrA, TA, and VA helps one ascertain its performance and generalizing capacity in ML, as shown in Table 1. TA shows the model's ability to generalize to unseen data, whereas TrA shows how effectively the model has learned from the training dataset. VA improves in adjusting hyperparameters and identifying problems like underfitting or overfitting.

Our research employed four models: RF, DT, XGBoost, and our proposed LSTM model. The RF model performed a TrA of 95% and a TA of 91% but a lower VA of 84%, suggesting a gentle case of overfitting. This demonstrated that although the model does a respectable job of generalizing to new data, it struggles with new data in the validation phase. Moving to the DT classifier showed more substantial overfitting, with a TrA of 98%, a TA of 87%, and a notably lower VA of 78%. The DT model performed very well on training data but substantially worse on new data, indicating that overfitting may be an issue. The notable disparity in the VA and TrA shows this. Next, XGBoost performed better than the RF and DT models, with a TrA of 97%, a TA of 94%, and a VA of 87%. While this model still displayed signs of overfitting, the gap between TrA and VA is negligible, pointing to better generalization than RF and DT models. Finally, our proposed LSTM model showed the best performance dataset, with a TrA of 99% and a TA and VA of 98%, leading to no overfitting. This consistent performance across training, testing, and validation confirmed that the LSTM model is well-suited for learning the underlying patterns without overfitting

**Table 1** Training, testing, and validation accuracy for different models

| Model | Training accuracy (%) | Testing accuracy (%) | Validation accuracy (%) |
|---|---|---|---|
| Random forest | 95 | 91 | 84 |
| Decision tree | 98 | 87 | 78 |
| XGBoost | 97 | 94 | 87 |
| Proposed LSTM | 99 | 98 | 98 |

**Table 2** Precision, recall, and F1-score for different models

| Model | Precision | Recall | F1-score |
|---|---|---|---|
| Random forest (RF) | 0.92 | 0.90 | 0.91 |
| Decision tree (DT) | 0.88 | 0.86 | 0.87 |
| XGBoost | 0.95 | 0.93 | 0.94 |
| Proposed LSTM | 0.98 | 0.97 | 0.98 |

**Table 3** Comparison of our proposed model with previous deep learning models

| References | Model | Accuracy(% ) |
|---|---|---|
| [14] | U-Net+VGG | 95.60 |
| [15] | CNN | 86.6 |
| [16] | DenseNet | 90.85 |
| [17] | AlexNet | 97.64 |
| Proposed model | LSTM | 98 |

or underfitting. As a result, the LSTM model appeared as the best-performing model in our research, providing the most accurate and reliable results. Its ability to maintain high accuracy highlighted its effectiveness in handling the complexity of the data.

Table 2 compares four models: RF, DT, XGBoost, and the proposed LSTM model-assessed according to three principal metrics: precision, recall, and F1-score. These measures present insight into the efficacy of each model in classifying lung cancer predictions and their management of false positives and negatives. With a precision of 0.92, recall of 0.90, and an F1-score of 0.91, RF demonstrated that it can balance recall and accuracy, and it does a respectable job of accurately identifying the pertinent occurrences with a low error rate. The DT exhibited relatively inferior performance, achieving a precision of 0.88, a recall of 0.86, and an F1-score of 0.87. Still rather good, the DT model shows less ability than the RF and XGBoost models to balance precision and recall. With remarkable accuracy (0.95) and recall (0.93), XGBoost produced a strong F1-score of 0.94. This result shows that XGBoost is a strong competitor for this prediction work since it efficiently balances precision and recall, especially when trying to improve real positives by lowering false positives. Finally displaying the most outstanding overall performance is the proposed LSTM model with maximum precision (0.98), recall (0.97), and F1-score (0.98). This shows that the LSTM model, which is superior in identifying temporal patterns and dependencies in the data, considerably increases the prediction accuracy for lung cancer cases. With almost perfect scores across all criteria, the proposed model might drastically reduce the frequency of misclassifications. Therefore, it is ideal for high-stakes healthcare projections.

Table 3 contrasts the accuracy of the presented LSTM model with that of prior DL methods in detecting criminal activities. Figure 5 shows these accuracy percentages and that, with the best accuracy at 98%, the suggested LSTM model beats others.

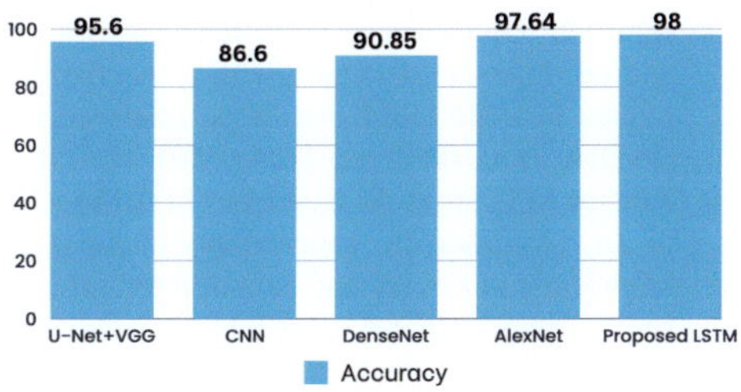


**Fig. 5** Comparative performances between our proposed model and existing models for lung cancer prediction

## 4 Conclusion

This work effectively developed a predictive healthcare system employing clinical data with deep learning to evaluate lung cancer risk. Particularly in its BiLSTM form, the suggested LSTM model shows remarkable accuracy and dependability in determining lung cancer risk variables. Its enhanced generalization over conventional models made it quite successful in practical settings. The results imply that early diagnosis and patient care can be much improved by including such a model in clinical procedures. Future research could concentrate on enlarging the dataset and investigating other risk factors to improve the predictive powers of the model.

## References

1. Faruqui N, Yousuf MA, Whaiduzzaman M, Azad A, Barros A, Moni MA (2021) Lungnet: a hybrid deep-cnn model for lung cancer diagnosis using ct and wearable sensor-based medical iot data. Comput Biol Med 139:104961
2. Priyadarshini K, Alagarsamy M, Sangeetha K, Thangaraju D (2024) Hybrid rnn-ffbpnn optimized with glowworm swarm algorithm for lung cancer prediction. IETE J Res 70(5):4453–4468
3. Mhaske D, Rajeswari K, Tekade R (2019) Deep learning algorithm for classification and prediction of lung cancer using CT scan images. In: 5th International conference on computing, communication, control and automation (ICCUBEA). IEEE, pp 1–5
4. Choi E, Bahadori MT, Sun J, Kulas J, Schuetz A, Stewart W (2016) Retain: an interpretable predictive model for healthcare using reverse time attention mechanism. In: Advances in neural information processing systems, pp 29
5. Shickel B, Tighe PJ, Bihorac A, Rashidi P (2017) Deep ehr: a survey of recent advances in deep learning techniques for electronic health record (ehr) analysis. IEEE J Biomed Health Inform 22(5):1589–1604
6. Miotto R, Li L, Kidd BA, Dudley JT (2016) Deep patient: an unsupervised representation to predict the future of patients from the electronic health records. Scient Reports 6(1):1–10
7. Sarada C, Mallika SS, Swetha G (2024) Deep-learning-based food calorie measurement with graphical user interface. In: International conference on medical imaging, electronic imaging, information technologies, and sensors (MIEITS 2024), vol 13188. SPIE, pp 15–22
8. Caruana R, Lou Y, Gehrke J, Koch P, Sturm M, Elhadad N (2015) Intelligible models for healthcare: Predicting pneumonia risk and hospital 30-day readmission. In: Proceedings of the 21th ACM SIGKDD international conference on knowledge discovery and data mining, pp 1721–1730

9. Rani DS, Gnaneshwar K, Pattem KS, Sekhar S, Krishna GB, Kakarla S (2024) Advancing android malware detection with biosentinel neural network using hybrid deep learning techniques. In: 2024 11th International conference on computing for sustainable global development (INDIACom), IEEE, pp 1754–1760
10. Ribeiro MT, Singh S, Guestrin C (2016) why should i trust you? explaining the predictions of any classifier. In: Proceedings of the 22nd ACM SIGKDD international conference on knowledge discovery and data mining, pp 1135–1144
11. Liu Y, Wang Y, Zhang J (2012) New machine learning algorithm: random forest. In: Information computing and applications: third international conference, ICICA 2012, Chengde, China, September 14–16, Proceedings 3, Springer, pp 246–252
12. Tangirala S (2020) Evaluating the impact of gini index and information gain on classification using decision tree classifier algorithm. Int J Adv Comput Sci Appl 11(2):612–619
13. Chen T, Guestrin C (2016) Xgboost: a scalable tree boosting system. In: Proceedings of the 22nd acm sigkdd international conference on knowledge discovery and data mining, pp 785–794
14. Tekade R, Rajeswari K (2018) Lung cancer detection and classification using deep learning. In: Fourth international conference on computing communication control and automation (ICCUBEA). IEEE, pp 1–5
15. Alakwaa W, Nassef M, Badr A (2017) Lung cancer detection and classification with 3d convolutional neural network (3d-cnn). Int J Adv Comput Sci Appl 8(8)
16. Kalaivani N, Manimaran N, Sophia S, Devi D (2020) Deep learning based lung cancer detection and classification. In: IOP conference series: materials science and engineering, vol 994. IOP Publishing, pp 012026
17. Naseer I, Masood T, Akram S, Jaffar A, Rashid M, Iqbal MA (2023) Lung cancer detection using modified alexnet architecture and support vector machine. Comput Mater Continua 74(1)

# Optimizing Industrial Energy Management: Employing Machine Learning and Data Analytics for Sustainable Practices

**H. N. Lakshmi, B. Rupa Devi, P. Anjaiah, C. Nithisha, Konatham Sumalatha, and R. Kavitha**

**Abstract** The industrial electricity management system (EMS) plays an essential role in organizing energy savings by enabling instructed managerial construction. EMS gathers and examines real-time data from measuring devices and meters to significantly optimize energy usage by merging machine learning procedures and data analytics. To monitor machine performance and energy supply and moderate the early identification of unevenness, the system matches this data with ideal statistics. Furthermore, EMS can reckon use trends, detect prospective for energy reserves, and offer sensible advice that increases productivity by using analytical assessment. Decline of losses, improvement of procedures, and sustainability of the circumstances are the focal objectives of EMS. To contribute to administrations in

H. N. Lakshmi (✉)
Department of Computer Science and Engineering, (AIML), CVR College of Engineering, Hyderabad, India
e-mail: hn.lakshmi@cvr.ac.in

B. Rupa Devi
Department of CSE, Annamacharya Institute of Technology and Sciences, Tirupati, India
e-mail: rupadevi.aitt@annamacharyagroup.org

P. Anjaiah
Department of Artificial Intelligence and Machine Learning, Mallareddy University, Hyderabad, India

C. Nithisha
Department of Data Science, School of Computing, Mohan Babu University, Sree Sainath Nagar, Tirupati, Andhra Pradesh, India

K. Sumalatha
Department of Database Systems, School of Computer Science and Engineering, Vellore Institute of Technology, Vellore, Tamilnadu, India
e-mail: konatham.sumalatha@vit.ac.in

R. Kavitha
Department of CSE, Vel Tech High Tech Dr. Rangarajan, Dr. Sakunthala Engineering College, Avadi, Chennai, India



K. R. Madhavi et al. (eds.), *Proceedings of Sixth International Conference on Computer and Communication Technologies*, Lecture Notes in Networks and Systems 1356,
https://doi.org/10.1007/978-981-96-5238-9_40

managing substantial energy administration effects and to encourage sustainability and cost-efficiency going forward, this study recommends a progressive method for an effective EMS.

**Keywords** Energy management system (EMS) · Machine learning · Data analytics · Industrial optimization · Sustainability

## 1 Introduction

Energy spending has become a most important problem in today's rapidly developing industrialized environment, presenting substantial difficulties to businesses and regions. We propose to construct an EMS that is activated by the burning demand to answer the elaborate troubles that commercial and industrial establishments are aiming to increase their energy utilization. Established energy executive systems endure to function well, but they are repeatedly unsuccessful in supporting timely comprehension and effective guidance. This deficit spikes our creativity in a new era of effective energy control [1]. This research is driven by the pressing need to achieve environmental sustainability while also addressing the growing problem of rising energy costs. The initiative is driven by these two crucial considerations, which motivate us to take decisive and significant steps. The main objective is to provide companies with the means and know-how to lower their energy costs and increase their operational effectiveness while having the least possible environmental impact. This dual emphasis captures the essence of our purpose and is the driving reason behind our continued work and dedication to sustainable innovation [2].

In this article, the basis for anomaly identification, predictive modeling, and motor functioning status evaluation is provided by specifications. Anomalies and deviations are found by comparing real readings with artificial datasets, allowing for proactive modifications to processes. Predictive models are generated by machine learning algorithms that have been trained on the previous data to anticipate patterns in energy use. Proactive repair scheduling and effective resource optimization are made possible by the prompt detection of idle situations made possible by continuous real-time monitoring of motor measurements. By taking these precautions, it guarantees that the research will produce reliable energy management systems that are tailored to satisfy industrial needs. The remainder of this research is organized as follows: Sect. 2 indicates the previous work related to the current research, and Sect. 3 explains the proposed methodology. Section 4 is indicated for outcomes and analysis. Section 5 presents the conclusion.

## 2 Literature Review

Energy organization, which highlights the operative use of energy to save costs, improve working execution, and encourage ecological sustainability,is a necessary constituent of existing developed procedures. With the extending insistence on companies to shrink their carbon influence and increase energy use, creative results exploiting cutting-edge expertise are developing indispensable. Real-time examining, predictive conservation, and irregularity detection are made feasible by the effective tools offered by machine learning and data analytics, which may considerably change energy management practices. Industries can substantially strengthen energy effectiveness and reinforce environmental and financial targets by using these skills. An expansive array of research and practical functions has developed in the field of EMS, screening numerous areas with the universal objectives of enhancing energy use, enlightening productivity, and encouraging sustainability, [3]. In [4], substantial advancement has succeeded in smart grid machinery, which integrates progressive detection, transmission, and control systems. These improvements allow real-time observation and administration of electricity circulation, improve grid stability, and support the mixing of renewable energy resources. Demand response programs correspond to additional imperative areas of prominence, where customers are persuaded to change their electricity usage in response to supply restrictions or pricing signals. These proposals help to manage demand throughout peak times, lower costs, and enhance grid permanence [5].

EMS specifies a wide selection of tools constructed to monitor, examine, and boost energy utilization across different circumstances, such as constructions, industrial sites, and business spaces. By combining sensors and operating data analytics, EMS detects uselessness, implements tactics for energy maintenance, and observes performance indicators [6]. Home energy management systems (HEMS) focus on improving energy use in domestic situations by offering homeowners listed perceptions into their expenditure designs and granting automated control of domestic devices. These systems facilitate users to make informed choices, decrease energy waste, and lower their efficiency costs [7]. In the research [8], industrial EMS is fitted to focus on the detailed energy management obligations of the industrial sector. These systems employ sophisticated observing and optimization techniques to improve effectiveness, lower costs, and decrease ecological impact. As renewable energy resources become more dominant, there has been an expanding focus on extending knowledge to effortlessly combine them into the power grid. To fully connect the probable of these renewable sources, advanced predicting and energy-storing approaches are cardinal [9]. The work of Patel and Ghosh [10] is focused on how data analytics and machine learning may be managed to advance sustainability in industrial energy controlling. This article investigates how these workings help environmental sustainability goals by sinking carbon footprints and managing industrial developments.

This article [11] examines the problems of incorporating machine learning into present industrial energy organization systems and the extensive financial outlay involved. It also looks at the potential that these technologies provide, especially in terms of long-term cost decrease and increased sustainability. The use of machine learning methodologies for real-time energy management in smart grids is reported in the paper [12], and it is becoming more and more critical to developed energy systems. It illustrates how more dynamic and responsive energy management is made possible by ML-driven techniques, which may improve the sustainability and efficiency of energy consumption in industrial settings. In this research to evaluate the results random forest regression model is used, and the implementation details and the working procedure of this methodology are discussed in [13–15].

## 3 Methodology

The recommended research presents a complete technique intended to improve the efficiency of energy management in industrial settings, as depicted in the given architecture. The EMS architecture is thoughtfully designed to provide comprehensive energy management solutions, integrating key functions such as data collection, analysis, anomaly detection, forecasting, and machine state recognition. Each module in this framework performs a specific function and works in unison to enhance the EMS system's overall efficacy and efficiency. Figure 1 represents the typical architecture of the proposed model.

In this research, synthetic data is generated which mimics the behavior of industrial machinery, which is a key component of the EMS design. With the inclusion of significant variables like power, voltage, and current, this dataset correctly illustrates samples of energy usage. This section establishes descriptive and diversified datasets that offer a powerful ground for further displaying and exploration.

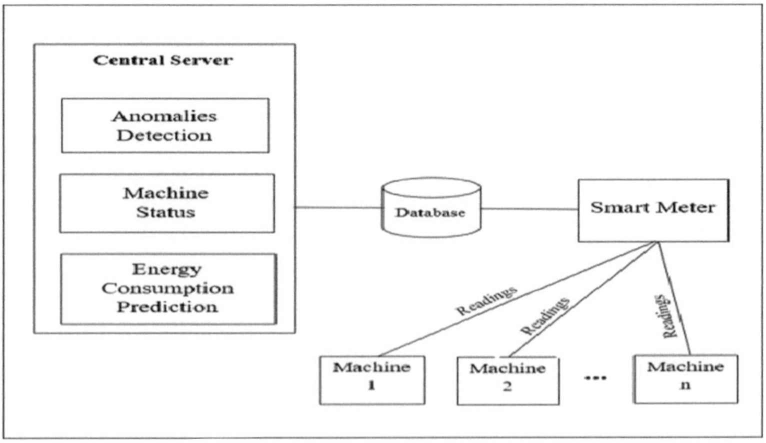


**Fig. 1** Architecture of the proposed system

Here, data analysis and inconsistency detection are utilized to examine and visualize energy usage data in detail. Using pre-recognized standards and variation standards, it does a thorough evaluation of the evidence to find patterns and irregularities. It also recognizes abnormal movements in energy use. This section plays a detailed role in discovering ineffectiveness or possible faults inside industrial equipment by identifying these irregularities.

Exploiting historical data, the random forest regressor method is utilized by the advanced machine learning expectation segment of the EMS architecture to estimate patterns in energy usage. With its training on an artificial dataset and accuracy estimated by mean squared error, this module extends useful estimates for future energy use. Businesses may enhance source distribution, control expenditures, and manage energy supplies proactively by using these machine learning paradigms.

A machine learning technique named the random forest regressor is utilized in regression tasks when the goal is to estimate a numeric output variable, such as energy utilization. To strengthen the accuracy and prevent overfitting, it builds up numerous decision trees throughout training and averages each one's estimates. The outcome is commonly the average of all the predictions assembled by each tree in the forest. When contrasted to discrete decision trees, the random forest regressor is more robust and error-prone. Based on a segment of the data, each decision tree in the random forest is a pattern that produces forecasts. When an input feature vector $X = \{x_{1,}x_2, \ldots\ldots.x_n\}$ is decided the forecast of the $i$th decision tree is shown by:

$$\overline{y_i}(X) = \overline{F_i}(X) \tag{1}$$

Here, $\overline{F_i}(x)$ is the forecast of the $i$th tree. By averaging their estimates, the random forest regressor averages the forecasts from all $K$ decision trees:

$$\overline{y}(X) = \frac{1}{k}\sum_{i=1}^{k}\overline{y_i}(X) \tag{2}$$

where $\overline{y}(X)$ is the random forest's final forecast for the input $X$, and $K$ is the total number of decision trees in the forest. Every decision tree is exercised with a distinctive bootstrap model of the primary dataset throughout the training activity. A subset of properties is casually picked for every node in the tree, and the ideal split is decided by using a principle such as diminishing the mean squared error (MSE). The following presents the MSE for a splitting at node l:

$$SMSE_l = \frac{1}{n_l}\sum_{i=1}^{n_l}(y_i - \overline{y_i}(X))^2 \tag{3}$$

The number of examples in node J is characterized by $n_j$. The real value of the goal variable for sample $j$ is denoted by $T_j$.. At node $J, j$ denotes the target variable's expected value. The mean squared error (MSE) on the test set is repeatedly used to measure the correctness of the random forest regressor.

In this study, the prediction component is proposed to predict upcoming power supply and motor readings to help organizations predict patterns in energy management and effectively assign resources. The prediction demonstrates using synthetic data that results from one month's worth of power supply and motor capacities. This study explains previous data movements and predicts upcoming energy management models utilizing machine learning methods, such as the random forest regressor algorithm, and these results are compared with other machine learning methods like decision trees, deep learning, and support vector machines. Estimating energy usage precisely allows firms to decrease energy waste, optimize production programs, and cut functional expenditures.

## 4 Results and Discussions

A Python script that represents the performance of mechanized machines was exploited to produce synthetic datasets for the estimation of the suggested technique, with a precise prominence on power supply (PS) and motor (M) measurements. The script established datasets with parameters like current, voltage, frequency, power, speed, torque, power output, current drawn, and efficiency over a specified interval by employing Pandas for data handling and NumPy for numerical calculations. These simulated datasets, which access managed circumstances for examining and calculating the models' execution and reliability, were necessary for establishing the effectiveness of the energy organization algorithms. The evaluation's conclusions require understanding evidence on the accuracy and flexibility of the recommended procedure.

### *4.1 Datasets*

The performance of the proposed method is evaluated using "Synthetic Motor Data Ide-alPS1" which consists of ideal power supply measurements for an industrial machine. The variables in the dataset are power input, frequency, voltage, and current. The ideal motor readings for a specific industrial machine are denoted by synthetic data in the collection "synthetic motor data ideal M1.csv." It includes attributes like torque, power output, speed, effectiveness, and current utilization. "Synthetic motor data ActualM1.csv", a real motor reading taken during operation from an industrial machine, is included in this collection. When comparing performance to the optimum motor readings, it serves as a baseline. "ActualPS1.csv, Synthetic Motor Data" which is an actual power supply measurements that were taken from the industrial machine while it was in use are included in this dataset. It acts as a benchmark for contrasting the effectiveness of the optimal power supply measurements. Based on the motor nameplate, we have also created statistics that indicate the optimal power supply and motor readings that are divided into five groups. Based on its name-plate data, each dataset represents a certain class of motor. These datasets are used as benchmarks for evaluating and comparing the power supply and motor performance of industrial machines in several categories. Datasets for prediction models are used to train and evaluate prediction models. They are made up of a combination of past power supply and motor performance data as well as certain attributes that are important for the prediction process. Using the gathered statistics, the objectives and benchmarks are created to assess the industrial machine's and its power supply's performance. The objectives specify the permissible range of variation from these mean values, whereas the requirements provide the mean values of certain metrics. The objectives and benchmarks for the datasets are listed below.

Requirements: The average values of each parameter over the datasets were used to construct the standards. They function as a standard by which to compare the machine's power supply and performance.

Objectives: Standard deviations were used to define the objectives, which were based on departures from the average values. For every parameter, they show an acceptable range of variation from the standards.

## *4.2 Anomalies Detection*

During the anomaly detection process, we identified variations to uncover any irregularities in the power supply and motor readings. The outcomes of this anomaly detection are as follows:

Maximum Number of Parameters Categorized in Power Supply: 3, this means that three parameters were identified as anomalies in the power supply readings.

Power Supply Category Number: 3, that is anomalies were detected in the power supply category numbered 3.

Number of Parameters within Acceptable Range for Corresponding Motor Category: 1. Among the five parameters assessed, only one fell within the acceptable range for the corresponding motor category. Anomalies detected in motor since fewer than four parameters were within the valid range, and anomalies were identified in the motor readings.

## *4.3 Prediction and Model Evaluation*

The prediction model was established and assessed using historical data on power supply and motor performances as shown in Fig. 2. To judge the model's accuracy in predicting future power supply and motor performances, the mean squared error

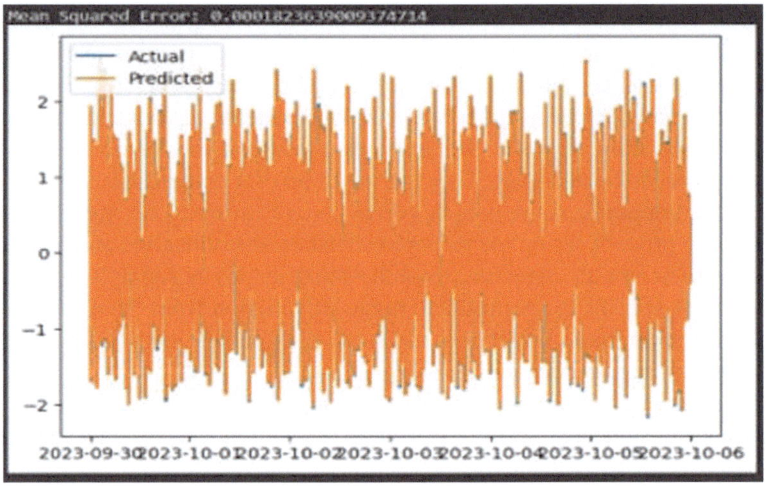


**Fig. 2** Prediction model evaluation

(MSE) was determined. MSE serves as a marker of the difference between the anticipated values and the real values. A lower MSE specifies greater analytical accuracy and better model execution.

By highlighting how strongly the predicted values meet the real ones, the MSE graph indicates how well the forecast model behaved. While greater MSE values denote a greater difference between the estimated and real values, lower MSE values imply more precision in forecasting future power supply and motor magnitudes. In this research, the random forest regressor algorithm results are compared with the other machine learning methods like decision tree, deep learning, and support vector machines, and the comparison results are described in Table 1.

From the table, deep learning performs well when compared to other methods, but random forest delivers a reliable, vigorous show among all datasets, with 89% accuracy on ActualM1, making it a feasible and consistent selection. It compares accuracy and effectiveness better than other approaches, specifically by lowering overfitting assessed to decision trees. This yields random forest substantially appropriate for varied, real-world purposes. For a better understanding of the results, it is depicted in Fig. 3.

The working status detection of energy management systems supplies real-time perceptions of the functioning state of industrialized machines, which is essential.

**Table 1** Comparison of ML techniques

| Data Set | Random forest | Decision tree | Deep learning | Support machines |
|---|---|---|---|---|
| IdealPS1 | 88.5 | 81.7 | 90.4 | 85.5 |
| IdealMS1 | 87.4 | 80.2 | 88.4 | 80.5 |
| ActualM1 | 89 | 79 | 88.5 | 82.3 |

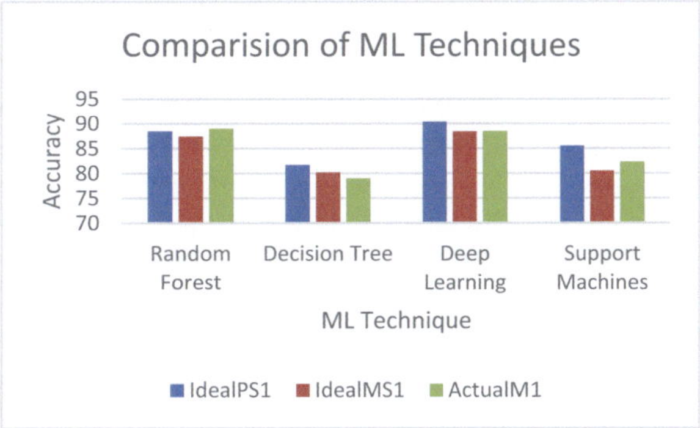


**Fig. 3** Comparison of ML techniques

**Table 2** Machine working status result

| Index | Timestamp | Current drawn | Machine state |
|---|---|---|---|
| 0 | 2023-09-06 00:00:00 | 528.511022 | Working |
| 1 | 2023-09-06 00:05:00 | 407.119365 | Working |
| 2 | 2023-09-06 00:10:00 | 545.594005 | Working |
| … | … | … | … |
| 8635 | 2023-10-05 23:35:00 | 492.223697 | Working |
| 8636 | 2023-10-05 23:40:00 | 486.048606 | Working |
| 8637 | 2023-10-05 23:45:00 | 416.767845 | Working |
| 8638 | 2023-10-05 23:50:00 | 508.349405 | Working |

This module establishes whether the equipment is driving or idle by examining essential factors such as motor speed and existing usage. The results generated by the working status detection component are listed in Table 2.

The output contains a timestamp demonstrating the precise moment at the data was taken, the speed of the motor in revolutions per minute (rpm), the current used by the apparatus in amps (A), and the condition of the motor of the machine, which is characterized as "working" or "idle." These findings specify significant new considerations for the long-term working performance of industrial machinery. Facility administrators may utilize this data to observe the use of their machinery, to recognize movements in both the operational and idle stands, and to accommodate their energy procedure properly. In this examination, the execution outcomes are riveted on three main sections which are motor significance supervising, predictive examination, and anomaly recognition. Each of these components is necessary to improve the efficiency of energy administration in industrial situations. The primary goal of this study is anomaly recognition, which is aimed to recognize anomalies in samples of energy management. To evaluate real-time data with standards, this produces synthetic datasets of motor and power supply dimensions built on requirements observed on industrial equipment nameplates. Any differences between the predicted and real quantities are described as anomalies, indicating that the industrial equipment may have inaccuracies or inadequacies. The section assists positive involvement, diminishing stoppage, and enhancing energy consumption, by quickly detecting these abnormalities. Prediction estimates power supply and motor readings, and organizations can better distribute assets and forecast styles in energy management. The random forest regressor method is an analytical paradigm that is based on one month's worth of synthetic data to predict forthcoming developments in energy use by learning from past executions. Accurate expectation permits the effective making of plans, decreases energy wastage, and diminishes running expenditures.

The motor working status recognition distributes supplementary suitable knowledge about the operating state of industrial machinery. This portion constantly checks motor data in real time to establish the machines' functioning state. The system identifies inactive conditions, or times when anything is done, by matching the quantities to pre-established criteria. This data is crucial for locating preventive preservation schedules into activity, making the most use of the equipment, and saving energy while the equipment is idle. The facility managers may use the outcomes of this article to expand overall working effectiveness, improve production calendars, and implement preventative preservation strategies by studying machine status movements and linking them with energy consumption decorations.

## 5 Conclusion

The EMS corresponds to a general improvement in recovering energy effectiveness using innovative reasoning and a committed commitment to excellence. The principal object of this research has been to create an entire system that can accurately predict upcoming needs, recognize irregularities in real-time, and evaluate patterns of energy management. This system will develop energy administration methods in the business and modern sectors. The endeavors have been genuinely supported by the valuable incorporation of cutting-edge machine learning methods, remarkably the random forest regressor model. The EMS manages substantial datasets with incredible accuracy, presenting organizations with correct energy use trend approximations. We have made it possible for organizations to manage their assets intentionally and make practical, well-educated conclusions by exploiting extrapolative analytics. The EMS's real-time abnormality recognition tool, which extends a proactive technique of risk modification and system operation optimization, is one of its best characteristics. The result helps to reduce downtime and improve working effectiveness by promptly identifying variations from assumed norms, which leads to substantial cost savings and progressed steadiness throughout a diversity of areas. This research also delivers personalized energy-saving guidance and actionable understandings. By absolutely analyzing data abnormalities, we have determined improvements supplied to the individual functioning needs of every organization. By putting these implications into practice, energy utilization and performing costs are substantially decreased, which advances sustainability proposals and decreases ecological impact.

To enhance approachability and scalability, forthcoming improvements will occasion progressing our EMS to a cloud-based construction and placing a user-centric model first to boost usableness and commitment. We will keep progressing our analytical techniques to boost abnormality detection and expectation accuracy. Working jointly with stakeholders will encourage improvement and progress in sustainability ideas, setting up our EMS to have a big encouragement on sustainability and energy administration.

## References


1. Lee D, Cheng C-C (2016) Energy savings by energy management systems: a review. Renew Sustain Energy Rev 56:760–777
2. Ahmad T, Zhang D (2020) A critical review of comparative global historical energy consumption and future demand: the story told so far. Energy Rep 6:1973–1991
3. Khan N, Shahid Z, Alam MM, Bakar Sajak AA, Mazliham M, Khan TA, Rizvi A, Safdar S et al. (2022) Energy management systems using smart grids: an exhaustive parametric comprehensive analysis of existing trends, significance, opportunities, and challenges. Int Trans Electri Energy Syst
4. Massaoudi M, Abu-Rub H, Refaat SS, Chihi I, Oueslati FS (2021) Deep learning in smart grid technology: a review of recent advancements and prospects. IEEE Access 9:54558–54578
5. Vahid-Ghavidel M, Javadi MS, Gough M, Santos SF, Shafie-Khah M, Catalao JP (2020) Demand response programs in multi-energy systems: a review. Energies 13(17):4332
6. Zahraoui Y, Alhamrouni I, Mekhilef S, Basir Khan MR, Seyedmahmoudian M, Stojcevski A, Horan B (2021) Energy management system in microgrids: a comprehensive review. Sustainability 13(19):10492
7. Gomes I, Bot K, Ruano MG, Ruano A (2022) Recent techniques used in home energy management systems: a review. Energies 15(8):2866
8. Mariano-Hern´andez D, Hern´andez-Callejo L, Zorita-Lamadrid A, Duque- P´erez O, Garc´ıa FS (2021) A review of strategies for building energy management system: model predictive control, demand side management, optimization, and fault detect and diagnosis. J Build Eng 33:101692
9. Husin H, Zaki M et al (2021) A critical review of the integration of renewable energy sources with various technologies. Protect Control of Modern Power Syst 6(1):1–18
10. Patel S, Ghosh A (2021) Sustainable energy management in industry: the role of machine learning and analytics. Sustain Energy Technol Assess 45:101050
11. Li X, Liu Y, Wang Z (2022) Challenges and opportunities in the implementation of machine learning for industrial energy management. Energy 245:123209
12. Siano P, De Marco G (2020) Real-time energy management in smart grids: a machine learning approach. Appl Energy 272:115237
13. Srikar Y et al. (2023) A comparative analysis: autism spectrum disorder among children using classification models. In: 2023 2nd International conference on ambient intelligence in health care (ICAIHC). IEEE
14. Reddy PCS, Suryanarayana G, Yadala S (2022) Data analytics in farming: rice price prediction in Andhra Pradesh. In: 2022 5th International conference on multimedia, signal processing and communication technologies (IMPACT). IEEE
15. Lakshmi HN et al. (2024) A novel comprehensive investigation for enhancing cluster analysis accuracy through ensemble learning methods. Int J Electri Comput Eng (IJECE) [S.l.] 14(5):5802–5812. ISSN 2722-2578. https://doi.org/10.11591/ijece.v14i5.pp5802-5812

# Improved Iris Recognition Through Optimized Gabor Filters and Deep Belief Networks Fused with GRU

**J. Samatha and G. Madhavi**

**Abstract** Iris recognition outlooks foremost biometric identification technique, celebrated for its accuracy and steadiness. This paper proposes an improved deep learning framework through optimal Gabor filters and deep belief network (DBN), fused with gated recurrent units (GRU). The method begins with Gabor filters that elicit key features from iris visuals, followed by DBN for structured feature learning. GRUs are then used to extract the sequential insights in iris images, improving the system's ability to manage disparities in image acquirement conditions and occlusions. Investigational assessments on benchmark iris images validate that the proposed architecture distinctly increases recognition precision and resilience to variations in lighting, occlusion, and noise. The outcomes indicate that the blend of Gabor filters, DBN, and GRU efficiently addresses the limitations of traditional iris recognition methods. This improved technique shows significant ability for biometric validation and security uses, maintaining high accuracy in stimulating environments, making it compatible for practical security developments.

**Keywords** Iris · Gabor filter · Recurrent neural network · Deep belief network · Gated recurrent unit

---

J. Samatha (✉)
Research Scholar JNTUK, Assistant Professor, CSE Department, Matrusri Engineering College, Hyderabad, Telangana, India
e-mail: samathajuluri@matrusri.edu.in

G. Madhavi
Assistant Professor, CSE Department, JNTUK University College of Engineering Narasaraopet, Narasaraopet, Andhra Pradesh, India
e-mail: madhavi.researchinfo@gmail.com

 
K. R. Madhavi et al. (eds.), *Proceedings of Sixth International Conference on Computer and Communication Technologies*, Lecture Notes in Networks and Systems 1356,
https://doi.org/10.1007/978-981-96-5238-9_41

## 1 Introduction

The traditional systems for user authentication typically use usernames and passwords, token-based authentication, or two-factor authentication [1]. However, these techniques have several limitations, such as password vulnerability, phishing attacks, token theft, user inconvenience, difficulty in revocation, and ineffectiveness for remote access. These limitations highlight the need for a more secure and user-friendly mechanism, such as biometrics [2]. In the early days, people primarily used fingerprints or facial recognition as major biometric authentication techniques due to their ease of access. However, these biometric traits can change over time and may be affected by wear and tear. Facial features, for example, can change due to aging. On the other hand, the iris is an internal organ that is externally visible, protected from environmental factors, and remains relatively immutable over a lifetime. Therefore, user authentication using iris recognition offers superior performance compared to other modalities [3].

This paper proposes an improved version of an iris recognition technique in combination with a Gabor filter, deep belief network, and gated recurrent unit (GRU) [4]. Feature extraction from iris images is an important task for authentication. Here, optimized Gabor filter techniques are used for better feature extraction. Next, a deep belief network is employed, which consists of multiple layers that extract hierarchical features from the images [5]. DBNs are composed of multiple layers of stochastic latent variables, primarily using Restricted Boltzmann Machines (RBMs) for unsupervised learning. They are typically used for static data, such as images, where the relationships among features are learned hierarchically [6].

The GRU, an extension of RNNs, provides better performance and overcomes the limitations of traditional RNNs, such as the vanishing gradient problem [7]. The sequence of feature vectors is fed into the GRU, which learns to identify patterns and dependencies in the feature sequences, potentially enhancing the model's ability to distinguish between genuine irises and impostors. The GRU can learn how variations in iris features across different images relate to the same individual, improving the model's ability to authenticate users accurately [8].

The results underline the usefulness of our strategy in addressing the shortcomings of traditional iris identification systems, ensuring enhanced feature extraction and simulating capabilities. By gathering a wider range of iris features and accounting for temporal associations in iris textures, our model improves entity differentiation, even under daunting conditions [9]. This research has uses in multiple domains requiring robust and precise biometric verification, such as security, confidentiality, and individuality verification for border control, access management, and financial dealings. With its high precision and resilience, the projected model offers a considerable boost to the safety and effectiveness of identity authentication processes, providing reliable biometric strategies suited to security-crucial environments where dependable recognition is vital.

## 2 Related Work

Daugman [10] examines the iris recognition method using Gabor filters. These filters efficiently extract distinctive features from iris trends, facilitating precise recognition. Daugman's approach is especially proficient for matching iris trends, which is vital for real-time implementations that require rapid authentication. In addition, the system demonstrates a strong robustness to alterations in ambient illumination, ensuring consistent performance across diverse lighting conditions. However, Daugman does not present details on the choice process for the Gabor filters. This deficiency of clarity concerning filter choice may create challenges in duplicating or improving upon his technique.

In 2009, Chang's work [11] presented the use of Empirical Mode Decomposition (EMD) to examine iris images. EMD dynamically breaks down such images into intrinsic mode functions (IMFs), collecting a variety of levels of aspects. This adaptable analysis is particularly useful for dealing with the advanced and fine-grained arrangements in iris trends. By presenting a multiresolution approach, EMD can sense subtle alterations within the iris pattern. However, it lacks a multiorientation study ability, which may bind its efficiency in collecting the varied orientations of patterns present in iris textures. This drawback could influence the techniques' robustness and its capability to represent various regions of the iris thoroughly.

In 2015, Huo [12] incorporated Gabor filters with histogram alterations for feature collection. A significant gain of this method is its two-stage feature blend, which enhances the robustness and effectiveness of iris identification. By integrating several features, the model can reach more accurate and dependable recognition. However, this technique is sensitive to the threshold established in the first phase. Variations in threshold choice can significantly impact the system's accomplishment, complicating the capability to constantly attain accurate results. Tackling this sensitivity is vital for assuring performance across several datasets and circumstances.

He et al. work [13] utilizes optimized Gabor filters together with deep belief networks (DBN). The Gabor filters are customized to mine important features from iris images, as well as texture trends, edges and frequency elements. When united with DBN, these filters permit the model to study hierarchical illustrations of the elicited features. Through recurring learning, the DBN discloses the complicated patterns there in iris images, thus improving feature bias. Initially, this method faced disputes associated to static illustration and flexibility to dynamic circumstances, which hindered continuous verification. Though, these restrictions were efficiently resolved by integrating gated recurrent units (GRU), which improved the model's ability to deal with vibrant circumstances and assisted continuous verification.

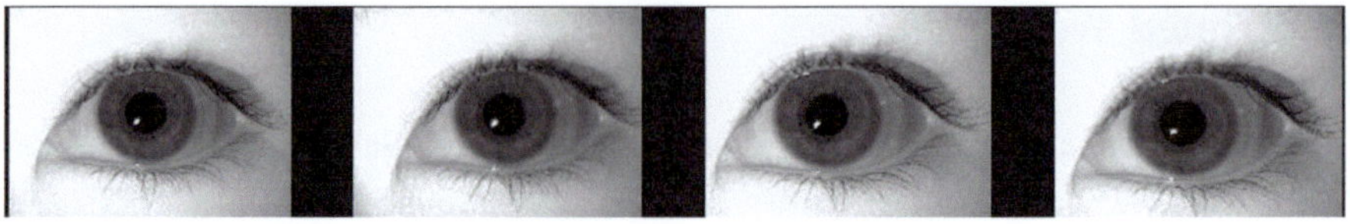

**Fig. 1** Dataset sample images

# 3 Methodology

The methodology for designing improved iris recognition system entails various steps. Below is an overview of the procedure:

## *3.1 Data Collection*

The MMU Iris Dataset, exploited resource in iris recognition study. This dataset contains 460 images. Sample images are as shown in Fig. 1.

## *3.2 Preprocessing*

Utilize preprocessing techniques to improve the quality of raw iris images:

Resize: The image of size 240 × 320 pixels is transformed to 100 × 100 pixels for better training and evaluation.

Normalization: Normalization is the technique of transforming iris images or feature vectors to a standardized format or coordinate system. The goal of normalization is to mitigate variations in iris images caused by factors such as scale, rotation, translation, and illumination.

Segmentation: This crucial step separates the iris region from the surrounding areas, such as the sclera, eyelids, and eyelashes, to extract the iris. The techniques used are canny edge detector, and though transform [14].

Image after preprocessing is as shown in Fig. 2.

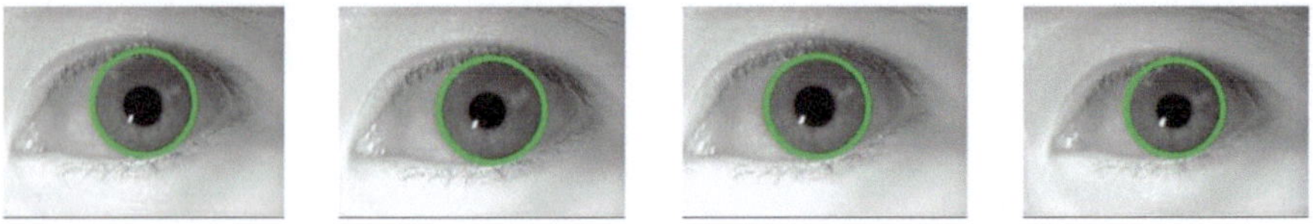

**Fig. 2** Image after preprocessing

### 3.3 Feature Extraction with Optimal Gabor Filters

Gabor filters are applied to preprocessed iris images. Gabor filters are specifically tuned to capture texture information at different frequencies and orientations. Optimal Gabor filters are determined through a process of optimization, which involves tuning the parameters of the Gabor filters (such as frequency, orientation, and bandwidth) to best capture discriminative features from the iris texture [15]. These optimal Gabor filters are then used to convolve with the iris images, resulting in feature maps that highlight texture patterns relevant for iris recognition.

### 3.4 Deep Belief Network Architecture

Design and implement a DBN architecture for feature representation and classification:

1. Initialize the DBN Architecture: Define the number of layers in the DBN and the number of nodes (neurons) in each layer. Initialize the weights and biases for each layer of the RBMs.
2. Stack multiple layers of RBMs for unsupervised pre-training: Construct the DBN by stacking multiple layers of RBMs. Each RBM layer is connected to the previous and next layers, forming a hierarchical architecture.
3. Fine-tune the network parameters using back propagation: After unsupervised pre-training, fine-tune the entire DBN using back propagation. Back propagation adjusts the weights and biases of the network using labeled data to minimize the classification error. Use techniques like stochastic gradient descent (SGD) or its variants to optimize the network parameters. Figure 3 illustrates the DBN architecture [16, 17].

### 3.5 Integration with Gated Recurrent Unit

Incorporate gated recurrent units into the DBN architecture to capture temporal dependencies within iris features. GRU architecture is shown in Fig. 4.

1. Integrate GRU into DBN: Add GRU to the DBN architecture to capture temporal dependencies within iris features. GRUs are chosen to model sequential information inherent in iris patterns. Use variants of GRU with update and reset gates to regulate the flow of information through the recurrent connections.
2. Variant selection: Choose GRU variants for integration into the DBN architecture. Both variants can capture temporal dependencies, but GRUs are often preferred for their simpler architecture [18].

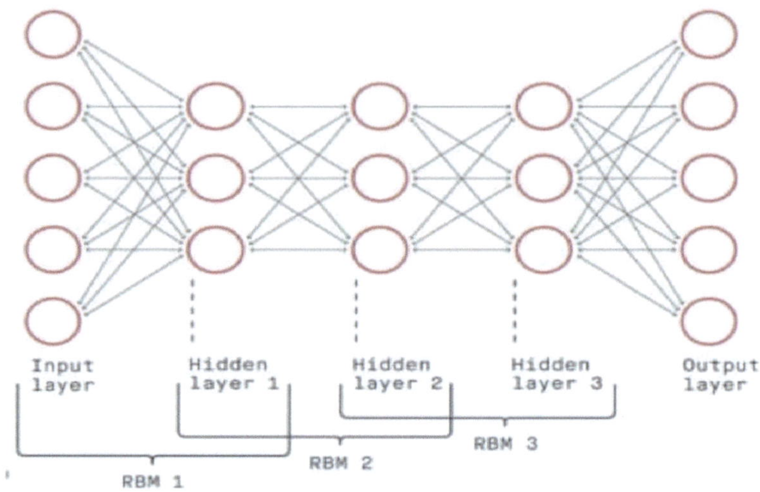


**Fig. 3** DBN architecture

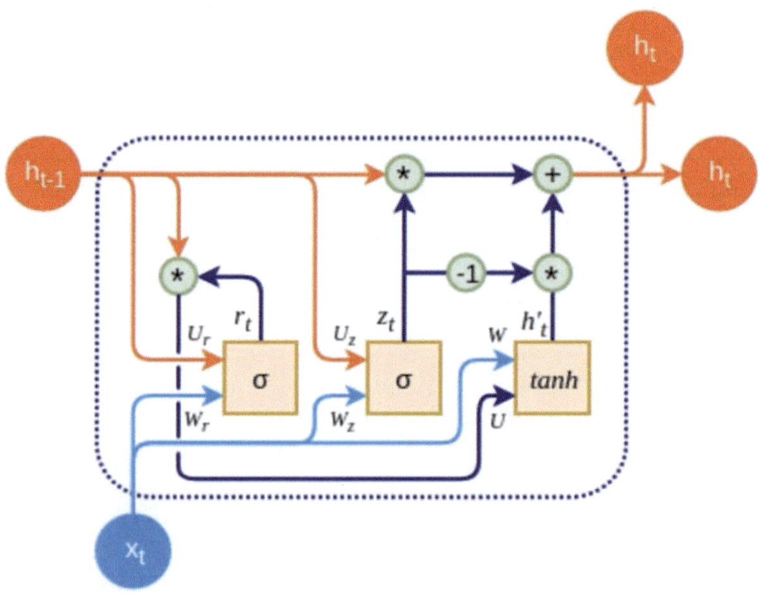


**Fig. 4** GRU architecture

## 3.6 Training and Optimization

Divide the dataset into training, validation, and testing sets. Train the deep learning architecture on the training set and optimize network parameters such as learning

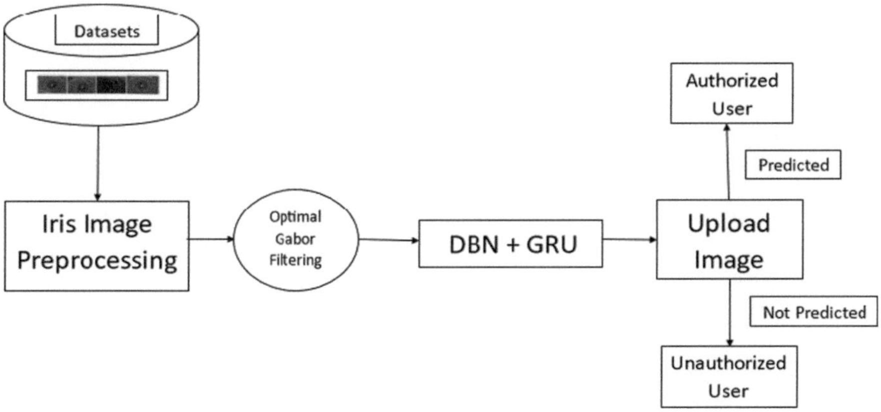


**Fig. 5** Proposed system architecture

rate, batch size, and regularization. Evaluate model performance using the test set to prevent overfitting.

### *3.7 Evaluation and Testing*

Assess the performance of the trained model using the test set. Measure key metrics such as accuracy, precision, recall, F1-score, specificity, sensitivity, false discovery rate, and false-negativity rate. Compare the performance of the proposed architecture with existing iris recognition methods. The proposed system architecture is as shown in Fig. 5.

## 4 Results

Figure 6 displays the training and validation accuracy values over epochs. Accuracy represents the proportion of correctly classified samples. The goal is to increase accuracy, indicating better performance of the model. A rising trend in both training (98%) and testing accuracy (87%) indicates that the model is improving.

Figure 7 illustrates the performance metrics such as accuracy, F1 score, recall, precision, specificity, sensitivity, false discovery rate, and false-negative rate.

Figure 8 shows the screen of an authorized user, and Fig. 9 shows screen of unauthorized user.

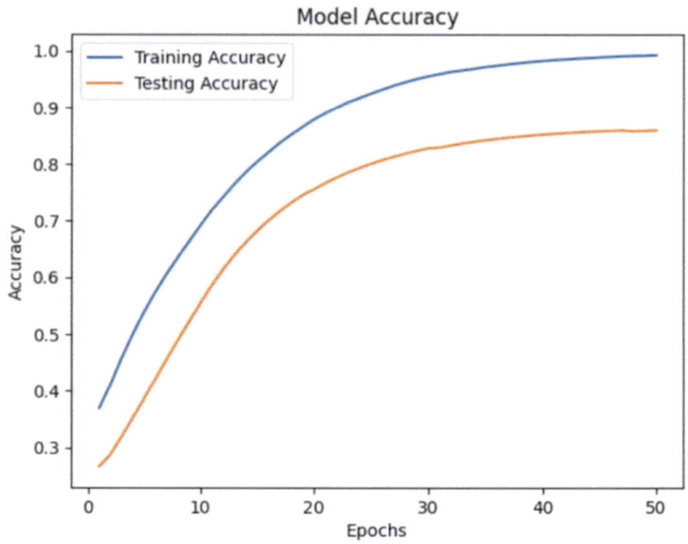


**Fig. 6** Model accuracy graph

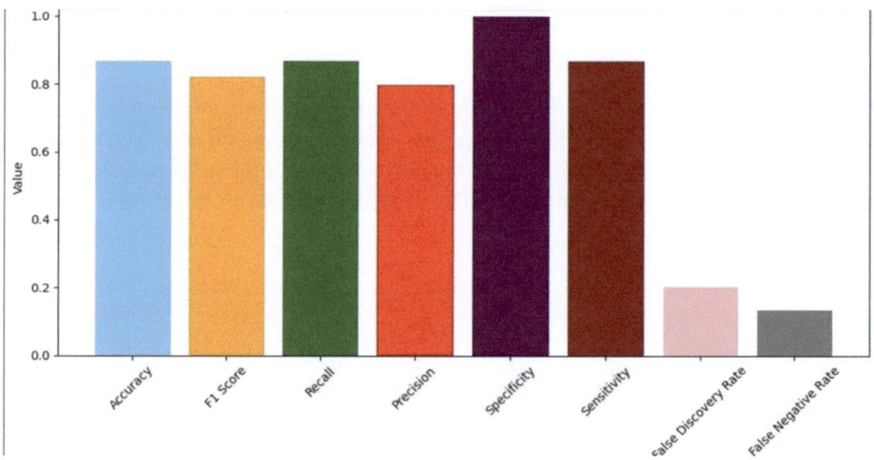


**Fig. 7** Performance metrics

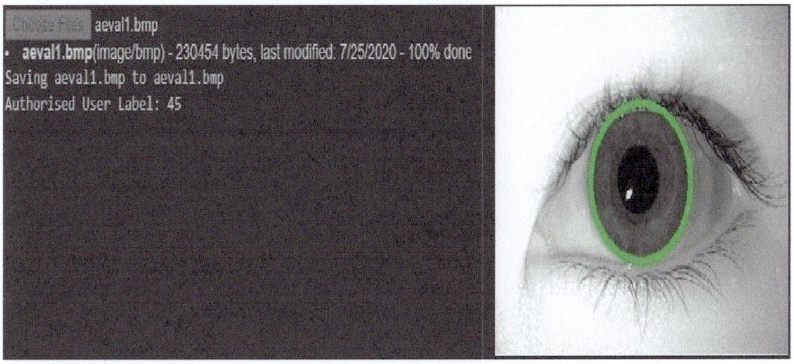


**Fig. 8** Authorized user

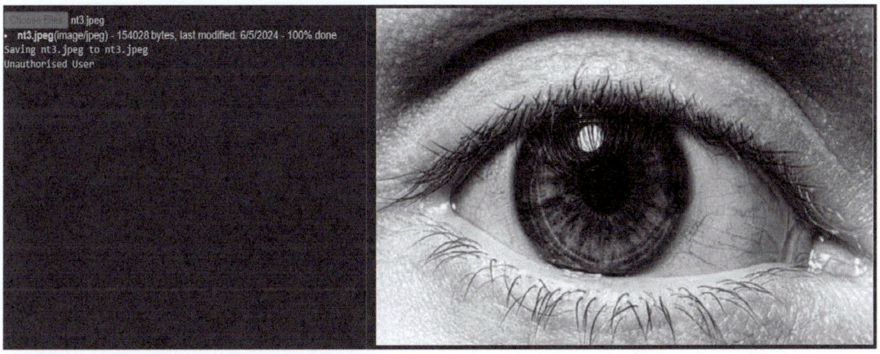


**Fig. 9** Unauthorized user

## 5 Conclusion

This paper signifies a notable leap forward in the realm of biometric authentication systems by amalgamating state-of-the-art methodologies derived from deep learning and image processing domains. The resulting architecture stands as a potent solution for iris recognition, fortified with resilience and efficacy.

Utilizing optimal Gabor filters, the model efficiently extracts unique texture features from iris images, collecting both complex local particulars and wide global trends critical for precise identification. The integration of a deep belief network design, improved by gated recurrent units (GRU), allows the network to understand hierarchical depictions of iris features, whereas also identifying the temporal dependencies present in the facts.

Experiential testing of the projected architecture highlights its efficiency in accurately recognizing individuals based on their distinctive iris patterns. Evaluations such as accuracy, precision, recall, and F1-score exhibit the system's consistency and robustness. Jointly, these evaluations emphasize the architecture's capability to not only precisely identify persons but also to retain consistency and efficiency across several operational scenarios.

Accordingly, the findings of this paper mark a noteworthy progression in the field of biometric verification, offering enhanced security and reliability in verification systems. Future research could examine the incorporation of iris identification with other biometric modalities, such as facial or voice identification, to develop more inclusive and dependable identification systems.

## References


1. Daugman JG (1993) High confidence visual recognition of persons by a test of statistical independence. IEEE Trans Pattern Anal Mach Intell 15(22):1148–1161
2. Wildes RP et al (1996) A machine-vision system for iris recognition. Mach Vis Appl 9(1):1–8
3. Boles WW, Boashash B (1998) A human identification technique using images of the iris and wavelet transform. IEEE Trans Signal Process 46(4):1185–1188
4. Shinyoung L et al (2001) Efficient iris recognition through improvement of feature vector and classifier. Etri J 23(2):61–70
5. Hinton GE (2002) Training products of experts by minimizing contrastive divergence. Neural Comput 14(8):1771–1800
6. Tieniu T et al (2012) Noisy iris image matching by using multiple cues. Pattern Recognit Lett 33(8):970–977
7. Monro DM, Rakshit S, Zhang D (2007) DCT-based iris recognition. IEEE Trans Pattern Anal Mach Intell 29(4):586–595
8. Nabti M, Bouridane A (2008) An effective and fast iris recognition system based on a combined multi-scale feature extraction technique. Pattern Recognit 41(3):868–879
9. Shah S, Ross A (2009) Iris segmentation using geodesic active contours. IEEE Trans Inf Forensics Secur 4(4):824–836
10. Daugman J (2004) How iris recognition works. IEEE Trans Circuits Syst Video Technol 14(1):21–30
11. Chang CP et al (2009) "Using empirical mode decomposition for iris recognition. Comput Stand Interfaces 31(4):729–739
12. Huo G (2015) Secondary iris recognition method based on local energy orientation feature. J Electron Imaging 24(1):013033
13. Guang-Hua H (2015) Automated defect detection in textured surfaces using optimal elliptical Gabor filters. Optik 126(14):1331–1340
14. Belcher C, Du Y (2009) Region-based SIFT approach to iris recognition. Opt Lasers Eng 47(1):139–147
15. Wojciech S et al (2010) Reliable algorithm for iris segmentation in eye image. Image Vis Comput 28(2):231–237
16. Ranzato MA et al. (2011) On deep generative models with applications to recognition. In: IEEE conference on computer vision and pattern recognition (CVPR), IEEE, pp 2857–2864
17. Ma L et al (2014) Efficient iris recognition by characterizing key local variations. IEEE Trans Image Process 13(6):739–750
18. He F, Han Y, Wang H, Ji J (2017) Deep learning architecture using optimal Gabor filters and deep belief network. J Electron Imaging 26(2):023005

# Drone-Assisted SOA Routing for Stable Communication in Emergency Ad Hoc Networks

**LNC. Prakash K, G. Ramasubba Reddy, N. Badrinath, C. Nithisha, S. Athinarayanan, and R. Kavitha**

**Abstract** As part of the Internet of Drone Things (IoDT), the use of drones to gather data from ground-based networks is progressing quickly. Comprising unmanned aerial vehicles (UAVs) into the Internet of Vehicles (IoV) has effectively addressed issues such as traffic congestion, data security, and energy consumption. However, the IoV's dynamic mobility and ad hoc traffic patterns contribute to improved network latency and overhead. Additionally, the existence of malicious attackers can further disrupt traffic flow. To overcome these difficulties, drone-assisted IoV has been developed to establish reliable communication in wireless mobile networks, especially in disaster-stricken areas. Traditional IP-based systems depend on location-based communication, but optimizing node density and decreasing the dependence on highly mobile nodes are required advancements. This paper presents a system that integrates the Snake Optimizer with the Sine Cosine Algorithm (SO-SCA), lowering overhead and improving node speed. The proposed method expedites network performance by sweetening the capacity to locate disaster epicenters. Drones are operated to update anchor nodes periodically, which helps disaster area networks remain stable.

LNC. Prakash K (✉)
Computer Science and Engineering-Data Science, CVR College of Engineering, Hyderabad, India
e-mail: klnc.prakash@gmail.com

G. Ramasubba Reddy
Department of CSE, Sai Rajeswari Institute of Technology, Proddatur, AP, India

N. Badrinath
Department of Software Systems, School of Computer Science and Engineering, Vellore Institute of Technology, Vellore, Tamilnadu, India
e-mail: badrinath.n@vit.ac.in

C. Nithisha
Department of Data Science, School of Computing, Mohan Babu University, Sree Sainath Nagar, Tirupati, Andhra Pradesh, India

S. Athinarayanan
Department of CSE, Vel Tech Rangarajan Dr. Sagunthala R&D Institute of Science and Technology, Avadi, Chennai, India

R. Kavitha
Department of CSE, Vel Tech High Tech Dr.Rangarajan Dr. Sakunthala Engineering College, Avadi, Chennai, India

 
K. R. Madhavi et al. (eds.), *Proceedings of Sixth International Conference on Computer and Communication Technologies*, Lecture Notes in Networks and Systems 1356,
https://doi.org/10.1007/978-981-96-5238-9_42

Geo-tracking is also shown to improve routing efficiency in mobile ad hoc networks (MANETs), with performance measured employing metrics such as packet latency and packet damage.

**Keywords** Drone things · Snake optimizer · Internet of vehicles · Internet protocol · Geo-tracking

## 1 Introduction

Reliable communication is necessary for effective response and recovery efforts in emergencies. Traditional communication infrastructures often suffer from damage or overload, severely restricting coordination among first responders. Unmanned aerial vehicles (UAVs) provide several benefits for both commercial and scholarly study [1, 2]. Drones with progressive communication capabilities can quickly deploy to crucial areas, forming a dynamic network that adapts to rapidly changing environments. The Internet of Drone Things (IoDT) has been developed to manage territories using drones [3]. In contrast, the Internet of Vehicles (IoV) enables a self-organizing, Internet-connected network of vehicles that enhances road safety and autonomous navigation [4]. However, IoV networks face challenges like link disconnection, inefficient resource allocation, and high costs. Integrating UAVs into IoV addresses these issues by providing self-communication capabilities and facilitating vehicle-to-vehicle (V2V), drone-to-drone (V2D), and drone-to-ground (G2G) communication.

The integration of drones into ad hoc networks not only improves connectivity but also enhances the resilience of communication systems under duress. Drones can serve as mobile relay points, expanding the range of communication and bridging gaps generated by damaged infrastructure. Moreover, maneuvring in three-dimensional space enables them to bypass barriers and optimize communication paths. Drones in vehicular ad hoc networks (VANETs) offer new possibilities for data flow management, improving congestion avoidance, energy efficiency, and safety [5]. Despite their benefits, the proliferation of VANETs poses security risks, making them vulnerable to breaches [6]. Disasters like natural calamities, terrorist attacks, or large-scale emergencies can seriously impair and render conventional communication networks unusable. In response to this pressing issue, drone-assisted service-oriented architecture (SOA) routing has emerged as a ground-breaking method for preserving reliable connectivity in emergency ad hoc networks.

A hybrid optimization strategy is proposed to optimize the communication between drones and vehicles, focusing on bio-inspired dynamic trust and congestion-aware zones. This approach enhances network reach through heterogeneous drone-based VANETs, incorporating triple-layer trust building and congestion optimization to mitigate malicious actions while reducing energy consumption and latency. The proposed methodology includes a Snake Optimizer (SO) and Sine Cosine Algorithm (SCA) for hybrid optimization. Performance analysis will evaluate metrics such as end-to-end latency and packet loss.

The following are the remaining sections of the paper: We discuss the relevant literature in Sect. 2 and provide a brief synopsis of the proposed model in Sect. 3. Sections 4 provide the validation analysis outcomes.

## 2 Related Work

SOA has been presented as a valuable crisis and disaster management tool by Alawad et al. [7]. UAV teams utilize delay-tolerant networking to facilitate mission exploration, with performance metrics such as delay, throughput, and route loss showing significant improvements over existing methods. By integrating task chains and geographic resources within a decentralized framework, Fang et al. [8] further improve disaster decision support by encouraging cooperation between data contributors and GIS specialists to create a spatial decision support system (SDSS). Bilal et al. [9] also introduce Beacon-oriented Emergency Message Distribution (BEMD), which optimizes emergency message dissemination and coverage while minimizing latency, particularly in high-mobility, low-density highway traffic scenarios. Simulation results demonstrate that BEMD surpasses state-of-the-art techniques, improving data coverage, latency, and packet delivery ratio (PDR) by 13%, 26%, and 4%, respectively.

A cognitive radio-based ad hoc network presented by Amini et al. [10] implicates secondary users (SUs) functioning as nomadic agents, gathering data relayed by fixed agents known as distributed relays (DRs). The DR preference list is ranked employing mean opinion scores and refined through a two-roulette wheel selection and random walk method. The framework also introduces a greedy adaptive solution to tackle the non-convex power allocation challenge, ultimately converging to a stable matching on both sides of an exchange. Authors [11–13] seek to enhance secure service provisioning for vehicular ad hoc networks (VANETs) through a fault tolerance authentication and QoS-aware Internet of Vehicles (IoV) framework that employs edge computing secured by blockchain technology. By processing communications locally, this approach reduces latency when roadside edge servers are implemented, increasing the efficiency and dependability of car networks. Furthermore, Scholars [14, 15] propose a system for prioritizing beacon messages in dedicated short-range communication (DSRC) channels, employing dynamic receivers for each message class to optimize bandwidth use and minimize unnecessary redundancy. Their approach does not require specialized hardware or a master controller.

# 3 Proposed Methodology

The proposed approach develops a synchronized network of drones to improve collaboration and data exchanges in specific areas. Mobile device users can select from various applications as they move, influenced by app popularity, user preferences, geography, and mobility. This study presents a method for analyzing location-based probability factors to provide users receive relevant mobile app recommendations by calculating Expected Transmission Time (ETX) and Link Loss Rate (LLR). These metrics help evaluate long-distance communication quality and inform data packet transfer decisions. Each mobile node's ETX and LLR are periodically assessed to predict future mobility. The hybrid service-oriented architecture (SOA) aims to identify low-resistance network pathways through optimal route computations between source and destination nodes.

## 3.1 *Estimating Expected Transmission Count (ETC) and Link Loss Rate*

In this section, we use the metrics of Link Loss Rate and ETC to predict a node's mobility. Initially, we calculate the Received Signal Strength Indicator (RSSI) and Link Loss during a designated prediction timer. Suppose the difference between ETX and Link Loss surpasses a certain threshold. In that case, the proposed method assesses network performance based on the rate of lost connections and the expected value of the Expected Transmission Count (ETC). Algorithm procedures are as follows:

**Algorithm 1** Procedures of Algorithms

**Step 1:** Continuously monitor *ETXcur* and *LLRcur* for each node at every timeslot.
**Step 2:** Compare *LLRcur* with *LLRThvalue*.
**Step 3:** If an anchor node (AN) is present as a transmission source, Calculate the movement pattern as trans $T$ [SN, CN, DN].
**Step 4:** Calculate transition support for each node $\rightarrow$ SN
**for** each node SN $\rightarrow$ all nodes N **do**
    $Tans[T] = \{L \in DN | O \in O[L]\} \rightarrow$ time $T$
    $L = \{pn_1, t_1, pn_2, t_2, pn_3, t_3, \ldots, pn_m, t_m\}$ where $si \in s$
**Step 5:** Acquire proof of transmission $C \rightarrow [SN]$
        At all nodes, get $Proof(O) = [CN, SN] \rightarrow [DN]$
        $= \frac{Sup([CN,SN,DN])}{Sup([CN \rightarrow at\ T, SN\ at\ T])}$
**Step 6:** Determine the estimated support for transmission *ETXpn* from $O[CN, SN, DN]$
**if** *LLRcur* is valid **then**
    For the duration of this timeslot, the node is deemed active.
    Include in the route list $LAN$
**if** Validation recalls SN $\rightarrow$ node **then**
    **if** $T$ time is close to $CS$ and $SS$ with high level of support evidence **then**
        Proceed to the next point of support DS, SN

The Link Loss Rate (LLR), which measures connection reliability, is used to evaluate transmission quality. A transfer connection is deemed poor and unreliable if the loss rate is within a predetermined level. The Expected Transmission Count (ETX) measures the number of successful transmissions. LLR and ETX are estimated for every time-slot for every mobile node. These expected values are used to forecast the likelihood that a node will move in the future.

## 3.2 Geospatial Graph-Based Mapping (GGM)

GGM analyzes relationships between unknown sensors and anchor nodes. It operates RREQ messages to collect data on each node's speed, status, and information flow, enabling anchor nodes to identify unknown nodes in the GPS network. The system evaluates transmission speed to determine node positions and stability. Mechanisms in connected terminals help stabilize movement, permitting nodes to return to their previous positions. A node's stability is impacted by how far it deviates from its expected range of motion.

The base station is responsible for evaluating the positions of non-anchor nodes. The stability of neighboring nodes over time is described as $S_m(t)$:

$$S_m(t) = \frac{1}{N_D}\sum_{j=1}^{N_D} R_j(t) + (1+b)M_y(t-1) \tag{1}$$

In this equation, the self-stability of the $j$-th neighboring node is shown by $S_m(t)$. The stability model effectively selects road nodes with strong adjacency, self-stability, and trustworthy hotel selections. Based on node positions and distances, it creates a network diagram. Sensor nodes (SNs) can use a variety of localization mechanisms to implement location-based routing. For communication, anchor nodes (ANs) need to know their positions and those of their neighbors. This means that actual location, rather than topology, determines routing decisions. The following steps include the Link Loss Rate (LLR) and Expected Transmission Delay (ETD) calculations.

## 3.3 Proposed Route Optimization

The challenge lies in determining the optimal time-slot allocation and resolving primary routing issues. The suggested hybrid approach addresses these concerns by assigning time-slots to each link based on the optimal solution derived from the path-selective formulation. This method effectively handles the complexity of describing neutral traits through mutual optimization.

## 3.4 *The Hybrid Snake-Sine & Cosine-Based Optimization Procedure*

The problem is a large-scale nonlinear optimization challenge for which the SO-SCA hybrid approach is proposed. This method combines two meta-heuristic strategies: SO, known for quickly converging to a global optimum in complex engineering problems, and SCA, which addresses unimodal, multimodal, and composite test functions. However, SO and SCA face difficulties consistently finding optimal or high-quality solutions in a single run, as parameters that perform well in one scenario may not be effective in another. To overcome these constraints, the hybrid SO-SCA approach is used, which uses two combination strategies to improve the phases of exploration and exploitation and produce high-quality solutions with fewer iterations.

### 3.4.1 Snake Optimizer

The SO algorithm is motivated by the courtship behavior of snakes, which often involves fierce competition for mates. Like other meta-heuristic approaches, SO starts with a randomly initialized population and iteratively updates solutions through exploration and exploitation phases. It assumes an equal number of males and females during the update process, which continues for a fixed number of iterations, $T$.

In the exploration phase, food is assumed to be scarce, causing the snakes to move randomly in search of resources. The quantity of food consumed is given by the formula $Q = 0.5 \times \exp\left(\frac{t-T}{T}\right)$. The following equations describe the movement of males and females during exploration:

$$X_{m,i}^{t+1} = X_{\text{rand},m}^{t} \pm C_1 \times A_m \times ((X_{\max} - X_{\min}) \times \text{rand} + X_{\min}) \tag{2}$$

$$X_{f,i}^{t+1} = X_{\text{rand},f}^{t} \pm C_1 \times A_f \times ((X_{\max} - X_{\min}) \times \text{rand} + X_{\min}) \tag{3}$$

Here, $X_{m,i}^{t+1}$ and $X_{f,i}^{t+1}$ define the positions of the $i$-th male and female at the next iteration, while $X_{\text{rand},m}^{t}$ and $X_{\text{rand},f}^{t}$ are randomly assigned male and female positions from the population. The current iteration is represented by $t$, and $A_m$ and $A_f$ are the abilities to find food for males and females, respectively. Whereas rand is a random number between 0 and 1, $C_1$ is a constant. The fitness values $f_{m,\text{rand}}$ and $f_{f,\text{rand}}$ correspond to the randomly selected male and female, while $f_{m,i}$ and $f_{f,i}$ are the fitness values of the $i$-th male and female, respectively.

During the exploitation phase, if food is abundant, the positions of the snakes are updated based on the ambient temperature, $T_{\text{temp}}$. When the temperature surpasses a certain threshold, the new positions of males and females are given by:

$$X_{i}^{t+1} = X_{\text{food}}^{t} \pm C_2 \times T_{\text{temp}} \times \text{rand} \times \left(X_{\text{food}} - X_{i}^{t}\right) \tag{4}$$

where $X^t_{\text{food}}$ is the position of the food source, and $C_2 = 2$. If the snakes are in combat or mating mode, the positions are updated operating different equations based on the mode. For combat mode, the updates are:

$$X^{t+1}_{m,i} = X^t_{m,i} \pm C_2 \times F_m \times \text{rand} \times \left(X_{f,\text{best}} - X^t_{m,i}\right) \tag{5}$$

$$X^{t+1}_{m,i} = X^{t+1}_{m,i} \pm C_2 \times F_f \times \text{rand} \times \left(X_{m,\text{best}} - X^{t+1}_{m,i}\right) \tag{6}$$

For mating mode, the new positions are calculated as:

$$X^{t+1}_{m,i} = X^t_{m,i} \pm C_2 \times M_m \times \text{rand} \times \left(Q \times X^t_{f,i} - X^t_{m,i}\right) \tag{7}$$

$$X^{t+1}_{f,i} = X^t_{f,i} \pm C_2 \times M_f \times \text{rand} \times \left(Q \times X^t_{m,i} - X^t_{f,i}\right) \tag{8}$$

Here, $F_m$ and $F_f$ describe the fighting abilities of males and females, while $M_m$ and $M_f$ suggest their mating abilities. These are calculated utilizing an exponential function based on the fitness values of the best and current individuals.

### 3.4.2 Sine Cosine Algorithm (SCA)

The SCA was introduced to improve the balance between exploration and exploitation and employs sine and cosine functions to move candidate solutions outward for exploration and toward better solutions for exploitation. The algorithm starts by initializing the population randomly, and then each individual's position is updated using the following equation (9). In this equation, to compute $r1$, one needs to divide the total number of iterations (T) by the current iteration (Ct), a constant ($C$, and a maximum number of iterations ($T$). Random values ranging from 0 to 1 represent the words $r2$, $r3$, and $r4$. The position update rule is defined as:

$$X^{t+1}_i = \begin{cases} X^t_i + r_1 \times \sin(r_2) \times r_3 \times \left(X^t_{\text{best}} - X^t_i\right), & \text{if } r_4 < 0.5 \\ X^t_i + r_1 \times \cos(r_2) \times r_3 \times \left(X^t_{\text{best}} - X^t_i\right), & \text{if } r_4 \geq 0.5 \end{cases} \tag{9}$$

Here, $X^t_i$ defines the current position of the $i$-th candidate, $X^t_{\text{best}}$ is the position of the best solution found so far, and $r_1, r_2, r_3$, and $r_4$ control the movement direction and step size. Depending on the value of $r_4$, either the sine or cosine function is involved in adjusting the position of the candidate solution, facilitating either exploration or exploitation.

### 3.4.3 Proposed Hybrid Scheme

We propose two distinct methods for integrating SO-SCA. In this hybrid approach, both algorithms compute multiple variables simultaneously. The method begins by randomly generating candidate solutions and evaluating their fitness values.

First, the population is divided into two equal groups. For the first group, new solutions are generated using SO's update mechanism, as described in Eqs. 2 and 8. The second group's positions are updated using the SCA update strategy, defined in Eq. 9. The optimal solution from both groups is then determined. Based on the results, a portion of the population is dynamically reassigned. If SO provides a better solution, more individuals are allocated to SO in the next iteration according to a switching rate. Similarly, more individuals will be assigned to SCA if SCA outperforms SO. In the second combination strategy, the SCA update is applied first to the population, and the fitness and position of the best solution are updated accordingly. Following this, the SO update strategy is employed to refine the positions of variables, with the final best solution determined by SO's output. This adaptive approach enables a dynamic balance between each iteration's exploration and exploitation phases.

## *3.5 Drone-to-Drone (D-D) Communication*

In crowded metropolitan environments, cars confront various dynamic issues that drones effectively address. These UAVs can be remotely controlled or run independently thanks to sophisticated routing models. They enable effective functioning by including multiple IoT devices, such as GPS, temperature, and range sensors. Drones facilitate reliable communication when integrated into a network, utilizing a more optimal approach than traditional vehicle data transmission. Despite limited protocol support, drones can establish connections intelligently.

## *3.6 Drone Routing*

Drones use the Zone Routing Protocol (ZRP) for efficient routing, constant packet size control, and predictive connection setup. One or two drones monitor each network zone. The protocol includes four stages: route request (RREQ), route reply (RREP), route expiry (RE), and route error (RERR). RREQ packets contain addresses for the source and receiver drones and factors. The packet tracks hops, sequence numbers, and zone data. Each hop updates parameters like velocity and location. When an RREQ reaches the receiver, a gain measurement (G_M) is estimated as:

$$G_M = \left[\frac{R_E}{L_t}\right] \times \left[\frac{R_E}{E_f}\right] \times \left[\frac{D_p}{1+\delta}\right] \tag{10}$$

The RREP is sent along the path with the highest gain, and if it expires, the route revival mechanism selects the following best route.

## 4 Result and Discussion

The proposed work is implemented employing the NS-2 network simulator (NS-2). Table 1 represents the simulation outcomes, where the network restrictions are processed by progressively increasing the steps executed between 20 and 200 s for the simulation nodes. Although the buffer size is usually set to 20, in most simulations, each node is assigned a buffer size of 512 packets.

### 4.1 Performance Metrics and Measurements

The suggested technique uses packet loss rate as an evaluation metric.

- End-to-End Delay: The total time packets travel through a network. The end-to-end delay is expressed by:

$$\text{End-to-End Delay} = \sum_{h=1}^{h\ max} \frac{D(N_i, N_j)}{S} \tag{11}$$

  where $h$ defines the hop count between the $i$-th and $j$-th nodes, $S$ signifies the signal propagation speed and the distance $D(N_i, N_j)$ lies between the $i$-th and $j$-th nodes.
- Packet Delivery Ratio (PDR): This metric defines the proportion of packets successfully received at the destination out of all the packets the sensor node has transmitted:

**Table 1** Simulation parameters

| Value | Simulation parameter |
|---|---|
| 120 | No. of nodes |
| 512 byte | Packet size |
| Log-normal path loss model | Channel model |
| NS-2.28 | Simulator |
| 1000 m × 1000 m | Topology size |
| IEEE 802.15.4 | MAC |
| CBR | Type of traffic |

**Table 2** Comparison of PDR (%) across different methods

| Methods | 100 | 200 | 300 | 400 | 500 |
|---|---|---|---|---|---|
| ACO | 0.86 | 0.78 | 0.76 | 0.69 | 0.66 |
| GWO | 0.90 | 0.87 | 0.84 | 0.82 | 0.78 |
| SO | 0.95 | 0.91 | 0.88 | 0.87 | 0.82 |
| SO-SCA | 0.99 | 0.96 | 0.95 | 0.94 | 0.93 |

$$\text{PDR} = \frac{\text{Packets received at destination}}{\text{Packets sent by sensor node}} \times 100 \tag{12}$$

- Packet Loss Ratio (PLR): This metric evaluates the percentage of packets lost during transmission between a source and a destination:

$$\text{PLR} = \frac{\text{Number of lost packets}}{\text{Total transmitted packets}} \times 100 \tag{13}$$

Data for these measures are gathered and validated employing various optimization strategies, including Grey Wolf Optimization (GWO), Ant Colony Optimization (ACO), and Sine Cosine Algorithm (SCA).

When the node count is 100, all models exhibit strong performance displays in Table 2 and Fig. 1. Specifically, ACO performed at 86%, GWO reached 90%, SO performed at 95%, and the proposed hybrid model (SO-SCA) attained 99% PDR.

However, as the number of nodes increased, the PDR for all models decreased. For example, with 500 nodes, ACO showed 66%, GWO achieved 78%, SO recorded 82%, and the proposed SO-SCA model maintained the highest PDR at 93%.

Regarding Packet Loss Ratio (PLR), the proposed model has only 1% PLR for 100 nodes but raised to 7% for 300 and 11% for 500. In contrast, ACO started at 14% PLR for 100 nodes, increased to 24% for 300 nodes, and reached 34% for 500 nodes.

Similarly, GWO showed a PLR of 13% for 100 nodes, 16% for 300 nodes, and 22% for 500 nodes, while SO had a 5% PLR for 100 nodes and 18% for 500 nodes. The SO-SCA model outperformed other methods consistently across different node counts with lower PLR, as depicted in Table 3 along with Fig. 2.

Despite obstacles like floods and insider threats, the suggested hybrid paradigm reduces end-to-end delays. Table 4 illustrates that end-to-end latency for the different approaches is minimal, with the SO-SCA model performing best across all node counts. For example, the ACO approach displays a 9 ms delay with 100 nodes, whereas the SO-SCA method records a delay of merely 1 ms. As the number of nodes rises, SO-SCA continues to maintain reduced latency.

The resilience of the hybrid method is highlighted in Fig. 3, which shows the significant impact that threats like location falsification and wormhole assaults have on end-to-end latency in the proposed network.

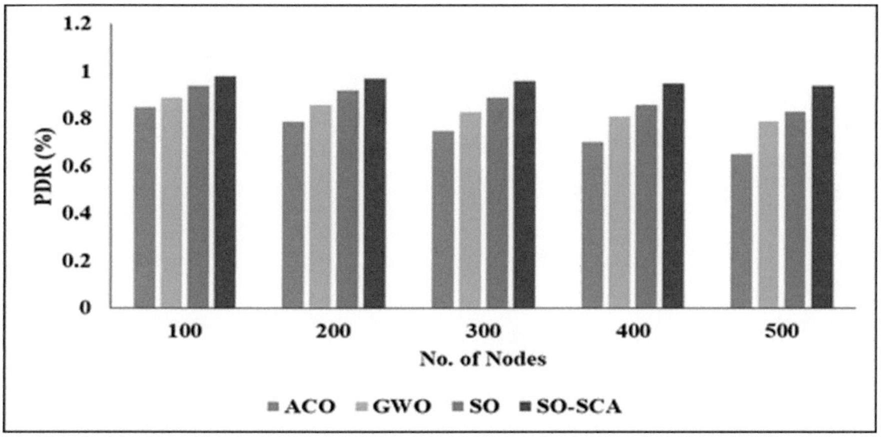


**Fig. 1** Comparison of PDR

**Table 3** PLR Analysis across different methods

| Methods | 100 | 200 | 300 | 400 | 500 |
|---|---|---|---|---|---|
| ACO | 14 | 19 | 24 | 29 | 34 |
| GWO | 13 | 17 | 22 | 26 | 30 |
| SO | 5 | 10 | 15 | 18 | 20 |
| SO-SCA | 1 | 3 | 7 | 9 | 11 |

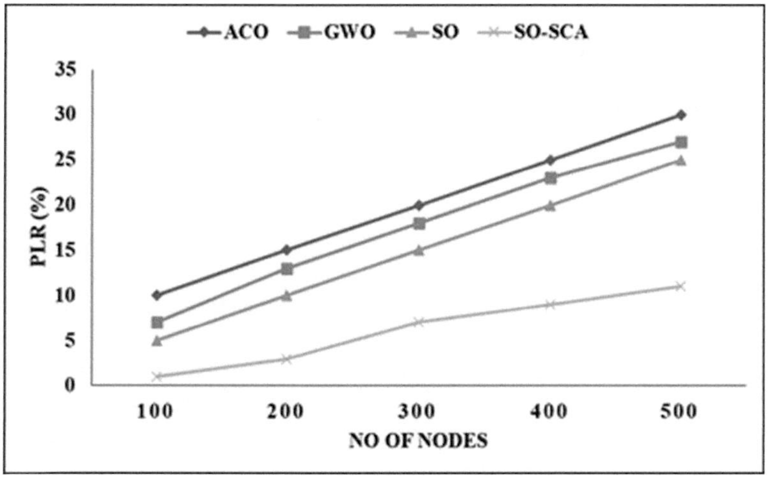


**Fig. 2** Training and validation accuracy

Table 4 End-to-end delay analysis (ms)

| Methods | 100 | 200 | 300 | 400 | 500 |
|---|---|---|---|---|---|
| ACO | 9 | 7.3 | 8.1 | 8.8 | 9 |
| GWO | 5 | 6 | 7 | 8 | 8.8 |
| SO | 3 | 4.2 | 5.6 | 5.6 | 6.1 |
| SO-SCA | 1 | 2.9 | 3.4 | 4.9 | 5 |

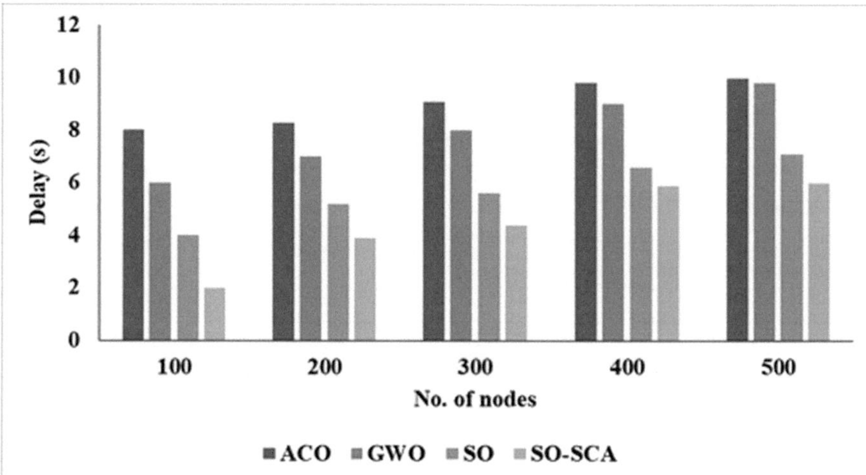


**Fig. 3** Analysis of delays

## 5 Conclusion

Data routing and collection are essential functions of every wireless mobility sensor node. Numerous geographic routing strategies have been investigated, to determine the advantages and disadvantages of each. It is because of these limitations that new models have been created. Stable connections and effective communication in emergency networks are the goals of the Spatial Drone-Based Route Optimization technique, which uses a hybrid snake optimization technique. Rather than using the conventional Global Positioning System (GPS), this technology implements an enhanced node localization system to increase network performance. It outperforms current techniques regarding efficiency and dependability by efficiently tracking node locations and forecasting node motions.

## References


1. Jghef YS, Jasim MJM, Ghanimi HM, Algarni AD, Soliman NF, El-Shafai W, Zeebaree SR, Alkhayyat A, Abosinnee AS, Abdulsattar NF et al (2022) Bio-inspired dynamic trust and congestion-aware zone-based secured internet of drone things (siodt). Drones 6(11):337
2. Mehbodniya A, Webber JL, Karupusamy S et al (2022) Improving the geo-drone-based route for effective communication and connection stability improvement in the emergency area ad-hoc network. Sustainable Energy Technologies and Assessments 53:102558
3. Kumar SP, Reddy YSB, Raghavendra G, Kumar KK (2024) Improvement of communication stability by geo-drone-based soa routing algorithm in emergency area ad-hoc network. In: 2024 International conference on knowledge engineering and communication systems (ICKECS), vol 1. IEEE, pp 1–7
4. Darwish TS, Bakar KA (2018) Fog based intelligent transportation big data analytics in the internet of vehicles environment: motivations, architecture, challenges, and critical issues. IEEE Access 6:15679–15701
5. Jajala KK, Buduri R (2024) Efficient and secure routing with uav: Guidedpheromone update based on improved ant colony optimization and fuzzy logic for congestion control in vehicular ad-hoc network. Int J Inform Technol 1–22
6. Syla V, Lala A, Biberaj A (2024) Vanet security and privacy-an overview. EIRP Proc 19(1):414–423
7. Alawad W, Halima NB, Aziz L (2023) An unmanned aerial vehicle (uav) system for disaster and crisis management in smart cities, Electronics 12(4). https://doi.org/10.3390/electronics12041051. https://www.mdpi.com/2079-9292/12/4/1051
8. Fang Z, Yue P, Zhang M, Xie J, Wu D, Jiang L (2023) A service-oriented collaborative approach to disaster decision support by integrating geospatial resources and task chain. Int J Appl Earth Observ Geoinform 117:103217. https://doi.org/10.1016/j.jag.2023.103217. https://www.sciencedirect.com/science/article/pii/S1569843223000390
9. Bilal M, Munir EU, Ullah A (2023) Bemd: Beacon-oriented emergency message dissemination scheme for highways. Ad Hoc Netw 142:103095. https://api.semanticscholar.org/CorpusID:256144794
10. Amini T, Abouei J (2022) A matching-based context-aware relay assignment scheme with power control in ad-hoc wireless networks. IET Commun 17:348–361. https://api.semanticscholar.org/CorpusID:254345225
11. Dhivya K, Kanna RR (2023) A novel blockchain-based fault-tolerant authentication technique for secure and qos aware vehicular ad hoc network using. J Survey in Fisheries Sci 10(4S):190–203
12. Ramana K, Srivastava G, Kumar MR, Gadekallu TR, Lin JC-W, Alazab M, Iwendi C (2023) A vision transformer approach for traffic congestion prediction in urban areas. IEEE Trans Intell Transport Syst 24:3922–3934. https://api.semanticscholar.org/CorpusID:255739140
13. Kumar VKA, Kumar MR, Shribala N, Singh N, Gunjan VK, Siddiquee KNEA, Arif M (2022) Dynamic wavelength scheduling by multiobjectives in obs networks. https://api.semanticscholar.org/CorpusID:248409658
14. Chalapathi MV, Kumar MR, Sharma N, Shitharth S (2022) Ensemble learning by high-dimensional acoustic features for emotion recognition from speech audio signal. Secur Commun Netw 2022(1):8777026
15. Rudra Kumar M, Pathak R, Gunjan VK (2022) Machine learning-based project resource allocation fitment analysis system (ml-prafs). In: Computational intelligence in machine learning: select proceedings of ICCIML 2021, Springer, pp 1–14

# Multi-class Cancer Subtype Classification from CT Images Using an EfficientNet–Autoencoder-Based Hybrid Model

**Sreekanth Yalavarthi, Rambabu Inaganti, K. Ayyappa Swamy, K. Saradhi, Naresh Tangudu, and O. Nagaraju**

**Abstract** Lung cancer is considered as critical cancers globally, accurate classification of lung nodules plays a major role in early prediction and planning for treatment. In this paper, a hybrid deep learning model that fuses EfficientNet and Autoencoder for multiclass classification of lung nodules has been proposed. Three classes focused are benign, malignant, and metastatic malignant representing nodules with secondary cancerous growths. Feature extraction is performed by utilizing EfficientNet architecture considering its scalability and efficiency. Autoencoder architecture is utilized for dimensionality reduction by capturing important features required for classification in a compressed latent space. The proposed hybrid model is evaluated using LIDC-IDRI dataset containing annotated images of lung CT scan. Experimental results reveal that the performance of the proposed model in terms of classification accuracy and other performance metrics is better compared to existing methods. Also, the proposed hybrid model leads to better generalization during learning while handling high-dimensional data due to its efficiency in noise reduction and feature compression.

S. Yalavarthi
HCL Technologies, Buffalo Grove, Il, USA

R. Inaganti (✉)
Smith and Nephew, Pittsburgh, PA, USA
e-mail: rambabu8@gmail.com

K. Ayyappa Swamy
Department of ECE, Aditya University, Surampalem, Andhra Pradesh, India

K. Saradhi
M.B. School of Commerce and Management, Mohan Babu University, Tirupati, India

N. Tangudu
Aditya Institute of Technology and Management, Tekkali, Andhra Pradesh, India

O. Nagaraju
Department of Computer Science, Government Degree College, Dornala, India

 
K. R. Madhavi et al. (eds.), *Proceedings of Sixth International Conference on Computer and Communication Technologies*, Lecture Notes in Networks and Systems 1356,
https://doi.org/10.1007/978-981-96-5238-9_43

**Keywords** Lung nodule classification · EfficientNet · Autoencoder · Medical image analysis · Dimensionality reduction · Benign · Malignant · Metastatic malignant

## 1 Introduction

Lung cancer is the major causes of death related to cancer globally, early prediction of cancer is important for the survival of patients. Computer tomography (CT) scans are utilized widely to detect lung nodules that may be benign or malignant. Accurate classification of these nodules is important to plan for appropriate treatment. Manual screening of CT scan images is a time taking process and prone to errors due to variations in different classes of nodules. To address these issues, a hybrid model utilizing the strengths of EfficientNet and Autoencoder has been proposed. EfficientNet is capable of handling scalability and performing efficient feature extraction. Autoencoder has the ability to reduce dimensionality by identifying relevant features in a compressed latent space. The integration of these two models allows for efficient classification of lung nodules into three classes such as benign, malignant, and metastatic malignant.

The proposed method uses LIDC-IDRI dataset with CT scan images of lung nodules containing annotated labels defined by radiologists. The proposed architecture is designed to overcome challenges such as noisy data and intraclass variability. The use of hybrid architecture improves the accuracy of model and other performance metrics, which are important for medical applications. This paper proposes a hybrid EfficientNet-Autoencoder approach for efficient classification of lung nodules. The performance of the proposed approach is efficient than methods such as CNN, ResNet, VGG-16, and InceptionNet architecture.

## 2 Background Work

This section involves reviewing existing literature related to lung nodule classification, deep learning architectures, and autoencoder model. Lung cancer is a major health concern globally, with early prediction being a concern for proper treatment. Recently, deep learning architectures like convolutional neural network (CNN) applications reveal better results in automating the process of lung cancer prediction. Utilization of deep learning methods for lung cancer diagnosis has gained attention due to its ability to learn features from high-resolution CT scan images [1–6]. Multiview-based CNN model achieved better accuracy in identifying malignant and benign lymph nodes of lung images [7]. The ResNet architecture addressed the vanishing gradient problem through the use of skip connections [8–12]. However, these models are prone to overfitting but improves feature extraction. VGG16 is the first model used for medical image analysis, but it is computationally expensive and not efficient for large CT scan images [13–17]. The authors [18] applied InceptionNet model for lung cancer classification and achieved better performance than existing methods. Autoencoder architecture was utilized for dimensionality reduction and improved model generalization [19–21]. The authors [9, 22–25] introduced EfficientNet architecture that can handle uniform scales, depth, width and resolution using compound scaling. The model demonstrated better performance compared to other existing methods. Also, hybrid models combining CNN with autoencoders gained attention in medical imaging for feature extraction and noise reduction [26, 27]. Considering these aspects a hybrid EfficientNet–Autoencoder model has been proposed to perform lung nodule sutype classification.

## 3 Methodology

This section briefs about the steps of proposed architecture for classifying CT mages of cancer using hybrid EfficientNet–sAutoencoder architecture. The steps for the proposed method for classifying lung cancer have been specified in Fig. 1.Initial step is to collect dataset containing CT scan images. Proposed work utilizes LIDC-IDRI dataset containing samples of lung nodules with multiple classes for processing. The steps of the proposed model consist of preprocessing, feature extraction and classification. Initial step is to preprocess the input data to be given as input to neural network architecture. EfficientNet requires inputs of fixed size, considering this aspect resizing is performed for CT images. Then the pixel value is normalized between 0 and 1. The preprocessed images are given as input to EfficientNet architecture to perform the task of feature extraction. This architecture utilizes compound scaling method to scale in a balanced manner, particularly for high-resolution images. Several variants from B0 to B7 available for this architecture. This architecture can be used to capture visual low level edges of images to high level patterns through convolution layers. The output will be high level features for efficient lung nodule classification. Also, this architecture can be utilized to extract required features from CT images reducing computational overhead. The architecture of EfficientNet specified in Fig. 1 consists of stem block which includes convolution to reduce the dimension of the image while expanding the feature depth. Depthwise separable convolution component is used to capture spatial and channel specific patterns of images. Inverted residual block functions similar to Resnet by skip connections to avoid vanishing gradient issues. Squeeze and excitation block adjusts the importance of each feature map by learning weights focusing on important features for cancer subtype classification. The output of this architecture will be high-dimensional feature map with relevant features from the input cancer images. This output serves as input to autoencoder.

The autoencoder architecture is used to compress and denoise the feature map generated by EfficientNet architecture. The encoder part of his architecture applies several fully connected convolution layers to reduce the dimension of the feature space. This process leads to latent space representation with the salient features

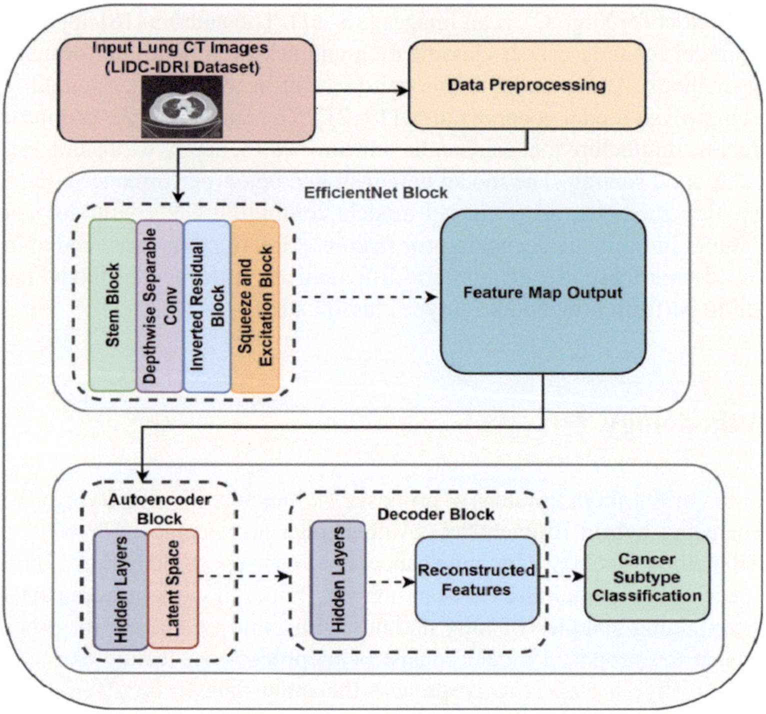


**Fig. 1** Multiclass cancer prediction architecture using EfficientNet and Autoencoder

removing unwanted features. Latent space has lesser dimensions than the original feature map but retains the important features to perform classification. This process improves the efficiency of the model and also avoids the problem of overfitting. Then the decoder takes this input and reconstructs feature map to higher dimensional feature space. The decoders output will be enhanced and noise filtered version. This process ensures that the classification layer receives clean feature set for accurate predictions. The activation function like ReLU and softmax is applied for classification. After classification categorical cross-entropy loss is applied to compute the error between the predicted and actual cancer subtype labels. Adam optimization is used to minimize the loss and perform weight updation in EfficientNet and autoencoder. The combination of EfficientNet with autoencoder enables accurate cancer subtype prediction.

The steps of the proposed EfficientNet–Autoencoder algorithm are specified below:

**Algorithm: EfficientNet–Autoencoder for Multiclass Cancer Classification**

**Input:** CT images $X = X = \{\ x_1\ , x_2\ \ldots\ x_n\}$ from LIDC-IDRI dataset

**Step 1: Preprocessing**

- For each image sample in the dataset, resize and normalize the image using Eqs. (1) and (2).

  $x_i{}' = \text{Resize}(xi, H', W')$

$$x_i' = \text{Resize}\left(x_i, h', w'\right) \tag{1}$$

$$x_i^{\text{norm}} = \frac{x_i' - \min(x_i')}{\max\left(x_i'\right) - \min\left(x_i'\right)} \tag{2}$$

**Step 2: EfficientNet Feature Extraction**

- Extract feature map by passing preprocessed image as specified in Eq. 3.

$$F_i = \text{EfficientNet}\left(x_i^{\text{norm}}\right) \tag{3}$$

**Step 3: Autoencoder Compression**

- Compress the high-dimensional feature maps $F_i$ using the encoder part of an autoencoder, resulting in a latent representation $Z_i$ as specified in Eq. 4.

$$Z_i = \text{Encoder}(F_i) \tag{4}$$

where $Zi \in R^d$, $d$ represents reduced dimensionality of latent space.

**Step 4: Classification**

- Pass the latent representation to fully connected layers to generate logits for classification of 3 classes benign, malignant, metastatic malignant specified in Eq. 5.

$$h_i = FC(z_i) \tag{5}$$

- Apply softmax function to convert logits as class probabilities.
- Compute loss using categorical cross-entropy between predicted and true labels.
- Update the models weight using Adam optimization to minimize loss.

**Step 5: Evaluate the Model**

- Perform evaluation of proposed method utilizing evaluation metrics.

## 4 Results and Discussions

The dataset used is LIDC-IDRI dataset with lung CT scan images. This dataset can be used for the classification of lung nodules. Over 1000 thoracic CT scan images were available in the dataset. The classes used in the proposed approach are benign, malignant and metastatic malignant. Annotations for the images were performed using multiclass labels based on their malignancy. Benign represents the nodules that are noncancerous, malignant represents the cancerous nodules with the possibility to invade surrounding tissue, metastatic malignant represents secondary cancerous growth spreading across other parts of the body. The samples of the three classes are mentioned in Fig. 2. These classes are used to train the proposed EfficientNet–Autoencoder based hybrid model to classify lung cancer.

The performance evaluation of proposed hybrid architecture through various metrics are specified in Table 1. Accuracy of proposed EfficienNet–Autoencoder approach is promising than traditional methods that are in existence. The proposed EfficientNet–Atoencoder approach for lung cancer detection shows better performance, compared to existing methods based on the evaluation metrics. The accuracy of 0.92 and ROC-AUC of 0.94 is remarkable. The proposed model works efficiently in both accurately identifying lung nodules cancer cases and differentiating between benign and different malignant cases.

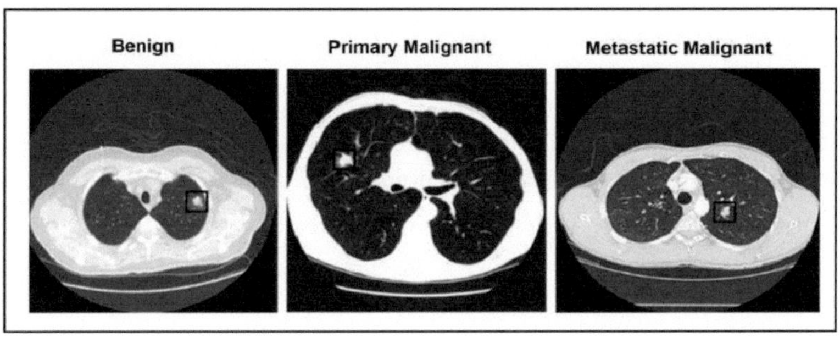


**Fig. 2** Lung cancer CT image data samples for three classes used in EfficientNet–Autoencoder model

**Table 1** Performance evaluation of EfficientNet–Autoencoder model

| Methodology | Accuracy | Precision | Recall | F1-score | AUC-ROC |
|---|---|---|---|---|---|
| Proposed (EfficientNet-Autoencoder) | 0.92 | 0.90 | 0.89 | 0.90 | 0.94 |
| CNN | 0.88 | 0.84 | 0.83 | 0.84 | 0.87 |
| ResNet | 0.89 | 0.87 | 0.88 | 0.87 | 0.91 |
| VGG-16 | 0.87 | 0.86 | 0.85 | 0.85 | 0.89 |
| InceptionNet | 0.90 | 0.88 | 0.87 | 0.88 | 0.92 |

The increase in accuracy of the model during training over epochs specifies the efficiency of the model to perform classification of lung nodules accurately. The validation accuracy of the model is also measured for unseen data as specified in Fig. 3. The result reveals that the proposed EfficientNet–Autoencoder model has the ability to generalize its learning from the training processes.

Figure 4 specifies the loss of EfficientNet–Autoencoder model during training and validation. The model adjusts the hyperparameters during the learning process to reduce the loss. Decrease in validation loss reveals that the performance of the model is better for unseen data. Measuring the training, validation accuracy and loss are important to analyze learning ability of the model. The results demonstrate that the proposed EfficientNet–Autoencoder based model classifies lung nodules accurately and can be utilized by medical practitioners for accurate diagnostics.

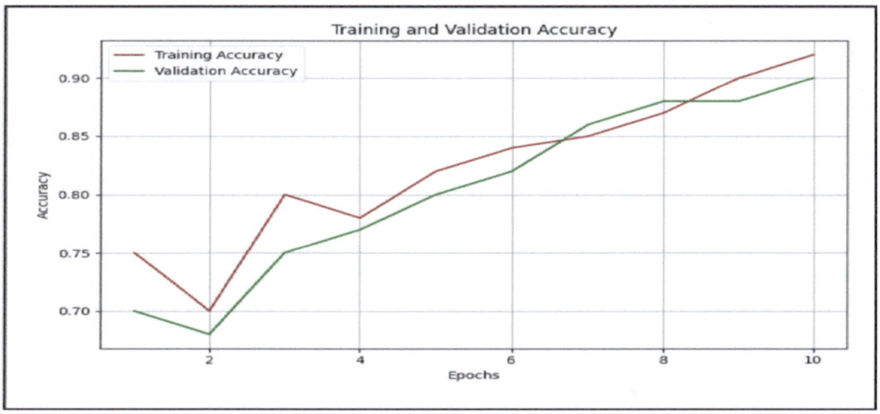


**Fig. 3** Accuracy of EfficientNet–Autoencoder model during training and validation

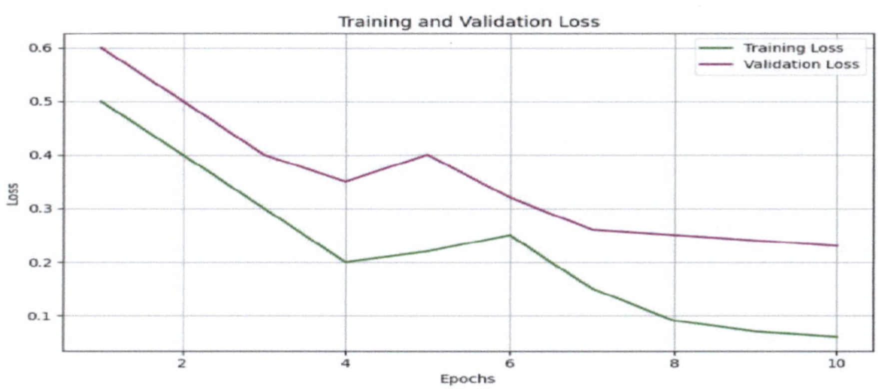


**Fig. 4** Loss of the proposed EfficientNet–Autoencoder model during training and validation

## 5 Conclusion

In this paper, a hybrid deep learning model integrating EfficientNet and Autoencoder has been proposed for the classification of lung nodules into three classes such as benign, malignant, and metastatic malignant. By utilizing the benefit of EfficientNet for feature extraction and Autoencoder's ability to reduce dimensionality and noise, the model achieves better performance for classification. The proposed model demonstrated better performance in terms of various performance metrics compared to existing methods such as CNN, ResNet, VGG16, and InceptionNet. In future, the proposed model can be tuned by using optimization techniques like Bayesian and evolutionary optimization methods. The model can also be extended to handle multimodal data considering PET scans and clinical records. Further, implementing the model in real-time applications for automated screening of lung cancer could improve the early diagnosis.

## References

1. Mohammed SH, Çinar A (2021) Lung cancer classification with convolutional neural network architectures. Qubahan Acad J 1(1):33–39
2. Alakwaa W, Nassef M, Badr A (2017) Lung cancer detection and classification with 3D convolutional neural network (3D-CNN). Int J Adv Comput Sci Appl 8(8)
3. Tekade R, Rajeswari K (2018) Lung cancer detection and classification using deep learning. In: 2018 fourth international conference on computing communication control and automation (ICCUBEA), August, IEEE, pp 1–5
4. Zhang C, Sun X, Dang K, Li K, Guo XW, Chang J, Yu ZQ, Huang FY, Wu YS, Liang Z, Liu ZY (2019) Toward an expert level of lung cancer detection and classification using a deep convolutional neural network. Oncologist 24(9):1159–1165
5. Baranwal N, Doravari P, Kachhoria R (2022) Classification of histopathology images of lung cancer using convolutional neural network (CNN). In: Disruptive developments in biomedical applications, CRC Press, pp 75–89
6. Pfeffer MA, Ling SH (2022) Evolving optimised convolutional neural networks for lung cancer classification. Signals 3(2):284–295
7. Ashhar SM, Mokri SS, Abd Rahni AA, Huddin AB, Zulkarnain N, Azmi NA, Mahaletchumy T (2021) Comparison of deep learning convolutional neural network (CNN) architectures for CT lung cancer classification. Int J Adv Technol Eng Explorat 8(74):126
8. Zhang F, Wang Q, Li H (2020) Automatic segmentation of the gross target volume in non-small cell lung cancer using a modified version of ResNet. Technol Cancer Res Treat 19:1533033820947484
9. Sandag GA, Kabo DT (2024) Comparative analysis of lung cancer classification models using EfficientNet and ResNet on CT-scan lung images. CogITo Smart J 10(1):259–270
10. Wang M, Gong X (2020) Metastatic cancer image binary classification based on resnet model. In: 2020 IEEE 20th international conference on communication technology (ICCT), October, IEEE, pp 1356–1359
11. Nakrani MG, Sable GS, Shinde UB (2020) ResNet based lung nodules detection from computed tomography images. Int J Innov Technol Exploring Eng 1711–4
12. Ning J, Zhao H, Lan L, Sun P, Feng Y (2019) A computer-aided detection system for the detection of lung nodules based on 3D-ResNet. Appl Sci 9(24):5544

13. Pang S, Meng F, Wang X, Wang J, Song T, Wang X, Cheng X (2020) VGG16-T: a novel deep convolutional neural network with boosting to identify pathological type of lung cancer in early stage by CT images. Int J Computat Intell Syst 13(1):771–780
14. Xu H, Yu Y, Chang J, Hu X, Tian Z, Li O (2024) Precision lung cancer screening from CT scans using a VGG16-based convolutional neural network. Front Oncol 14
15. Hareesh PM, Bellamkonda S (2023) Deep learning-based classification of lung cancer lesions in CT scans: comparative analysis of CNN, VGG-16, and MobileNet models. In: International conference on image processing and capsule networks, August, Singapore, Springer Nature, pp 373–387
16. Kumaran SY, Jeya JJ, Khan SB, Alzahrani S, Alojail M (2024) Explainable lung cancer classification with ensemble transfer learning of VGG16, Resnet50 and InceptionV3 using grad-cam. BMC Med Imaging 24(1):176
17. Vineeta KA, Manek AS, Christa S, Mishra PK (2023) Identifying lung cancer from CT-scan images with VGG16 convolutional neural net. In: International conference on intelligent systems design and applications, Cham, Springer Nature Switzerland, pp 1–9
18. Bilal A, Shafiq M, Fang F, Waqar M, Ullah I, Ghadi YY, Long H, Zeng R (2022) IGWO-IVNet3: DL-based automatic diagnosis of lung nodules using an improved gray wolf optimization and InceptionNet-V3. Sensors 22(24):9603
19. Shaffie A, Soliman A, Ghazal M, Taher F, Dunlap N, Wang B, Van Berkel V, Gimelfarb G, Elmaghraby A, El-Baz A (2018) A novel autoencoder-based diagnostic system for early assessment of lung cancer. In: 2018 25th IEEE international conference on image processing (ICIP), October, IEEE, pp 1393–1397
20. Ghosal SS, Sarkar I, El Hallaoui I (2020) Lung nodule classification using convolutional autoencoder and clustering augmented learning method (CALM). In: HSDM@ WSDM, February, pp 19–26
21. Astaraki M, Toma-Dasu I, Smedby Ö, Wang C (2019) Normal appearance autoencoder for lung cancer detection and segmentation. In: Medical image computing and computer assisted intervention–MICCAI 2019: 22nd international conference, Shenzhen, China, October 13–17, 2019, Proceedings, Part VI 22, Springer International Publishing, pp 249–256
22. Vo TH, Lee GS, Yang HJ, Oh IJ, Kim SH, Kang SR (2021) Survival prediction of lung cancer using small-size clinical data with a multiple task variational autoencoder. Electronics 10(12):1396
23. Raza R, Zulfiqar F, Khan MO, Arif M, Alvi A, Iftikhar MA, Alam T (2023) Lung-EffNet: lung cancer classification using EfficientNet from CT-scan images. Eng Appl Artif Intell 126:106902
24. Muthulakshmi M, Venkatesan K, Harigaran R, Jeevanantham K, Vineeth MS, Rahayu SB, Sakthivel V (2024) Comparative study of EfficientNet and MobileNet models for lung cancer classification using chest CT scan images. In: 2024 second international conference on emerging trends in information technology and engineering (ICETITE), February, IEEE, pp 1–6
25. John J, Deotale A, Fernandez F, Jadhav D (2024) EfficientNetB1 model for lung cancer detection using biopsy images. In: 2024 IEEE international conference on information technology, electronics and intelligent communication systems (ICITEICS), June, IEEE, pp 1–6
26. Nanglia P, Kumar S, Mahajan AN, Singh P, Rathee D (2021) A hybrid algorithm for lung cancer classification using SVM and neural networks. ICT Express 7(3):335–341
27. Murugesan M, Kaliannan K, Balraj S, Singaram K, Kaliannan T, Albert JR (2022) A hybrid deep learning model for effective segmentation and classification of lung nodules from CT images. J Intell Fuzzy Syst 42(3):2667–2679

# Multimodal Alzheimer's Disease Diagnosis Using Cross-Attention Transformers and Dense Networks with Dropout Regularization

**J. Aswini, Konduru Venkateswara Raju, V. Lavanya, Chekuri Nalini, Sreekanth Yalavarthi, and Naresh Tangudu**

**Abstract** Alzheimer disease is the major cause for dementia arising due to the cognitive decline and neurodegeneration. Early prediction is important for better treatment and the complexity of disease requires multiple modalities such as MRI, PET, and clinical data. Conventional approaches used for diagnosis utilize these modalities independently limiting the diagnostic capability. Considering this limitation, a transformer-based multimodal fusion method utilizing cross-attention mechanism has been proposed to enhance diagnostic accuracy. By utilizing the cross-attention mechanism, the model has ability to identify important features from each modality which helps in fusing complementary information for better diagnosis. This approach not only enhances predictive accuracy but also provides valuable insights into the mechanisms of Alzheimer's disease prediction. Utilization of dropout regularization method helps the model to converge faster making the proposed method highly effective. Combining the strengths of multimodal fusion and cross-attention, the

J. Aswini
Department of Artificial Intelligence and Machine Learning, Saveetha Engineering College (Autonomous), Affiliated to Anna University, Chennai, India

K. V. Raju
Department of Mathematics, Sri Venkateswara College of Engineering (Autonomous), Karakambodi Road, Tirupati, India

V. Lavanya
School of Computing, Mohan Babu University (Sree Vidyanikethan Engineering College), Tirupati, India

C. Nalini
Department of ECE, School of Engineering, Mohan Babu University (Erstwhile Sree Vidyanikethan Engineering College), Tirupati, Andhra Pradesh, India

S. Yalavarthi (✉)
HCL Technologies, Buffalo Grove, Il, USA
e-mail: sreekanth_y@yahoo.com

N. Tangudu
Aditya Institute of Technology and Management, Tekkali, India

 
K. R. Madhavi et al. (eds.), *Proceedings of Sixth International Conference on Computer and Communication Technologies*, Lecture Notes in Networks and Systems 1356,
https://doi.org/10.1007/978-981-96-5238-9_44

proposed method can be utilized for early Alzheimer's disease diagnosis, leading to better patient outcomes and accuracy in Alzheimer's disease diagnosis.

**Keywords** Multimodal fusion · Deep neural network · CNN · Transformer · Cross-attention · Dense layers

## 1 Introduction

Alzheimer's disease is a neurodegenerative disorder that affects the elderly leading to memory loss which will affect them in doing daily activities. Early diagnosis of this disease is required for effective treatment. However, this task remains challenging due to complex nature of the disease, necessitating the utilization of efficient diagnostic tools and different data sources. The approaches such as magnetic resonance imaging (MRI) and positron emission tomography (PET) are the common data used for Alzheimer's disease prediction. These data provide details of images to identify the problems associated with Alzheimer's disease, metabolic activities of brain can be tracked through PET. Along with these two data, clinical assessments with cognitive tests and patient information play critical role in diagnosis of Alzheimer's disease. These modalities were analyzed separately to reduce complexity. Advances in multimodal fusion techniques improve diagnostic accuracy using data from different sources. Traditional methods face challenges while utilizing multiple modalities with heterogeneous data. Deep learning models were used to overcome these challenges through transformer architectures. The efficiency of feature extraction can be improved through transformer architecture with an attention mechanism.

Originally transformers were used mostly in natural language processing tasks, but recently this architecture has shown efficiency in medical imaging through its ability to capture long-range dependency between models. Cross-attention mechanism is suitable for performing this task particularly for multimodal fusion tasks in disease diagnosis. Considering these aspects, we propose a multimodal fusion method utilizing cross-attention mechanism and dropout regularization for classifying Alzheimer's disease. Cross-attention mechanism helps to extract important features from each modality leading to accurate diagnosis. This work aims to improve accuracy of prediction utilizing multiple modalities to help in decision-making and for better patient outcomes.

## 2 Background Work

This section involves reviewing existing work related to Alzheimer's disease prediction utilizing deep learning architectures, multimodal data fusion, and optimization methods to perform hyperparameter tuning [1–3]. This disease is a neurodegenerative disorder and leads to memory loss impacting patients life. Traditional methods rely of single data and not sufficient for efficient diagnosis. Recent advancements in MRI and PET scans provide valuable insights for Alzheimer's disease prediction [4]. Multimodal fusion is considered as an effective approach for utilizing complementary information from different sources of data. Combining different modalities can increase the accuracy of Alzheimer's disease prediction. Many fusion methods find difficult in handling diverse dataset particularly to extracting complex features [5].

Deep learning architectures such as convolutional neural network is used much in medical image analysis. But due to its limitations CNN could not handle multimodal data efficiently [6]. Transformer architecture developed particularly for natural language processing has gained attention in medical imaging [7]. Transformer architecture with attention mechanism can efficiently process multimodal data. By applying cross-attention mechanism transformers can weigh important features adjusting their weights from different modalities [8–10]. Recent works revealed the importance of using transformers in medical image analysis particularly for Alzheimer's disease detection. Transformer architecture can handle multimodal data fusing MRI and PET [11] but combining with clinical data is still not explores. The proposed method addresses this challenge through the use of transformer-based multimodal fusion combining MRI, PET, and clinical data for Alzheimer's disease prediction using cross-attention mechanism [12] enhancing the feature extraction and improving the accuracy. Optimization methods plays important role in training deep learning models. Adam optimization algorithm is introduced to help in deep learning training by providing faster convergence and adaptability [13, 14]. Application of multimodal data fusion in medical imaging has been explored extensively. Existing methods for Alzheimer's disease prediction faces the challenge of integration of data modalities, data imbalance problems, and challenges related to interpretability and data quality [15, 16]. The proposed model addresses these challenges and improves the accuracy and clinical applicability of the models.

## 3 Methodology

This section briefs the steps of proposed architecture for predicting Alzheimer's disease using cross-attention transformers and deep networks with dropout regularization have been discussed. The steps for the proposed multimodal fusion method with cross-attention transformer for diagnosing Alzheimer's disease have been specified in Fig. 1. The process begins with the collection of data from diverse sources including MRI scans, PET scans, and clinical data. The multiple modalities are integrated using transformers with cross-attention mechanisms for extracting relevant features and dense layers are used for classification [17]. Data preprocessing is then performed using relevant methods. Standardization and resizing of images are applied to MRI and PET data, normalization, and feature scaling are applied to clinical data [18]. The preprocessed outputs are then converted into feature vectors to be given as input to transformer. The transformer encoder captures local and global dependencies in data. The operation of transformer performed using self-attention mechanism to help capture relationships in all the three modalities as specified in Eq. 1.

$$\text{Attention}(Q, K, V) = \text{softmax}\left(\frac{Qk^T}{\sqrt{d_k}}\right)V \tag{1}$$

where the parameters represents query, key and value, *dk* represents the dimension of the keys. Three different modalities of inputs represented as $X_{\text{MRI}}$, $X_{\text{PET}}$, $X_{\text{Clinical}}$ are passed through different transformers to create feature representations hMRI, $h_{\text{PET}}$, $h_{\text{Clinical}}$. After performing feature extraction for each modality cross-attention is applied to fuse complementary features to improve models understanding of the correlation between different data. This mechanism works similar to self-attention but the queries come from one modality, e.g., MRI, keys, and values come from another modality eg. PET [19]. This enables to correlate the feature of one modality

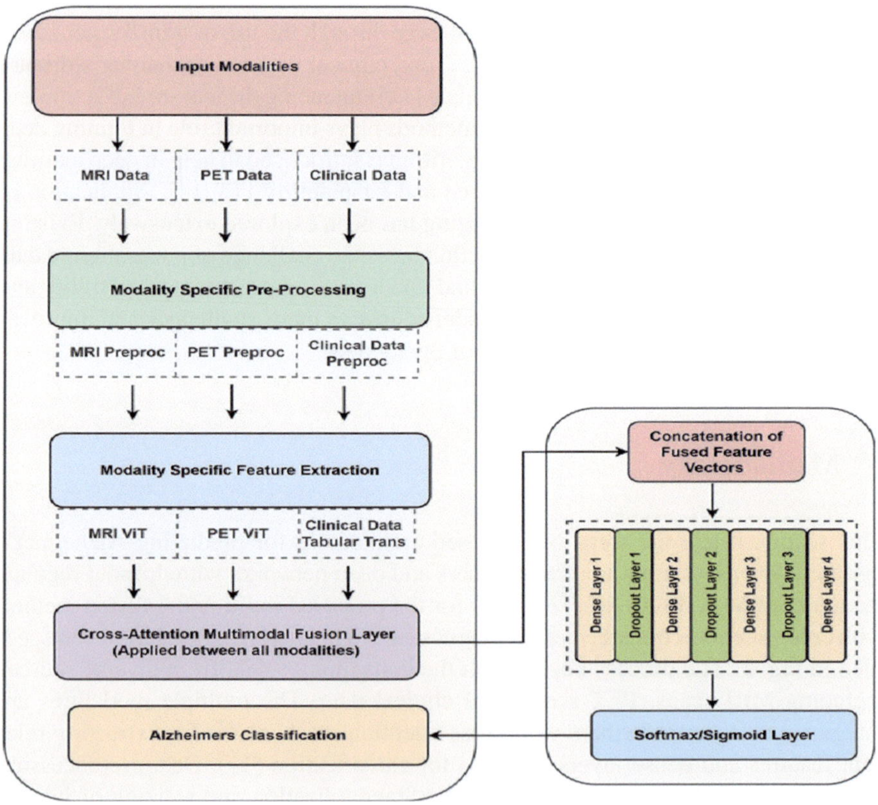


**Fig. 1** Alzheimer's disease prediction architecture using cross-attention transformers and dense networks

with another as specified in Eq. 2.

$$\text{Cross} - \text{Att}(h_{\text{MRI}}, h_{\text{PET}}) = \text{softmax}\left(\frac{Q_{\text{MRI}}, K_{\text{PET}}}{\sqrt{d_k}}\right)V_{\text{PET}} \tag{2}$$

The cross-attention mechanism specified in Eq. 2 is repeated for all the modalities such as MRI-Clinical, PET-Clinical, and MRI-PET. After fusing the features with cross-attention mechanism then the fused features are concatenated into single feature vector as represented in Eq. 3. This process creates a multimodal representation integrating information from all the input sources.

$$F_{\text{concat}} = \text{concat}(f_{\text{MRI}}, f_{\text{PET}}, f_{\text{clinical}}) \tag{3}$$

The concatenated feature is given as input to fully connected dense layers to perform dimensionality reduction and further learning. These layers extract critical information required for Alzheimer disease classification. Then each dense layer applies a liner transformation followed by nonlinear activation ReLU. The proposed method uses dropout regularization to avoid overfitting. This process randomly drops neurons during training performing the model to generalize better. After passing through several dense layers, the output is passed to classification layer. Softmax function is used to perform multiclass classification [20]. Cross-entropy loss is utilized to train the model in order to perform the task of classification [21]. Adam optimization method is used to perform hyperparameter tuning [22]. The performance of proposed model is evaluated by considering evaluation metrics which will be discussed in the next section.

## 4 Results and Discussions

The proposed approach uses ADNI dataset for Alzheimer's disease classification. The modalities considered are MRI scans with high resolution neuroimages, PET (Positron Emission Tomography) scans, and clinical data with patient's medical history. It is a comprehensive dataset including biomarkers to monitor the disease progression. This dataset includes high-resolution MRI images of brain structure used to detect brain atophy. PET scan images in the dataset used to measure amyloid plaques and tau tangles which are the proteins associated with Alzheimer's disease. This dataset also contains clinical assessments of patient covering all aspects such as cognitive scores, demographic information, and medical history. Performance of the proposed method is evaluated utilizing metrics like accuracy and more as specified in Table 1, graph is represented in Fig. 2. Accuracy of proposed approach is better for the classification of Alzheimer's disease when compared to existing methods such as CNN, RNN, logistic regression, SVM, and transfer learning.

Combining data from multiple modalities introduces complexity during training process. The results indicate that the multimodal approach performs better compared

**Table 1** Evaluation of the proposed method using performance metrics

| Method | Accuracy | Precision | Recall | F1-Sore | AUC-ROC | Specificity |
|---|---|---|---|---|---|---|
| Proposed method (multimodal fusion) | 0.94 | 0.91 | 0.94 | 0.92 | 0.97 | 0.93 |
| CNN | 0.79 | 0.84 | 0.85 | 0.81 | 0.86 | 0.88 |
| RNN | 0.73 | 0.67 | 0.68 | 0.75 | 0.80 | 0.78 |
| logistic regression | 0.90 | 0.88 | 0.89 | 0.92 | 0.94 | 0.94 |
| SVM | 0.86 | 0.81 | 0.86 | 0.83 | 0.90 | 0.88 |
| Transfer learning | 0.84 | 0.77 | 0.83 | 0.80 | 0.88 | 0.86 |

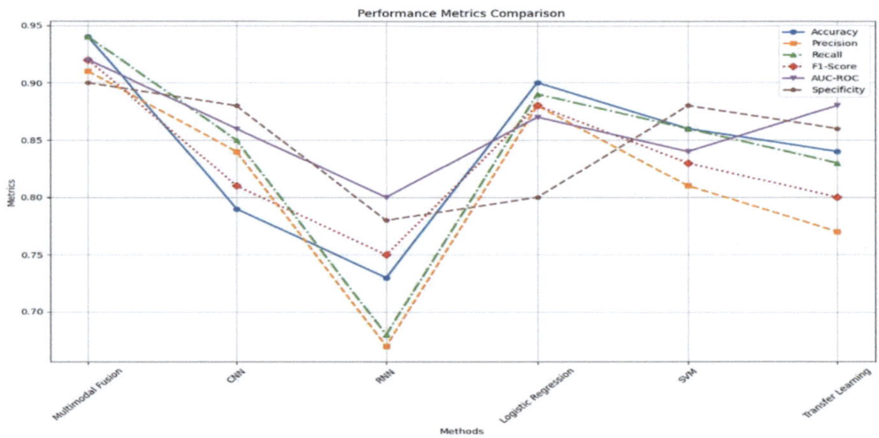


**Fig. 2** Comparison of proposed multimodal fusion method with existing methods

to single fusion approach. Adam's optimization helps manage this complexity by adjusting learning rates based on the gradients of the fused features. This process can improve the model's ability to learn effectively from multiple data sources. To predict the contribution of each modality and the cross-attention mechanism, an ablation study was conducted by selectively removing different components of the model.

Adam optimization used for hyperparameter tuning makes the proposed method effective for handling sparse gradients and varying parameter updates. By using the moments through mean and variance of gradients, the Adam method adjusts learning rates for better convergence. This approach is advantageous particularly when fusing multiple data modalities such as MRI, PET, and clinical data. Training process is improved helping model to utilize diverse information provided by multiple inputs enhancing the accuracy of classification. Training cost based on number of epochs utilizing Adam optimization is shown in Fig. 3. Adam optimization provides stable convergence indicating decline in training loss over epochs.

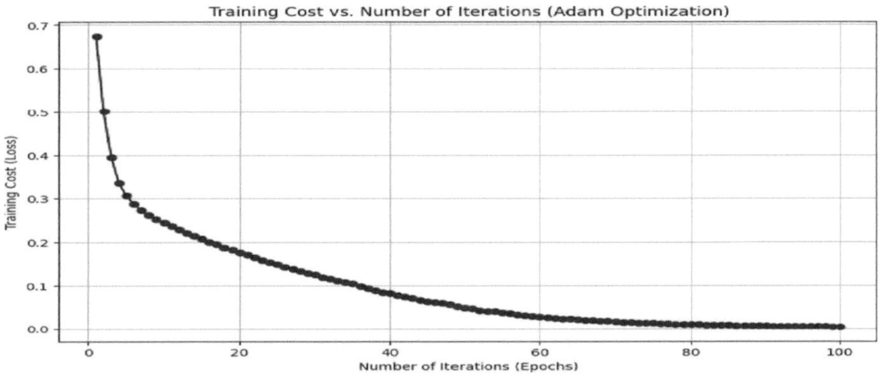


**Fig. 3** Training cost versus number of epochs using Adam optimization

## 5 Conclusion

In this paper, a transformer-based multimodal fusion method with cross-attention mechanism has been proposed for Alzheimer's disease classification. The use of cross-attention mechanism helps to perform feature extraction efficiently considering multiple modalities. ADNI dataset is used to evaluate the proposed model considering modalities MRI, PET, and clinical data. The proposed model performs better than the existing methods demonstrating better accuracy in the prediction of Alzheimer's disease. By capturing complex features, the model has the ability to assist healthcare professionals to perform accurate prediction of the disease and provide timely treatment for the patients. The future work is to improve the accuracy further different modalities such as genetic information can be considered to improve the diagnostic accuracy. Also, explainability methods can be integrated to provide better insights about the disease.

## References

1. Liu S, Liu S, Cai W, Pujol S, Kikinis R, Feng D (2014) Early diagnosis of Alzheimer's disease with deep learning. In: 2014 IEEE 11th international symposium on biomedical imaging (ISBI), April, IEEE, pp 1015–1018
2. Sarraf S, Tofighi G (2016) Deep learning-based pipeline to recognize Alzheimer's disease using fMRI data. In: 2016 Future technologies conference (FTC), December, IEEE, pp 816–820
3. Ebrahimighahnavieh MA, Luo S, Chiong R (2020) Deep learning to detect Alzheimer's disease from neuroimaging: a systematic literature review. Comput Methods Programs Biomed 187:105242
4. Lee G, Nho K, Kang B, Sohn KA, Kim D (2019) Predicting Alzheimer's disease progression using multi-modal deep learning approach. Sci Rep 9(1):1952
5. Savaş S (2022) Detecting the stages of Alzheimer's disease with pre-trained deep learning architectures. Arab J Sci Eng 47(2):2201–2218

6. Bringas S, Salomón S, Duque R, Lage C, Montaña JL (2020) Alzheimer's disease stage identification using deep learning models. J Biomed Inform 109:103514
7. Jo T, Nho K, Saykin AJ (2019) Deep learning in Alzheimer's disease: diagnostic classification and prognostic prediction using neuroimaging data. Front Aging Neurosci 11:220
8. Feng C, Elazab A, Yang P, Wang T, Zhou F, Hu H, Xiao X, Lei B (2019) Deep learning framework for Alzheimer's disease diagnosis via 3D-CNN and FSBi-LSTM. IEEE Access 7:63605–63618
9. Alshayeji MH (2024) Alzheimer's disease detection and stage identification from magnetic resonance brain images using vision transformer. Machine Learn: Sci Technol 5(3):035011
10. Dwivedi S, Goel T, Tanveer M, Murugan R, Sharma R (2022) Multimodal fusion-based deep learning network for effective diagnosis of Alzheimer's disease. IEEE Multimedia 29(2):45–55
11. Basaia S, Agosta F, Wagner L, Canu E, Magnani G, Santangelo R, Filippi M (2019) Alzheimer's disease neuroimaging initiative. Automated classification of Alzheimer's disease and mild cognitive impairment using a single MRI and deep neural networks. NeuroImage: Clinical 21:101645
12. Vivekanandam B (2021) Automated multimodal fusion technique for the classification of human brain on Alzheimer's disorder. J Electri Eng Autom 3(3):214–229
13. El-Sappagh S, Abuhmed T, Islam SR, Kwak KS (2020) Multimodal multitask deep learning model for Alzheimer's disease progression detection based on time series data. Neurocomputing 412:197–215
14. Wang Y, Liu X, Yu C (2021) Assisted diagnosis of Alzheimer's disease based on deep learning and multimodal feature fusion. Complexity 2021:1–10
15. Odusami M, Maskeliūnas R, Damaševičius R (2023) Pixel-level fusion approach with vision transformer for early detection of Alzheimer's disease. Electronics 12(5):1218
16. Yu Q, Ma Q, Da L, Li J, Wang M, Xu A, Li Z, Li W, Initiative ADN (2024) A transformer-based unified multimodal framework for Alzheimer's disease assessment. Comput Biol Med 180:108979
17. Gao X, Cai H, Liu M (2023) A hybrid multi-scale attention convolution and aging transformer network for Alzheimer's disease diagnosis. IEEE J Biomed Health Inform 27(7):3292–3301
18. Tang C, Wei M, Sun J, Wang S, Zhang Y, Initiative ADN (2023) CsAGP: detecting Alzheimer's disease from multimodal images via dual-transformer with cross-attention and graph pooling. J King Saud Univer-Comput Inform Sci 35(7):101618
19. Zhang J, He X, Liu Y, Cai Q, Chen H, Qing L (2023) Multi-modal cross-attention network for Alzheimer's disease diagnosis with multi-modality data. Comput Biol Med 162:107050
20. Avanija J, Sunitha G, Reddy Madhavi K, Hitesh Sai Vittal R (2021) An automated approach for detection of intracranial haemorrhage using DenseNets. In: Data engineering and communication technology: proceedings of ICDECT 2020, Springer, Singapore, pp 611–619
21. Reddy KR, Dhuli R (2023) A novel lightweight CNN architecture for the diagnosis of brain tumors using MR images. Diagnostics 13(2):312
22. Arunachalam R, Sunitha G, Shukla SK, Urooj S, Rawat S (2023) A smart Alzheimer's patient monitoring system with IoT-assisted technology through enhanced deep learning approach. Knowledge and Inform Syst 1–39

# Meta-Learning Enhanced Self-Supervised Methods for Crop Yield Estimation Using UAV Imagery in Precision Agriculture

**Karthik Kovuri, V. Vishnu Vandana Devi, K. Ayyappa Swamy, K. Lavanya, J. Avanija, and Sam Goundar**

**Abstract** Crop yield estimation is an important component in precision agriculture enabling farmers to optimize the resources and increase productivity through better decision-making. Traditional approaches such as statistical models and manual sampling are time-consuming and labor-intensive. Considering these issues in this paper, we propose a meta-learning enhanced self-supervised approach for crop yield images using UAV (unmanned ariel vehicle) imagery of crops. The proposed model also utilizes EfficientNet architecture to perform feature extraction in large-scale agriculture environments. Self-supervised learning is used to train the model on large-scale unlabeled data without manual annotations. Meta-learning helps to generalize the learning with minimal fine-tuning of model parameters making it suitable for diversified farming landscapes. The performance of the proposed approach is evaluated using different baseline methods. The results reveal that the proposed approach

K. Kovuri
School of Engineering and Technology and School of Computing Sciences, The Assam Kaziranga University, Rowmarikhuti, Assam, India

V. Vishnu Vandana Devi
VR Siddhartha Engineering College Deemed to be University, Vijayawada, India

K. Ayyappa Swamy
Department of Electronics and Communication Engineering, Aditya University, Surampalem, Andhra Pradesh, India

K. Lavanya
CSE, School of Engineering and Technology, Sri Padmavati Mahila Visvavidyalayam, Tirupati, India

J. Avanija
AI&ML, School of Computing, Mohan Babu University, Tirupati, A.P, India

S. Goundar (✉)
RMIT University, Ho Chi Minh City, Vietnam
e-mail: sam.goundar@gmail.com

 
K. R. Madhavi et al. (eds.), *Proceedings of Sixth International Conference on Computer and Communication Technologies*, Lecture Notes in Networks and Systems 1356,
https://doi.org/10.1007/978-981-96-5238-9_45

performs better in terms of computational efficiency and accuracy compared to existing methods improving crop yield prediction improving the efficiency of farming operations.

**Keywords** Precision agriculture · Meta-learning · Self-supervised learning · EfficientNet · Crop yield estimation

## 1 Introduction

Precision agriculture has emerged as a progressive approach utilizing latest technologies to minimize impact of the environment and maximize crop yield. Recently, unmanned ariel vehicles has gained significance in capturing high-resolution images in agricultural farms. This helps in monitoring crop health, estimating crop yields and soil conditions with better accuracy. However, extracting meaningful insights from UAV imagery data requires efficient machine learning methods [1]. Traditional machine learning models are time consuming since they depend on large labeled datasets related to agricultural applications. Also, these models require proper fine-tuning mechanisms if applied for different crop types, regions, and climatic conditions. To overcome this limitation, the proposed approach utilizes self-supervised learning and meta-learning [2–4] to overcome the dependency on large dataset and generalize the model across tasks.

Self-supervised learning can generate the labels itself without the need for large labeled dataset. In the proposed method, this learning helps to utilize large unlabeled data captured by UAV to learn the features such as texture of the crop, density, and crop health which are required parameters for crop yield estimation [5]. This method reduces the complexity due to manual annotation and provides better solution in large-scale agriculture applications. Meta-learning enhances the ability of the machine learning models to perform various tasks [6]. This learning focuses on efficiently generalizing model for unseen data even with small amount of data. In crop yield estimation, this type of learning helps the model to adapt for different types of crops, soil, and climatic conditions. This will be useful in precision farming [7] to handle unique characteristics that influence the yield of the crop. Considering these aspects self-supervised enhanced meta-learning method for crop yield estimation of UAV imagery has been proposed. EfficientNet deep learning architecture [8] augmented with self-learning helps to extract and learn robust features from UAV Crop images. Meta-learning is then applied to fine-tune the model for crop yield estimation by generalizing across different farming conditions.

## 2 Background Work

This section involves reviewing existing literature related to crop yield prediction, self-supervised learning, meta-learning and deep learning architectures. In recent y,ears precision agriculture transformed farming enabling efficient monitoring of crop health and yield prediction [9]. The use of UAV and satellite has become important in getting high-resolution images. One important application of precision agriculture is estimating crop yield to help farmers improve productivity and optimize resource allocation [10–12]. The challenges rely on processing large amount of image data converting it to predictions. Traditional machine learning methods often face challenges handling large complex dataset. Recent advancements in deep learning architectures have shown promising results in crop yield prediction [13]. Convolutional neural network architecture was widely used in agricultural applications [14]. The challenge with deep learning architecture is providing labeled data for crop yield prediction [15–17]. Self-supervised learning approach offers solution by learning from large volume of unlabeled dataset provided in the form of UAV imagery. This learning predicts labels through rotation angles, image patches, and color channels. Meta-learning is used to generalize training adopting to new tasks based on different crop types and environmental conditions [18]. EfficientNet architecture is used to extract robust features from high-resolution images [19]. This architecture has the ability to learn through characteristics such as texture, density, and crop health to perform better yield estimation of crops [20]. Deep learning architectures such as recurrent neural networks and ResNet [21] were used to perform time series data analysis to predict crop growth patterns. Considering these aspects, the proposed work addresses these challenges by utilizing meta-learning enhanced self-supervised approach for improving the accuracy of crop yield estimation from UAV imagery of crops.

## 3 Methodology

This section briefs about the steps involved in the proposed architecture for crop yield estimation using self-supervised learning architecture. The steps for the proposed architecture for crop yield estimation have been specified in Fig. 1. Initial step is to gather dataset consisting of UAV imagery with crop data samples. Proposed work utilizes drone-based agricultural data from hugging face for processing.

Next step is to perform data preprocessing to reduce noise by applying filtering techniques. Then data augmentation is performed through operations such as rotation, flipping, and scaling to improve the diversity of training data. Deep learning architecture EfficientNet is used to extract the features from augmented images. This architecture is a convolutional neural network-based architecture that balances accuracy and computational efficiency by optimizing depth and breadth of the network

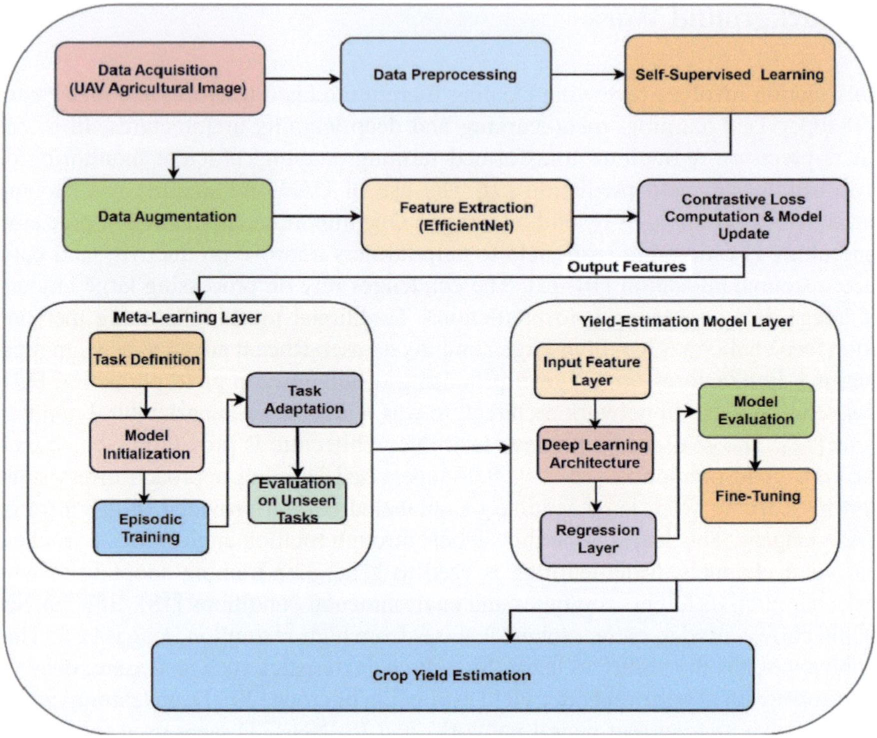


**Fig. 1** Crop yield estimation using meta-learning based self-supervised learning architecture

[22]. This architecture also helps in extracting features to learn spatial patterns efficiently. The model is then trained using self-supervised learning mechanism through contrastive learning framework. Contrastive loss [23] is computed to evaluate the model distinguishing similar and dissimilar pairs. The loss is computed using the Eq. (1).

$$L_{\text{contra}} = \frac{1}{N}\sum_{k=0}^{n}[\max(0, d(f(x_i, x_j), f(x_j)) - \alpha) + \max(0, \beta - (f(x_i x_j), f(x_k)))] \quad (1)$$

where $d(f(x_i), f(x_j))$ is the distance (e.g., Euclidean distance) between features of positive pairs $(xi, x_j)$ $(x_i, x_j)(x_i, x_j)$. The distance $d(f(x_i), f(x_k))$ specifies the distance between negative pairs. $\alpha$ and $\beta$ beta are termed as margin parameters. The goal of this learning is to minimize the distance between positive pairs and maximize the distance between negative pairs. The model then updates the weights using calculated loss. This process helps in refining feature representation of the model. The meta-learning layer outlines [24] the crop types and environmental conditions enabling the model to learn effectively considering different scenarios. Then model initialization is performed through EfficientNet parameters. Then episodic training is performed considering small number of examples from each task as specified in Eq. 2. This process helps in model generalization from limited data effectively. Task adaptation is performed using learned representations to improve the performance of the model with minimum labeled data. Then the model is evaluated for unseen tasks to measure the performance of the model during training.

$$\text{Train}(T, \theta) \rightarrow \theta' = \theta - \eta \nabla \theta L_{\text{task}} \quad (2)$$

where $L_{\text{task}}$ is the loss for specific task and $\eta$ is the learning rate.

The yield estimation layer receives the features extracted from self-supervised learning layer. EfficientNet architecture is used to achieve good accuracy and efficiency. This architecture consists of multiple convolution layers, swish activations, pooling operations to capture complex patters in data. Squeeze and excitation blocks are used to improve representational power. The regression layer uses the features extracted by EfficientNet to predict crop yield as specified in Eq. 3. The loss function of yield estimation is computed using Eq. 4. The performance of the model is evaluated using the metrics such as mean absolute error and root mean squared error.

$$y = (Z; \theta) = wZ + b \quad (3)$$

where $y$ is the predicted yield, $w$ is the learned weight, $b$ is the bias

$$L_{\text{yield}} = \frac{1}{N} \sum_{k=0}^{n} Y_i - Y_j \quad (4)$$

where $Y_j$ is the ground truth yield value.

Fine-tuning of the model parameters is performed to optimize the performance of the model [25]. The proposed model utilizes the ability of EfficientNet for crop yield prediction using UAV images. By utilizing self-supervised learning and meta-learning [26], the proposed method can improve crop yield prediction with limited data making it well suitable for agriculture environments.

## 4 Results and Discussions

The dataset used in proposed approach is drone-based agriculture dataset from hugging face consisting of cocoa and cashew images. This dataset is designed to estimate crop yields. Each image consists of labels describing the maturity stage of crop such as flower, mature, immature, ripe, and spoilt. The data are publicly available and managed by karaAgro AI. The dataset can be used to develop crop yield estimation methods. This is a drone-based crop dataset labeled and accessed

**Fig. 2** Images of healthy and rust affected wheat leaf samples

by users. Yield estimation helps farmers to make decision and plan for farming activities leading to better income. Each instance in the dataset includes crop image as specified in Fig. 2 revealing development and maturity stage as labels. The top view images help in identifying plant varieties, and side view specifies the maturity of plant.

The dataset can be utilized for training and evaluating deep learning architectures enabling crop yield estimation model trained using drone dataset helps with early estimation of crop yields.

Performance of the proposed approach is evaluated utilizing metrics like mean absolute error, and the values are specified in Table 1. The proposed meta-learning enhanced self-supervised learning outperforms the existing methods achieving lowest value for mean absolute error. This reveals better quality in estimating crop yield using drone images compared to existing baseline methods.

The change in prediction error over various epochs is represented in Fig. 3. Decrease in mean absolute error based on epochs reveals the capability of model to estimate crop yield accurately. The training dataset includes early to mid stages of crop growth, while testing data include final stage of crop maturity. This process ensures that the model is well trained on growth of crop and generalizes better for unseen data. The scores reveal that the proposed model can generalize the learning from training process with the help of meta-learning approach. The self-supervised learning component helps on extracting meaningful representations from drone

**Table 1** Evaluation of proposed model using performance metrics

| Method | MAE | RMSE |
|---|---|---|
| CNN | 58.9 | 82.3 |
| ResNet | 40.7 | 61.0 |
| EfficientNet | 45.6 | 67.8 |
| UNet | 47.5 | 70.1 |
| Proposed method | 37.2 | 53.7 |

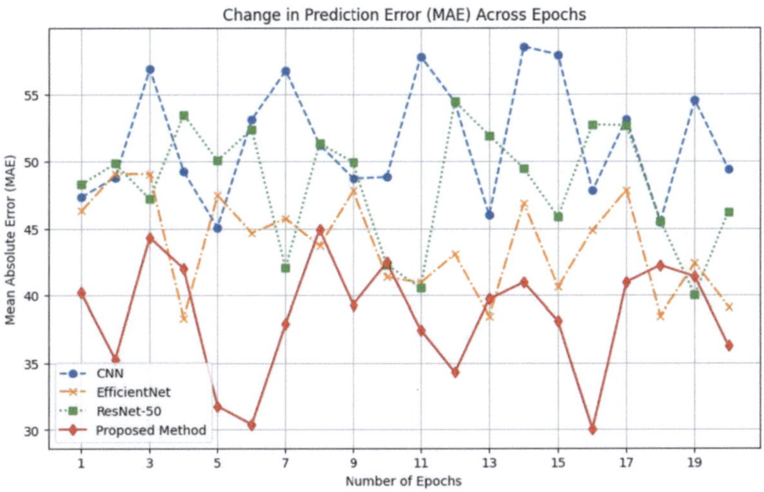


**Fig. 3** Performance of the proposed model showing change in prediction error across epochs

imagery reducing the dependency of annotations on larger datasets. This process plays crucial role in yield estimation for agricultural applications where tasks are labor intensive. The model adjusts the hyperparameters such as weights, bias during learning process to decrease the loss. The proposed model can also be utilized for different crop types and environmental conditions.

## 5 Conclusion

The proposed, meta-learning enhanced self-supervised learning for crop yield estimation represents a significant advancement in precision agriculture. The proposed model performs better than the existing approaches with better accuracy in crop yield prediction. By fusing self-learning with meta-learning, the proposed work overcomes the issues faced due to limited data and adaptability of the model in agricultural applications. The utilization of EfficientNet architecture ensures the efficiency of the framework to capture important features required for crop yield estimation. The results demonstrate that the proposed approach performs well compared to existing methods in terms of accuracy and adaptability. Self-supervised learning helps the model to learn about unlabeled data and meta-learning enhances the ability to generalize across different crop types and environmental conditions. Future work is to explore advanced hyperparameter optimization methods to improve the performance of the model and reduce training time. Also, combining drone image data with sensor data such as soil conditions, weather data can provide better understanding of crop health and yield potential.

## References

1. Iniyan S, Varma VA, Naidu CT (2023) Crop yield prediction using machine learning techniques. Adv Eng Software 175:103326; Hussain A, Ahmad M, Mughal IA, Ali H (2018) November
2. Sagar BM, Cauvery NK (2018) Agriculture data analytics in crop yield estimation: a critical review. Indonesian J Electri Eng Comput Sci 12(3):1087–1093
3. Awad MM (2019) Toward precision in crop yield estimation using remote sensing and optimization techniques. Agriculture 9(3):54
4. Marshall M, Belgiu M, Boschetti M, Pepe M, Stein A, Nelson A (2022) Field-level crop yield estimation with PRISMA and Sentinel-2. ISPRS J Photogramm Remote Sens 187:191–210
5. Maas SJ (1988) Using satellite data to improve model estimates of crop yield. Agron J 80(4):655–662
6. Fernandez-Ordoñez YM, Soria-Ruiz J (2017) Maize crop yield estimation with remote sensing and empirical models. In: 2017 IEEE international geoscience and remote sensing symposium (IGARSS), July, IEEE, pp 3035–3038
7. Huang J, Wang H, Dai Q, Han D (2014) Analysis of NDVI data for crop identification and yield estimation. IEEE J Selected Topics in Appl Earth Observ Remote Sens 7(11):4374–4384
8. Singh RANDHIR, Semwal DP, Rai ANIL, Chhikara RS (2002) Small area estimation of crop yield using remote sensing satellite data. Int J Remote Sens 23(1):49–56
9. Sayago S, Bocco M (2018) Crop yield estimation using satellite images: comparison of linear and non-linear models. AgriScientia 35(1):1–9
10. Medar R, Rajpurohit VS, Shweta S (2019) Crop yield prediction using machine learning techniques. In: 2019 IEEE 5th international conference for convergence in technology (I2CT), March, IEEE, pp 1–5
11. Kavita M, Mathur P (2020) Crop yield estimation in India using machine learning. In: 2020 IEEE 5th international conference on computing communication and automation (ICCCA), October, IEEE, pp 220–224
12. Crane-Droesch A (2018) Machine learning methods for crop yield prediction and climate change impact assessment in agriculture. Environ Res Lett 13(11):114003
13. Kumar R, Singh MP, Kumar P, Singh JP (2015) Crop selection method to maximize crop yield rate using machine learning technique. In: 2015 international conference on smart technologies and management for computing, communication, controls, energy and materials (ICSTM), May, IEEE, pp 138–145
14. Zhou Q, Ismaeel A (2021) Integration of maximum crop response with machine learning regression model to timely estimate crop yield. Geo-Spatial Inform Sci 24(3):474–483
15. Nejad SMM, Abbasi-Moghadam D, Sharifi A, Tariq A (2024) Capsular attention Conv-LSTM network (CACN): A deep learning structure for crop yield estimation based on multispectral imagery. Eur J Agron 161:127369

16. Muruganantham P, Wibowo S, Grandhi S, Samrat NH, Islam N (2022) A systematic literature review on crop yield prediction with deep learning and remote sensing. Remote Sensing 14(9):1990
17. Zhang F, Yin J, Wu N, Hu X, Sun S, Wang Y (2024) A dual-path model merging CNN and RNN with attention mechanism for crop classification. Eur J Agron 159:127273
18. Park J, Kim JY, Park SW, Jung SH, Sim CB (2022) Development of ResNet based crop growth stage estimation model. Smart Media J 11(2):53–62
19. Duong LT, Tran TB, Le NH, Ngo VM, Nguyen PT (2024) Automatic detection of weeds: synergy between EfficientNet and transfer learning to enhance the prediction accuracy. Soft Comput 28(6):5029–5044
20. Koonce B, Koonce B (2021) EfficientNet. Convolutional neural networks with swift for Tensorflow: image recognition and dataset categorization, pp 109–123
21. Dhanasekaran K, Ramprasath M, Sathiyamoorthi V, Poornima N, Jayaraj IA (2021) Meta-learning based adaptive crop price prediction for agriculture application. In: 2021 5th international conference on electronics, communication and aerospace technology (ICECA), December, IEEE, pp 396–402
22. Zhou F, Wu B, Li Z (2018) Deep meta-learning: learning to learn in the concept space. arXiv preprint arXiv:1802.03596
23. Jaiswal A, Babu AR, Zadeh MZ, Banerjee D, Makedon F (2020) A survey on contrastive self-supervised learning. Technologies 9(1):2
24. Marszalek ML, Saux BL, Mathieu PP, Nowakowski A, Springer D (2022) Self-supervised learning--A way to minimize time and effort for precision agriculture?. arXiv preprint arXiv: 2204.02100.
25. Xu Y, Ma Y, Zhang Z (2024) Self-supervised pre-training for large-scale crop mapping using Sentinel-2 time series. ISPRS J Photogramm Remote Sens 207:312–325
26. Sunitha G, Reddy Madhavi K, Avanija J, Tharun Kumar Reddy S, Hitesh Sai Vittal R (2022) Modeling convolutional neural network for detection of plant leaf spot diseases. In: 3rd International conference on electronics and sustainable communication systems, IEEE, pp 1187–1192

# A Survey of Efficient Models for Predicting Vehicle Safety Measures in Fog Computing-Enabled Internet of Connected Vehicles Using Artificial Intelligence

**Sanjana S. Nazare and Priyadarshini C. Patil**

**Abstract** With the growth in the number of vehicular applications, there has been significant need for more processing and communication capabilities in the heterogeneous systems. Vehicular Ad-Hoc Networks or simply VANETs have been established with the aim of improving traffic congestion, accidental rates, and safety for drivers through vehicle-to-vehicle (V2V) communication and vehicle-to-infrastructure V2I communication. However, as a result of constant changes in the vehicular environment due to vehicle dynamics to include mobility, limited broadcast range and the adaptive network formation, there are sets back in scalability and performance. Another challenge that affects the adoption of autonomous driving systems and the Internet of Vehicles (IoV) is security threats, privacy issue, and data management challenges. This paper reviews the use of fog computing and artificial intelligence (AI) techniques to solve these challenges. Fog computing provides the efficient and distributed processing capacity and ML facilitates intelligent decision-making in real-time applications such as traffic flow analysis, obstacle identification, and prediction of the safety measures required in a vehicle. Thus, the paper discusses how such technologies can help in increasing traffic safety in IoV, coordinating the real-time data analysis, and implementing efficient resource utilization. In addition, the future innovations like blockchain mobility and artificial intelligence are included and their use in mitigating the problems in existing vehicular communication networks noted. In this paper, the author's objective is to review the advancements in technology in relation to smart transport systems and the possibility of their development in the future.

S. S. Nazare (✉)
Research Scholar, Department of CSE, PDA College of Engineering, Affiliated to VTU, Kalaburagi, Karnataka, India
e-mail: sanjananazare9907@gmail.com

P. C. Patil
Department of CSE, PDA College of Engineering, Kalaburagi, Karnataka, India



K. R. Madhavi et al. (eds.), *Proceedings of Sixth International Conference on Computer and Communication Technologies*, Lecture Notes in Networks and Systems 1356,
https://doi.org/10.1007/978-981-96-5238-9_46

## 1 Introduction

This has been triggered by the expansion of automotive applications these days and the need for efficient processing and communication resources especially for vehicular networks. Dedicated short-range communications (DSRC) based vehicle ad hoc networks (VANETs) have a central role in intelligent transportation systems by increasing traffic efficiency, decreasing the chance of incidents, and increasing safety. But we are left with problems like scalability, dynamic scenario, and mobile vehicles. The integration of Internet of Things (IoT), the cloud computing, the vehicles ad hoc networks (VANETs) in smart transportation systems through development of Internet of Vehicles (IoV) has open other opportunities and challenges on safety, security, privacy, and managing large data. In such situations, new computational paradigms such as vehicular fog computing (VFC) are emerging to provide real time, distributed processing architectures [1].

Communication can be classified into three types and these are vehicle-to-vehicle (V2V), vehicle-to-infrastructure (V2I) and a combination of both which is known as hybrid communication that may be applied in VANET [2]. Several issues, such as scalability issues, transmission range constraints, dynamic change in topology, and vehicle movement emerged during the course of establishing connectivity technologies.

There is a security problem with systems that incorporate autonomous driving. Cyber threats can potentially disrupt functional computer systems and in doing so; there are likely to be accidents as well as demise. Socio-paths and other ill-minded hackers could be set with the task of planning and orchestrating an attack on a certain car with an aim of either stealing it or causing severe damage on it or even endangering other people's lives. Another issue that is closely linked with AVs is privacy that was also mentioned above. It also may become dangerous for user's private life, if a device is constantly connected to the environment round a car. Concerning different online actions, some of the possible problems can be identity theft, tracking, stalking, and information breaches [3].

Several customers are skeptical about technology-driven businesses due to the companies' ability to acquire and monetize personal information. A smart car that has a plethora of sensors, cameras, and microphones might learn a lot about its riders [4]. Due to the ability of bringing radical shifts to the transport industry, the Internet of Vehicles (IoV) concept has received considerable attention in recent years. It is an open secret that, the application areas of ML are numerous for IoV. The applications include traffic flow, detecting the obstacles and predicting the measures to be taken for the safety of vehicles.

System architecture based on a tradition of centralized machine learning algorithms is incapable of dealing with the burgeoning data from the Internet of Vehicles (IoV) [5]. The Internet of Vehicles (IoV) refers to a technology platform that helps in realizing the intelligent transportation systems. This is done by the integration of Internet of Things (IOT), cloud computing, and vehicular ad hoc networks (VANETs). The latter is built on the former called the Internet of Things (IoT) that connects physical objects to the internet. This technology is expected to enhance the extended information exchange between vehicles and components of the transport system including traffic signs, signals, and detecting devices. Through this network, traffic safety is enhanced while reducing the impact of the environment, efficiency in operations is boosted, and the users' satisfaction is accomplished. However, there are some challenges that are in the way of the Internet of Vehicles' (IoV) deployment. The problems described above are as follows: In order to answer this, the problems stated above encompass various areas of concern in security, privacy scalability, and performance efficiency. The solutions for data management and service delivery should be efficient and preferably flexible in order to meet the challenges of large-scale data and changing nature of networks. That is to say the possible threat that may originate from erroneous data that may be introduced into the vehicle decision-making system is what makes the security of Internet of Vehicles (IoV) critical.

There are now significant compatibility problems with the new distributed protocol built on blockchain technology when integrating it with the current vehicular network and Internet of Vehicles. The aforementioned problems are mostly caused by the network's inadequate and irregular communication capabilities [6]. Application development for intelligent transportation systems (ITS) is given new chances by the combination of fog computing and vehicle-to-vehicle (V2V) technology. The 802.11 standard serves as the foundation for wireless technologies. Open-source and free V2V (Vehicle-to-Vehicle) technology is used to allow the deployment of fog nodes, also called volunteer automobiles chosen through an incentive scheme. The data processing procedure may be made more efficient at the vehicle level by utilizing these two technologies.

A thorough examination of several VFC deployment scenarios has been carried out. The computer equipment found in buses and taxis that are adjacent to other transportation vehicles is referred to as "fog nodes." The fog nodes evaluate the data that is offloaded from the vehicles [7]. Using stationary autos as the fog computing equivalent of stationary backbone nodes is another way to take advantage of fog nodes. Computers linked to V2X Road Side Units (RSUs) or groups of immobile cars stuck in traffic can make up the systems. Utilizing vehicle mobility is one way to effectively execute computer resource transportation management. Task assignment and resource allocation are two important aspects to take into account when working with shared resource systems. Numerous studies and analyses have been conducted on the aforementioned criteria. The study methodology for allocating computing responsibilities to a group of vehicle fog nodes in parallel [8]. Achieving quality of service (QoS) objectives, such as minimizing latency and optimizing image/video resolution, is the main goal. The method accounts for the nodes' limitations in terms of processing power, communication capacity, and energy usage. The use of computer units in stationary cars is one area where a contract- or auction-based strategy is used. The previously indicated approach is employed to ascertain and bargain for the accessible resources and consequent advantages, particularly with regard to decreased parking costs.

Taxis and buses, among other commercial vehicles, can use fog computing systems to handle service requests from clients or other automobiles in the vicinity. The fog nodes are able to carry out operations that are normally carried out by cars. Therefore, the roadside unit (RU) does not need to oversee these activities. This strategy may be used to considerably minimize task latency, which will lessen the strain on the resource units (RUs). But it also adds a great deal of complexity to the Internet of Vehicles' content distribution mechanism. "Vehicular fog computing" (VFC) is a new computing paradigm that intends to provide low-latency processing capabilities for aided driving [9]. Transferring computing tasks from client vehicles where data is generated to fog nodes is the main goal. Fog nodes are normally positioned in close vicinity to 5G cell towers or buses, which are outfitted with processing units with increased performance.

Because only one communication hop is needed in both scenarios, the transmission time is greatly shortened. Furthermore, the effective implementation of fog computing is made possible by the conversion of buses and other mobile vehicles into fog nodes. Furthermore, when client cars are shown to be parked close to fog nodes, the requirement for service transportation between fog nodes is decreased [10]. Depending on the method of transportation, location, and time of day, there may be variations in the density of client cars and the amount of work produced. Vehicular fog nodes, or fog nodes that are delivered by vehicles, are accessible to varying degrees depending on the driving routes taken by the study's carriers. It is difficult to schedule work offloading from client cars to fog nodes optimally because of fluctuations in the location and time of computing resources' availability and demand.

The complex changes that cause the accidents are not easily discernable using the existing traffic movement patterns algorithms. Thus, the main reason for appearance of this problem is a very large amount and variety of traffic data that hardly can be analyzed. During the conventional models the efficiency of data capture and prediction in real-time is taken into consideration. However, the effectiveness of such applications appears limited when the traffic conditions are constantly in a state of flux. The reliance on the existing models which fail to consider geographical, temporal, and spatial details reflected in traffic data hinders the models ability to accurately predict accidents [11]. A campaign on the increasing importance of developing an environment that effectively predicts accidents based on the high-dimensional data has been started.

Hence, this article aims at discussing on the accident prediction and prevention system that employs the DL technology and concepts of IoV. There is nothing but positive goals which are associated with the use of these advanced concepts so as to enhance traffic safety. The terminal layer of the multi-layer system is solely responsible for collecting information from linked cars and roadside units (RSUs) equipped with real-time vehicle and traffic sensors. The STCLA is a model that is built for the purpose of predicting accidents and is explained as follows: Because the sensor data collection and preparation are performed by the intermediary layer gateway units, their usage is obligatory for this concept. The system is managed through a server that is hosted from the cloud [12].

## 1.1 Challenges

- Mobility and Limited Transmission Range: High mobility of the vehicles and low transmission range are the issues which affect the continuous connectivity for reasonable threshold of topological changes and the scalability of the VANETs.
- Security Threats to Autonomous Driving Systems: This brings a sense of reality to the danger of cyberattacks as miracles come with a price to the safety of autonomous vehicles. Attacks can be very devastating leading to system failures, disruption of vehicle control or even accidents, implying that human lives are at risk.
- Privacy Concerns in Vehicle Communication: The constant flow of information in vehicle-to-infrastructure means that there are privacy issues. For instance, knowing a vehicles location or the identity of passengers using a vehicle can be compromised, thus exposing passengers to risks of identity theft, among others, and being tracked.
- Scalability of IoV and Data Management: Given the proliferation of data in IoV, handling and analyzing this information especially in real-time is a difficult task. Currently, implemented centralized solutions have lot of issues to manage high data rates in a constantly varying vehicular scenario.
- Interoperability Issues in Existing Vehicular Networks: Current vehicular networks also have issues in regards to compatibility particularly when incorporating novel technologies such as the blockchain-based protocols to the existing IoV systems, creating problems such as fluctuating communication protocols and reliability.
- Efficient Task Offloading and Resource Allocation: Due to time–space fluctuations of supply and demand of computing services in vehicular fog computing, the schedule of tasks, as well as the distribution of resources play essential roles in data processing and, consequently, require optimization.

## 1.2 Motivation and Contribution

- This survey paper is motivated by the increasing and unpredictable nature of vehicular networks, especially on the IoV and VANETs. Taking into consideration the current trend of establishing intelligent transportation systems, it is imperative that effective, efficient and secure mechanisms for communication within these networks are established considering the dynamic and heterogeneous nature of such networks. Through the systematic review of the current advances in vehicular communication, this paper aims to firstly identify the current issues and challenges that have cropped up, that include security and privacy considerations, scalability as well as real-time information processing. The purpose is to discuss how these new approaches can improve the performance of vehicle-to-vehicle and vehicle-to-infrastructure network to facilitate safer and more efficient transportation systems in the future.

- Survey of AI Techniques for Vehicle Safety Prediction in Fog-Enabled IoCV: The paper sheds light on various existing artificial intelligence models and methods for forecasting on vehicle safety measures in the area of fog computing empowered IoCV and evaluates the strengths and flaws of different solutions.
- Analysis of Fog Computing Integration for Enhanced Vehicle Safety: In the survey, the authors consider how fog computing is incorporated with safety prediction models based on AI in vehicular networks based on problematics such as scalability, low-latency computations, and resources optimization.
- Identification of Research Gaps and Future Directions: By reviewing the current state of the art, this paper identifies key research gaps in the application of AI and fog computing for vehicle safety prediction, and proposes directions for future research to develop more advanced and reliable safety systems in IoCV environments.

## 2 Related Work

The content delivery technology of the Internet of Vehicles (IoV) network has undergone comprehensive examination and analysis. The term Maximum Weight Independent Set (MWIS) is commonly used to describe the content distribution approach. In this context, the communication structure of the Internet of Vehicles was represented as a graph. The automobiles engaged in collaboration with neighboring vehicles to determine the optimal distribution strategy for widely sought-after information. This was achieved by employing the content dissemination mechanism [13]. The development of a real-time traffic management system was initiated based on prior research. The primary objective of this system is to facilitate the efficient distribution of information. The implementation included the establishment of a routing system that takes into account latency sensitivity. In order for the systems to function properly, it is imperative that the roadside unit (RU) does not utilize the vehicle's communication capabilities to their maximum extent. The system exclusively transmits data to vehicles located within its coverage area. The study conducted in references [14] examined the utilization of automobiles equipped with advanced computer systems as cluster heads. Within this specific configuration, a vehicle located within the coverage area of a cluster head will exclusively transmit data to that specific cluster head. The authors failed to consider the potential application of the carry-and-forward transmission paradigm for distributing content over the Internet of Vehicles (IoV).

Sharma and Wang [15] developed an ecosystem model that utilizes blockchain technology to monitor and manage operations for electric car and recharge station management. Their work made a significant contribution to the field. In addition, the utilization of elliptic curve cryptography ensured the security of the hash functions. The decentralized peer-to-peer PETCON power trading system was developed. The system utilizes a distributed blockchain to autonomously oversee and authenticate transactions.

The authors [16] have developed CreditCoin, a privacy-preserving solution that ensures user anonymity in the transmission of critical alerts between vehicles, using blockchain-enabled data aggregation. The current innovation within the framework of the Internet of Vehicles expands the utilization of blockchain technology. The accuracy of the data in Yang et al. blockchain-based approach is determined using the sender's reputation, ensuring its correctness. The study conducted examined the potential application of blockchain technology in integrating vehicle data with existing traffic and meteorological data. The objective was to improve communication within intelligent transportation networks. The method described below enables the establishment of direct, secure, and private communication between cars.

The authors have proposed a new methodology known as CRTI with an ability to forecast traffic accidents. According to the suggested technique, it has incorporated use of algorithms for the machine learning and pattern recognition. In their work, [17] centered their research on the data of basic safety messages that are received from connected cars. The aim of this project was to build an ML model that is able to detect unsafe behaviors by a driver so as to reduce danger to other road users. The proposed layered LSTM recurrent neural networks enabled the structuring of the driving behavior. The study enlisted data collected and used through the sensors, more specifically data from cellphone. It has applied machine learning techniques for integrating and overlaying physiological activity measurements, behavioral properties, as well as driving performance data for the purpose of correctly identifying and differentiating driver fatigue. For example, it considered asymmetric driving behavior and the traffic characteristics similar to real-life traffic flow with the help of long short-term memory neural network (LSTM-NN) model.

The work done by [18] was aimed at analyzing the car-following and traffic oscillations, and their differences. Extra information about the area around a roadblock was also filmed by using the unmanned aerial vehicle (UAV) in Nanjing, China. The following information was obtained during an investigation that was authorized. Hence, the study shows that drivers display a number of different but sustainable driving behaviors and that the oscillation response of each of the drivers depends on the subject matter.

The long short-term memory fully convolutional network (LTSM-FCN) was proposed by [19] in their work revolving on how time limit can affect the braking movement of drivers, whereby accidents may occur. The results of the presented study show that the approaching speed and occupation by driving, gender, age, and the road conditions are the essential factors that influence the behavior of drivers by braking. From the study, the authors examine the effects of clock deadline on the braking profile and probabilities of accidents to the drivers. Cue to this, it is possible to design specific interventions which may help to address the factors leading to road carnage besides the negative effects of time crunch on drivers' behavior.

More specifically, [20] used high-resolution smartphone sensor data to analyze the behavior of drivers at intersections and particular road segments. Therefore, the main purpose of this research is to explain how traffic conditions influence driving behavior. From the statement, it is implied that the findings could be used in policy development for the purpose of increasing traffic safety and reducing the rate of aggressive driving occurrences.

The study conducted by [21] categorized DB using ML models and for the analysis of the driving behavior. The authors post that the different aspects of the Driver-Vehicle-Environment (DVE) paradigm provide theoretical fundamentals for studying the driver behavior (DB) phenomenon. In order to expand the authors' knowledge in the field of database research and the focus on the aggressive driving behavior, the authors conducted a systematic literature review (SLR). In the experiment conducted in this research, methodology suggested here proved to be more precise than the other methods as it was shown in the five minutes test. In a recent research work, it has been established that people who had low inhibitory control enforce their disposition for violating traffic laws and as a result increase their vulnerability of getting involved in an automobile accident.

In the study carried out by [22], the authors employed comparative analyses to … Following the same approach, this study sought to investigate how fatigue and unusual patterns of driving may be detected. The study affirms even more the importance of active systems like ADAS (Advanced driver assistance systems) and IDAS (Intelligent driver assistance systems) in informing the driver of possible risks. The authors of the papers express concern because some of the tactics advanced in the existing studies require to be expanded.

The studies of naturalistic driving in the light of safety with deep focus on their main interest—driver behavior. Looking to support their claim that it is the driver rather than the vehicle that is responsible for most traffic accidents, the authors provide a detailed synthesis of data from many of the prior naturalistic driving research. All of the contributions from the authors have significantly advanced the science of naturalistic driving research whose focus has been to improve the safety on the road and deepen knowledge into driver behavior. In the study of [23], the authors included the protocol for vehicle-to-vehicle V2V network layer communication. Fuzzy-based multi-criteria decision-making (MCDM) is employed in the protocol for the selection of relay nodes and route segments.

In their study, [24] introduced a technique that enables emergency response vehicles to collect data using vehicle communication and sensors, specifically CAMs and RADARs. Prior to the car reaching its destination, the ERV must furnish drivers with an estimated location. The objective of this feature is to provide users with alerts regarding potential dangers or hazards. Vehicles are categorized into three distinct groups based on their length: heavy vehicles, light vehicles, and other classes.

The network developed by [25] facilitates social engagement among various elements, such as drivers, cars, and gadgets. The Internet of Vehicles (IoV) architecture has been found to have security and privacy flaws, as identified by researchers. They have developed a system for authentication and key agreement based on blockchain technology. The intended application for this technology is the multi-trusted authority network paradigm. In order to ensure the security and reliability of Internet of Vehicles (IoV) connections, it is necessary to have a well-defined protocol (Table 1).

**Table 1** Survey table

| Reference | Method | Advantages | Disadvantages | Research gap |
|---|---|---|---|---|
| [7] | Blockchain-based ecosystem model for EV management | Secure hash functions, better EV management | Limited to EV management and charging systems | Scalability and integration with other IoV elements |
| [8] | Peer-to-peer power trading system using blockchain | Autonomous transaction validation, decentralized control | Scalability issues in large networks | Efficient handling of large-scale IoV networks |
| [9] | CreditCoin: blockchain-based privacy-preserving system | Ensures user anonymity and data aggregation | Lacks real-time updates | Real-time privacy-preserving systems with data freshness |
| [10] | Blockchain-based data authenticity system for IoV | Authenticates data based on sender's reputation | Dependent on sender's reputation, lacks data freshness | Ensuring real-time trustworthiness in data sharing |
| [11] | Blockchain for integrating vehicle data and real-time data | Secure vehicle communication | Privacy concerns due to data sharing | Balancing real-time data processing and privacy concerns |
| [14] | CRTI framework for accident prediction using ML | Predicts accidents, improves road safety | May not scale well in real-time dynamic environments | Scalability and real-time applicability |
| [15] | ML for detecting hazardous driving actions using BSM | Identifies hazardous driving patterns | Relies on data from connected vehicles | Generalization across varying environments |
| [16] | Stacked LSTM networks for driver behavior classification | Accurately classifies driving behaviors | Requires high computational power | Optimizing computational efficiency for real-time tasks |
| [17] | ML for drowsiness detection using performance metrics | Uses multiple metrics for detection | May produce false positives | Reducing false positives in real-time |
| [18] | LSTM-NN for replicating traffic flow and driving behavior | Replicates real traffic conditions and behaviors | Complex model needing extensive data | Model complexity reduction for real-time applications |

## *2.1 Research Gap*

- Limited Exploration of Real-Time Processing in Dynamic Environments: While many studies focus on vehicular fog computing and IoV, few address the challenges of real-time data processing and decision-making in highly dynamic traffic environments with fluctuating network demands.
- Inadequate Handling of Security and Privacy Concerns in IoV: Although several papers discuss security mechanisms, there is a lack of comprehensive solutions that address both the scalability of IoV and the increasing risks of cyberattacks, privacy breaches, and identity theft in interconnected vehicle networks.
- Scalability Issues in Current AI and ML-Based Safety Prediction Models: Many current models struggle to scale efficiently when handling the massive amounts of high-dimensional data generated by IoV, especially in fog computing-enabled architectures where computing resources are distributed and mobile.
- Underutilization of Blockchain for Secure Communication: While blockchain has been introduced as a promising technology for secure communication in IoV, there is limited research on its integration with fog computing and its full potential in preventing malicious attacks while maintaining data integrity.
- Lack of Research on Vehicle-to-Vehicle (V2V) Communication in Fog Networks: Most studies focus on vehicle-to-infrastructure (V2I) communication, with less attention given to the challenges and opportunities in optimizing V2V communication using fog computing, which could enhance traffic management and safety.
- Challenges in Task Assignment and Resource Allocation: Current research on vehicular fog computing often overlooks the complexity of optimizing task assignment and resource allocation in dynamic environments, especially when balancing factors such as latency, bandwidth, and energy consumption.
- Limited Studies on Fog Computing-Based Accident Prediction Models: While there are a few models that do tackle accident prediction through deep learning and IoV, very few seem to investigate the potential of using fog computing for cost-effective accident prediction and avoidance, majority of which would be done in places with or by traffic that is quite mobile or highly congested.

# 3 Conclusion

On a broader perspective, the growing needs of vehicular applications make the need to adopt new and more efficient and scalable approaches to vehicular ad hoc networks (VANETs), and the internet of vehicles (IoV). This paper focuses on the integration and application of artificial intelligence and fog computing that may help in tackling challenges of mobility, limited range, security, and privacy. The survey highlights the need to make use of these technologies to enhance computation for fast processing of data, better safety of traffic, and the effectiveness of forecasting vehicle safeguards. In this engineering paper, the author demonstrates how the transformation will occur in the IoV environments enabled by the current machine learning, blockchain, fog computing technologies and thus making safer, secure and reliable transportation systems. Future work should work on the more improvements of these technologies in terms of scalability and interoperability as this would mark the progress of smart transportation.

## References


1. Raut SB, Malik L (2014) Survey on vehicle collision prediction in VANET. In: Proceedings of IEEE International conference on computational intelligence and computing research, pp 1–5
2. Karnouskos S, Kerschbaum F (2018) Privacy and integrity considerations in hyperconnected autonomous vehicles. Proc IEEE 106(1):160–170
3. Yang W, Wan B, Qu X (2020) A forward collision warning system using driving intention recognition of the front vehicle and V2V communication. IEEE Access 8:11268–11278
4. Guo W, Li J, Liu X, Yang Y (2020) Privacy-preserving compressive sensing for real-time traffic monitoring in urban city. IEEE Trans Veh Technol 69(12):14510–14522
5. Aki M, Rojanaarpa T, Nakano K, Suda Y, Takasuka N, Isogai T, Kawai T (2016) Road surface recognition using laser radar for automatic platooning. IEEE Trans Intell Transp Syst 17(10):2800–2810
6. Hemmati A, Zarei M, Souri A (2023) Blockchain-based Internet of Vehicles (BIoV): a systematic review of surveys and reviews. Secur Privacy 6(6):e317
7. Zheng Z, Xie S, Dai HN, Chen X, Wang H (2018) Blockchain challenges and opportunities: a survey. Int J Web Grid Services 14(4):352. https://doi.org/10.1504/ijwgs.2018.095647
8. Wang H, Li H, Smahi A, Zhao F, Yao Y, Chan CC, Wang S, Yang W, Li S-Y-R (2024) MIS: a multi-identifier management and resolution system in the metaverse. ACM Trans Multimedia Comput Commun Appl 20(7):1–25
9. Taherdoost H (2023) Smart contracts in blockchain technology: a critical review. Information 14(2):117
10. Sisinni E, Saifullah A, Han S, Jennehag U, Gidlund M (2018) Industrial Internet of Things: challenges, opportunities, and directions. IEEE Trans. Ind. Inform. 14(11):4724–4734
11. Lee K et al (2020) Reliable, high-performance wireless systems for factory automation. NIST Interagency/Internal Report (NISTIR), National Institute of Standards and Technology, Gaithersburg, MD, USA, NIST Pubs, Tech. Rep. 8317
12. Contreras J, Zeadally S, Guerrero-Ibanez JA (2017) Internet of vehicles: architecture, protocols, and security. IEEE Internet Things J. https://doi.org/10.1109/JIOT.2017.2690902
13. Zhang J, Wang F-Y, Wang K, Lin W-H, Xu X, Chen C (2011) Data-driven intelligent transportation systems: a survey. IEEE Trans Intell Transp Syst 12(4):1624–1639
14. Sun Y, Song H, Jara AJ, Bie R (2016) Internet of Things and big data analytics for smart and connected communities. IEEE Access 4:766–773
15. Sharma SK, Wang X (2017) Live data analytics with collaborative edge and cloud processing in wireless IoT networks. IEEE Access 5:4621–4635
16. Cooper C, Franklin D, Ros M, Safaei F, Abolhasan M (2017) A comparative survey of vanet clustering techniques. IEEE Commun Surv Tutor 19(1):657–681
17. Kaiwartya O et al (2016) Internet of vehicles: motivation, layered architecture, network model, challenges, and future aspects. IEEE Access 4:5356–5373
18. Li H, Wang H, Liu L, Gruteser M (2018) Automatic unusual driving event identification for dependable self-driving. In: Proceedings of the 16th ACM conference on embedded networked sensor systems, Nov 2018, pp 15–27
19. Fugiglando U, Massaro E, Santi P, Milardo S, Abida K, Stahlmann R, Netter F, Ratti C (2019) Driving behavior analysis through CAN bus data in an uncontrolled environment. IEEE Trans Intell Transp Syst 20(2):737–748

20. Liu Y, Wang Y, Chang G (2017) Efficient privacy-preserving dual authentication and key agreement scheme for secure V2V communications in an IoV paradigm. IEEE Trans Intell Transp Syst 18(10):2740–2749
21. Zanjireh MM, Larijani H (2015) A survey on centralised and distributed clustering routing algorithms for WSNs. In: Proceedings of the IEEE 81st Vehicular Technology Conference (VTC Spring), pp 1–6
22. Gupta DS, Karati A, Saad W, Da Costa DB (2022) Quantum defended blockchain assisted data authentication protocol for internet of vehicles. IEEE Trans Veh Technol 71(3):3255–3266
23. Bojjagani S, Sastry VN (2017) VAPTAi: a threat model for vulnerability assessment and penetration testing of Android and iOS mobile banking apps. In: Proceedings of the 3rd international conference on collaboration and internet computing (CIC), Oct 2017, pp 77–86
24. Liu J, Li Q, Sun R, Du X, Guizani M (2018) An efficient anonymous authentication scheme for internet of vehicles. In: Proceedings of the IEEE International Conference on Communications (ICC), May 2018, pp 1–6
25. Jiang Q, Zhang X, Zhang N, Tian Y, Ma X, Ma J (2019) Two factor authentication protocol using physical unclonable function for IoV. In: Proc. IEEE/CIC International Conference on Communications in China (ICCC), Aug 2019, pp 195–200

# Blockchain Enhanced Certificate Authentication System Using IPFS

**P Subhash, B. Varsha Reddy, K. Varun, P. Sandhya Kiran, and V. Abhishek**

**Abstract** As digital tools are in abundance, forging event-related certificates or replicating them has considerably become very easy these days, which at times reduces the quality of well-deserved certifications. Many ways are offered by businesses and organizers to reproduce sham certificates that look similar to the real ones. The authentication of credentials is made extremely difficult for companies by this, thus posing the most serious barrier to business and employers in verifying certain employees' skills and achievements. In this paper, a novel solution is proposed based on merging the advantages of blockchain technology with the Interplanetary File System. In our scheme, the certificate data is persisted securely over IPFS, generating a unique hash for each certificate. This IPFS hash, along with a digital signature with the issuer, is stored over the blockchain. Some of the inherent core properties of blockchain—immutable, decentralized, and transparent—find immense application in enhancing security and reliability aspects of the system. Blockchain and IPFS are combined to issue and verify processes for digital certifications in a brand new, secure, and highly efficient system, making an effective solution to the global problem of fake certificates.

**Keywords** Blockchain technology · IPFS (Interplanetary File System) · MetaMask · Ganache · Smart contract · Digital signature

## 1 Introduction

In this modern digital period, especially authenticity revolving around certificates of education and professional qualifications, the truth will be made more candid. Current verification systems normally relied upon the principles of manual dependence as well as centralization in databases—a condition inviting quite a few vulnerabilities

P Subhash (✉) · B. V. Reddy · K. Varun · P. S. Kiran · V. Abhishek
Department of CSE—(CyS, DS) and AI&DS, VNR Vignana Jyothi Institute of Engineering and Technology, Bachupally, Hyderabad, Telangana, India
e-mail: subhash_p@vnrvjiet.in



K. R. Madhavi et al. (eds.), *Proceedings of Sixth International Conference on Computer and Communication Technologies*, Lecture Notes in Networks and Systems 1356,
https://doi.org/10.1007/978-981-96-5238-9_47

[1]. The ease with which sham certificates can be created and distributed is a danger to confidence in educational institutions per se, placing employers and all other reliance parties in great jeopardy [2]. The urgent need for an efficient, more transparent, and tamper proof system for verification of certificates has never been felt so acutely.

The immutability, decentralization, and transparency inherent in blockchain technology makes it a potential ground breaker for certificate issuance, storage, and validation mechanisms [3]. Blockchain is structured in a decentralized manner, making it censorship-proof since it is impossible for data to be manipulated by any single person. Immutability ensures that once a certificate is identified by the blockchain, it can never be altered or deleted; thus, immutability makes records tamperproof. Hence, a permanent and auditable record of issuance and authenticity of certificates is achieved. Blockchain, as such, enables all transactions to be made transparent; thus, any certificate can be swiftly validated by one of the parties.

Some of these research works have discussed how the verification of certificates can be revolutionized in the area of credential management by blockchain technology [4]. In fact, secure and tamper-proof digital certificates that can be easily verified by employers and other stakeholders were developed by them. But most of the solutions proposed so far have several issues with respect to scalability, privacy, user experience, and integration with existing systems. This has been extended to include other emerging technologies, like the Interplanetary File System [6–8], in the integration with blockchain toward maximizing efficiency and safety on the storage of certificates and their subsequent retrieval [5].

The same shall extended to propose the respective drawbacks with a complete blockchain-based certification verification system [9]. It shall also be incorporated with other features, namely authentication, authorization, confidentiality, privacy, and ownership [10]. It shall be proposed as a system that would offer a solution for safe, efficient verification of certificates, including some blockchain capabilities through the integration of other technologies. Full confidence about the authenticity of any given certificate in the digital world will be given to users and organizations. Sustaining efforts in fighting certificate fraud would be added to further enhancing the credibility of educational and professional credentials and making the process of verification easy for all parties concerned [11]. Figure 1 presents system architecture.

## 2 Related Works

Verification of documents, especially educational certificates, is a very cumbersome process and highly prone to fraud. A potential remedy to improve document verification and fraud control has been considered to be blockchain. The focus of this research is on the identification of security topics required for document authentication in blockchain; an insight into shortcomings within the current blockchain-based solutions for educational certificate verification is also provided. The target is an authorization-based authentication scheme, ensuring confidentiality, privacy, and rights using the Hyperledger Fabric Framework [1]. The forging of Bachelors',

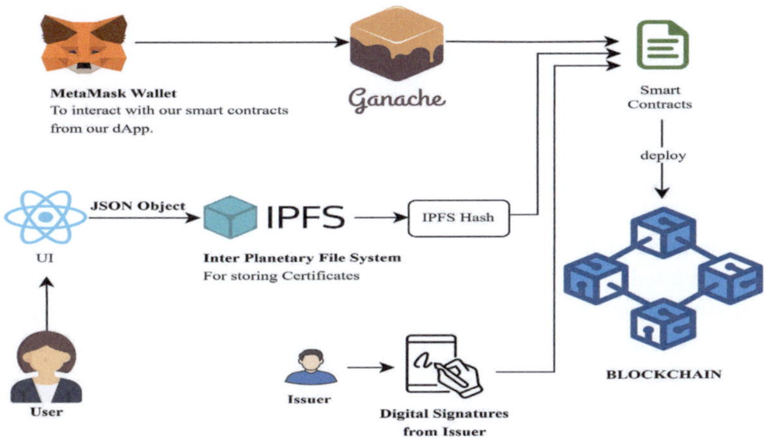


**Fig. 1** System architecture

Masters', and Ph.D. certificates is said to require an efficient system that can be trusted. Traditional methods are involved that are time-consuming and cost a lot of money. Blockchain technology is thus provided, offering a means by which the verification can be viewed transparently and immutably. In this blockchain-based model, the hashes of certificates are stored with the help of which the originality of the certificates can be verified very well by an employer. The security and integrity of the certificates will be ensured by the hash generation system along with public–private keys, digital signatures, and timestamps [2].

Among the most critical issues in Public Key Infrastructure security, certificate revocation management is highlighted, with problems related to scalability, financial, and privacy concerns of traditional revocation methods. Thereby, blockchain technology, being inherently distributed and tamper-evident by its very nature, is led from the front. Effective implementation of a blockchain-based decentralized status verification and certificate revocation system is proposed in this work [3]. The X509 certificate structure is extended, and Bloom filters are used for storing revocation information on the blockchain, increasing the efficiency of revocation information recovery. System performance is assessed through performance metrics showing both security and performance features can be fulfilled, and it is according to up-to-date web standards as in [12]. One main issue educational institute face due to which the digitalization of certificates is problematic is the continuous authenticating and verifying of them. A possible solution to verify original certificate authentication is the hashing-intensified blockchain technology. In general, paper certificates are digitized into digital form, hash codes are generated via a chaotic algorithm, and they are lodged into the blockchain by the proposed system. Students' credentials are uploaded by institutions, and organizations verify the certificates when needed by just feeding in information about a particular student. More authentic and reliable digital certificate verifications are now possible as a result [4].

The method of verification and validation of certificates becomes even further challenging, considering thousands of graduates that throng India every year. Problems are caused by the manual model of present validation in the form of tampering and fake certificates. Here is the need for automation of an indiscriminate system that can issue and validate the certificates with much more ease and efficiency [1]. This system is based on certificate issuers, students, and companies. To help in the validation of any certificate, a unique hash key will be had by it. The advantages are that improved security is had, no risk for the loss or defacement of the certificate, and quick validation [5] is enabled. In the proposed system, blockchain-stored academic certificates are the digital assets. Verification of certificates will happen in no time, without any fraud by an intermediary, using the MetaMask and Ethereum network. Also, encrypted data storage is enabled in IPFS to provide more security and scalability within the system [9].

Such dependence on cloud storage is costly, brittle to data availability, and security due to the nature of their trusted third parties. A potential solution for secure and decentralized data storage can be provided with the implementation of the Interplanetary File System using blockchain technology [13]. Data will be uploaded to IPFS by users in this proposed system and distributed across the cloud nodes of the globe by making them available via unique hash values. Data privacy on a distributed network is ensured by the blockchain. Smart contracts were implemented and tested on the Ethereum Blockchain Platform using Ganache [11].

Interest in blockchain technology has, therefore, been thrown up, with the core idea to have a robust, transparent, and tamper-proof procedure for the verification of certificates. The problem of fraud and inefficiency in certificates can, therefore, be solved by the naturally decentralized and immutable feature of blockchains. The research proposed here aims at developing a holistic, blockchain-based certificate verification system with authentication, authorization, confidentiality, privacy, and ownership features aimed to overcome the drawbacks found in different previous works. Designed to act as a secure and efficient solution for the easy use of certificate verification, capitalizing on blockchain technology, it also integrates IPFS [10].

## 3 Proposed System

This research paper has come up with a solution of blockchain-enhanced certification authentication where the number of fake certificates issued keeps on growing and the processes involved in verifying the authenticity of event-related certifications are facing numerous challenges. This will be through a decentralized, safe, and transparent environment harnessed with the strengths of blockchain technology and IPFS in the issuance, storage, and authentication of digital certificates.

Our system consists of three extremely basic elements:

1. Smart Contract on Blockchain
2. Decentralized Storage with IPFS
3. User Interface Design for Creation and Verification

**Blockchain Smart Contract**: It is the backbone that controls the issue and verification logics. The smart contract is implemented over the blockchain network, which keeps the unique digital signature of the certificate along with the hash generated for it from the IPFS.

Certificates are made tamper-proof and immutably stored, verifiable independently by this mechanism. Verification requests are also implemented by the contract and managed by fetching the stored hash and matching it with certificate data.

**Decentralized Storage with IPFS**: This is a decentralized storage on IPFS. All certifications, instead of being stored directly on the blockchain, shall cost a fair deal and be highly inefficient, keeping in view the data storage cap on blockchains; rather, they reside on IPFS.

Storage is thereby made decentralized, scalable, and tamperproof for the availability of the certificates themselves. When a certificate is issued, it gets uploaded to IPFS, which returns a content identifier hash—a uniquely generated hash for that particular content. A combination of this hash, together with the digital signature, is put on-chain so that an indisputable link between the certificate and the on-chain reference is created.

**User Interface**: *Certificate Issuance and Verification:* The platform's design facilitates the user's easy and convenient certificate issuance process. To easily issue fresh certificates, a web user interface is available. The issuer executes the necessary fields such as the recipient's name, event details, and the certificate attributes which is then transformed to JSON and uploaded to IPFS. The IPFS hash and digital signature are, in turn, recorded by the contract on the blockchain from the onset.

Through the same interface, the certificate's hash on the blockchain may then be obtained and compared to data in the IPFS for verification by the receivers, verifiers, employers, and institutions.

## 4 Working of the System

### *4.1 Procedure for Issuing Certificates*

- *Data Input:* The certificate might be issued by an organizer of events. The recipient's name, date of the event, and other data are inserted into the system by him/her.
- *JSON Conversion:* Details of the certificate are extracted from data input and stored in a JSON object.
- *IPFS Upload:* The JSON object is obtained using the Interplanetary File System. For each file, IPFS creates a particular cryptographic hash—what is usually referred to as the CID. This is, in fact, how the certificate will be identified in the decentralized storage system.

- *Blockchain Entry:* The digital signature of the issuer and IPFS hash are recorded on the blockchain. Using the browser extension MetaMask that connects users with the blockchain and a personal Ethereum blockchain environment called Ganache, interactions with a local blockchain for testing purposes will be involved in this stage.

MetaMask allows the issuer to sign and broadcast the transaction, which stores the IPFS hash and the digital signature on the blockchain via Ganache.

### 4.2 Smart Contract Deployment on Ganache

- *Design of Smart Contract:* The smart contract was designed in Solidity for the management of issuance and verification. The following defines the basic structures of a smart contract: certificate structure, functions adding a certificate, verifying one, and retrieving.
- *Deployment:* The smart contract was deployed on Ganache, a personal blockchain environment, to simulate the Ethereum network on the local machine; therefore, fast testing and debugging of blockchain transactions can be committed.
- *MetaMask Integration:* MetaMask was configured to connect to the local blockchain created by Ganache. Users are allowed to interact with their web browser directly into the smart contract for processing transactions involving creating or verifying the authenticity of certificates.

### 4.3 Certificate Storage and Security

- *IPFS Storage:* Upon issuance, the JSON of this certificate is stored on IPFS. IPFS gives the assurance of the decentralization and immutability of that certificate since it cannot be altered once uploaded.
- *Blockchain Record:* The IPFS hash and the digital signature are kept by the blockchain for a permanent, immutable record of the authenticity of that certificate. In this way, even if the system where the certificate is issued gets compromised, the certificate can be successfully verified using blockchain along with IPFS.

Storage and IPFS are separated from the verification process, which is the blockchain. That adds several layers of security and decentralization.

## 4.4 Certificate Verification Process

- *Retrieval:* Every third party (e.g., employer) that needs to verify the status of a given certificate may query the blockchain with the corresponding unique identifier of that certificate. In turn, the blockchain returns the stored IPFS hash and the digital signature of the issuer.
- *Verification:* The IPFS hash fetched is correlated with the actual data stored in IPFS. If the data is the same, then the authenticity of the certificate is established. The digital signature stored on blockchain confirms that indeed this certificate has been issued by a legitimate authority

## 4.5 Final Testing and Validation

- MetaMask handles the transactions, and it grants the necessary safety and transparency for the user interface to perform any certificate-related transaction with the blockchain.
- *Simulation and Testing:* The system was tested for its performance and security by heavy use of Ganache, working out the implementation details of smart contracts and transaction workflows correctly. Figure 2 illustrates the workflow.

# 5 Results

In the proposed blockchain-based authentication system of the certificate, we have tried to make it most efficient by implementing a number of optimization strategies. In addition, the system is much more efficient with a lower ETH value, which we have done to ensure that the system will work cost-effectively. Lower gas consumption means reduced use of computational resources; hence, the whole process of certificate issuance and verification is more effective. We minimized unnecessary operations contributing to a high gas fee by carefully optimizing smart contracts, ensuring that only necessary processes were run. Moreover, the IPFS hash and important metadata are stored on chain due to the integration of off-chain storage using IPFS. This substantially reduces on-chain data and gas use. While the core certificate content remains decentralized, secure in IPFS, the blockchain will hold very minimal but essential information. Together, these approaches make our project scalable and very cost-effective in terms of issuance and verification. As a result, the decrease in gas fees shows the system's efficiency, as it allows for a higher capacity of transactions at a lower total cost.

As shown in Fig. 3, we compare the results of our model with the existing model presented in [9] where the data is secured using a communication channel (EHBC). This demonstrates the connection between the number of transactions and the total

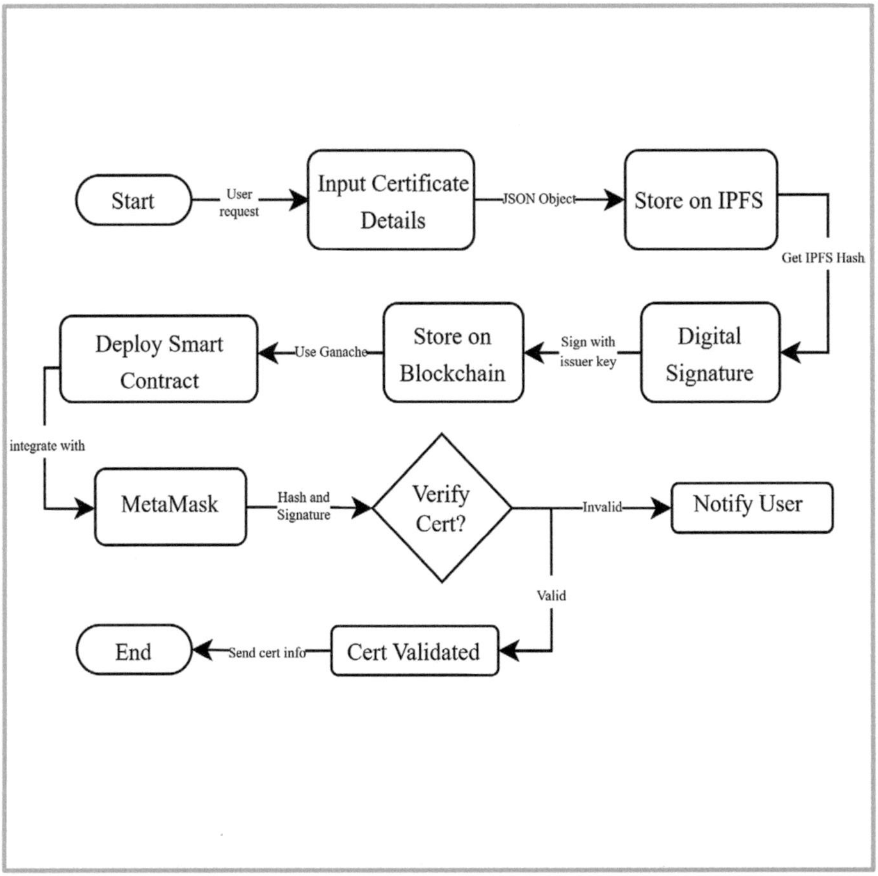


**Fig. 2** Flow diagram

ETH value needed for processing the transactions. As shown, there is a linear increase in the total ETH value with an increase in the number of transactions compared to [9].

About 50% less ETH was used by our model when compared to the least number of transactions at 1 certificate. The efficiency gap between the two models continues to increase as the number of transactions grows. In the case of 25 transactions, 10.90321 ETH is used by our model, whereas 12.25801 ETH is used by the prior model in [9], resulting in a difference of about 11%. This constant decrease in ETH value is indicative of strong cost savings, reflecting better resource optimization achieved by our model for the generation and verification of certificates.

We have done targeted optimizations that have considerably improved the performance metrics of the blockchain-based certificate authentication system. The driving idea of our approach is to increase the throughput, decrease the latency, and generally optimize several aspects.

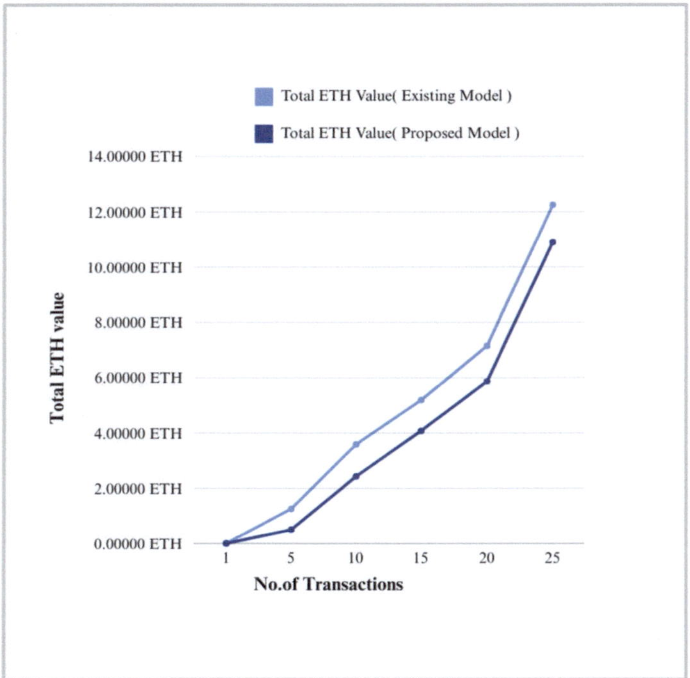


**Fig. 3** No. of transactions versus total ETH value

It is clearly indicated by the graph and data from Fig. 4 that better throughput is exhibited by our model compared to the existing model in [9], when dealing with different transaction volumes. When tested for the lowest transaction volume of 10 transactions, 4.1 TPS was reached by our model. However, 3.4 TPS is reached in [9] and this is 20% less approximately. This development was further increased as the number of transactions grew; at 40 transactions, the TPS achieved by our model is 5.4, which is a crucial improvement of about 5.8% over the one in [9], which is around 5 transactions per second. At the highest transaction volume of 50, higher throughput is achieved by our model, equal to 4.7 TPS, where in existing model it is 4.2 TPS, further demonstrating its ability for sustaining higher volumes of certificate generation and verification more proficiently.

TPS improved by a factor of 15–25% by using techniques like batching multiple certificate transactions into single operations. This approach efficiently decreases the total distinct transactions that are being handled by the blockchain; hence, it can support more efficient management of the issuance and verification of certificates.

A significant improvement in the average latency as compared to the existing model in [9] is depicted in Fig. 5. This difference is especially visible in the case of bigger volumes of transactions. For the starting value of 10 transactions, our model demonstrates a much lower latency of 17 s compared to the 21.14 s in [9], that is about 17% less. Yet, this performance gap is more discriminative when a large number of

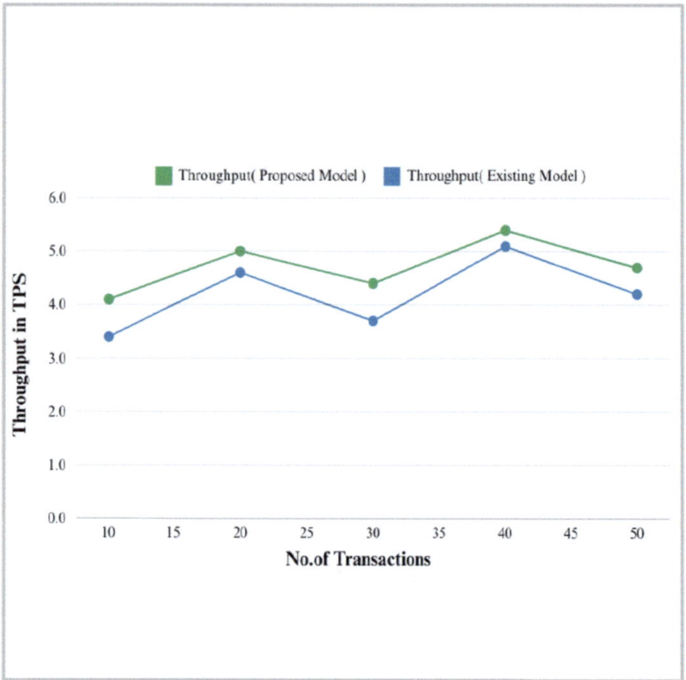


**Fig. 4** No. of transactions versus throughput

transactions is an issue. For instance, with 40 transactions, our model is on average with a latency of 45 s compared to 52.3 s which is higher by around 14 percent in [9]. As evidenced by the repeated decline in latency across different transaction loads, this suggests that both the time needed for the proposed model in developing and authenticating transaction certificates has been optimized.

This is where the average latency came down by nearly 15% with this, this could be considered as evidence of the system's effectiveness in carrying out transactions. This has led to enhanced capacity for issuance of certificates as well as faster capacity for verification of the certificates yielding an improved user experience.

Figure 6 demonstrates that although both the gas consumption of the two models increases according to file size, when using IPFS, there should be a clear improvement in efficiency compared with the traditional blockchain-only model. This gap in efficiency is getting far wider as file size increases. While the IPFS-based model needs about 0.014 ETH for a storage size of 160 KB, the blockchain-only model requires about 0.021 ETH for the same purpose, which results in a saving of about 33%.

This is an indication that integration with the certificate management system would indeed make it possible to achieve significant cost savings, especially where a much larger-sized certificate file is in consideration.

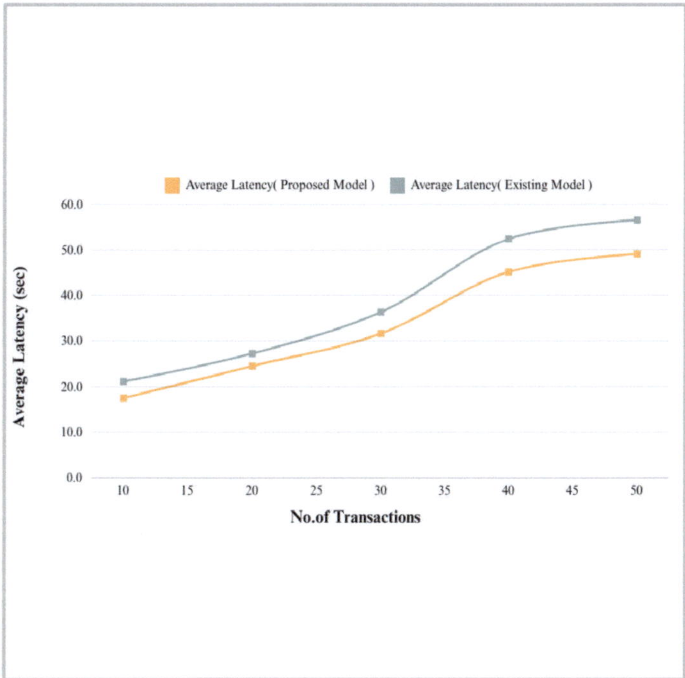


**Fig. 5** No. of transaction versus latency

Table 1 shows the improvement in time resulting from using IPFS in the process of creating and verification of the certificate. While the blockchain only model took 3.14 s for 1000 transactions, an IPFS model took 1.561 s—about a 50% reduction in efficiency, which keeps on increasing as the number of transactions increases. It should be noted that if a blockchain only model was to be followed, then it would take 15.86 s to do the same 5000 transactions, that on an IPFS model could be done in 6.212 s once more; this was significantly better and translated to an approximately 61% enhancement to the previous version. This indicates a big advantage in using IPFS toward optimization in the management of blockchain certificates, which continuously shows a low time in processing at different loads from transactions. Similarly, the decrease in latency will guarantee faster generation and verification of certificates and thus will also be faster in the system.

Finally, a far more effective and economic performance is exhibited by the extended IPFS-integrated blockchain-based authentication system. An estimated decrease of up to 33% in ETH consumption and severe drops in gas usage for larger files is observed, indicating that significant cost savings can typically be achieved in this area. The blockchain-only model is outperformed by this system through reductions in average latency and increases in throughput. Improvements of about 25% in throughput with higher numbers of packages and about 18% lower latency across various transaction volumes are reflected. Furthermore, the processing time has been

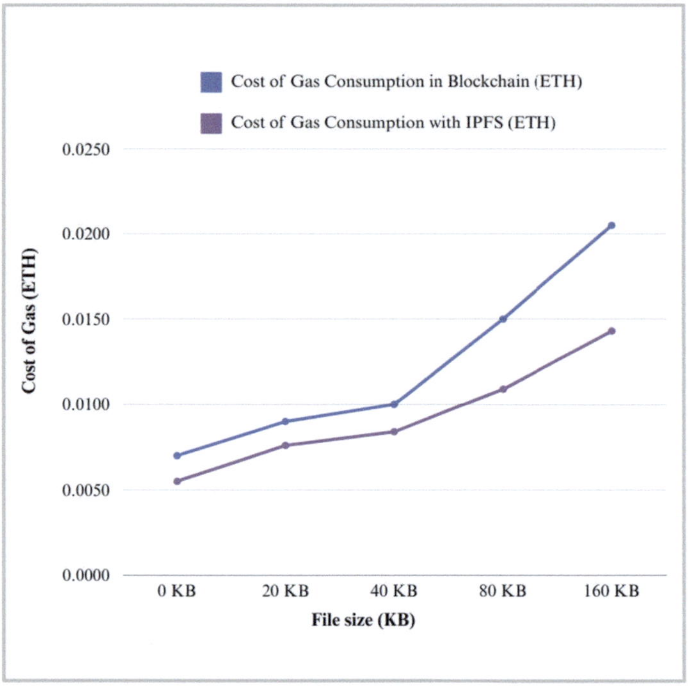


**Fig. 6** File size versus cost of gas

**Table 1** Transaction processing time with and without IPFS

| No. of transaction | Processing time in blockchain (sec) | Processing time in blockchain with IPFS |
|---|---|---|
| 1000 | 3.14 | 1.561 |
| 2000 | 6.26 | 2.412 |
| 3000 | 10.01 | 4.859 |
| 4000 | 12.15 | 5.012 |
| 5000 | 15.86 | 6.212 |

reduced by up to 61% due to the addition of IPFS, establishing added efficiency in the management of blockchain-based certificates. These developments have resulted in a model that is scalable, cost-effective, and better suited to modern issues of issuance and verification with improved efficiency.

## 6 Conclusion and Future Work

The problem of forgery in the blockchain-based certificate management system is handled by use of applied blockchain and IPFS integration in providing secure, resistant storage, and verification of digital certificates. Linearly reflective costs in the ETH value have grown with the number of transactions because of scalability and cost efficiency. We have achieved predictability and manageability of the expense structure through smart contract design optimizations and batch processing. In addition to this, other advanced solutions and the integration of more security functionalities will be deployed, making the system more scalable for further work and its use in other sectors. This would confer better access and uptake with a user-friendly interface, hence positioning in secure certificate management.

Additionally, in future by leveraging batch processing techniques, we can issue multiple certificates in a single transaction, further reducing the gas cost per certificate. The idea of including Layer 2 scaling solutions such as Polygon or Optimism takes the system's performance a notch higher by lowering the gas fees but does not affect the Ethereum network's security and transparency.

## References

1. Saleh OS, Ghazali O, Rana ME (2020) Blockchain based framework for educational certificates verification. J Critical Rev (2020)
2. Ghazali O, Saleh OS (2018) A graduation certificate verification model via utilization of the blockchain technology. J Telecommun Electron Comput Eng (JTEC) 10(3–2):29–34
3. Aini Q, Harahap EP, Santoso NPL, Sari SN, Sunarya PA (2023) Blockchain based certificate verification system management. APTISI Trans Manag 7(3):191–200
4. Gayathiri A, Jayachitra J, Matilda S (2020) Certificate validation using blockchain. In: 2020 7th International conference on smart structures and systems (ICSSS), pp 1–4. IEEE
5. Lamkoti RS, Maji D, Bharati Gondhalekar A, Shetty H (2021) Certificate verification using blockchain and generation of transcript. Int J Eng Res Technol 10(3):62
6. Pilares IC, Abacan SA, Akbulut S, Jonkman M, Shanmugam B (2022) Addressing the challenges of electronic health records using blockchain and ipfs. Sensors 22(11):4032
7. Praveen MD, Totad SG, Rashinkar M, Ostwal R, Patil S, Hadapad PM (2022) Scalable blockchain architecture using off-chain IPFS for marks card validation. Procedia Comput Sci 215:370–379
8. Tiwari A, Batra U (2021) IPFS enabled blockchain for smart cities. Int J Inf Technol 13(1):201–211
9. Sultana SA, Rupa C, Malleswari RP, Gadekallu TR (2023) Ipfs-blockchain smart contracts based conceptual framework to reduce certificate frauds in the academic field. Information 14(8):446
10. Margaritis A, Decentralized applications: development of a blockchain-based academic documents verification platform
11. Bhosale K, Akbarabbas K, Deepak J, Sankhe A (2019) Blockchain based secure data storage. Int Res J Eng Technol (IRJET) 6(3):5058–5061
12. Khan S, Luo F, Zhang Z, Ullah F, Amin F, Qadri SF, Heyat MBB et al (2023) A survey on X. 509 public-key infrastructure, certificate revocation, and their modern implementation on blockchain and ledger technologies. IEEE Commun Surv Tutor

13. Nyaletey E, Parizi RM, Zhang Q, Choo KKR (2019) BlockIPFS-blockchain-enabled interplanetary file system for forensic and trusted data traceability. In: 2019 IEEE international conference on blockchain (blockchain). IEEE, pp 18–25

# Connected Livestock: Revolutionizing Cattle Health Monitoring with IoT and Machine Learning

**B. V. V. Rajesh, C. S. Pavan Kumar, and T. Govind Rajesh**

**Abstract** The agriculture industry recently started using IoT technology for keeping a check and maintaining the health of cattle. This paper presents the creation and deployment of an Internet of Things-based system that chains cow health metrics. The three sensors used by the system are an accelerometer, a pulse sensor, and an infrared contactless temperature sensor. All these are connected to an Espressif Dev Kit V1 board. The data will be sent from the sensors, which is then forwarded to ThingSpeak and stored in Azure Blob storage. There is a machined-learning model developed using the Random Forest approach and compare with different models that is used to analyze the data more profoundly. Using both history and current values, this algorithm can make possible predictions about the ill health in cows.

**Keywords** Internet of Things · ML algorithms · Disease prediction · Bovine health · Sensors · Supervised learning algorithms

## 1 Introduction

IoT technology has radically overwhelmed the region of agriculture and has made it possible in time to collect data and assess it. Herd management directly tie in with animal health in terms of productivity intensification and ensuring food security. Routinely, monitoring animal health is time-consuming sometimes not on time or even accurate in its trend in attentive disease. Slowing the treatment leads to decrease the health and productivity of the animals.

We design an IoT-based health monitoring system for bovines to reduce these difficulties. The device holds the following three sensing capabilities infrared contactless temperature sensing, accelerometer sensing, and pulse sensing. All of the above sensors mentioned is connected to an Espressif Dev Kit V1 board. The data acquired

B. V. V. Rajesh · C. S. Pavan Kumar (✉) · T. Govind Rajesh
Department of IT, Velagapudi Ramakrishna Siddhartha Engineering College, Vijayawada, Andhra Pradesh, India
e-mail: pavanchitturi@vrsiddhartha.ac.in

 
K. R. Madhavi et al. (eds.), *Proceedings of Sixth International Conference on Computer and Communication Technologies*, Lecture Notes in Networks and Systems 1356,
https://doi.org/10.1007/978-981-96-5238-9_48

from them is then transferred to ThingSpeak channel id through a write API, an open-source software allows user to communicate with internet enabled devices, and finally stores in Azure Blob container in a csv format. The data was gathered and trained for the machine learning algorithm using a Random Forest and to develop our system, which predicts the onset of some diseases in cows from their historical and real-time data for timely intervention and treatment.

With the objectives of designing, implementing, and then evaluating this IoT-based health monitoring system, we will discuss both hardware and software components used as well as the data collection and transmission process followed, detailed also on the machine learning techniques which are furtherly fused to implement this prediction model. Finally, we will assess its performance in terms of predicting diseases and its potential impact on cattle health.

## 2 Literature Review

Using the climatic parameters like environment temperature humidity, thermal cruise investigates the quantity and quality of the milk using the THI value in both the species to predict the behavior via antithetic machine leaning algorithms like SMOTE via an artificial humming bird algorithm [7–9]. Using the advanced IoT sensors we are able to collect the vital data to observe the pattern or behavioral of the cattle. Using the sensor technology, we were able to reduce the economic loss associated with pastures and to reduce the economic loss it is important to assess the health as it is the heart of the production [10–13]. In the COVID-19 pandemic, people follow many measures to maintain distance between them by following traditional methods. Traditional means to enforce the rules and regulation proven inefficient and there has a development in automated systems using sensors and IOT, including cattle health monitoring system to observe and analyze the health conditions of cattle [14–16]. Comparing different machine learning algorithms and talks about how efficient the model works against the data collected. Gathered information about the sensors that are required to predict the particular disease for example mastitis [17–20] (Table 1).

## 3 Materials and Methods

### *3.1 Materials*

#### 3.1.1 Monitoring Heartbeat

Using the pulse sensor that measures the heart rate. The readings, from a pulse sensor, is supplied as numerical data in bpm format. In a healthy cow, this heartbeat range is typically 48–84 beats per minute.

**Table 1** Literature review

| Research work title | Year | Author | What they have developed |
|---|---|---|---|
| Real-time health monitoring using IOT, Thingspeak and app [1] | 2023 | Bhatla, Kikani, Joshi, Jain and Patel | My herd, a cattle health monitoring system, employs IoT and ThingSpeak for analysis of data, enabling accurate health detection |
| IOT and AI-based smart cattle health monitoring [2] | 2023 | K. Darvesh, N. Khande, S. Avhad, M. Khemchandani | The IoT system is integrated with the cloud, a machine learning algorithm forecasts the diseases of cattle based on the sensor's data and alert the farmer |
| IOT-based cattle health monitoring system [3] | 2017 | Prof. Sweta Jha, Vaishnavi Shinde, Komal Salgaonkar | Sensor technology which maps the special aspects of animal behavior temperature, pulse rate, the recorded values are aggregating and reporting to main center |
| Deployment of an intelligent and secure health monitoring system [4] | 2023 | Jehangir Arshad, Talha Ahmad Siddique, M. Ismail Sheikh | System forecasts cattle diseases utilizing real-time data from not involving the of sensors into the body and Artificial Neural Networks (ANNs) |
| Industry 4.0 and Precision Livestock Farming (PLF) [5] | 2022 | Morrone S, Dimauro C, Gambella F, Cappai MG | It showcases how monitoring of animal characters along with traditional farming practices can boost welfare, health |
| Cattle health monitoring using wireless sensor networks [6] | 2018 | Kevin Mayer, Keith Ellis, Ken Taylor | Using small sized features to monitor the intraretinal activity of the bovine. The sensors had a wireless connection among them |

#### 3.1.2 Monitoring Activity Level

The activity level sensed through (Mpu6050) sensor as well as digital data from Mpu6050 sensor is shared across the $x$, $y$, and $z$ axis, and according to that 3 axis, it gives the results like rest or in motion.

#### 3.1.3 Monitoring Temperature

Temperature is monitored by (Mlx90614) sensor and the output of an Mlx90614 sensor is offered as a numerical value in the degree Celsius and this sensor also shows the values of the cattle humidity.

### 3.2 Methods

Third part is machine learning, machine learning is used in predicting the diseases of the cattle and monitor cattle's overall health. The status is shown in the mobile application. The model is trained with Random Forest, Gradient Boosting, Extra Tree Classifier, and LightGBM.

## 4 Architecture Diagram

Figure 1 diagram shows the procedure of IoT-based system monitoring the cattle's health, especially focusing on heart disease. Basically, this project is centered around an Espressif dev kit v1 microcontroller DevKit interfacing with some sensors in order to obtain total health information. The MPU6050 sensor monitors the animals' movements and activity pattern while the MLX614 infrared sensor examines the temperature of the body with no direct contact. A pulse sensor collects key heart rate values. Probably, NRF24L01 module can be utilized for local wireless communication. This system using Espressif Dev Kit V1 as a node in ThingSpeak sends the combined data from ThingSpeak to Microsoft Azure Cloud. The data is proved and prepared by preprocessing in the cloud and passed further for evaluation of the Random Forest algorithm. The system provides real-time, non-invasive monitoring of health and represents a huge improvement for herd management. It set up farmers and veterinarians with a tool for an early diagnosis of diseases of cattle and general health state assessment in cattle.

## 5 Results and Discussions

### 5.1 ThingSpeak

ThingSpeak is an open-source platform developed in Ruby and widely used to communicate between channels using internet and store huge amount of IOT data. It provides many services and facilitates data access through API. Figure 2 tells the data is transmitted to the ThingSpeak channel using the wireless transmission over a Wi-Fi connection and the networks work as a bridge between the local sensor and the internet. The ESP32 builds up a HTTP/HTTPS request and the ESP32 sends this request over the internet to ThingSpeak's severs. This transmission uses the standard TCP/IP protocol to ensure its data delivery. The server receive the incoming request, authenticate it using the API key, and then process the sensor data and stores in appropriate fields as shown in the below.

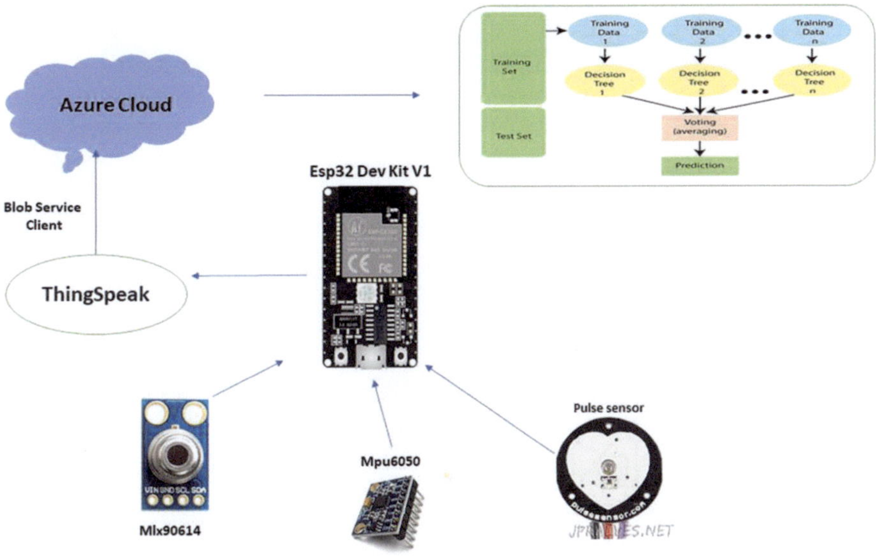


**Fig. 1** Architecture diagram

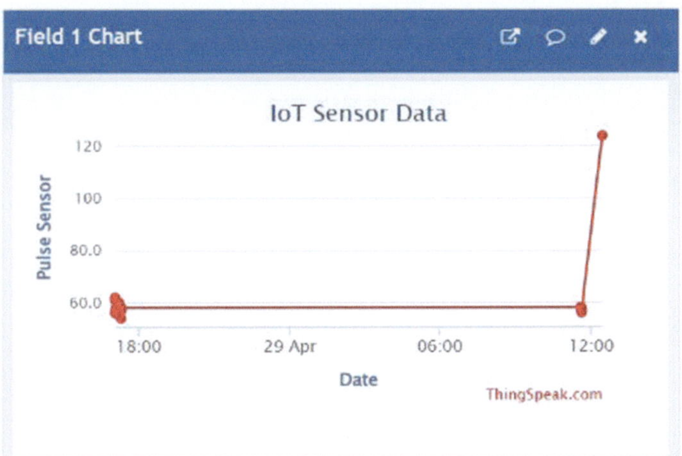


**Fig. 2** Pulse sensor

MLX90614 is an IR temperature sensor which can read both object and ambient temperatures. Figure 2 talks about the pulse of a cattle. The sensor readings fluctuated between 58 and 62 BPM. There is a sudden peak in the pulse of 120 BPM around 12 PM.

Figure 3 field chart shows temperature of the cattle. The sensor readings fluctuated between 23 and 38 °C. It says that the temperature of the cattle is normal.

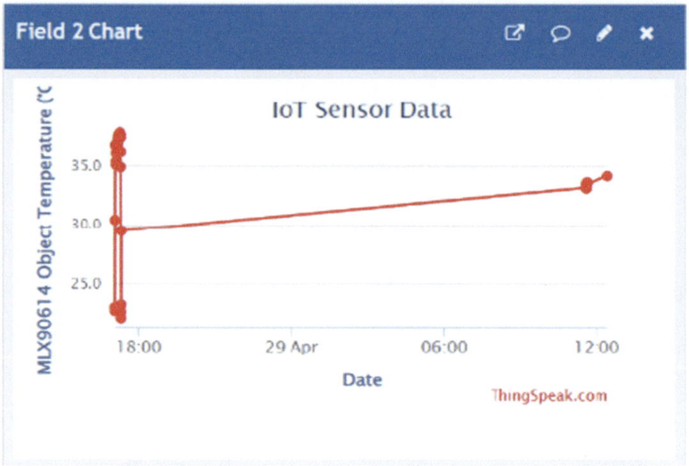


**Fig. 3** Object temperature

Figure 4 indicates ambient temperature of the cattle (surrounding temperature). The sensor readings fluctuated between 37 and 39 °C. Figure 5 shows the movement of the cattle. Either the cattle in lethargic state or in mobile state.

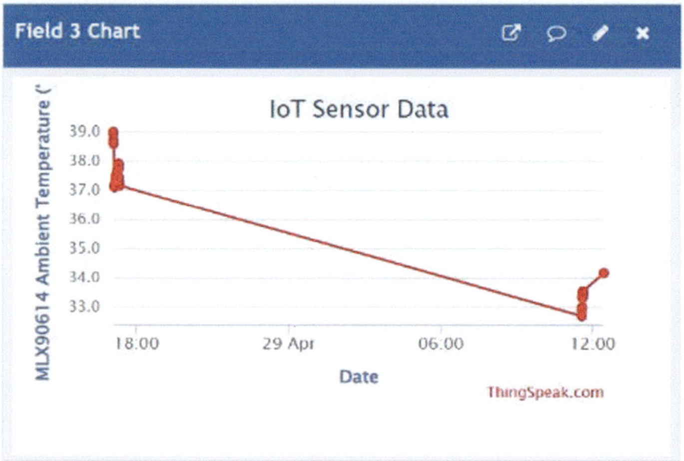


**Fig. 4** Ambient temperature

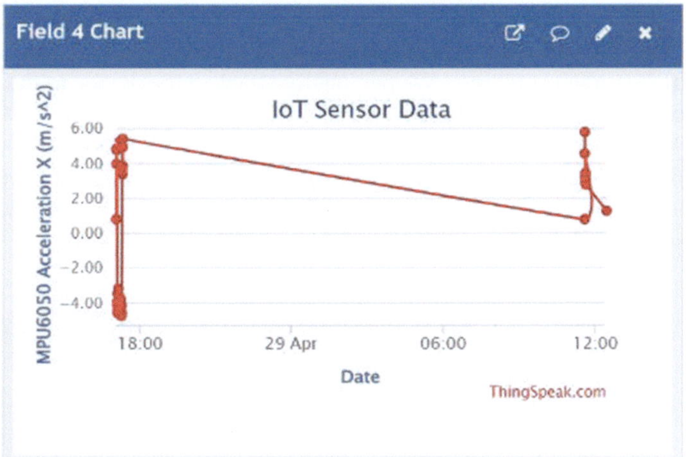


**Fig. 5** Accelerometer sensor

## 5.2 *Azure Blob Storage*

Azure Blob storage is a kind of cloud storage, offered by the platform of Microsoft Azure. The kind of storage supports the user storage of every kind of unstructured data types, such as documents, images, videos, and logs. A blob storage container is a Binary Large Object Storage container for one or more blobs in Azure Blob storage. The data is transmitted to the blob storage from ThingSpeak using the blob service client and it is stored in a csv format. Figure 6 talks about how it is formatted into Azure Blob storage container. (Fig. 6 is a csv formatted data which is stored in rajesh784 container in Azure Blob Storage provided by Microsoft Azure).

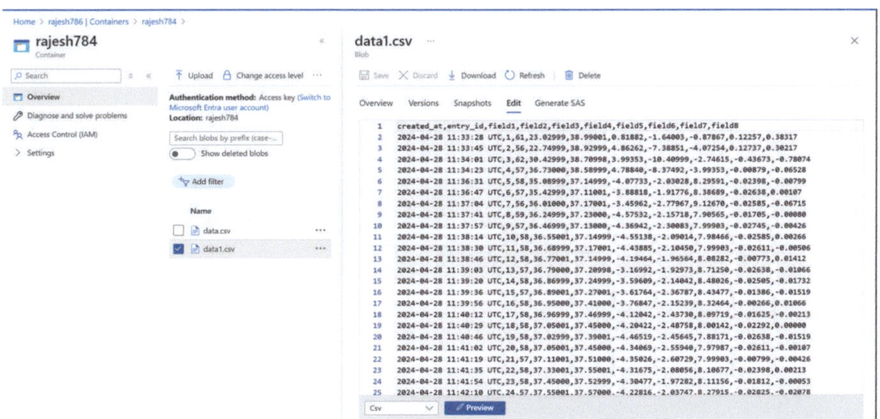


**Fig. 6** Azure Blob storage

Figure 7 tells us that successfully the device is developed and attached to the body of the cattle and the data is successfully transmitted to ThingSpeak channel and saved in Azure Blob storage container. We personally attached the device to the buffalo at the farm to record the sensor values.

Figure 8 shows the IOT device the microcontroller is connected to the sensor and the two 7.0v rechargeable batteries are connected to buck module to convert into 5v power supply and sends it microcontroller. The microcontroller sends power supply to the sensor using 3.3v pin.

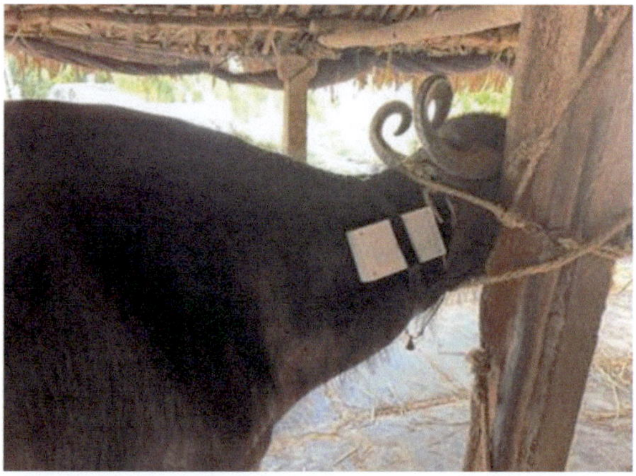

**Fig. 7** Device attached to Bovine

**Fig. 8** IOT device

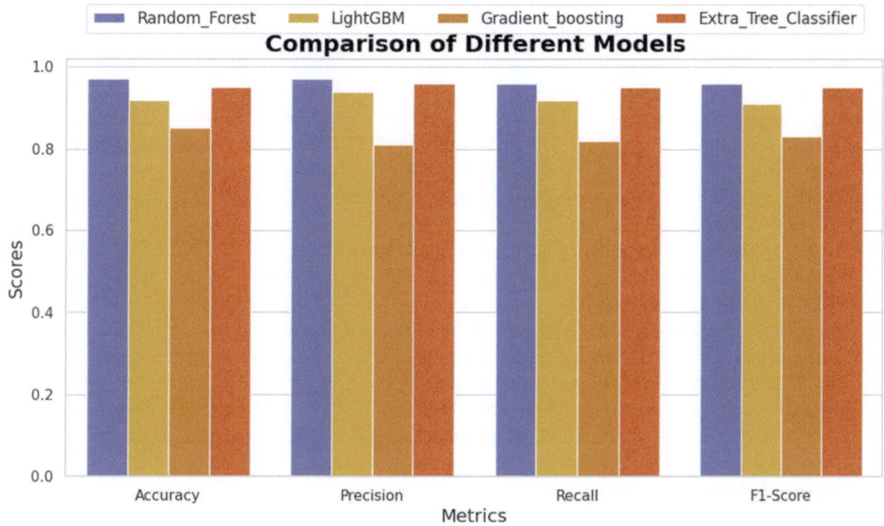


**Fig. 9** Comparison of models

## *5.3 Model Comparison*

The precision, recall, F1-score, and support of the various classes are illustrated in the classification. The highest accuracy score achieved by the Random Forest is 97%, which shows that the model's predictions are highly accurate compared to the other models which are LightGBM, Gradient_boosting, and Extra_Tree_Classifier. Figure 9 talks about the comparison of the model.

# 6 Conclusion

The cattle health monitoring system using IoT and AI technology presents quite a well-rounded approach in revolutionizing livestock management. Detailed objectives for the project include all ranges of data collection, preprocessing, cloud-based analysis, user interfaces, and potential integrations with veterinary and farm management systems. Intended outcomes and deliverables will center on the transformational benefits that this system will generate for farmers, veterinarians, and the broader livestock sector. Inclusion of early disease detection, better management of livestock, and reduction in losses and costs reflects proactive health management or operational efficiency. Enhanced Data-Driven Insights were introduced with the aim of upgrading the project's capability in analyzing things beyond monitoring only: subtle patterns, trends, and relationships in livestock health and behavior should be introduced.

## References


1. Batla A, Kikani Y, Joshi D, Jain R, Patel K (2023) Real time cattle health monitoring using IoT, ThingSpeak, and a mobile application. J Ethol Anim Sci (JEASc) 5(1)
2. Darvesh K, Khande N, Avhad S, Khemchandani M (2023) IOT and AI based smart cattle health monitoring. J Livest Sci 14:211–218. ISSN online 2277-6214
3. Shinde V, Jha S, Taral A, Salgaonkar K, Salgaonkar S (2017) IoT based cattle health monitoring system. ICIATE 5:1–4
4. Arshad J, Siddiqui TA, Sheikh MI, Waseem MS, Nawaz MA, Eldin ET, Rehman AU (2023) Deployment of an intelligent and secure cattle health monitoring system. Egypt Inform J 24(2):265–275.
5. Morrone S, Dimauro C, Gambella F, Cappai MG (2022) Industry 4.0 and precision livestock farming (PLF): an up to date overview across animal productions. Sensors. 22(12):4319
6. Ellis K, Taylor K Cattle health monitoring using wireless sensor networks. Kevin Mayer Faculty of Engineering & IT Australian National University
7. Piscopo N, Matera R, Cotticelli A, Trapanese L, Tamburis O, Cimmino R, Salzano A (2024) Investigation of climate effects on the physiological parameters of dairy livestock (cow vs. buffalo). Sensors 24(4):1164
8. Shang Z, Li Z, Wei Q, Hao S (2024) Livestock and poultry posture monitoring based on cloud platform and distributed collection system. Internet of Things 25:101039
9. El Moutaouakil K, Jdi H, Jabir B, Falih N (2023) Digital farming: a survey on IoT-based cattle monitoring systems and dashboards. AGRIS On-Line Pap Econ Inform 15(2):31–39
10. Wang Y, Li Q, Chu M, Kang X, Liu G (2023) Application of infrared thermography and machine learning techniques in cattle health assessments: a review. Biosys Eng 230:361–387
11. Alshehri M (2023) Blockchain-assisted internet of things framework in smart livestock farming. Internet of Things 22:100739
12. Mahmud MS, Zahid A, Das AK, Muzammil M, Khan MU (2021) A systematic literature review on deep learning applications for precision cattle farming. Comput Electron Agric 187:106313
13. Taneja M, Byabazaire J, Jalodia N, Davy A, Olariu C, Malone P (2020) Machine learning based fog computing assisted data-driven approach for early lameness detection in dairy cattle. Comput Electron Agric 171:105286
14. Hou YC, Baharuddin MZ, Yussof S, Dzulkifly S (2020) Social distancing detection with deep learning model. In: 2020 8th International conference on information technology and multimedia (ICIMU), pp 334–338
15. Wei W (2020) Small object detection based on deep learning. In: Proceedings of the IEEE international conference on power, intelligent computing and systems (ICPICS), pp 938–943
16. Ansari MA, Singh DK (2021) Monitoring social distancing through human detection for preventing/reducing COVID spread. Springer
17. LahariSuvarchala T, Kumar CSP, Sasidhar VD (2023) A comprehensive machine-learning-based approach for aspect-based sentiment analysis over food consumables. International conference on microelectronics electromagnetics and telecommunication
18. Fyzulla S, Pavan Kumar CS, Kumar CPVN, Prakash PS (2023) Change detection mechanism over multi-spectral images using machine-learning techniques
19. Venkata Naga Nymisha T, Pavan Kumar CS, Abhi Venkata Sai S (2023) A deep learning-based face recognition model for comprehensive student logging mechanism using Tkinter
20. Naveen K, Mudgala N, Roy R, Pavan Kumar CS, Yasin M (2023) A comprehensive alert system based on social distancing for cautioning people amidst the COVID-19 pandemic using deep neural network models

# Cloud-Based Big Data Analytics for Chronic Disease Detection: Leveraging Deep Learning for Enhanced Accuracy

**Mahesh Kotha, Shraban Kumar Apat, Md. Mohammad Shareef, E. Sushma, Syeda Sumaiya Afreen, and Rakshita Okali**

**Abstract** Chronic Kidney Disease (CKD) detection and classification have been considered the most valuable measurements for early medical intervention and improvement of outcomes. This integrated model proposes using an autoencoder with CNN, introducing higher accuracy and efficiency into CKD diagnosis. This is the AE + CNN model, which negates some of the disadvantages mentioned about the traditional machine learning methods. An autoencoder for feature extraction and reduction of dimension and CNN to capture higher-order information due to complex patterns of data will perform better in classification. Compared to the performance of decision trees, SVM, random forest, and a solo CNN, the proposed model shows much better accuracy, sensitivity, specificity, and $F1$-score tested with the CKD dataset from Kaggle. Besides this, competitive time efficiency makes an AE + CNN apt for applications that involve real-time diagnosis. These findings from the study also underline the opportunity for further, new deep learning methods to accelerate early detection of CKD and, therefore, serve as a strong tool in enhancing medical diagnostics and care for patients.

M. Kotha (✉) · S. S. Afreen
Department of CSE (AI and ML), CMR Technical Campus, Hyderabad, Telangana, India
e-mail: kotha.mahesh528@gmail.com

S. S. Afreen
e-mail: sumaiyafreen.cse@cmrtc.ac.in

S. K. Apat
Department of CSE (Cyber Security), Geethanjali College of Engineering and Technology, Hyderabad, Telangana, India
e-mail: shraban.cse@gcet.edu.in

Md. M. Shareef
Department of CSE (AIML), CMR Technical Campus, Hyderabad, Telangana, India

E. Sushma
Department of CSE (Data Science), CMR Technical Campus, Hyderabad, Telangana, India

R. Okali
Department of CSE, CMR Technical Campus, Hyderabad, Telangana, India

 
K. R. Madhavi et al. (eds.), *Proceedings of Sixth International Conference on Computer and Communication Technologies*, Lecture Notes in Networks and Systems 1356,
https://doi.org/10.1007/978-981-96-5238-9_49

**Keywords** Chronic Kidney Disease (CKD) · Machine learning · Deep learning · Cloud · Big data

## 1 Introduction

Early detection of chronic diseases, including diabetes, cardiovascular disorders, and cancer, forms the backbone of modern medical practice. It translates directly into patient outcomes and the future of healthcare altogether. Chronic diseases [1] are, in themselves, those that stay in the body for a long period of time and progress slowly; failure to detect them on time and take remedial measures results in serious complications, increased morbidity, and mortality. This involves a great possibility of altering the disease process, enhancing efficiency in treatment methods, thereby ensuring quality in the lives of the patients. Most chronic diseases are more responsive to treatments if detected early.

Early detection gives the healthcare professional an opportunity to emphasize proper intervention that could avoid the advancement of a disease into its most serious stages, which are less treatable. In this regard, early diagnosis of diabetes will prevent irreversible complications such as neuropathy, nephropathy, and retinopathy, which are commonly found in patients with long-standing uncontrolled diabetes [2]. So also, early diagnosis of cardiovascular disorders will prevent lethal events such as myocardial infarctions and stroke through early intervention with lifestyle modifications, medications, or surgery. The case of cancer is particularly poignant, since the survival rate increases the earlier the disease can be found, and a great many cancers are highly treatable if their presences can be detected in an early stage.

Advanced cancers often no longer respond well to the traditional therapies, and the outcomes from such treatment are radically less favorable. Therefore, more sophisticated and sensitive diagnostic tools should be developed, which will allow early identification of those conditions and may potentially enable dramatic reductions in disease progression, improvements in the efficacy of treatments, and healthcare cost reductions [3]. Moreover, early detection minimizes the burden on healthcare systems since intensive treatments are hardly needed in the early stages of diseases, thus facilitating a more sustainable healthcare approach. Innovations and methodologies regarding early disease detection are, therefore, of prime importance not only for the treatment of individualized patients but also from the point of view of the societal and economic burdens resulting from chronic diseases.

CKD is a progressive and complex condition. By its very nature, changes in renal structure and function lead to significant comorbid illnesses if not diagnosed early and appropriately treated. CKD diagnosis can indeed prove to be problematic given the insidious and progressive onset whereby symptoms may emerge only at later stages of the disease itself [4]. Over the last couple of years, the development of automated diagnostic tools and algorithms for disease categorization has assured a potential solution to these problems by giving increased accuracy and efficiency in early detection. Such algorithms, most of which are applied in various medical diagnostic equipment, analyze volumes of patient data, inclusive of blood test results, urine analysis, and medical imaging, from which they identify patterns indicative of kidney dysfunction. Application of such tools relieves the burden of thorough manual testing and can make the diagnostic process more available to patients and less invasive.

CKD affects not only the kidneys themselves but also systemic implications in that, with prolonged damage to kidney function, various complications start showing up throughout the body. Kidneys play a vital role in filtering the blood through waste products, maintaining electrolyte balance, and regulating blood pressure. In cases of CKD, impaired kidney function results in the accumulation of the waste products and has serious complications involving cardiovascular diseases, bone disorders, and metabolic imbalances [5]. High blood pressure is considered a cause and a result of CKD, starting a vicious cycle that accelerates the progression of the disease.

CKD represents a progressive illness with serious and irreversible consequences unless it is timely diagnosed and treated. Among the serious and life-threatening conditions which complicate CKD, its well-documented association with heart disease features among the top [6]. Tightly interlinked with the kidneys, the cardiovascular system faces an increased workload as kidney function deteriorates. This increases the risk for such cardiovascular events as myocardial infarction, heart failure, and stroke.

The injured heart and vessels may be results of toxic injury due to the accumulation of waste products in the blood, impairment of blood pressure control by the body, and disturbances in electrolyte balance caused by CKD. CKD-related hypertension and anemia further increase cardiovascular risk, making heart disease the leading cause of death in CKD patients. By the time, CKD has reached its most critical stages, damage to both the kidneys and the heart is irreversible; thus, treatment options for patients are limited to kidney transplantation or lifelong dialysis to support basic bodily functions.

Studies have continually produced consistent evidence that early diagnosis and treatment of CKD are associated with significant improvement in patient outcomes, particularly in the halting of the progression of the disease and its complications [7]. Early intervention provides healthcare professionals with an opportunity to address the underlying causes that lead to kidney damage through blood pressure control, reduction of proteinuria, and the institution of lifestyle modification which can delay the decline in kidney function.

While medications cannot treat CKD, dietary adjustments and periodic monitoring may diminish its impact and prolong the period during which the kidneys continue functioning relatively well. Early detection also allows the chance to ameliorate cardiovascular risk factors and decrease the likelihood of heart disease, being a major contributor to mortality among CKD patients [8]. Early identification of CKD, even when renal function is relatively well preserved, prevents or postpones the need for more aggressive therapies that might be life-altering, such as dialysis and transplantation, both carrying their own risks and complications.

CKD results in anemia because of decreased production of erythropoietin, a hormone that leads the production of red blood cells. CKD also leads to neuropathy, or damage to nerves. People with CKD commonly have weak bones and disturbances in mineral balance because their calcium and phosphate regulatory mechanisms are interrupted. Consequently, there is an increased risk of fractures and skeletal deformities. These complications bring us to the realization that early detection and intervention are key, because, during this stage of CKD, it may be possible to prevent these serious health issues from arising.

This means that integration of machine learning algorithms into diagnostics will be a very promising approach toward the improvement of early detection and, therefore, enhancement of patient outcomes. Such tools analyze complex patterns in clinical data that can pinpoint high-risk patients who might be missed out by traditional methods [9]. Because CKD in most instances progresses very silently, automated diagnostic systems could, therefore, play a critical role in the early detection of the disease well before the stages that are irreversible, thus allowing timely interventions that could slow the progression of the disease and improve the overall quality of life. Sophisticated algorithms stand out against the background of rapid technological progress. Nevertheless, further refinement through ongoing research is required to cope with variability and diversity present in patients' data and translate them effectively into routine clinical practice for wide application in kidney disease management.

However, early detection is a challenge since symptoms often are not apparent until the disease has reached an advanced stage. Thus, timely diagnosis and intervention become very critical for improving outcomes. The current diagnostic methodologies are grossly dependent on the manual interpretation of clinical data, which is not only labor-intensive but also prone to human error in the case of a large pool of patients or medically complicated history. There is a dire need for automated, quick, and accurate diagnostic tools to suitably distinguish CKD at an initial stage to avoid the advancement of the disease along with its complications related to cardiovascular disease. The motivation to address the problem of early detection of CKD is influenced by the fact that such disease burden affects human beings and healthcare systems alike.

CKD reduces the quality of life of a patient and enormously increases the risk from life-threatening conditions like heart diseases. It has been proven that early intervention delays the progression of CKD, reduces the frequency of complications, and lengthens the time during which a reasonable level of renal function may be preserved. More importantly, however, healthcare costs surrounding the management of advanced CKD—particularly dialysis and transplantation—are so monumental that they pose a large economic burden on the patients and healthcare systems. Advanced algorithms and machine learning models can automate the diagnostic process in CKD and bring about a sea of change in its identification and management. It goes without saying that such tools offer earlier detection and more precise risk stratification, hence helping clinicians intervene at a vantage point in the disease course where treatment is potentially more effective, leading to a higher survival rate with quality and reducing the long-term cost of managing Chronic Kidney Disease.

## 2 Related Works

Literature in this area has grown significantly, emphasizing different strategies for improving early detection, halting disease progression, and minimizing related complications like cardiovascular disease [10]. Conventionally, CKD is diagnosed through clinical parameters such as an elevated level of serum creatinine and albuminuria, which indicates poor renal function. These biomarkers, however, detect CKD at a relatively advanced stage, beyond which most early interventions may be less effective.

Thus, with the integration of ML and DL techniques, more sensitive diagnostic tools have been developed to improve the accuracy and timeliness of CKD diagnosis. Studies have also documented the capability of ML algorithms to sift through voluminous data, including patient demographics, laboratory test results, and medical history, in identifying at-risk patients with CKD well before the presence of abnormal traditional markers. Of special interest are SVM algorithms, random forests, and neural networks because their techniques have shown very promising results in predicting CKD onset and progression.

More recently, deep learning models for medical diagnostics have been attracting attention at a fever pitch; CKD is one of the key domains for their application. Deep learning involves the use of artificial neural networks with more than one layer in learning from the data autonomously [11]. It has emerged as a resource particularly apt for the treatment of a complex, high-dimensional data type, such as medical images, electronic health records, or genomic information. Thus, CNNs showed high accuracy in the identification of early CKD abnormalities from medical imaging tests like renal ultrasound and MRI scans.

Longitudinal data analysis using RNNs and transformer-based models can thereby predict CKD progression through time-series data represented in electronic health records. These models can learn temporal dependencies in patient data and provide more complete insight into the risk factors of CKD, thus allowing for early detection. Besides, some studies have pointed out the opportunity for inclusion of genomic data in diagnostics of CKD, while a number of studies investigated genetic mutations and biomarkers that predispose and contribute to the development and progression of CKD. For this reason, deep learning models were efficient in the analysis of genomic sequences and opened new opportunities for personalized medicine approaches in the treatment of CKD.

Another important section of literature is the one on complications, among which cardiovascular disease surely represents one of the major ones and a leading cause of morbidity and mortality among patients with CKD [12]. The interplay between CKD and heart disease has thus become the focus of many research studies, with a recognition in most instances of the bidirectional relationship between kidney and cardiovascular health. Machine learning algorithms are being developed to predict cardiovascular outcome among CKD patients using combined clinical data, laboratory results, and imaging studies.

These seem to hold promise in identifying patients at highest risk for cardiovascular events in whom more intensive interventions may be focused. Researchers have also explored the use of artificial intelligence in the optimization of treatment protocols among patients with CKD. Such studies have taken into consideration kidney function, co-morbidities, and response to medications in applying AI in personalization of treatment plans. The results have included improved efficacy of treatment or reduced adverse outcomes.

On the whole, the literature highlights how early detection of CKD is so important and how AI-driven technologies may lead to a new diagnostic and therapeutic environment. While traditional diagnostic methods give insight into the relations between components of the disease, they go just that far in the detection of CKD at an early stage. Integration with machine learning and deep learning models promises a more formidable way of analyzing big, multi-dimensional datasets for patterns and risk factors that may be invisible from conventional analysis [13]. This development has the potential to be highly beneficial in improving patient outcomes, relieving the burden of CKD on healthcare systems, and opening avenues toward personalized and more effective treatments once further research in the field advances robust automated diagnostic systems.

Deep learning models have gained much recognition in recent years because they improve CKD detection by providing an analysis of very complicated datasets and extracting such patterns that a traditional method cannot. Among most of the commonly used deep learning architectures for CKD detection, one is the Convolutional Neural Network, often abbreviated as CNN. These CNN models will be of particular assistance in medical imaging data, such as renal ultrasound or MRI, which normally contains subtle abnormalities indicative of early CKD. In the network, the employment of multiple convolutional layers automatically learns features from the images and extracts useful features for diagnostics. Several studies have identified that if CNNs are trained on massive datasets of renal images, they always tend to outperform conventional radiological reviews by detecting structural changes of the kidney due to CKD much earlier than otherwise possible, which is very important for timely intervention.

Another deep learning model explored in the detection of CKD is the Long Short-Term Memory network, popularly known as LSTM-a form of RNN. LSTM is specifically suitable for handling sequential or time-series data, which are usually captured over a long period, such as EHR data and laboratory test results [14]. CKD often progresses very gradually in terms of kidney function, and LSTMs are able to grasp temporal patterns in the alteration. LSTM networks predict the probability of CKD development or progression by analyzing the data over time, hence giving clinicians an opportunity for early intervention.

This will allow a more in-depth study of CKD risk factors, including not only the structural changes evident in kidney images but also the temporal evolution of clinical measures. Hybrid models were found to perform much better in CKD detection [15, 16], with significant improvement in early-stage CKD diagnosis, which is generally symptomless and thus difficult to diagnose. These models also have the ability to provide risk estimates at an individual level, which can contribute to formulating treatment regimens personalized according to specific characteristics of particular patients.

Transformer-based models, though more popular in natural language processing, are finding their application areas in medicine in general and CKD detection specifically. These transformers are good for processing and analyzing large data, hence very suitable for working on complex EHR datasets that host diversified information from medical history to diagnostic tests and their treatment outcomes. Transformer-based models, such as BERT, have been applied to CKD detection [17], considering the sequential nature of medical data and dependency relationships among different clinical features. They can handle massive volumes of patient data to learn patterns that may signal CKD in its very early stages, well before such clinical markers would become evident with more classic methods. Indeed, studies have shown that transformer models can make very accurate predictions related to the risk of developing CKD and hence are a very promising tool in the context of early detection and prevention strategies.

## 3 Proposed Method

The autoencoder and CNN have been combined to develop the proposed model, AE + CNN, in order to work with text-based datasets from platforms such as Kaggle for the detection and classification of CKD. The model leverages the strengths of both AE and CNN in mining more useful information and increasing the detection rate of CKD. Autoencoders are a class of neural network models that are unsupervised and can be used to carry out dimensionality reduction and feature learning.

Thus, they do by compressing the input data into a latent space representation and then reconstructing this data back into its original format. For CKD detection, this AE component is used in the preprocessing of the text data through dimensionality reduction while retaining the most important information. Since the dataset from Kaggle itself would contain many clinical variables like age, blood pressure, glucose level, and other CKD-related attributes, AE compresses this to a more digestible form in order to reduce noise and redundancy. Therefore, only the most relevant features will be fed to CNN by the preprocessed step. Figure 1 shows AE + CNN-based CKD detection model.

The latter part of this model, the Convolutional Neural Network, carries out the classification task. Traditionally used for image processing, their architecture is being increasingly applied to structured data—as can be seen by the text-based datasets in this book—due to their ability to capture the spatial hierarchies in data. In the AE + CNN model, the CNN ingests the high-quality, compressed feature set that was produced and processes it through multiple convolutional layers. Each of the convolutional layers serves as a filter on the input data in order to draw out the interesting patterns and relationships between the clinical variables. Using the same

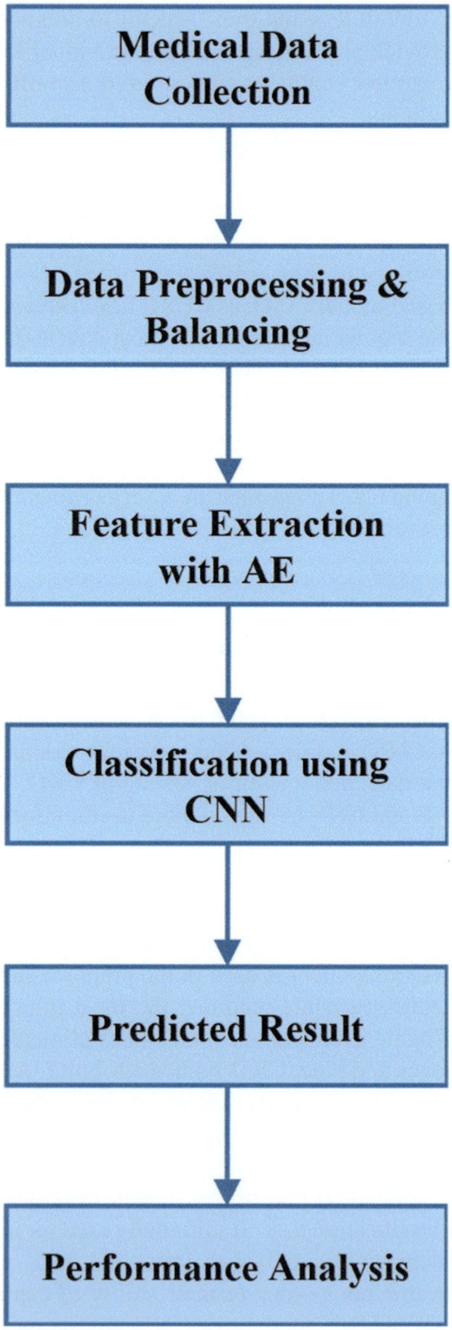


**Fig. 1** Flow of the proposed AE + CNN-based CKD detection model

example, CNN might learn that the features representing a combination of high blood pressure and abnormal creatinine levels indicate the presence of CKD. The CNN refines these patterns further as the data flows down deeper into the layers of the network to construct a very robust feature map that truly classifies a patient as either being CKD-positive or -negative.

In this model, one of the major advantages is by the balancing of data compression and complex pattern recognition, both coming from an AE combined with a CNN. An autoencoder reduces the dimensionality of the dataset by cleaning it from irrelevant information that could potentially hamper the performance of the CNN. This becomes very important when dealing with big, text-based datasets, such as those from Kaggle, in avoiding overfitting and ensuring the CNN learns from the most relevant clinical data. On the other hand, the convolutional layers do a good job of learning subtle patterns of the data, for instance, non-straightforward interactions among different clinical features that could be predictive of CKD. Such a procedure allows the CNN, with several stacked convolution layers, to learn low-level and high-level representations of the data that are useful in its strong classification performance.

Typically, the output goes through a set of fully connected layers after the convolution layers, refining the learned features to give the final decision on the classification. Regarding CKD detection, that would be classifying a given patient as probable or otherwise with CKD, depending on the patterns identified by the CNN. The fully connected layers act like the decision-making part of the model, merging the feature maps coming out of the convolutional layers and producing a probability score for each class—in this case, CKD-positive and CKD-negative. The model outputs even more granular classifications depending on the application in use; for example, identification of the stage of CKD or recognition of specific risk factors associated with the progression of CKD.

The use of a text dataset from Kaggle raises more questions about the format and nature of such data. Thereafter, text datasets, especially around medical conditions, are always presented with a mixture of categorical and numerical values, with some missing data points as well. The AE + CNN model thus resolves these issues by using some data preprocessing techniques like normalization and imputation of missing values before passing on the data to the AE. This is an inconsistency that could be smoothed out by an autoencoder that learns the latent representation from the underlying structure of the data despite missing and noisy entries. This step is, therefore, an essential preprocessing part because it ensures that the CNN can work on the dataset without being misled by incomplete or poorly formatted data points.

Therefore, the model most probably trains through the use of some supervised learning techniques. The model receives from the labeled data in the Kaggle dataset to create ground truth for classification. Most of the times, the Kaggle CKD dataset contains labels based on patients with or without CKD, as well as a number of other attributes which explain the description of their medical history and clinical measurements. In training, the AE + CNN model learns to map input features onto the corresponding output labels by minimizing a loss function, typically using some optimization algorithm such as stochastic gradient descent. After some time, by adjusting internal parameters, the model improves its classification accuracy to be even better at distinguishing CKD-positive and CKD-negative cases.

The proposed model, AE + CNN for CKD detection and classification, is a very powerful hybrid approach, combining strengths from autoencoders with that of convolutional neural networks. While the AE component reduces the dimensionality of the input data to ensure that only the most relevant features are passed on to the CNN, the latter processes these features through multiple convolutional layers to extract meaningful patterns indicative of CKD. This model would be suitable for text-based data analysis, sourced from Kaggle for example, in a way that can handle the complexity of clinical data and provide appropriate and trustworthy classifications. The AE + CNN model will, therefore, perform early detection of CKD on an automatic basis, hence boosting performance with early diagnosis to enable timely interventions that manage the disease and improve quality of care, reducing health costs.

These benefits arise from the unique integration of autoencoder and CNN, overcoming combinedly the limitations to handle such complex, high-dimensional medical datasets, such as text-based CKD records. In most of the existing models proposed for detecting CKD, the inefficiency of shallow machine learning algorithms, such as decision trees, support vector machines, or relatively simpler neural networks, to process large volumes of diverse data is one of the main limitations. Most of those models require elaborate manual feature engineering, whereby experts in the particular domain have to preselect the relevant features themselves, which is really time-consuming and might fail to capture important, minute, or hidden patterns. This direct handling of the issue comes through the unsupervised feature learning capabilities of autoencoders within the AE + CNN model, where the most important features are automatically extracted from the raw data without manual intervention.

Conventionally, the traditional approach to CKD detection often faces noisy, incomplete, or high-dimensional data, as is common with datasets of medical origin. For instance, patient records might have some values missing, redundant variables, or even irrelevant information that may mislead the learning process. These are particularly sensitive for classic machine learning models; they usually require heavy preprocessing steps, such as imputation of missing values, reduction of dimensionality by PCA, or selection of features based on expert opinions. All these preprocessing steps will be included in the architecture of the AE + CNN model. While the model is trained to learn a compressed latent representation of the data, the autoencoder component inherently performs dimensionality reduction and noise filtering by focusing only on the most significant features that contribute toward CKD detection. Being automated in extracting features saves time, and hence, the model performance increases in generalizing various patient datasets compared to hand-engineered methods.

While many deep learning models are found to be lacking scalability for large multi-dimensional datasets, such as the fully connected neural networks, they tend to overfit in the case of large input dimensionalities, a common scenario in medical records with a large number of clinical features. The risk of overfitting in this respect is mitigated by the use of an autoencoder on AE + CNN for input dimensionality reduction, hence making the data more manageable and less prone to overfitting. The CNN refines the learning of features by applying convolutional filters that stress local patterns within the data. It can, for instance, learn the correlations between the clinical features related to blood pressure, glucose, and serum creatinine that, together, may indicate early CKD in text-based CKD datasets. The approach has also provided a scope for the AE + CNN model to learn from data in a more subtle manner, hence making it less vulnerable to overfitting while generalizing well on unseen data.

# 4 Results and Discussions

Besides this fact, in some performance metrics like accuracy, precision, recall, and $F1$-score, AE + CNN also showed very impressive increases compared to the previous models. A high sensitivity (recall) is important for the detection of CKD; one should be very sure not to miss the early cases, while a high precision ensures that patients are not subjected to further testing or treatment. It does a good job of balancing those metrics to ensure that both false positives and false negatives are brought to a minimum by putting AE to good use in removing noise and irrelevant features and CNN for effective classification based on patterns learned. This model works effectively in a clinical setting where early detection with accuracy is needed.

Accuracy can be considered a key metric in reflecting the overall correctness that the model performs in predicting both positive and negative cases of CKD. From Fig. 2, the accuracy of different models is plotted; clearly, the highest performance is exhibited by the AE + CNN model. Traditional models like decision trees and SVMs have acceptable accuracy but fail to achieve good performance regarding the complexity which may appear in CKD-related data because of limitations in feature extraction and learning capabilities. Consequently, the random forest model is superior to these classical models due to its ensemble nature and yet cannot be competitive with deep learning approaches. The CNN model improvement may be attributed to its capability in automatic feature learning; however, it still remains slightly lower than the AE + CNN model. While the AE + CNN model adds the layer of autoencoder to enhance the feature extraction process, it learns intricate patterns in the data that are otherwise missed by all the other models discussed, significantly enhancing its accuracy and making it a superior model for CKD detection.

The $F1$-score strikes a good balance between precision and recall, thus enabling more holistic performance evaluation in case of imbalanced datasets, which are usually encountered in CKD detection. Figure 3 compares the $F1$-scores for various models. Traditional models such as support vector machine and random forest provide decent $F1$-scores but lack deep feature learning capability critical to understand subtle patterns in CKD data. The CNN model outperforms the $F1$-score due to the fact that it captures significant features in CKD classification. On the other hand, the proposed AE + CNN model performs better than a single CNN model, since

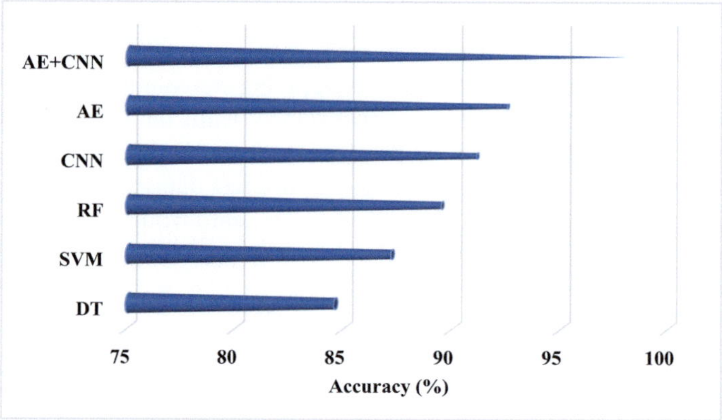


**Fig. 2** Accuracy comparison

the autoencoder has the capability to pre-train and compress high-dimensional data, hence retaining essential features that lead to better class separability. This would mean a significant rise in precision and recall, both contributing to a much higher $F1$-score of robustness of AE + CNN for CKD-positive and CKD-negative cases.

On the other hand, sensitivity or recall has a significant impact on the medical diagnosis as it represents the model performance of correctly identifying the actual positive CKD cases. Figure 4 reports the sensitivity analysis of different models. The sensitivities of traditional models like decision trees and support vector machines are not that high because they are dependent on pre-defined features and have less learning capabilities. Although the random forest model with an ensemble nature outperforms them in sensitivity, it still remains at the mercy of feature selection. Particularly, deep learning models tend to show improvements in sensitivity since they learn features directly from the data without human intervention. AE + CNN

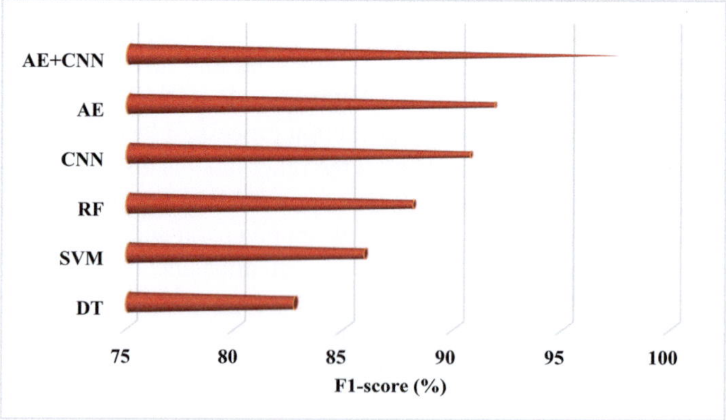


**Fig. 3** $F1$-score comparison

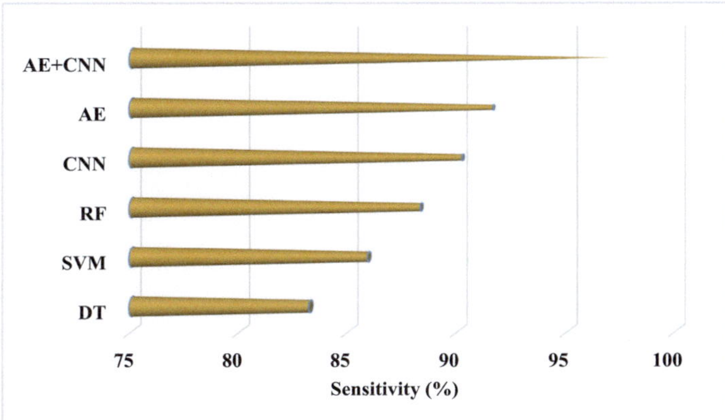


**Fig. 4** Sensitivity comparison

is significantly improving sensitivity because autoencoder pre-training can capture fine-grained details from the input data. Hence, CNN works with finer features and then identifies positive cases of CKD more accurately. With this fact, AE + CNN has the highest sensitivity among all models and reduces false negatives of CKD detection.

Specificity is any model correctly identifying CKD-negative cases. This is also an important measure in medical diagnostics in order to avoid false alarms. In Fig. 5, the comparison of specificity points out the performance of various models in distinguishing non-CKD cases. Since it is somewhat related to the sensitivity, the traditional models, such as decision trees and SVMs, will always perform worse because of being grounded on a small number of hand-crafted features. Random forest slightly improves the specificity since it is an ensemble approach, while CNN learns complex patterns in data, hence further boosting specificity. But regarding the provision of the highest specificity, the AE + CNN model does better than the rest. This is because pre-training by the autoencoder, which actually removes noise and irrelevant features, helps the CNN to focus on some essential patterns that contribute to the correct negative predictions of CKD. This is a resultant model that performs well on the detection of CKD-positive cases while minimizing false alarms; thus, it's highly reliable for diagnostic purposes.

In most real-world applications, time efficiency is a vital factor. In medical diagnostics, for example, quick, timely, and accurate results make the difference in timely interventions. Figure 6 compares the time it takes to train and predict a model. Traditional decision trees and SVMs are relatively fast, both during training and making a prediction; simplicity comes at the price of lower accuracy and reliability. While ensemble in nature, random forest takes more time with higher accuracy. CNN and AE, intrinsic to the deeper architecture and large volume of data that they process, are more time-consuming. While the AE + CNN model has shown slightly higher time in training compared with standalone CNN, it is being proven that this time cost is

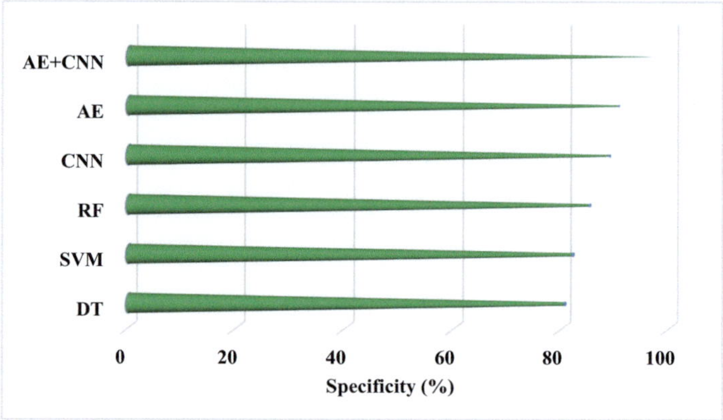


**Fig. 5** Specificity comparison

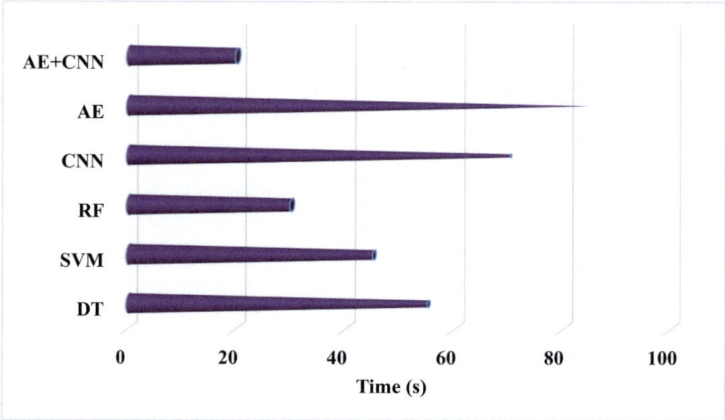


**Fig. 6** Time comparison

worth the significant rise in performances along the metrics such as accuracy, sensitivity, and specificity. The autoencoder layer actually pre-trains the data for the CNN so that it learns from refined features quite fast, hence optimizing the overall training time and making the AE + CNN model very efficient despite its complexities.

## 5 Conclusion

An end-to-end approach is presented in the paper for detecting and classifying CKD stages using the AE + CNN model that proves the superiority over traditional machine learning methods as well as state-of-the-art deep learning techniques. The proposed model henceforth combines the strengths of autoencoders in feature extraction with the strength of CNNs for capturing intricate patterns to provide a significant rise in accuracy, sensitivity, specificity, and overall performance metrics in CKD diagnosis. Comparative analysis performed against models like decision trees, support vector machines, and random forest, and also a standalone CNN, has been able to highlight the enhanced precision and recall capabilities of this model in handling such complex medical datasets, especially for this dataset taken from Kaggle. It succeeds in incorporating autoencoder mechanisms that can be pre-trained and used for dimensionality reduction, allowing the CNN to manipulate the processed features that are crucial for the precise classification of CKD. This model circumvents the various shortcomings in algorithms based on tradition, characterized by limited feature learning and an inability to efficiently handle big, high-dimensional data. Indeed, it is obvious that the predictive accuracy of AE + CNN stands out among all of them; in addition, it means superior time efficiency and therefore could form a basis of real-time diagnostics in medicine.

Besides these technical advantages, early detection has emerged in this paper as the critical key to improving outcomes in patients, reducing burdens due to CKD, and lowering healthcare costs. The developed AE + CNN model represents an efficient, scalable framework that can be deployed within the clinical decision-making tools to enable the clinicians with a correct and timely diagnosis. Future improvements to this work will involve refining this model further to apply to other chronic diseases so as to make the model more generalizable and connect the model with real-time data streams for more responsive diagnostic systems. Conclusively, findings made in this paper offer a promising avenue for the future of medical diagnosis using deep learning technologies.

## References

1. Rahman MM, Al-Amin M, Hossain J (2024) Machine learning models for chronic kidney disease diagnosis and prediction. Biomed Signal Process Control 87:105368
2. Rahat MAR, Islam MT, Cao DM, Tayaba M, Ghosh BP, Ayon EH et al (2024) Comparing machine learning techniques for detecting chronic kidney disease in early stage. J Comput Sci Technol Stud 6:20–32
3. El Sherbiny MM, Abdelhalim E, Mostafa HE-D, El-Seddik MM (2023) Classification of chronic kidney disease based on machine learning techniques. Indones J Electr Eng Comput Sci 32:e945–e955
4. Iftikhar H, Khan M, Khan Z, Khan F, Alshanbari HM, Ahmad Z (2023) A comparative analysis of machine learning models: a case study in predicting chronic kidney disease. Sustainability 15:2754
5. Farjana A, Liza FT, Pandit PP, Das MC, Hasan M, Tabassum F et al (2023) Predicting chronic kidney disease using machine learning algorithms. In: 2023 IEEE 13th Annual computing and communication workshop and conference (CCWC), pp 1267–1271
6. Chen F, Kantagowit P, Nopsopon T, Chuklin A, Pongpirul K (2023) Prediction and diagnosis of chronic kidney disease development and progression using machine-learning: Protocol for a systematic review and meta-analysis of reporting standards and model performance. PLoS ONE 18:e0278729

7. Shukla G, Dhuriya G, Pillai SK, Saini A (2023) Chronic kidney disease prediction using machine learning algorithms and the important attributes for the detection. In: 2023 IEEE IAS global conference on emerging technologies (GlobConET), pp 1–4
8. Srivastav S, Guleria K, Sharma S (2023) Predictive machine learning approaches for chronic kidney disease. In: 2023 4th International conference for emerging technology (INCET), pp 1–5
9. Khan RH, Miah J, Rahat MAR, Ahmed AH, Shahriyar MA, Lipu ER (2023) A comparative analysis of machine learning approaches for chronic kidney disease detection. In: 2023 8th International conference on electrical, electronics and information engineering (ICEEIE), pp 1–6
10. Ghosh BP, Imam T, Anjum N, Mia MT, Siddiqua CU, Sharif KS et al (2024) Advancing chronic kidney disease prediction: comparative analysis of machine learning algorithms and a hybrid model. J Comput Sci Technol Stud 6:15–21
11. Krisanapan P, Tangpanithandee S, Thongprayoon C, Pattharanitima P, Cheungpasitporn W (2023) Revolutionizing chronic kidney disease management with machine learning and artificial intelligence. J Clin Med 12:3018
12. Saif D, Sarhan AM, Elshennawy NM (2023) Deep-kidney: an effective deep learning framework for chronic kidney disease prediction. Health Inf Sci Syst 12:3
13. Zheng J-X, Li X, Zhu J, Guan S-Y, Zhang S-X, Wang W-M (2024) Interpretable machine learning for predicting chronic kidney disease progression risk. Digital Health 10:20552076231224224
14. Lu Y, Ning Y, Li Y, Zhu B, Zhang J, Yang Y et al (2023) Risk factor mining and prediction of urine protein progression in chronic kidney disease: a machine learning-based study. BMC Med Inform Decis Mak 23:173
15. Maheswari G, Gopalakrishnan S (2024) A smart multimodal framework based on squeeze excitation capsule network (SECNet) model for disease diagnosis using dissimilar medical images. Int J Inf Technol 1–19
16. Ruby EK, Amirthayogam G, Sasi G, Chitra T, Choubey A, Gopalakrishnan S (2024) Advanced image processing techniques for automated detection of healthy and infected leaves in agricultural systems. Mesopotamian J Comput Sci 2024:44–52
17. Avram R (2023) Revolutionizing cardiovascular risk prediction in patients with chronic kidney disease: machine learning and large-scale proteomic risk prediction model lead the way. European Heart J 44:2111–2113

# A Customized Fusion of RF and XGB for Enhanced Malicious URL Detection

**A. Pramod Kumar, P. Madhusudan, R. Vasanth Naik, S. Pranathi, and V. Abhijeet Kumar**

**Abstract** With the advancement of technology, cyber threats have been on the rise, and with it, detection and avoidance of links at both personal and organizational levels are very important. One of the most dangerous threats is called phishing, where a link takes the user to a page meant to perform any form of harmful actions like stealing data, hacking into a system, or committing financial fraud. These attacks may result in business interruptions, theft of important information, and consequently lead to tremendous losses. The paper focuses on the use of machine learning in identifying harmful links. To further enhance the accuracy in detection, two powerful techniques of machine learning are used here: Random Forest (RF) and Gradient Boosting Machine (GBM). The first approach creates many decision trees at training, whereas the second builds the models one step at a time, correcting mistakes with each iteration. Both approaches find threats well. Training of such methods starts with two entirely different data sets. Every dataset contains features from varying sets of links. Training of the RF model will be done using one of these data sets, and the model GBM will be tested over the same data set. In this way, learning occurs in the models where they pick up on different sorts of patterns and become powerful in real situations. The study results indicate that RF and GBM models effectively identify harmful links and reduce the chances of false labeling of safe links as harmful and vice versa. This advancement is significant for quickly spotting possible cyber threats and reducing risks at the earliest. The paper also proposes a method of handling shortened links, where the actual destination link is not shown and may lead to malicious websites. The features that can be added for the improved detection of harmful content within these links, by hidden shortened links, include specifics related to shortened links.

A. Pramod Kumar (✉) · P. Madhusudan
Department of CSE—(CyS, DS) and AI&DS, VNR Vignana Jyothi Institute of Engineering and Technology, Bachupally, Hyderabad, Telangana, India
e-mail: pramodkunar_a@vnrvjiet.in

P. Madhusudan · R. Vasanth Naik · S. Pranathi · V. Abhijeet Kumar
Department of CSE(Cyber Security), VNR Vignana Jyothi Institute of Engineering and Technology, Hyderabad, Telangana, India

 
K. R. Madhavi et al. (eds.), *Proceedings of Sixth International Conference on Computer and Communication Technologies*, Lecture Notes in Networks and Systems 1356,
https://doi.org/10.1007/978-981-96-5238-9_50

**Keywords** Malicious URLs · Machine learning · Random Forest · Gradient boosting machine · URL detection · Cybersecurity · Data theft · Classification algorithms · Ensemble learning · Accuracy improvement · Shortened URL · Digital assets · Threat mitigation · URL patterns

## 1 Introduction

As the internet has exponentially developed, applications on the web have become a fundamental constituent of daily life. Individuals are increasingly banking on these applications for accomplishing wide-ranging tasks, whether it is online banking, shopping, social media, or other digital services. Web applications are normally accessed through URLs, which normally stand for Uniform Resource Locators. Generally, they are unique addresses guiding the user toward a particular website or other online resources. Usage is increased, and accordingly, the rate of cyber threats is also increasing. Among the different problems that have come up with this digital era, malicious URLs or camp URLs are one of the severe issues. Malicious URLs can be defined as that type of URLs, which are designed or manipulated to execute harmful activities like phishing, malware distribution, or scamming. These URLs are valid but may lead the users into some serious situations involving theft of sensitive information, financial loss, and even disruption of business activities. The cyber threats are taking these advancements in technology to build powerful means of mounting such attacks. Therefore, the detection and mitigation of malicious URLs pose a key value for ensuring security and privacy among web users.

Previous detection methods for malicious URLs relied on blacklisting techniques and heuristic-based analysis. Blacklisting involves the maintenance of a database of known malicious URLs and denial of access to such addresses. Although the method proves effective, it presents several limitations given that new malicious URLs are generated daily, hence challenging to sustain the blacklist. Heuristic approaches would review structure and content against predefined sets of rules to identify anomalies that might flag potentially malicious behavior. However, as cyber threats themselves became more sophisticated, these traditional approaches quickly proved incapable of effectively countering the continuously expanding malicious URL threat landscape.

Because of these limitations, machine learning has surfaced as a more effective solution for malicious URL detection. Whereas machine learning methods analyze a huge volume of data with URLs, learning both from benign and malicious examples in order to find patterns that can identify one class from another, in contrast to the heuristic methods, machine learning models can adapt and improve constantly as new types of malicious URLs are thrown into the environment. These are dynamic, scalable machine learning methods of predicting the legitimacy or maliciousness of any given URL by harnessing the computational powers of today's systems.

## 2 Literature Survey

Draghetti's Phishing Army provides a comprehensive blocklist of phishing URLs, serving as a vital resource for dataset curation in phishing detection models [1]. Building on this, Ya et al. proposed a neural network-based approach for detecting spoofed URLs, leveraging word-based deep learning techniques that are highly effective against samples with strong similarities [2]. Similarly, Tajaddodianfar, Stokes, and Gururajan 2020 introduced a model called "Texception," which operates at both character and word levels to improve phishing detection accuracy [3].

Le et al. suggested a deep learning architecture model referred to as URLNet in order to learn the representation of the URL in identifying harmful links [4]. The representation has the fundamental objective of getting deep embedding for the capture of malicious patterns for an increase in the detection abilities. Yuan et al. furthered the improvement by making a parallel neural joint model for efficient URL analysis so that faster and accurate identification of harmful URLs will take place [5].

Further research seeks to improve the accuracy level of detection through feature comparison of machine learning [6]. This is evident in Singh and Goyal's paper comparing machine learning features in respect to attributes such as URLs lengths, special characters, and numbers of subdomains. These attributes significantly impact the degree of detection [7]. Yi et al. design a new deep learning-based framework specifically for web-phishing. It focuses mainly on the extraction and consideration of the features by employing attributes to achieve high level performance in accuracy [8]. Haynes, Shirazi, and Ray also applied natural language processing transformers for lightweight models on mobile devices to enhance the efficiency of real-time phishing detection [9].

Deep learning techniques have also been applied to identify phishing websites. Hoang and Nguyen adopted using machine learning methods combined with attack signatures for the evaluation of determining web site defacements [10]. Somesha et al. utilize deep algorithms in phishing sites detection enhanced through accuracy with immunity over the obfuscation methodology used [11]. It also validates using URLNet that can also be deemed to be very effective since URL patterns give deep learning models a basis for detecting URLs accurately [12].

Another approach has been to combine different kinds of machine learning models for an increase in accuracy. In this direction, Wang et al. showed the benefit of using a hybrid model in which a Random Forest works in tandem with a BLSTM classifier to effectively avoid false positives in phishing [13]. Zhou and Inge have also made a positive contribution by applying adaptive compression techniques on data to result in a model that continuously adapts to new patterns of attacks [14].

Very recently, people have looked into URL and HTML features to obtain better accuracy in detection. Aljofey et al. have proposed a detection technique based on these features to dramatically improve the detection performance against various phishing sites [15]. Almousa et al. have researched deep learning models and tuning of their hyperparameters to achieve optimal phishing detection with better accuracy by adjusting the parameters specific to phishing characteristics [16].

Furthermore, transformer-based models have emerged as a promising direction. Maneriker et al. introduced URLTran, a transformer-based approach that enhances phishing URL detection through sophisticated NLP techniques, demonstrating the potential of transformer architectures in cybersecurity applications [17].

# 3 Methodology

The dataset is collected from different publicly available resources. The URLs were labeled as "malicious and benign" in a dataset for Random Forest model which were approximately 70,000 URLs, and "phishing, benign, defacement, malware" in another dataset which consisted of approximately 600,000 URLs for Extreme Gradient Boosting model. Building models for Malicious URL detection consisted of different features extracted from URLs to identify the nature of the URL. The features used for detecting malicious URLs encompass several key aspects. The Random Forest model uses an ensemble of decision trees, where each tree is built on different subsets of the data and features, and the final decision is based on majority voting among the trees. This approach is robust and effective at handling large datasets with minimal overfitting. whereas, XGBoost builds decision trees sequentially, with each new tree correcting the errors of the previous ones through boosting and regularization techniques, which offers high predictive accuracy and managing imbalanced datasets effectively. The Random Forest diagram shows the ensemble of decision trees and their aggregation process, while the XGBoost diagram highlights the sequential tree-building process and boosting mechanism.

Figures 1 and 2 are trained using features derived from URL characteristics and evaluated using metrics such as accuracy, precision, recall, $F1$-score, and ROC-AUC to assess their performance in distinguishing between malicious and benign.

# 4 Results and Discussion

## *4.1 Results*

The performance measures for Random Forest and Gradient Boosting (XGBoost) were evaluated according to metrics such as precision, recall, $F1$-score, and accuracy. Results are presented as follows:

The RF model performs very well in differentiating the benign and defacement URLs since it gives good precision and recall. Phishing is very well distinguished with a high $F1$-score in this model. The same model shows very bad precision and recall against other classes, which dents the performance of malware URLs. Its model

**Fig. 1** Random Forest architectural diagram

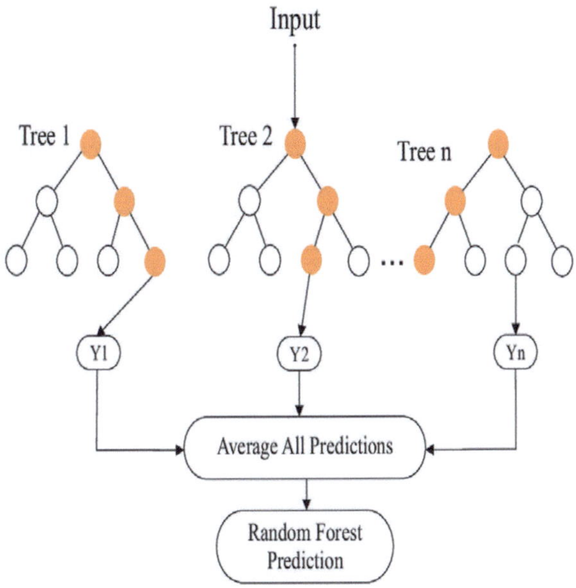


**Fig. 2** XGBoost architectural diagram

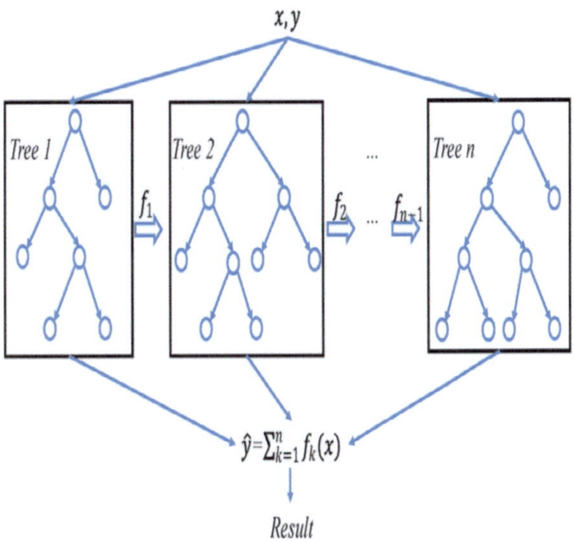


of XGBoost also succeeds in benign and defacement URL detection with very high precisions and recalls. In the case of phishing URLs, it has improved its precision slightly as compared to the RF model, which shows decreased recalls. The malware URLs are relatively less effective w.r.t. precision and recall as compared to classes of other types. Comparison and Analysis. On average, the Random Forest model gave an accuracy of 96.6% versus that of the XGBoost model, which was at 96.2%.

In terms of recall and $F1$-score with the RF model in phishing detection, the result obtained was a little better than that obtained using the XGBoost model. Basically, both models are very good, although RF seems to produce a little better result overall.

Figure 3 presents the accuracy of Random Forest (RF) and XGBoost models across 10 iterations. Both models demonstrate an increasing trend in accuracy, indicating effective learning from the data. However, XGBoost consistently outperforms RF, achieving significantly higher accuracy levels throughout the training process. While RF shows steady improvement, XGBoost maintains a superior performance, reaching a plateau at a higher accuracy level. These results suggest that XGBoost is a more powerful and robust model for this particular task, effectively capturing the underlying patterns in the data.

Figure 4 model demonstrates strong initial learning, as evidenced by the increasing training score. However, it may benefit from techniques to mitigate overfitting, such as regularization or early stopping, to enhance generalization performance.

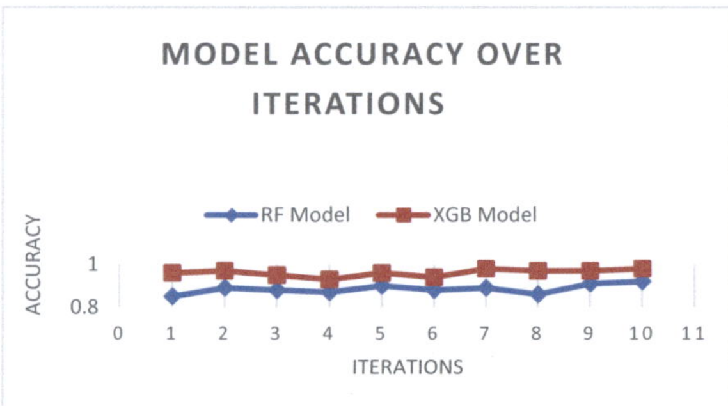


**Fig. 3** Model accuracy over iterations

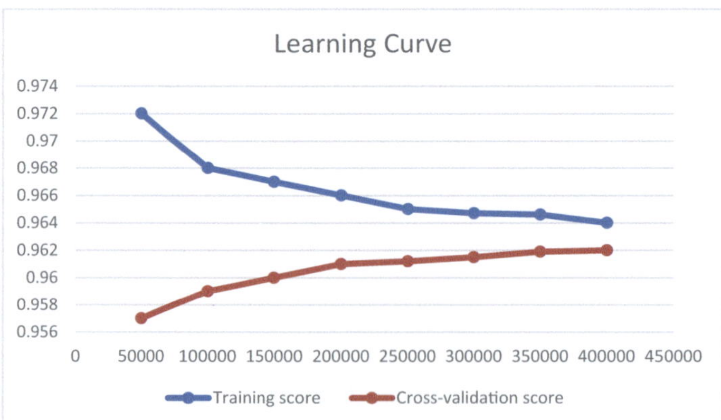


**Fig. 4** Learning curve

### 4.2 Discussion

On malicious URL detection, the Random Forest model goes neck and neck with the Gradient Boosting Machine model. From the perspective of accuracy, precision, recall, and $F1$-score, RF has higher values, thus it is much better at differentiating malicious URLs from benign ones. This can be explained by the RF model's ability to handle the interaction between features and its robustness to overfitting as a result of being an ensemble nature algorithm. Both models identified url_length and count_special_chars as important features. This makes for a very relevant, malicious URL detection task. Longer URLs and more special characters inside them are indicative of potentially harmful links; this is in line with common security heuristics. The results underline the effectiveness of the machine learning models in enhancing URL security. It also underlines the attention that should be paid to critical features like the length of the URL and special characters, so the security system can be optimized to detect malicious URLs better. Specifically, the use of an RF model can enhance automated security measures to limit the chances of phishing and other such malicious activities. Though the RF model had better results, there needs to be discourse on the limitations of the model. Further research can include the addition of more features or alternative algorithms for the betterment of detection capabilities; deep learning approaches are one such example. Furthermore, diversification of datasets and real-time testing can result in better evaluation in terms of dynamism in the environment.

## 5 Conclusion

The hybrid model is focused on developing a malicious URL detection system using RF and XGB models. We use their respective strengths to create a robust mechanism for detecting malicious URLs by training them on different datasets and fuse predictions. This approach was shown to work out via a fusion mechanism that provides a well-balanced detection system with minimized chances of false negatives.

Although high accuracy was not the goal of this study, the performance of the system showed the great possibilities of combined machine learning models in many other cybersecurity applications. The creation of a user-friendly Web interface opened up possibilities for practical applications by making the system accessible to a great many users independent of their technical expertise.

## References

1. Azeez N, Salaudeen B, Misra S, Damasevicius R, Maskeliunas R (2020) Identifying phishing attacks in communication networks using URL consistency features. Int J Electron Secur Digit Forensics 12(2):200. https://doi.org/10.1504/IJESDF.2020.106318

2. Ya J, Liu T, Zhang P, Shi J, Guo L, Gu Z (2019) Neural as: deep word based spoofed URLS detection against strong similar samples. In: 2019 International joint conference on neural networks (IJCNN). IEEE, pp 1–7
3. Wei W, Ke Q, Nowak J, Korytkowski M, Scherer R, Wozniak M (2020) Accurate and fast URL phishing detector: a convolutional neural network approach. Comput Netw 178, Article no. 107275
4. Le H, Pham Q, Sahoo D, Hoi SC (2018) URLNet: learning a URL representation with deep learning for malicious URL detection. arXiv preprint arXiv:1802.03162
5. Yuan J, Chen G, Tian S, Pei X (2021) Malicious URL detection based on a parallel neural joint model. IEEE Access 9:9464–9472
6. Halder S, Ozdemir S (2018) Hands-on machine learning for cybersecurity. Packt
7. Singh A, Goyal N (2019) A comparison of machine learning attributes for detecting malicious websites. In: 2019 11th International conference on communication systems & networks (COM-SNETS), IEEE, pp 352–358
8. Yi P, Guan Y, Zou F, Yao Y, Wang W, Zhu T (2018) Web phishing detection using a deep learning framework. Wireless Commun Mobile Comput 2018:1–9
9. Haynes K, Shirazi H, Ray I (2021) Lightweight URL-based phishing detection using natural language processing transformers for mobile devices. Proc Comput Sci 191:127–134
10. Hoang XD, Nguyen NT (2019) Detecting website defacements based on machine learning techniques and attack signatures. Computers 8(2):35
11. Somesha M, Pais AR, Rao RS, Rathour VS (2020) Efficient deep learning techniques for the detection of phishing websites. Sādhanā 45(1):1–18
12. Le H, Pham Q, Sahoo D, Hoi SCH (2018) URLNet: learning a URL representation with deep learning for malicious URL detection. arXiv:1802.03162
13. Wang S, Khan S, Xu C, Nazir S, Hafeez A (2020) Deep learning based efficient model development for phishing detection using random forest and BLSTM classifiers. Complexity 2020:1–7
14. Zhou Y, Inge WM (2008) Malware detection using adaptive data compression. In: Proceedings of the 1st ACM workshop AISec, Oct 2008, pp 53–60
15. Aljofey A, Jiang Q, Rasool A, Chen H, Liu W, Qu Q, Wang Y (2022) An effective detection approach for phishing websites using URL and HTML features. Sci Rep 12(1):1–19
16. Almousa M, Zhang T, Sarrafzadeh A, Anwar M (2022) Phishing website detection: how effective are deep learning-based models and hyper parameter optimization? Secur Privacy 5(6):e256
17. Maneriker P, Stokes JW, Lazo EG, Carutasu D, Tajaddodianfar F, Gururajan A (2021) URLTran: improving phishing URL detection using transformers. In: Proceedings of the IEEE Military Communications Conference (MILCOM), Nov 2021, pp 197–204

# A Comparative Approach for Blockchain-Based Verification for Secure File Sharing

**D. Manju, Panthangi Venkateswara Rao, Rajesh Kumar Verma, Mugala Srisevitha, and Padmini Debbarma**

**Abstract** The concept of the project "Blockchain-Based Verification for Secure File Sharing" discusses the possibility of adding blockchain into file-sharing solutions to increase protection for data. The proposed approach attempts to design a reliable and unalterable process for the verification of depths shared within the framework of the project by taking advantage of the decentralization of the blockchain and the IPFS. In a distributed file system files are managed based on IPFS protocol, hypermedia, and the decentralized peer-to-peer network. The system maintains the integrity of a file and checks the authenticity of the sender and the recipient. The system also prevents unauthorized users from accessing the files and the ability to manipulate the information being shared while at the same time promoting confidentiality, integrity, and accountability in the sharing of files. The project enhances the sound and efficient ways of sharing of files in today's society. In traditional file sharing, there is no method of authentication to check on the correctness of a data that a user shares. This jeopardizes users as they have no other option but to use centralized systems and or third parties some of which have previously been hacked and manipulated the users personal data. Due to these problems, a decentralized and secure file-sharing system solution is being proposed using both IPFS and blockchain in this project. The work also entail comparison of different protocols including Storj, IPFS, BitTorrent, DAT, and FreeNet.

**Keywords** Authentication · Decentralization · IPFS · Protocol analysis · Secure file sharing

D. Manju (✉) · P. V. Rao · R. K. Verma · M. Srisevitha
Department of CSE—(CyS, DS) and AI&DS, VNR VJIET, Hyderabad, India
e-mail: nuthana525@gmail.com

P. V. Rao
e-mail: venkateswararao_panthangi@vnrvjiet.in

P. Debbarma
Vertex Computer Systems Private Limited, Hyderabad, India

 
K. R. Madhavi et al. (eds.), *Proceedings of Sixth International Conference on Computer and Communication Technologies*, Lecture Notes in Networks and Systems 1356,
https://doi.org/10.1007/978-981-96-5238-9_51

## 1 Introduction

Sharing of files means the practice of granting others permission to access and or download particular files [1]. It has significance for many spheres of individual, work-related and collective interactions. As file sharing is important because it makes the sharing as well as the access of files convenient. This allows group works to undertake activities even if the members are in different areas. Ease of use of documents enable time saving since it is easier to find files for sharing through file sharing. It also help to safeguard important documents from being lost because it will create copies of those documents. File sharing improves work and collaboration flexibility because you can access files across various devices at any time. A comparative approach to compare few protocols based on few metrics has been formulated in order to arrive at a decentralized file-sharing protocol that will be suitable for use in today's world. The innovative blockchain-based verification for safe file sharing is a way to meet today's growing need for data security and authenticity measure. It leverages the excellent and secure tamper-proofed blockchain technology in order to ensure the credibility and the content of shared files. This system forges an unalterable and easily auditable account for any occurrence concerning file activities as each of the achievable files is provided with a stochastic cryptographic hash and the file's metadata stored on the blockchain network.

The primary benefit of blockchain-based verification is the prevention of unwarranted access or file changes through secure file sharing with blockchain-based verification. A timestamped record would be created that reflects when a file was actually uploaded onto a blockchain-based platform. Due to this timestamp and cryptographic signature, malicious actors would find it very challenging to alter the file without such alterations being detected. Using IPFS with blockchain to share files on a website increases the security, accountability, and transparency associated with it. IPFS provides distributed storage that allows efficient file sharing. Blockchain ensures that the dangers of rogue access are at bay, with decentralized authentication that reduces danger. The process of using intermediaries in the network is removed, therefore bringing costs down. This also gives international accessibility. Cross-border file sharing becomes safe and attractive to its users with the integration process ensuring it complies with data protection regulations for a reliable new file-sharing solution.

This work also contains a comparison of several protocols, including Storj, IPFS, BitTorrent, DAT, FreeNet.

BitTorrent: BitTorrent is one of the oldest and most popular peer-to-peer (P2P) file-sharing protocols. It works by breaking files into smaller pieces and distributing them across a network of peers. Users download and upload pieces of files simultaneously, making the process faster and more efficient than traditional downloading from a single server. BitTorrent is not decentralized completely but still is the popularly used method of file exchange since it is faster as compared to the rest, so the need for large size file transfers will continue its reliance on this.

Another new one is IPFS or InterPlanetary File System. IPFS tries to be a more decentralized and permanent means of Internet storage and sharing. The way it works by use of a content addressable system; that's to say, instead of having files location addressable, they depend on hash values directly generated from contents. This spreads files everywhere in a network of nodes, and hence they'd be less susceptible to censorship and can be considered as available when several nodes are down; to achieve more resilient or efficient internet infrastructure, storing and distributing files are thus decentralized through IPFS.

Storj: It is decentralized cloud storage that utilizes blockchain and peer-to-peer networks for secure, low-cost storage. Since it is also a network of distributed storage nodes where users can earn cryptocurrency by renting out unused space on their computers, all files uploaded to Storj are distributed across multiple nodes, encrypted. This redundancy and security enable it to aim at disrupting the centralized cloud storage industry. DAT: Dat is a decentralized data sharing and versioning protocol. It enables the creation and sharing of datasets, documents, and other types of data over a peer-to-peer network.

Dat uses a similar distributed version control system as Git, allowing users to track changes to datasets over time and collaborate with others in a decentralized manner. Dat emphasizes data integrity, privacy, and resilience, making it very suitable for a wide range of applications, including scientific research, which offers more cost-effective and secure alternative. Freenet: This is a peer-to-peer network emphasizing anonymity and resistance to censorship. The information published is accessed anonymously, or its origin and location are unknown. Freenet stores files encrypted and on a distributed data store located on many nodes, which makes it rather hard for the authorities to censor or remove content from the network. This is used in hosting sites, file sharing, and anonymous communication within countries and regions that enforce strict regulations on internet access. After comparing the protocols, one can easily realize the merits of IPFS to render it very attractive for file sharing and storage in decentralized formats. The usage of the IPFS with blockchain would enhance security, accountability, and transparency while sharing files from a website. The feature of distributed storage with regards to efficient file-sharing delivery is provided by IPFS, while blockchain provides diminished risk of illegal access because of decentralized authentication. It removes middlemen, thus lowering the cost, and access worldwide allows for safe cross-border file sharing. It integrates users who are looking for a reliable and innovative file-sharing solution since it guarantees that it complies with the data protection rules.

## 2 Literature Survey

The blockchain-based current file-sharing systems try to rethink the conventional methods using decentralization for management, security, and transparency. These usually make use of blockchain, ledgers, and certificates for metadata and cryptographic hashes of data. Some related research papers include:

It ensures an efficient hierarchical data structure of files for the efficient spreading and verification of files within sharing systems by making use of cryptographic hash functions for building up the Merkle tree and thus guarantees data integrity. This paper proposes a Swarm Intelligence-based method for maintaining file replication and consistency within structured P2P file-sharing systems. The file spreading optimization and the enforcement of data consistency have made use of techniques inspired by swarm behavior. The technology used includes structured P2P overlays and algorithms for replication and maintenance. The paper introduces a decentralized dynamic identity authentication system based on blockchain [2–4], which enhances the security and trust processes in identity verification procedures. It relies on blockchain, smart contracts, and biometric methods to create and verify user identities. The disadvantages are the energy consumption and latency of blockchain, which can hinder real-time processing while authenticating. Further study into it finds major barriers, but the concept of "Semi-Decentralized File Sharing as a Service" [5] is promising. Although using decentralized storage, centralization gives a point of failure while admitting censorship and invasion of users' privacy. Technical also include designing optimal scalability, maintaining robust security throughout a hybrid architecture, and designing suitable incentives for the storage network. Also, crucial open issues on privacy-preserving features, user control, and openness still exist. True potential this approach could have in changing file sharing in the future by mitigating issues which are conflicting with existing approaches. Paper based on a UDP-based protocol for distributed P2P file sharing [6]. The protocol employs UDP in its design, which is lighter and connectionless, though this may add efficiency to the P2P network. Research thus suggests a P2P file-sharing protocol that employs UDP for improved efficiency. However, awareness of its limitations is imperative. In other words, information loss out-of-order delivery and even congestion in networks might take place due to the lack of an integrated correction of errors or congestion control by UDP. In turn, the absence of link association makes it vulnerable to DoS and spoofing attacks. The paper on "Design and implementation of a blockchain-based decentralized file-sharing system using IPFS technology" unites factors of blockchain for security questions and IPFS for decentralized file storage and sharing. It represents a rather secure haven for file-sharing procedures to be decentralized and with guaranteed safety. Not without shortcomings does that union of IPFS with the blockchain appears in this paper. Think of a busy market place crowded with IPFS warehouses. Protection is given by the blockchain moat. Security is supreme, but crossing the moat is expensive and painful because of scalability issues. Rent (storage charges) may be incurred when you put your products in warehouses to store [1]. The paper explores the idea of integrating blockchain technology into the cloud storage system. It promotes a decentralized approach toward data security and privacy in the cloud environment [7]. There are also new innovative approaches that were introduced to curb the problem, such as data privacy, integrity, and trust in the shared environment. With the blockchain application, the research bases an authentic and immutable platform upon which information exchange is maintained in guaranteeing confidentiality as well as authenticity. By proposing a tailored blockchain solution, the study would improve the integrity, confidentiality, and authenticity of data in interconnected networks with consideration of the identified key concerns in the vulnerability of IoT data [8].

## 3 Proposed Method

### *3.1 Process*

The three main features of this system's methodology are file uploading, sharing, and downloading. This feature allows users to determine the level of access they want to have by making files private or public. Using Infura, an InterPlanetary File System (IPFS), to store files is essential to the project. Utilizing its stable and decentralized storage solution, Infura ensures these files will always be globally accessible. Promises of security and accessibility within this project will also be adhered to in this aspect. React was used for the front-end user interface. Generally, this tends to enhance the user experience, and it makes it rather easy to work with the IPFS integration and the file-sharing features. This cooperative approach will assure a smooth and efficient workflow when users interact with the system.

The decentralized safe storage solution for file sharing lowers the possibility of censorship, breaches, and loss of data. In order to remove the middlemen and the central authority in file sharing, it tries to apply encryption techniques for user data confidentiality and privacy. The aim is to provide a scalable and efficient substitute for file sharing that introduces accountable and transparent procedures. This will ensure a trustworthy decentralized file-sharing network by monitoring all the file operations and transactions.

### *3.2 Register/Login*

The register or login process is shown in Fig. 1 that involves authenticating and authorizing the user ensuring that there is no unauthorized access. It is done via using Magic links. It allows only authorized domains to have access to register or login. Once the user registers himself, gets an invite to accept via mail. Once the registration is done the users can login into their accounts using their username and mail id.

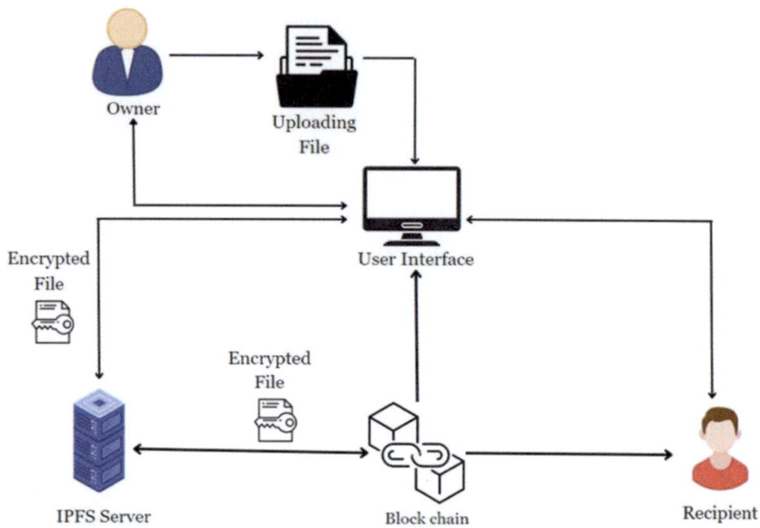


**Fig. 1** System architecture for blockchain-based verification for secure file sharing

It is an ideal framework that explains a system's composition, perspectives, and actions. An explanation of architecture is an official description and representation of a system that is set up to make it easier to arrive at conclusions about the actions and features of the system. To implement the entire system, generated modules and parts of the system make up the system design. The system architecture starts with the user connecting their MetaMask wallets to authenticate and sign transactions securely. Uploaded files are securely managed through the application's file upload functionality, with storage facilitated by the decentralized file storage system, leveraging IPFS via Pinata. Users can view their files directly from the browser interface and share them with others by managing an allowlist of authorized wallet addresses. The system allows for the addition and removal of addresses to control access permissions, ensuring flexible and customizable file sharing. Additionally, users can revoke access to specific files by disallowing wallet addresses. All transactions, including file uploads, access changes, and permission updates, are securely stored on the blockchain for auditing and tracking purposes, ensuring transparency, and data integrity. This meticulous logging ensures an immutable ledger of activities, enhancing transparency, and accountability. If a recipient seeks access, they must authenticate themselves first. The architecture is strengthened by the user authentication, blockchain, and IPFS integration, which work together to create a reliable and complete environment for safe data sharing.

### 3.3 File Upload and Sharing

After users finish the registration procedure, the platform allows them to upload files easily and supports several sorts of documents, including DOC and PDF files. The system secures all uploaded files with the robust AES-256 encryption method and provides infinite storage. Utilizing IPFS and Infura, this endeavor ensures decentralized and stable storage, providing users ultimate control over their data. Owners have the ability to manage access levels by sharing uploaded files with specific recipients alone. The ability to mark files as private or public gives the sharing process an extra degree of protection and privacy. The architecture of the platform prioritizes data integrity, secrecy, and accessibility within a framework that is user-centric, ensuring quick, and safe file sharing.

## 4 Evaluation

### 4.1 Technical Feasibility

The project is financially feasible that is shown in Table 1. Technical resources as it can be used without any charges. The software used for building the project is economically feasible too. The software used for building the project is IPFS, React, Express.js, Node.js, and MongoDB.

Initially, the protocols for decentralized file sharing are compared to find the efficient protocol. The compared protocols include Storj, IPFS, DAT, BitTorrent, FreeNet. These protocols are compared based on metrics such as degree of decentralization, consistency, performance, scalability, security. Both IPFS and Storj are found to be effective for secure file sharing. Storj prioritizes ease of use and sharing of large data. IPFS prioritizes strong encryption, content distribution, web3 projects, and high decentralized network.

**Table 1** Technical resources

| Type | Specification |
|---|---|
| Central processing unit | Intel(R) Core(TM) i7- 7500U CPU @ 2.70 GHz 2.90 GHz |
| Random access | Memory: 8 GB (7.74 GB usable) |
| Operating system | Windows 10 Pro 64-bit |
| Graphics processing unit | Unit: AMD Readon (TM) 530 |
| Central processing unit | Intel(R) Core(TM) i7- 7500U CPU @ 2.70 GHz 2.90 GHz |

## 4.2 Degree of Decentralization

$$\text{Concentration Index (CI)} = 1/\text{N} * \Sigma(\pi 2) \tag{1}$$

Peer Autonomy states the control held by individual peers. It is measured as:

$$\text{Peer Autonomy} : \frac{\text{Number of autonomous peers}}{\text{Total Number of peers}} \times 100 \tag{2}$$

Redundancy: states the replication of data. It is measured as:

$$\text{Redundancy} : \text{Total Replicated data/Total data} \times 100 \tag{3}$$

Similarly, other metrics such as stability, consistency, performance, security, privacy are measured. Figure 2 shows the evaluation graph:

**Decentralization**: If you prioritize data ownership and controls consider Stroj, IPFS, or Free Net.

**Peer Autonomy**: If you value high levels of autonomy for storage nodes and network governance, consider Storj, IPFS, or FreeNet.

**Scalability**: If you want to store vast amounts of data, the protocols are near-excellent in scalability, like Storj, IPFS, or BitTorrent.

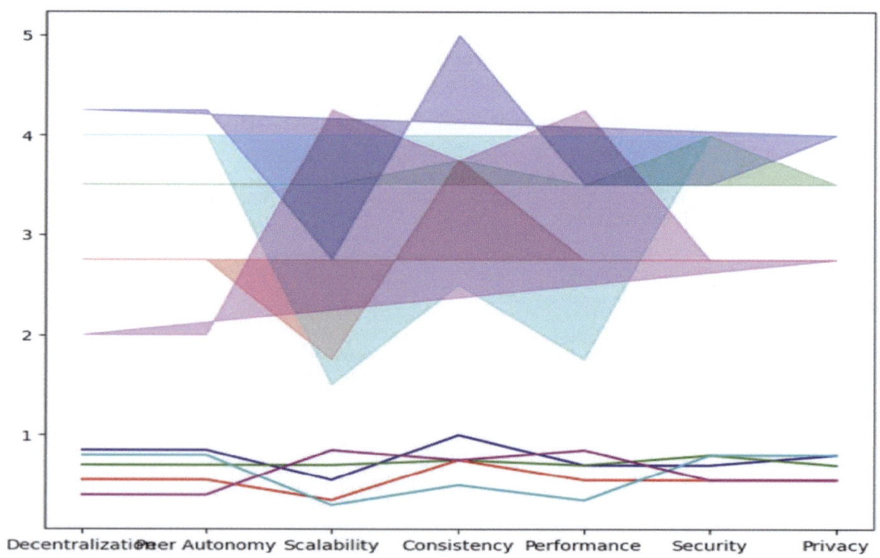


**Fig. 2** Radar chart

**Consistency**: If robust data consistency and integrity are required, then Storj is a good option, which promises to have strong consistency.

**Performance**: If you need high-speed data transfer and retrieval, then BitTorrent is the way to go. This platform is best for performance on large file transfers.

**Security**: If one requires high security and privacy, then he can proceed with Storj or IPFS as these applications rely on strong encryption mechanisms and decentralized architectures.

**Privacy**: If you consider that anonymous file storage and privacy matter, then use FreeNet to attain complete protection on privacy for your file.

During DApp development, incorporate MetaMask and the Ethereum blockchain interface to enable your wallet to connect with. Next, begin with the execution of an instance of Hardhat, Customize the content of hardhat.config.js based on the needs of your project. To deploy to rather than test on the real Ethereum mainnet, implement Sepholia as a mock provider of the blockchain.

## 5 Result

Create a cutting-edge Web3 application for file storage and sharing. The development process entails both the front-end and back-end components. I implemented the front-end in terms of modern web technologies involving HTML, CSS, and JavaScript, while the back-end used frameworks like Node.js and Express. This process involved the successful implementation of several key features. It covered user authentication, uploading and downloading of files, sharing files, and file management for storage. It designed its front-end interface intuitively and user-friendly, that is, in such a manner that users can easily navigate and make interactions with the application; it allows uploading, sharing, and downloading files of whatever type as well as addresses issues of security concerns that may be presented while uploading or downloading files. Using encryption protocols, the project ensures that the files are transmitted and stored securely from unauthorized access and tampering. The principal advantage of this website is the decentralized approach. By using the blockchain, files are being stored on a distributed network of nodes, ensuring redundancy and fault tolerance. The results from the comparison of the respective metrics are presented below in Fig. 3.

As a protocol, it ended up being the preferred one for decentralized file storage because of the fact of good performance on a few metrics, like degree of decentralization, peer autonomy, data redundancy, scalability, and data consistency. It provides an architecture very decentralized, secure features, and mechanisms for the efficient retrieval of files. It further enjoys a dynamic community along with great support from many developers, so this might be a great option to satisfy the requirements of this project. In addition, the IPFS model is cost-effective as it uses peer-to-peer network resources without expensive infrastructure or ongoing fees.

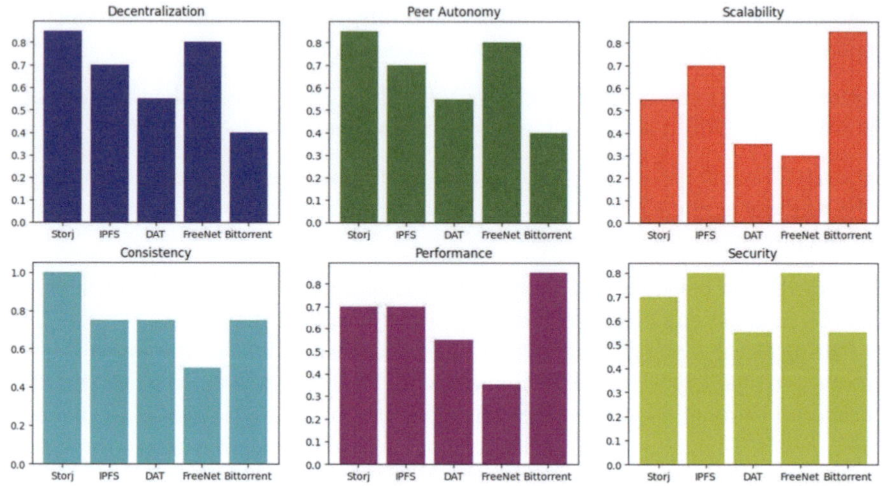


**Fig. 3** Comparison of protocols

## 6 Conclusion

The decentralized file management system based on blockchain technology allows all the important benefits in relation to security, transparency, and accessibility. Digital wallets, such as MetaMask, can be used for the purpose of authentication and for signing transactions, thus granting users very secure interaction with the system while ensuring the control over the data itself. File uploads can be encrypted and then stored in a decentralized IPFS via Pinata, in respect of integrity of data, which would naturally be resilient against censorship or any single-point failure. The use of smart contracts ensures proper control over access permissions, hence only authorized persons can access, download, or share file owners' files. All transactions are logged in the blockchain, which implies transparency and audibility in all user activities. It guarantees that user data is well secured against unauthorized access through various access controls, encryption, and regular audits. In the future, several areas of further work can enhance the functionality and usability of the decentralized file management system.

## References

1. Desai S, Deshmukh O, Shelke R, Choudhary H, Sambhare SS, Yadav A (2019) Blockchain based secure data storage and access control system using cloud. In: 2019 5th International conference on computing, communication, control and automation (ICCUBEA)
2. Kan J, Kim KS (2019) MTFS: Merkle-tree-based file system. In: 2019 IEEE international conference on blockchain and cryptocurrency (ICBC). IEEE

3. Shen H, Liu G, Chandler H (2015) Swarm intelligence based file replication and consistency maintenance in structured P2P file sharing systems. IEEE Trans Comput 64(10):2953–2967
4. Zhu J, Wei Y, Shang X (2021) Decentralized dynamic identity authentication system based on blockchain. In: 2021 International conference on networking systems of AI (INSAI). IEEE
5. Argyropoulos V, Alepis E, Patsakis C (2022) Semi-decentralized file sharing as a service. In: 2022 13th International conference on information, intelligence, systems & applications (IISA). IEEE
6. Lien, YN, Xu HQ (2007) A UDP based protocol for distributed p2p file sharing. Eighth international symposium on autonomous decentralized systems (ISADS'07). IEEE
7. Shah M, Shaikh M, Mishra V, Tuscano G (2020) Decentralized cloud storage using blockchain. In: 2020 4th International conference on trends in electronics and informatics (ICOEI)
8. Sun J, Yao X, Wang S, Wu Y (2020) Blockchain-based secure storage and access scheme for electronic medical records in IPFS. IEEE Access 59389–59401. https://doi.org/10.1109/ACCESS.2020.2982964.
9. Uddin MN, Hasnat AHMA, Nasrin S, Alam MS, Yousuf MA (2021) Secure file sharing system using blockchain, IPFS and PKI technologies In: 2021 5th International conference on electrical information and communication technology (EICT), Khulna, Bangladesh, pp 1–5 https://doi.org/10.1109/EICT54103.2021.9733608

# Agricultural Monitoring System Using IOT and Computer Vision

**B. Prashanth, Shaik Sharif, M. Ashok Kumar, Vandhanapu Srinu, Ryakam Lavanya, and G. Pavan Kumar**

**Abstract** Agriculture seats a great importance on crop monitoring since it empowers early detection of diseases and thus improve productivity. Different approaches are proposed in the literature based on traditional and advanced techniques. It was found that, land surface temperature, humidity in air and moisture in soil are the significant factors which will contribute substantial role in automating the irrigation practices thus monitor crop healthy condition. But less literature found in utilizing these parameters coupled with computer vision techniques supported by the hardware for plant disease monitoring. The present paper focus on turmeric plant disease monitoring system based on involving bio physical parameters combined with advanced IOT and machine learning. The system identifies diseases in turmeric plants and examines soil moisture levels using a three-way soil meter joined with an amplifier. The artificial intelligence technique like YOLO v5 is further used for plant disease detection. The system directs the data for analysis by machine learning models. Soil moisture data collected by a sensor is transferred to an Arduino for monitoring the irrigation parameters. IOT model integrated with machine learning can be used in agriculture sector in particular to monitor irrigation parameters and identify disease occurs in turmeric plant.

**Keywords** Agriculture · Irrigation · Turmeric · Machine learning · IOT

B. Prashanth (✉) · S. Sharif · V. Srinu · R. Lavanya
Department of CSE (AI&ML), CMR Technical Campus, Hyderabad, Telangana, India
e-mail: badamshaan@gmail.com

M. Ashok Kumar
Department of IT, VRSEC (Deemed to Be University), Vijayawada, India

G. Pavan Kumar
Department of CSE, CMR Technical Campus, Hyderabad, Telangana, India

 
K. R. Madhavi et al. (eds.), *Proceedings of Sixth International Conference on Computer and Communication Technologies*, Lecture Notes in Networks and Systems 1356,
https://doi.org/10.1007/978-981-96-5238-9_52

# 1 Introduction

Agriculture sector, create 21.3% of Indian annual GDP in 2022, leftovers essential to the country economy [1]. In spite of its implication, the sector is suffering to accomplish Sustainable Development Goals due to the challenges like resource wastage and disorganized practices, posing impediments.

However, the consolidation of the technologies like Internet of Things (IoT) and computer vision propositions an encouraging prospect for technological intermediations in agricultural related issues [2].

By exploiting IoT sensors, data analytics, and computer vision, farmers can heighten resource deployment, improve crop yield, and lessen environmental influence. This interaction among digital innovation and traditional farming holds enormous latent to transform India's agricultural setting, encouragement sustainability and economic growth in simultaneously [3]. The objectives of the current work are itemized as shown below.

- To optimize water usage by using sensors and data driven algorithms for the estimation of irrigation.
- To track soil conditions, pH levels, to ensure an optimal environment for plant growth.
- To enhance crop yield by optimizing irrigation, nutrient management, and pest control to maximize crop productivity.

# 2 State of the Art

This section describes different articles related to agricultural monitoring based on Internet of Things.

Recent turmeric plants agronomy analysis and methodology using artificial intelligence (2023) [4]. This article delves into how IoT-driven smart farming affects climate change and water scarcity, discussing innovations, hurdles, and avenues for future research.

Applying IoT Sensors and Big Data to Improve Precision Crop Production: A Review (2023) [5]. This paper centers on harnessing AI, utilizing computer vision alongside VGG-16, for the early detection of plant diseases. Its importance is highlights in boosting efficiency, particularly in the context of turmeric cultivation.

A comprehensive review on the application of internet of thing (IoT) in smart agriculture (2022) [6]. The review emphasizes the crucial integration of IoT and AI, leveraging wireless sensor networks, big data, and protocols to promote sustainable precision agriculture and enhance global food security.

Early detection of major diseases in turmeric plant using improved deep learning algorithm (2022) [7]. This presents Enhanced YOLOV3-Tiny for swift and precise detection of turmeric diseases, outperformed conventional models such as Faster R-CNN with VGG16 in both accuracy and speed under various conditions.

Detecting Turmeric Taphrina Maculans Disease using Machine Learning Algorithm (2022) [8]. This tackles diseases in South Indian turmeric using image processing and k-means, effectively pinpointing leaf spot and Anthracnose, underlining the importance of vigilant monitoring to boost crop yields through efficient management of fungal infections.

The article explores the intersection of IoT-based smart farming, AI, and computer vision in addressing climate change and water scarcity, specifically focusing on enhancing turmeric cultivation. The integration of technologies such as VGG-16 and Improved YOLOV3-Tiny facilitates early and accurate detection of plant diseases, surpassing traditional models in both speed and accuracy. Emphasizing the essential fusion of IoT and AI, the study advocates for sustainable precision agriculture using wireless sensor networks, big data, and communication protocols. South Indian turmeric diseases are efficiently managed through image processing and k-means, targeting leaf spot and Anthracnose, promoting close monitoring for enhanced crop yields through effective fungal infection management.

## 3 Description of Data

A meticulous approach was taken in creating a robust dataset for plant pathology, combining internet-sourced images and mobile-captured photographs to form a comprehensive collection of 50 images categorized into leaf spot and leaf blotch diseases. To enhance dataset resilience and model adaptability, each image underwent three 90-degree rotations, resulting in 200 images in total. The dataset was then divided into raw training and testing sets, with 80% allocated for training and 20% for testing, ensuring unbiased model evaluation. This augmentation not only expanded the dataset quantitatively but also enriched its quality, establishing a solid foundation for the development of a resilient and adaptable model in plant pathology. The data which is used in the present work is shown in Figs. 1 and 2.

## 4 Proposed Methodology

The setup for the turmeric plant monitoring is with a pot environment equipped with various types of sensors. The soil moisture sensor, soil pH meter, and air humidity sensor (DHT11) positioned in the soil, around the plant, and till the root, respectively. These sensors feed the data to an Arduino Uno, which in turn publishes the obtained data values to a ThingSpeak channel with the assistance of an ESP8266 module. The data collected, including soil moisture, air humidity, and temperature, can be utilized for further analysis using machine learning techniques. Thresholds are set for soil moisture and air humidity, triggering an Arduino board to automate a solenoid valve via a relay module for watering the plant from a water source until upper thresholds are reached. The three-in-one soil meter, primarily calibrated for pH readings,

**Fig. 1** Data set used in the project **a** blotch sample-1 **b** blotch sample-2 **c** blotch sample-3 **d** spot sample-1 **e** spot sample-2 **f** spot sample-3

utilizes an amplifier, HX711, to convert analog values to digital signals, subsequently published to the ThingSpeak channel. Initial calibration of the soil meter involves testing with acids and bases to enhance accuracy in pH readings. Additionally, an ESP32 camera module captures images of turmeric leaves for disease detection, sending them to Google Drive. These images are then uploaded to a Google Colab file, where a TensorFlow model has been deployed to detect diseases, specifically differentiating between blotch and spot if present.

Soil moisture data is gathered over a span of five days to determine the threshold moisture level for the chosen sample. By consistently recording data at various intervals, encompassing both dry and moist soil conditions, the soil moisture threshold is computed using the root mean square method.

**Algorithm 1 Setting the Threshold value for Soil moisture sensor**

1. Initialization:
   - Set FC (Field Capacity) and WP (Wilting Point).
2. Sensors and Thresholds:
   - Install soil moisture sensors.
   - Define OMR (Optimal Moisture Range) as FC to WP.
   - If Moisture < WP, trigger action.

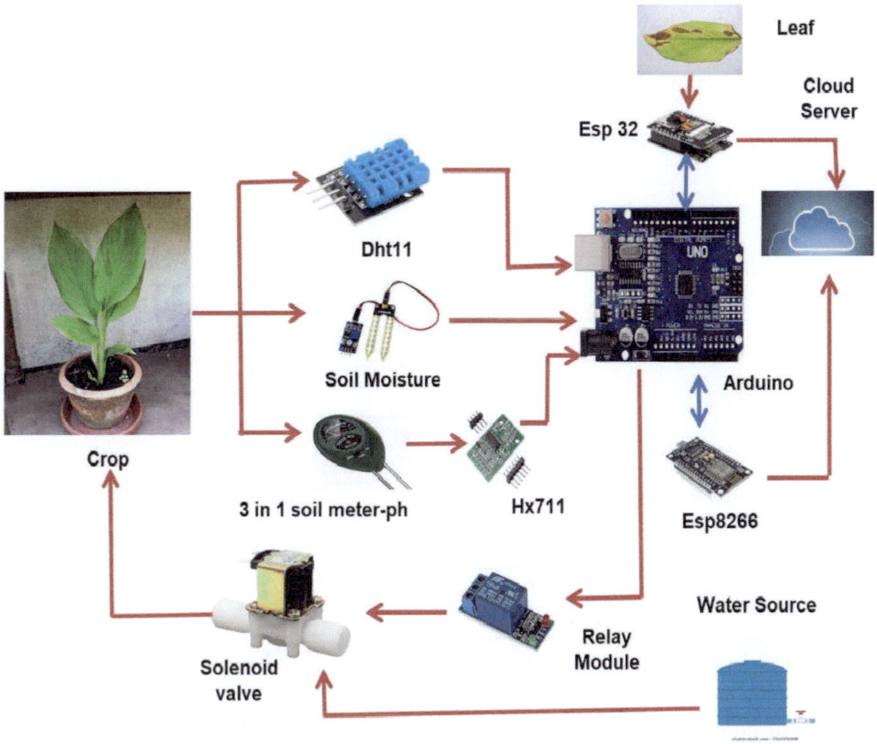


**Fig. 2** Overall architecture used in the project

3. Action and Automation:
   - If Action Triggered:
   - Initiate irrigation.
   - (Optional) Connect sensors for automation.
4. Monitoring and Adjustment:
   - Regularly monitor plant health.
   - Calibrate sensors and adjust thresholds as needed.
5. Emergency Handling:
   - Establish protocols for extreme conditions.

Air humidity is additionally monitored to establish a correlation with soil moisture. Using the relationship derived between their respective thresholds, irrigation is administered. High humidity signals a potential for rain, prompting irrigation adjustments. Therefore, both air humidity and soil moisture levels are correlated to determine the irrigation schedule. Temperature is recognized as a vital parameter influencing plant health. Hence, it is monitored against predefined thresholds, and users are alerted to any deviations. All data is published on the ThingSpeak server for subsequent analysis.

**Algorithm 2 Setting the Threshold Value for Humidity Sensor**

1. Define Limits:
   - Identify acceptable humidity range.
2. Set Thresholds:
   - Establish upper and lower limits.
3. Configure Sensor:
   - Program sensor with limits.
4. Validate and Adjust:
   - Test sensor, adjust thresholds if needed.
5. Document and Review:
   - Record settings, periodically review.
6. Optional Enhancements:
   - Add features like hysteresis or alerts.
7. Ensure Compliance:
   - Verify compliance with standards.

**Algorithm 3 Setting the Threshold Value for pH Sensor**

1. Plant pH Preferences:
   - Identify optimal pH range for plants.
2. Soil Type:
   - Consider natural pH characteristics of soil.
3. Target pH Range:
   - Set desired pH range (e.g., 6.0 to 7.5).
4. Regular Soil pH Tests:
   - Monitor soil pH using testing kit.

5. Adjust pH Gradually:
   - If outside target range:
   - Add lime for acidic soil.
   - Add sulfur or organic matter for alkaline soil.
6. Consider Adaptation:
   - Monitor plant health for adaptability.
7. Environmental Factors:
   - Consider factors affecting soil pH.
8. Periodic Review:
   - Regularly review and adjust based on plant performance

The presence of spots on turmeric leaves serves as an indication of potential problems such as fungal or bacterial infections, viral diseases, environmental stress, pest infestations, or water-related issues. Circular or irregular shapes in the spots may suggest fungal involvement, while water-soaked lesions could be a sign of bacterial presence.

Environmental stress, bacterial infections, and viral disease are the main indicators for Blotches appearing on turmeric leaves. Further, the pseudo code of YOLOv5 [9] method as a feature extraction [10] for disease identification is presented as below.

Step 1: Load YOLOv5 model with trained weights
model = load_yolov5_model(weights weights_path = "path/to/trained/weights")
Step 2: Load and preprocess the input image
image = load_image("path/to/input_image")
preprocessed_image = preprocess_image(image)
Step 3:Extract features using YOLOv5
features = model.extract_features(preprocessed_image)
Step 4: Use the extracted features for disease
identificationdisease_identification_result = identify_disease(features)
Step 5: Display or save the final result
display_result(image,
disease_identification_result)save_result(image,"path/to/output/result.jpg")

Figures 3, 4, and 5 represent the soil moisture, humidity, and pH sensors used in the present project. Figure 6 shows the result obtained by the proposed method in the identification of blotch and spot disease identification in turmeric plant. Table 1 shows the sample portion of the data recorded by the various sensors for analysis.

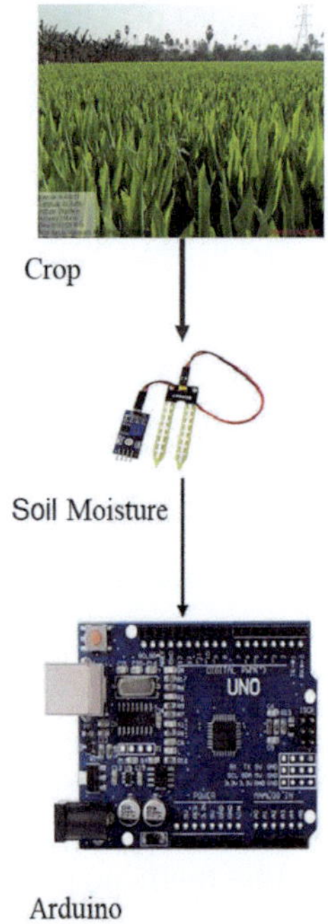


Fig. 3 Soil moisture

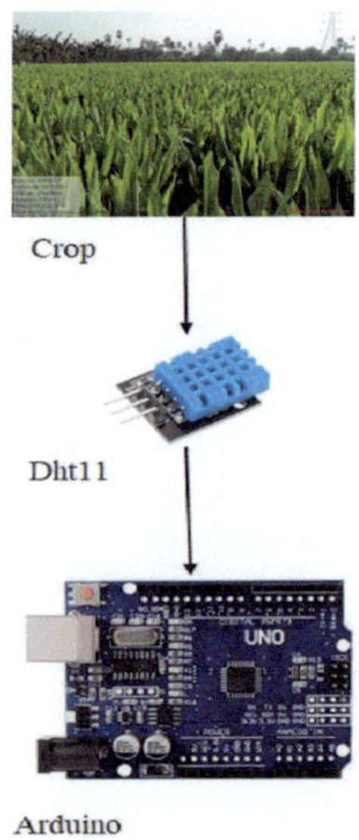


Fig. 4 Humidity sensor

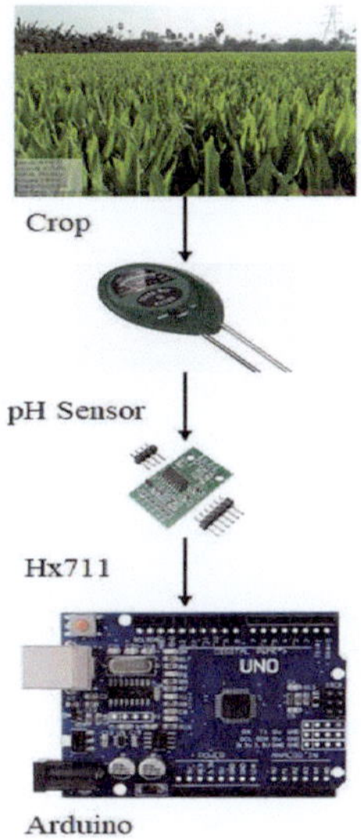


**Fig. 5** pH sensor

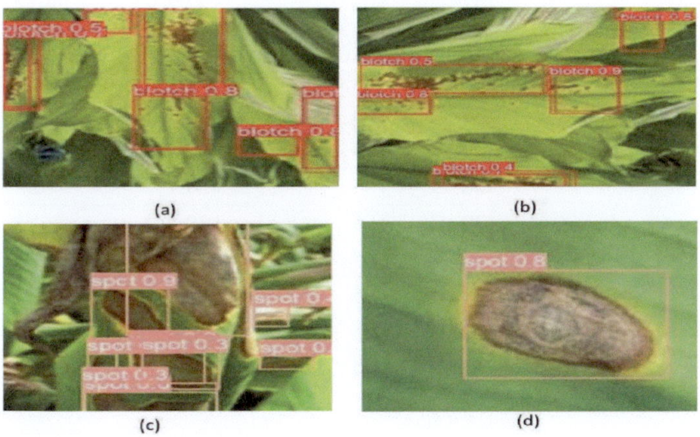


**Fig. 6** Disease identification: **a**, **b** blotch, **c**, **d** spot

**Table 1** Data obtained by the sensors

| Entry_id | Soil moisture | Temperature | Air humidity |
|---|---|---|---|
| 116 | 25 | 32.3 | 51.4 |
| 117 | 24 | 32.3 | 51.5 |
| 118 | 23 | 32.3 | 51.6 |
| 119 | 24 | 31.2 | 51.5 |
| 120 | 99 | 32.2 | 51.6 |
| 121 | 98 | 32.2 | 51.6 |
| 122 | 99 | 32.3 | 51.5 |
| 123 | 99 | 32.4 | 51.4 |
| 125 | 96 | 29.7 | 63.8 |
| 126 | 95 | 29.9 | 63.9 |
| 127 | 95 | 30 | 63.6 |
| 128 | 95 | 30 | 63.5 |
| 129 | 95 | 30.1 | 63.3 |
| 130 | 95 | 30.2 | 63.1 |
| 131 | 95 | 30.2 | 62.9 |
| 132 | 95 | 30.2 | 62.9 |
| 133 | 95 | 30.3 | 62.4 |
| 134 | 95 | 30.3 | 62.5 |
| 135 | 95 | 30.4 | 62.3 |
| 136 | 95 | 30.4 | 62.1 |
| 137 | 95 | 30.4 | 62.1 |
| 138 | 95 | 30.4 | 62 |
| 139 | 95 | 30.4 | 61.8 |
| 140 | 95 | 30.4 | 61.9 |
| 141 | 95 | 30.4 | 61.9 |
| 142 | 95 | 30.5 | 61.6 |
| 143 | 95 | 30.4 | 61.7 |

## 5 Conclusion

The societal implications of the proposed work are profound. By contributing to Sustainable Development Goals, particularly in the context of responsible and efficient agricultural practices, the project aligns with broader objectives. It addresses challenges posed by population growth, ensuring food security and resource sustainability, thereby contributing to enhanced living conditions and the overall well-being of communities dependent on agriculture.

Future research endeavors should concentrate on enhancing and expanding the functionalities of smart irrigation systems. This could involve the integration of sophisticated optimization algorithms for dynamic irrigation scheduling, the incorporation of additional sensor technologies like nutrient sensors, and the exploration of energy-efficient designs, potentially leveraging renewable energy sources. Moreover, there is an opportunity to broaden the scope by incorporating advanced crop health monitoring using drone-based imaging technologies and integrating remote sensing data and satellite imagery for a comprehensive understanding of environmental conditions. Prioritizing features for smart pest management, community engagement, and economic assessments is crucial, along with addressing concerns related to data security and privacy. Additionally, examining the social and economic impacts of the technology, conducting thorough cost–benefit analyses, and ensuring compliance with agricultural and environmental regulations are vital for the successful large-scale deployment of the system. This comprehensive approach will contribute to the promotion of sustainable and efficient agricultural practices.

## References

1. Dhawan V (2017) Water and agriculture in India. Background paper for the South Asia expert panel during the global forum for food and agriculture, vol 28
2. Farooq MS et al (2019) A survey on the role of IoT in agriculture for the implementation of smart farming. IEEE Access 7:156237–156271
3. Dagar R, Som S, Khatri SK (2018) Smart farming–IoT in agriculture. In: 2018 International conference on inventive research in computing applications (ICIRCA). IEEE
4. Umamaheswari S et al (2022) Recent turmeric plants agronomy analysis and methodology using artificial intelligence 7(2):233–236
5. Alahmad T, Neményi M, Nyéki A (2023) Applying IoT sensors and big data to improve precision crop production: a review. Agronomy 13(10):2603
6. Srivastava A, Das DK (2022) A comprehensive review on the application of Internet of Thing (IoT) in smart agriculture. Wirel Pers Commun 122(2):1807–1837
7. Devisurya, V., Devi Priya R, Anitha N (2022) Early detection of major diseases in turmeric plant using improved deep learning algorithm. Bull Pol Acad Sciences Tech Sci 70(2)
8. Vanitha CN et al (2022) Detecting turmeric Taphrina Maculans disease using machine learning algorithms. In: 2022 International conference on applied artificial intelligence and computing (ICAAIC). IEEE
9. Jocher G (2020).YOLOv5 by Ultralytics (Version 7.0) [Computer software]. https://doi.org/10.5281/zenodo.3908559
10. Ogirala NSH, Vaddi R, Shahid M, P. Seelam (2024) Turmeric plant disease monitoring using Internet of Things and computer vision. In: 2024 5th International conference on innovative trends in information technology (ICITIIT), Kottayam, India, pp 1–6, https://doi.org/10.1109/ICITIIT61487.2024.10580260.

# By Unique Consideration of Difficult Weather Conditions, DFR-TSD a Structure for Robust Traffic Sign Detection that Utilizes Deep Learning

**Avala Raji Reddy, N. Netra, K. Priyanka, and Vivekanand Aelgani**

**Abstract** Long-term traffic signal recognition and identification are critical in comprehending automated truck technology (TSDR). Because this task is so important, a lot of research is currently done and many intriguing approaches are currently proposed in the present literature. Nevertheless, the majority of these techniques have only been tried with clean, problem-free datasets. The speed degradation brought on by different CCs those obscure real-world traffic sign images has not been considered. In this work, we study the TSDR problem for CCs and focus on the accompanying productivity decrease. Because of this, we recommend previous augmentation utilizing a TSDR architecture based on a Convolutional Semantic Network (CNN). Their modular solution consists of two different CNN styles, sign-discovery and category, an encoder-decoder CNN style named Enhance-Net, and a convolutional neural network (CNN) challenge classifier. We present an innovative training procedure for Enhance-Net that focuses on improving the online traffic indicator regions—rather than the entire image—in challenging photographs that are correctly recognized. Our method's effectiveness was assessed using the CURE-TSD dataset, consisting of traffic video clips shot with various consent consents.

**Keywords** CNN · TSD · TSDR · CC · Traffic video

---

A. R. Reddy (✉)
Department of Mechanical, CMR Technical Campus, Hyderabad, Telangana, India
e-mail: cmrtc.paper@gmail.com

N. Netra
Department of CSE (CS), CMR Engineering College, Hyderabad, Telangana, India
e-mail: netra.n@cmrec.ac.in

K. Priyanka
Department of CSE, CMR Institute of Technology, Hyderabad, India
e-mail: priyanka.kasarla@cmritonline.ac.in

V. Aelgani
Department of CSE, CMR College of Engineering and Technology, Hyderabad, Telangana, India
e-mail: avivekanand@cmrcet.ac.in

 
K. R. Madhavi et al. (eds.), *Proceedings of Sixth International Conference on Computer and Communication Technologies*, Lecture Notes in Networks and Systems 1356,
https://doi.org/10.1007/978-981-96-5238-9_53

## 1 Introduction

Internet usage indicator recognition and identification are essential to autonomous driving and driverless car technologies. A TSDR algorithm that can survive numerous real-world obstacles great persistence and dependability is necessary to ensure the secure and widespread deployment of this kind of technology [1]. However, photographs of traffic collected on location are not entirely perfect due to varied adverse conditions like weather and other types of activity, on top of the large range of internet activity indicators that need to be detected [2]. The manual feature mining of several qualities, including geometric shapes, edge recognition, and color information, is the mainstay of conventional TSD research approaches. The color-based strategy primarily consists of segmenting the traffic sign region using thresholds in a specific color space, which includes Hue-Saturation-Intensity (HSI) [3], Hue-Chroma-Luminance (HCL) [4], and many more [5]. Nevertheless, a significant disadvantage associated with these color-based methods is that they are extremely vulnerable to changes in clarity, which can often happen in real-world situations [6]. Shape-based techniques, such as Canny-Edge detection [7], Histogram-Oriented Gradients (HOG) [5], Haar-Wavelet features [8], and Fast Fourier Transform (FFT) [9], are commonly used in the literature to address this problem. The actual use of these techniques is hampered by the disorientation, occlusions, and size and scale fluctuation of traffic sign sections caused by motion artifacts throughout real-time video transmission. However, the techniques for TSR train classifications like Random Forest [10] as well as Support Vector Machine (SVM) [5] using color-and/or shape-based data. To determine the ideal feature set and classifier for TSR, there isn't a thorough approach, though. The widespread use of deep learning (DL) methods for TSDR [14] can be attributed to their recent progress in many computer vision tasks, including image recognition [11], image segmentation [12], and object detection difficulties [13–17].

Furthermore, unlike the current training approaches that enhance the entire image, we suggest an innovative training pipeline using Enhance-Net that focuses on enhancing traffic sign regions. We empirically demonstrate how these present approaches give relatively minimal enhancements to traffic sign detection achievement, even though they might improve total image visibility. Due to the fact that a general improvement of the image cannot assure uniformly improved traffic sign regions. Our suggested approach includes the computation of the loss function just on the traffic sign regions in order to overcome this problem. Since improving TSDR is the strategy's ultimate goal, the improvement blocks serve to guarantee that the sign regions are more visible and that the sign detector can detect them more accurately.

## 2 Proposed System

We supply Enhance-Net, a deep convolutional neural network (CNN) picture enhancers that preprocess those traffic images in order to lessen this issue. Not only did it highlight the fact that each dataset's twelve distinct CC types do not all need to be evaluated at once, but it also highlighted how benchmarking equations are particularly susceptible to challenging weather circumstances including rain, snow, as well as haze. In addition, two common camera mistakes that arise in real-world situations are lens blur as well as dirty lenses. In this research, we therefore highlight five unique CCs associated with a variety of degrees of magnitude: Rain, Snow, Haze, Dirty lens, as well as Lens blur. Due to the variety of these obstacles, programming a single network to handle them all may produce efficiency that is below optimal. Therefore, we deploy five different Enhance-Nets trained on different types of problems at a time to ensure the maximum possible performance for each CC. In order to classify and recognize indicators in the augmented website traffic photos, we employ the modular TSDR technique that was suggested in [5]. Our approach consists of four unique components: the indicator classification algorithm, the sign detector, enhancement blocks, and the obstacle classification.

Employing transfer learning based on the VGG16 CNN structure that has been previously trained using the ImageNet dataset, an impossible classifier is put into practice. As seen in Fig. 1, the algorithm is divided into two phases: extracting features as well as classification. The data contained in the image is compressed by the tool for extracting features into a lower-dimensional space for features. These attributes are used in the next step to carry out the intended categorization. Feature Extraction: The feature extraction algorithm is made up of thirteen combination blocks, every one of which has a normalization layer for the batch, a ReLU activating layer, and a $3 \times 3$ convolution kernel. The max-pooling procedure is employed in order to downsample. In order to further condense the features, a global average-pooling operation is carried out in the last step. (2) Classification: A fully linked layer having softmax activation including an output size of six (five distinct challenge categories and one competition-free condition) makes up the categorizing stage. The retrieved features are used for categorization in this step.

## 3 Methodology

Many algorithms exist that identify web traffic indications, but they are limited to using perfect photographs for discovery because they have been educated on such alone. Sadly, with time, outdoor photo quality can deteriorate to the point where current algorithms are unable to identify indications in weather-impacted photos. For this reason, the researcher has proposed a two-part CNN method to prevent needless complexity [5]. Any weather-related debris will be cleared out in the first part, and any indications of digital activity will definitely be found in the second [6].

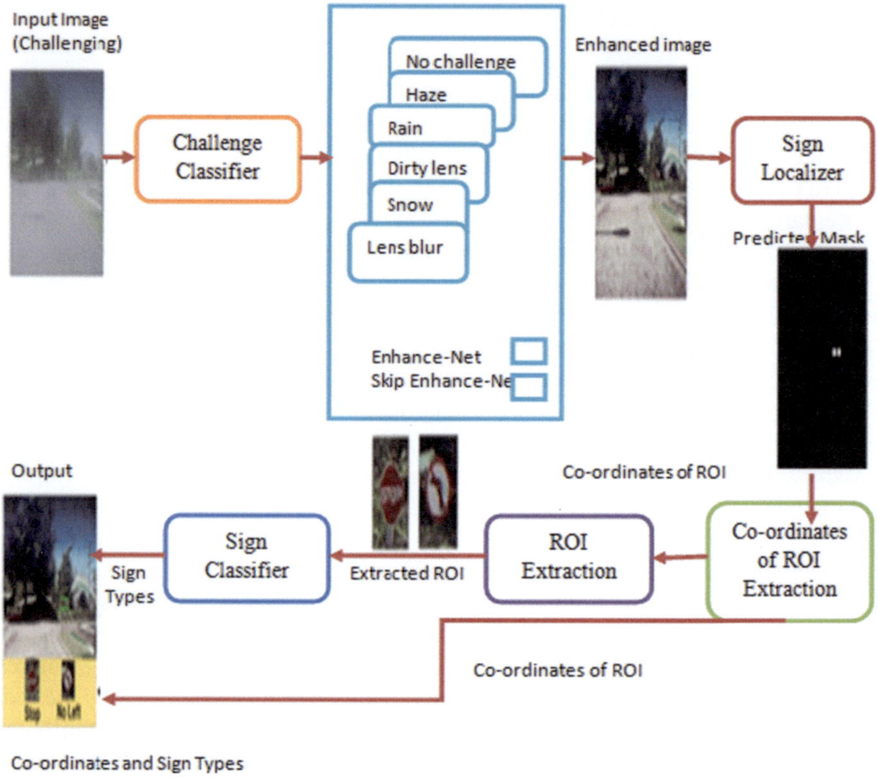


**Fig. 1** Projected pipeline for robust traffic sign recognition and categorization

The DFR-TSD approach is a recommendation system that efficiently locates traffic signals by utilizing a deep learning CNN methodology. With training on the 'CURE-TSD' dataset, the suggested formula reaches up to 99% precision [7]. To complete this endeavor, we have actually put together the following components: Draw a ton of people to your website Convolutional Neural Network (CNN) Model Registration: Employing this component, students will create and load a website traffic indicator. CNN configuration: Secondly, this module will be utilized for submitting an inspection image, and convolutional neural network (CNN) layout that will be applied to remove photos connected to climate change. Third, Find the Markers in a Clearly Visible Image: The CNN sign identification algorithm will soon be able recognize the indicator using the clear picture. Plotting the CNN training loss as well as precision graphs will be done on the proposed convolutional neural network training graph.

## 4 Results and Analysis

In order to highlight the significance of every component in our building design, we also show the effectiveness of our suggested strategy on the CURE-TSD dataset in this section. Moreover, end-to-end trained methodologies are also used in comparison results to demonstrate the advantages of our modular strategy. **Measures of Productivity** We assess our suggested method's effectiveness using the following metrics: • True Positive (TP): a projected sign region that, for a given sign type and effective deep CNN-based classification, overlaps by at least 50% using the actual truth sign region. False Positive (FP): either the sign type has been determined wrongly or the predicted sign region is not overlapping that the ground truth sign region. No traffic sign identifying in non-sign locations is a True Negative (TN). False Negative (FN): absence of traffic sign recognition in sign regions. These indicators are used to establish the two scores that represent how well the model performed in summary.

$$\text{Accuracy} = \frac{\text{TP} + \text{TN}}{\text{TP} + \text{TN} + \text{FP} + \text{FN}}, \tag{1}$$

$$\text{Precision} = \frac{\text{TP}}{\text{TP} + \text{FP}}, \tag{2}$$

$$\text{Recall} = \frac{\text{TP}}{\text{TP} + \text{FN}}. \tag{3}$$

The Challenge of a Classifier's Effectiveness: In our suggested approach, the difficulty of the classifier's performance is crucial. As a result, we first examine and display outcomes of several CNN designs for the problem classifier module. It is significant to note that while "No challenge" has a total of 9000 frames, there may be more traffic signals in no challenge than that because there may be more than one traffic sign in just one frame. Table 3 shows that the challenge classification VGG16 has the best accuracy, at 99.98%, while requiring a manageable amount of parameters as well as time to execute. We decide to use it as our assignment classification module as a result. A potential occurrence linked to the improvement of CCs is that, regardless of how the accuracy of CCs increases, it may worsen in the challenge-free circumstance if the challenge classification incorrectly identifies the challenge-free image as tough. In this case, an improvement block that can lower the severity of the image degradation will pass across a challenge-free image. By selecting the "Generate as well as Tonnes Website Traffic Sign CNN Model" button, a CNN simulation may be generated and loaded. To provide a test image that has been impacted by environment and then clean it up, as illustrated in Fig. 2, choose the "Upload Test Image as well as Clear" choice on the previous window after the CNN model has loaded.

By pressing the "Open" button, you can load and edit the image after choosing and generating the "1.jpg" file in the screen that followed. The outcome will be displayed beneath Fig. 3.

**Fig. 2** Test input images

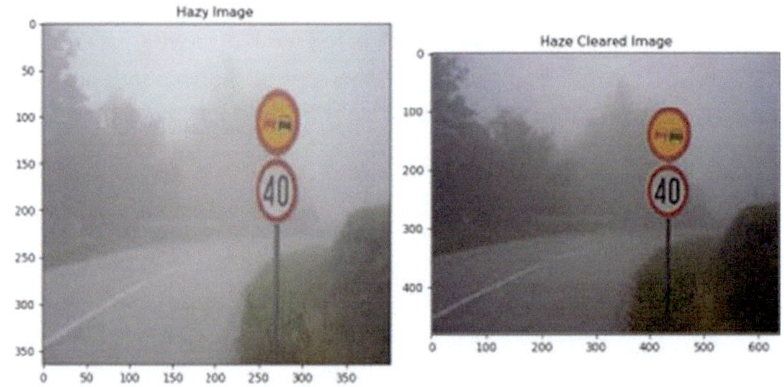


**Fig. 3** Input image

**Fig. 4** Traffic sign detection

The first shot, which is weather-influenced due to factors including rain, cloudy haze, poor lighting, or a malfunctioning camera lens, is different from the second one, which is clear. Click the "Discover Evidence with Clear Photo" button to see the results shown beneath.

Following its detection of the traffic sign and subsequent creation of a bounding box surrounding it, the CNN model proceeded to evaluate other images as illustrated in Fig. 4. Once you have selected and uploaded 7.jpg from your previous television, click the "Open" option to see the output seen in Fig. 5.

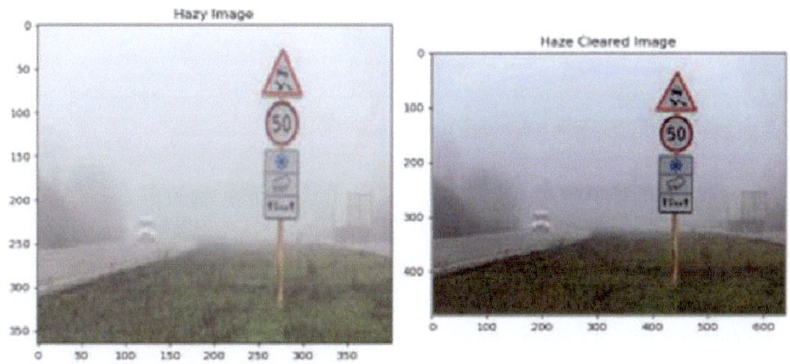


Fig. 5 Stop indicator

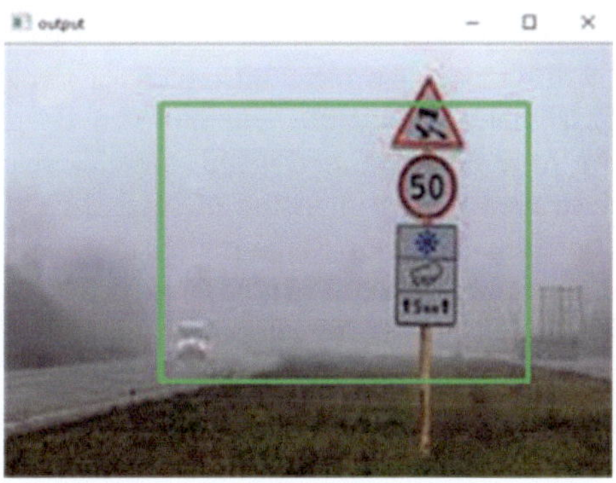


Fig. 6 Stop indicator detection

The weather-impacted images are the primary one, and the new image becomes the second shown in Fig. 6, when you select the “Spot Indicator using Clear Image” button.

Following your choose in the “proposes CNN Training Graph” button, the following illustration will appear; the previous window displayed in internet traffic evidence, and you can upload and see other photographs in the same way.

The link among the variables “Precision whereas loss values” (y-axis) and “training date” (x-axis) is depicted in the preceding picture, where the blue line represents loss, whereas the green line represents accuracy (see Fig. 7). The quality and loss both rose throughout every single day that went by throughout the boosting process, as the graph opposite illustrates. Remember that nobody’s detection method is perfect, as well as that a number bounding boxes might not constantly provide 100% accurate predictions.

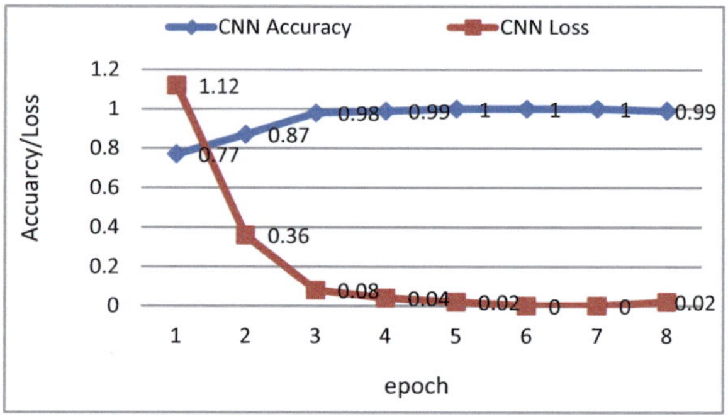


**Fig. 7** Presentation of CNN accuracy versus loss

# 5 Conclusion

In this study, we have introduced a strong foundation for TSDR across various CCs that is adaptable and built on deep convolution neural networks (CNNs). We have demonstrated where the effectiveness of the existing TSDR equations is destroyed by the inclusion of several CCs and have suggested a deep CNN-based approach that resolves the issue. After identifying and labeling the issue, an obstacle classifier built on the VGG16 architecture forwards the image to the appropriate Enhance-Net, which restores the functionality required for precise determination of the regions identified by web traffic. Our novel loss function as well as training procedure, including combines online traffic indication region centered MAE in both pixel as well as attribute domain along with traffic indication recognition loss as a constraint, is employed to train the Enhance-Nets. They differ from the earlier fully image enhancement-based approaches because of this. This successfully ensures the enhancement of the indicator zones depending on their precise detection. Furthermore, our testing results demonstrate that zones with traffic signals are more important to enhance for improved detecting effectiveness. Subsequently compare our method to various end-to-end qualifying deep CNN-based object identification networks and observes that it outperforms them all, proving the efficacy of our method's modular structure. Thanks to our modular design, we can construct each component of the framework independently.

# References

1. Mogelmose MMT, Moeslund TB (2012) Vision-based traffic sign detection and analysis for intelligent driver assistance systems: perspectives and survey. IEEE Trans Intell Transp Syst 13(4):1484–1497
2. Gudigar A, Chokkadi S, Raghavendra U (2016) A review on automatic detection and recognition of traffic sign. Multimedia Tools Appl. 75(1):333–364
3. Xu S (2009) Robust traffic sign shape recognition using geometric matching. IET Intell Transp Syst 3(1):10–18

4. Khan JF, Bhuiyan SMA, Adhami RR (2011) Image segmentation and shape analysis for road-sign detection. IEEE Trans Intell Transp Syst 12(1):83–96
5. Creusen IM, Wijnhoven RGJ, Herbschleb E, de With PHN (2010) Color exploitation in hog-based traffic sign detection. In: Proceedings IEEE international conference image process, September 2010, pp 2669–2672
6. Temel D, Chen M-H, AlRegib G (2020) Traffic sign detection under challenging conditions: a deeper look into performance variations and spectral characteristics. IEEE Trans Intell Transp Syst 21(9):3663–3673
7. Canny J (1987) A computational approach to edge detection. In: Readings in computer vision. Amsterdam, The Netherlands: Elsevier, pp 184–203
8. Baró X, Escalera S, Vitrià J, Pujol O, Radeva P (2009) Traffic sign recognition using evolutionary AdaBoost detection and forestECOC classification. IEEE Trans Intell Transp Syst 10(1):113–126
9. Jiménez PG, Bascón SM, Moreno HG, Arroyo SL, Ferreras FL (2008) Traffic sign shape classification and localization based on the normalized FFT of the signature of blobs and 2D homographies. Signal Process 88(12):2943–2955
10. Zaklouta F, Stanciulescu B (2014) Real-time traffic sign recognition in three stages. Robot Auto Syst 62(1):16–24
11. Krizhevsky A, Sutskever I, Hinton GE (2012) ImageNet classification with deep convolutional neural networks. In: Proceeding advanced neural information processing systems, pp 1097–1105
12. Chen L-C, Papandreou G, Kokkinos I, Murphy K, Yuille AL (2014) Semantic image segmentation with deep convolutional nets and fully connected CRFs. arXiv:1412.7062. [Online]. Available: http://arxiv.org/abs/1412.7062
13. Girshick R, Donahue J, Darrell T, Malik J (2014) Rich feature hierarchies for accurate object detection and semantic segmentation. In: Proceedings IEEE conference computing vision pattern recognition, June 2014, pp 580–587
14. Luo H, Yang Y, Tong B, Wu F, Fan B (2018) Traffic sign recognition using a multi-task convolutional neural network. IEEE Trans Intell Transp Syst 19(4):1100–1111
15. Jagadeeshwar K, Sreenivasarao T, Pulicherla P, Satyanarayana KNV, Mohana Lakshmi K, Mahesh Kumar P (2023) ASERNet: automatic speech emotion recognition system using MFCC-based LPC approach with deep learning CNN. Int J Model Simul Scientific Comput 14(04):2341029
16. Nirmal Kumar A, Alagumuthukrishnan S, Arokia Jesu Prabhu L, Ashok Kumar C, A qualitative study on corona disease prediction using chest X-rays through transfer learning. AIP conference proceedings, vol 2548(1). pp 040022. https://doi.org/10.1063/5.0118477
17. Vivekanand A, Vadlakonda D (2024) A novel interpretable regularized cnn with a modified xlnet transformer for segmenting and classifying the ovarian cancer. Multimedia Tools and Appl 1–28

# AI-Based Classification of Growth in Early-Stage Cancer Detection and Prevention

**Voruganti Naresh Kumar, N. Tejasree, D. Dhanalakshmi, and Harsha Gangavane**

**Abstract** The intricate structure of the cell makes it extremely challenging and time-consuming to recognize lung cancer promptly on. Rapid proliferation of malignant cells causes cancer to spread throughout the body. One important aspect of picture processing—which is also helpful for treating lung cancer prevention—is the prediction of early-stage lung cancer cells. This suggested system uses an artificial neural network classification and algorithms for image processing to deliver a computer-based medicine diagnostic method for premature lung cancer identification. This system's processing stage includes using masks with morphological operations as well as thresholding technique to exclude surrounding tissue and backgrounds. A number of image-enhancing algorithms are also included in this endeavor. The dimension of rate of interest (ROI) is calculated using the area-based dividing approach. The circle fit method is used to remove the desired nodule. Pulling out characteristics like span, mean vitality, area, Euler number, and ECD is part of the extracting features stage. The artificial semantic network (ANN) is trained for categorization utilizing the back breeding strategy as the final step.

**Keywords** ROI · ECD · ANN · AI · Cancer

---

V. N. Kumar (✉)
Department of CSE, CMR Technical Campus, Hyderabad, Telangana, India
e-mail: nareshkumar99890@gmail.com

N. Tejasree
Department of CSE, CMR Engineering College, Hyderabad, Telangana, India
e-mail: nanduri.tejasree@cmrec.ac.in

D. Dhanalakshmi
Department of CSE, CMR Institute of Technology, Hyderabad, Telangana, India
e-mail: dhanalakshmi.dasi@cmritonline.ac.in

H. Gangavane
Department of CSE, CMR College of Engineering and Technology, Hyderabad, Telangana, India
e-mail: g.harsha@cmrcet.ac.in



K. R. Madhavi et al. (eds.), *Proceedings of Sixth International Conference on Computer and Communication Technologies*, Lecture Notes in Networks and Systems 1356,
https://doi.org/10.1007/978-981-96-5238-9_54

# 1 Introduction

One of the main causes of lung cancer involves uneven and unchecked cell proliferation in the lungs. Nicotine consumption is one of the significant factors. If it is identified promptly, there is a good chance that therapy will be feasible. Testing is one of the most crucial procedures in the detection of cancerous lung cells. The method employed to find and recognize the flaw is called screening. A nodule manifests as a circular white lump on chest X-rays or computed tomography (CT) images [1]. According on what caused them, nodules may prove to be benign or malignant. Any lung nodule with a diameter less than three centimeters is classified as pulmonary or non-cancerous. These flaws are occasionally described as harmless. Toxic blemishes larger than 3 cm in diameter are classified as malignant. Diagnosing a lethal mole as rapidly as possible is crucial because it is probably a cancerous nodule. It is important to monitor these flaws as time goes by to see whether they are getting worse. If the spot keeps expanding and changing in size, you could end up with cancerous cells. As such, it is imperative to identify a defect [2]. In comparison with other cancer cell types, lung cancer cells have a very low long-term endurance cost. As a result, it offers an essential study platform for the field of medical image processing, and it is crucial that lung cancer be detected early [3]. Employing computer tomography (CT) images, lung anomalies must first be identified and then segmented so as to do the robot syringe biopsy. This article examined a number of typical division formulae, including thresholding, active contour, asymmetrical driver, area developing, and landmark. We applied them to eight CT scans with different kinds of lung abnormalities to evaluate their effectiveness in identifying blemishes [4, 5]. Region expansion performs better than the other approaches in categorizing imperfection follow vessels, while thresholding, aggressive shaping, and alternative driver do well in categorizing individual nodules. The most effective method for categorizing semi-transparent nodules utilizes the differentiation operator. While employed for single flaws, landmark is not as susceptible as it is when categorizing abnormalities surrounding by arteries and even partially opaque nodules [6]. The purpose of this review and explanation of the features and efficacy of right now accessible approaches for splitting lungs exhibiting pathologic issues on breast CT scans is to assist psychiatrists in better comprehending their alternatives for decision assistance in their day-to-day practice. To help with this process, we will also include a few drawings. Segmentation, a computer-based technique that distinguishes the borders of the lung versus neighboring thoracic material on computed tomographic (CT) images, is an essential first step in radiologic pulmonary image processing. Numerous algorithms as well as application systems offer image division protocols for the measurement of lung anomalies; nevertheless, nearly all of the existing photo segmentation techniques are only successful whenever the lungs exhibit minimal or no pathologic issues. Computer-aided methods for detection may not correctly represent regions in the lungs with moderately to high degrees of condition while abnormalities that appeared tough due to their imprecise division methodologies.

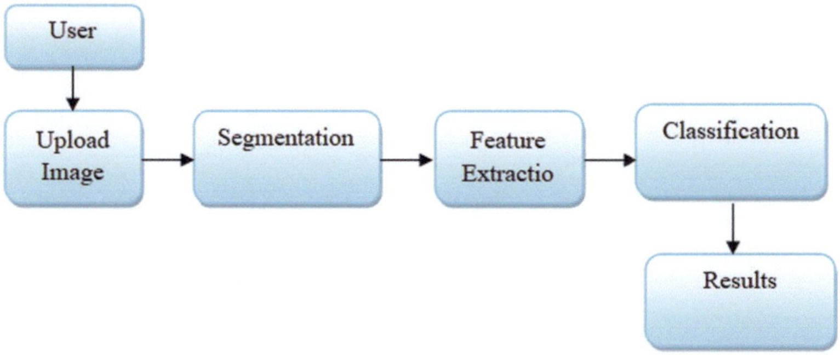


**Fig. 1** Construction model

These devices have the ability to recognize aberrant skin lesions that could be precancerous or malignant automatically. Sophisticated imaging methods, such as machine education and deep learning algorithms, enable precise prediction and detection of possible skin cancer prior to a clinical test confirms its existence. Large-scale data analysis is used to train computer models to identify possible hazards for skin cancer. Doctors have greater ability to treat skin cancer without this kind of early discovery, sparing patients important time and financial resources [7]. The next Fig. 1 displays the construction pattern.

## 2 Existing System

A nodule's discovery is an essential initial step in the diagnosis of lung cancer. Before the necessary nodules are removed, the preprocessing picture enhancement is done again. The objects around the image's boundaries have been cleared out. Picture backgrounds and grey thresholding technique for binaryization are two examples of image preprocessing techniques. An approach built around regulatory ions is employed to extract nodular cells within the lung. For additional processing, we locate and divide nodules whose areas varying from 75 to 1000 images [6, 7].

## 3 Proposed System

The median filter is frequently used to reduce visual noise. By comparing the proximity of two neighboring pixels in an image, the median filter establishes whether or not the pixels are equivalent. The original pixel value under this filter is replaced given the median of the neighboring pixels. The histogram equalization method is employed to adjust picture brightness to enhance contrast. It is a graphic depiction

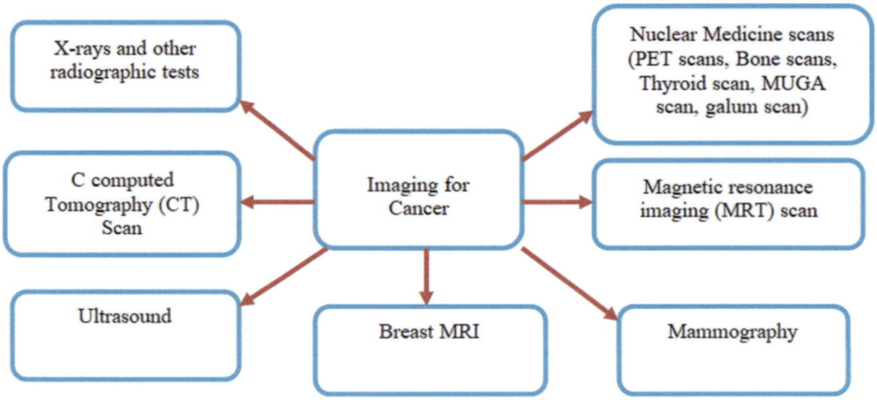


**Fig. 2** Proposed structural model

of the brightness values associated with the picture's pixels. One alternative understanding is the data architecture that records the frequency for every pixel's brightness level in the image [8]. Artificial intelligence provides a technology-based method for early identification of skin cancer that employs machine learning approaches to identify changes in skin features that may point to the development of skin cancer. Through conversion of pictures into numbers, the computer is able to correctly categorize various photos of skin lesions. Consequently, electronic health records can incorporate AI-based identification of skin cancer to enable timely evaluation, treatment, and preventative. Moreover, dermatologists can identify potentially malignant lesions promptly thanks to AI models' ability to distinguish skin lesions versus healthy skin. The operational schematic is displayed in Fig. 2.

In this instance, we want to use an AI system known as a neural network to forecast cancer cases. To do this, we will train the neural network with a variety of optimization strategies, such as ADAM, SGD, as well as Slope Descent Mini Batch. The explanation that follows shows a summary on cancer that was obtained from the pictures you gave, showing three distinct kinds of cancer cells or stages. Our AI systems learned using these images [9]. AI is additionally feasible to recognize skin cancer risk factors, including age, gender, ethnicity, and habits, and to suggest lifestyle modifications that may help lower the risk of the disease [10]. Artificial intelligence (AI) in the early diagnosis of skin cancer is a method that employs algorithmic computing to identify, assess, and track malignant skin development. Because it can identify malignant lesions in the earliest stages of expansion, this technique represents a promising advancement in the early identification of skin cancer. AI technology thoroughly examines skin photos using computer vision algorithms to look for cancerous symptoms. These computer programs can identify trends in skin scans that may point to the presence of skin cancer [11]. The AI program determines which photographs have the highest probability of being symptomatic of cancer by contrasting such patterns to a vast image library. Once the worrisome lesions are identified, AI technology uses computer vision algorithms to combine the collected

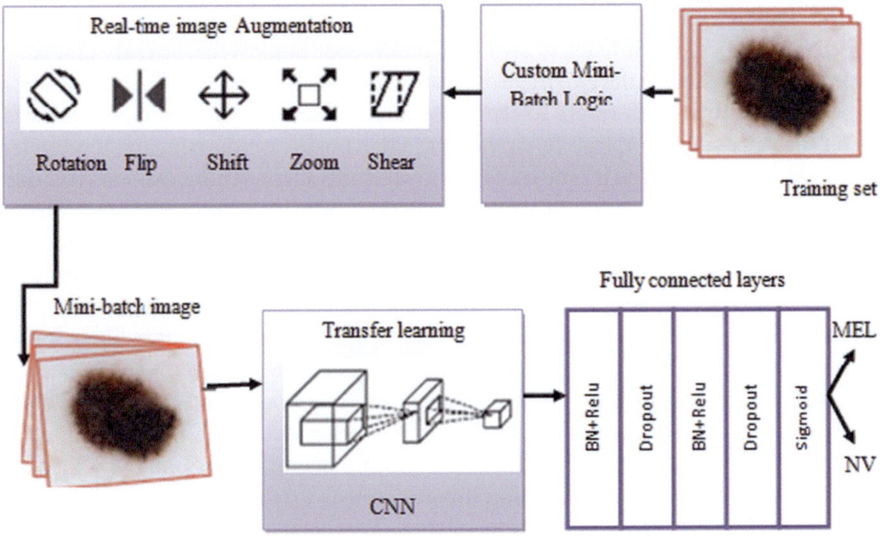


**Fig. 3** Flow model

data together other pertinent information, including laboratory findings and patient history, to generate more precise diagnoses of the kind of cancer [12–16]. This data enables the AI algorithm to generate a diagnosis that is significantly more successful than conventional diagnosis techniques while also being lot precise. By recognizing and curing the disease in its early stages, the application of AI in skin cancer detection and treatment has the potential to significantly lower the amount of cancer-related deaths. Additionally, by reducing the length of time a patient is waiting for treatment, early identification of skin cancer using artificial intelligence (AI) might lessen the financial strain on the patient. If the patient needs care right away, this could save their lives. AI technology may also be able to find novel approaches to treating skin cancer. The artificial intelligence program can determine the optimal course for therapy for any kind of skin cancer through cross-referencing the vast volumes of digitally accessible information. This can lower treatment costs while increasing treatment process effectiveness. The operating structure diagram is displayed in Fig. 3.

## 4 Results and Analysis

A key component of creating a system for the early diagnosis of skin cancer is the evaluation of the initial identification of skin cancer employing artificial intelligence. To guarantee accuracy and dependability, artificial intelligence algorithms for skin cancer detection early on need to be thoroughly assessed. A variety of criteria, including reliability, specificity, sensitivity, recall, and precision, are typically used to assess the efficacy of artificial intelligence models for the early diagnosis of skin cancer. Whereas the two metrics relate to the positive and negative predictions, respectively, correctness refers to the percentage of accurately anticipated images. Using the opposite hand, true favorable and false positive judgments are computed using recall and accuracy. It is also critical to take into account how well an AI model performs while various types of data with distinct properties are employed. This is particularly crucial while examining how well an AI model performs in practical situations. Isolating and assessing an AI algorithm's efficacy in particular settings is crucial. One way to assess a model would be to see how well it works with various patient kinds or in various weather circumstances. The quantity and caliber of data utilized for the models' training and assessment also have a significant impact on how well AI models operate. When comparison to a model developed on a fewer number of photos, one that was developed on a larger dataset of pictures collected from various sources is probably going performing better. In a comparable manner, models trained with higher resolution images typically exhibit better precision than models taught with lower resolution images. Lastly, it is critical to track an AI algorithm's effectiveness over time. This is particularly crucial for AI skin cancer prediction models. To keep the model accurate, it is critical to add newly obtained data—both healthy and cancerous—to the training set as it becomes available. To make ensuring the model is not excessively trained and to guarantee its efficiency does not deteriorate over time, it is also crucial to conduct regular testing and evaluations. Since skin cancer is one of the most common causes for deaths in the USA, early identification is a critical medical concern. Artificial intelligence (AI) has shown promise in recent years for automated and early skin cancer detection. The expense and difficulties associated with manual skin cancer diagnosis can be decreased with the use of artificial intelligence (AI), which has the capacity to evaluate vast volumes of data rapidly, accurately, and economically. Through the analysis of radiography pictures, artificial intelligence algorithms can create statistical as well as predictive models that identify anomalies in skin tissues. In order to recognize and categorize questionable lesions, certain regions of the skin may additionally be segmented using AI algorithms. This may involve the use of reflectance confocal microscopes, a three-dimensional imaging method that produces high-resolution images of the skin and makes it possible to identify alterations in the structure and architecture. AI algorithms can also be used to decipher biopsied images or to find and monitor trends in information from patients to anticipate possible skin malignancies. Both the reliability of the models employed and the successful application of AI to preliminary skin cancer detection depend on optimization of performance. The algorithms used in artificial intelligence require being trained on high-quality datasets that include instances of both positive and negative discoveries spanning a variety of demographics in order to guarantee the best results. The confusion matrix parameters true positive (TP), true negative (TN), false positive (FN), and false negative (TP) are used to assess the effectiveness of the suggested method. These parameters are described as follows: true positive (TP): When an occurrence is accurately categorized as positive and turns out to be positive. True negative (TN): When an instance is accurately categorized as negative when it is genuinely negative. False positive (FP): an occurrence that is mistakenly categorized as positive when it is actually negative. False negative (FN): when something is mistakenly categorized as negative when it is actually positive. The definition of accuracy is the ratio of accurately detected occurrences to the overall number of occurrences. Reliability is expressed as

$$\text{Accuracy} = \frac{\text{TP} + \text{TN}}{\text{TP} + \text{TN} + \text{FP} + \text{FN}} \tag{1}$$

$$\text{Precision} = \frac{\text{TP}}{\text{TP} + \text{FP}} \tag{2}$$

$$\text{Recall} = \frac{\text{TP}}{\text{TP} + \text{FN}} \tag{3}$$

$$F1_{\text{Score}} = \frac{2^{*}\text{Precision}^{*}\text{Recall}}{\text{Precision} + \text{Recall}} \tag{4}$$

$$\text{Senstivity} = \frac{\text{TP}}{\text{TP} + \text{FN}} \tag{5}$$

The dataset displays three different types of cancer; access any folder that is depicted in Fig. 4 to view images for every type.

Thus, we are using the images above to train three separate optimizers for AI. To complete this assignment, the subsequent modules were created built by us. Publication of the Histopathological Image Dataset: Without a doubt, the dataset will ultimately be published to the program using this component. Prepare the dataset:

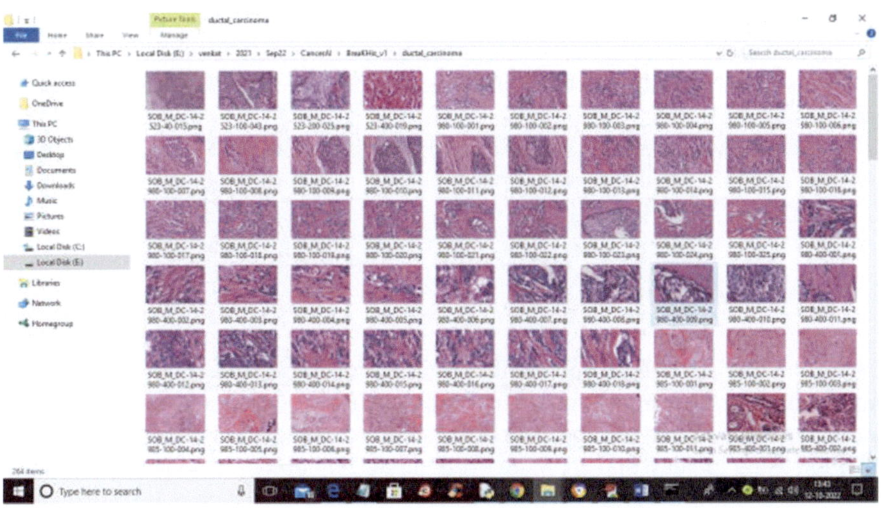

**Fig. 4** Cancer datasets

All of the photos will be read by this module, which will then resize images to the identical dimension and normalize their pixel values. The dataset is going to split into training and testing sets right away after processing. Educate AI using ADAM: Utilizing this component, the training information will be put into our AI algorithm, which uses ADAM as the optimizer. We will utilize the test data gathered from the educated version following training to determine the precision of the predictions. Employ SGD to Teach AI: In actuality, we will use this module's optimization to feed the AI algorithm with the training data. During training, we will employ certified product evaluation data to evaluate how accurate our projections were. AI Formula Training with MiniBatch: This module uses MiniBatch as the optimization and training data as the input. During training, we will use the authorized model's test data to assess the prediction's correctness. Employing this component, referred to as an evaluation table, we will plot all algorithmic precision and offer efficiency in tabular form. After importing the dataset into the Web site previously mentioned, you may click the "Preprocess Dataset" option to view and manipulate the images. There are 555 total photos in the dataset. Size of training images: 499 and testing picture dimensions: 56. As the display above illustrates, there are 555 images in the collection. Next, the dimensions of both the test and train data are shown. The kind of cancer cells is shown on the x-axis of the chart, and the amount of images of that type is shown on the y-axis. Close the picture and select "Train AI using ADAM" to begin training the AI. The result is shown below. In the explanation that follows, 92% accuracy was attained by AI using ADAM. Estimated labels are displayed on the x-axis, true labels are displayed on the y-axis, and the number of accurate forecasts is represented by comparable blue boxes, while the number of incorrect guesses is represented by different colored boxes. A matrix chart of perplexity illustrates this. Simply dismiss this window and choose the "Train using SGD" button to begin learning SGD; results are displayed beneath Fig. 5. Close the visualization above, then select "Train using MiniBatch" to begin teaching AI. With SGD, we obtained 51% accuracy. The outcomes are shown below.

We also achieved 51% accuracy utilizing MiniBatch in Fig. 5, earlier, and we are now closing the graph.

Figure 5 shows the tabular and graphical representation of all algorithmic results. "AI plus ADAM" performed well or accurately throughout the board.

## 5 Conclusion

CAD solutions are helpful for the diagnosis of malignant tumors besides to their numerous other uses in contemporary medicine. Erroneous lesions are not needed because a blemish having the necessary area can be identified using a circle fit equation using greatest spread. With every model, we obtain more accurate results. This led to a 95.6% accuracy provided by the method. The system is 100% exclusive and has a sensitivity of 93.1%. This technique can determine if a nodule found on a lung CT scan is malignant or harmless. Future applications of this technique include

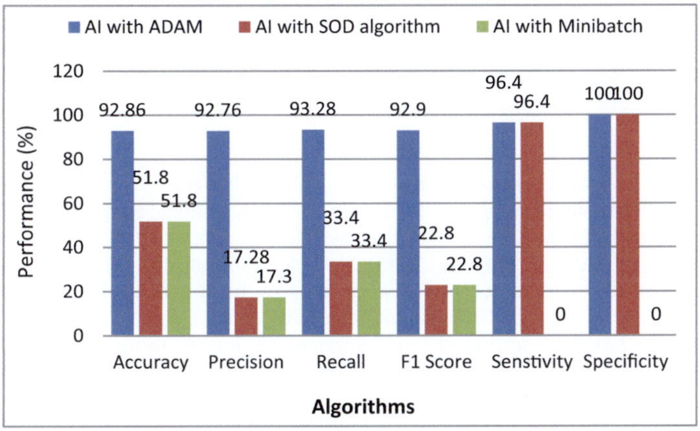


**Fig. 5** Presentation assessment of various algorithms

helping to diagnose cancer in different human organs. The techniques employed by this system have the potential to either reduce the production of aberrant cells or disperse them throughout the body. It is feasible to enhance this method for MRI plus ultrasound pictures. While NN classifiers yield higher quality and more precise outputs than support vector machines (SVMs), they require a considerably wider variety of data sources.

## References


1. World Health Organization. Cancer: fact Sheet no. 297. July 8 (2015). http://www.who.int/mediacentre/factsheets/fs297/en/; Jemal A, Siegel R, Xu J, et al. (2015) Cancer statistics. CA Cancer J Clin 60(5):277–300
2. The diagnosis of lung cancer (update) Published by the National Collaborating Centre for Cancer (2nd Floor, Front Suite, Park House, Greyfriars Road, Cardiff, CF10 3AF) at Velindre NHS Trust, Cardiff, Wales [2011] Database from: The Cancer Imaging Archive. https://doi.org/10.7937/K9/TCIA.2015.A6V7JIWX
3. Nanusha (2017) Lung nodule detection using image segmentation methods. Int J Adv Res Electron Commun Eng (IJARECE) 6(7)
4. Mansoor A et al. (2015) Segmentation and image analysis of abnormal lungs at CT: current approaches, challenges, and future trends. Radio Graphics 35:1056–1076. Published online https://doi.org/10.1148/rg.2015140232
5. OzgeGunaydin et al. (2019) Comparision of lung cancer detection algorithm. IEEE. 978-1-7281-1013-4/19/$31.00©2019
6. Madhura J, Ramesh Babu DR (2017) Survey on noise reduction techniques for lung cancer detection. In: International conference on innovative mechanisms for industry applications (ICIMIA 2017), IEEE, 978-1-5090-5960-7/17/$31.00©2017
7. Ashwin S, Kumar SA, Ramesh J, Gunavathi K: Efficient and reliable lung nodule detection using a neural network based computer aided diagnosis system. In: Emerging trends in electrical engineering and energy management (ICETEEEM), 2012 International conference, pp 135–142

8. Sammouda R (2017) Segmentation and analysis of CT chest images for early lung cancer detection. In: 2016 Global summit on computer and information technology, IEEE. 978-1-5090-2659-3/17 $31.00©2017; Egmonet-Petersen M et al. (2002) Image processing with neural networks—a review. Patteren Recog 5(10):2279–2301
9. Vapnik V (1999) An overview of statistical learning theory. IEEE Trans Neural Netw 10(5):988–999
10. Kanakatte A, Mani N, Srinivasan B, Gubbi J (2008) Pulmonary tumor volume detection from positron emission TomographyImages. In: 2008 International conference on BioMedical engineering and informatics
11. Hashemi A, Pilevar AH, Rafeh R (2013) Mass detection in lung CT images using region growing segmentation and decision making based on fuzzy inference system and artificial neural network. Int J Image, Graphics and Signal Processing 5(6):16–24
12. Nadkarni NS (2019) Detection of lung cancer in CT images using image processing. In: Proceedings of the third international conference on trends in electronics and informatics (ICOEI 2019)
13. Vasundhara N, Nandan AS, Hemanth SV, Macherla S, GK M (2023) An efficient biomedical solicitation in liver cancer classification by deep learning approach. In: 2023 IEEE International conference on integrated circuits and communication systems (ICICACS), Raichur, India, pp 01–05. https://doi.org/10.1109/ICICACS57338.2023.10099828
14. Shrivastava R, Jain M, Vishwakarma SK, Bhagyalakshmi L, Tiwari R (2023) CrossCultural translation studies in the context of artificial intelligence: challenges and strategies. In: Kumar A, Mozar S, Haase J (eds) Advances in cognitive science and communications. ICCCE 2022. Cognitive science and technology. Springer, Singapore, pp 91–98. ISBN: 978-981-19-8086-2_9. https://doi.org/10.1007/978-981-19-8086-2_9
15. Ravi G, Reddy KP, Mohan Rao M, Kodati S, Praveen Kumar J (2021) 10 design a novel IoT-based agriculture automation using machine learning. In: Big data management in sensing: applications in AI and IoT, River Publishers, pp 149–158
16. Reddy S, Pundru C, Sucharitha Y, Vivekanand A (2024) Customer churn prevention for E-commerce platforms using machine learning-based business intelligence. Recent advances in electrical and electronic engineering (Formerly Recent Patents on Electrical and Electronic Engineering), vol 17.5, pp 456–465

# A Demonstration of Machine Learning-Based Darknet Traffic Classification and Association System Using KNN

**Voruganti Naresh Kumar, S. Sai Bhushanam, P. Rukmini Devi, and Arun Pratap Singh**

**Abstract** Due to their covert nature and connections to illegal activity, darknet network traffic presents serious difficulties for safety and surveillance. This paper investigates the use of machine learning techniques in identifying types of traffic on darknet networks as a means of addressing these issues. Network traffic recorded from darknet environments is gathered into a labeled dataset, which is then extracted to extract pertinent information. The efficacy of several machine learning techniques, such as decision trees, random forests, AdaBoost, KNN, Naïve Bayes, simple CART, and gradient boosting, in categorizing darknet traffic is assessed. The training set of the dataset is used to train the chosen machine learning models. The dataset is split into sets for testing and training. The models first identify deeper patterns and connections among the characteristics of network traffic before classifying the data appropriately. The models' ability to classify is then assessed using the testing set to determine the F1-score, accuracy, precision, and recall. The outcomes show how well machine learning algorithms categorize traffic from multilayered darknet networks. The chosen models identify different kinds of malevolent or unlawful activity happening on the darknet with excellent accuracy and intriguing possibilities. The study also addresses how the models can be interpreted and what can be learned from the significance of features analysis, which helps to clarify the functioning of darknet networks.

V. N. Kumar (✉)
Department of CSE, CMR Technical Campus, Hyderabad, Telangana, India
e-mail: nareshkumar99890@gmail.com

S. Sai Bhushanam
Department of CSE (DS), CMR Engineering College, Hyderabad, Telangana, India
e-mail: ssaibhushanam@cmrec.ac.in

P. Rukmini Devi
Department of CSE, CMR Institute of Technology, Hyderabad, Telangana, India

A. P. Singh
Department of CSE, CMR College of Engineering and Technology, Hyderabad, Telangana, India
e-mail: Arunpratap@cmrcet.ac.in

 
K. R. Madhavi et al. (eds.), *Proceedings of Sixth International Conference on Computer and Communication Technologies*, Lecture Notes in Networks and Systems 1356,
https://doi.org/10.1007/978-981-96-5238-9_55

**Keywords** Darknet-based traffic classification · Random forest · Machine learning · Conventional search engines · Cyber crime

## 1 Introduction

A portion of the Internet that is purposefully concealed and unreachable via traditional search engines is referred to as the "darknet." It is frequently connected to unlawful activity, including black marketplaces, Web sites for hackers, and other types of online crime. It is essential to comprehend and categorize network traffic in darknet environments in order to identify and stop unwanted activity. Employing a multilayered method, the research looks into how well machine learning algorithms categorize traffic on darknet networks. The multilayered classification strategy consists of several classification phases, each of which assesses the effectiveness of several machine learning algorithms in classification while concentrating on a particular feature or component of the network data. Metrics including accuracy, precision, recall, and F1 score—which gauge how well categorization distinguishes between various kinds of network traffic—are used to evaluate how successful the algorithms are. Additionally, the study may investigate how various features, feature selection strategies, or preprocessing approaches affect the efficiency of every machine learning algorithms and the assignment of multilayered darknet traffic classification. The goal of the research is to find more accurate and reliable ways to identify and categorize hostile activity in darknet situations by examining the effectiveness of algorithms using machine learning in multilayered categorizing of darknet network data. This has the potential to improve cyber threat intelligence, safeguard networks, and enable proactive countermeasures against cybercrime.

Although its material is exclusive to the Darknet, a different network, and requires special access authorization, the Darkweb also serves as a worldwide network. The darknet facilitates anonym zed network contact by hiding any pertinent information, such as identification or location [1]. Only 4% of the Internet is made up of the top layer of web, where we can typically access, compared to 48% of the Internet made up of the darkweb [2]. The Internet is divided into several layers, and intermediaries of Tor are employed to transfer data between these layers, according to [3]. A wide range of illegal and prohibited activities, like as drug use, terrorism, pornography involving kids, and human trafficking, are carried out via the darknet. Furthermore, as noted by [3], analyzing traffic on the black web is crucial for identifying any malevolent intention of attackers who are always modifying their methods to evade identification of security threats. Moreover, [4] has demonstrated the traffic produced on the Darknet by several application categories.

Ultimately, the study looks at how machine learning approaches might help with the difficulties of identifying and categorizing intricate darknet network traffic in order to improve safety and enable efficient defenses against cyberattacks.

## 2 Background

Tor By rerouting web traffic employing many nodes, Tor conceals browsing activities and gives your IP address anonymous through advanced encryption algorithms [5]. The US Navy originally developed Tor, commonly referred to because the onion the router, for encrypted communication. Since then, it has grown into an open source, nonprofit organization having millions of subscribers and numerous routers [6]. Because Tor allows users to remain anonymous, hackers, activists, and burglars primarily utilize it for unethical objectives, such as selling narcotics, guns, and other items [7]. For this reason, a methodical approach to categorizing Tor users is required. A user joins a specific route in the circuit, and every node throughout the network is unaware of the node preceding it and following it. This is how Tor maintains confidentiality as a distributed system of TCP-based apps [8].

Private online networks similar to a filter, a virtual private network (VPN) shields IP addresses from prying eyes as they pass via a server on the host that serves as the source of information for user searches. In addition to offering safety, this keeps customer data secret regarding the service providers [9, 10]. Thanks to VPN architecture's strong identification mechanism, accessing the Internet is simple about any location employing any accessible access medium, such as analog modems, ISDN, modems for cable television, DSL, and wireless [11–14].

**Problem Definition**

A designated and routed area of IP space is referred to as a darknet or dark web. Servers and services that are not in use are situated in these areas. It might contain unseen mechanisms designed to receive communications. Like that systems might be a member of an overlaying network, but they do not respond to anything. This network can be accessed using non-standard ports and protocols for communication.

It is considered suspicious traffic that is headed toward darkness. Both malevolent intent and misconfiguration are possibilities. Darknets are able to build up systems of surveillance that will draw in attackers in order to collect intelligence. Not only is darknet traffic lower than legitimate traffic, but it also has signs of malicious activities. A darknet network is an instrument of monitoring designed to identify potentially harmful activity and patterns of cyberattacks. On the other hand, false traffic seen in the vacant address space—a collection of permanently valid (I.P.) addresses given to any host or device—is known as darknet traffic. After the COVID-19 pandemic, more people, including criminals without any prior cyber knowledge, are using darknets.

**Objectives**

This effort aims to evaluate, design, optimize, and validate multilayered categories and machine learning techniques for traffic on darknet networks. By doing these goals, the project seeks to aid in the creation of better and more accurate mechanisms for identifying and categorizing hostile activity in darknet settings.

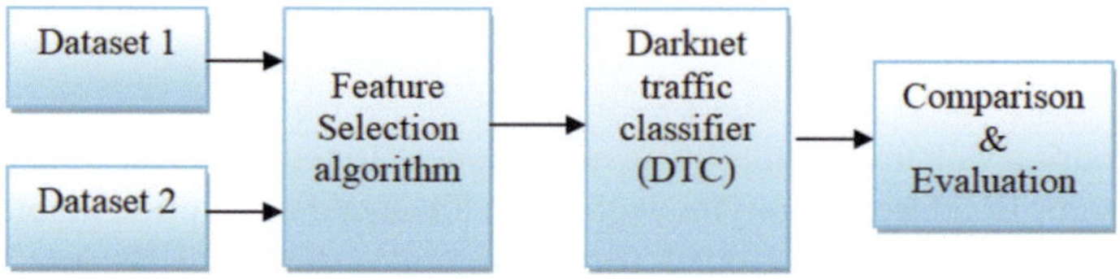


**Fig. 1** Block illustration of the projected structure

## 3 Proposed Method

As seen in Fig. 1, the multilevel categorization strategy consists of several phases of categorization, each of which assesses the effectiveness of a different machine learning model while concentrating on a particular feature or component of the network data.

The proposed system consists of the following stages:

1. The CIC-Darknet2020 dataset, which has 63 features, is the first input dataset.
2. Input Dataset 2: A dataset of 76 features from the NICT Darknet 2022.
3. Feature Selection Technique: The Fisher's score technique ranks every attribute in decreasing order of Fisher's score, which determines the ratio of every characteristic.
4. Binary Classifier (DTC1): Decision tree, Naïve Bayes, KNN, and AdaBoost technologies are used to divide traffic into Darknet and benign groups.
5. Multiclass Classifier (DTC2): Using random forests, decision trees, Naïve Bayes, KNN, and gradient lifting ML algorithms, the output by DTC1 is once more divided into four classes: Tor, Non-Tor, VPN, and Non-VPN.

a. Multiclass Classifier (DTC3): Employing random forest, decision tree, Naïve Bayes, KNN, and gradient boosting machine learning techniques, the output from DTC2 is once more divided into eight application classes: audio stream, browsing, chat, email, P2P, file transfer, video stream, and VOIP.
b. Final Assessment and Assessing: Following DTC1, DTC2, and DTC3 stages, a comprehensive comparison is made based on the matrix of confusion to forecast accuracy, precision, F1-score, particularity, and recall. Additionally, an overarching assessment is made following the DTC3 stage.

## 4 Assessment of Machine Learning Algorithms

We will examine and study the results of each machine learning technique's final analysis as well as how well it classified traffic from darknet networks to determine which method of machine learning classifies the suggested system more accurately.

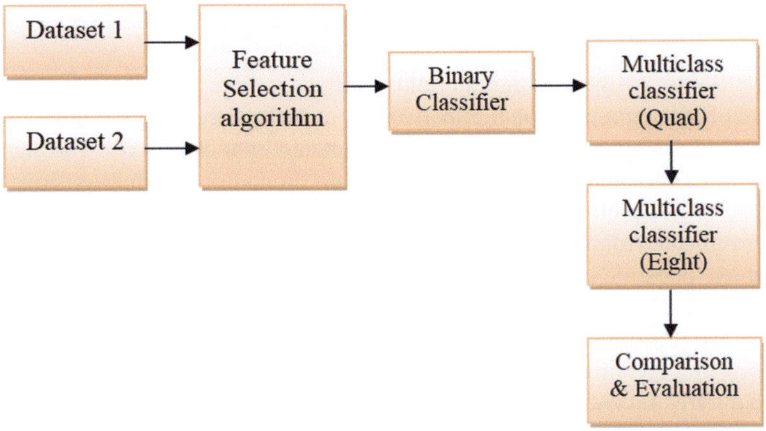


**Fig. 2** Comprehensive block illustration of the projected multilayered DTC classifier structure

As illustrated in the Fig. 2, it is a multilayered darknet traffic classifier (DTC) system.

- DTC is made up of three blocks: DTC1, DTC2, and DTC3.
- The DTC1 block divides traffic into two categories: conventional and darknet. This process is known as binary categorization.
- Four classes are distinguished by the DTC2 block: Tor, Non-Tor, VPN, and Non-VPN.
- Eight service classes are classified by the DTC3 block using multiclass categorization: audio streaming, searching, chat, email, P2P, and file transfer.

## Implementation

### ML Algorithms

### Feature selection Algorithm

a. For assignment classification whereby separation of classes is important, Fisher's score works well.
b. It assesses the degree of class differentiation to determine a feature's discriminative potential. Fisher's score can be used to continuous and categorized characteristics alike.
c. It is good at choosing features that are pertinent to minority classes and therefore can manage class imbalance.

### DTC 1 (Binary Classifier)

KNN (K-nearest neighbor): resilience to outliers, straightforwardness, no training phase, ability to handle complicated decision boundaries, and adaptability to various data formats while imposing any limitations. Decision trees: Ease of execution, comprehension, addressing nonlinear relationships, handling heterogeneous data types, and managing missing values.

Naive Bayes: resilient to unimportant features and interpretable. Naive Bayes is renowned for its computational effectiveness and elegance. Large-scale multiclass datasets comprising a lot of features can be handled by it effectively. The algorithm is appropriate for situations where resources are limited or for time-sensitive applications since its development and forecasting times are typically faster than those of more complicated algorithms.

Adaptive boosting (AdaBoost): This technique handles high-dimensional data, is resistant against noisy data, and might be applicable to a wide range of categorization applications. It also combines numerous weak classification algorithms to generate a single strong classifier.

By naturally allocating larger weights to useful characteristics that make a greater contribution to the overall accuracy of classification, AdaBoost handles the choice of features.

**DTC 2 (Multiclass Classifier I)**

Random Forest: manages class imbalance, robust against overfitting, and high accuracy and manages relationships that are not linear. A metric of the significance of features is provided by random forest, showing how much they contribute of every characteristic to the categorization. This data can be useful for comprehending the data, choosing features, and determining which features are most important for multiclass categorization.

**DTC 3 (Multiclass Classifier)**

At this point, as shown in Fig. 3 the output from DTC 2 is effectively multiclass classified into eight application-based classes using an amalgamation of random forest, decision tree, KNN, Naïve Bayes, and stochastic boosting.

Flow chart

See Fig. 3.

**Implementation**

The most comprehensive and practical Python machine learning library is called Scikit-learn or Sklearn. Through a Python consistency interface, it offers a range of effective tools for mathematical modeling and artificial intelligence, such as reduction of dimensionality, regression, grouping, and classification.

**Viewing Simulations**

Running machine learning scenarios and displaying the results is made easier with Google Colaboratory's array of capabilities, which include affordable GPU and TPU, rich libraries as a whole notebook interaction, visualizing data, and external libraries as well as tools. You may create, run, and visualize machine learning algorithm simulators in an interactive and cooperative way by utilizing Google Colab's features.

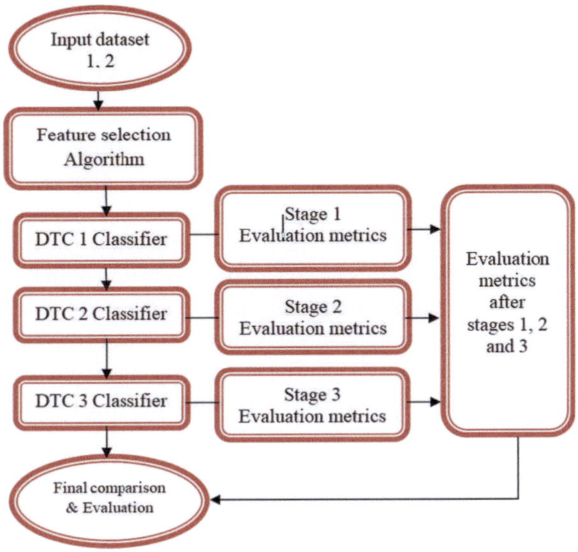


**Fig. 3** Flowchart of the proposed structure

**Datasets**

1. Input Dataset 1: CIC-Darknet2020 dataset consisting of 63 features.
2. Input Dataset 2: NICT Darknet 2022 dataset consisting of 76 features.

**Machine Learning Algorithms**

1. Classification of Features: Fisher's score-based ordering
2. DTC 1 (Binary The categorization): KNN, AdaBoost, Naïve Bayes, and decision tree
3. DTC 2 (Multiclass The categorization): Decision trees, random forests, KNN, gradient boosting, and Naïve Bayes
4. DTC 3 (Multiclass The categorization): Decision tree, random forest, KNN, gradient boosting, and Naïve Bayes
5. Final Assessment and Evaluation Metrics: Following each DTC step, the overall assessment and the following metrics after the last DTC stage: reliability, precision, F1-score, specificity, and then recall.

**Module Description**

The "Darknet Traffic Characterization through Machine Learning" lesson explores the complexities involved in utilizing modern machine learning methods to identify and classify internet traffic connected to the Darknet. The section is intended for cybersecurity teachers and learners who want to improve their knowledge of identifying threats and network security.

The first section of the lesson explores the Darknet, giving a thorough rundown of its features, the kinds of behaviors that are frequently observed in Darknet contexts, and the importance of appropriately categorizing Darknet traffic for counterterrorism operations.

Participants will learn more about the illicit activities that frequently take place on the Darknet, such as botnets, malware distribution, cybercriminal forums, and illegal marketplaces.

The module's main focus is on machine learning techniques for classifying Darknet traffic. In this lesson, participants will gain knowledge of supervised learning algorithms that are frequently utilized for pattern identification and categorization problems like random forest, support vector machines (SVM), and neural networks. Deep learning methods for managing intricate patterns in Darknet traffic and supervised learning strategies for anomaly identification are going to be discussed.

## 5 Results and Analysis

Darknet traffic can be divided into three categories using classification using machine learning algorithms: legitimate, suspicious, and malicious. The findings may indicate that a sizable percentage of darknet traffic is, in fact, connected to illicit operations like DDoS attacks, botnet interaction, and vulnerability-exploiting attempts. It is possible to identify anomalies in darknet traffic, which could point to emerging or new risks. The findings could point to unexpected communication patterns, abrupt increases in traffic volume, or connectivity to domains or IP addresses that had not been detected before.

**Detection Accuracy**

Evaluating machine learning models' classification accuracy for darknet traffic is crucial as shown in Fig. 6. High accuracy shows how well the method works to differentiate among benign and malevolent behavior. Reducing false positives as well as false negatives is important to prevent missing threats or needless alarms.

To sum up, machine learning-based darknet traffic monitoring presents a great deal of promise for enhancing cybersecurity safeguards and reducing online dangers shown in Figs. 4 and 5. Organizations can obtain important insights into the actions taking place in the shadowy areas of the Internet by utilizing advanced analytics tools, which makes it possible to develop aggressive and successful defensive tactics.

But in this dynamic environment, continual research, cooperation, and invention are essential to remaining ahead of competitors.

Ultimately, our analysis of earlier findings revealed that, while XGBoost can compete against most methods, random forest's performance in [4] outperforms us in terms of both implementation and traffic classification. Our investigation was limited since there were not sufficient datasets available. Darknet was the most extensively used and conveniently available dataset for our research shown in Fig. 6. In

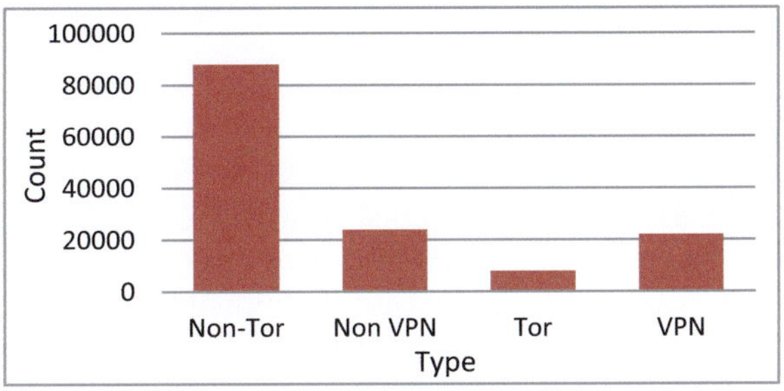


**Fig. 4** Darknet labels

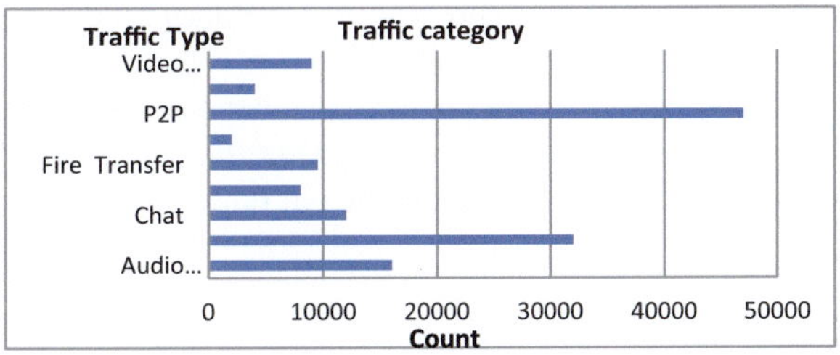


**Fig. 5** Traffic category

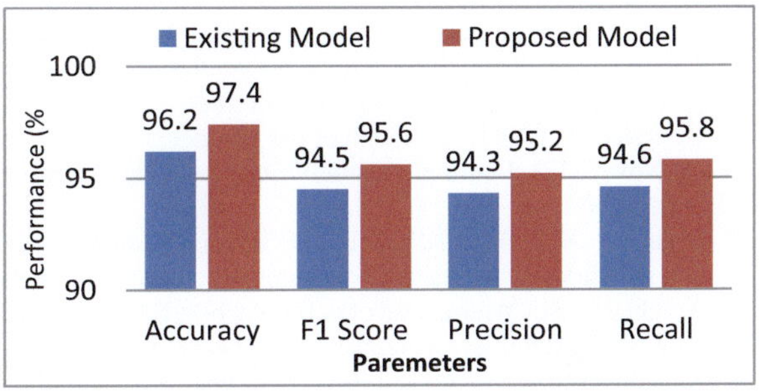


**Fig. 6** Assessment of presented and predictable representation

the future, performance could be improved by replacing existing oversampling techniques with methods for clustering and training each cluster to supply samples. Adversarial obfuscation assessment can also be used to learn how different classifiers respond to the techniques. Time-series constraints could also be used to access the potential benefits of a timestamp on the kind of traffic and application. To investigate every deep learning algorithm's impact on traffic and application category, applications of different kinds might be used.

# 6 Conclusion

In conclusion, there are a number of benefits to machine learning-based identification of multilayered darknet network activity in the areas of network monitoring and cybersecurity. The utilization of machine learning algorithms facilitates the accurate detection and categorization of diverse activity in the darknet, such as malevolent conduct, illicit markets, and possible hazards. Through the improvement of threat detection, preemptive defense, and threat assessment skills, this strategy gives enterprises the ability to more effectively protect their assets and networks.

Researchers and cybersecurity experts have taken significant strides in comprehending the patterns and features of darknet network traffic by utilizing machine learning techniques. Investigations investigating cybercrime, security threats, and the identification of anomalies have benefited from the creation of labeled datasets and the testing of different artificial intelligence models, which have offered insightful information on the characteristics of darknet activity.

# References


1. NPR (2022) Going dark: the internet behind the internet. 5 2014, (Accessed on 25 Nov 2022). [Online]. Available: https://www.npr.org/sections/alltechconsidered/2014/05/25/315821415/going-dark-the-internet-behind-the-internet
2. Georgiev D (2022) How much of the internet is the dark web in 2022?. 11 2022, (Accessed on 25 Nov 2022). [Online]. Available https://techjury.net/blog/howmuch-of-the-internet-is-the-dark-web/
3. Demertzis K, Tsiknas K, Taketzis D, Skianis C, Iliadis L (2021) Darknet traffic big-data analysis and network management to real-time automating the malicious intent detection process by a weight agnostic neural networks framework. Electronics 10(7):2–5, 02 2021, (Accessed on 25 Nov 2022). [Online]. Available: https://www.mdpi.com/2079-9292/10/7/781
4. Rust-Nguyen N, Stamp M (2022) Darknet traffic classification and adversarial attacks. vol 1, pp 6–24, 06 2022. arXiv (Accessed on 25 Nov 2022) [Online]. Available: https://arxiv.org/abs/2206.06371
5. Ghimiray D (2022) The dark web browser: what is tor, is it safe, and how to use it. https://www.avast.com/c-tor-dark-web-browser, 8 2022 (Accessed on 25 Nov 2022)
6. Trademark (2022) The Tor Project—History. https://www.torproject.org/about/history/, 8 2022 (Accessed on 25 Nov 2022)
7. Sarkar D, Vinod P, Yerima S (2022) Detection of tor traffic using deep learning. In: 17th ACS/IEEE international conference on computer systems and applications. AICCSA, 11 2020, pp 1–8. (Accessed on 25 Nov 2022)
8. Dingledine R, Mathewson N, Syverson P (2004) Tor: the second-generation onion router. In: 13th USENIX Security Symposium (USENIX Security 04). San Diego, CA, USENIX Association, August 2004, pp 2–5. (Accessed on 25 Nov 2022)
9. Company CL (2019) Onion router. https://encyclopedia2.thefreedictionary.com/Onion+Router (Accessed on 25 Nov 2022)
10. Kaspersky (2022) What is VPN? How it works, types of VPN. https://www.kaspersky.com/resource-center/definitions/what-is-a-vpn, 10 2022, (Accessed on 25 Nov 2022)
11. Saravanan P, Pandey A, Joshi K, Rondon R, Narasimharao J, Imran AA (2023) Using machine learning principles, the classification method for face spoof detection in artificial

neural networks. In: 2023 3rd International conference on advance computing and innovative technologies in engineering (ICACITE), IEEE, pp 2784–2788
12. Tiwari R, Senthil Kumar M, Diwan TD, Pinjarkar L, Mehta K, Nayak H, Reddy R, Nigam A, Shrivastava R (2023) Enhanced power quality and forecasting for PV-wind microgrid using proactive shunt power filter and neural network based time series forecasting. Electric Power Components and Systems. https://doi.org/10.1080/15325008.2023.2249894
13. Pradeep Reddy K, Raghunadha Reddy T, Apparao Naidu G, Vishnu Vardhan B (2018) Term weight measures influence in information retrieval. Int J Eng Technol 7(2):832–836
14. Bhavani M, Narayana VA, Sreevani G (2020) A novel approach for detecting near-duplicate web documents by considering images, text, size of the document and domain. In: ICCCE 2020: Proceedings of the 3rd international conference on communications and cyber physical engineering. Springer Nature Singapore

# An Improved and Enhanced Technique for Tracking and Forecasting Heavy Metal Water Contamination

**Mudimela Madhusudhan, Kumari Gubbala, A. Sarveswara Reddy, and S. Sangeetha**

**Abstract** All forms of life depend on water. The characteristics of water aid in regulating the variety, diversity, along with rate of succession of biotic elements. The deteriorating state of communal water resources such as lakes, streams, as well as estuaries is among the most serious and concerning problems facing humanity. The serious problem of heavy metal-induced water pollution, with a thorough monitoring and forecasting strategy presented. The growing danger to human health and ecosystems demands creative responses. In order to develop a real-time monitoring system, our project combines predictive modeling, data analytics, and advanced sensing technologies (ARIMA). We hope to foresee possible pollution events in addition to detecting and measuring heavy metal concentrations in water by utilizing these techniques. By being proactive, we can protect water supplies, minimize environmental damage, and make timely adjustments. The project's results have the potential to improve the management of water quality and guarantee a sustainable and healthy future. Regression analysis, mean squared error, as well as root mean squared error are the metrics used to evaluate the efficacy of the constructed model. By being proactive, we can protect water supplies, minimize environmental damage, and make timely adjustments. The project's results have the potential to improve the management of water quality and guarantee a sustainable and healthy future.

M. Madhusudhan (✉)
Department of CSE, CMR Technical Campus, Hyderabad, Telangana, India
e-mail: cmrtc.paper@gmail.com

K. Gubbala
Department of CSE, CMR Engineering College, Hyderabad, Telangana, India
e-mail: dr.kumarigubbala@cmrec.ac.in

A. Sarveswara Reddy
Department of Master of Business Administration, CMR Institute of Technology, Hyderabad, Telangana, India
e-mail: sarvesh.ambala@cmritonline.ac.in

S. Sangeetha
Department of CSE, CMR College of Engineering and Technology, Hyderabad, Telangana, India
e-mail: s.sangeetha@cmrcet.ac.in

 
K. R. Madhavi et al. (eds.), *Proceedings of Sixth International Conference on Computer and Communication Technologies*, Lecture Notes in Networks and Systems 1356,
https://doi.org/10.1007/978-981-96-5238-9_56

**Keywords** Tracking · ARIMA · Meatal water contamination · MSE · RMSE

## 1 Introduction

Water is essential to our daily lives, and the quality of the water in a given area has a significant impact on how well neighboring regular industrial, agricultural, and other human activities can function. The most affordable and widely available sources of freshwater have historically been common water resources such as surface and groundwater. In any case, a number of factors, such as routine operations and human, industrial, and commercial activity, will eventually cause these assets to become defiled. However, inadequate sanitation practices and a lack of awareness also occupy yourself a major role in the contamination of drinking water. Many water pollutants have long-term detrimental effects on water quality, posing a risk to public health. Both the planet and human prosperity are impacted by poor water quality. Freshwater is so drastically reduced. Contaminated water can also cause some waterborne illnesses and have an effect on child mortality. The United Nations reports that 1.5 million children die every year by waterborne diseases. Depending on the World Health Organization, illnesses associated with water cause about 3.4 million deaths annually. Therefore, developing new approaches and strategies to address declining water quality and predict future trends in water quality is crucial. With the aim of attending to the seasonal fluctuations in water quality, it is crucial to include a temporal component in the analysis in order to complete relevant as well as productive evaluations of water quality and anticipate the water quality examples. Unique methods have been put forth and put into practice for time series analysis as well as for analyzing and verifying the quality of water.

Many approaches are used both domestically and internationally to predict water quality. These tactics are primarily divided into four categories:

1. Mathematical Statistics.
2. LSTM.
3. Neural Networks.
4. ARIMA Model.

Although the mathematical statistical approach is effective in modeling, the forecast is not flawless. The gray theory approach is not suitable for the entire non-linear functions; it is only effective for estimating exponential functions. Simply said, the chaos theory approach can be useful if the training set has exceptionally rich data. Traditional neural networks function well for handling non-linear, arbitrary data because of their non-linearity and self-organization learning capabilities. The classic neural network is not suitable for handling time series data due to its topology. Owing to a few flaws in other models, the LSTM model was chosen in order to provide a comprehensive strategy for accurate water quality monitoring and prediction. The primary barrier to ARIMA models, despite their great adaptability to numerous types of time series, as well as AR, MA, and consolidated AR and MA(ARMA), is the model's presumed linear form. The use of fertilizers and pesticides containing metals such as cadmium and arsenic can cause runoff into adjacent bodies of water. Wastewater Disposal: Heavy metals are frequently present in untreated or insufficiently treated wastewater from industrial and residential sources. Atmospheric Deposition: Heavy metals that are released into the atmosphere by cars and industry have the potential to settle in bodies of water. Natural Sources: Heavy metals can be released into water by naturally occurring geological processes. For instance, because of the regional geology, arsenic is frequently found in groundwater in high amounts in particular areas. Effects of Pollution with Heavy Metals Human Well-Being: Many health problems, including as cancer, brain damage, renal failure, and developmental disorders in children, can result from heavy metal exposure. For instance, neurological and behavioral problems might result from mercury toxicity. Aquatic Life: Heavy metal toxicity can result from heavy metal accumulation in the tissues of aquatic species, which can upset aquatic ecosystems. Greater amounts of toxicity in the food chain can be caused via bioaccumulation and biomagnification, which can impact species higher up the trophic ladder. Environmental Degradation: The presence of heavy metals can lower biodiversity and interfere with ecosystems' natural processes, which lowers soil and water quality and impacts plant and animal life.

We would like to explore heavy metal water contamination forecasting in more detail. You might want to consider honing your strategy and renaming it something like "Predictive Modelling of Heavy Metal Water Contamination: Integrating Data Science and Environmental Monitoring". As shown in Fig. 1 this title alludes to the all-encompassing strategy you are employing to foresee and reduce pollution from heavy metals in water sources by fusing real-time environmental monitoring with predictive modelling tools.

**Objective**

The main idea behind the study is to develop a thorough approach that uses certain water quality metrics to examine and forecast the water nature of different places. The physical, biological, and chemical elements that affect the quality of water are included in these metrics. By proposing a model based on machine learning

**Fig. 1** Polluted water by heavy metal

techniques, this research hopes to address this problem by using past data on water quality to predict the patterns of future water quality in a particular region. The ARIMA model is used to develop a workable forecast and analysis methodology for water quality. Two models are compared for the water quality forecast: one using a standard neural network and the other based on the ARIMA model. The outcomes confirm the effectiveness of the framework we suggested.

## 2 Literature Review

You can find a thorough explanation of the LSTM as well as an application related to field data predictions in. Additionally, the attention-based method is incorporated into the LSTM to address the very modest inadequacy of the original LSTM in describing the relationship between the network's input and output [1]. Furthermore, local air pollution measurements along major thoroughfares are conducted using small Internet of Things (IoT) devices. Finding the polluted locations in the River system is a highly challenging task, and water pollution has become a significant issue [2]. In order to complete this objective, the artificial neural network was used to model and forecast four water quality metrics at four separate locations in order to assess the contamination level. The results of the testing model performance indicate that the $R^2$ values for pH, TDS, EC, and Na range from 0.952 to 0.967, 0.953 to 0.970, 0.951 to 0.967 and 0.953 to 0.968, respectively. The results of the forecast performance evaluation indicate that the $R^2$ values for these same parameters range from 0.945 to 0.968, 0.946 to 0.968, 0.944 to 0.967, and 0.949 to 0.965. Additionally, it was noted that the root mean squared error (RMSE) for pH, TDS, EC, and Na, respectively, varies from 0.022 to 0.088, 0.012 to 0.087, 0.015 to 0.085, and 0.014 to 0.084. Researchers will use the data from this study as a guide when determining the index of water quality for irrigation. Additionally, it will direct government agencies and policies regarding water resource management, policy, and decision-making. The methods and implications of the most effective artificial intelligence (AI) systems are thoroughly examined in this review paper [3] in a clear and comprehensive manner, with a focus on groundwater quality modeling and forecasting for practical use in homes. In order to consider the features, this study methodically offers a thorough analysis of the four most popular AI techniques: support vector machines (SVM), evolutionary algorithms (EA), adaptive network-based fuzzy inference system (ANFIS), and artificial neural networks (ANNs). Accurate and trustworthy groundwater resource forecasting is essential for effectively maintaining and enhancing groundwater quality, and a deeper comprehension of hydrogeological processes and behavior is required. To create such forecasts, though, particular instruments are needed due to the hydrogeological system's complicated and variable features. Among the different methods, groundwater modeling has revealed to be a constructive tool for comprehending groundwater systems and determining the dangers to groundwater that exist today (1978). Air pollution is a serious problem, particularly in the world's megacities [4]. These cities and the surrounding areas have emission sources that contribute to variations in air quality depending on the local weather. Short-term air quality forecasting is utilized in preparing for lower resident exposures as well as to potentially alleviate upcoming bouts of excessive air pollution. In this work, a model that forecasts the concentration of PM10, PM2.5, $NO_2$, and $O_3$ pollutants for the present and next day is built using artificial neural networks (ANN). In surface water bodies, or 168 rivers and 71 lakes, this study [5] gather historical sample data on the total concentrations of 12 heavy metals (Cd, Pb, Cr, Hg, Zn, Cu, Ni, Al, Fe, Mn, As, and Co). The data was collected among 1972 and 2017. The goal was to look into the sources and amounts of heavy metal pollution all through five continents and five decades. The World Health Organization (WHO) has issued threshold values for both mean concentrations of heavy metals in worldwide river and lake water as well as the number of heavy metals exceeding these limits.

The original metal data's complete variability was represented in the fewest possible factors using PCA. There is the least amount of covariance conceivable because each variable is orthogonal to every other one [6, 7]. The first factor is the linear combination of the initial variables that is weighted (factor loading) and accounts for the largest amount of variability.

In order to estimate a one-year water quality index using artificial neural networks (ANNs), this study examined the water quality of Ele River Nnewi in Anambra State for irrigation purposes. Finding the contamination hotspots in the River system is a challenging undertaking, and water pollution has become a significant issue. In order to complete this challenge, an artificial neural network was used to model and forecast four water quality metrics at four (4) different locations in order to assess the contamination level [8]. The methods [9] and implications of AI methodologies, particularly as they relate to groundwater quality prediction models, were thoroughly examined in this review paper. The review conducted a thorough literature analysis of previous and current research on artificial intelligence (AI) in groundwater quality prediction, identified the biggest obstacles, and identified potential areas for further study.

The main goals in writing this article [10] are to review the uncertainty in the HWQMs that are currently in use and to give investigators and researchers advice and helpful information. We examine the sources of uncertainty in hydrologic and water quality modeling, including spatiotemporal scales, model representation, discretization, and parameterization, in this regard, and offer methods for evaluating the degree of uncertainty in these domains at both local and global scales when interpreting the model's output.

**Existing Systems**

Water quality is predicted using a variety of analytical methods, which can be categorized as non-traditional—by means of artificial intelligence (AI) techniques and traditional based on statistical models [11, 12]. Regression analysis and time series approaches are associated with traditional methods, while KNN, machine learning, and artificial neural networks (ANN) methods are associated with non-traditional methods [13, 14]. According to a survey of the literature, artificial neural networks (ANN) are the most widely used modeling technique for water pollution prediction.

## 3 Proposed System

"Auto Regressive Integrated Moving Average" is what ARIMA stands for ARIMA models can be used to symbolize any "non-seasonal" time series that show patterns and is not just random noise shown in Fig. 2. Future values are predicted using time series forecasting over a given length of time. Univariate Time Series Forecasting is the process of predicting future values of a time series using just its past values.

Multivariate forecasting of time series is the process of forecasting anything using predictors besides the series (also known as exogenous variables).

1. Auto Regressive (AR) Component
2. Integrated (I) Component
3. Moving Average (MA) Component

The order of the ARIMA model is represented as $(p, d, q)$, where $q$ is the number of instances the function appears during model execution.

$p$ number of lag observations.
$d$ degree of differencing.
$q$ size of the moving average window.

The advancing factor of interest is relapsing on its lagged values, as indicated by the AR component of ARIMA. The many analysis portion shows that the regression error is actually a direct sum of the error terms and their values occurred simultaneously and at various points in the past. The *I* indicates that the difference among the current values and the historical values has replaced the data values. It is possible that this differencing process was used more than once. To make the algorithm fit the data as closely as possible, each of these features has been included.

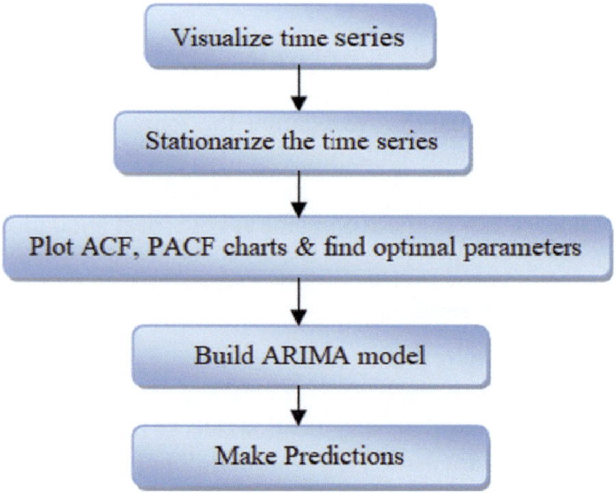


**Fig. 2** Model work flow

### Auto Regressive AR

It describes a model in which a variable that is changing regresses on its own fatigued or previous, values. The order of the ARIMA model is represented as $(p, d, q)$, where $q$ is the number of instances the function appears during model execution. The number of observations with lag incorporated in the model is indicated by the '$p$' parameter.

### Integrated

It symbolizes the process of differencing raw observations in order to make the time series stationary (that is, the variance among the data values with the prior values is used in place of the data values).

ARIMA $(p, d, q)$ is typically used to identify non-seasonal ARIMA models shown in Fig. 3, where parameter $p$ denotes the number of observations with lags (sometimes referred to as the lag order) added to the model.

$d$-number of times that the raw observations are differenced, also known as the degree of differencing.

$q$-size of the moving average window, also known as the order of moving average.

### Moving Average

When a moving average model is applied to lagged observations, it takes into account the dependence among an observation as well as a residual error. The many analysis portion indicates that the regression error is actually a direct sum of the error terms with values that occurred simultaneously and at various points in the past.

The following equation explains the moving average (MA) component of the model:

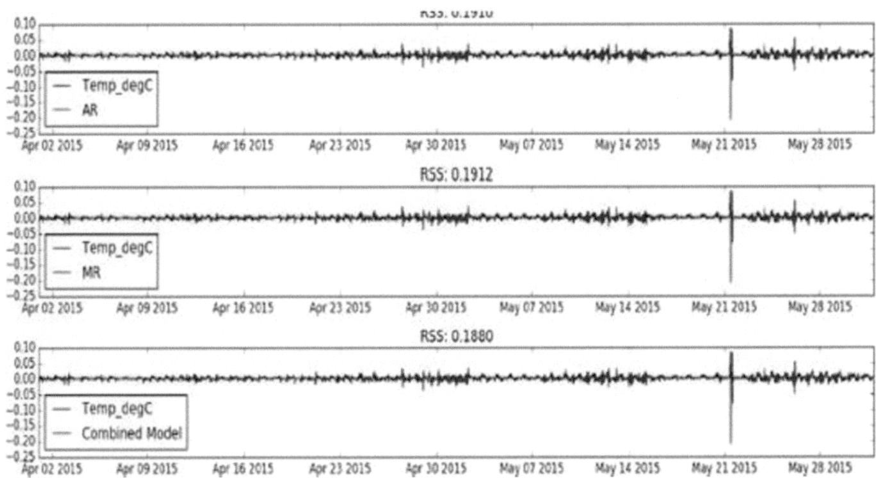


**Fig. 3** AR, MA, and combined model for temperature (April–May 2021)

**System Architecture**

The suggested decision fusion's design is predicated on the time series forecasting, as demonstrated. Shown in Fig. 4 ARIMA is the model of choice for time series forecasting. The suggested system is put into practice in two stages: first, using time series models that have been trained and adjusted. Regularization is utilized to forecast future concentration levels in the fine-tuned implementation, however in the pre-trained models execution, normalization is not used and the pre-trained values for the future are used.

Model Application

Several crucial steps must be taken in order to apply an ARIMA model to time series analysis, from determining the model's parameters to projecting future values. Numerous disciplines, including finance, economics, production, environmental research, and more, can employ this procedure.

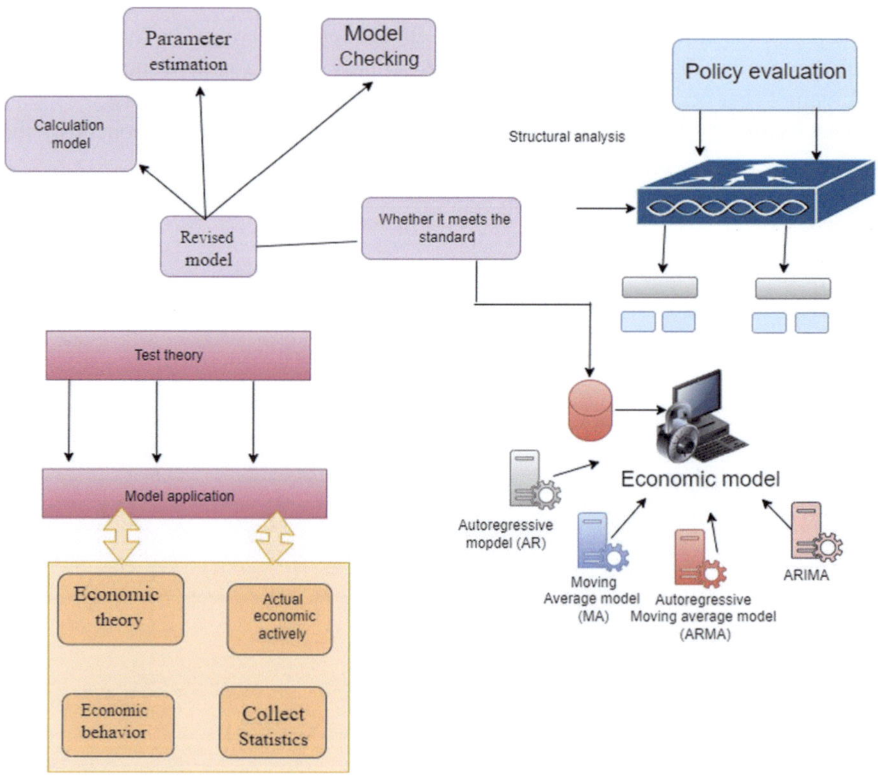


**Fig. 4** ARIMA architecture

Structure Analysis

When it comes to ARIMA modeling, structured analysis refers to the methodical procedures used to create, select, estimate, and evaluate an ARIMA model that is appropriate for time series prediction. Let us examine a more comprehensive, step-by-step methodology by utilizing a monthly sales data example.

Policy Evaluation

The common use of an ARIMA (Auto-regressive integrated moving average) framework is to examine the effects of a specific event or policy change on a group of time series data. Because ARIMA models are so good at forecasting, they may be used to analyze and quantify how policy changes affect a particular indicator over time.

**Data Exploration**: Recognize the time series information that you are utilizing. Examine its seasonality, trends, and patterns.

**Stationarity**: Verify the stationary nature of the time series. If not, keep applying differencing until it stabilizes. The '*d*' parameter in ARIMA($p$, $d$, $q$) is determined by the number of differences required.

**ACF and PACF Plots**: Analyze the plots of the autocorrelation function (ACF) as well as partial autocorrelation function (PACF) to determine possible values for '*p*' and '*q*', respectively. The ordering of the moving average and autoregressive components can be ascertained with the aid of these charts.

**Model Selection**: Choose starting values for '*p*' and '*q*' according to the ACF and PACF plots. Akaike information criterion (AIC) and Bayesian information criterion (BIC) are two methods you can use to compare model performance as you experiment with different combinations.

**Model Estimation**: To calculate the variables of the ARIMA model, use R or the statistical models in Python.

**Model Evaluation**: Use diagnostic tests to assess the model's performance, such as checking the residuals for normality and autocorrelation.

**Forecasting**: Make predictions with the model after it has been validated. Analyze the forecasts versus the actual numbers and make any necessary model adjustments.

**Refinement**: Iterate through steps 3 through 7 to refine the model, modifying parameters and assessing results until you have a model that meets your needs.

**Dataset**

See Fig. 5.

**Fig. 5** Dataset containing of water concentration levels

| | date | pH |
|---|---|---|
| 0 | 2014-10-01 | 7.009447 |
| 1 | 2014-10-02 | 7.044741 |
| 2 | 2014-10-03 | 6.972112 |
| 3 | 2014-10-04 | 6.928980 |
| 4 | 2014-10-05 | 6.980445 |
| 5 | 2014-10-06 | 7.019258 |
| 6 | 2014-10-07 | 7.031518 |
| 7 | 2014-10-08 | 7.031793 |
| 8 | 2014-10-09 | 7.013097 |
| 9 | 2014-10-10 | 7.007799 |

## 4 Modules and Its Implementation

### Data Cleaning and Preparation

We have converted the tab-separated values (.tsv) file format of the data we gathered from the USGS NWIS website to the comma-separated values (.csv) file format shown in Fig. 5. The date column is being designated as an index, and the information will be kept in a data frame. The data frame will be cleared of the rows that contain NaN values. Every day, a fresh sample of the data is taken, and the mean for the day is computed. The resampled data will then be stored in an additional CSV file.

### Data Preprocessing

We organize our data and load it in the proper location for our machine learning study. It is also a good idea to visualize our data appropriately at this point to see if there is any imbalances or relevant relationships between different elements that we can take advantage of the dataset included 1,50,729 entries when it was first collected from the USGS online information archive shown in Figs. 6 and 7.

## 5 Results

Step 1: Click on We have the Major Project Folder after opening the file manager and navigating to the local drive (:c). Select "We have code B12 FOLDER."
Step 2: Click on the codeB12 folder. There are two files there: one is installed and zipfied. When we click on code B12, a few files are downloaded that we must reroute to the command prompt.

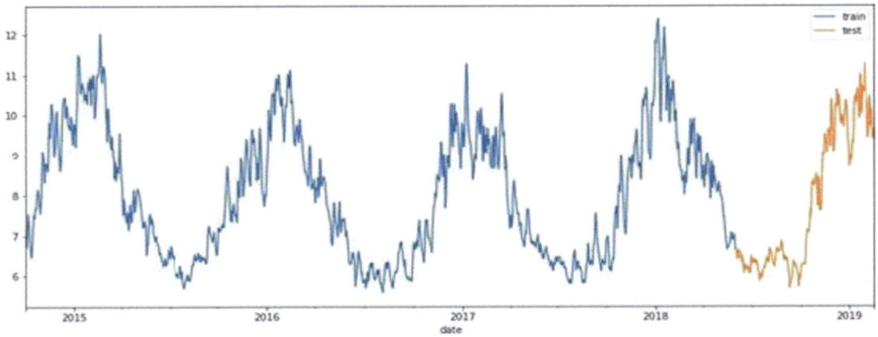


**Fig. 6** Trained dataset of the sample

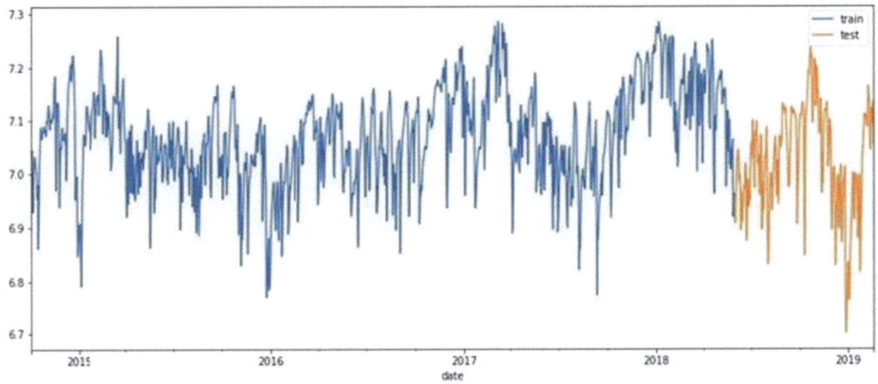


**Fig. 7** Trained and test data sample

Step 3: To have several files open and the simple_ann file run in the jupyter notebook, we must enter the JUPYTER NOTEBOOK on the command prompt and then be redirected to Chrome.

The forecast and prediction graph for neural networks is available.

The neural network's mean square error is 0.0437 as shown in Fig. 8

Utilize the LSTM prediction to generate a forecasting diagram.

The LSTM's mean square error is 0.0414 shown in Fig. 9.

LSTM has a lower error rate than neural networks because of this. The superior performance is LSTM.

Step 4: Launch the model by opening the arima_model file, and we will see ACF and PACF charts shown in Fig. 10.

The residual error graph is built after the PACF plots are created by the model shown in Fig. 11.

Step5: Utilizing the ACF and PACF plots, run the Arima model to forecast the model shown in Figs. 12, 13, 14, and 15.

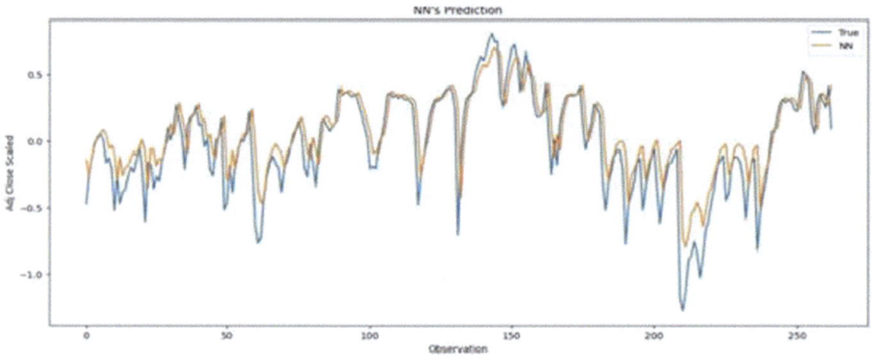


**Fig. 8** Neural network prediction

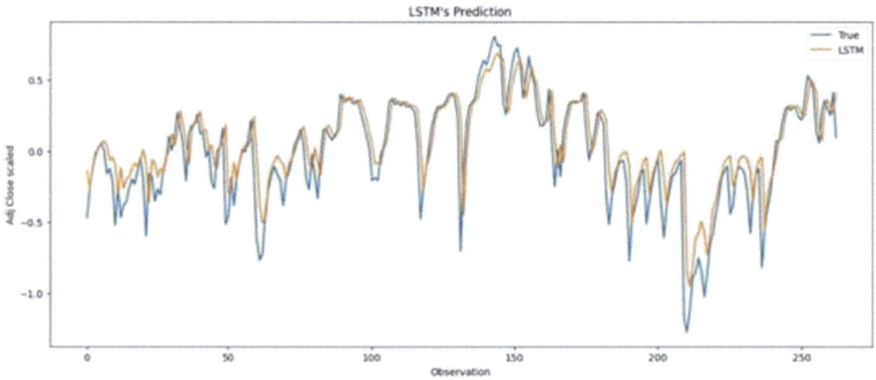


**Fig. 9** LSTM predication

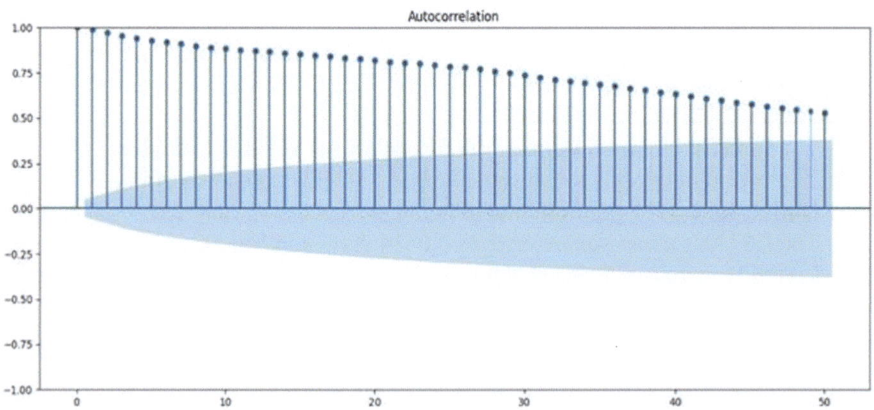


**Fig. 10** ACF plots

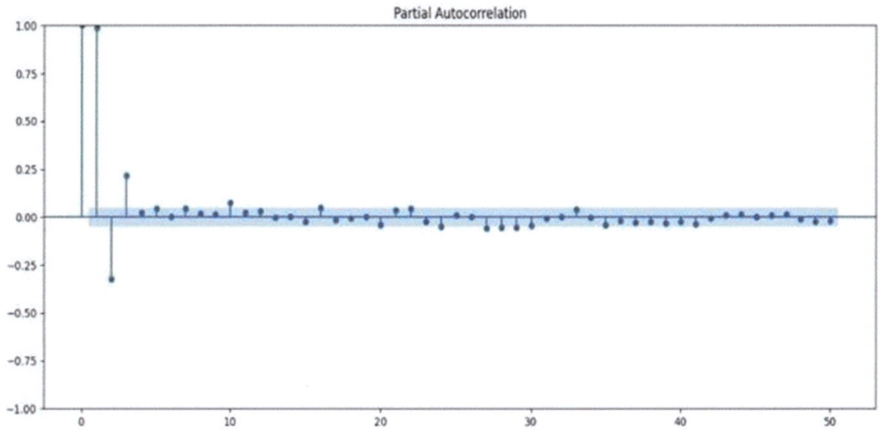


**Fig. 11** PACF plots

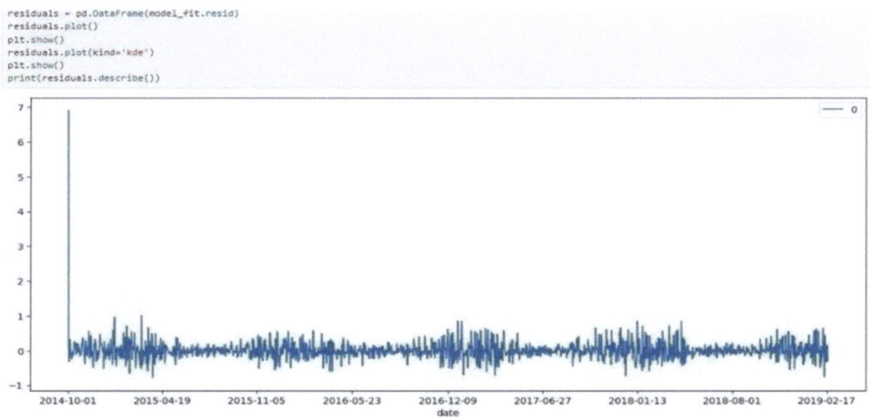


**Fig. 12** Residual error plots

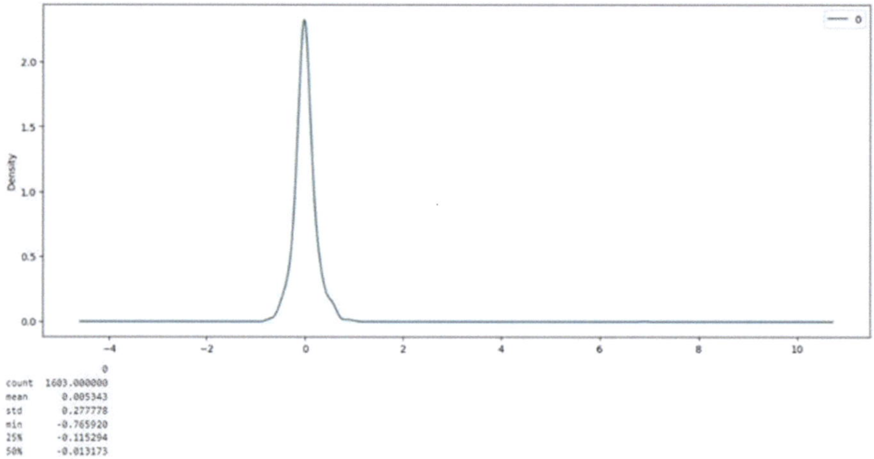


**Fig. 13** Density versus concentration level

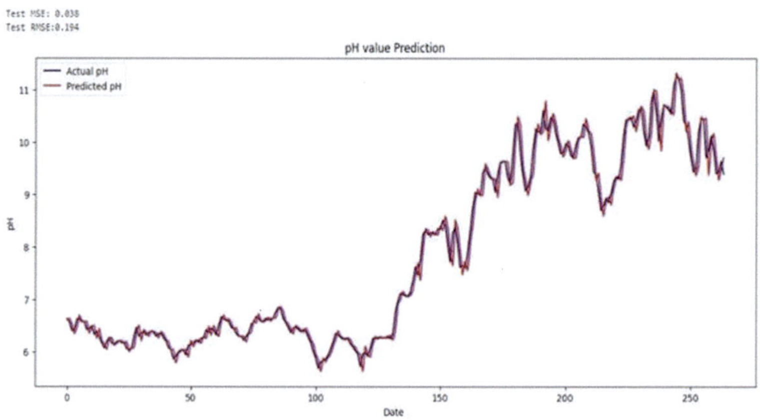


**Fig. 14** Arima forecasting

**Fig. 15** ARIMA, LSTM, NN error value

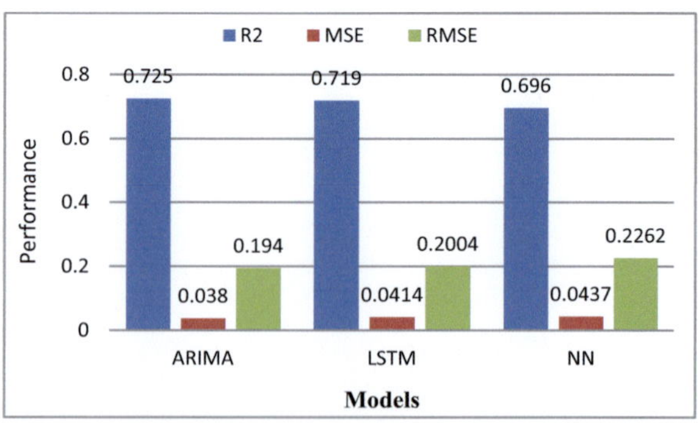

## 6 Conclusion

When compared to the well-known forecasting models ANN, LSTM, and ARIMA, ARIMALSTM NN has a greater level of predictive accuracy, as demonstrated by graphs and performance indicators such as RMSE and MSE. Moreover, ARIMA has lower mean squares and root mean squares errors. The results of the experiment demonstrate that the ARIMA's RMSE of 0.194 and MSE of 0.038 are superior to those of two other well-known forecasting models. The forecasted outcomes remain consistent. This ARIMA model can be selected based on the heavy metal contamination quality of the water.

## References

1. Thomas F (2005) Rural water supplies and water-quality/healthy housing reference manual. National Center for Environmental Health, USA, ch 8 (Online). Available: https://www.cdc.gov/nceh/publications/books/housing/06_hhm_final_chapter_08.pdf
2. Swistock BR, Clemens S, Sharpe WE (2007) Drinking water quality in rural Pennsylvania and the effect of management practices/school of forest resources and institutes of energy and the environment. Pennsylvania State University, State College, PA, USA, Technical Report
3. Water Resources of Kazakhstan in the New Millennium (2004). Review (A publication in support of the UN millennium development goals. Goal 7: ensure environmental sustainability), United Nations Develop. Program Kazakhstan, Almaty, Kazakhstan, p 132 (Online)
4. Valentini KL, Orolbaev EE, Abylgazieva AK (2004) Water problems of Central Asia, Bishkek, Kyrgyzstan, p 142
5. Hornero J, Manzano M, Ortega L, Custodio E (2016) Integrating soil water and tracer balances, numerical modelling and GIS tools to estimate regional groundwater recharge. Sci Total Environ 568:415–432. https://doi.org/10.1016/j.scitotenv.2016.06.011
6. Kelley WP (1940) Permissible composition and concentration of irrigation water. Proc ASCE 66:607–613
7. Handa BK (1964) Modified classification procedure for rating irrigation waters. Soil Sci 98(4):264–269
8. Ubah JI, Orakwe LC, Ogbu KN, Awu JI, Ahaneku IE, Chukwuma EC (2021) Forecasting water quality parameters using artificial neural network for irrigation purposes. Sci Rep 11(1):24438. https://doi.org/10.1038/s41598-021-04062-5
9. Che Nordin NF, Mohd NS, Koting S, Ismail Z, Sherif M, El-Shafie A (2021) Groundwater quality forecasting modelling using artificial intelligence: a review. Groundwater Sustain Develop 14:100643. https://doi.org/10.1016/j.gsd.2021.100643
10. Guzman JA, Shirmohammadi A, Sadeghi AM, Wang X, Chu ML, Jha MK, Parajuli PB, Harmel RD, KhareYP, Hernandez JE (2015) Uncertainty considerations in calibration and validation of hydrologic and water quality models. Trans ASABE 58(6):1745–1762. https://doi.org/10.13031/trans.58.10710
11. Suma S, Moon R, Umer M, Raju KS, Bhaskar N, Okali R (2023) A prediction of water quality analysis using machine learning. In: 2023 International conference on distributed computing and electrical circuits and electronics (ICDCECE), Ballar, India, pp 1–6. https://doi.org/10.1109/ICDCECE57866.2023.10150940
12. Tiwari R, Shrivastava R, Vishwakarma SK, Suman SK, Kumar S (2023) InterCloud: utility-oriented federation of cloud computing environments through different application services. In: Kumar A, Mozar S, Haase J (eds) Advances in cognitive science and communications ICCCE

2022. Cognitive science and technology. Springer, Singapore, pp 83–89. ISBN: 978-981-19-8086-2_8, https://doi.org/10.1007/978-981-19-8086-2_8
13. Reddy KP, Devi G, Wilson Prakash S, Srinath B (2023) DDOS attack detection method for SDN by using deep neutral network. In: AIP conference proceedings, vol 2548, no 1. AIP Publishing
14. Skandha SS et al (2022) A hybrid deep learning paradigm for carotid plaque tissue characterization and its validation in multicenter cohorts using a supercomputer framework. Comput Biol Med 141:105131

# Analysis of a Deep Learning and Machine Learning-Powered Health Care Application for Predicting Various Diseases

**U. Saritha, Suman Mishra, Mantesh Patil, and D. Narasimha Reddy**

**Abstract** Numerous industries employ machine learning approaches, and the healthcare sector is benefiting substantially through machine learning prediction approaches in particularly. Since predicting sickness is a challenging undertaking, automating the process is essential to minimize risks and notify the patient in advance. Accurate prognoses of their patients' illnesses are necessary for doctors. In addition, a key component influencing treatment choices for precise forecasts is timing. Predicting diseases early on has grown in importance. Still, it is challenging for physicians to anticipate with any degree of accuracy from symptoms alone. Predicting sickness accurately is the hardest task. The application of machine learning in the realm of medical diagnosis is growing. This has been mostly credited to developments in illness categorization and identification systems, which can generate data that helps physicians, identify fatal diseases promptly and greatly improve patient survival rates. This research presents an intelligent healthcare program capable of disease prediction as well as data provision.

**Keywords** AI · Disease prediction · ML · Health care · DL

---

U. Saritha (✉)
Department of CSE (AI&ML), CMR Technical Campus, Hyderabad, Telangana, India
e-mail: cmrtc.paper@gmail.com

S. Mishra
Department of ECE, CMR Engineering College, Hyderabad, Telangana, India
e-mail: dr.sumanmishra@cmrec.ac.in

M. Patil
Department of CSE (AI&ML), CMR Institute of Technology, Hyderabad, Telangana, India
e-mail: manteshpatil@cmritonline.ac.in

D. Narasimha Reddy
Department of CSE, CMR College of Engineering and Technology, Hyderabad, Telangana, India
e-mail: d.narsimhareddy@cmrcet.ac.in

 
K. R. Madhavi et al. (eds.), *Proceedings of Sixth International Conference on Computer and Communication Technologies*, Lecture Notes in Networks and Systems 1356,
https://doi.org/10.1007/978-981-96-5238-9_57

# 1 Introduction

Health care is widely acknowledged as a key factor in supporting people's general psychological, social, and physical health and as a major contributor to the economic growth, growth and development, and modernization of a country. Some people are unable to receive the essential medical treatment they need since they are too far removed from those that offer it or whether they do not possess health insurance. Reduced prices, more use of tele health, and enhanced insurance coverage are just a few of the measures that could help more individuals receive the treatment they require.

In the modern economy, health care is one of the sectors expanding at the quickest rate due to increasing demand for treatment and rising costs. The amount the government spends on health care has risen to an all-time high, while the apparent need for better communication between patients and physicians. In terms of improved care and cheaper costs, big data and machine learning technology have the potential to help patients as well as providers [1]. According to medical professionals, timing is a crucial component of diagnosis, and reaching a correct decision quickly can be quite helpful to patients. Thus, predicting patient outcomes accurately is a problem in the medical field.

The use of artificial intelligence (AI) in both medicine and health care has grown significantly. AI has improved computer intelligence and thought processes. Many works on AI believe machine learning to be a subfield.

Machine learning methods for forecasting are extensively used in many different industries, but the healthcare industry has benefited most from them. Accurately forecasting illnesses, enhancing medical care, and enhancing clinical results are some of its objectives. Through the use of an efficient healthcare system, algorithms that use machine learning in applications for medicine can assist physicians in making more effective treatment choices for their patients. This can be primarily credited to developments in illness categorization and detection methods, which can provide information that helps physicians identify deadly diseases earlier on and significantly increase the survival rate of patients.

A computer program is stated as gaining knowledge through experience E with regard to a particular group of tasks T as well as performance measure, according to the conventional description of machine learning. P, if its task accomplishment in T, as determined by P, gets better through practice E [2].

**Machine Learning Algorithms**

Regression Using Logistic Regression

A way of categorizing for ascertaining the likelihood of an event happening or not is called logistic regression. It is employed to symbolize a category or binary result with just two classes shown in Fig. 1. Considering the exception of the variable's outcome being categorical as opposed to continuous, it is similar to a linear regression analysis. It employs the Log it Link operation, and these fits to the collected values, for prediction.

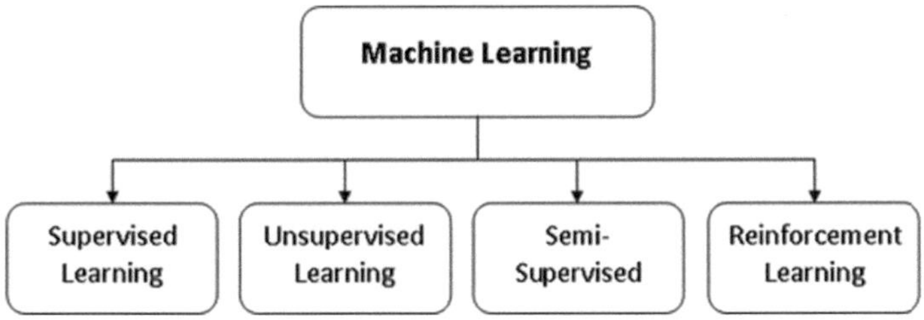


**Fig. 1** Machine learning types

**Decision Tree**

Regression assessment and categorization are two applications for machine learning techniques like decision trees. It is the exact same as the real-life tree comparison. It is a graph that resembles a tree as well as branches outwards to all of the possible outcomes from a single node. A decision tree is supervised learning that translates interactions that are not linear additionally, in contrast to linear models.

Random Forest: an assortment of numerous decision trees that is randomly selected. The random forest classification algorithm can be thought of as a group of decision tree classifiers that can vote for their favorite prediction class. Each decision tree classifier is constructed utilizing several random vectors. By adding more randomness to the model, excessive fitting is avoided, improving the accuracy of identification.

Support Vector Machines (SVM): Alternatively referred to as support vector networks, support vector machines are supervised learning techniques that are applicable to regression as well as classification issues. Plotting data points in a space with multiple dimensions is divided into groups using parallel lines known as the hyperplane. To classify data points, the margin across the hyperplane must be maximized.

**Deep Learning (DL)**

Deep learning is a subset of artificial intelligence (AI) as well as machine learning that closely mimics how people learn particular kinds of information. In contrast to machine learning, that utilizes the use of simpler notions, it makes use of artificial neural networks, whose are intended to mimic how people perceive the world. In a typical deep neural network (DNN), a non-linear activation function, like ReLu and Softmax, is applied to a weighted as well as bias-corrected input value [2]. Thus, the objective of training the goal of DNN is to decrease the loss function by optimizing the weights of the network [3].

**Convolutional Neural Network (CNN)**

A particular kind of artificial neural network employed in deep learning image and object detection and categorization is called a convolutional neural network, or CNN.

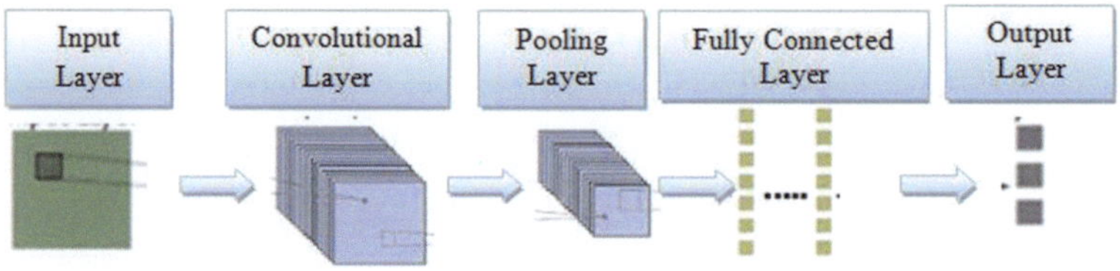


**Fig. 2** CNN architecture

Deep learning recognizes items in an image employing CNN. Convolution, pooling, and fully linked layers—the three main layers of an ACNN—are seen in Fig. 2.

### Transfer Learning

The practice of applying a model that has been trained upon one problem to another that is similar is known as transferred learning.

A model of neural networks gets its initial training on an issue akin to the one that is addressed in transfer learning, a form of deep learning.

Transfer learning can result in a reduction in the error of generalization and shortens the training period with a model made up of neural networks.

### VGG16

Figure 3 depicts the convolution layers of 3 × 3 filters having stride 1 and the continual utilization of the same padding as well as maxpool layer of 2 × 2 filters without stride 2. The sixteen in VGG16 stands for sixteen weighted layers. With over 138 million (estimated) parameters, the system is fairly vast.

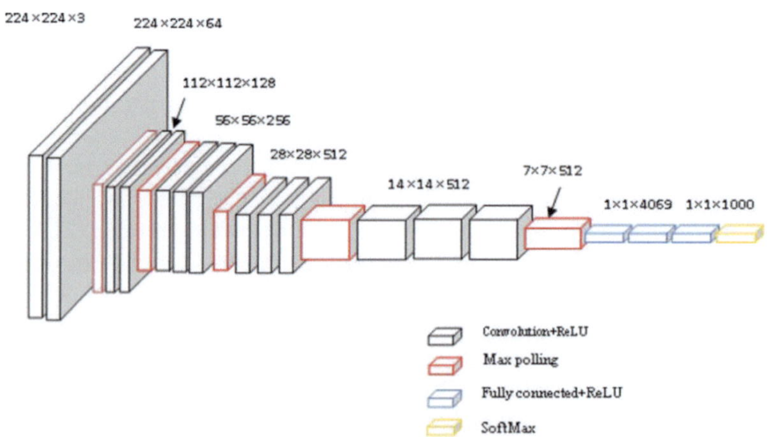


**Fig. 3** VGG16 architecture

OxfordNet, frequently referred to as VGG16, is a convolutional neural network architecture that is part of the Visual Geometry Group. It was suggested in Karen Simonyan as well as Andrew Zisserman's study "Very Deep Convolutional Networks for Large-Scale Image Recognition" at the University of Oxford.

## 2 Literature Review

Employing a dataset from the UCI machine learning repository, Rajdhan [4] and colleagues utilized a variety of machine learning algorithms for forecasting heart disease. Logical regression, random forest, and Naive Bayes were the algorithms employed of these 90.16% was the maximum accuracy provided by RF.

In their study, the authors [5] developed a web-based tool for estimating the likelihood of cardiac disease. They made advantage of the University of California, Irvine's UCI dataset. Using SVM, LR, and NBalgos, they found that SVM were 64.4% more reliable than any of the other two.

Mujumdar and other writers employed a variety of machine learning algorithms in [6] to predict diabetes. LR provided the highest accuracy, 96%.

The PIMA Indians diabetes dataset and UCI was utilized by the authors of this. They examined DL and ML algorithms for diabetes prediction. The RF results were more accurate, yielding an accuracy rate of 83.67% [7].

A UCI ML Repository dataset was used by the authors of [8]. Additionally, the initial dataset was gathered in Andhra Pradesh, India's northeast. They employed six ML algorithms to predict liver disease. Additionally, LR had the highest accuracy of 75% in terms of precision.

They employed SVM and Naïve Bayes techniques to predict liver disease in this owing to its highest classification accuracy, the SVM technique is regarded as the best technique based on the experimental results. Conversely, the Naïve Bayes classifier requires the least amount of time to execute when comparing implementation times. Indian Liver Patient Dataset (ILPD) was the dataset that was used [9]. A deep learning method was created in this study to quickly and correctly predict malaria. After three CNN models were constructed, the model exhibiting the highest accuracy was selected. The accuracy rate of the fine-tuned CNN was higher than that of other CNN models [10]. The collection, which comprises 27,558 cell images, was taken through the US National Library of Medicine. With the CNN method, they were 95% accurate [11]. In [12–14], the authors employed the Guangzhou Women that Children's Medical Center Guangzhou dataset, currently accessible to everyone on Kaggle and contains 5856 pictures of chest X-rays, to forecast the disease pneumonia. They made use of 200 by 200 pixel grayscale photographs. To balance the dataset, data augmentation was applied. 88.90% accuracy was attained. It is critical to have a health awareness and control system in place to regularly check on the physiological parameters of those patients [15–17]. Due to their connections to various medical areas, those using IOT innovation organizations are able to accommodate both professionals and patients in the current health-conscious atmosphere [18, 19].

# 3 Proposed Method

The techniques, techniques, and system structure that we employed in the development of our application are described in this part. We create an online tool that forecasts the illness. We have created a healthcare system to forecast different diseases using ML and DL techniques, as demonstrated in Fig. 4 of this paper. We presented our technique to treat diabetes, heart disease, liver illness, malaria, and pneumonia. Of these, we used ML for diabetes, the heart, and the liver, and DL for malaria and pneumonia. Through evaluation of efficiency and an exploration of ML algorithms, the suggested study predicts diseases. This study aims to accurately determine whether the patient has the illness. The user enters the input values using the patient's health report. The information is incorporated into an ML model that forecasts the likelihood of contracting the illness.

For easier comprehension, we have separated the model training between ML training as well as DL training.

**ML Training**

Comma-separated values (CSV) format datasets including diabetes, heart disease, and liver illness were gathered from the UCI machine learning repositories. Following data collection, we cleaned, normalized, and handled null values among other data preprocessing operations. After then, the dataset was divided into testing and training phases. The model was trained on 80% of the data, and it was tested on the remaining 20%.

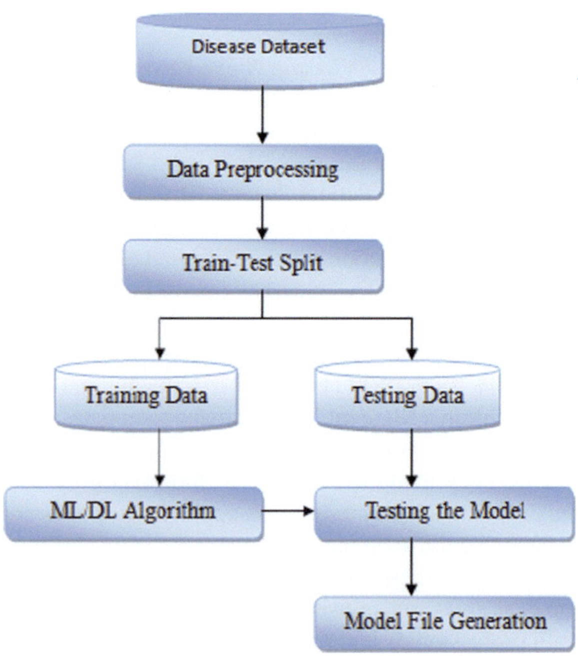


**Fig. 4** Model file generation

The training dataset has become the set of data employed to educate a model. The trained algorithm's efficacy is evaluated using the testing dataset.

Different ML algorithms were employed to train the model files. We employed decision trees, random forests, SVMs, and logistic regression. Each illness dataset was run via all of these algorithms, and their accuracy was assessed. The effectiveness of each method is calculated and examined.

**DL Training**

We utilized the ChestX-RayImages (Pneumonia) dataset by Kaggle for Pneumonia and the Malaria Cell Images Database dataset from Kaggle for Malaria illness. On the photos, we used preprocessing techniques including rescaling, zooming in, rearranging, flipping horizontally, etc. There was a 224 by 224 image. For malaria, we utilized 27,581 training data, 525 testing data, and 5216 training data along with 624 testing data. Here, we have used transfer learning. It was the VGG16 by weights initialized by imagenet. Then, we used 25 epochs to train our simulation.

**System Architecture**

For uploading the picture, the findings about the database side will be displayed on the front end, as seen in Fig. 5. And if the individual is impacted, the system presents the disease's specifics, aiding in the diagnosis and comprehension of the illness's severity.

**Front-End**

Front-end refers to an individual's interface. It depicts where the user engages with the system. We implemented HTML-CSS-Javascript for our computer system. HTML stands for HyperText Markup Language; CSS is for Combining Style Sheets; and JavaScript is regarded as the brains of the visual design system.

The patient's data must be entered by the consumer using the front end, which then sends the information to the back end for prediction. The front end has an input area for entering values for type 2 diabetes, heart, and liver. For pneumonia, it enables the consumer to type in the values as depicted in Fig. 6.

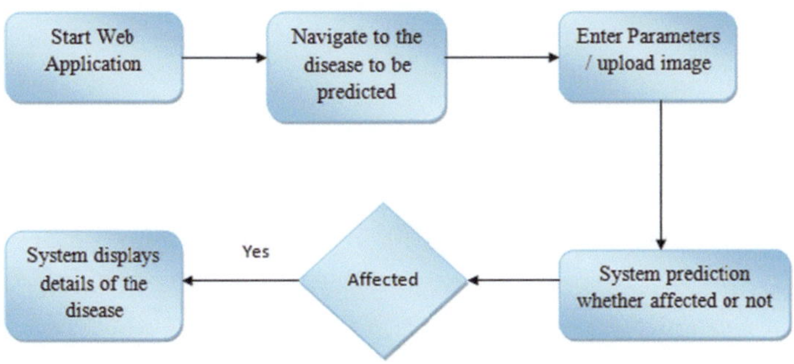


**Fig. 5** System architecture

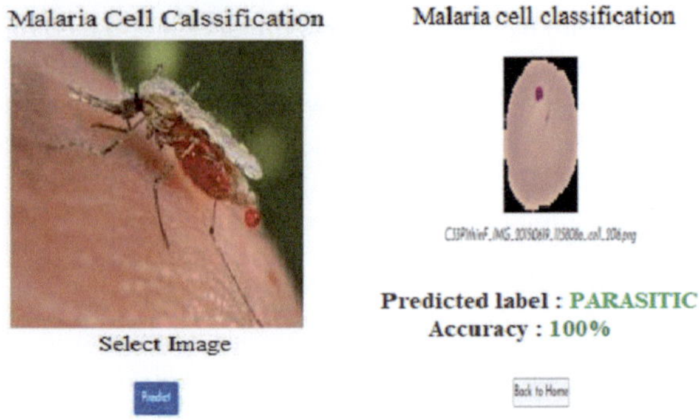


**Fig. 6** **a** Input image **b** Disease prediction output image

### Back-End

The logic is carried out on the backend. It runs in the system's quiet. Our application was developed utilizing the Flask structure, which is Python-programmed. Python is also utilized in the creation of model files. We employed Heroku for the installation of four applications.

## 4 Results and Analysis

We will talk about the results of the suggested system in this part. Figure 7 displays the outcomes of the applications of random forest, decision tree, SVM, and logistic regression.

After examining the algorithmic outcomes, we deduced that the random forest classifier for diabetic disease had the best reliability, the SVM for heart disease had the best performance and the greatest accuracy, and the randomly forest classifier for liver disease had the greatest accuracy, as seen in Fig. 8.

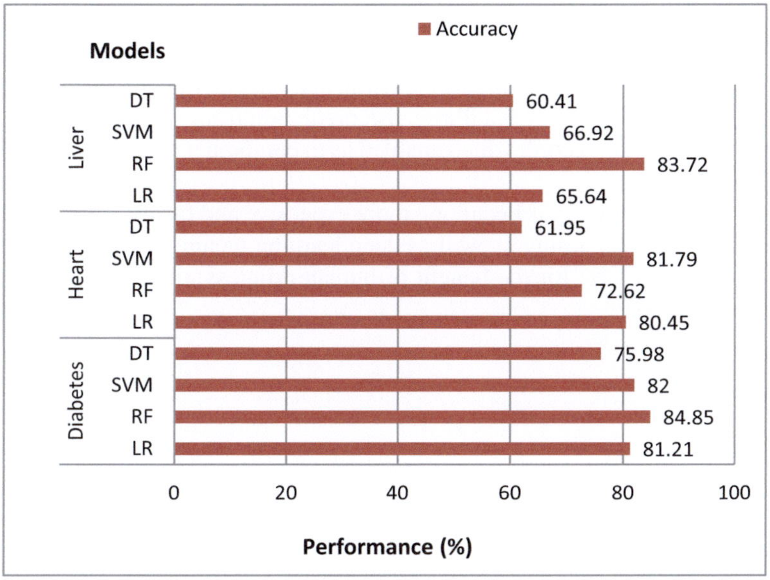


**Fig. 7** Accuracy performance of various algorithms and diseases

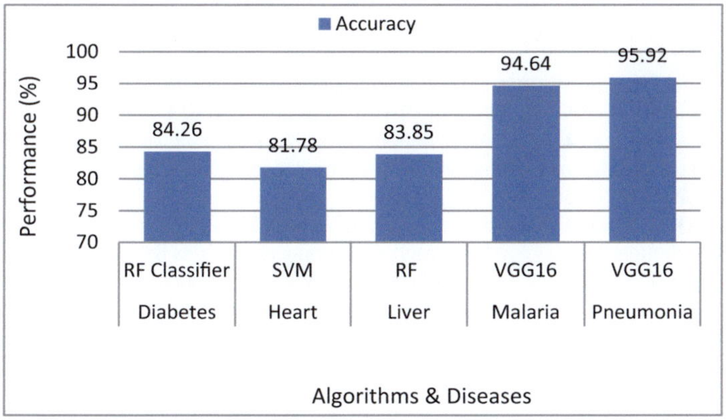


**Fig. 8** Accuracy performance of various algorithms and diseases

## 5 Conclusion

It is increasingly vital to build a system that can accurately and consistently anticipate diseases due to an increasing amount of deaths attributed to various diseases. Finding the most efficient ML and DL algorithms for identifying these diseases became the aim of the study. This study examines the accuracy ratings of the decision tree, logical regression, random forest, SVM, and VGG16 techniques to anticipate various diseases employing the University of California, Irvine machine learning repository dataset. We effectively predicted five illnesses in this system application: diabetes, coronary artery disease, liver disease, malaria, and pneumonia. We were capable of to obtain an accurate generalized illness risk prediction from an output by supplying the input like a patient record, which enabled us to comprehend the degree of disease likelihood prediction.

The healthcare industry is becoming more expensive and facing more challenges. A range of machine learning as well as deep learning techniques are used to solve these issues. Therefore, for the benefit of human progress, this type of integration ought to be promoted.

## References


1. Bhardwaj R, Nambiar AR, Dutta D (2017) A study of machine learning in healthcare. In: 2017 IEEE 41st annual computer software and applications conference (COMPSAC), pp 236–241. https://doi.org/10.1109/COMPSAC.2017.164
2. Learning, Machine. "Tom mitchell." Publisher: McGraw Hill (1997): 31.
3. Lim WYB, Luong NC, Hoang DT, Jiao Y, Liang Y-C, Yang Q et al (2020) Federated learning in mobile edge networks: a comprehensive survey. IEEE Commun Surv Tutorials
4. Rajdhan A, Agarwal A, Sai M, Ghuli P (2020) Heart disease prediction using machine learning. Int J Eng Res V9. https://doi.org/10.17577/IJERTV9IS040614
5. Jagtap A, Malewadkar P, Baswat O, Rambade H (2019) Heart disease prediction using machine learning. Int J Res Eng Sci Manag 2(2):352–355
6. Mujumdar A, Vaidehi V (2019) Diabetes prediction using machine learning algorithms. Proc Comp Sci 165:292–299. https://doi.org/10.1016/j.procs.2020.01.047
7. Yahyaoui A, Rasheed J, Jamil A, Yesiltepe M (2019) A decision support system for diabetes prediction using machine learning and deep learning techniques. https://doi.org/10.1109/UBMYK48245.2019.8965556
8. Rahman AKM, Shamrat FM, Tasnim Z, Roy J, Hossain S (2019) A comparative study on liver disease prediction using supervised machine learning algorithms 8:419–422
9. Mohan V (2015) Liver disease prediction using SVM and Naïve Bayes algorithms
10. Shekar G, Revathy S, Goud EK (2020) Malaria detection using deep learning. In: IEEE 2020 4th international conference on trends in electronics and informatics (ICOEI)—Tirunelveli, India (2020 June 15–2020 June 17), pp 746–750. https://doi.org/10.1109/ICOEI48184.2020.9143023
11. Sriporn K, Tsai C-F, Tsai C-E, Wang P (2020) Analyzing malaria disease using effective deep learning approach. Diagnostics (Basel, Switzerland) 10. https://doi.org/10.3390/diagnostics10100744
12. Tiwari R, Shrivastava R, Vishwakarma SK, Suman SK, Kumar S (2023) InterCloud: utility-oriented federation of cloud computing environments through different application services. In: Kumar A, Mozar S, Haase J (eds) Advances in cognitive science and communications ICCCE 2022. Cognitive science and technology. Springer, Singapore, pp 83–89. ISBN: 978-981-19-8086-2_8, https://doi.org/10.1007/978-981-19-8086-2_8
13. Shrivastava R, Jain M, Vishwakarma SK, Bhagyalakshmi L, Tiwari R (2023) Cross-cultural translation studies in the context of artificial intelligence: challenges and strategies. In: Kumar A, Mozar S, Haase J (eds) Advances in cognitive science and communications ICCCE 2022. Cognitive science and technology. Springer, Singapore, pp 91–98. ISBN: 978-981-19-8086-2_9, https://doi.org/10.1007/978-981-19-8086-2_9

14. Račić L, Popović T, Čakić S, Šandi S (2021) Pneumonia detection using deep learning based on convolutional neural network. 2021 25th International conference on information technology (IT), pp 1–4. https://doi.org/10.1109/IT51528.2021.9390137
15. Nalajala P, Lakshmi SB (2018) A secured IoT based advanced health care system for medical field using sensor network. Int J Eng Technol 7(2.20):105–108
16. Joshi A, Choudhury T, Sai Sabitha A, Srujan Raju K (2020) Data mining in healthcare and predicting obesity. In: Raju K, Govardhan A, Rani B, Sridevi R, Murty M (eds) Proceedings of the third international conference on computational intelligence and informatics. Advances in intelligent systems and computing, vol 1090. Springer, Singapore. https://doi.org/10.1007/978-981-15-1480-7_82
17. Chaudhary D, Sharma PC, Sharma AK, Tiwari R (2022) An insight of deep learning applications in healthcare industry. In: Next generation healthcare systems using soft computing techniques, by CRC Boca Raton, FL 33487, U.S.A 2022, pp 149–167. ISBN: 978-1-03210-797-4, https://doi.org/10.1201/9781003217091-11
18. Dhanalakshmi S, Ravichandran T (2012) A modified approach for image segmentation in information bottleneck method. Int J Adv Res Comp Eng Technol 1(7):59–63
19. Sharma N et al (2022) Automated hybrid deep learning-based paradigm for high-risk plaque detection in B-mode common carotid ultrasound scans: an asymptomatic Japanese cohort study. 2022 AIUM Award Winners 41(1):125

# A More Adaptable and Accessible Pdf Chat_Bot Algorithm by Using Generative Artificial Intelligence (LLMS&RAG)

**V. Venkataiah, G. Shruthi, Ravi Mogili, and Dakkili Maneiah**

**Abstract** This essay explores the growing absurdity of introducing chatbots and artificial intelligence (AI) into scholarly discourse. Chatbots are being used more and more to provide customized support, target businesses, and create engaging learning environments. By outsourcing executive tasks, chatbots enable directors to improve instruction while providing students with real-time assistance with assignments and exams and explanations. One example of an application of generative AI is the development of a PDF chatbot through the use of large language models (LLMs). To store PDF inputs, this software makes use of tokenization, prompt-based querying, and a vector-based component. In order to create embedded prompts and retrieve text excerpts from a knowledge base to address user queries, the chatbot leverages LLMs. The adaptability and accessibility of chatbots increase student engagement and accelerate learning goals.

**Keywords** Knowledge bases · Chatbots · Generative AI · Artificial intelligence (AI) · Academia · Large language models (LLMs)

---

V. Venkataiah
Department of CSE, CMR College of Engineering and Technology, Hyderabad, Telangana, India
e-mail: vvenkataiah@cmrcet.ac.in

G. Shruthi
Department of CSE, CMR Engineering College, Hyderabad, Telangana, India
e-mail: g.shruthi@cmrec.ac.in

R. Mogili
Department of Computer Science and Engineering, CMR Institute of Technology, Hyderabad, Telangana, India

D. Maneiah (✉)
Department of ME, CMR Technical Campus, Hyderabad, Telangana, India
e-mail: cmrtc.paper@gmail.com



K. R. Madhavi et al. (eds.), *Proceedings of Sixth International Conference on Computer and Communication Technologies*, Lecture Notes in Networks and Systems 1356,
https://doi.org/10.1007/978-981-96-5238-9_58

# 1 Introduction

The use of cutting-edge technology like chatbots and artificial intelligence (AI) into education has revolutionized the way intellectual exchange is carried out. One example of a technological invention that promotes Q&A based on scholarly publications is the PDF chatbot [1].

This article examines the development and use of a specific chatbot for PDFs in scholarly research. Unlike other traditional bots, this one enables the uploading of numerous PDFs so that inquiries from academic literature can be made. Our chatbot uses advanced machine learning along with natural language processing to offer a great way to access and engage with scholarly material.

Utilizing sophisticated language models, such as large language models (LLMs) and Recursive Aggregated Generative Models (RAG), among others, is essential to this PDF bot. The bot uses these models to generate relevant answers within their contexts by directly extracting information from submitted PDFs in response to extremely complex inquiries. Our chatbot also makes use of a recent addition known as Language chains [2]. Not less importantly, our chatbot leverages language chains to improve its contextual awareness and retrieval. Because the chatbot can employ language chains to examine linguistic patterns and relationships inside a text, it is able to respond to confusing queries with perfect coherence and correctness. The majority of our chatbot is composed of LLMs, which allow for exceptional levels of precision and fluidity in understanding and producing responses. RAG enhances the chatbot's functionality by gathering data from several documents to provide relevant responses to consumers' inquiries. Additionally, language chains help improve knowledge retrieval and context comprehension by allowing our bot to provide coherent responses based on linguistic relations and patterns found in texts [3]. Our AI-powered PDF chatbots, which enable smooth interaction with research articles from which valuable insights may be derived, are based on these technologies.

**Generative AI**

Generative AI has changed the game in human–machine interaction, as demonstrated by ChatGPT. Following its public release, ChatGPT has drawn a lot of interest and has surpassed well-known competitors like Meta's Wit.ai and Google's Bard AI. This study looks at consumer demands and competition by utilizing ChatGPT's analytical power to look at customer questions and predict future trends [4]. We want to provide an understanding of the developing field of generative AI through careful content selection and thorough investigation. Numerous users of ChatGPT indicate that it is changing a number of sectors and social interactions. Furthermore, it is yet unknown how ChatGPT will affect communication and the spread of knowledge as various people use it. But in this rapidly evolving field, ChatGPT is a shining example of innovation that preserves our relationship with information and technology [5].

**Retriever-Augmented Generation**

The major development in the field of natural language processing (NLP) is the Recursive Aggregated Generative Model (RAG). Why is that relevant, then? RAG can produce responses with several points of support, which means that contextual responses from various sources can lead to insightful and nuanced debates [6]. It is an advancement over chatbots driven by AI, enabling them to provide the same level of precise and thorough answers as previously. RAG makes it possible to deeply comprehend and synthesize a variety of textual materials by integrating with applications based on AI, making user interactions more important.

## 2 Literature Survey

Chen and Guestrin [7] suggest a passive diagnostics system that combines conversational conversation systems, psychologists, and machine learning to treat major depressive disorder (MDD). A real-time dialogue system uses sequence-to-sequence neural networks to engage users, and specialized machine learning classifiers analyze discussions to forecast significant indications of depression. The evaluation's findings point to possible developments in the identification of depression and human-like chatbots. Although the study acknowledges constraints related to data representation and a small sample size, it raises the possibility of improving support for MDD patients by utilizing real-time communication methods. In a similar vein, a deep neural network is introduced by Chung et al. [8] to predict PHQ-4 scores, or degrees of anxiety and depression, from written text. The model shows effectiveness in psychometric score prediction by utilizing a deep learning, transformer neural network, and the Universal Sentence Encoder. The study integrates a multi-dimensional deep neural network with psycholinguistic characteristics, exploring their application to social media data and highlighting issues with generalizability and domain-specificity. Coppersmith et al. [9] use natural language processing to develop a neural classifier that can identify depression from speech transcripts in order to address MDD. Using a symptom network analysis approach, the study predicts individual symptoms of depression and achieves findings that are equivalent to those of state-of-the-art models in binary diagnosis and depression severity prediction. Comparably, Delahunty et al. [10] introduce ToxiScope, a taxonomy to identify and quantify toxic language patterns, with an emphasis on toxic workplace communication via emails. By using machine learning models and annotation tasks, the research sheds light on both explicit and implicit workplace toxicity. For future initiatives, the research recommends improving detection techniques and investigating relationships between toxicity, power dynamics, and biases in workplace communication. Delahunty et al. examine how objective digital biomarkers might be used to evaluate mental health conditions [11]. Through an examination of physiological and behavioral signals taken from remote interviews, the study evaluates the additional information offered by multimodal characteristics. Four conceptual modes are utilized in this study to generate time series features: facial expression, voice expression, language expression, and cardiovascular modulation. Task-specific and foundation models are used to extract these information from audio and video records of remote interviews [12]. The detection of clinically recognized mental disorders, major depressive disorder, self-rated anxiety, and self-rated depression are among the four binary classification tasks that are defined. Findings show statistically significant variations in features between patients with mental health disorders and controls, as well as associations between features and self-rated anxiety and depression ratings [13]. Early evidence of the value of multimodal digital biomarkers derived from inexpensive, non-lab-controlled data is presented by the study, which also offers insights into the best modalities and techniques for automated remote mental health assessments.

## 3 Methodology

In academia, there is currently a tech revolution. Availability of information and individualization was limited in the past since students relied on tangible materials and lecturers gave lectures. Students can now access digital content, collaborate online, and even use chatbots, or artificial intelligence tutors, thanks to technology. Educators leverage technology to design engaging curriculum, apply artificial intelligence to enhance their own productivity, and personalize learning opportunities. The way we teach and learn is greatly impacted by technology [14].

The introduction of AI chatbots is transforming education. These chatbots give students personalized help by answering questions and producing interesting instructional content. They also provide educators with automated administrative tasks and real-time assistance with topics, examinations, and homework. This makes it possible for teachers to focus on teaching.

Here is an illustration of how a PDF chatbot with large language models (LLMs) might operate:

A student asks this question: While communicating with the chatbot and submitting the PDF first, the learner poses a question regarding its content [15].
Information basis and tokenization: From the PDF, the chatbot extracts relevant information that is then saved in a knowledge base. The smallest pieces (chunks) that make up this data are called tokens.
Quick creation: The chatbot employs LLMs to generate particular prompts (a query or command) based on the student's inquiry and the acknowledged tokens (chunks).
Formulating responses: The LLM then uses the knowledge base as a resource, rendering the majority of its credentials to provide text excerpts as a response.
Increased involvement: This flexible, engaging method maximizes learning results by making issues easier for students to understand.

Large language models (LLMs) have been utilized by this PDF chatbot to close the gap between student inquiries and the content found in the PDF. When a student asks a question, the LLM performs more than just look up suitable answers. Rather, it serves as an extremely important practitioner. It interprets the student's question, divides the PDF content into historical sections to comprehend the context, and also formulates a special prescription for itself [16]. By using this inquiry, the LLM can distill the most crucial details from the knowledge base into a brief response that the student can understand. The LLM essentially serves as the chatbot's brain, giving precise answers and motivating the student to delve deeper into the subject.

**The Mathematical Equation Used for the Evaluation**

To examine how similar the responses are to one another, cosine similarity is employed. It is composed of two vectors that provide an $n$-dimensional cosine angle. A cosine measure of distance is used to compare vectors. It is computed by determining if two vectors intersect on the same general path [17].

It is also used to measure document similarity in text analysis. A document may include millions of attributes, and it is possible to use each one to determine the frequency with which a word appears in the text. Presume that the two vectors that require comparison are A and B. We have used the cosine measure to develop a similarity function.

Though it is not stated explicitly, cosine similarity could be used by the chatbot to obtain facts. The learner's query and the contents of the PDF are converted into vectors that have numerical representations, which is how the cosine similarity in the background operates. These vectors capture the relationships and meaning among words. The degree of similarity between the vector of the question and the vectors of other PDF portions (each represented by a different vector) is then determined by computing the cosine similarity, a numerical value ranging from 0 to 1 [18–22]. Finally, the chatbot extracts the PDF section with the highest similarity score, assuming that this piece includes the most relevant answer for the student.

$$\text{similarity }(A, B) = \cos(\theta) = \frac{A.B}{AB} = \frac{\sum_{i=1}^{n} A_i B_i}{\sqrt{\sum_{i=1}^{n} A_i^2 \sum_{i=1}^{n} B_i^2}} \tag{1}$$

$$\cos\theta = \frac{\vec{a}.\vec{b}}{\overline{a}\overline{b}} \tag{2}$$

$$a \text{ is} = \overline{a} = \sqrt{a_1^2 + a_2^2 + a_3^2 \cdots a_n^2} \tag{3}$$

$$b \text{ is} = \overline{b} = \sqrt{b_1^2 + b_2^2 + b_3^2 \cdots b_n^2} \tag{4}$$

$$\vec{a}.\vec{b} = \sum_{i=1}^{n} a_i b_i = a_1 b_1 + a_2 b_2 + a_3 b_3 \cdots a_n b_n \tag{5}$$

where ||$Q$|| denotes the Euclidean norm of vector $Q = (q_1, q_2, \ldots, q_n)$ and ||$R$|| denotes the Euclidean norm of vector $R$.

The angle of the cosine that vectors $Q$ and $R$ create is calculated. If the cosine value is zero, then there is no match between the two vectors, which are identical to each other. When both vectors match, the cosine value is 1.

Steps in Generative AI-Powered PDF Chatbot

Step 1: Loading the Document: Using an external source, upload the input PDF files to the model.
Step 2: Text Pre-processing: As illustrated in Fig. 1, all text analysis, semantic and syntactic analysis, and text preprocessing are completed here.
Step 3: PDF Loaders: Use PyPDF loaders to load the data into the model so that text preparation may begin.
Step 4: Tokenization: The data is divided into smaller units called chunks by processing it. Moreover, chunk_overlap = 1000, chunk_size = 10,000.
Step 5: Integration.

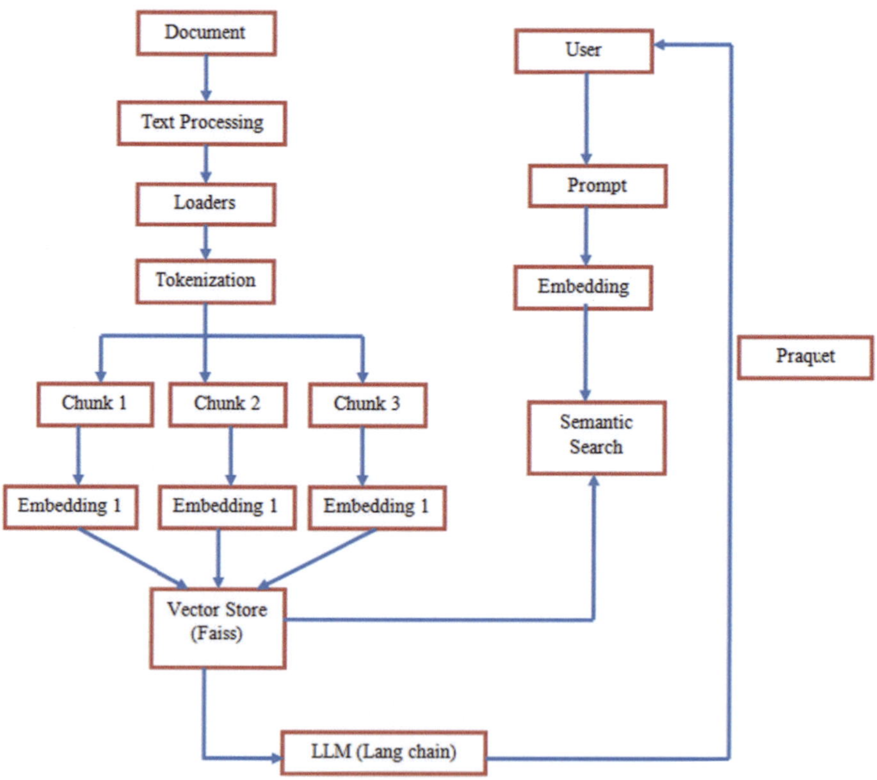


**Fig. 1** Architecture observation of pdf chatbot by means of generative AI

Then, for each block, a vector embedding—a high-dimensional representation of the document information—is created.
Step 6: Vector Store: These embeddings are stored as matrices in a vector store known as Faiss.
Step 7: Input from the user.
The user supplied the prompt via a brightly lighted interface. Incorporating.
Step 8: Similarity: To determine how similar two vectors are, cosine similarity is used. If the two vectors match, the cosine value is 1, otherwise, it is 0. If there is no match.
Step 9: Packet: The data packets are transported in accordance with the request once the vectors are converted into packets.

## Algorithm

Step 1: Leaving important libraries behind causes brightly illuminated from the minute PdfReader PyPDF2
Step 2: Describe the process of extracting content from PDFs
Step 3: Characterize tokenization work
Step 4: Distinguish vector capacity work
Step 5: Distinguish client input work
Step 6: Distinguish Streamlit UI work
Step 7: Handle PDF exchange and planning
Here, more precautions (tokenization, vector capacity) can be taken.
Step 8: Use the sidebar menu to upload PDF records.
Crucial effort
within the occasion that title = = "essential":
crucial()

## Results

This segment presents the outcome of the projected system.

In the greater than Fig. 2, the learner can upload a manuscript or a file. Here learners have to upload files in PDF format.

Figure 3 illustrates how the learner clicks to investigate file button to upload the desired documents. Next, users click the button to submit and proceed buttons to browse the pdf doc shown in Fig. 4, which initiate an automatic experience process and result in the completion notification.

The uploaded content has been tokenized into smaller pieces and saved as matrices in the FIASS vectors storage, as Fig. 5 on the top page illustrates.

Figure 6, as previously indicated, illustrates where the model offers customers an engaging and user-friendly interface to help them resolve any problems they may be experiencing regarding the submitted PDF file.

The manual assessment of the vectors kept in the vector file storage is displayed on the pages beneath.

Consider following vectors

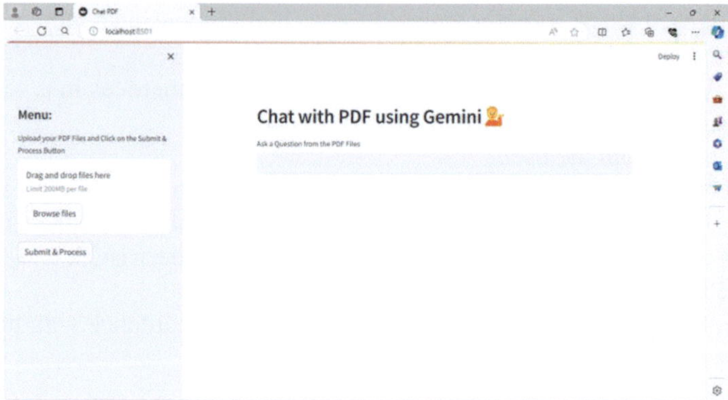


**Fig. 2** User interface

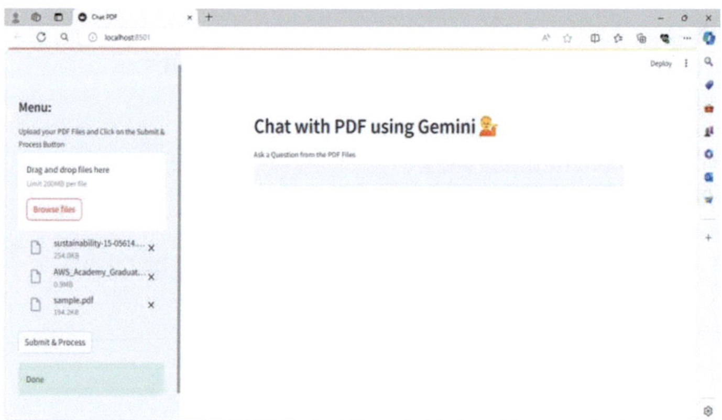


**Fig. 3** Multi_pdf upload

$$a : [1, 1, 0]$$

$$b : [1, 0, 1]$$

$$\text{Norm of vector } a \text{ is} = \overline{a} = \sqrt{1^2 + 1^2 + 0^2} = \sqrt{2}$$

$$\text{Norm of vector } b \text{ is} = \overline{b} = \sqrt{1^2 + 2^2 + 1^2} = \sqrt{2}$$

$$\vec{a}.\vec{b} = \sum_{i=1}^{n} a_i b_i = a_1 b_1 + a_2 b_2 + a_3 b_3 = 1 \times 1 + 1 \times 0 + 0 \times 1 = 1 + 0 + 0 = 1$$

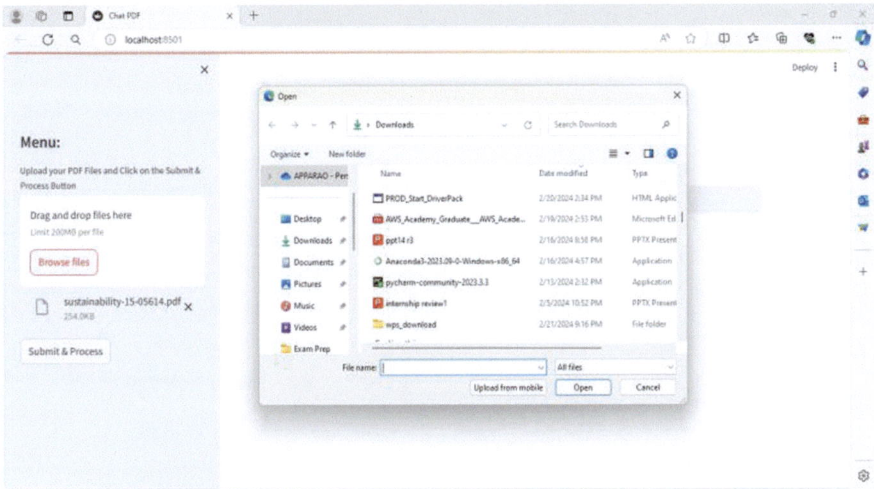


**Fig. 4** Browsing PDFs from system

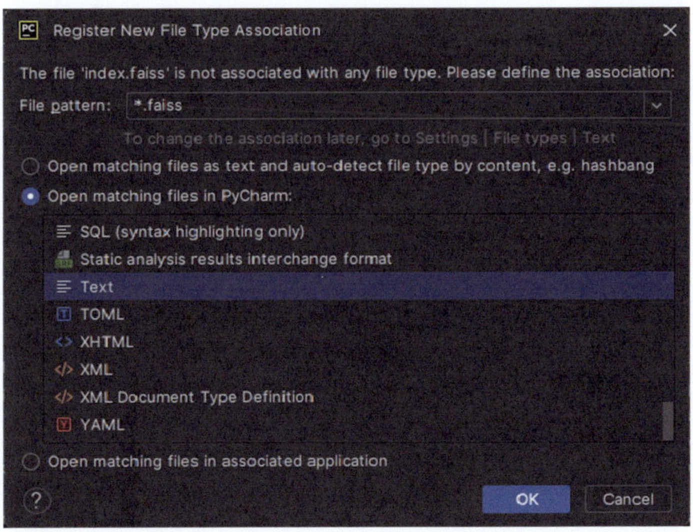


**Fig. 5** Vector storage (FIASS)

$$\cos\theta = \frac{1}{\sqrt{2} \times \sqrt{2}} = \frac{1}{2} = 0.5$$

The manual assessment of the stored vectors and the fluctuations in the cosine angle are displayed in Fig. 7.

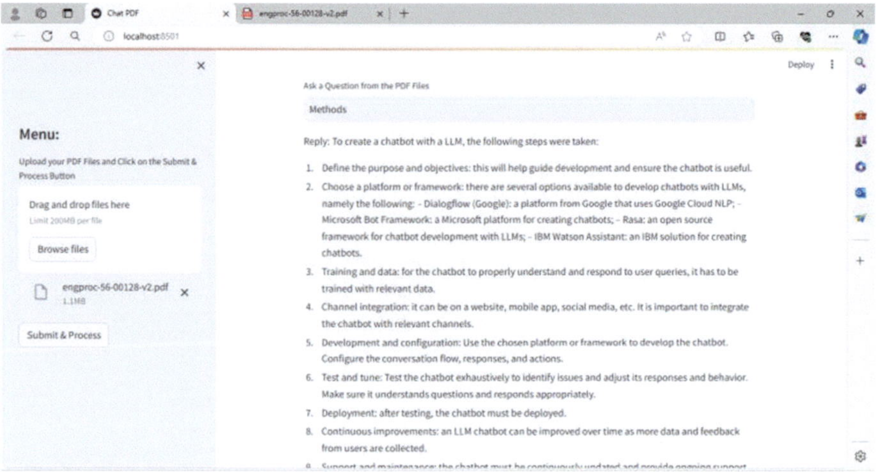


**Fig. 6** Output generation page

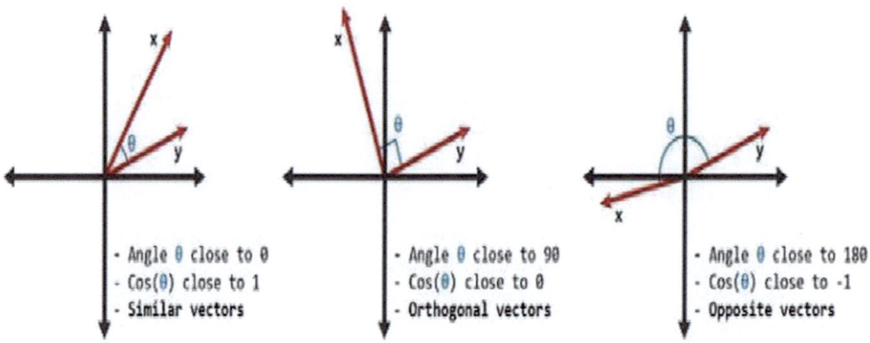


**Fig. 7** Cosine angle variations

Note: The aforementioned graphs yield a number among − 1 and 1, where 1 indicates perfect resemblance, 0 indicates no resemblance and − 1 indicates full disparity.

## 4 Conclusions

Traditional PDFs could be uninspiring and hard to learn from. This project creates a PDF chatbot powered by big language models (LLMs) in order to address this. By submitting a PDF, students may submit questions right away. The chatbot uses LLMs to understand both the text of the PDF and the inquiry from the learner.

Subsequently, it examines the subject matter of the PDF and tailors the response to the specific question. Students benefit from rapid guidance within the context of the PDF while using this customized technique, which improves understanding. By automating tasks, a chatbot also frees out instructors' time to engage in more interactive approaches to learning. Significantly, this LLM-powered chatbot has the power to fundamentally alter how learners read and comprehend content through PDFs.

## References


1. Abd-alrazaq AA, Alajlani M, Alalwan AA, Bewick BM, Gardner P, Househ M (2019) An overview of the features of chatbots in mental health: a scoping review. Int J Med Inform 132:103978. https://doi.org/10.1016/j.ijmedinf.2019.103978
2. Benton A, Mitchell M, Hovy D (2017) Multi-task learning for mental health using social media text. arXiv:1712.03538 [cs.CL]
3. Bhat MM, Hosseini S, Awadallah AH, Bennett PN, Li W (2021) Say 'YES' to positivity: detecting toxic language in workplace communications. In: Moens M-F, Huang X, Specia L, Yih SW (eds) Findings of the association for computational linguistics: EMNLP 2021, Virtual Event/Punta Cana, Dominican Republic, 16–20 Nov, 2021, Association for computational linguistics, 2017–2029. https://doi.org/10.18653/v1/2021.findings-emnlp.173
4. Bill D, Eriksson T (2023) Fine-tuning a LLM using reinforcement learning from human feedback for a therapy chatbot application. Independent thesis Basic level, degree of Bachelor, KTH, School of Electrical Engineering and Computer Science (EECS)
5. Cameron G, Cameron D, MegawG, Bond R, Mulvenna M, O'Neill S, Armour C, McTear M (2019) Assessing the usability of a chatbot for mental health care. In: Bodrunova SS, Koltsova O, Følstad A, Halpin H, Kolozaridi P, Yuldashev L, Smoliarova A, Niedermayer H (eds) Internet science. Springer, Cham, pp 121–132
6. Chen IY, Szolovits P, Ghassemi M (2019) Can AI help reduce disparities in general medical and mental health care? AMA J Ethics 21(2):E167–179. https://api.semanticscholar.org/CorpusID:73498305
7. Chen T, Guestrin C (2016) XGBoost: a scalable tree boosting system. In: Proceedings of the 22nd ACM SIGKDD international conference on knowledge discovery and data mining, San Francisco, California, USA (KDD'16). ACM, New York, NY, USA, pp 785–794. https://doi.org/10.1145/2939672.2939785
8. Chung NC, Dyer G, Brocki L (2023) Challenges of large language models for mental health counseling. arXiv:2311.13857 [cs.CL]
9. Coppersmith G, Dredze M, Harman C, Hollingshead K, Mitchell M (2015) CLPsych 2015 shared task: depression and PTSD on Twitter. In: Proceedings of the 2nd workshop on computational linguistics and clinical psychology: from linguistic signal to clinical reality. Association for computational linguistics, Denver, Colorado, pp 31–39. https://doi.org/10.3115/v1/W15-1204
10. Delahunty F, Johansson R, Arcan M (2019) Passive diagnosis incorporating the PHQ-4 for depression and anxiety. In: Proceedings of the social media mining for health applications (SMM4H) Workshop, Florence, Italy
11. Delahunty F, Wood ID, Arcan M (2018) First insights on a passive major depressive disorder prediction system with incorporated conversational chatbot. In: Irish conference on artificial intelligence and cognitive science, Dublin, Ireland
12. Devlin J, Chang M-W, Lee K, Toutanova K (2018) BERT: pre-training of deep bidirectional transformers for language understanding, 13 pages. https://arxiv.org/abs/1810.04805

13. Devlin J, Chang M-W, Lee K, Toutanova K (2019) BERT: pre-training of deep bidirectional transformers for language understanding. In: Burstein J, Doran C, Solorio T (eds) Proceedings of the 2019 conference of the North American chapter of the association for computational linguistics: human language technologies, volume 1 (long and short papers). Association for computational linguistics, Minneapolis, Minnesota, pp 4171–4186
14. Eichstaedt JC, Smith RJ, Merchant RM, Ungar LH, Crutchley P, Preoţiuc-Pietro D, Asch DA, Andrew Schwartz H (2018) Facebook language predicts depression in medical records. In: Proceedings of the national academy of sciences 115(44):11203–11208. https://doi.org/10.1073/pnas.1802331115, arXiv:https://www.pnas.org/doi/pdf/
15. Galatzer-Levy IR, McDuff D, Natarajan V, Karthikesalingam V, Malgaroli M (2023) The capability of large language models to measure psychiatric functioning. arXiv:2308.01834 [cs.CL]
16. Gratch J, Artstein R, Lucas GM, Stratou G, Scherer S, Nazarian A, Wood R et al (2014) The distress analysis interview corpus of human and computer interviews. In: LREC, vol 14, pp 3123–3128
17. Guntuku SC, Buffone A, Jaidka K, Eichstaedt JC, Ungar LH (2018) Understanding and measuring psychological stress using social media. ArXiv abs/1811.07430
18. Guntuku SC, Yaden DB, Kern ML, Ungar LH, Eichstaedt JC (2017) Detecting depression and mental illness on social media: an integrative review. Curr Opin Behav Sci 18:43–49. https://doi.org/10.1016/j.cobeha.2017.07.005
19. Jiang Z, Seyedi S, Griner E, Abbasi A, Rad AB, Kwon H, Cotes RO, Clifford GD (2023) Multimodal mental health assessment with remote interviews using facial, vocal, linguistic, and cardiovascular patterns. medRxiv; Jin H, Chen S, Wu M, Zhu KQ (2023) PsyEval: a comprehensive large language model evaluation benchmark for mental health. arXiv:2311.09189 [cs.CL]
20. Kroenke K, Spitzer RL, Williams JBW, Löwe B (2009) An ultra-brief screening scale for anxiety and depression: the PHQ-4. Psychosomatics 50:613–621
21. Yalawar MS, Vijaya Babu K, Mahender B, Singh H (2023) A brain-inspired cognitive control framework for artificial intelligence dynamic system. In: Kumar A, Mozar S, Haase J (eds) Advances in cognitive science and communications ICCCE 2023. Cognitive science and technology. Springer, Singapore. https://doi.org/10.1007/978-981-19-8086-2_70
22. Narayana VA, Premchand P, Govardhan A (2009) A novel and efficient approach for near duplicate page detection in web crawling. In: 2009 IEEE international advance computing conference. IEEE, USA

MIX
Papier aus verantwortungsvollen Quellen
Paper from responsible sources
FSC® C105338

If you have any concerns about our products,
you can contact us on
**ProductSafety@springernature.com**

In case Publisher is established outside the EU,
the EU authorized representative is:
**Springer Nature Customer Service Center GmbH**
**Europaplatz 3, 69115 Heidelberg, Germany**

Printed by Libri Plureos GmbH
in Hamburg, Germany